COMMERCIAL FRUIT PROCESSING

Second Edition

COMMERCIAL FRUIT PROCESSING

Second Edition

Edited by

JASPER GUY WOODROOF
Department of Food Science
University of Georgia
Experiment, Georgia

BOR SHIUN LUH
Department of Food Science and Technology
University of California
Davis, California

AVI PUBLISHING COMPANY, INC.
WESTPORT, CONNECTICUT

ISBN-0-87055-502-2

Printed in the United States of America

A B C D E 5 4 3 2 1 0 9 8 7 6

Contents

Contributors

Larry R. Beuchat, Department of Food Science, University of Georgia, Agricultural Experiment Station, Experiment, GA 30212
J. I. Chung, California Food Institute, Mountainview, CA 94041
James E. Epperson, Department of Agricultural Economics, University of Georgia, Athens, GA 30602
B. Feinberg, Food Technology Consultant, Berkeley, CA 94708
James S. L. How,* North Carolina State University, Raleigh, North Carolina 27695
Edward J. Hsu, Department of Biology, University of Missouri—Kansas City, Kansas City, MO 64110
C. E. Kean, California and Hawaiian Sugar Company, Crockett Refinery, Crockett, CA 94525
B. S. Luh, Department of Food Science and Technology, University of California, Davis, CA 95616
Nancy J. Moon, Pioneer Hi-Bred International, Genetics Division, Johnston, IA 50131
Stanley E. Prussia, University of Georgia, Agricultural Experiment Station, Experiment, GA 30212
Robert Lorne Shewfelt, Department of Food Science, University of Georgia, Agricultural Experiment Station, Experiment, GA 30212
L. P. Somogyi, Etel, Inc., San Rafael, CA 94662
Warren K. Trotter, Agricultural Research Service, U.S. Department of Agriculture, Richard B. Russell Research Laboratory, Athens, GA 30603
Glenn G. Watters, Fruit and Vegetable Chemistry, Western Regional Research Laboratory, U.S. Department of Agriculture, Berkeley, CA 94708
J. G. Woodroof, Department of Food Science, University of Georgia, Experiment, GA 30212
Clyde T. Young, Department of Food Science, North Carolina State University, Raleigh, NC 27695-7624

*Present address: Hershey Foods Corp., Technical Center, P.O. Box 805, Hershey, PA 17033-0805.

Preface to Second Edition

Although some of the chapter titles of this second edition of *Commercial Fruit Processing* are similar to those of the first edition, Chapters 2, 8, 11, 15, and 16 have been completely rewritten, and the remaining 11 chapters have been extensively revised and updated. New material and topics added to this edition include expanded discussion of the history of fruit processing; extended list of fruits and fruit products; list of specialty fruit products made from 27 fruits; contract packing; breeding fruits for processing; list of molds commonly associated with fresh fruits; list of antimetabolites produced by molds; development of cryogenic railcar; freeze-dried fruits; pitting dates; table of opening and closing dates for canning fruits; aseptic processing of fruit puree; fruit irradiation; fruit leather; processing of bananas, mango, and kiwifruit; exploding blueberries; U.S. Food and Drug Administration regulations and location of FDA consumer affairs officers; list of products covered by Food and Drug Standards of Identity, Quality, and Fill; list of products covered by USDA Grade Standards for Fresh and Processing Fruits; changes in regulations for U.S. wine labels; and storage of raisins.

Since about 1950 commercial fruit processing operations have been getting larger and fewer in number. The industry is diversifying away from single commodities, away from single geographical areas, away from seasonal operations, away from standardized materials and size of packages, and away from specific markets. The trend is toward different year-round operations, products, processes, and markets that can be changed and adjusted to short-term as well as long-term discoveries and demands. Every segment of the industry is subject to alteration to accommodate changes in domestic and international conditions. Several earlier trends have continued during the 1980s:

- decline in freezing some of the one-time leaders, such as peaches and strawberries
- use of less sweeteners with natural sugars in order to reduce calories

- use of fewer additives containing sodium, spices, artificial colors and flavors, and "energy"
- continued use of fruits in cereals, salads, cakes, pies, and other combinations, as a source of minerals, vitamins, fiber, and natural flavors and colors

An important recent innovation is low-moisture processing, in which fruit, with no added sugar, preservative, or carrier, is converted into convenient dehydrated forms. Development of this technology has been stimulated by high transportation rates, improvements in technology, and revolutionary new packages. In addition to raisins, prunes, and dehydrated apples, pears, peaches, and apricots, bananas are available in flakes, slices, and granules; pineapple and other tropical fruits also are available in new forms. Another low-moisture product is apple fiber solids, consisting of cell wall material (cellulose, hemicellulose, lignin, and pectin) and apple sugars. Low-moisture forms of other fruits are becoming more common.

Commercial Fruit Processing is a companion volume to *Commercial Vegetable Processing,* also edited by B. S. Luh and J. G. Woodroof; both are being updated and revised simultaneously.

Grateful acknowledgments and thanks go to contributors who wrote in their own area of expertise on commercial fruit processing. Credit also goes to more than a dozen commercial companies and individuals who supplied photographs, charts, tables, and data from commercial operations.

Thanks also to Ann Autry who typed, corrected, and edited the manuscript; and to Naomi C. Woodroof, my wife, for assisting in research.

Jasper Guy Woodroof
Bor Shiun Luh

1

History and Growth of Fruit Processing

J. G. Woodroof

The oldest method of processing fruits is sun-drying. Figs, grapes, and dates grown in ancient Canaan and Egypt were among the first fruits dried, requiring no peeling or special treatment. Formation of molds, fermentation, and insect attacks were among the problems associated with man's initial efforts to preserve fruits.

It's likely that wine and vinegar were by-products formed during fermentations that accompanied early attempts to preserve grapes and grape juice for sacramental purposes. Yeasts were naturally present on the skins of the fruit, and the rate and extent of fermentation were controlled by the addition of honey. The temperature was varied by storing containers in caves, and exposure to air (oxygen) was reduced by use of special ceramic containers. For the past seven centuries fermentation has been an important area of research and development. Even today, viticulture, enology, and microbiology continue to produce more and better wines, brandy, and related beverages.

Spices have also been used as a fruit preservative from the time of antiquity. They contain alkaloids, which act as antioxidants, and add flavor, color, and bouquet. When added to fruit along with vinegar and honey (sugar), they provide the foundation for spiced fruits and fruit pickles.

Chemical preservation of fruit, by the use of vinegar, wine, and sugars

Commercial Fruit Processing, 2nd Edition

ISBN 0-87055-502-2

also was practiced before biblical times. Pickles and even fruit preserves are partially or wholly preserved by chemicals. In pickles the preservative may be sugar, salt, vinegar (acetic acid), alcohol (wine or liquors), and by-products. Chemicals preserve foods in several ways. The chemical may be a poison to the microorganism that causes spoilage; it may provide an environment in which the microorganism cannot grow, even though it is not killed; or it may react with the fruit to form a new product. A high sugar content also serves another purpose: it ties up moisture, which is then not available to microorganisms. Many procedures have been improved and products bettered as the use of undesirable chemicals have been eliminated, and others have been developed that are effective in preserving food without harmful side effects to people. Among these are ascorbic acid, which is added to frozen fruits and canned apple juice to preserve a light color, and benzoic acid and sorbic acid, which are used to prevent mold and yeast growth. Calcium chloride is added to canned fruits for firmness.

The greatest breakthrough in fruit processing, however, was development of thermal processing, the technique of preserving foods by heat. Nicholas Appert, who perfected thermal processing of fruits, vegetables, and meat and published his methods in 1810 (Goldblith 1972), received a prize of 12,000 francs from the French government for his work. Appert's process consisted of four steps:

1. Enclosing the food to be preserved in bottles
2. Corking the bottles carefully
3. Submerging the bottles of food in boiling water for a time, depending upon the nature of the food
4. Removing the bottles and cooling them

Since Appert's early work, the technology of thermal processing has advanced and become quite sophisticated. The details of this important technique are discussed in later chapters, but several significant developments are listed here:

- Tin, aluminum, steel, or containers that can be hermetically sealed can be used instead of bottles.
- Some fruits retain their color, flavor, and texture better in containers of special metals; and containers of certain metals require a protective lacquer to prevent color or flavor reactions between the fruit and container.
- Some fruit products may be aseptically packed by heat-sterilizing the product and conveying it aseptically into containers that have been presterilized, followed by aseptic closure of the containers.
- High-temperature short-time (HTST) processing can reduce treatment

time for many fruit products and is especially suited for purées, juices, and other products with heat-sensitive color and flavors.

- Most fruit products with pH 5.5–6.0 can be processed in boiling water and retain acceptable color, flavor, and texture, but some products with a pH $\geqslant$ 7.0 require such a long treatment time to kill microorganisms that the texture of the fruit is destroyed.
- Because high-temperature short-time heat-sterilizing reduces the extent of products denaturization, high-speed heat exchangers and turnover containers are being developed, resulting in constantly improved processed fruit products.

MAJOR TECHNIQUES IN FRUIT PROCESSING

The 10 major techniques used in commercial fruit processing, in their approximate order of importance, are

1. Canning at high temperatures
2. Freezing at low temperatures
3. Drying by reducing the moisture content
4. Preserving with high sugar concentrations (e.g., preserves, syrups)
5. Concentration by removing a high portion of water
6. Preserving with chemicals such as sulfur dioxide, sorbic acid, sodium benzoate, ascorbic acid, acetic acid, citric acid, and alcohol
7. Fermentation with yeasts and bacteria
8. Pickling with sugar, spices, and vinegar
9. Reduction of oxidation by the use of vacuum, antioxidants, and reducing agents
10. Reduction of visible light by the use of opaque containers and dark storage

Often, a combination of the two or more of these techniques is used in processing a particular product. Other materials may be added to achieve specific effects: calcium salts to improve texture; ascorbic acid to retain red colors; sugars to enhance flavors; spices to add pungency; tannins and caffeine as stimulants and to add bitterness.

There are at least 35 commercially manufactured fruit products (see Table 1.1) and an equal number of minor or specialty fruit products. Practically every major fruit processing operation such as a canning, freezing, drying, or juice extraction plant can be adapted for the manufacture of minor or specialty products from fruit that is overripe, hail damaged, irregularly formed, bruised by mechanical handling, from a little known cultivar, off-season, or produced to meet a special market. Sometimes the manufacture of specialty products provides substantial revenue and furnishes employment in out-of-season months. It also provides a way to utilize agricultural products more completely. As more fruits are mechanically harvested, graded, and sorted, bruising is increased, and a smaller percentage of choice product results. This increases the need for developing by-products to help finance a commercial processing operation. In many instances manufacture of specialty products is economically attractive even if the operation only breaks even because it eliminates the cost of disposing of otherwise waste materials such as peels, seed, cores, and windfall, immature, or overripe fruit.

The leading methods of processing fruits have shifted continuously during the centuries from sun-drying, sugar preserving, and chemical preserving to canning, artificial dehydration, concentrating, radiation, and freezing, and combinations of one or more of these methods; but as long as records have been kept, the percentage of fruits processed, compared with that eaten fresh, has steadily increased. This general trend has been due to one or more improvements in the cultivars used for processing; horticultural practices and production of fruits especially suitable for processing; mechanical harvesting of fruits for processing; techniques for processing large volumes of fruits; containers that extend the shelf-life of processed fruits; the standards and nutritional value of processed fruits; and the marketing of processed fruits, especially imported ones.

Some processing methods are more suitable for some fruits than for others, although some shifts have occurred over the years. For example, strawberries have not been extensively dehydrated or canned, but are the leading fruit for freezing; peaches and pineapples have remained leading canned items in the face of all other methods of processing. On the other hand, canned fruit juices have been available in moderate amounts for many years but have been largely superseded by frozen, concentrated products in recent years. Powdered fruits and fruit juices were very poor products until freeze-drying improved such products almost beyond expectations. Changes in cultivars, methods of handling, processing techniques, eating habits, packaging, transportation, and export demand will produce other shifts in commercial fruit processing in the future.

DEVELOPMENTS IN THERMAL PROCESSING

Advances in the science and technology of canning during the twentieth century have been mainly in three areas: applied bacteriology, processing technology, and processing equipment. Many improvements stem from research at the National Canners Association (NCA) laboratories by W. D. Bigelow and Catheart on the effect of acidity in lowering the processing requirements of canned foods.[1] Bigelow developed a series of processing studies that resulted in the publication of a number of NCA bulletins that helped canners to establish processing parameters for different sizes of cans and various types of foods and to thus plan their manufacturing operations on a sound technological basis. The first in the series was *Some Safety Measures in Canning Factories* by A. W. Bitting, which became the bible of the canning industry.

The rate of heat penetration has been the basis of applied bacteriology and the development of processing techniques. In 1917, Bigelow conceived the idea of using thermocouples to measure the temperature of the interior of pressure kettles and cans of food, and to determine the distribution of heat. This technique permitted a continuous measurement of the heat within a can inside the cooker, and was thus a great improvement over the previous instrument that measured only specific isolated temperatures. Today, thermocouple measurements have become the basis for all of our present-day knowledge of time-temperature processes for canning food (Goldblith 1972).

From the field of thermal bacteriology, which was largely developed through the auspices of the NCA, came the 12-D process. This refers to the combination of time and temperature necessary to reduce a colony of *Clostridium botulinum* spores from 10^{12} to 1. (This reduction of the decimal number 10 (the D) from an exponent of 12 is what gave rise to the name 12-D.)

Since the control of *C. botulinum* and other microorganisms was not as difficult a problem in acid fruits as it was in nonacid vegetables and meats, research emphasis was shifted to improving the retention of nutrients, chiefly vitamin C. This was done chiefly by developing techniques of uniformly cooking the product throughout the container. In this way overcooking near the walls of the container, in order to adequately heat the center, was avoided. Uniform heating was relatively easy to attain for canned or pickled fruits in light syrup that could be rolled, turned end-over-end, or vibrated during the heating to cause the liquid to "circulate"; but for purees, butters, preserves, and very viscous mixtures this was not possible. One way to maintain a relatively uniform cooking temperature of heavy products, it was found, is to use containers 4 in. or less in diameter; another method is to precook the product and can it aseptically.

Since about 1950 numerous commercial sterilizing methods and equip-

[1]Now known as the National Food Processors Association (NFPA), the National Canners Association was established in 1907, began supporting research in 1910, and organized an industrial laboratory in 1913. No organization in the United States has been as helpful to canners of all products as the NCA (Goldblith 1972).

ment have been developed for fruits. Vertical and horizontal "still" retorts have been replaced by "continuous" methods of sterilization; and practically all "hand" batch operations have become "mechanized" around-the-clock operations. These new designs possess one or more of the following advantages: improved product quality; higher capacity per plant surface unit area; continuous operation; less variation in product quality; fewer package failures; and lower expenditures for steam, cooling water, and labor per case of product. Among the disadvantages are relatively large capitalization costs, especially for smaller processors with limited production; short seasonal operations; and less sophisticated quality control equipment. Some of the newer techniques that are fast becoming common are described in the following sections (see Chapter 12).

Continuous Agitating Cookers

Agitating continuous retorts and water-bath cookers have become common for canning fruits. They permit a considerable reduction in processing times because agitation permits higher rates of heat penetration into the food. Agitation may be accomplished by rolling containers, turning containers end-over-end, or vibrating the product (similar to paint mixing) in containers.

Production costs with this type of equipment are reduced through savings in labor, steam, venting, cooling water, and plant space. More product uniformity and higher quality result from the shorter processing time and better control of time and temperature, both in heating and cooling. Furthermore, can damage and product losses are less.

Aseptic Packing

Aseptic canning was first used in commercial operations in 1950, in a system known as the Martin Aseptic Canning Process. The system is basically a high-temperature short-time (HTST) sterilizing process. It combines flash sterilization and cooling with aseptic methods of packaging for fluid and semifluid products, thus eliminating the retorting and subsequent cooling phases. This system consists of four separate operations carried out simultaneously in a closed interconnected apparatus using a mechanically synchronized process: (1) sterilization of the product under appropriate quick heating, holding, and cooling; (2) sterilization of the containers and covers with superheated steam; (3) aseptic filling of the cooled, sterile product into the sterile containers; and (4) aseptic sealing of the containers with sterile covers.

The heat exchange process used with aseptic canning is suitable for packing preserves and jams and requires that the liquid or semiliquid product be pumped through the heating section of the sterilizer where the food is quickly brought to the desired temperature. Holding the product at this temperature for a predetermined period assures complete sterilization. A cooling system completes the processing, which is automatically regulated by a controller-recorder. Process time is controlled by the rate of flow of the product through the system.

The use of chemicals (e.g., hydrogen peroxide), instead of heat, for presterilization of containers offers the possibility of aseptic packaging in paper and foil containers. The FDA has approved use of hydrogen peroxide for presterilization of containers for aseptic packaging. Data from the FDA Food Technology Engineering Center in Cincinnati, Ohio, has provided information on potential critical points in this technique. For example, when hydrogen peroxide is sprayed into a container from a vertical nozzle, shutoff droplets may drip into the filled container, resulting in a residual hydrogen peroxide concentration higher than the FDA's legal tolerance of 0.10 ppm. One solution to this problem would be to position the nozzle so that the shutoff droplet does not fall into the container.

An even more important container for aseptic packing may be the retort pouch. The retort pouch has been under development for many years, but the need for high-speed filling and processing equipment has slowed the commercialization of these products. The FDA contributed to the delay because of concerns about migration of the adhesive component used in the laminations. The National Cancer Institute has identified 2,4-toluenediamine, a component of the urethane-type adhesive used in early pouches, as a carcinogen, and it has recently been confirmed that this chemical migrates out of pouch adhesives. Fortunately, newer types of pouches do not need this type of adhesive and can meet FDA safety requirements.

The inclusion of air in a retort pouch has been shown to have a significant effect on heating rates and sterilizing F_0 values. For example, sterilizing F_0 values as high as 9.7 can be obtained with good removal of air (approximately 5 ml of air remaining in a 5-lb pouch); but if as much as 250 ml of air are present in a pouch, the F_0 value would drop to 2.5, which is not sufficient for commercial sterilization. The FDA is also concerned about the integrity of these retort pouches, and various methods for testing this have been investigated (Schaffner 1982).

Aseptic Drum Processing

Pumpable fruit products, such as purees and concentrates of peaches, apricots, pears, and berries of a wide variety of densities are commonly packed in 55-gal containers in aseptic drum processing systems. This process typically involves sterilizing a food product with pressurized steam to 60 psig at approximately 300°F, cooling under 20–26 in. of vacuum; filling under vacuum; and sealing aseptically.Sterilization takes place in a heat exchanger as the product flows through. The temperature varies from 212° to 300°F depending upon the pH, viscosity, and sensitivity of the product.

The containers are sterilized inside and out with 100 psig steam and filled under vacuum with 1-1/4-in. headspace. An interlock control system prevents sterile product from entering the container until the sterilizing cycle is completed. While it is not necessary for the product to be held under refrigeration, storage life is greatly expanded by refrigerated storage.

In comparison with other packing methods, aseptic drum processing of purées and concentrates has several advantages: reduced handling and transport costs; reduced storage space; improved quality; lower labor costs in filling, opening, and emptying containers; and salvage value of the larger containers after use.

Hydrostatic Cookers

In hydrostatic sterilization, developed in Europe about 1950 and now used in the United States, steam pressure is maintained by water pressure. Hydrostatic cookers are made up of water chambers and steam chambers. The water temperature in the water chambers or legs ranges from about 60° to 260°F. The temperature of the steam in the steam chamber is controlled by the pressure produced by the water legs, and it can be regulated by moving the overflow level up or down in the down-traveling leg. Steam temperatures between 240° and 265°F are generally used. Several models of hydrostatic cooker are manufactured in this country and in Europe. Although the basic principle of all models is the same, their design and the details, such as water and steam temperatures, vary greatly.

Cans are conveyed through a hydrostatic cooker by means of carriers connected to heavy-duty chains which produce positive can travel control. As cans enter the down-traveling water leg, where the temperature is about 180°F, the can temperature begins to increase. As cans move down this leg, they encounter progressively hotter water. In the lower part of this leg the water temperature reaches 225°–245°F; then, near the water seal area next to the steam chamber, the water temperature becomes progressively closer to that of the steam. In the steam chamber cans are exposed to a temperature of 240°–265°F, the steam temperature being set to suit the product undergoing sterilization. In some hydrostatic cookers cans make two passes, one up and one down the steam chamber; other models are designed for four or six passes. After leaving the steam chamber, cans again go through a water seal into water at a temperature of 225°–245°F where the cooling cycle commences. Cans are then conveyed through progressively cooler water to the top of the up-traveling water leg where the water temperature is between 190° and 200°F. In other models the temperature of the water in the up-traveling leg is as low as 60°F and can cooling is completed within this leg.

An important advantage of hydrostatic cookers is the progressive variation in temperature and pressure to which cans are exposed. Rapid changes in pressure, which may distort and weaken can seams, are eliminated. The amount of heat and water required for processing can be reduced by recirculating the cooling water in counterflow to the cans. The heat that has been put into the cans during sterilization is then transferred from the outgoing to the incoming cans, greatly decreasing steam requirements. Lower water consumption can also be significant in reducing waste disposal problems. Labor requirements with hydrostatic sterilization are low; and automatic control is possible, providing improved process uniformity. Some models enable a processor to run two can sizes or two products simultaneously. Retention time at sterilization temperature is about 10% less than with still retorts.

TRENDS IN PROCESSED FRUIT PRODUCTS

There was an abundance of fruits to choose from in colonial America. Native fruits included blackberries, cranberries, raspberries, gooseberies, elderberries, huckleberries, blueberries, strawberries, crab apples, wild cherries, fox grapes, muscadine grapes, persimmons, plums, figs, and others. These were processed by the Indians, English, Spanish, and French settlers according to their national background. To these were added introduced fruits such as apples, peaches, nectarines, apricots, Italian plums, Japanese persimmons, European grapes, citrus fruits, strawberries, quinces, and others.

Fermented beverages, including beer, wine, cider, brandy, cordials, and nectars, topped the list of early fruit products. These have since given way, in part, to fruit juices, punches, nectars, ades, and concentrates. The products are now canned, frozen, dried, powdered, concentrated, and fortified.

Drying whole, sliced, pureed, or mashed fruits was another early method of preservation. This was the forerunner of the present dried, dehydrated, dehydrofrozen, and granulated fruit industry.

A third method of preserving whole, sliced, or sectioned fruits was to soak them in honey or sugar syrup, or to boil the juice down until it was heavy syrup and treat the fruit with it. Sometimes the fruits or juices were mixed, and occasionally spices were added. This was the forerunner of the present methods of making preserves, jams, marmalades, glaze, jelly, butters, and sauces.

In Table 1.1 are listed the known commercial products made from fruits grown and processed in tropical and temperate areas of Central and South

Table 1.1. List of 52 Fruits and 36 of Their Commercial Products

	Processed fruit products																																			
Name of Fruit	Alcoholic cocktails	Brandy	Brined	Cakes	Canned	Cereals	Champagne	Chunks	Citric acid	Cocktail	Confections	Cookies	Crushed	Diet spread	Dried	Essence	Frozen	Glace	Jam	Jelly	Juice	Leather	Marmalade	Nectar	Pectin	Pickles	Pie	Pie filling	Preserves	Purée	Sauce	Snack bars	Strained purée	Syrup	Vinegar	Wine
Apple (*Pyrus malus*)	•	•		•	•	•	•	•	•		•	•	•		•	•	•	•		•	•				•		•	•		•	•	•			•	
Apricot (*Prunus armeniaca*)		•		•	•	•		•			•	•	•	•	•		•		•					•			•	•	•	•		•	•			
Avocado (*Persea americana*)													•																	•						
Banana																																				
(*Musa sapientum*)															•															•						
(*M. paradisiaca*)															•															•						
(*M. cavendeishii*)						•						•	•		•		•													•			•			
Baobab (*Actansonia*)													•														•			•						
Blackberries (*Rubus* alleghaniensis)	•	•		•	•					•		•	•			•			•	•	•			•			•	•	•	•		•	•	•		•
Boysenberries (*Rubus*)		•		•	•	•				•	•	•				•			•	•	•			•			•	•	•				•	•		•
Breadfruit (*Artocarpus altilis*)								•																						•						
Cherries																																				
sour (*Prunus cerasus*)			•	•	•				•	•	•	•		•			•		•	•	•			•			•	•	•	•						•
sweet (*P. avium*)			•	•	•					•	•	•					•							•			•	•	•	•						
Coconut (*Livistona chinensis*)	•	•		•	•			•			•	•	•		•		•				•			•			•			•		•	•	•		
Crabapple (*Malus pumila*)																				•					•	•										
Cranberries (*Vaccinium macrocarpum*)					•						•		•		•			•	•	•	•									•	•		•	•		
Currants																																				
(*Ribes vulgare*, rubrum)					•							•	•			•		•	•											•						
(*R. nigrum*)					•									•																•						
Elderberries (*Sambuca nigra*)	•	•															•			•	•			•						•						•
Figs (*Ficus carica*)				•	•	•				•	•	•	•	•	•		•	•				•							•	•		•				
Gooseberries (*Ribes hirtellum*)					•					•	•									•	•								•							•

Grapefruit (*Citrus paradise*, pommelo)					•			•			•		•				•	•			•		•													
Grapes																																				
(*Vitis labrusca*)	•						•						•	•					•		•										•				•	•
(*V. rotundifolia*)							•												•	•	•										•				•	•
(*V. vinifera*)	•				•	•	•			•					•		•		•	•	•														•	•
Guava (*Psidium guajava*)													•	•					•	•		•								•		•				
Honeydew melon (*Cucumic melo*)										•							•	•																		
Kiwifruit (*Acitidia chinensis*)					•					•			•		•																					
Kumquat (*Fortunella margarita*)																			•																	
Lemon (*Citrus limon*)	•								•				•			•	•	•			•		•		•											
Lime (*Citrus aurantifolia*)	•								•				•			•	•				•		•		•											
Longan (*Euphoria*)									•				•																							
Loquat (*Eriobotrya japonica*)										•																				•						
Lychee (*Litchi chinensis*)					•										•																					
Mango (*Mangifera indica*)													•																	•						
Nectarine (*Prunus nectarina*)					•					•	•	•	•	•			•		•											•						
Olive (*Olea europala*)			•		•																															
Orange (*Citrus qurantium*, sinensis)						•	•				•	•	•	•		•	•	•			•		•		•											•
Papaya (*Carica papaya*)													•								•	•								•		•				
Sapote (*Pouteria sapota*)													•								•									•						
Passion fruit (*Passiflora*)													•								•			•						•			•			
Peaches (*Prunus persica*)		•	•	•	•	•		•		•	•	•	•	•	•	•	•		•		•	•				•	•	•	•	•	•	•	•	•		•
Pear (*Pyrus communis*)		•			•	•		•		•	•	•	•	•	•		•																			
Persimmon																																				
(*Diospyros kaki*)								•					•																							•
(*D. virginiana*)		•											•						•		•		•		•		•	•	•					•		
Pineapple (*Ananas cormosus*)				•	•	•		•		•	•	•	•	•			•	•	•			•							•	•			•			
Plum (*Prunus domestica*)						•				•	•		•	•							•									•			•			
Pomegranate (*Punica granatum*)																			•		•	•								•		•	•			
Prune (*Prunus domestica*)				•	•	•				•	•	•	•	•	•				•	•					•				•	•					•	•
Quince (*Cydonia vulgaris*)																		•	•	•	•			•		•			•	•		•	•			•
Raspberry (*Rubus idueus* or *R. stigosus*)				•	•	•					•	•	•	•		•	•		•	•	•			•		•		•	•			•	•	•		
Strawberry (*Fragaria chiloensis*)		•		•		•					•	•	•				•				•															
Tangerine (*Citrus reticulata*)										•							•	•			•									•						
Tucuma (*Astrocaryum tucuma*)													•																							

America, and South Asia. Other specialty products made from fruits include butters; beverage bases; bread, muffin, and cookie mixes; frozen, canned, and dried chunks; breakfast and cocktail drinks; diet liquids and spreads; and by-products such as pectin, citric acid, ascorbic acid, colors, and flavors.

Worldwide, most types of fruit can be eaten raw with little preparation. Possibly half are graded and prepared for immediate consumption by hundreds of recipes; the choice of the crop is processed, packaged, and transported for year-round consumption. Every fruit-growing area has its own fruit types, cultivars, and favorite products. While most fruit products are used directly as food, others go into confections, bakery goods, cosmetic items, pharmaceuticals, diet foods, and other products. They provide a "natural" source of colors, flavors, vitamins, minerals, and texturizers to scores of food and nonfood products (see Chapter 10).

Apples[2]

Apples are the leading deciduous fruit in the United States. Although annual apple production fluctuates somewhat, it increased from about 150,000,000 (42 lb) units in 1965 to 184,000,000 units in 1981. About 80,450,000 units in 1965 and about 105,564,000 units in 1981 were consumed fresh, while about 60,000,000 and about 90,000,000 units, respectively, were processed. (Fruit not eaten fresh or processed are discarded as "culls.")

In 1970, the numbers of units made into different products were as follows: 27,190,000—canned; 23,810,000—juice; 4,188,000—dried; 5,094,000—frozen; and 3,540,000—other products. In 1981 the corresponding figures were 24,021,000; 42,674,000; 4,693,000; 4,112,000; and 2,173,000. These data show that between 1970 and 1981 the quantities of canned, dried, and frozen apples remained fairly constant; apple juice production increased by about 40%; and the quantity of apples going into other products dropped by about half.

From 1965 to 1981 the average number of units of apples processed annually in different areas was as follows: New York, 13,357,000; New England, 1,293,000; Virginia–Pennsylvania, 16,129,000; Michigan, 9,381,000; Washington, 13,092,000; and California, 9,214,000.

Peaches[2]

The quantity of peaches canned in the United States per year increased from 7,624,000 cases in 1923 to 34,744,000 cases in 1980 (see Chapter 16). Of this latter quantity, 32,624,000 cases were California clingstone, 1,832,000 were California freestone, and 288,000 were from other states.

[2]Data on apples obtained from International Apple Institute; on peaches from National Food Packers Association; and on frozen fruits from American Frozen Food Institute.

In addition to domestic retail sales, 1,196,000 cases were for the armed forces, 2,828,000 were for export, and 38,400 were for the Veterans Administration during 1980.

Frozen Fruits[2]

The annual frozen strawberry pack in the United States increased from 63,776,000 lb in 1942 to 312,293,000 lb in 1956, but declined to 210,558 lb in 1981. About 60% of frozen strawberries are sliced, of which 68% are packed in California, 28% in the Northwest, and 3.5% in the South and Midwest.

The frozen cherry pack increased from 41,820,000 lb in 1942 to 85,848,000 lb in 1981, when 70% of frozen cherries were packed in the Midwest, 17.3% in the West, and 11.3% in the Northwest. However, the year-to-year fluctuation in the frozen cherry pack is considerable. For example, in 1958, 86,195,000 lb were frozen; in 1964, 202,522,000 lb were frozen.

The frozen red raspberry pack was 14,206,000 lb in 1942 and 22,841,000 lb in 1981, a rather small change in comparison to other fruits. In 1981, 80% were frozen in bulk for later processing, 16.8% were frozen for retail trade, and 2.2% were frozen for foodservice use.

The frozen blueberry pack steadily increased from 1,716,000 lb in 1942 to 50,141,000 lb in 1981. In 1981, 21,372,000 lb were from the Northeast, 25,927,000 lb from the Midwest, and 2,842,000 lb from the West; 80% of the pack was frozen in 30 lb containers.

The frozen blackberry pack increased from 8,229,000 lb in 1942 to 26,970,000 lb in 1960, and reached a high of 29,186,000 lb in 1970. By 1981 the pack had fallen to 16,997,000 lb. In most years more than half of the pack is in barrels.

In 1960 the following quantities (in gallons) of frozen concentrated citrus juices were produced; orange—77,999,000; grapefruit—1,639,000; lemon—1,150,000; grapefruit/orange—284,000; lemonade—14,750,000; tangerine—324,000; and limeade—849,000. In 1980 frozen orange juice 174,537,000; grapefruit juice 20,763,000; tangerine juice 1,175,000; and lemonade 1,851,000.

Summary of Fruit Packs

Each country's agriculture produces fruits that are prepared, processed, and consumed according to local tastes, technology, and economy. These native fruits are always part of the country's staple diet, but the same fruits—processed for and consumed in another country—may be novelties (for example, the recently popular kiwifruit in the United States). Data on the packs of major fruits in specified countries of the world from 1974 to 1981 are presented in Table 1.2.

Table 1.2. Fruit Packs in Specified Countries by Commodity

Commodity and country	1974	1975	1976	1977	1978	1979	1980	1981
Apples & applesauce								
Canada	1,018	802	611	737	823	1,177	710	n.a.
Germany, West	1,406	1,548	1,419	1,251	2,090	1,910	1,985	1,985
South Africa	278	187	247	262	438	284	238	204
United States	18,730	10,705	10,763	12,679	13,651	15,155	12,284	10,300
Apricots								
Australia	553	647	423	602	338	640	341	545
Canada	42	49	49	34	46	47	66	31
Greece	1,427	1,301	1,599	2,186	500	1,250	1,665	2,500
South Africa	478	717	480	732	794	925	1,098	707
Spain	1,400	700	700	300	900	300	225	735
United States	1,987	4,421	2,387	2,269	2,127	2,887	2,994	1,208
Cherries								
France	558	337	408	242	516	570	495	520
Germany, West	1,066	1,058	1,258	786	1,183	1,383	1,730	1,460
Italy	297	254	210	210	165	230	375	415
Japan	398	256	613	630	551	669	450	380
United States								
red pitted	1,188	1,273	438	605	582	526	545	213
sweet	623	412	464	500	485	651	428	316
Fruit mixtures								
Argentina	637	607	588	539	284	318	132	118
Australia	1,369	1,659	1,070	1,039	1,374	1,366	1,511	1,732
France[a]	1,224	958	1,220	895	1,029	1,036	990	1,165
Italy	1,272	1,187	2,500	3,330	3,540	2,705	2,915	2,790
Japan	337	227	334	451	214	387	295	295
South Africa	1,819	1,643	1,728	1,974	1,744	2,130	2,454	1,935
United States	16,742	14,968	14,854	15,139	13,857	16,960	18,041	13,928

Peaches								
Argentina	1,764	1,837	3,919	3,184	1,702	1,960	1,911	1,370
Australia	2,519	3,739	2,879	2,656	2,313	2,589	2,645	3,405
Canada	375	326	256	232	151	192	238	142
Chile	421	460	381	394	461	805	563	612
France[a]	436	145	520	320	529	532	465	540
Greece	2,016	2,428	4,223	3,602	5,414	5,414	5,650	4,560
Italy	1,229	933	1,250	1,000	1,085	2,295	2,165	1,875
Japan	3,595	2,775	2,453	2,272	1,886	2,222	2,020	1,615
South Africa	6,105	6,250	5,890	5,443	4,750	5,347	6,443	4,158
Spain[a]	1,300	700	1,200	600	1,000	700	600	980
United States	32,431	29,196	24,983	29,414	21,529	25,031	26,841	21,763
Pears								
Australia	3,102	2,869	1,739	1,654	2,233	2,923	2,705	2,275
Canada	634	450	388	393	458	519	487	398
France[a]	487	383	346	150	358	650	555	665
Italy	2,330	1,993	2,130	1,295	1,315	2,130	2,250	2,580
Japan	181	125	143	122	54	55	43	42
South Africa	1,506	1,559	1,828	1,712	1,456	1,595	1,782	1,415
United States	10,692	9,776	11,518	9,614	9,026	10,568	10,928	9,700
Plums & prunes								
Canada	111	134	73[d]	62[d]	118[d]	91[d]	71[d]	57[d]
Germany, West	717	320	1,019	649	659	710	700	290
United Kingdom	1,165	762	801	742	921	789	495	378
United States	1,166	1,447	1,042	817	923	711	1,066	940
Pineapple								
Australia	1,822	1,402	1,923	1,379	1,591	2,064	1,587	n.a.
Ivory Coast	3,429	4,164	3,550	2,613	2,744	2,954	2,758	2,352
Kenya[b]	424	984	1,465	2,221	2,375	2,300	1,884	n.a.
Malaysia	2,971	2,156	2,333	2,446	2,379	2,289	2,096	1,752
Martinique[c]	389	455	503	333	245	376	393	n.a.

(continued)

Table 1.2. *(Continued)*

Commodity and country	1974	1975	1976	1977	1978	1979	1980	1981
Mexico	2,148	2,090	2,368	3,035	3,118	3,464	3,545	n.a.
Philippines	6,810	6,908	8,573	12,602	12,623	12,993	13,522	13,154
South Africa	2,702	2,559	2,477	2,402	2,652	2,693	2,666	2,888
Taiwan	2,179	1,663	1,136	908	1,265	1,500	1,817	226
Thailand	1,608	2,033	3,295	4,617	5,328	5,825	6,393	7,819
Hawaii	8,110	8,200	8,270	8,490	7,620	7,470	6,940	n.a.

Source: USDA, Foreign Agricultural Service Fruit and Vegetable Division Commodity Analysis Branch.

Note: The annual canned fruit packs in the Southern Hemisphere countries (Australia, South Africa, etc.) commence in November or December of each year and become seasonally heavy in the following January or February. For the purposes of this table such packs are aligned with the U.S. packing marketing season, which commences the following summer. For example, the Southern Hemisphere packs completed in early 1975 are shown under the column for 1975 (marketing season 1975–1976). Figures are ×1000 45-lb cases (equivalent to 24 #2-1/2 cans/case).

[a] Syrup pack only.

[b] Exports only.

[c] Imports into France.

[d] Plums only.

In Table 1.3 are data for imports of canned and frozen fruits into the United States; U.S. exports of canned and frozen fruits are summarized in Table 1.4.

CONTRACT PACKING

Contract processing and packing involves the processing and/or packing of a product for a company or individual who in turn sells the product under a brand name or private or generic label. This type of arrangement has been used in processing meats, vegetables, and a few fruits. Usually Company A contracts with Company B to pack a product in order to broaden its line of items on a trial basis, with no capital investment and with minimum risk and financial exposure.

Contract packing may be a means by which a new and inexperienced company can benefit from the expertise of old and successful companies. Company A may have contracts with more than one packer; also a large packer may pack for several distributor companies. By contracting to pack for other companies, an established firm can utilize existing equipment, experience, and expertise to the fullest without having to maintain a far-flung sales force. On the other hand, contract packaging permits a small or new firm to try out many areas of food processing without committing itself to long-term commitments and expenses, and enables a new firm to get acquainted with new processes such as aseptic packaging, freeze-drying or flexible retort packaging.

FRUIT PROCESSING IN LATIN AMERICA

In much of Latin America very few citrus fruits, cherries, plums, or small fruits including various kinds of berries are processed in any form. Frozen fruits are not served even in the most fashionable hotels and other eating places; and ice cream containing fruits is very rare. Candies and bakery products (cookies, cakes, pies, and pastries) seldom contain fruits; and fruit cakes and fruit rolls are not seen. Fresh fruit is widely served in season and is often unpeeled. Fruit processing in Argentina is fairly representative of that in other Latin American countries and is discussed here for that reason.

The equipment for handling and processing fruits in Argentina is very modern. Conveyors, washers, peelers, blanchers, concentrators, filling and closing machines, cookers, coolers, and labeling machines are equal to those used in the United States. However, labor-saving devices and automation are less developed than in the United States because one of the primary aims of Argentine industries is to give employment to local people.

Table 1.3. U.S. Imports of Canned and Frozen Fruits

	1978		1979		1980		1981	
	Volume	Value ($1000)	Volume	Value ($1000)	Volume	Value ($1000)	Volume	Value ($1000)
Fruits, prepared or preserved (1000 lb)								
Apples and sauce	7,783	1,013	2,018	578	2,733	725	3,696	906
Apricots	548	215	1,020	408	237	193	427	231
Berries, other, canned, frozen	10,941	3,545	3,780	1,923	5,951	3,041	3,714	1,887
Blueberries, frozen	11,365	8,813	4,464	3,284	6,803	3,869	6,204	3,768
Cherries								
in brine	154	104	1	1	35	28	626	403
canned or brandied	359	278	1,190	836	265	247	349	293
Oranges, Mandarin	97,653	34,670	75,205	30,163	85,886	35,826	87,042	38,725
Other citrus fruits	10,374	2,405	12,816	3,616	11,968	3,705	11,881	3,713
Figs	73	49	16	57	91	63	87	70
Grapes	46	22	48	41	189	49	389	75
Guavas	1,201	366	1,534	545	1,089	379	1,114	423

Olives (1000 gal)	18,160	76,377	14,673	74,898	14,689	77,904	15,036	76,986
Papayas	2,336	768	1,389	520	2,979	1,054	2,283	997
Peaches	84	33	560	180	465	95	502	127
Pears	50	31	46	26	32	22	34	31
Pineapples	387,256	88,506	457,000	108,000	472,624	116,302	435,011	118,729
Prunes, plums	1,954	1,712	1,885	2,024	1,976	2,690	1,661	2,642
Strawberries, frozen	97,649	22,038	112,160	34,408	83,472	26,812	60,123	25,225
Fruits, other	14,720	4,264	19,239	11,287	20,093	7,017	22,993	9,297
Fruit mixtures	18,899	4,758	16,418	5,099	22,857	7,299	14,386	5,131
Juices (1000 gal)								
Apple, pear	44,394	36,990	66,569	66,939	43,550	40,066	81,603	60,227
Grape	724	1,028	835	1,388	865	1,325	1,373	2,006
Lime	1,009	1,342	2,191	2,089	1,396	1,192	917	953
Orange concentrate	150,741	109,238	160,018	109,967	100,014	66,584	214,231	178,271
Citrus, other	305	624	2,124	2,456	2,951	4,132	11,393	16,774
Pineapple	24,658	12,642	33,992	20,081	39,775	25,988	37,329	26,211
Other fruit	3,763	8,286	2,618	6,663	2,014	4,318	3,640	7,387
Vegetable	121	270	404	624	423	590	240	509

Source: Bureau of the Census, U.S. Department of Commerce. Courtesy and cooperation of the National Food Processors Association.

Table 1.4. U.S. Exports of Canned and Frozen Fruits

	1978		1979		1980		1981	
	Volume (1000 gal)	Value ($1000)	Volume (1000 gal)	Value ($1000)	Volume (1000 gal)	Value ($1000)	Volume (1000 gal)	Value ($1000)
Fruit juices								
Not concentrated								
Grapefruit	3,907	6,049	3,803	6,576	4,061	7,460	3,650	7,412
Orange	9,203	16,971	9,263	19,382	10,030	19,633	9,080	19,193
Other citrus, citrus mix	2,261	2,905	1,901	3,135	2,001	3,143	1,824	2,717
Grape	1,304	2,151	1,829	3,854	2,785	3,590	1,906	3,119
Pineapple	2,203	3,427	1,810	3,386	2,864	4,827	2,866	6,242
Other	4,994	8,812	4,868	8,023	6,468	10,851	6,679	13,824
Total not concentrated	23,871	40,315	23,474	44,356	28,209	49,504	26,004	52,507
Frozen concentrate								
Grapefruit	4,537	9,886	5,727	14,804	12,496	20,292	14,238	20,873
Orange								
containers < 32 oz.	26,898	45,773	30,508	51,778	34,668	57,155	34,170	58,978
containers 32 oz–1 gal	3,815	7,915	5,115	9,581	7,012	10,424	8,725	13,906
containers > 1 gal	5,879	15,838	7,341	19,543	24,559	32,337	32,370	38,746
Other citrus	4,178	6,220	4,514	7,070	6,506	7,939	13,311	14,954
Grape	256	748	243	1,136	688	1,732	882	1,903
Other	1,802	3,659	2,407	4,537	4,404	6,405	4,961	6,962
Total frozen concentrate	47,366	90,309	55,856	108,447	90,332	136,284	108,658	156,322
Concentrate, not frozen (hot pack)								
Grapefruit	1,305	2,656	1,290	2,626	1,791	3,716	2,014	3,317
Orange	4,092	12,284	3,428	11,793	6,681	11,647	6,715	9,728
Other citrus, citrus mix	807	2,275	828	2,609	1,356	2,457	1,388	1,875
Grape	2,053	2,709	2,461	4,492	2,220	3,323	4,310	8,550
Pineapple	608	1,229	565	1,134	885	953	1,229	1,273
Other	2,380	5,598	3,260	4,839	8,865	9,969	12,339	12,711

Total concentrate, not frozen	11,245	26,751	11,832	27,493	21,798	32,064	27,994	37,454
Canned fruits								
Apricots	4,907	1,491	3,998	1,247	2,292	1,114	2,659	
Cherries								
maraschino	4,120	2,594	7,239	4,102	4,797	3,414	5,346	
sweet	4,830	2,781	6,429	3,282	3,718	2,339	4,357	
tart	11,126	6,597	5,160	3,400	14,763	7,234	2,886	
Berries	3,283	1,360	3,352	1,463	3,089	1,660	4,073	
Grapefruit	2,234	619	1,979	623	3,432	1,371	2,125	
Other citrus, citrus mix	20,572	5,002	2,995	867	2,113	824	2,674	
Olives								
canned	2,466	2,266	4,156	3,694	3,016	2,606	4,144	
other	1,374	704	1,577	797	1,187	742	974	
Peaches	188,112	44,831	117,996	35,714	126,829	42,408	116,953	
Pears	7,097	2,673	6,814	2,533	5,586	2,222	6,340	
Pineapple	31,022	10,115	25,400	8,461	20,240	6,929	23,534	
Other canned fruits	14,825	5,677	13,533	5,063	15,092	6,741	15,663	
Mixtures, noncitrus	95,423	31,107	102,732	37,584	118,910	46,624	110,305	
Total canned fruits	391,390	117,818	303,358	108,829	325,066	126,227	302,033	
Baby food fruits	12,313	5,440	9,736	4,985	11,175	5,249	16,966	
Other fruit pastes and pulps	14,971	4,539	15,959	5,912	18,924	7,338	24,168	
Jams, marmalade, and fruit butter	8,088	4,495	8,503	5,185	8,634	5,275	8,678	
Frozen fruits								
Blueberries	1,147	415	1,871	924	12,415	6,258	20,278	
Strawberries	9,535	3,586	5,159	2,443	4,364	2,037	6,561	
Other frozen berries	4,597	1,404	5,174	1,726	10,762	3,577	8,035	
Cherries								
sweet	5,276	2,381	4,510	2,361	5,411	2,787	9,091	
tart	1,735	848	1,487	914	993	415	1,654	
Other frozen fruits	2,891	1,155	11,793	3,012	6,624	2,506	8,040	
Total frozen fruits	26,182	9,789	29,993	11,381	40,569	17,581	53,663	

Source: Bureau of the Census, U.S. Department of Commerce. Courtesy of the National Food Processors Association.

Most fruit processing plants are large enough to provide separate buildings for manufacturing equipment; storage of sugar, salt, and other ingredients; storage of empty packages and containers; storage of finished products; and administrative offices.

These plants can types and cultivars of fruits that have successive ripening dates. For example, a plant may begin the season with several cultivars of plums, cherries, apricots, and operate on these for 30 days; then can peaches from early, midseason, and late ripening areas for 45 days; followed by pears from early and late ripening cultivars and areas for 30 days.

At the same time, fruits or portions of fruits that are not of canning quality are made into jams, marmalade, or jelly. Due largely to the high cost of electricity no fruits, except grapes for fruit cocktails, are refrigerated for the purpose of extending the processing season. By staggering ripening dates, a well-located plant will operate on fresh fruits 60–70 days per year. Many of the fruit processing plants also make tomato paste, concentrate, and catsup. A typical day's run might be 2000 bu of peaches, 1700 bu of tomatoes, and 500 bu of pears.

The same plant may brine sufficient cherries and store white seedless grapes for fruit cocktails, and dry apricot pits for oil.

Fruit processing in Latin America differs from that in the United States in the selection of the raw materials and labeling the finished product. In Argentina, for example, local, provincial, and national laws with respect to cultivars and quality of raw fruits for processing are almost nonexistent. Thus, the government sets no standards concerning fruits that are underripe, overripe, bruised, sour, molded, spotted with disease, hail damaged, or contaminated with dust, insect frass, or spray residue. Individual commercial companies set their own specifications and grades for products and standards for operation; the consuming public is the ultimate judge of quality control.

In contrast to processed food products made in the United States, the labels on Argentine processed fruits and other food products do not carry a list of the ingredients. While the labels are beautifully illustrated pictorially, the kind, cultivar, and grade of fruit used and the kind and amount of additives are not indicated. This simplified labeling is a barrier to exportation of Argentine food products into the United States and certain other countries.

Tree Fruits

Tree fruits in Argentina are grown almost exclusively in three provinces where river water is available for irrigation. The great plains (or pampas) are too dry for fruit growing.

Although a wide variety of tree fruits are grown along the east slopes of the Andes Mountains, where melting snow provides water in the form of hundreds of small streams most of the year, the larger fruit-growing and processing units are along the larger rivers. The demand for water from the rivers for irrigation is so great that most of the rivers are "used up" within a hundred miles of the mountains. Year by year, fruit growers and processors are forced to use more underground water from wells; and, year by year, 8-in. pipes are being sunk deeper into the ground to obtain nonbrackish water.

River water was ample a few years ago, and water from 100-ft wells was used only as a supplement in the event of light snow on the mountains. Now, most growers and processors use both river water and well water; and new developments use water almost exclusively from a series of 700- to 800-ft wells.

The Buenos Aires Province, along the 3-mi-wide Panara River has sufficient rainfall for most fruits without irrigation. In this almost subtropical region bananas, pineapples, citrus fruits, strawberries, nectarines, plums, and early peaches are grown for the vast Buenos Aires fresh fruit market. Efforts are being made to produce yellow-flesh, clingstone peaches on the higher elevations for canning.

The southern fruit-growing area is the Rio Negro Province and is watered by the Rio Negro river from snow on the Andes Mountains. Large quantities of apples, pears, cherries, plums, and a few peaches are grown here, largely for the fresh export market. Efforts are being made to increase the production of yellow-flesh, clingstone peaches in this area for canning.

In the Mendoza area, along the Mendoza and other rivers, and foothills of the Andes Mountains, are produced a large number of tree fruits.

The manufacture of jam and marmalade from apricots, peaches, pears, prunes, quince, grapes, and citrus fruits is the chief method of processing. These products are served extensively for "continental breakfast" and sweets on other occasions. Jelly is also made from apples, quinces, grapes, oranges, lemons, and other fruits, and used similarly as jam.

Fruits that are canned in the Mendoza Province include apricots, pears, peaches, prunes, strawberries, and cocktails (pears, peaches, cherries, white grapes, and apricots).

Dried fruits include prunes, apricots, figs, peaches, apples, and pears.

Brined fruits include cherries and olives. A large quantity of olive oil is manufactured in Argentina.

Grapes

Growing and processing grapes into wine is highly specialized in certain areas of Argentina. For example, 68% of the cultivated area in Mendoza Province is planted to 5,250,000 acres of grapes; of the 186,571,000 tons of grapes produced on the average from 1970 to 1975, 85% are made into 315,000,000 gal. of wine—40% red, 31% rose, 15% white, and 14% champagne and others. This province has approximately 2000 wineries and 1385 wine storage cellars. Several wineries employ more than 1100 people and make in excess of 5,200,000 gal. of wine. Reportedly, the biggest winery in the world, with an annual capacity of 200,000,000 liters, is located at Maipu, Mendoza. Much of this wine is transported in tank trucks to other cellars for bottling.

As a wine-producing country, Argentina is in fourth place. In 1970, Italy produced 65,237,000 hectoliters (100 liters); France, 65,120,000; Spain, 23,902,304; and Argentina, 19,623,783. Italy and France each produce about 23% of the world wine production; Spain, 8%; Argentina, 7%; Russia, 6% Rumania, 3%; and other countries together, the remaining 30%.

By-products of the wineries are brandy from the hulls, grape seed oil, alcohol, and tartaric acid. The Mendoza area, for example, produces $170,000,000 of tartaric acid annually (Anon. 1972).

REFERENCES

ANON. 1967. Fruits, non-citrus—Production, Use, Value. USDA Statistical Reporting Service Bull. 407.

ANON. 1972. International Annual Income. Industrialization of fruits in Mendoza 13. Dept. Fruit Industries, Mendoza, Argentina. (Spanish).

GOLDBLITH, S. A. 1972. Science and technology of thermal processing. Part 2. Food Technol. *26*(1) 64–69.

SCHAFFNER, R. M. 1982. The impact of technology assessment on food regulations. Food Technol. *36*(9) 82–84.

2

Harvesting, Handling, and Holding Fruit

S. E. Prussia and J. G. Woodroof

The quality of processed fruit products depends on the quality of the incoming fruit, which in turn depends in large part on how the fruit is harvested, handled, and stored. Although processors can use fruit that is unacceptable as a fresh product, the costs of sorting, trimming, and related operations must be considered. Thus, an understanding of the operations necessary to deliver quality fruit to a processing plant is important to maintaining economic processing operations.

The purpose of this chapter is to provide sufficient background on harvesting, handling, and holding equipment and methods to help identify steps in the system that affect fruit quality. Representative practices, unique approaches, and methods with potential are emphasized. More comprehensive information on specific operations is available in the books by O'Brien *et al.* (1983), Ryall and Pentzer (1982), Sarig (1984) and the other literature cited.

HARVESTING FRUIT

Because the quality of fruit delivered to a processing plant depends largely on factors related to harvesting, an understanding of harvest principles can help decision makers plan operations that result in the highest

Commercial Fruit Processing, 2nd Edition

ISBN 0-87055-502-2

fruit quality possible within specified economic constraints. Research and development efforts around the world are resulting in methods to improve productivity and efficiency for both manually and machine-harvested fruit.

Manual harvesting continues to be the predominant method of detaching and delivering fruit to a centralized container. Relatively new scientific approaches are focusing on methods to improve human performance. Several crops can now be successfully harvested mechanically by mass-removal techniques, but each crop presents special problems that must be overcome to allow mechanical harvest. A common problem is excessive fruit damage caused by impacts with tree limbs and other fruits. Many of the current research approaches to minimizing fruit damage require significant changes in production practices.

Manual Harvest

Most of the world's production of fruit is harvested by hand. The primary advantage of manual harvest is the versatility and adaptability of humans. Despite phenomenal technological advances, it will be many years before computer-controlled robots can equal the ability of humans to detect and select fruit based on color, maturity, defects, size, shape, and other physical and biological factors. Even optimistic estimates (Orrock and Fisher 1983) indicate it will be several years before robots can economically harvest oranges, the crop most suitable for robot picking of individual fruits. Consequently, most mechanized harvest techniques rely on mass removal of fruit, which usually results in more damage than is commonly found with manually harvested fruit.

Manual harvesting of fruit involves low overhead (fixed costs) compared with the high expenses associated with mechanical harvesting. When workers are available, harvest capacity can be increased with little or no increase in per-unit costs. Likewise, when yields are low, growers are not burdened with the costs of unused or underutilized equipment, which is a problem with mechanical harvesting. For these reasons, manual harvest methods are preferable to mechanical methods for growers with low production volumes. A major disadvantage of manual harvest is the limited output of workers. In addition, worker output has experienced a 30% decline in the past 30 years. Worker output is affected by human endurance, skill level, physical strength and speed, environmental conditions, fruit accessibility, fruit density, and worker incentive. Even when optimum conditions are present, the output possible with manual harvesting is far less than is common with mechanical-harvesting equipment.

Another disadvantage of manual harvest is variability in worker performance. Worker output measured in terms of harvest rate and product quality varies from crew to crew, person to person, and even hour to hour for a specific individual. Financial incentives to increase output require time-consumming accounting procedures and can force time-saving short-cuts that increase damage. Unnoticeable variability in product damage is especially critical since most fruits do not exhibit evidence of damage for several hours or days.

Unpredictable availability of trained workers can be a problem for both immediate needs and long-term planning. Worker availability is influenced by government policies for social programs, immigration, and finances. The number of workers available for a crop can be sharply reduced when other crops are mechanized to the point that migrant workers find it is no longer advantageous to remain in the area at harvest time.

The costs for manual harvest are increasing and in some circumstances may make an operation economically unsound. Costs for improved benefits and services and compliance with new government regulations often are more important than wage increases.

Human Factors Principles. Human factors principles are now being applied to many fruit-harvesting situations. Human factors is a rapidly expanding discipline that formally addresses the relationships between people and technology (Kantowitz and Sorkin 1983). In Europe the discipline is called ergonomics. Ergonomics focuses on how work affects people with emphasis on ways to reduce fatigue, while human factors addresses the man–machine interface with emphasis on designs that reduce the potential for human error (Eastman Kodak Company 1983). A bibliography of references on the subject is available (Wilkinson and Baker 1984).

From the human factors viewpoint, manual harvest operations can be divided into four operations: *detecting* fruit suitable for harvest, *directing* the hand to its location, *detaching* it with specific motions, and *delivering* the fruit to a container (Fig. 2.1). Also addressed in this section are endurance and strength considerations when harvesting and handling fruit.

Detecting Fruit. Visual detection of fruit for harvest is primarily a visual search effort. Detection of ripe fruit requires rapid searching of leaves, limbs, immature fruit, and other materials. Search time is less when selection is based on color than when it is based on size or shape. Search theory also indicates that search times decrease as luminance from the object increases.

Visual detection of fruit is affected by individual differences in acuity, color sensitivity, and perception of size and distance. There are several types of visual acuity; however, most deal with the resolution of black and white detail. Visual targets (eye charts) measure minimum separable

FIG. 2.1. The primary assets of workers are their ability to rapidly detect fruit suitable for harvest, direct their hand to it, detach it with complex movements, and deliver it to a container. The human factors discipline provides principles needed to help improve manual harvest.

acuity, that is, the smallest feature or smallest space between features that can be detected. Vernier acuity, stereoscopic acuity, and other acuity measurements also help evaluate a person's ability to sense visual stimuli.

Color is one of the primary characteristics used by workers in deciding which fruits to harvest. However, approximately 8% of the U.S. male population of European ancestry exhibits deficiencies in color vision, although less than 1% of women have color deficiencies. Color vision abnormalities usually involve red and green, the critical colors for distinguishing fruit ripeness. Even workers with normal vision cannot be expected to differentiate between more than two dozen colors on an absolute basis, with no opportunity for comparison. This means, for example, that only two or three hues of red can be distinguished by matching a mental standard. However, 100,000–300,000 different colors can be differentiated when comparing two objects side by side. Color detection by harvest workers can be improved by periodic reference to appropriate color standards or true-size color replicas of product ranging from unacceptable to perfect. Harvest operations under artificial lighting require careful attention to the ability of workers to discern colors.

Size and distance are perceived by combining visual cues. For example, a large object at great distance projects a smaller image on the eye's retina than a small object that is nearby. Good size discrimination helps workers estimate fruit maturity and distance perception helps them rapidly perform tasks such as ladder positioning, bin placing, and fruit grasping.

Normal stereoscopic vision provides two sources of information not available with monocular vision—disparity and convergence angle, which are primarily beneficial for perceiving nearby objects. Several monocular distance and size cues assist people with normal vision and provide the only input for people with vision in only one eye. Human factors and psychology textbooks provide details on the use of monocular cues such as interposition, movement parallax, linear perspective, and size consistency.

Visual discrimination of absolute sizes is rather poor (five to seven identifiable sizes). For example, a 20% difference in fruit size is required for distinction on an absolute basis. As with color, however, a much larger number of sizes can be distinguished on a relative basis. When fruit size is important, periodic referral to a simple go-nogo template could minimize the unnecessary handling of fruit later rejected because of incorrect size.

Directing Hand Movement. Movement of the hand to the location of detected fruit requires speed and accuracy for a person to be an efficient worker. Both can be improved by applying principles involving response time, which is composed of reaction time and movement time.

Reaction time is proportional to the amount of information that is processed in making the decision to make a movement (Hick's law). Consequently, it is best for workers to have a clear mental picture of fruit suitable for harvest. Practice also reduces reaction time. Therefore, experienced workers are more valuable, especially when payment is by the hour.

Movement time depends on factors such as body member moved, distance traveled, direction of motion, accuracy required, and termination conditions. Harvest speeds are greatest when fruit is within easy reach of workers; harvesting rate is decreased when stooping, leaning, and other body maneuvers are required.

Direction of motion also influences movement times. A simple experiment shows that you can move your finger more quickly when tapping it than when moving it sideways. In general, arm movements that are primarily a pivoting of the elbow tend to take less time than those with a greater degree of upper-arm and shoulder action.

Movement accuracy is related to movement time (MT) by Fitt's law:

$$MT = a + b \log (2D/W)$$

where D is the distance from a starting point to a target that has width W. The terms a and b are constants. This equation shows that movement time increases as the target gets smaller (W decreases). Stated another way, it takes longer to direct your hand to a cherry than to an apple.

Termination conditions for movements to objects can be classified as (1) fixed location, (2) variable location, and (3) mixed with other objects. Reach times are considerably less for fixed objects. Nearly 20% more time is needed when visual control is necessary to direct movement to an object; even more time is required when the object is mixed with similar ones. Most workers that harvest fruit face the worst combination of termination conditions: item location varies at each site and the desired item is mixed with other items exhibiting a wide range of shapes, colors, and other characteristics.

Detaching Fruit. Detachment of the fruit after the hand is directed to it requires high levels of manual dexterity and tactile sensitivity. Human manipulative skills can be very complex. Fruit ranging in size from grapes to grapefruit can be quickly grasped by using two fingers or all five fingers; workers almost automatically compensate for the infinite number of shapes encountered. Then, movements in several degrees of freedom such as rotation, translation, and twisting are often combined in a smooth motion to detach the fruit. In comparison, a typical robot has about six joints that require complex control programs to provide only limited ranges of motion.

Tactual sensitivity plays a predominant role when detaching fruit. Finger tips are the most sensitive body surface for distinguishing pressure, and wearing gloves can reduce tactile sensitivity by 50%. Low temperatures also have a detrimental effect on touch. Workers who harvest fruit manually most likely rely on the sense of touch more than is generally realized.

Delivering Fruit. Many of the same motions discussed for directing the hand to fruit are required to deliver it to a container. Reaction time is probably delayed by the need to mentally process tactual signals before sending the neural signal to retract the hand. Visual and tactual information is obtained about fruit quality as it is delivered to the container. Rotation of the fruit gives the first input on the quality of the opposite side of the fruit. Occasionally, the item is studied for a slightly longer time than normally required to deliver it to the container. Subsequent selection decisions are often based on the feedback obtained.

Movement time is related to travel distance and accuracy as previously described by Fitt's law. Speed is greater for short distances and large openings in the container or bag. The container opening should be placed at a height that puts it close to the most fruits. The opening should also give the largest possible terminal position (width in Fitt's law). Container depth should allow hand-placing fruit on the bottom when necessary.

Endurance and Strength. The endurance of workers hand-picking fruit is primarily a function of their energy expenditure rate, total working time, and duration of rest periods. Moving a ladder requires nearly twice the energy as general picking; carrying a full bag also is energy intensive. Pickers are 25% more efficient when picking fruit that can be reached from the ground than when using ladders.

Energy expenditure should be limited to a daily maximum of 2000 kcal or 4.2 kcal/min for an 8-hr work day. McCormick and Sanders (1982) give the following formula for estimating the total amount of rest necessary to stay under a maximum daily energy expenditure:

$$\text{rest} = \frac{\text{total working time (avg kcal/min} - 4.2)}{\text{avg kcal/min} - 1.5}$$

Based on the above formula and 4.6-kcal/min energy expenditure, workers doing general harvesting should have slightly over 5 min of rest/hr.

Understanding human factors principles can help increase the total amount of work accomplished per unit of energy expenditure. Energy efficiency depends on how a load is carried. For example, carrying a double pack over the shoulder is far better than carrying the same mass in each hand. The principle is to minimize the number of muscle groups in use and to keep the load close to the body's center of gravity. Hand holds are very helpful when lifting boxes.

The maximum load that should be lifted on a continuing basis depends on age, sex, personal differences, and other considerations. There is wide variation in maximum recommended loads stated by various researchers and government agencies. Generally, 25 kg is considered an acceptable maximum for men 20–35 years old (highest strength range). The maximum recommended load decreases progressively at lower and higher ages and reaches 15 kg for persons 14–16 years old and those over 50. Exercise can increase strength from 30 to 50%. Women have about two-thirds the strength of men.

Improving Human Performance. Efforts to improve the daily output of workers have typically focused on one or more of the following areas: testing worker capability, worker positioners, fruit accessibility, equipment to aid workers, personal comfort, training, and financial incentives.

Testing. Substantial improvements in the efficiency of workers who harvest fruit manually can be realized through testing. Visual acuity and color tests are very simple to administer by employers. Standard field-administered tests could also be developed for other sensory abilities such as touch sensitivity and motion control, as well as strength and endurance. Based on such tests, employers could assign workers to those tasks they are most suited for. Such an approach would improve overall worker efficiency and probably contribute to worker safety by having workers avoid situations where their impaired condition may increase the possibilities of accidents.

Worker Positioners. Ladders and multiworker mobile platforms are examples of worker positioners, which are used to position a worker in a convenient and comfortable location relative to fruiting regions of the plant. Methods to elevate workers are needed for tree fruits; methods to lower workers are needed for berry crops.

Ladders with three legs, which provide stability on uneven terrain, are the most common worker positioners for tree fruits. Although ladders are quite versatile and inexpensive, from 20 to 30% of a worker's time is typically devoted to moving the ladder, climbing up or down the ladder, moving fruit to field containers, and other activities other than picking fruit. Since the primary asset of manual harvesting is the ability of workers to detect and detach fruit, there is considerable interest in developing both single and multiworker mobile positioners.

Single-worker mobile positioners for fruit trees are generally designed to substitute for ladders. Most early positioners were mounted on a self-propelled, three-wheel chassis. The operator usually had three-dimensional control of the platform. The most efficient machines positioned a boom rather than the chassis. With such equipment, worker productivity can be increased up to 20% compared with ladders.

Multiworker mobile positioners can theoretically improve the cost-to-benefit ratio by spreading machine cost over the output from more than one worker. Most multiworker positioners are on a four-wheel chassis designed to travel parallel to rows of trees. Operation in conventional plantings requires lateral movement mechanisms to enable access to the entire tree. An approach gaining in popularity uses high-density trees pruned in the shape of narrow hedgerows. A multilevel positioner travels along the row while workers pick the fruit accessible from their level. Conveyors and other materials-handling methods deliver the fruit from each worker to a centralized container.

Several machines have been developed to position workers near the ground for berry harvesting. A multiworker positioner developed for cucumbers provides head rests for workers in the prone position (Walton *et al.* 1983). The three-wheeled vehicle increased harvest rates up to 46% (Fig. 2.2).

Economic considerations explain why neither single-worker nor multiworker mobile positioners have displaced ladders on a large-scale basis. In general, the time saved by positioners does not compare favorably to the capital costs and operating costs of the machines. The need for additional management also reduces the attractiveness of positioners. The productivity of workers using positioners can be substantially less than

FIG. 2.2. Multiworker positioners relieve workers of materials-handling efforts and can provide comforts such as shade. (Courtesy L. R. Walton.)

expected due to factors such as low yields and the need to travel as slow as required by the slowest picker.

Fruit Accessibility. The use of dwarf trees in high-density tree-wall (hedgerow) plantings makes the fruit accessible to workers standing on the ground as an alternative to positioning manual harvesters near the fruit. Worker productivity can be increased by this approach with no investment for mechanical equipment. Elevating fruit that grows near the ground to an accessible height is more difficult than growing dwarf trees. However, it may be possible to raise standard growing beds enough to enhance the performance of machines developed to position seated workers at a convenient height between beds.

Equipment. Manual fruit detachment and handling aids have been developed for some fruits. An ergonomically designed tool for clipping orange stems increases worker output up to 8%. Detachment of strawberries is aided by a clipper blade attached to a finger. Other detachment aids include reach-extending poles to shake fruit loose, scoops with tines for combing through vines with small berries, and hand-held mechanical vibrators.

Handling aids range from picking bags to elaborate machines to convey and containerize fruit. Many types of picking bags are in widespread use for ladder and bag systems. Successful bags are light, durable, and inexpensive. Rigid pails are required for some delicate fruits. The first layer of fruit should be hand-placed to prevent damage which can be twice as much for fruit-to-pail impact as for fruit-on-fruit impact. Caution is also needed to prevent damage that results to fruit in the bag due to impact with the ladder as the worker descends.

In the Netherlands a machine to eliminate the use of bags by ground-level workers was developed. Workers in each row walk behind a hopper that guides the fruit to an inclined conveyor. The conveyor elevates the fruit to a cross conveyor that gently transfers it to a lowering device that places it in pallet bins. The bin-handling equipment and the conveyors are mounted on a power unit that travels in a row wider than the others.

Personal Comfort. Workers need comfortable conditions both for their well-being and to encourage high productivity. Long-term benefits result from attention to worker needs such as sanitation facilities, refreshment centers, and field transport equipment. Proper clothing is required to protect workers from low temperatures and sharp, toxic, or otherwise harmful plant materials. Heat stress when temperatures are high can be reduced in some cases by the use of shades and fans. All equipment should be well maintained and serviced to assure safe operation, high reliability, and proper function.

Training. Proper training of workers typically results in more production with less effort and the need for fewer workers and less management. Training programs should be designed to focus on specific steps important for the crop. For many crops, it is important to train workers in ways to reduce bruise damage, which is undetectable for several hours or days. Significant time can be saved by workers who plan ladder placement according to their optimum reach. Likewise, the ability to arrange for a full bag when at the bottom of the ladder is useful. Fruit selection depends on a clear understanding of acceptance standards. Proper removal technique is needed for some fruit crops to reduce removal force and prevent plugging and other damage. Careful placement of fruit in picking containers is more critical for some fruit than others. The proper procedure for dumping picking bags is to lower them to the bottom of the bin before releasing their contents. A good training program needs periodic followup and evaluations. Well-trained workers are needed to preserve the advantage manual harvest has in terms of high-quality fruit.

Financial Incentives. Well-planned financial incentive programs are designed to fit various crops and other circumstances. Hourly pay rates (by individual or group) allow the most flexibility when performing special harvesting but require supervision to maintain high levels of output. Piecework payment systems encourage high productivity but require records and the monitoring of quality. A common approach is to not pay for a unit (bag, bucket, container) of fruit unless it meets a minimum grade.

Group piecework payment systems are necessary when fruit is pooled from several workers (e.g., family members, organized crews, or multi-worker positioners with centralized containerization). Special payment incentives are often given for difficult working conditions such as extremely tall trees and for times when fruit is sparse.

Mechanical Harvest

Labor costs account for 30–60% of the total production costs for most manually harvested fruit crops. Harvest mechanization can reduce labor costs by greatly increasing fruit removal rates, but may require additional preharvest and postharvest operations besides the substantial equipment cost. Successful machine harvest of several crops has required an interdisciplinary research effort involving engineers, plant scientists, food scientists, physiologists, pathologists, and economists.

The initial stimulus for developing mechanical harvesters was the shortage of labor brought about by World War II. Early mechanization efforts focused on the mechanical design of equipment to remove fruit from trees, vines, and bushes that had shapes suitable for manual harvesting. It was quickly realized that changes in cultural practices were needed to facilitate the development of efficient mechanical harvesters. Analyses of harvest operations led to design criteria for both the harvester and the plants. New training and pruning practices were developed to accommodate mechanical harvesters. Maturity variability was reduced for some species through genetic developments and a better understanding of cultural influences on maturity.

Prunes, figs, dates, and tart cherries were among the first fruit crops to be harvested mostly by machines. For several years now commercial equipment has been available to harvest cranberries, blackberries, raspberries, and blueberries. Most harvesting machines depend on indirect detection and selection methods. Therefore, processors receive fruit with wider variations in maturity than is normal for hand-harvested fruit.

About 11% of the U.S. fruit production for processing was mechanically harvested in 1978 (ASAE 1983). Difficulty in maintaining product quality was identified as the major constraint that prevents more use of machines, although the high costs of mechanical harvesters have also impeded their adoption.

Production Methods for Improving Quality. Most production operations undertaken by fruit growers have the improvement of fruit quality as the goal. Realistic profits depend in part on the ability to market a crop with a large percentage of high-quality fruits. Genetic factors, planting and training methods, and pruning and thinning techniques help to improve fruit quality (Claypool 1983). Other factors that affect fruit quality include plant nutrition, pest control, and harvest considerations.

Genetic improvements aid in mechanical harvest through yield improvements and the uniformity in maturation needed for once-over harvesting. Genetic reduction of tree size has enabled the development of practical over-the-row harvesters. Dwarf apple trees have been the most successful.

Planting and training methods must be especially tailored for trees that are to be mechanically harvested. Conventional densities are 87–125 trees/ha. Ultrahigh densities fof 10,000 plants/ha have been demonstrated in "meadow orchard" systems (Alper *et al.* 1980). Several training systems have been developed to reduce the number and force of impacts as fruit falls through the canopy during mechanical shaking. Special designs include a butterfly shape (Diener *et al.* 1974), the Tatura Trellis, and the Lincoln Canopy. These thin-canopy approaches enable mechanical pruning and harvesting but tend to suffer from low yields. Freestanding trees also benefit from training to accommodate the operation of mechanical harvesters and to reduce the number of impacts.

Pruning and thinning improve crop yield, fruit quality, and maintenance of annual cropping. Mechanical pruning is practiced on peaches, citrus, apples, grapes, and other fruit crops. Fruit thinning is used to balance the number and distribution of fruits on a tree with the ability of the tree to produce fruit with proper size and maturity. The stone fruits and pome fruits are the main species thinned. Thinning methods include caustic sprays, mechanical shaking, and manual removal of blooms or fruit.

Proper levels of plant nutrients are necessary to give the potential for the high yields that are needed when mechanical harvesting equipment is employed. Soil and leaf tests can reveal nutrients that might be deficient. Annual applications of nitrogen are required to maintain adequate levels. Low levels may cause some fruit species to set small numbers of fruit buds for the next year's crop even if the current crop load is normal. Excessive levels of nitrogen can cause poor color, soft fruit, small sized fruit, irregular ripening of fruits on the same tree, and uneven ripening on a single fruit. Soil moisture levels need to be adequate for proper movement of nutrients.

Control of pests (insects, birds, animals, and weeds) is important when fruit crops are mechanically harvested. Pest-damaged crops often have poor yields, and the harvested fruit may be unacceptable to processors. Uncontrolled weeds can make harvesting difficult and increase the cost of separating undesirable material.

Harvest considerations that affect fruit quality include fruit maturity, harvest scheduling, and harvest method. A mature fruit is one that has developed to the point that it is considered ripe or has the potential to ripen after harvest. During ripening—the final phase of maturation—the fruit develops desirable flavor, texture, and aroma. Nonclimacteric fruit (blackberry, cherry, citrus, fig, grape, pineapple, strawberry, etc.) are best when ripened on the plant. Climacteric fruit (apple, apricot, peach, pear, plum, etc.) develop optimum quality when allowed to ripen after harvest.

Adoption of mechanical harvesting can be seriously limited by variations in fruit maturity. Because of inherent genetic factors, some cultivars mature over long periods or remain in an acceptable range of maturity for different time periods. Soil variability also can cause maturity variations even when fertilizer is applied uniformly. Fruits on major branches can vary in maturity due to uneven crop load. Differences in chronological age of fruits can result from weather conditions that cause an extended bloom period.

Variations in maturity can be reduced by modifying cultural practices. Crop control methods such as pruning and thinning help to give uniform crops. Applications of chemicals can improve uniformity and hasten maturity. Growth-regulating chemicals can also extend the harvest season and reduce the number of harvests needed. More hectares of a crop can be harvested per machine when control of maturity allows an extended harvest season.

Harvest timing is extremely important. A once-over harvest can be scheduled for some fruit crops. Due to variations in maturity, an early harvest results in reduced yields and high percentages of immature fruit. A late harvest creates problems for the processor when dealing with fruits of advanced maturity. Quantitative methods have been developed to estimate the optimum number of harvests and the interval between harvests (Topham and Mason 1981). A more uniform flow of material through a processing plant can be achieved by proper selection of cultivars and diverse geographical distribution of growing regions.

Economic Considerations. A primary reason for machine harvesting is to reduce production costs. Large savings are possible because hand harvesting commonly represents 30–60% of total production costs. Economic analyses to determine the advisability of converting from manual to machine harvest would be easier if both systems yielded fruit of equal quality. However, mechanical harvesting often requires special handling, sorting, grading, and processing methods, and the costs of these must be considered when evaluating the economics of the entire system. The costs for postharvest and processing changes are often reflected in lower prices for mechanically harvested fruit.

Manual or Machine Harvest Decisions. A grower who is evaluating harvest methods needs to compare the net income from hand harvest with estimated income from machine harvest. The risks associated with hand harvest, such as the unavailability of labor, can influence decisions, but machine systems are usually not adopted until there is a clear economic advantage.

Mechanically harvested crops are often less valuable on a per unit basis than are hand-harvested crops. This difference must be compensated for by greater yields and/or lower harvesting costs possible with machine harvest. Kepner *et al.* (1978) given an equation for calculating the "break-even crop-value ratio" (R) to help growers estimate the price a mechanically harvested crop must bring to break even with higher prices from manually harvested fruit:

$$R = 1 - \frac{H - M}{G}$$

where H is the total hand-harvesting costs in dollars/crop ha/year; M the total mechanical-harvesting costs in dollars/year; and G the gross income or crop value for hand-harvested crop in dollars/ha. The value for M includes machine fixed costs, machine operating costs, labor costs, and any postharvest costs not required for fruit harvested manually.

High yields and uniform crop distribution can significantly lower overhead costs per unit of production. Machine harvesting is best for high yields, whereas hand harvesting is superior for low yields. Methods to reduce mechanical-harvesting costs for small holdings include custom harvest, cooperative ownership, and adaptability of equipment to multiple crops. Difficult-to-analyze factors are technological obsolescence, potential for reduced life of mechanically harvested orchards, and possible increases in diseases and pests. A slight advantage for machine harvest is the possibility of favorable income tax regulations.

Impact on the Industry. Mechanization of harvest operations leads to changes throughout the fruit industry. An immediate change is often the need for increased processing capacity. This is due to the high harvest rates and reduced storage life of most fruits that are mechanically harvested. New operations may also be required to clean, sort, grade, cool, or otherwise prepare the product for processing.

The size and number of farms tend to change when a crop is mechanically harvested. Economics of size help make large-scale operations more efficient. As the size of farms increases, there is a decrease in the number of farms unless demand for the crop increases. Mechanical harvesting can also be expected to favor grower specialization in one or a combination of crops adaptable to a particular system.

Conversion to mechanical harvesting also can alter utilization patterns. High consumer prices for products made from crops difficult to mechanize causes a shift to substitute products with lower prices. Thus, a reduction of product selection can result.

Social and Political Considerations. Social and political developments substantially influence the availability and cost of manual labor. The seasonal labor needs for agricultural work have traditionally been filled by migratory workers or local workers who rely on several crops or nonfarm work to provide ample yearly income.

It is interesting to note that farm employment in California has remained relatively constant during the period when many crops were converted to mechanical harvest. Some laws and regulations encourage farm employers to have fewer workers for longer periods. A higher percentage of the workers need specific skills as higher technology production practices are adapted. Some labor developments such as unionization give incentive to use machines even when manual-harvesting costs are lower.

Mechanical Detachment Methods

Many devices have been developed to detach fruit mechanically. Commercially successful machines must remove a high percentage of the mature crop, have high harvest rates, cause minimum damage to fruits and tree, and be economically attractive. Economic mass removal has been achieved for several crops by use of direct contact devices or vibrational techniques. Many direct contact devices cause very little damage but are presently unacceptable because they have low harvest rates. Delayed evidence of bark damage is a possibility with methods that vibrate tree trunks or limbs. Also discussed in this section is the vibration of tree canopies with pulsating blasts of air.

Direct Contact Detachment. Many ingenious mechanical devices have been devised to remove fruit by direct contact of machine elements with the fruit or nearby supporting stems. However, development of new concepts and machines is slowed by high costs and limited markets. Detachment methods are usually limited to the specific crop for which they were developed and sometimes to a specific production region. The following discussion summarizes existing direct contact detachment methods and describes methods that could lead to improved detachment concepts.

Combing, raking, and stripping are commonly used terms to describe a method of detaching berries such as lowbush blueberries, cranberries, and strawberries. The basic concept in combing is to pass through the fruit-bearing plant with closely spaced tines that trap fruits while allowing vines, stems, leaves, and other material to remain on the plant. Clearance between tines is set to prevent passage of mature fruits between tines.

The tines can be stationary relative to the machine or mounted on rotating drums.

The combing concept has also been applied to apples grown on trees trained in a "fruit wall." An over-the-row harvester described by LeFlufy (1982) is based on two sets of harvesting arms that enter the bottom of the tree row and move upward through both sides of the trees as the machine travels continuously over the wall.

Clipping or cutting is a method of harvesting based on the detachment of the entire portion of the plant that contains fruit. The fruit is separated from unwanted material either on the harvester or at a processing plant. Separation efficiencies can be improved by mowing excess foliage ahead of the cutter. Strawberries and grapes are the two crops that have benefited most by the cutting methods. Meadow orchard harvesting is a novel approach that has been investigated for apples, peaches, and other crops. Hudson (1971) proposed that extremely high-density apple plantings (70,000 trees/ha) could be harvested by cutting the entire plant every other year. The harvester would then separate the fruit and shred the shoots before returning them to the ground. Thus, prunning and harvesting would be accomplished in one pass. Chemicals could cause fruit set on 1-year-old trees. Researchers in Israel have developed a harvester for meadow orchard systems (Alper *et al.* 1984).

Impacting or slapping the canopy is a successful mass-removal technique for many fruit crops, especially caneberries (bramble berry or bushberries), cranberries, and grapes. Detachment is based on direct impact with a berry or acceleration of nearby canes, stems, or supporting structures. Vibration by impact is distinguished from the trunk and limb vibration methods (discussed in the next section) by the proximity of the moving element to the fruit.

Caneberry and cranberry harvesters based on impact allow multiple harvest because the force required for detachment decreases as berries reach optimum maturity. Maturity selection based on detachment force is normally superior to that based on color, which can be nearly the same for berries with wide variations in maturity. Therefore, fruits detached by shaking have higher soluble solids and are more uniform in maturity than manually harvested berries.

Impacting frequency, amplitude, and duration are the main variables that affect performance of impact harvesters (Smith and Ramsay 1983). Poor removal results from low frequency, low applitude, and short duration shakers. Immature fruit, leaves, and other material are removed when impacters operate at excessive frequency, amplitude, or duration. Proper combinations of these variables depend on the type of impacter, plant cultivar, cultural practices, production season, crop maturity, and even time of day.

Machine elements that are used to detach fruit by impact include panels, rods, and fingers mounted on drums or wheels. Both panel and rod impacters cause lateral bush movement by contact near the surface of the canopy. Finger impacters on drums are designed to penetrate the bush to cause vibration throughout the fruit-bearing region. A method reviewed by Cargill and Kirk (1983) was a finger impacter developed for harvesting apples and raspberries grown on a horizontal Lincoln Canopy System. Two horizontal shafts, each with four vibrating finger wheels, engage the top of the canopy. Detached berries fall through the canopy to a collection conveyor. Cranberries harvested in flooded bogs can be detached with a "water reel" or "beater."

Harvesters for grapes are designed to detach the maximum amount of fruit possible in a once-over harvest. The detachment is a function of stroke length, frequency of vibration, orientation of vibration, and path of motion (vertical or elliptical). Frequency has been found to be less important than stroke. Chemicals are needed to prevent skin-fracturing at the point of detachment.

Commercial grape-harvesting machines can be classified into three categories: cane shakers, trunk shakes, and combined cane/truck shakers. With a cane shaker, bunches and individual grapes are removed by direct contact between the vibrating rods and the vine canopy. With trunk shakers, which have recently been introduced for grapes, bars in the shape of skiis vibrate back and forth in unison as they pass both sides of the trunk. Grapes are detached by movements transferred through the vine. Trunk shakers cause less damage to the grapes and remove fewer leaves than rod shakers. Commercial machines that combine cane and trunk shaking offer many advantages. The trunk shaker portion removes fruit from the region near the trellis stake, while the rod shaker detaches the other grapes.

Spinning, twisting, and pulling are direct contact methods that have been used to detach citrus fruits because mass removal by trunk and limb shaking is not yet suitable in most cases. A straight pull parallel to the axis of the fruit tends to remove part of the peel with the stem (plugging). Therefore, manual harvesters detach citrus with a combined rotation and sharp jerk at an angle to the fruit's major axis. Similar motions are not easily duplicated by machines. Poor fruit removal and low harvest rates have prevented commercial success for the devices.

Apples and peaches have been detached with a pushing device (Peterson 1982A). A bank of equally spaced rods is pushed through shallow fruiting canopies such as those typical of the Tatura and the Lincoln canopy trellis (Fig. 2.3). Padded tips on the rods apply sufficient force directly to the fruit to cause detachment. A simple cam mechanism on each rod releases when a preset spring force is exceeded due to contact of

FIG. 2.3. Shallow-canopy fruit trees harvested with an experimental direct contact "rod press fruit removal mechanism." (Courtesy D. L. Peterson.)

a rod with a non-yielding object. Clever design of the mechanism allows it to automatically reset when the unit is withdrawn from the tree. Fruit removal rates averaged 85% when the device was tested on 'Red' and 'Golden Delicious' fruit. The apples averaged 80% Fancy–Extra Fancy. Similar results have been obtained with peaches.

Vibrational Detachment. Vibration of tree limbs and trunks by shaking is probably the oldest and most common method to mechanically detach tree fruits. Vibrational detachment differs from the direct contact vibration devices by the need to transmit forces to more distant fruit. Fridley (1983) gives an excellent history and description of vibrational harvest methods and equipment. Cable shakers represented an early attempt to mechanize walnut harvest in the mid-1940s. Shaking frequency was limited by the time it took for the limb to return because only tension could be transmitted by the cable.

Boom shakers were commercially available by the mid-1950s. Replacement of the cable with a solid boom clamped to the tree allowed sustained vibrations at frequencies higher than the natural frequency of the tree. Amplitude of the shake was essentially the same as the stroke of the eccentric due to the rigid connection. Other boom shakers delivered an impulse by knocking the tree with a rubber pad. Engineering research was undertaken to determine empirical and theoretical relationships among stroke, frequency, fruit removal, power requirements, duration of shake, direction of shake, clamping method, and other parameters. This section discusses vibrational detachment from the standpoint of vibrational reactions of the fruit and tree, vibration transfer from the shaker to the tree, and vibration generation equipment.

Vibration Reactions. Efficient fruit detachment depends on reaction forces at the fruit–stem juncture and good force transmission through the tree. Fruit detached with low-frequency vibrations tend to have longer stems and higher levels of damage. At high frequencies the fruits tend to remain more stationary, which results in detachment at the natural abscission zone. Creation of the necessary vibration at the location of the fruit depends on good vibration transmission within the tree.

Vibration transmission varies depending on tree species, pruning practices, vibration characteristics, and other factors. Transmission is poor in limbs that grow upward and then curve downward and in limbs that are long and willowy. Transmissibility can be improved by pruning to give shorter and stiffer branches. Studies have verified the advantage of shaking at several frequencies.

Direction of vibration input is also important. A 1-sec vertical shake under limbs results in the removal of more fruit than a similar horizontal shake. Trunk shakers are limited to linear or multidirectional horizontal motions. Only a small portion of the limbs receive a longitudinal vibration from a linear shaker.

The nature of the vibration also influences fruit removal. Several short bursts are more effective than a long-duration shake. Short bursts also have the advantage of giving the fruit collection and removal equipment time to clear fruit to prevent fruit-on-fruit impacts. The nature of the vibration refers to the wave form of the input. Most shakers deliver a sinusoidal force. Results have been encouraging for impact and recoil-impact shakers. An advantage of impact shaking is that branches tend to move away from the fruit. Very little fruit movement takes place before the detached fruit falls straight down. Minimum damage is achieved when fruit is removed with several small impacts. Continuous shakers with wave forms other than sinusoidal shapes have not been investigated to any extent.

Vibration Transfer. The transfer of vibration is a critical link between the vibration source and the tree trunk or limb. Hydraulically actuated clamps are the main method used to transfer vibratory forces. Serious bark injury from a canker disease may result if the cambium layer is damaged. Because the obvious evidence of damage appears only later, careful operation and periodic bark inspection is imperative during harvest operations. Bark injury is usually caused by excessive tangential (longitudinal) stresses. These shear stresses can be minimized by positioning the shaker perpendicular to the limb or trunk and by clamp designs that deliver only radial force.

Radial stress on the bark results from clamping pressure and the transfer of vibrational forces. Clamping pressure must be set high enough to prevent any relative motion between the clamp and the tree during shaking, but less than allowable radial stresses at the cambium layer. The allowable stresses increase with tree age, decrease with turgidity, and increase from spring to fall.

Clamping action is intermittent when the shaker is manually directed to a limb or tree to be harvested. High-density plantings such as trees in hedge rows require automated or continuous clamping action. A continuously moving over-the-row shake-and-catch-type harvester described by Peterson (1982B) can automatically engage over 200 trees/hr. Tennes and Brown (1981) and van der Werken (1981) have also developed continuous shakers.

Vibration Generators. Mechanical devices create the vibratory force that is transferred to the tree where reaction forces cause fruit detachment. Three types of mechanisms generate the vibrations. Inertia-type tree-shakers are based on slider-crank and eccentric rotating mass devices. The long slender shape of the slider-crank unit makes it desirable for reaching over catching frames to shake limbs. Rotating mass units have found widespread use on trunk-shaking machines. Various multidirectional shake patterns are obtained by changing the relative angular velocities and masses of two rotating eccentrics.

Impact shakers deliver an impulsive force to the tree trunk or limb by causing a moving mass to strike a member in contact with the tree. Double-impact and recoil-impact devices have been developed to improve fruit removal. The double-impact unit delivers impacts from opposite sides of the tree which are timed to cause maximum trunk movement. Millier *et al.* (1983) describe a recoil-impact unit designed to pull the tree towards the shaker (recoil) then push it away (impact). Recoil force is initiated when a compressed spring is released to accelerate a mass that strikes an impact transfer tube to cause an impact to the tree. Sensors and controls allow either automatic or manual operation of this well-designed machine.

Air-Blast Vibration. For several years oscillating air blasts have been evaluated as a noncontact method to detach fruit through vibration of the foliage. Air velocities of 40–70 m/sec have been pulsated at 1.0–1.5 Hz by mechanical devices such as rotating louvers, oscillating louvers, and wobble plates placed in the air discharge. The high rate of harvest (176 trees/hr) attained by a conical-air-scan shaker developed in Florida overcame previous limitations from low harvest rates. A pulsating conical pattern is caused by a fixed-vane assembly that rotates at 1.2 Hz in front of the 1.4-m diameter axial fan. The conical pattern opens up the foliage rather than packing it toward the center of the tree. However, data available in 1984 indicate harvest costs for air-blast vibrations are nearly twice the costs of manual harvest (Whitney and Coppock 1984).

Mechanical Collection Methods

The advent of shaking as a viable fruit detachment method greatly increased the need for mechanical collection equipment. Neither direct hand pickup from the ground nor manually handled canvas catch sheets or frames provided adequate collection rates. Consequently, mechanical pickup methods were developed for fruit crops that could be allowed to fall on the ground. Interest in applying shaking methods to soft fruit crops required the development of catching equipment that cushioned impact and prevented contact with the ground. Results from continuing development efforts are improving the equipment for both pickup and catching collection methods.

Both pickup and catching collection methods have several advantages and disadvantages depending on the crop. Studies have shown that the capacity of limb shakers can be nearly doubled when shaking is independent of collection operations. Catching systems also limit a grower's flexibility by the need to mechanize both removal and collection at the same time. Pickup systems also have the ability to collect fruit that falls before harvest, maneuver easily under trees with limited skirt clearance, and operate efficiently in native stand plantations with irregular tree spacings. Excessive damage to soft fruits and the need for expensive preharvest operations limit the use of pickup systems to a few fruit species.

Advantages of the shake and catch method include the ability to prevent fruit contact with ground contamination, single-pass harvest, and reduced impact damage on specially designed catching frames. However, catching systems are expensive and in many cases require extensive pruning, training, and other cultural practices. Reduced yields are a disadvantage of some training systems. Fruit damage continues to be a problem for many shake and catch systems.

Ground Collection. Oranges and grapefruit are probably the most suitable fruit crops for collection by pickup machines. Similar machines have also been developed for figs, prunes, and apples. Collection by pickup requires proper orchard conditions and preharvest operations. The soil must be able to withstand smoothing operations without excessive water runoff, soil erosion, or reduction in water infiltration. A smooth surface is required to give high fruit recovery. Soil preparation also influences the amount of damage to the fruit. Recently tilled soil can cushion the impact of oranges enough to yield 95–99% undamaged fruit. Damage to prunes is minimized by soil that has been rolled smooth to eliminate clods. Pickup collection also requires preharvest removal of fallen limbs, sticks, and decayed fruit.

Sweeping Methods. A sweeping machine is commonly used to concentrate fruit at the tree skirt line or the row center. Concentration is needed to improve efficiency of the pickup machine (harvester) and to remove fruit located under low tree skirts that make direct pickup difficult or impossible. Oblique rakes are commercially available for citrus (Sumner and Churchill 1977). Cylindrical augers with rubber fingers are also used with citrus (Ruff and LePori 1983). Wheel sweeps and end sweeps are commonly made from horizontal brushes.

Pickup Methods. Fruit harvest by pickup involves elevating the fruit and placing it in containers. Pickup harvesters often include concentrating rakes to eliminate the need for a separate sweeping operation. A recently developed pickup machine uses a double-sided rake to collect citrus from two rows (Churchill and Hedden 1983).

Direct-loading and side-loading rod draper chain techniques have been successful with citrus mainly due to their high volume capability (1.3 kg/sec/m of width). Another benefit is that soil is separated from the fruit as the chain lifts from the ground.

Several pickup concepts are based on rotating reels or drums. A reel with forward rotation uses a drum with rubber fingers to push fruit onto a conveyor. Several commercial pickup machines have utilized the reverse-rotation reel concept. Rubber fingers on a reel move the fruit forward and upward to cross-conveyors.

Attempts to pick up fruit by vacuum have not been successful mainly due to excessive noise, dust, and power requirements. A vacuum pickup machine developed for oranges was limited by low capacity (0.54–0.82 ton/hr).

Shake and Catch Collection. Catching surfaces have emerged from over 40 years of research as the most reliable method of collecting most fruit crops removed by mechanical shaking. They have evolved from manually moved canvas surfaces to highly sophisticated catch frames that are mechanically activated and include collection and containerization equipment.

With current technology shake and catch harvesting is limited to fruits for processing and to fruits that can withstand some impact. Tree shakers and catching frames became practical for the harvest of soft fruits with the development of the inertia principle of tree shaking and the application of decelerator strips that minimize impact injury by slowing the fall of fruits. The primary advantage of inertia shakers is that the shaking forces are isolated from the carrying vehicle and therefore they can be mounted directly on catching frames. With shake and catch harvesting, some type of deceleration device is essential to minimize injury caused by fruit falling onto other fruits, particularly over conveyors.

Tree Fruit-Catching Frames. Diener and Fridley (1983) give an excellent description of past, present, and potential future catching equipment. Early extension machines worked by pulling out two canvas surfaces from a machine that moved straight down the aisle. Later machines had a powered roll-out to extend a catching frame for each side of the tree. Typical extension machines could harvest about 30 trees/hr.

Harvest rate was increased by machines that consisted of a powered catch frame on both sides of the tree. Extension and retraction time was eliminated because each half moved forward near the tree trunk line. Some machines have identical halves with a shaker, conveyor, and handling equipment. Other machines have the shaker on one half and the remainder of the equipment on the other half (Diener *et al.* 1982). Two-unit harvesters have been successful for both standard and dwarf trees. Overlapping flexible materials are commonly used to prevent the loss of fruit near the trunk and between the two machine halves.

Commercial machines with a catching surface that resembles an inverted umbrella or funnel are available. The machines are driven to the tree to be harvested and the shaker is clamped to the trunk. One design extends cantilevered arms that support the catching surface as it forms an umbrella shape under the tree. Another design uses ground wheels to support the frame as it extends.

Over-the-row machines eliminate the operational difficulties of two-unit harvesters. Much of the current research and development on tree fruit harvesting focuses on hedge row plantings of dwarf trees which are suitable for over-the-row harvesters. A wide variety of both single- and multi-level catching surfaces have been mounted on machines.

Peterson (1982B) developed a continuously moving over-the-row harvester for tree crops (Fig. 2.4). Fruit-catching surfaces consist of two inclined panels on the shaker side and an inclined panel and horizontal conveyor on the other side of the harvester. One of the inclined panels on the shaker side moves with the shaker mechanism as it sequences through the lateral and longitudinal motions needed for continuous machine movement. An open-cell urethane foam was used to make a tight seal around the trunk, but tests showed that a more durable material was needed.

FIG. 2.4. Continuously moving over-the-row shake and catch harvester can harvest over 200 trees per hr. (Courtesy D. L. Peterson.)

Padded rods were placed over the conveyor to reduce fruit-on-fruit impacts. The machine was designed so two men could easily reduce its width for road travel or prepare it for harvesting.

Most over-the-row cane and bushberry harvesters have a catching frame consisting of a series of overlapping plates. The hinged plates are individually spring-loaded such that plant stems or vines automatically open each plate the required amount as the machine moves forward. The plates deflect the berries to conveyors on each side of the harvester.

A unique system was developed for black currants that deflects the entire bush to a 45° angle. The detached berries then fall to a parallel collecting surface without the need for a seal near the base of the plants. Another alternative to reduce berry loss is a device that squeezes the bush bases between two belts that move at the same speed as the machine but in the opposite direction.

Cushioning Materials and Decelerators. Early efforts to reduce fruit damage addressed the construction of the catching surface. According to Diener and Fridley (1983) a well-designed catching surface should (1) prevent rebound by adsorbing most of the impact energy; (2) be soft enough to prevent bruises on soft fruit without danger of "bottoming";

(3) be stiff enough and flat enough to allow fruit to roll to the conveyors; and (4) be able to resist rubbing and abrasion. These authors state that the rebound height of an object such as a steel ball should not exceed 10% of the initial drop height. Quantitative measures of rebound, softness, and stiffness are also described. Backing materials with openings, such as wire mesh and perforated panels, have been found to be superior to solid sheets of material.

Many cushioning materials have been found satisfactory for catching surfaces. Primary catching surfaces are usually a taut fabric or a composite surface made up of cushioning material over plywood. Lightweight saran (220 g) has been effective for tart cherries. However, rebounding was a problem when the saran was stretched too tight. If tension was too loose, the fruits tended to create pockets that prevented rolling of the cherries to the conveyor. The problem of pocket formation is more severe with larger fruits. Neoprene-impregnated nylon has also been an effective material when stretched across catching frames.

Padding materials placed over solid surfaces are typically closed-cell or open-cell foams. Vinyl sheets or neoprene canvas are used to cover the foam to protect it from abrasion and from deterioration due to ultraviolet light. The covering tends to stiffen the surface and reduce penetration depth. Firm closed-cell foams are best for padding conveyors, sharp edges, and transfer points. Open-cell materials such as polyurethane and polyether (38–50 mm thick) are best for primary catching surfaces.

Decelerators are devices placed in the path of fruits to slow them down before they reach the catching surface. Decelerators are not as necessary over sloping catching surfaces as they are over conveyors. Strips made from nylon fabric have been arranged in two or three offset layers. Fruits strike a strip and roll to the next layer or the catching surface. The number of fruit-on-fruit impacts is reduced because the first fruits to reach the catching surface are protected from direct impact from fruits with high velocity. The need for attention to strip tension is eliminated by the use of circular tubes made from or covered with cushioning material.

Padded flaps are needed on conveyor sides to slow fruits rolling down inclined catching surfaces. Basically, padding is needed on any surface when there is relative velocity between the fruit and the surface.

Tree and Fruit Modifications. A major cause of fruit damage during shake and catch harvesting is the impact of fruits on limbs. Several pruning and training techniques have been developed or proposed that reduce the probability of such impacts.

An encouraging result from field studies of apple trees with a horizontal butterfly shape was that total production of fancy fruit was nearly as great as from conventionally shaped trees. The Lincoln Canopy system gives similar results. Reduced amounts of damage have been found when open center trees were harvested with conventional shake and catch equipment. Peaches harvested by a machine developed for trees trained on a Tatura trellis have shown less damage than hand-harvested fruits.

Sarig *et al.* (1976) modified the fruit to make it more bruise resistant. Apple trees were sprayed with a urea–formaldehyde to coat both the fruit and the limbs. Harvesting the foam-covered fruit with conventional equipment resulted in significant reductions in bruise damage. The use of foam could be successful if an easy method could be found to remove it from the fruit.

A similar approach, reported by Johnson *et al.* (1983), overcame the removal problem. A water-soluble dextrin adhesive was first sprayed on the fruit; then, preformed polystyrene particles were blown onto the fruit. A difficulty was the formation of imprint bruises under the particles.

Catching Surfaces in the Tree. Damage to fruit can be reduced by placing decelerators or catching surfaces near the point of detachment to reduce the frequency and velocity of impacts.

One development was called the plateau system because the tree was divided into two levels. One catching surface was placed in an open space at mid-height on the tree; the other catching surface was placed in the conventional location. Bruising was reduced to levels similar to those found with manual harvest.

The space-fill concept is a unique method to reduce the speed of fruit falling through the tree. Berlage and Langmo (1976) describe tests with an over-the-row system using low-density polyethylene spheres (75 mm diameter) that fill voids in the canopy before shaking is started. After an initial shake, the fruit and spheres were conveyed to a separation system. Undetached fruit was removed by additional shakes as the spheres were lowered. Decreases in damage to 'Golden Delicious' apples did not justify the 9- to 13-fold increase in harvest time.

Tines have been inserted into the tree canopy to both decelerate the fruit and to direct it out of the canopy. Fridley *et al.* (1975) developed a device with inflatable tapered tines made from rubber-covered nylon. Eight levels of tines directed fruit to conveyors on four levels. 'Golden Delicious' apples harvested with the device had 71% Extra Fancy compared to 88% for manual harvest.

Johnson *et al.* (1983) describe the use of insertable tines made from flat foam wheels attached to 2.5-cm conduit. The wheels had lobes that encouraged them to rotate when branches were contacted. The tines were arranged into surfaces at three levels. Apples harvested with this device compared favorably with hand-harvested fruit.

A unique harvester developed in the Netherlands (van de Werken 1981) has two levels of insertable tines on each side of a continuously moving over-the-row machine (Fig. 2.5). The innovative aspect of this machine is

FIG. 2.5. Two levels of tines are inserted into the tree canopy by this over-the-row harvester. The harvester is unique in its ability to keep the tines stationary relative to the trees while the machine moves continuously forward. (Courtesy J. van der Werken.)

that the inclined tines remain stationary relative to the tree as the harvester moves forward. Padded troughs on the tines deliver the fruit to conveyors when the tines retract. Two people could operate the machine at a speed of 13 cm/sec to give a capacity of 1 ha/day. Harvested fruit ranged from 60% Class 1 for 'Golden Delicious' to 93% Class 1 for 'Cox' apples.

HANDLING FRUIT

An understanding of the mechanical properties of fruit enables the design of conveyors, lowerators, containers, and other equipment that minimize bruising. Operations such as separating trash and sorting of fruit by quality increase possible holding periods and add value to the product. Bulk handling methods are rapidly being adopted to reduce costs and to enable increased mechanical-harvesting rates. Quality standards are available for some fruit crops. Bulk load-sampling equipment delivers randomly selected samples for quality evaluations. A key to retaining fruit quality is the analysis of the entire handling system using systems analysis techniques.

Fruit Properties

Mechanical damage to fruit during handling is the result of injury to individual cells and tissues. The amount of injury caused by a given load depends on the microstructure of the skin, cell size, cell wall thickness, and intercellular space. For example, thick-walled cells in the skin of apples and pears can remain free of discernible injury after transferring loads that damage softer tissues below the skin. Damage to fruit flesh is also related to differences in the strength of adjacent structures such as pits in stone fruits and cores in pome fruits. High turgor pressure makes fruit sensitive to damage.

Other variables that affect the resistance of fruit to bruising include maturity, temperature, and cultivar. The influence of these variables can depend on the method of loading (static, impact, or vibrational). In general, fruits become softer as they mature. For impact loading of apricots, peaches, and pears, the amount of impact energy needed to cause injury decreases as the fruit becomes softer. Consequently, allowable drop heights decrease.

Fruits such as strawberries, raspberries, blackberries, and cherries exhibit increased resistance to bruising as temperature is lowered. Under static conditions of loading, apples are more resistant to damage at higher temperatures than at lower ones. An explanation is that high temperatures produce less turgid cells that enable greater deformation without rupturing. In actual practice apples bruise more easily from impact when they are cold than when they are warm.

Some cultivars are more resistant to bruising than others. This difference offers an opportunity for plant breeders to develop new cultivars that can withstand mechanical forces imposed during harvesting, handling, and holding. Differences in bruise resistance also points out the need for special care with some cultivars. It is important for reports giving results of bruise tests to give a complete description of the test conditions including maturity, temperature, cultivar, and loading method.

Damage to fruit flesh can occur from either internal or external forces. Internal forces can be caused by changes in temperature, moisture content, chemical balances, or biological factors. Skin cracks in sweet cherries are an example of damage due to internal forces. However, the predominant source of damage to fruit is from either dynamic or static external forces.

Impact Damage. Impacts during the handling of fruit are one of the most common causes of bruises and result in large economic losses. In the apple processing industry, studies have shown that bruised flesh represents 2.8% of the total weight of the apples and manual removal of bruises accounts for 85% of the total materials and labor costs for apple slices.

Handpicked cherries normally lose 12–14% of their total mass when pitted, with about half of the loss attributed to pit mass. Mechanically harvested cherries lose 14–24% of their total mass due to bruises received during normal handling. Bruised cherries also result in a final product with less red color and a softer texture.

Impacts occur when the fruit is harvested (either manually or mechanically), during movement to containers, while in transit, upon container dumping, and during handling at the processing plant. Reductions in bruising would increase processing yields and permit longer storage periods for the raw material. Then, processing schedules could be more flexible and the need for extra equipment to handle peak flows could be reduced.

Impacts are defined as collisions with very short durations that produce stress waves which travel back and forth inside the fruit several times before being dissipated. The following discussion summarizes the Hertz theory and some other approaches for quantifying fruit reactions during impact

Hertz Theory of Contact Stresses. The collision of two spheres or the impact of a sphere on a thick plate can be described mathematically if the materials are hard (elastic) and the initial velocity is low. Theories that account for plastic deformation more accurately describe the collisions when materials are soft or initial velocity is high. Although fruit impacts do not meet the requirements of the Hertz law, extensive use has been made of the approach because calculation methods are relatively straightforward and correlations with experimental results have been good.

Using the theory to find the maximum shear stress shows that changes in fruit weight and drop height have less influence on maximum shear stress than the moduli of elasticity or the radii of the fruit and the impacting surface (Horsfield *et al.* 1972).

A drop height of only 2.5 cm can initiate damage when a peach impacts a very hard (steel) object with a 2.54-cm radius. Changing from the 2.54-cm radius to a flat plate increases allowable drop height to 28 cm. Padding on the 2.54 radius can give a drop height of 41 cm and the same padding on a flat surface allows a height of 457 cm. Changing to stiffer padding on a flat surface reduces allowable drop height to 140 cm. Therefore, impact damage is a serious problem during mechanical harvesting when fruit falls through a tree canopy with small diameter, hard limbs.

The depth of bruises below the skin of fruit depends on the radii of the contact surfaces and the modulus of the fruit. Small radii result in bruises near the surface. Large radii cause bruises well below the surface. Deep bruises are often undetected unless they are severe enough to cause surface damage. Fruits with a high modulus of elasticity (stiff) tend to bruise nearer the surface than fruits with a low modulus (soft).

Other Impact Theories. In addition to the application of the Hertz theory of contact, Mohsenin (1978) provides extensive coverage of other theories in a section on impact damage. Many alternative theories have been developed because the assumptions of the Hertz theory of contact do not hold for most applications. Some failure theories consider the fact that plastic yielding of tissues takes place in addition to elastic deformation. The size of permanent indentations can be measured to indicate the portion of the impact energy that gave plastic deformation.

Dynamic yield pressure (P_d) is defined as the impact pressure needed to cause plastic flow. Dynamic yield pressure is not necessarily the same as the static pressure required to cause plastic flow. Equations have been developed for predicting P_d based on the impact of a spherical indenter dropped on a fruit. Due to very high impulsive forces, extremely light impacts have been shown to start plastic deformations.

Viscoelastic impact theories account for variation in mechanical properties due to the rate of deformation. Equations to describe viscoelastic materials include parameters for both the viscous and elastic nature of fruit. Encouraging results have been published that indicate viscoelastic theory could lead to a generalized failure criterion.

Vibrational Damage. Nearly all fruit receives some damage due to vibrations during the journey from farm to processor. Serious damage from vibrations can occur on harvesters, transport vehicles, and conveyors. The top layer of fruit in a bulk container receive vibrational damage because certain combinations of amplitudes and frequencies of vertical vibrations cause bouncing of individual fruits. While weightless the fruit is free to rotate which causes subsequent impacts to occur over the entire surface. The resulting surface discoloration, removal of bloom, and cell wall fatigue is termed "roller bruising" by O'Brien and Gaffney (1983). Roller bruising can also affect the second, third, and deeper layers when vibration is severe. The amount of damage depends on fruit properties, vibrational characteristics (duration, amplitude, and frequency), and container design.

Vibrational Properties of Fruit. The response of fruit to vibrational inputs is influenced by cultivar, maturity, and other factors. Consequently, a vibrating table can cause one fruit to bounce while another one remains in contact with the surface. Vibrating systems have a specific frequency (natural frequency) that causes the greatest accelerations. An example is an unbalanced automobile wheel. At a specific speed the wheel and suspension system vibrate with greater force than at slightly lower or higher speeds. The unbalanced, rotating mass is delivering an input to the suspension of the system (spring and shock absorber) at the natural frequency of the system. Vibration does not result if the input force occurs when the spring is moving towards the wheel (out-of-phase).

The natural frequency of fruits can be measured or computed from theoretical equations. For example, whole apples placed on electromechanical vibrators have been used to determine natural frequencies of the fruit (Hz).

Vibrational Input Characteristics. Rotating equipment on mechanical harvesters, uneven road surfaces, and periodic displacements on conveyors can cause vibrations of individual fruits and fruit in containers. Vibrations of a truck bed result from random diplacements of the truck wheels that are caused by road roughness. The frequency of the bed vibrations tend to be near the natural frequency of the truck suspension system. Unfortunately, trucks with conventional leaf springs have natural frequencies in the same range as many fruits. Air-ride trucks have a lower natural frequency and cause less bruising.

Input energy is a parameter that correlates well with bruise volume (Schoorl and Holt 1982). Comparisons of acceleration data (a measure of vibration intensity) and bruise measurements at various levels in apple containers indicate that substantial forces can occur without measurable movement. An interesting result of other studies is that laboratory tests with single vertical columns of fruit predict bruising nearly as well as tests with full cartons.

Container Design. Container depth, cover arrangement, and other design features have been found to affect the amount of damage in fruit. The shallow depth of lug boxes means a higher percentage of the fruit experiences roller bruising than is experienced by the single top layer in a pallet box. Roller bruising can be reduced by covers that restrain the top layer. Better arrival quality is the objective of bulge packing of fresh market produce. Roller damage is reduced because the top layer is restrained by pressure from the cover. The increased pressure can also change the product's natural frequency to a point outside the natural frequency of typical trucks. However, bulge-packed fruit tends to settle enough during shipment to become free of the cover.

An alternative is tight-filling of containers. This is accomplished by random-filling followed by a brief vibration to settle the contents. The fruit is then held in place by pressure from the cover. A 10% reduction in the height of grape bunches has been successful in preventing roller damage without evidence of damage due to the increased pressure. It might be possible to incorporate similar approaches with pallet bins and bulk loads of fruit for processing.

Static Loads. Bulk transportation and storage has become an economic necessity in recent years. Distortions, cracks, internal bruises, and other injuries result from excessive pressures caused by increased depths of fruit. Major considerations in static loads are creep characteristics of fruit and the container dimensions.

Fruits subjected to static loads undergo viscoelastic deformation described by creep phenomenon. A basic creep model can be represented by a spring placed in parallel with a damper (Kelvin model). A constant load results in a gradual increase in deformation that reaches a maximum of S/E, where S is the constant stress and E is the elasticity of the fruit. The Burgers model provides better prediction of actual fruit behavior by adding two more elements—a spring above the Kelvin model and a damper below it. Values for the parameters of each element are found experimentally. Placing the values in an equation based on the Burgers model makes it possible to predict the amount of deformation that will occur for a specified load. Mohsenin (1978) gives extensive coverage of equation development, parameter determinations, and examples for determining maximum container depth.

Conveying and Lowering

A basic rule for minimizing damage to free-flowing fruit during movement is to reduce the magnitude and frequency of impacts. Properly designed and operated conveyors and lowerators are extremely important components in low-damage handling systems. Both conveyors and lowerators find many applications on mechanical harvesters and handling systems. It is also very important to provide adequate padding on impacted surfaces.

Conveying Fruit. Conveyors are a common method of moving fruit. Applications of conveyors range from the direct movement of fruit on harvesters to rollers for pallet bins. Conveyors often elevate or lower fruit in addition to providing horizontal movements. Damage can be reduced by attention to several design and operating principles.

Damage due to overloading and plugging can be avoided with properly sized conveyors that have a uniform input flow. Receiving devices should operate at higher capacity than discharge devices. Operating speeds should be minimized by using wide conveyors. Padding should be used on all surfaces that are impacted by fruit. Drop heights at the discharge must be minimized by conveyor placement or with appropriate lowering devices. Deflectors on receiving conveyors prevent fruit-on-fruit impacts by clearing a fruit-free surface on part of the conveyor. When possible, it is beneficial to direct oncoming fruit in the direction of conveyor movement. Live sides can be designed to prevent scraping of the fruit as it moves past stationary sides.

Other conveyor design and selection factors include crop contamination, fuel efficiency, capacity, reliability, and convenience. Energy use increases with increased capacity and height of elevation and with the square of conveyor velocity. Greater speed, width, and product depth give increased capacity. Several types of conveyors are discussed in the following paragraphs and are summarized in Table 2.1.

Belt conveyors consisting of an endless rubber belt placed around two pulleys are one of the simplest types of conveyors. A flat or trough shaped support is needed under all except very short belt conveyors. The support prevents deflection of the belt but limits capacity due to sliding friction. Some conveyors have rollers for supports.

Chain conveyors move materials with metal, stiff rubber, or other kinds of flights pulled along a support surface by chains. Apron conveyors are similar to belt conveyors except the carrying surface is self-supporting. Some belts are made from flexible metal mesh that engages sprockets directly. Mesh size can be designed to eliminate small fruit and trash.

Rod conveyors consist of rigid members placed between chains traveling around sprockets. The rods, bars, tubes, or slats are placed close together to form a moving apron surface. The spacing can be designed to eliminate unwanted small material.

Containers attached to single or double chains hold fruits conveyed by bucket conveyors. One method of loading the buckets is to lift them through a hopper of material; otherwise, the fruit is delivered by another conveyor.

The advantages of pneumatic conveyors include simplicity, compactness, flexibility of use, and the ability to separate light material from the fruit. High power requirements are a principle disadvantage.

Both flumes and pipes are used to hydraulically convey products such as apples, cherries, peaches, and other fruit that can be placed in water. The factors affecting open-flume conveying include slope of the bottom, cross-sectional shape, specific gravity of the produce, concentration of produce, and water velocity. Fungicides are needed in the water to control diseases and rot in the produce.

Diener *et al.* (1981) described a system developed for field grinding of juice apples. The slurry was pumped from the harvester-towed grinder to a tanker trailer (Fig. 2.6). The same pump was used to unload the tanker at the processing plant. Primary advantages of the system: a 56% reduction in volume, ability to handle apples at the same capacity as the harvester, and elimination of pallet bins, which cost approximately $6/year/bin.

Some fruit can be conveyed without damage on oscillatory (vibratory) conveyors. The conveyor consists of a pan that is forced to move with sufficient amplitude and frequency to make the fruit periodically airborne. While the material is airborne the pan is moved down and back to cause the fruit to land slightly forward of its previous position.

Cartons, pallet bins, and, in some cases, individual fruits can be moved by roller conveyors. Roller conveyors have a full width rotating rod or several small wheels on a common axle. Low-friction bearings enable

Table 2.1. Summary of Conveyor Types and Features

Type	Elevation angle	Capacity	Ease of cleaning	Gentleness	Special considerations	References
Belt	45° with flights	Medium	Low	High	Simple	Monroe and O'Brien 1983
Chain	45°	Medium	Medium	Low		Monroe and O'Brien 1983
Rod or apron	45°	Medium	Medium	Medium high	Separation ability	Monroe and O'Brien 1983
Bucket	Vertical	Low	Medium	Medium	Complex	Tennes 1981
Pneumatic	Vertical	Medium	High	Low	High-power, air lock needed	Christensen and Ellis 1984
Hydraulic pipes	Vertical	High	High	Medium		Diener *et al.* 1981
Flumes	Horizontal	High	High	High	Water treatment required	Tennes *et al.* 1978
Oscillating	Horizontal	Low	High	Low	Uniform flow	Sweetwood and Persson 1982
Roller	Horizontal	Medium	Medium	Medium	Containers	Monroe and O'Brien 1983

FIG. 2.6. Apple-grinding line on the conveyor half of the WVU tree fruit harvester includes left to right, cleaning conveyor and blower, sorting conveyor, brush washer, and hammermill. The tanker is shown in the rear of the photo. (Courtesy R. G. Diener.)

easy movement of the container when the conveyor is level. Balls in place of the wheels enable the container to turn or rotate. Many fruit harvesters have roller conveyors for moving pallet bins. Roller conveyors are also an efficient method to move containers at processing plants.

Lowering Fruit. Special equipment is necessary to prevent bruising when fruit has the potential to fall more than a few inches. Handling operations that benefit from lowering devices include container filling in the field and conveyor-to-conveyor transfers. The primary purpose of a lowering device or filler is to gently decelerate falling fruit to cause a uniform, low-velocity descent. A secondary purpose is to distribute the fruit evenly in containers or on conveyors. Low velocities are necessary to keep energy levels low at the time of impact. Extreme care is necessary during the initial fill phase because the fruit is impacting the hard container. The main filling phase can result in fruit damage due to initial fruit-on-fruit impact and from several impacts of decreasing magnitude as the fruits roll down the side of the pile. Secondary bruising is reduced by maintaining a level fruit surface as the container fills. Only one or two impacts are typical during the topping-off phase, which accounts for 5–15% of the total fill. An inflatable air bag has been tested for reducing damage in the bottom of bulk trucks (Diener *et al.* 1983).

An inclined shute is a very simple, passive lowerator that is effective for conveyor-to-conveyor transfer and container filling. Decelerator strips (described in the section on catching surfaces) can be considered lowerators. Similar strips have been used to decelerate fruit falling from conveyors into pallet bins on mechanical harvesters. Four of the eight filler designs analyzed by O'Brien *et al.* (1980) were passive. The most effective passive device ranked fourth overall. It consisted of baffles and a combined decelerator and spreader.

Many active lowerators are based on conveyor techniques. An apron conveyor can be designed to lower fruit into a container or onto another conveyor after an elevating leg is complete. The best methods are swing conveyors and cam-controlled devices that alternately clamp and release the fruit.

Drop height can be reduced from a conveyor to pallet bin by tilting the bin. As the bin fills, the angle is reduced and the bin is lowered. Other lowerators have been based on holding fruit between padded belt conveyors. A bin filler developed in the Netherlands used a belt conveyor with flexible fingers to lower fruit. The conveyor raised as the bin filled. Fruit was distributed by rotating the pallet bin. Others have distributed fruit in pallet bins with a horizontal disk with openings. Apples have been distributed in a truck with a movable deflector mounted on an overhead conveyor.

Separating, Singulating, Sorting, and Special Operations

Efficient separating, singulating, and sorting equipment is needed for both farm and processing plant operations. Crops that are harvested mechanically require high-capacity equipment because acceptable holding periods are often reduced due to bruising and the presence of over-mature fruit and high levels of trash.

Mechanical harvesters usually have equipment for separating much of the unwanted material. Sorting of desirable from undesirable fruit is commonly done at centralized facilities because controlled conditions are required. Singulating is often needed to improve sorting accuracy. Separating and sorting accuracy is judged by the amount of good product with the bad and the amount of bad material with the good.

Although widely practiced for both manually harvested and machine-harvested fruit, manual separating and sorting is a tedious, imperfect operation. However, attention to human factors principles can greatly improve manual operations. Manual and mechanical separating, singulating, or sorting operations are often combined to capitalize on the strenghts of each.

Separating Methods. Unwanted material such as sticks, trash, and overmature fruit occupies valuable space and is expensive to transport to the processing plant where it is a burden to dispose of properly. An equally important reason for early separation is the harm unwanted material can have on good fruit: punctures from sticks; contamination from bacteria, fungi, and soil; heat from respiring leaves; and trash caused reductions in airflow that is needed for cooling. Both manual and mechanical separation requires at least one physical parameter of the wanted material to differ from the unwanted material.

Workers continue to perform the primary separation of trash and other unwanted material for many crops. Manual separation is often a secondary operation when mechanical devices remove most of the unwanted material. Manual separation can be based on a large number of physical parameters such as size, shape, color, and light reflectance. When workers are separating trash on a harvester, special attention should be given to safety, vibration, postures needed, temperature, dust, lighting, etc.

A popular method of separating trash on mechanical harvesters is the use of high-velocity airflows. Separation depends on differences in terminal velocity (flow rate needed to suspend an object). The best separation for pressure systems occurs when the flow of material is in a thin layer such as occurs near the discharge of a conveyor or elevator. Another common approach is to blow air through a conveyor with openings in the apron.

Suction systems are designed as either open or closed devices. Open devices usually pull air through a narrow opening extending over the width of a porous conveyor. Airflow due to the suction causes light material to enter the nozzle while the fruit remains. The principle is similar to the operation of a household vacuum sweeper. Closed suction devices carry a mixture of materials in a pipe or duct that suddenly increases in size. The resulting reduction in airflow velocity allows the fruit to fall while light material is blown through the fan. The fruit is removed in batches or an airlock door arrangement enables nearly continuous removal.

A difference in the specific gravity (density) of trash and fruit also can be used to separate materials. Fruit that sinks in water is readily separated from trash that floats. The specific gravity of water can be adjusted by adding salt, alcohol, or other materials, when needed, to cause the desired objects to float. Fluidized beds of granular material also separate products by flotation. The density of the bed is adjusted by selection of material and the airflow rate that causes the material to float.

Many devices have been developed to separate unwanted material that is larger or smaller than the desired fruits. Apron conveyors with rods or mesh belts successfully separate fruit such as oranges from soil and other small material. In other cases, the fruit falls through the conveyor while the trash is conveyed to a discharge. Improved separation usually results when the conveyor is vibrated. Harrison and Blecha (1983) provide a detailed analysis of separation by oscillating screens.

Mechanical separation of unwanted materials has been accomplished by exploiting many physical parameters other than pneumatic drag, density, and size. Magnets are effective for separating iron and steel items such as bolts, nails, staples, and wire. Walnut shell pieces have been successfully separated from the meat by coating the uncracked shells with magnetic fluid, cracking the nuts, and using magnets to cause a change in trajectory of coated shell pieces. (Krishnan and Berlage 1983).

Rotating cylinders separate material by using screens or pockets. Appropriate size screens on the cylinder surface retain only the desired material. Pockets on the inside of other designs catch small items near the bottom of the cylinder and lift them to a conveyor running longitudinally inside the cylinder. Meanwhile, the trash travels along the bottom of the cylinder until it discharges at the end. In both designs cylinder rotation provides the agitation needed for good separation. Rotation combined with an inclined cylinder causes the material to flow.

An inclined belt can separate round objects such as berries, apples, and peaches from leaves, sticks, and other trash less likely to roll. The angle is adjusted to cause the fruit to roll down the incline while the relatively flat trash is conveyed to the top of the conveyor.

Singulating Methods. It is often necessary to present individual items to subsequent operations such as sorting, pitting, and peeling. Creating a uniform flow of items at regular intervals (singulating) is a difficult task when the incoming items are randomly distributed in a single layer or in three dimensions. Variations in size, shape, and other characteristics of fruit make the singulating operation especially difficult to perform mechanically. Manual singularization provides high levels of versatility but is tedious work, is not 100% accurate, and has limited capacity.

Furman and Henderson (1982) identify four subproblems that need to be addressed when developing a singulator: agitation, gating, pocket formation, and exit. A partial prototype machine they developed could singulate approximately 2000 times/min.

Monroe and O'Brien (1983) describe several singularizing devices that have been reported. An interesting method for converting grape clusters to individual berries uses intermeshed, counterrotating brushes. Clusters fed from the top are held back by one brush with relatively slow speed, while the other brush strips the berries from the stem. Adjustable variables include brush speed, speed ratio, brush overlap, and bristle stiffness.

Sorting Methods. Sorting is performed to segregate material into lots based on fruit characteristics such as maturity, color, weight, size, defects, and firmness. Sorting fruit enhances the value of the crop by providing a more uniform product. Sorting by maturity enables mature fruit to be processed immediately and less mature fruit to be held for later use. Removal of decayed and overmature fruit extends storage life of the good product by minimizing decay organisms and excessive ethylene gas. There are many benefits to sorting as soon as possible after harvest. Sorting methods can be categorized as manual, mechanical, or electronic.

Manual Sorting. Sight and touch are the primary senses that give workers unique sorting ability. Probably more important is the ability of humans to process visual and other inputs and make intelligent decisions. Inputs related to several fruit characteristics can be rapidly integrated with standards, instructions, and historical data when deciding if a fruit is acceptable or should be removed. However, it is commonly known that manual sorting is far from perfect.

The human factors discipline has made great strides in understanding and predicting human performance at tasks such as inspection (Clare and Sinclair 1979). Most of the results for inspection can be applied to sorting tasks. Previously, error rates in the range of 20–30% were simply categorized as being due to "human nature." Now efforts are underway to develop improved theoretical models of inspector performance and to better understand factors affecting performance.

Theoretical models have focused on visual search, decision making, vigilance, perception, and other fundamental elements of the inspection process. Factors affecting inspection performance include individual and group differences, lighting, object dynamics, and training to improve sensitivity and response strategies.

Presentation rate is one of the most important variables affecting inspection performance. The required exposure time per item is a function of variables such as number of characteristics checked, item size, item density on conveyor, and acceptable error rate. Additional exposure time should be allowed when items rotate during inspection. A practical way to determine maximum belt speed (minimum exposure time) is to increase the speed until output quality approaches unacceptable limits.

Caplan *et al.* (1983) give over a dozen guidelines for improving inspection performance. They indicate that a moving product is most accurately inspected when it moves towards the operator rather than from side to side. A uniform flow of items is also extremely important. Directing a separate flow of items to each inspector eliminates repeated inspections of the same items and makes it possible to establish responsibility for the output. Visual acuity and color tests should be conducted with actual fruit samples. Inspectors should have the ability to vary belt speed within a small range when needed. Initial and continuing training periods are important. Regular checks (audits) of the sorting operation output enable feedback to inspectors about errors or missed defects and aids in identifying training needs. The need for accurate sorting can be emphasized by periodically reminding inspectors about the impact of errors. A 5-min rest period every 30 min improves performance. Photographic or simulated fruit specimens are valuable aids for comparative evaluations. (Improvements of 100% can result when aids are used for borderline cases.)

Mechanical and Related Sorting Methods. Many mechanical sorting devices are based on fruit size or resilience. Apron conveyors have been used for many years to remove undersized fruit on harvesters and during later sorting operations. Some sizers use stationary, powered rollers with increasing distances between the rollers to both convey and provide multiple size segregations. The rotating rollers also help to separate individual items. Another sizer design uses an expanding pitch lead screw to increase the opening between rotating rollers. Rotating tapered rollers have also been used to create an increasing opening size for fruit moving parallel to the axes of the rollers. The rollers usually slope toward the discharge end to move the fruit. Diverging cord-type belt conveyors also provide an increasing opening size in the direction of fruit travel.

Important operating characteristics of sizers include sorting efficiency, capacity, gentle handling, and ability to easily change size fractions. Sizer efficiency is improved by providing a uniform, single layer of fruit that is evenly distributed.

Differences in the resilience of fruit have been used primarily to sort mature from immature fruit (e.g., blueberries, cranberries, coffee cherries, oranges, dates, and peaches). Resilience principles can be applied to sorting by causing the fruit to bounce on an inclined, flat plate or to be deflected by a rotating drum. The less resilient, mature fruit falls into containers or conveyors close to the line of initial fall due to absorption of energy during impact. In some cases the more resilient (firm) fruit that bounces to more distant containers is the desired product. Consistant orientation of nonspherical fruit is important to prevent large variation in bounce depending on location of impact on the fruit surface. Other considerations for bounce sorting include drop height, impact surface rigidity, angle of impact surface, and container or partition height and location.

Impact force analysis is a relatively new sorting method similar to bounce sorting. Individual fruits are impacted on a hard, flat surface supported by a force transducer. The voltage output is analyzed in both the time and frequency domain. McDonald and Delwiche (1983) found a linear relationship between peach firmness and impact characteristics; however, significant differences existed depending on location of impact (cheek, suture, or top).

A related sorting concept is based on vibrational characteristics of fruit. Resilient fruit bounces over the side of a conveyor inclined perpendicular to the direction of travel when a vibrating plate is placed under part of the belt. Improvements in the technique are needed to overcome differences in resilience at various locations on the fruit.

Resonance characteristics of fruit have been studied as a possible sorting method. Stephenson *et al.* (1979) found that hard, green peaches and firm, colored peaches had resonant bands of 1700–1800 and 500–600 Hz, respectively, and this difference could be used to separate them. However, soft, colored peaches could not be separated from firm, colored ones.

Fruit firmness has been used to sort soft blueberries from high-quality firm berries with a belt conveyor positioned at an angle normal to the direction of travel. Incoming material entered the upper edge of the conveyor at one end. Firm berries rapidly rolled down the incline as the belt moved. The slowly rolling soft fruit was deflected off the belt before reaching the lower portion of the conveyor.

Many mechanical designs for sorting fruit have been based on principles other than size and resilience. Pneumatic sorters are available to separate immature and ripe blueberries. Density sorting in liquids can also separate fruit into maturity classes. Brine or alcohol solutions can be used when a specific gravity other than 1.0 is needed. Vacuum pockets on a rotating drum have been used to sort dates based on moisture content. A soft retaining belt on the outside of a rotating drum has enabled the separation of cherries with stems from stemless cherries.

Optical and Electronic Sorting. Many sorting methods rely on electronic processing of signals from optical, mechanical, and other types of sensors. Electronic circuits can analyze sensor output and activate systems to separate fruit into several categories. Electronic sorting equipment allows rapid changes in acceptance criteria, enables detection of small differences, has high speed operation, allows noncontact sensing, and provides data for other purposes.

Optical sorting equipment can be based on light reflectance, transmission, or delayed emission (Birth 1983). Several companies market color sorters for products ranging from cherries to lemons. Over 1 MT of tomatoes/hr can be color-sorted by each of the 24 channels on mechanical tomato harvesters. Optical sorting by light reflectance is based on measurement of the percentage of incident light at a particular wavelength that is reflected from a fruit surface compared to the amount reflected from a reference standard. Increased sorting sensitivity is often obtained by forming a reflectance index such as the difference between the reflectance at two wavelengths divided by the reflectance at one of the wavelengths.

Instruments are available to measure color based on tristimulus colorimetry principles. The Hunter *L, a,* and *b* values are determined in a three-dimensional color reference system (see Chapter 11 for a more complete description). The *a* value has been shown to be the most sensitive to changes in ground color as peaches mature (Delwiche and Baumgardner 1983). Reflectance principles are also useful for detecting surface flaws, mold, bruises, and other defects.

Light transmittance sorters are based on the measurement of the intensity of light that is transmitted through a fruit at specific wavelengths. Separation is possible when there are differences in the optical density (OD) which is defined as $OD = \log_{10} (I_1/I_2)$ where I_1 is the incident radiant energy and I_2 is the radiant energy transmitted through the fruit. Light transmittance equipment can detect differences in color, water core in apples, scald damage in cherries, blueberry ripeness, and other characteristics.

Delayed-light emission equipment is based on measurement of light emitted from a sample after the source has been removed. Again, the intensity at specific wavelengths is important. Applications include measurement of color, chlorophyll content, anthocyanin content, maturity, and mold levels.

Digital cameras and image-processing capabilities make it possible to sort fruit based on silhouettes or analysis of a full image. Volume and shape can be determined by using a computer to analyze silhouettes taken from several angles. Slaughter and Rohrback (1983) used an electrooptical device to measure the radius of curvatures needed when calculating contact stresses in blueberries.

Cornell University researchers Taylor *et al.* (1984) detected apple bruises by using digital imaging systems similar to those developed for interplanetary explorations. The experimental system worked as well or better than human graders in detecting bruises on ‘Red Delicious’ and ‘McIntosh,’ but did not work as well on ‘Golden Delicious’.

A produce-sorting system developed in England combines human skills with electronic capabilities to grade nearly 3 MT of potatoes/hr (Carlow 1983). The operator places a hand-held wand over defective items moving by on a conveyor. Magnetic and electronic components allow the system to determine when an ejector mechanism should be activated to reject the identified item as it exits the conveyor. The sorting machine effectively relieves the inspector of the time-consuming chore of handling rejected items.

FIG. 2.7. Computer-based system sorts fruit by weighing each item. The computer enables collection and analysis of extensive data. (Courtesy D. E. Hammett, Durand-Wayland, Inc.)

Electronic weighing has become an important method of sorting fruit by weight (Fig. 2.7). Microcomputers control the weighing of individual fruits located in cups on a conveyor. The computer also enables the collection and storage of important data.

Materials Handling and Transportation

The type of container selected for a fruit crop largely determines the most suitable materials-handling and transportation equipment. The functions of a container are to protect the fruit and make handling more efficient. The three major types of fruit containers are lug boxes, pallet bins, and bulk trailers (gondolas). The best type of container and amount of mechanical handling equipment depends on the fruit species, quantity handled, yearly usage, and intended uses.

Lug-Boxes. Containers suitable for manual handling are commonly called lug boxes or hampers. Lug boxes hold approximately 22.7 kg of fruit and are usually constructed from either wood or plastic. Lug boxes are still in use for some fruit (primarily berries) because they provide good protection and can be handled without expensive equipment.

At harvest time a quantity of lug boxes are typically distributed to workers or placed on a mechanical harvester. When full they are stacked on a transport vehicle or placed on a pallet for later loading on a truck with a forklift. Handling stacks of lug boxes on pallets with a forklift lowers labor requirements and reduces the amount of bruising (Holt and Schoorl 1981).

Pallet Bins. A container much larger than a lug box is needed for efficient mechanical handling of fruit. A standard is available (ASAE 1984) that specifies one square size, one rectangular size, and two overall heights for agricultural bins.

Pallet bins are also called bulk bins, bin pallets, bulk boxes, pallet boxes, tote bins, and tote boxes. Empty bins weigh 45–70 kg and when filled weight up to 500 kg. Most have two-way entry pallets for use by standard forklift trucks. Openings can be placed in the floor and sides to aid in ventilation and cooling.

Initial container costs per ton of fruit favor bins compared to lug boxes by a ratio of 2 to 1. Bins also result in reduced hauling charges because they hold 33% more fruit per volume. Fruit handled in bins have less roller bruising than in lug boxes. Empty bins are distributed in the orchard where mechanical harvesters empty the fruits into bins or hand pickers empty their picking buckets or bags directly into bins. Filled bins are moved with tractor forklifts or special machines to loading areas. From there they are hauled to a processing plant by highway trucks.

At the plants, bins are handled with industrial forklift units. The bins are emptied by mechanical dumpers of three general types. In one type of dumper, bins are fed from a supply track or conveyor into a bin-inverting section, where they are held between an upper and lower conveyor belt. There each bin is slowly and completely inverted, and as it moves forward is slowly lifted off the fruit, which is being carried on the lower conveyor belt. In a second type of dumper, bins are successively pushed into a tip-up section under a padded cover and are tipped up with a turn of about 110°–120°; the fruit is then slowly fed onto a conveyor at the start of the processing line through a hinged door at the lower edge of the cover. A rotating frame can be attached to forklifts in a third type of dumper.

Bulk Trailers. Citrus, apples, peaches, cranberries, and other fruits are usually loaded in pallet bins in the field and then dumped into 2.4 ×

12.2 m bulk trailers. Some fruit is loaded directly from mechanical harvesters into trailers.

Special watertight gondolas have been developed for grapes. For some fruit a trailer is used that has standard sides with no top. Often the walls have openings to provide ventilation. Fruit is also transported in truck-sized tubs that are carried on flatbed trailers. A tank trailer similar to a bulk milk transport trailer has been used for crushed apples because handling pallet bins delayed mechanical harvest. Cherries in brine have also been transported in tank trailers.

Special unloading methods and equipment are needed for bulk truck loads of fruit. One method for dumping a bulk load is to hydraulically elevate the fifth wheel on a tractor to cause the trailer to tilt from front to back. Some trailers are emptied by parking them on a pad that slopes to the side; then hinged doors are opened along the length of the trailer. Similar results are obtained by hoisting one side of containers that are hinged on the opposite side. Bulk loads can also be dumped by a self-contained hydraulic system that tilts the bed (dump truck). In other cases the entire truck is tilted. Another unloading method uses a live bottom consisting of conveyor belts or chains.

Many unloading methods are assisted by using hoses to flush the product into water tanks that cushion the falling fruit. Dry dumping can be done if bruising is not severe. Damage can be reduced by mechanically helping the product to move in a uniform flow. It is also important to reduce the drop height by slowly increasing the tilt angle when using dump beds or tilting platforms.

Fruit Quality

One component of consumer satisfaction with a product is "conformance to expectation" which implies a certain correspondence between price and quality. To illustrate, the purchaser of an inexpensive can of generic peaches is satisfied with the quality even if the size and color are nonuniform; the same product sold as a brand name would be considered unsatisfactory. However, buyers have definite expectations about the minimum level of performance regardless of purchase price.

It is difficult to define the quality acceptable to consumers because buyer expectations depend on location in the food distribution chain, quality available in the past, possibility for substitute products, economic circumstances, cultural background, and other variable factors.

Establishment of quality standards and the definition of product grades help growers and processors define quality (see Chapter 13). Conformance to the requirements of a standard is verified through proper sampling and quality evaluation methods. Improving the quality of processed fruit products requires consideration of the entire production, harvesting, handling, and holding system.

Sampling Fruit. Sampling of fruit delivered by growers is standard practice for determining grades on which to establish prices. Mechanical harvesting and bulk handling of fruit introduced the need for an efficient method of obtaining representative samples from large containers. A bulk load fruit sampler should (1) maximize randomness, (2) minimize damage to sample and load, (3) need only one operator, (4) eliminate human judgment, and (5) accommodate both bins and trailer gondolas.

Samples of deciduous fruits (peaches, pears, prunes, apples) in bins are obtained by lowering a lid fitted with a compartment chamber onto the top of a bin of fruit and partially inverting it to permit the fruit above a preselected door to flow into a chamber. The sample, consisting of approximately 5% of each layer of fruit in the bin, is removed through the arcuate lid of the chamber and placed on the inspection table for grading.

In many fruit-growing regions the standard bin-sampling method is based on a conveyor belt with randomly located holes. Fruit from randomly selected bins is emptied onto the sampling belt and made to flow in a single layer. A portion of the fruit falls through when holes in the belt match holes in the table.

Wine grape samples are obtained from a 180 × 180 mm core of undisturbed fruit from top to bottom at a randomly selected location of the gondola. Fruit in this core is from different vines, rows, and pickers. Grapes that are damaged are excluded from evaluation. It requires 45 sec to sample a gondola loaded with 4000 kg of grapes. The sample weighs approximately 20 kg after the cut grapes have been discarded. Only 6 min are required for inspecting the sample.

It is extremely important to obtain a completely random sample. It is best to take several small samples rather than one large sample. Randomness can be achieved by having several selection guides that indicate the location of the sample. Later sample locations are identified by subsequent pages in the selection guides. Labor and equipment must be available to gain access to lug boxes and bins not on top. Sample location for bulk trailers is changed by moving the core sampler and/or the trailer.

Evaluating Quality. Accurate evaluation of fruit quality is important at all harvesting, handling, and holding points in the system. Decisions to harvest an individual fruit, a specific tree or bush, a certain orchard, or a growing region are based on one or more quality evaluations. During harvest, field measurements of quality provide the information needed for instructions to workers and adjustments to machines. Transportation and holding operations benefit from feedback from subsequent quality evaluations.

Accurate quality evaluations of fruit samples are extremely important because the disposition of an entire shipment often depends on a sample that is less than 1% of the load. The results of quality evaluations also provide processors with valuable data concerning maturity, storage requirements, blending possibilities, and other factors that help improve utilization of a shipment.

Evaluation methods range from direct visual examinations to sophisticated laboratory procedures. Fruit quality is often judged by the human senses of sight, touch, smell, and taste. Industry and government standards for quality rely heavily on the skills of trained graders and inspectors whose performance is enhanced through training, testing, clear standards, physical examples, and proper lighting.

Instrumental evaluations of quality provide quantitative data that may identify fruit properties undetectable by direct inspection. Many low-cost instruments with reasonable accuracy have been developed for use outside the laboratory. Dependable results from instruments and other inspection equipment require clear operating instructions, adequate training, careful handling, and continuing supervision. A periodic calibration and repair schedule is necessary to assure accurate measurements. Calibrations should be made with standards traceable to nationally recognized standards.

Fruit condition and maturity are two of the most important characteristics that are evaluated when a shipment is delivered to a processor.

The condition of fruit is judged primarily by visual examination. Fruit in good condition is fresh, free of decay, and at the proper stage of ripeness. Shrivelled fruit has poor condition due to loss of moisture.

Delivered fruit is also evaluated for defects. Excessive bruising and punctures reduce possible holding time and can cause reduced processing yields. Large amounts of trash also hasten deterioration and require extra cleaning effort. The value of a shipment is inversely proportional to the costs required to prepare it for processing.

Fruit maturity at the time of harvest is a major determinate of final product quality. Immature fruit ships well but may lack desired texture, sweetness, and other important characteristics. Fruit picked when fully ripe requires extra care in handling and may be overripe by the time it is processed. Several methods are available for evaluating maturity.

Manual inspection of fruit gives an indication of maturity through subjective evaluation of color, size, shape, firmness, and possibly aroma. The accuracy of maturity judgments depends on the type of fruit, examiner skill, ambient conditions, and a host of other factors.

Instruments offer the opportunity to make better maturity decisions. Accurate measurement of color is now possible with reflectance or transmittance spectrophotometers. Visual color assessment is improved by using standardized light sources.

Fruit firmness is a good indicator of maturity. Manual squeezing of a fruit is commonly practiced to estimate firmness. A more quantitative measure results when a penetrometer is used. Instruments such as the Magness-Taylor, UC Fruit Firmness, and Effegi pressure testers (penetrometers) indicate the maximum force required to insert a plunger a specified depth into the fruit flesh. A patch of skin must be removed before testing some fruits.

Fruit maturity can also be evaluated by chemical and electrical measurements. Iodine placed on a sliced apple reacts with starch not yet converted to sugar, and the intensity of the color indicates the amount of starch remaining. The maturity of melons, grapes, and citrus are commonly estimated by the level of soluble solids measured with a refractometer or hydrometer. The ratio of soluble solids to acid is a good measure of maturity for many fruits. Titratable acidity and pH are also useful measures.

Some methods with potential are based on electrical resistance or capacitance measurements, vibration, bounce and resonant frequency techniques, and viscoelastic properties.

Retaining Quality through Systems Analysis. An optimum system is not necessarily obtained by optimizing each component. Conflicting requirements often require decisions that are not obvious until the entire system is considered. Problem-solving techniques associated with systems analysis provide a scientific basis for analyzing problems and making management decisions. Early work in this discipline (1940–1960) was known as operations research. Systems analysis techniques were rapidly adopted by many industries in developed countries. However, the food industry has been slow to apply the approach. Many national and international organizations are encouraging expanded use of the systems approach to help reduce food losses through distribution systems. The study of a complete system usually requires a multidisciplinary team.

The purpose of systems analysis efforts is to either analyze, identify, or synthesize a system. Analysis applications allow prediction of future system behavior for specified inputs. Identification is the essence of scientific experimentation—a system is identified that consistantly converts controlled inputs into the outputs observed. Synthesis is the design of a system that produces the desired output from a specified set of inputs. After identifying a system, it is common to synthesize alternate systems that are analyzed for diverse input conditions.

Systems Approach. In the systems approach, components and activities are viewed as interrelated parts of a larger entity called the system. Conceptual boundaries are placed around the system to separate it from the rest of the world. Interactions of the system with the outside world (inputs and outputs) are identified and quantified.

An important benefit of a systems approach is the necessity to carefully select the components and activities that belong in the system, the ones that interact with it, and the infinite number that are unrelated. For example, a fruit transportation system could be defined as a shipment of strawberries, the containers holding them, and the insulated truck. Interactions across the system boundary could include inputs such as initial fruit quality, road roughness, and weather conditions. An important output is the condition of the berries at the destination. The number of telephone poles passed, a parade 100 mi away, or the yield per hectare in a nearby corn field are examples of unrelated components or activities in the world outside the system. Including road roughness as an input enables analysis of the effect of decreasing truck speed. Such a change might improve quality by reducing bruising from shocks and vibration, but final berry quality could be lowered due to the delay in delivery.

The components in a system can often be defined and analyzed as a separate system. The truck in the previous example could be studied as a system with components (subsystems) consisting of the suspension system, ventilation system, drive train system, etc. Likewise, the transportation system could be considered a component of a super-system defined as the handling system for strawberries that also included the harvester, holding facilities at the processor, and preprocessing operations.

Skill and experience are needed when defining the boundaries of the system to be studied. An overly complex model results when exhaustive details are included. However, an extremely simple model may not represent the true system with sufficient accuracy.

Models and Simulations. Models usually represent components of a system. Simulations typically address the interactions of a number of components in large systems. Models can be classified as either iconic, analog, or abstract. Iconic models look identical to the system they represent (e.g., scale model of aircraft in wind tunnels or processing plant piping layout). Analog models act like the systems they represent even though they may appear quite different (e.g., electrical models of mechanical systems, flowcharts, block diagrams). Abstract models are sets of mathematical relationships that mimic the behavior of the system (e.g., distance = velocity × time; new balance = old balance + old balance × interest rate).

Simulation provides a framework for studying the interactions of the components that comprise a system. Mathematical models for the components enable synthesis of a new system by simply rearranging components or changing the parameters of the models.

Simulations of very simple systems can be performed by manual calculations. However, most simulations are done on mainframe digital computers. An exciting development is the availability of simulation programs for personal computers. Pratt (1984) lists 19 simulation packages available for a range of personal computers. He also lists nine international professional societies and associations devoted to computer simulation.

Large systems often contain both discrete and continuous components that are interrelated. For example, bulk loads of fruit could be undergoing continuous changes in temperature while waiting in an unloading queue. At least one computer simulation package is available to handle both types of models as part of one system (Pritsker and Pegden 1979).

Systems Analysis Techniques. Flow charts and block diagrams are useful techniques (tools) when using the systems approach. The construction of a chart or diagram helps clarify the boundaries of the system. Identifying the flows from one component to the next provides the opportunity to consider alternative paths. Recent advances in strawberry mechanization resulted from taking a systems approach (Hansen *et al.* 1984). The key to success was the concurrent development of the solid-set culture, the concept of mowing the crop, and the specialized processing equipment necessary for decapping the fruit.

Researchers at the University of Georgia have enhanced their ability to use a systems approach by developing a mobile research laboratory (Fig. 2.8). A multidisciplinary team travels with shipments of products from the farm to the processing plant. On-site quality measurements are made on samples collected at major handling points. The mobile lab also facilitates data collection and measurements needed to construct flow charts and simulation models.

Bender *et al.* (1976) discuss how some systems analysis techniques can be applied to the food industry. They give several examples that use linear programming to optimize the allocation of limited resources among competing demands. An example with apples showed that harvest should be delayed as long as possible, apple slices should be made first, and that sauce production is not profitable until after a certain harvest data.

Linear programming also is a popular mathematical technique for optimizing transportation problems involving production centers, storage locations, and consumption points. Integer programming is a technique similar to linear programming but with special provisions for discrete problems.

Dynamic programming is a mathematical technique that provides a systematic procedure for choosing the best policy (strategy) when the

FIG. 2.8. This mobile research laboratory (a) assists a multidisciplinary team when making the on-site quality evaluations (b) needed for a systems approach to postharvest handling.

system can be represented in several stages with more than one state (condition) possible at each stage (Davis and McKeown 1984).

Response surface techniques are useful for finding the optimum combination of variables in a process. The technique is illustrated by developing a three-dimensional surface for taste panel scores as a function of salt and sucrose levels in cured ham.

Techniques developed for production control include methods for assigning workers to tasks, selecting the optimum number of grading belt inspectors, machine sequencing, and optimizing machine performance. Other techniques help determine the quantity and quality of raw material to purchase in order to meet budgeted requirements for the finished product of a given quality. Scheduling and planning of the individual activities that comprise a complete project can be improved by using PERT (Program Evaluation and Review Technique) or CPM (Critical Path Method).

Game theory provides a framework for making decisions when there is an opponent (competitor, weather, etc.). Possible strategies are analyzed for minimum and maximum gain or loss based on a payoff matrix.

Bond graphs (Rosenberg and Karnopp 1983) are especially useful approach for representing a physical system in abstract terms useful in models and simulations. Bond graphs provide a step-by-step approach for converting a dynamic system into a diagram that gives insight into the system. It is also possible to use bond graphs to write equations for the system or to provide the input needed by a special computer program (ENPORT).

HOLDING FRUIT

Fruit for processing is normally held the minimum period possible. Unavoidable holding occurs while fruit is in harvesting containers, during transport, and when unloading is delayed at the processing plant. Deliberate holding is practiced in some cases to provide a uniform flow of material or to allow the fruit to mellow. Holding periods can be extended by cooling the fruit. Conventional or controlled-atmosphere storage facilities are used when long holding periods are planned. Aseptic bulk storage of semiprocessed grapes and other fruits has been successful.

Fruit Ripening

An important step in commercial fruit processing is the harvest of fruit at the proper stage of development. In order to avoid bruising during handling, most fruits are harvested before they reach optimum flavor, color, and nutritional qualities. Practically all fruits, except bananas, pears, and some apples, are harvested in the "firm-ripe" stage. At this time they have attained full shape and size, but are still firm enough to withstand reasonable handling without bruising. For several hours to a few weeks the fruit mellows, as evidenced by intensifying ground color, increasing flavors and aroma, and loss of crisp texture. This is the "mellow-ripe" stage when many fruits (including peaches, apricots, berries, grapes, plums, and kiwifruits) are in prime condition for eating fresh. Beyond this point catabolic changes result in a loss of firm texture and decrease in flavor, aroma, sweetness, and certain nutrients. Finally, the "soft-ripe" stage is reached when bananas become mushy, apples become mealy, and peaches, berries, cherries, plums, and kiwifruit turn soft and very juicy.

Ideally, fruits at the proper maturity for commercial processing are best held on the tree or vine until wanted for processing. Then they are harvested and moved into processing as quickly as possible without introducing fruit that is cut, bruised, overripe, or underripe. Since this ideal is not possible, alternatives are resorted to (Table 2.2).

Some fruits such as olives, avocados, pears, and apples do not have a definite maturation pattern. In such cases, maturation may be hastened by spraying with 2000 ppm of ethephon, or be delayed by holding under refrigeration.

Subtropical fruits (pineapple, avocado, grapefruit, limes) are subject to chilling injury when held at temperatures of 12°C or lower. They may be held at controlled-atmosphere (CA) conditions specific for each fruit. Waxes containing 1200–2400 ppm of benomyl are typically applied for control of fruit rot. While being held, tangerines absorb diphenyl in the peel oil (more than oranges, grapefruit, lemons, or limes do); and apples release aromas that may be absorbed by other products.

A few fruits are processed in the "firm-ripe" stage, when it is desired to retain the natural form of the fresh fruit (e.g., whole peach pickles, marachino cherries, whole olives, gooseberries, plums, grapes, and brandied peaches). To compensate for the undeveloped full natural flavor and aroma, other ingredients such as vinegar, spices, salt, sugar, calcium, colors, and flavors are added to such products.

Most fruits are canned, frozen, dehydrated, pickled, or preserved in the "mellow-ripe" stage to capture the maximum natural nutrients, flavor, aroma, and color. Flavors in canned and frozen fruits are enhanced with added sugar; colors are retained in dehydrated fruits with sulfites; and nutrients may be fortified with added ascorbic acid (vitamin C).

Fruits for juice, puree, wine, brandy, vinegar, leather, jelly, preserves, marmalade, butter, nectar, sauce, or essence recovery often are pro-

Table 2.2 Recommended Ripeness of Fruit at Harvest and Suggested Holding Conditions

Ripeness at harvest			Holding environment			Holding duration			
Mature green	Fully ripe	Firm ripe	Chill	Controlled atmosphere	Ambient	One day	Until ripe	Special	Minimum
Bananas	Citrus	All others	Cherries	Apples	All others	Mango	Bananas	Apples—up to 6 months	All others
Pears	Figs		Currants	Bananas		Nectarine	Kiwifruit	Apricots—1 to 2 days	
	Grapes		Figs	Pears		Peaches	Kumquat	Avocado—1 to 7 days	
	Olives		Gooseberries			Plum	Lemon	Blackberries, boysenberries—12 hr	
	Passion fruit		Raspberries				Lime	Cranberries—until mellow	
	Pineapple						Pears		
	Prune						Quince		
	Raspberries						Strawberries		

cessed in the "soft-ripe" stage because flavor and aroma are more important than texture.

For maximum use of product, and to extend plant operations, the same kind of fruit may be processed in more than one stage of ripeness. For example, firm-ripe, unbruised, mechanically harvested apples may be peeled, sliced and canned or frozen, and the bruised and misshaped fruit may be used in cider, butter, jelly, or vinegar. Most of the fruit juices, jellies, and preserves on the market are by-products of other major operations.

Many of the physiological and biochemical changes associated with fruit ripening have been extensively studied since 1945. These studies have provided qualitative and quantitative information on several important constituents that undergo change during ripening and senescence in relation to a wide range of postharvest environmental conditions. Many anabolic processes, such as the synthesis of pigments, lipids, nucleic acids, and other constituents, have been shown to occur during ripening. Although ripening normally occurs in fruit after growth ceases, it is a natural consequence of developmental changes during growth. A fruit is physiologically "ready" to ripen when it has gained the required potential.

The activity of many enzymes produces changes in fruits as they ripen. The increase in protein synthesis rate, the net increase in protein content, and the increased activity of several enzymes during the respiratory climacteric is the result of the synthesis of enzymes involved in various ripening reactions. The synthesis of enzymes takes place during the early stages of ripening before marked physical changes become apparent in the tissue. Ethylene is required in physiologically active concentrations in order to initiate ripening. Without sufficient ethylene, fruits that otherwise are ready to ripen continue to synthesize proteins for which capacity already exists and the fruit remains unripe.

Subatmospheric pressure (hypobaric) treatments delay the ripening of apricots by 15–37 days, peaches by 7–27 days, sweet cherries by 16–33 days, pears by 1.5–4.5 months, and apples by 2.5–3.5 months. In general, degradation of chlorophyll and starch, losses of sugars and titratable acidity, and formation of carotenoids are delayed by subatmospheric pressure treatments.

The rate of stone fruit ripening can be controlled by lowering fruit temperature through rapid removal of field heat. Hydrocooling and air cooling are the principal methods used.

Treating Fruit for Extended Life

Transporting and holding harvested fruit has many hazards, such as decay due to green mold and sour rot. Standard practice for chemical treatments of citrus fruits involves two sequential steps: (1) cleaning and disinfecting the fruit by washing in a soak tank containing either sodium *o*-phenylphenate (SOPP) or sodium carbonate (soda ash) or by brushing the fruit in a foam washer with SOPP, and (2) applying a wax that may contain SOPP, thiabendazole, 2-aminobutane (2-AB), or benomyl.

The common commercial application of postharvest chemical treatments is done during hydrocooling. Chlorine, for example, has been used extensively to reduce decay development. Other fungicides used in this water are captan, botran, and benomyl. With the introduction of botran and benomyl, preharvest sprays have shown real promise for postharvest decay control.

Water recirculation in hydrocoolers causes an accumulation of decay-producing organisms. The principal problem in connection with the use of approved chemicals is the maintenance of uniform concentrations, particularly in ice-refrigerated equipment, because of the constant dilution from the melting ice. The addition of mold disinfectants such as choline or approved phenol compounds reduces the buildup of bacteria and fungus spores, but will not kill infections already in the fruit or sterilize either the water or product surfaces. Brown rot and rhizopus decay spores on the surface and under the skin of peaches can be effectively destroyed by soaking the fruit in water at 52°C for 2–3 min. However, hot water treatment offers no protection against future contamination. The spores can also be destroyed by treating the fruit with a commercially available chemical fungicide which is usually put directly into the hydrocooling water.

Machine-harvested strawberries for freezing, jam, jelly, purée, or juice retain more natural color when treated with certain chemicals, especially SO_2, citric acid, or $SnCl_2$ (Sistrunk *et al.* 1982). It is recommended that machine harvesting of blackberries and raspberries be done in the morning, when ambient temperatures are low, and that field heat be removed immediately by holding the berries under refrigeration in an atmosphere of 20–40% CO_2. Spraying sweet cherries after harvest with 3% ethyl oleate can reduce the incidence of cracking from 29 to 11%. Preharvest spraying with AVG delays ripening of pears (Romani *et al.* 1983).

Lemons may be held for processing or eating fresh for up to 6 months under an atmosphere of 10% oxygen and nil CO_2 at 10°C, with continuous removal of ethylene. Before storage the lemons should be treated with GA, to reduce the rate of loss of chlorophyll from the ring; 2,4-D to maintain the button; and fungicide to restrict wastage. A combination treatment of sodium o-phenylphenate and benomyl is recommended to prevent growth of strains of *Penicillium* resistant to benomyl and related fungicides.

Cooling Fruit

Fruit for processing should be cooled to delay ripening if holding is necessary. Strawberries are normally cooled overnight to increase firmness to improve yield from decapping equipment. Some processors precool berries to increase the capacity of their freezer (Anon. 1981). Excellent references are available on thermal properties of fruit (Mohsenin 1980), cooling methods (ASHRAE 1982), and techniques (Ryall and Pentzer 1982). Therefore, this section focuses on applications of cooling to the processing crops most commonly cooled.

Peaches. Hydrocooling is the primary method for cooling peaches. Chilled water is flooded over fruit in bins conveyed through a tunnel. Water is sprayed on top of the fruit through nozzles or allowed to flow through openings in the bottom of an overhead flood pan. A water flow rate of 400–600 liter/min/m^2 is required when peaches are in pallet bins. Bulk fruit in layers less than 20 cm deep can be cooled with 200 liter/min/m^2 (Fig. 2.9). Resident time in the cooler is approximately 25–30 min for both bins and bulk systems.

The capacity of the refrigeration system can be improved by insulating the hydrocooler unit and by using ice to supplement cooling needs during peak periods. Monroe and O'Brien (1983) provide several factors to check if hydrocooler performance is inadequate. Inability to cool the water may be due to (1) floating ice that allows water to bypass it, (2) low water velocity past the ice, and (3) inadequate ice surface area due to small ice compartment or insufficient cooling coil surface area. Poor cooling of the fruit may be due to (1) inadequate refrigeration, (2) inadequate water circulation as indicated by a water temperature rise greater than 1.1°C, (3) excessive conveyor speed, (4) poor water distribution through the fruit, and (5) excessive heat gain through the hydrocooler walls.

Apples. Although apples may be hydrocooled (placed in flowing, chilled water), room cooling is the most common method. Room cooling involves placing pallet bins of fruit in a refrigerated room maintained at 0°C. Channels between bins and airflow rates should be sufficient to cool the apples to 3°C within 7 days. Ventilation openings in the bins can double cooling rates. Moisture should be added to keep the relative humidity near 90%. Moisture loss through condensation on evaporator coils can be reduced by providing a cooling surface area large enough to cause less than a 3°C temperature change.

Apples of a quality comparable to those obtained with conventional room cooling have been obtained by forcing outside air through apples in a partially buried silo (Burton *et al.* 1979). The exposed portion of the

Fig. 2.9. Hydrocooler designed for loose bulk cherries. Other designs handle pallet bins. (Courtesy J. M. Cline, Clarksville Machine Works.)

45,400-liter steel tank was insulated. Exhaust fans pulled air up through the apples if outside air was at least 2°C cooler than the apples. The fans automatically stopped if the outside air dropped below 0°C. In China, a related approach is to place fruit in underground caverns (Show-Chun 1982). Cooling is supplemented with mechanical refrigeration equipment.

Berries. Forced-air (pressure) cooling is the most suitable method of cooling berries. Cool air is forced through the berries by causing a pressure difference of 25 to 100 mm of water on opposite sides of vented containers. Airflows from 0.06 to 0.19 m^3/min/kg of fruit are normally sufficient. Forced-air cooling is usually four to ten times faster than room cooling. Therefore, the refrigeration system must have an equivalent increase in capacity. However, hydrocooling remains at least twice as fast as forced-air cooling.

At least 4% of the container side should be vented area. Elongated vents reduce obstruction by product. Air path length through the containers should be less than 1.0–1.2 m to prevent excessive air pressure.

Containers are typically stacked in rows in a cool room equipped with baffles and fans arranged to force air through the row. Equations are available for calculating pressure drop as a function of airflow rate and packing porosity (Chau *et al.* 1983).

Special Cooling Methods. Forced-air cooling methods and equipment developed for field-cooling broccoli and cauliflower (Freeman 1984) could have applications with fruit. Entire cooling systems are mounted on trailers to enable movement to production sites (Fig. 2.10). Shortly after harvest the field-packed product is at the desired holding temperature. A fleet of 14 mobile coolers was in operation the first year the units were designed.

Blueberries and muscadine grapes have been cooled with cryogenic gas (Rohrback *et al.* 1983). A primary advantage is rapid removal of field heat. The basic approach should be adaptable for most berry crops held for processing.

Field-freezing has been demonstrated for boysenberries. Nitrogen,

FIG. 2.10. Mobile forced-air cooling units enable field-cooling immediately after harvest. (Courtesy C. D. Freeman, Bud Antle Co.)

Freon, and other materials have been evaluated along with freezing techniques. Machine-harvested berries frozen in the field had less damage than hand-harvested berries when delivered to the processor. The field-frozen berries also yielded 95.1% combined grade A and B, while hand-harvested berries only yielded 83.7% of these grades. Unfortunately, the higher quality did not offset the higher costs for field-freezing.

Conventional Fruit Storage

Many fruits—especially apples, pears, berries, peaches, and grapes—are frequently stored for a few days to more than a month to prolong the processing season and to allow the fruit to mellow. The most obvious losses during holding are caused by mechanical injury, decay, and aging. Losses of moisture, vitamins, sugars, and starches are less noticeable but adversely affect the quality and nutritional value of fruit.

Rough handling and holding at undesirably high or low temperatures increase losses. On the other hand, during short holding periods most fruits become more uniformly colored, mellow, and more flavorsome. Pears and plums become sweeter, more highly colored, and smoother in texture.

Apples. A larger quantity of apples is stored before processing, and for a longer period of time, than for any other fruit. The length of storage may be relatively short for fall cultivars going into processing, and longer for winter cultivars.

For most apple cultivars, 30°–32°F is the recommended storage temperature. This is 1.5°–2.5°F above the average freezing point of apples. If held at this temperature, some cultivars develop physiological disorders that impair processing qualities. However, if temperatures are elevated to 38°–40°F, such storage disorders may not occur during the period involved; consequently, 36°F is often used. The relative humidity (RH) should be near 90%.

Chilling injury may occur at low temperatures even if freezing does not occur. The principal disorders classed as chilling injury in apples are soft scald, soggy breakdown, brown core and internal browning.

Apples have been successfully held in silos designed for hydraulic handling (Fig. 2.11). The water used to move the fruit into the silo is drained at the start of the storage period.

Pears. Pears do not ripen on the tree. 'Bartlett' is the most important cultivar for processing and is grown on the Pacific Coast. A limited quantity of the 'Kieffer' cultivars is grown in the eastern states. 'Winter Nelis' is also canned to a limited extent. A 10-day to 2-week cold storage period

FIG. 2.11. Storage silo for apples in which fruit is conveyed hydraulically. (Courtesy B. R. Tennes.)

is commonly used by canners because it improves the uniformity of ripening. Substantial quantities are stored for 2–4 months to better utilize processing facilities. Maturity at harvest has a very important bearing upon the subsequent storage life and processing quality of pears (Chen *et al.* 1983). If harvested too early, pears are subject to excessive water loss in storage; if harvested overmature their storage life is shortened by scald and core breakdown.

Since pears do not ripen when held at 30°–31°F with about 90% RH, they are held "dormant" until a specified number of days before processing. Each day the quantity required for one day's processing is removed from cold storage to room temperature and allowed to properly ripen for processing.

Flesh firmness, as measured by a pressure tester, is the best measure of the potential storage life of pears from any single orchard. For 'Bartlett' pears, a firmness of 17–19 lb measured on a Magnus-Taylor pressure tester or similar device, using a 3/16-in. plunger head, indicates best storage quality. Pressure test information for each lot going into storage is helpful. Controlled-atmosphere storage is used for pears to be stored for more than 1 month. Storage disorders for pears are similar to those for apples. No processing pears should be allowed to remain in storage for any of these disorders to develop.

Stone Fruits, Grapes, and Berries. Peaches, plums, apricots, and nectarines do not benefit from storage; however, they are frequently hydrocooled and stored at 31°F for a few days, to better withstand transportation and prolong the processing season.

Grapes and berries should ripen on the vine. When mature, they begin a slow process of breakdown, which can be slowed by refrigerated storage to prolong the period for processing. Storage is confined entirely to the European or California (*Vitis vinifera*) type of grapes, which includes about 90% of U.S. production. Of the 2.75 million tons produced, about 2 million tons are made into wine, juice, champagne, and minor grape products. Even though they ripen in the hottest part of the year, grapes can be stored for a month for processing if kept in refrigerated storage.

The American or Eastern (*V. labrusca*) type of grapes, of which 'Concord' is the leading cultivar, are taken directly to the juice or wine plant for immediate washing and debunching for processing. The skins are thin and much shattering occurs and fermentation begins within a few hours.

The Southern Muscadine (*V. rotundifolia*) type of grapes is borne in small bunches, shatters badly, and is difficult to store for over 7 days (Ballinger and McClure 1983). These grapes are highly colored and flavored, and the thick hull makes them highly suitable for wine, preserves, sauces, and jelly.

Strawberries, raspberries, loganberries, youngberries, blackberries, elderberries, gooseberries, and currants are frequently precooled and stored for 1–2 days to prolong the ripening season and allow the flavor and color to intensify.

Controlled-Atmosphere Storage

Early research that demonstrated the advantages of controlling the atmosphere, temperature, and humidity of a storage space for fruit was performed in England. Since 1930, there has been a growing fund of information on the effects of controlled atmosphere (CA) on the keeping qualities of fruits. Much of the knowledge on the influence of trace quantities of gases on harvested fruit has been obtained since 1960. Routine measurements of various gases are now possible because of major improvements in gas analysis techniques, especially in gas chromatography.

Utilization of CA storage has been largely confined to apples because this hardy fruit is particularly adapted to such storage. But the technique is also being applied to citrus, pears, and other fresh fruits.

The optimum atmospheric conditions for apples vary with cultivar. For example, 'MacIntosh' should be stored at 38°F and 3% oxygen (O_2) with carbon dioxide (CO_2) at 2.5% for the first month and then 5%; 'Delicious' apples at 32°F, 3% O_2, and 2.5% CO_2. The duration that apples can retain good quality under various conditions of storage and handling depends on maturity.

Assuming that the fruit is harvested at optimum color, flavor, and texture conditions, a processor can predict the life of apples. If apples are removed rapidly from orchards to a storage room, cooled quickly and sealed, and if the oxygen level is reduced to an acceptable level in 2–3 days and the buildup of CO_2 limited, apples can be removed and processed as desired.

Types of CA Systems. Three techniques are generally recognized for providing controlled-atmosphere conditions in a fruit storage room: natural, liquid nitrogen, and nitrogen generator systems.

Natural CA storage depends on the respiration of the fruit to generate CO_2 by reaction of storage room O_2 with carbohydrates in the fruit. The CO_2 concentration is controlled by the use of wet scrubbers, hydrated lime, or by an inert gas generator.

Successful operation of natural CA storage requires that the room be airtight, since leakage can cause the concentration of O_2 to increase beyond the allowable limits. Aside from the considerable cost of making storage rooms nearly airtight, natural CA storage requires relatively little capital investment; however, this method lacks flexibility and the ability to control CO_2 and O_2 levels.

Liquid nitrogen has been used to a limited extent for simultaneous refrigeration and control of atmosphere by flushing the ambient atmosphere out of the room. In general, though, compressed gaseous nitrogen is used to correct temporary faults in atmospheric conditions due to excessive leakage of air into a storage room. It is unlikely that the use of either gaseous or liquid nitrogen will ever be competitive with other systems for providing artificial atmospheres since nitrogen is less costly when obtained from inert gas generators.

The use of controlled-atmosphere generators has become increasingly popular, largely because they provide a means for rapidly reducing O_2 levels and for controlling CO_2 and O_2 levels throughout the storage period.

Two methods have been used for generating nitrogen to supply a controlled atmosphere in fruit storage rooms. In the older method, the room is continuously flushed with a gas mixture that consists of about 96% nitrogen, 2–3% O_2, and 1–2% CO_2. To produce this mixture, ambient air is passed through a blower, mixed with fuel (usually liquid propane or natural gas), and admitted to a burner in which essentially all the fuel and all but 1–3% of O_2 are converted to CO_2 and water vapor. Products from the burner go through a water-cooled condenser to remove most of the water vapor, then to a scrubber that removes most of the CO_2. A bypass valve around the scrubber helps to regulate the flow, the pressure drop, and the composition of the gas that is admitted to the storage space. Several factors influence operation of the system and the storage atmosphere produced: air leakage into the storage area; composition of the atmosphere supplied by the generator; respiration of the stored fruit; and leakage of the atmosphere out of the storage area. A flame-type burner can be used because the mixture fed to it contains the normal atmospheric complement of oxygen, i.e., 21% by volume. A catalytic burner is usually provided, however, because of the greater assurance it provides against the generation of partial oxidation products, which should be excluded from the storage room.

In the second major method of supplying an inert gas for CA storage, the room atmosphere is recirculated continuously through the generator equipment.

A controlled-atmosphere generator should (1) reduce the operator's work load; (2) be trouble-free and safe to use; (3) produce an atmosphere good for stored fruit; (4) be economical and reliable; (5) not create any air or water pollution problems, and (6) perform atmosphere-purification functions not otherwise performed in the storage facility.

For optimum CA storage of fruits the relative humidity should be 90–100%. The absorbent that removes CO_2 from the atmosphere also removes water vapor, so before the nitrogen–oxygen mixture is returned to the storage room it passes through a humidifier.

Some generators start and stop automatically in response to a signal keyed to the concentration of CO_2 in the storage room, but many operators prefer less costly equipment that can be manually started when measurements of CO_2 concentration indicate the need for its removal.

One type of unit converts the O_2 originally in the CA storage room into CO_2 and water vapor by reacting it with LP gas. A more direct means would be to react hydrogen with the O_2 because the only by-product, water, could be disposed of by cooling and condensation. There are disadvantages, however, to the use of hydrogen. The CO_2 produced by the common fuel gases can be removed either by the techniques used in conventional CA storage or by absorption on a solid.

In some CA systems, large plastic bags are connected to the storage room to provide a "flexible wall" that can react to changes in air pressure that occur with changes in barometric pressure and temperature. The bags could be retrofitted to the many storages that do not use them. It is important to use bags of adequate size.

In each system the inert gases fed to the storage room impose additional load on the refrigeration system since they are at a temperature 10°–15°F above that of the available cooling water and because they are nearly saturated with water vapor.

An inert gas (propane) generator for CA storage has been developed that is cheaper than a nitrogen generator (Scott *et al.* 1968). The unit is a simple, economical, flame-type generator using standard commercial components, modified to ensure safe, reliable operation and elimination of harmful impurities. The operating cost of the unit including fuel, utilities, depreciation, repairs, and interest on the investment is acceptable for a 25,000-bu storage. One advantage is that this unit can be used for smaller storages.

CA Storage of Apples. Apples are harvested "firm-ripe" and mellow on holding until optimum texture, flavor, color and aroma develop for the products made from them. When held under CA, the ideal temperature, relative humidity, O_2 level, and CO_2 level vary with each cultivar.

Apples stored for periods up to 6 months at 0°C in 1.5% CO_2 + 1.0% O_2 are firmer than those held in conventional CA storage (Lidster *et al.* 1983). The rate of firmness loss of apples declined with extended duration up to 120 days. The rate of loss of titratable acidity was much slower at low O_2 and CO_2 levels. The evolution of ethylene (C_2H_4) from apples increased to an asymptotic maximum, which was lower for apples stored in very low O_2. The respiration rate of apples was suppressed by 1.0% O_2. The accumulation of C_2H_4 in storage was increased by increasing O_2 levels.

Controlled-atmosphere storage is a method that may be adapted by warehouses to preserve the firm ripe qualities of apples for 6 months or more. While more applicable for apples for the fresh market, it may be used for processing apples, when the cost can be justified.

Aseptic Bulk Storage

The seasonal nature of fruit processing has contributed to economic inefficiency within the industry. Restricted processing seasons precipitate high costs through inefficiency in raw product scheduling, equipment, and finished product marketing. Most fruits, especially grapes, peaches, pears, apples, and berries, are suitable for holding in a sterile, semi-processed condition for later manufacture into puree, juice, sauce, butter, jelly, preserves, marmalade, and other products.

Bulk storage processing consists of holding large quantities of semi-processed fruit under aseptic conditions, in barrels, bags, or silo-like tanks, for postseason manufacture or final processing. The raw fruit is washed, sorted, chopped, and preheated. Next the pulp is deareated and sterilized at a high temperature for a short time. The sterile pulp is cooled and transferred to presterilized bulk containers. After filling, an inert gas is maintained in the headspace. The product is sterile and can be kept for several months. The tanks of product can be partially emptied with the remaining material kept sterile.

The containers of sterile, semiprocessed fruit can be shipped to distant plants, or exported to other countries. The process enables excess raw product supplies to be held until slack periods for final processing. In some cases it is more economical to process the final product at locations other than the production areas.

Bulk storage has been practiced in the Concord grape juice industry since about 1930. The 300,000 tons of these grapes grown around the Great Lakes and along the Atlantic seaboard, are used for juice, frozen concentrate, jelly, jams, drinks, and various noncarbonated beverages.

There have been three major technological revolutions in the Concord grape juice industry within the last 50 years. The first was that of bulk storage of juice. The second was continuous processing. Until 1960, the juice from grapes was extracted conventionally in a manner similar to apples, by building alternate layers of wooden racks and blankets containing the heated grapes. The advent of continuous pressing was very fortuitous since labor for the tedious cheese building and stripping operation was becoming difficult to obtain. Without the arrival of continuous pressing, it would be a very difficult chore to press 2000 tons in a plant in a single day. The third milestone was that of mechanical harvesting because of labor storage. Approximately 75–80% of the grapes in the Chatauqua–Erie grape belt are mechanically harvested.

Bulk Storage Processing System. Grapes, whether hand or mechanically harvested, are crushed and destemmed, heated to approximately 60°C, and treated for 30–40 min with a depectinizing enzyme to disintegrate the pulp and release juice. The juice is separated from skins and seeds in a continuous press with the addition of a purified cellulose additive. The extracted juice is then partially clarified to remove the gross insoluble solids and is then flash-pasteurized, cooled, and pumped directly to storage tanks. (See Chapter 6 for more details.)

Bulk storage processing of grape juice has several advantages. A substantial pooling of juice takes place with a resulting beneficial standardization and uniformity. Juice may be stored for 12 months or more and used as needed. The flavor of juice stored at 18°C or lower for 12 months does not deteriorate; some think it even improves with age. Shifts in marketing and sales patterns can be readily accommodated from the substantial pool available. Although the grape harvest season of approximately 30 days is a busy one, steady production and employment are a result of the reserve juice. Products leaving the plant are packaged frequently and throughout the year assuring freshness to the consumer.

Although a substantial amount of grape juice may be concentrated during the harvest season and stored in drums or in tankage, 40–50 million gal of grape juice are stored single-strength in tanks of one sort or another.

Bulk Storage Vessels. The earliest storage tanks were of wood and the problems of sterilization can be readily recognized. Some were left uncovered and an ice cap produced by freezing became the cover. The first tanks especially built for grape juice storage were made of concrete and lined with paraffin or asphaltic material and later with vinyl coatings. They had a capacity of about 50,000 gal.

In 1946, the first large stainless clad tanks, of 150,000 gal each, were fabricated. In that year, 30 tanks with a total capacity of 4,500,000 gal were erected. These tanks were approximately 32 ft in diameter and 25 ft high; five were contained in an insulated, refrigerated room, approximately 160 × 40 ft.

Subsequent tanks were of mild steel and suitably coated with a phenolic resin that required rather high bake temperatures. All these tanks were cylinders with slightly sloped tops and bottoms, with a bottom manhole on the shell placed about 24 in. from the ground level and a top manhole on the high side of the roof. A catwalk system allowed for access to these top manholes.

Storage tanks built recently have been 33 ft in diameter and 48 ft high and contain approximately 320,000 gal each. These tanks are lined with a phenolic resin (epoxy) requiring a somewhat milder bake. They have rounded radium where the shell meets the sloped bottom and hemispherical top; this latter feature allows for a complete fill into the top manhole, assuring the absence of entrapped air. An ultraviolet lamp is attached to the top manhole cover to prevent mold growth on the top surface in the manhole area.

Bulk storage of grape juice has been used successfully for many years. However, vigilance must be exercised in the construction of storage vessels and pasteurizing and cooling equipment to ensure a sanitary installation and process. Good quality control is needed to assure that appropriate steps are taken to maintain sterility. Cold storage of grape juice affords many advantages by increasing uniformity, maintaining quality, keeping the freshness of product, providing production continuity, and allowing for sales sensitivity.

FUTURE POSSIBILITIES

Microelectronics and biotechnology will certainly play an increasingly important role in the fruit processing industry. Technological innovations applicable to harvesting, handling, and holding fruit will result from targeted research as well as from spinoff from other endeavors. Systems analysis techniques also have the potential to make major impacts through modeling and simulation of entire systems. Although it is not possible to predict specific results, there are trends that indicate some general directions that are likely.

A general trend has been towards mechanization of not only harvest but also production and handling operations. The crops most suitable for mechanical harvest were addressed in the past few decades. Yet only 11% of the fruit in the United States was mechanically harvested in 1979 (ASAE 1983). Poor quality is the major problem that has limited the addition of crops to the list of ones mechanically harvested (Drake 1983). Reliance on illegal immigrant labor could be postponing recognition of the need to support research that would enable mechanical harvest of other crops (Martin 1983). A possible result is movement of fruit processing industries to countries with lower labor costs. Another possible outcome is a reduction of fruit products with higher labor requirements.

Microelectronics and computers offer tremendous opportunities for improved fruit quality. Harvester controls could be developed for more precise operation of harvesting mechanisms. Handling systems on harvester and transport vehicles could be controlled by signals from sensors that indicate incipient damage. A change from mass removal (direct contact or vibratory methods) to removal of individual fruits is a possibility with computer-based robotics. A major obstacle is the ability to detect and select fruit surrounded by leaves and other material. The development of successful visual-imaging equipment for sorting fruit with defects indicates that similar equipment for a robot could be available shortly.

Microelectronics and computers also enable improved communications and greatly enhance research capabilities.

Biotechnology has the potential to revolutionize agriculture. Several possibilities are based on genetic engineering. The development of fruit suitable for once-over harvest would be extremely beneficial. Delivery of a product with a more uniform maturity would enhance the quality of the final product. Other fruit characteristics such as bruise and decay resistance, size, shape, and shelf life could be altered by genetic engineering. A major breakthrough would be the ability to control plant form to facilitate mechanical harvest. It may be possible to modify tropical fruits for production in temperate zones.

Systems analysis techniques offer the ability to model and simulate complex systems. A growth model for fruit crops could predict estimates of yield, harvest date, and even fruit quality. Computer simulations could be developed for analyzing the harvesting, handling, and holding system. Before making changes in the physical system, the proposed change would be simulated to determine if the desired output would result. Simulations are also useful for scheduling, determining product mix, locating facilities, evaluating bottlenecks, and many other purposes.

People are ultimately the element that determines the success of a system. The importance of human decisions and actions is amplified when an operation is mechanized. One miscalculation by the operator of a mechanical harvester can cause a much greater loss than an incorrect decision by a worker picking fruit by hand. Consequently, it is imperative that future systems be operated by highly trained and motivated personnel. It is equally important for the equipment to be designed to compliment the strengths and weaknesses of the operator.

REFERENCES

Harvesting Fruit

ALPER, Y., EREZ, A., and BEN-ARIE, R. 1980. A new approach to mechanical harvesting of fresh market peaches grown in a meadow orchard. Trans. ASAE *23* (5) 1084–1088.

ALPER, Y., WOLF, I., BEN-ARIE, R., ELKIN, I., MIHAI, G., and ANTLER, A. 1984. A mechanical harvesting system for fresh-market peaches grown in a meadow orchard. pp. 52–57. *In* Fruit, Nut, and Vegetable Harvesting Mechanization. ASAE Publ. 5–84.Amer. Soc. Agric. Engr., St. Joseph, MI.

ASAE (American Society of Agricultural Engineers). 1983. Status of Harvest Mechanization of Horticultural Crops. ASAE Special Publ. 3–83. Amer. Soc. Agric. Engr., St. Joseph, MI.

BERLAGE, A. G. and LANGMO, R. D. 1976. Trunk shaker harvesting of apples surrounded by plastic spheres. Trans. ASAE *19* (2) 209–212.

CARGILL, B. F., AND KIRK, D. E. 1983. Detachment of fruits by direct contact devices. *In* Principles and Practices for Harvesting and Hanlding Fruits and Nuts. M. O'Brien, B. F. Cargill, and R. B. Fridley (Editors). AVI Publishing Co., Westport, CT.

CHURCHHILL, D. B. and HEDDEN, S. L. 1983. A double-sided rake-pickup machine for citrus. Trans. ASAE *26* (4) 1034–1036.

CLAYPOOL, L. L. 1983. Biology and culture of fruit production. *In* Principles and Practices for Harvesting and Handling Fruits and Nuts. M. O'Brien, B. F. Cargill, and R. B. Fridley (Editors). AVI Publishing Co., Westport, CT.

DIENER, R. G. and FRIDLEY, R. B. 1983. Collection by catching. *In* Principles and Practices for Harvesting and Handling Fruits and Nuts. M. O'Brien, B. F. Cargill, and R. B. Fridley (Editors). AVI Publishing Co., Westport, CT.

DIENER, R. G., ADAMS, R. E., and NESSELROAD, P. E. 1974. Economic Evaluation of the Butterfly Orchard Design for Mechanical Harvesting. West Va. Univ. Agric. Exp. Sta. Curr. Rpt. No. *64*.

DIENER, R. G., ELLIOTT, K. C., NESSELROAD, P. E., ADAMS, R. E. BLIZZARD, S. H., INGLE, M., and SINGHA, S. 1982. The West Virginia University tree fruit harvester. J. Agric. Engr. Res. *27,* 191–200.

EASTMAN KODAK COMPANY. 1983. Ergonomic Design for People at Work. Vol. 1. Lifetime Learning Publications, Belmont, CA.

FRIDLEY, R. B. 1983. Vibration and Vibratory Mechanisms for the Harvest of Tree Fruits. *In* Principles and Practices for Harvesting and Handling Fruits and Nuts. M. O'Brien, B. F. Cargill, and R. B. Fridley (Editors). AVI Publishing Co., Westport, CT.

FRIDLEY, R. B., CLAYPOOL, L. L., and MEHLSCHAU, J. J. 1975. A new approach to tree fruit collection. Trans. ASAE *18* (5) 859–863.

HUDSON, J. P. 1971. Meadow orchards. Agriculture (London) *78,* 157–160.

JOHNSON, J. C., DIENER, R. G., ELLIOTT, K. C., NESSELROAD, P. E., SINGHA, S., BLIZZARD, S. H., ADAMS, R. E., and INGLE, M. 1983. Two experimental bruise reduction methods for mechanically harvested apples. Trans. ASAE *26* (4) 1037–1040.

KANTOWITZ, B. H. and SORKIN, R. D. 1983. Human Factors: Understanding People–System Relationships. John Wiley & Sons, New York.

KEPNER, R. A., BAINER, R., and BARGER, E. L. 1978. Fruit and vegetable harvesting and handling. *In* Principles of Farm Machinery, 3rd ed. AVI Publishing Co., Westport, CT.

LEFLUFY, M. J. 1982. Apple harvesting by a combing technique. ASAE Paper 82–1065.

McCORMICK, E. J. and SANDERS, M. S. 1982. Human Factors in Engineering and Design. 4th ed. McGraw-Hill, New York.

MILLIER, W. F., VAN DER WERKEN, J., and THROOP, J. A. 1983. A recoil-impact shaker for semi-dwarf apple trees. ASAE Paper 83–1080.

O'BRIEN, M., CARGILL, B. F. and FRIDLEY, R. B. (Editors). 1983. Principles and Practices for Harvesting and Handling Fruits and Nuts. AVI Publishing Co., Westport, CT.

ORROCK, J. E. and FISHER, J. K. 1983. Intelligent robot systems—potential agricultural applications. *In* Robotics and Intelligent Machines in Agriculture. ASAE, Publ. 4–84 Amer. Soc. Agr. Engr., St. Joseph, MI.

PETERSON, D. L. 1982A. Rod press fruit removal mechanism. Trans. ASAE *25* (5) 1185–1188.

PETERSON, D. L. 1982B. Continuously moving over-the-row harvester for tree crops. Trans. ASAE *25* (6) 1478–1483.

RUFF, J. H. and LEPORI, W. A. 1983. Fruit and nut collection by pickup. *In* Principles and Practices for Harvesting and Handling Fruits and Nuts. M. O'Brien, B. F. Cargill, and R. B. Fridley (Editors). AVI Publishing Co., Westport, CT.

RYALL, A. L. and PENTZER, W. T. 1982. Handling, Transportation, and Storage of Fruits and Vegetables. Vol. 2. 2nd ed. AVI Publishing Co., Westport, CT.

SARIG, Y. (Editor). 1984. Fruit, Nut, and Vegetable Harvesting Mechanization. Proc. Intern. Symp. on Fruit, Nut, and Vegetable Mechanization, October 5–12, 1983, The Volcani Center, Bet Dagan, Israel. ASAE, Publ. 5–84. Amer. Soc. Agric. Engr., St. Joseph, MI.

SARIG, Y., SEGERLING, L. J., and LITTLE, R. W. 1976. Deformation analysis of foam-encapsulated apple under impact loading. ASAE Paper 76–1041.

SMITH, E. A. and RAMSAY, A. M. 1983. Forces during fruit removal by a mechanical raspberry harvester. J. Agric. Engr. Res. *28,* 21–32.

SUMMER, H. R. and CHURCHILL, D. B. 1977. Collecting and handling mechanically removed citrus fruit. Proc. Int. Soc. Citricuture *2,* 413–418.

TENNES, B. R. and BROWN, G. H. 1981. Design, development and testing of a sway-bar shaker for horticultural crops—a progress report. ASAE Paper 81–1059.

TOPHAM, P. B. and MASON, D. T. 1981. Modelling a raspberry harvest: Effects of changing the starting date of, and the interval between, machine harvest. Hort. Res. *21,* 29–39.

VAN DER WERKEN, J. 1981. Possibilities of harvest mechanization in H.D.P. hedgerow systems. ACTA Horticulturae *114,* 221–231.

WALTON, L. R., SWETNAM, L. D., and CASADA, J. H. 1983. A harvesting aid for bell peppers and pickling cucumbers. ASAE Paper 83–1076.

WHITNEY, J. D. and CAPPOCK, G. E. 1984. Harvest mechanization of Florida oranges destined for processing. pp. 149–155. *In* Fruit, Nut, and Vegetable Harvesting Mechanization. ASAE Publ. 5–84. Amer. Soc. Agr. Engr., St. Joseph, MI.

WILKINSON, R. H. and BAKER, L. D. 1984. Selected references on human factors engineering. ASAE Paper 84–1647.

Handling Fruit

ASAE. 1984. Agricultural pallet bins. S337. *In* 1984–1985 Agricultural Engineers Yearbook of Standards. Amer. Soc. Agric. Engr., St. Joseph, MI.

BENDER, F. E., KRAMER, A., AND KAHAN, G. 1976. Systems Analysis for the Food Industry. AVI Publishing Co., Westport, CT.

BIRTH, G. A. 1983. Optical Radiation. *In* Instrumentation and Measurement for Environmental Sciences. 2nd ed. B. W. Mitchell (Editor). ASAE Special Publ. 13–82. Amer. Soc. Agric. Engr., St. Joseph, MI.

CAPLAN, S. H., LUCAS, R. L., and MURPHY, T. J. 1983. Information transfer. *In* Ergonomic Design for People at Work. Vol. I. Eastman Kodak Company. Lifetime Learning Publications, Belmont, CA.

CARLOW, C. A. 1983. An instructible rejection system for quality grading of potatoes and other produce. J. Agric. Engr. Res. *28,* 373–383.

CHRISTENSEN, G. W. and ELLIS, R. F. 1984. Pneumatic tube system conveys potatoes at 20–30 tons/hour. Food Processing *45* (1) 174–175.

CLARE, J. N. and SINCLAIR, M. A. 1979. Search and the Human Observer. Taylor & Francis, Ltd., London.

DAVIS, K. R. and McKEOWN, P. G. 1984. Quantitative Models for Management. 2nd ed. Kent Publishing Co., Boston.

DELWICHE, M. J. and BAUMGARDNER, R. A. 1983. Development of ground color references as a maturity index for Southeastern peaches. ASAE Paper 83–6537.

DIENER, R. G., ELLIOTT, N. C., NESSELROAD, P. E., BLIZZARD, S. H., ADAMS, R. E., SINGHA, S., and INGLE, M. 1981. Field grinding of juice apples. Trans. ASAE *24* (6) 1429–1431.

DIENER, R. G. ELLIOTT, K. C., NESSELROAD, P. E., BLIZZARD, S. H., ADAMS, R. E., and SINGHA, S. 1983. Bulk handling of processing apples in the orchard. ASAE Paper 83–1083.

FURMAN, B. J. and HENDERSON, J. M. 1982. An improved rotary singulator. ASAE Paper 82–6013.

HANSEN, C. M., LEDEBUHR, R. L., VANEE, G., and FRIESEN, O. 1984. Systems approach to strawberry harvest mechanization. pp. 325–331. *In* Fruit, Nut, and Vegetable Harvesting Mechanization. ASAE Publ. 5–84. Amer. Soc. Agric. Engr., St. Joseph, MI.

HARRISON, H. P. and BLECHA, A. 1983. Screen oscillation and apperature size-sliding only. Trans. ASAE *26* (2) 343–348.

HOLT, J. E. and SCHOORL, D. 1981. Fruit packaging and handling distribution systems: An evaluation method. Agricultural Systems *7,* 209–218.

HORSFIELD, B. C., FRIDLEY, R. B., and CLAYPOOL, L. L. 1972. Application of theory of elasticity to the design of fruit harvesting and handling equipment for minimum bruising. Trans. ASAE *15* (4) 746–750.

KRISHNAN, P. and BERLAGE, A. G. 1983. Separation of shells from walnut meat using magnetic methods. ASAE Paper 83–6538.

McDONALD, T. and DELWICHE, M. J. 1983. Non-destructive sensing of peach flesh firmness using impact force analysis. ASAE Paper 83–6540.

MOHSENIN, N. N. 1978. Physical Properties of Plant and Animal Materials. Gordon and Breach Science Publishers, New York.

MONROE, G. E. and O'BRIEN, M. 1983. Postharvest functions. *In* Principles and Practices for Harvesting and Handling Fruits and Vegetables. M. O'Brien, B. F. Cargill, and R. B. Fridley (Editors). AVI Publishing Co., Westport, CT.

O'BRIEN, M. and GAFFNEY, J. J. 1983. Handling and transportation operations. *In* Principles and Practices for Harvesting and Handling Fruits and Vegetables. M. O'Brien, B. F. Cargill, and R. B. Fridley (Editors). AVI Publishing Co., Westport, CT.

O'BRIEN, M., PAASCH, R. K., and GARRETT, R. E. 1980. Fillers for fruit and vegetable damage reduction. Trans. ASAE *23* (1) 71–73.

PRATT, C. A. 1984. Going further. Byte *9* (3) 204–208.

PRITSKER, A. A. B. and PEGDEN, C. D. 1979. Introduction to Simulation and SLAM. John Wiley and Sons, New York.

ROSENBERG, R. C. and KARNOPP, D. C. 1983. Introduction to Physical System Dynamics. McGraw-Hill, New York.

SCHOORL, D. and HOLT, S. E. 1982. Road-vehicle-load interactions for transport of fruit and vegetables. Agricultural Systems *8,* 143–155.

SLAUGHTER, D. C. and ROHRBACH, R. P. 1983. Quick method for determining the principle radii of curvature in blueberries. ASAE Paper 83–6539.

STEPHENSON, K. Q., ROTZ, C. A., and SINGH, M. 1979. Selective sorting by resonance technqiues. Trans. ASAE *22* (2) 279–282.

SWEETWOOD, P. C. and PERSSON, S. P. 1982. Vibratory feeder for vacuum-type seed-metering device. ASAE Paper 82–1075.

TAYLOR, R. W., REHKUGLER, G. E., and THROOP, J. A. 1984. Apple bruise detection using a digital line scan camera system. *In* Agricultural Electronics—1983 and Beyond, Volume II, pp. 652–662. American Society of Agricultural Engineers, St. Joseph, Michigan.

TENNES, B. R. 1981. Design and development of a bucket conveyor for apples. ASAE Paper 81–1065.

TENNES, B. R., BURTON, C. L., and BROWN, G. K. 1978. A bulk handling and storing system for apples. Trans. ASAE. *21* (6) 1088–1091.

Holding and Future Possibilities

ANON. 1981. Precooling conserves energy, safeguards quality, ups capacity for organ packer. Quick Frozen Food. *Dec.*, 36–39, 59.

ASHRAE. 1982. Methods of precooling fruits and vegetables. *In* ASHRAE Applications Guide Book. Amer. Soc. for Heating, Refrigeration, Air Conditioning Engineers, Atlanta, GA.

BALLINGER, W. E. and McCLURE, W. F. 1983. The effect of ripeness on storage of 'Carlos' muscadine grapes. Scientia Horticulturae *18*, 241–245.

BURTON, C. L., TENNES, B. R., and BROWN, G. K. 1979. Air-cooled storage of bulk apples in a silo—a progress report. Trans. ASAE *22* (3) 510–512.

CHAU, K. V., GAFFNEY, J. J., BAIRD, C. D., and CHURCH, G. A. 1983. Resistance to air flow of oranges in bulk and in cartons. ASAE Paper 83–6007.

CHEN, P. M., MELLENTHIN, W. M., and KELLY, S. B. 1983. Fruit quality of 'Bosc' pears (*Pyrus communis* L.) stored in air or one percent oxygen as influenced by maturity. Scientia Horticulturae *21*, 45–52.

DRAKE, S. R. 1983. Introduction to the symposium on the influence of mechanical harvesting on the quality of horticultural crops. HortScience *18*, 406.

FREEMAN, C. D. 1984. Cost reducing technologies in cooling fresh vegetables. ASAE Paper 84–1074.

LIDSTER, P. D., LIGHTFOOT, H. J., and MCRAE, K. B. 1983. Fruit quality and respiration of 'McIntosh' apples in response to ethylene, very low oxygen and carbon dioxide storage atmospheres. Scientia Horticulturae *20*, 71–83.

MARTIN, P. L. 1983. Labor-intensive agriculture. Scientific American *249* (4), 54–59.

MOHSENIN, N. N. 1980. Thermal Properties of Foods and Agricultural Materials. Gordon and Breach, New York.

MONROE, G. E. and O'BRIEN, M. 1983. Postharvest functions. *In* Principles and Practices for Harvesting and Handling Fruits and Nuts. M. O'Brien, B. F. Cargill, and R. B. Fridley (Editors). AVI Publishing Co., Westport, CT.

PEARL, R. C. 1971. The concept of bulk storage processing and its implications to industry. HortScience *6*, 221–222.

ROHRBACK, R. P., FERRELL, R., BEASLEY, E. O., and FOWLER, J. R. 1983. Precooling blueberries and muscadine grapes with liquid carbon dioxide. ASAE Paper 83–6008.

ROMANI, R., LABAVITCH, J., YAMASHITA, T., HESS, B., and RAE, H. 1983. Preharvest AVG treatment of 'Bartlett' pear fruits: Effects of ripening, color change, and volatiles. J. Amer. Soc. Hort. Sci. *108*, 1046–1049.

SCOTT, K. R., PHILLIPS, W. R., and JOHNSON, F. B. 1968. Controlled-atmosphere generator for apple storage. Trans. Am. Soc. Agr. Engrs. *11*, 120–122.

SISTRUNK, W. A., MORRIS, J. R., and HOZUP, J. 1982. The effect of chemical treatments and heat on color stability of frozen machine-harvested strawberries for jam. J. Amer. Hort. Sci. *107*, 693–697.

SHOW-CHUN, Qi. 1982. The application of mechanical refrigeration to cave storage of fruit. Intern. J. Refrigeration *5* (4) 235–237.

3

Fruit Washing, Peeling, and Preparation for Processing

J. G. Woodroof

Fruit processing is continuous from the time the fruit leaves the growing plant till it is protected securely from microorganisms or macroorganisms. Harvesting, sorting, handling, storing, washing, peeling, preparation, heating, refrigerating, dehydrating, concentrating, treating with chemicals, and packaging are steps in the chain of events.

Detached fruit may be ripened by exposure to alternating periods of light and darkness, the light being confined to the blue, red, and far red regions of the light spectrum, with temperature being maintained lower during the dark periods. Equipment for processing, transporting, and dispensing of processed fruits is more adequate now than ever before and is improving each year.

This chapter treats three of the above steps—washing, peeling, and preparation for processing.

WASHING

Fruit is washed to remove dust, dirt, insect frass, mold spores, plant parts, and filth that might contaminate or affect the color, aroma, or flavor of the fruit. Washing certain fruits, as strawberries or raspberries for freezing, that are not peeled or heated is almost the only means of "clean-

Commercial Fruit Processing, 2nd Edition

ISBN 0-87055-502-2

FIG. 3.1. Processing oranges into frozen concentrate. Oranges at frozen orange concentrate plant entering scrubbing station. (Courtesy of USDA, Office of Information.)

ing.'' With other fruits, as cherries for brining, washing is a supplemental method of cleaning. Fruits that are peeled, such as peaches, pears, apples and apricots, are seldom washed before peeling.

Washing by flooding enormous quantities of water over the fruit is less common now than formerly, due to the increased necessity to conserve water. To be most effective and economical, washing with water must be accompanied with brushing, rubbing, and forcing the water against the fruit and into crevices. Detergents are frequently used in the wash or rinse water. Washing with water is often supplemented with air currents to remove light materials and perforated or rod screens to remove small, heavy materials.

Washing after peeling, pitting, or slicing removes sugars, vitamins, minerals, and other soluble materials and should be reduced to a minimum. Cold water keeps fruit firm and reduces leaching. Fluming in cold water reduces contact with the air and reduces bruising of soft fruit and is a retardant to oxidation if an antioxidant is used.

Of the losses in fruit quality attributed to processing, possibly as much occurs in the washing operation as in all the other steps combined. The effectiveness of the washing operation depends on the amount, temperature, acidity, hardness, and mineral content of the water and the force at which it is applied.

Cleaning Berries

Machine cleaning of berries is a delicate job. Berries are too fragile to survive heavy blasts of air, and water-cleaning systems have never been able to completely remove small particles that cling tenaciously to the curvilinear berry surfaces.

The McLauchlan air-and-water cleaner includes an air cushion dump, an air and vacuum chamber cleaner, a water wash, a dewaterer, and size grader. The machine performs all the functions of a dump belt, a dry cleaner, a washer, and a grader in a distance of 8 ft. In operation, berry flats are dumped directly into the first air chamber. Product damage at this point is minimized by an upblast of air, which cushions the fall of the delicate fruit. As the berries fall, they are actually suspended in air. After entering the machine, the berries pass to a vacuum chamber where vibrating air currents shake out the fruit and air-scrub the berries to remove all loose debris. Berries are then moved to a water bath, sprayed, dewatered, and graded by size. The bulk of the cleaning, including removal of insects and dry matter, occurs by air, while dust and dirt is removed by water.

Electronic Grading of Color

Today, apples and certain other fruits are merchandized by color as much as by taste, size, and shape. It has been established that the skin color of fruits bears a direct relation with fruit quality. Hence, grading on the basis of color has become a key factor in the sorting of fruit.

Colosie and Breeze (1973) developed an electronic apple redness meter for measuring color with a sensitivity to differences in redness closely equivalent to that of the human eye. It was found suitable for defining the minimum acceptable color for red apples ('McIntosh', 'Red Delicious', 'Spartan') without the necessity for separate color calibration for each cultivar.

The USDA color grade of hand-harvested grapes is determined visually. Such a technique is not satisfactory for grading the color of mechanically harvested grapes because the skins are often separated from the pulp and the harvest consists of a mixture of skin, pulp, whole fruit, and juice in a 4 × 4 × 4 ft plastic-lined bin. Fresh grapes must have adequate color if they are to produce concentrated frozen juice of satisfactory color. The USDA color grade of frozen concentrated sweetened juice is based on the absorbance of the juice (pH 3.2) at 340 and 520 nm. Absorbance at 340 nm indicates degree of degradation and at 520 nm indicates the anthocyanin (Acy) content.

Electronic color grading is used extensively with oranges, lemons, and limes, as well as with peanuts, pecans, beverages, and printed labels. Improvements are made constantly in accuracy, speed, and color bands, and fruit processors should keep abreast of new developments.

Air versus Water Cleaning

Mechanical harvesting of seasonal fruits has caused great changes in the cleaning requirements for raw materials. The mobility of mechanical harvesters precludes the use of large amounts of water for in-field cleaning and trash removal. Yet mechanical harvesting can result in large amounts of trash, broken product, and dirt being delivered to the processing plant.

Air cleaning has proven effective for trash removal and separation of loose soil and other low bulk density materials. Air separation is basically a form of pneumatic transport, and is effective only if the physical properties of the unwanted material differ significantly from those of the fruit material being cleaned.

Pneumatic transport of high moisture particulate food presents several major materials-handling problems not encountered in conventional water fluming. Water cushions, suspends, and continuously cleans food particles and conveying equipment. Moist food particles can be damaged during pneumatic transport by hitting the transport pipe at bends and during discharge. Each time a particle contacts the pipe surface, a layer of food material may be deposited. This food material is subject to microbial growth which will further contaminate the food and the downstream transport tube.

The removal of smear soil, insect fragments, plant exudates, or other incidental contaminants by means of air scouring alone does not seem feasible at the present time. The energy needed to release this type of soil from the surface of food can best be supplied by mechanical means combined with a limited amount of water or possibly another liquid. Although air is effective for dry cleaning, it is a poor medium for cooling, blanching, sterilizing, and removing stains or foreign flavors and odors from fruits being processed (Heldman 1974).

Enzymatic browning of peaches, nectarines, apricots, plums, and other fruits that are to be preserved by freezing, drying, and puréeing occurs during harvesting, handling, and processing. This reaction results in deleterious changes in color, aroma, flavor, and nutritional value.

Enzymatic browning can be prevented by several techniques: 1. heat denaturation of the browning enzyme, polyphenoloxidase; 2. exclusion of oxygen by packaging; 3. addition of reducing, antioxidative agents such as ascorbic acid; 4. lowering the temperature to the freezing point of water or lower; 5. lowering the pH with citric or other acids: 6. spraying the fruit during the growing period with plant growth regulators, such as gibberellic acid or ethephon (Paulson *et al.* 1980).

PEACHES, NECTARINES, AND APRICOTS

Peaches, nectarines, and apricots may be peeled by hand; with boiling water or steam; with lye or alkalies (chiefly sodium hydroxide and potassium hydroxide); by dry-caustic peeling with infrared heat; with high-pressure steam; by freezing; and even with acids.

Hand Peeling

Peeling peaches by hand has a few advantages: Little equipment is needed; enzyme stimulation (e.g., heat and alkalies) which increase browning, are not introduced; and the hand-removed peels may be pulped and used for wine, vinegar, and animal feed. Furthermore, a smaller amount of water is used than in other peeling methods, and the wash water is not contaminated with chemicals. However, all of these advantages are more than offset by high labor costs in most areas of the world. As late as 1945, peach peels in the Southeast could be sold for enough to pay for the operation. Hand peeling multiplies the chances of contaminating the prepared peaches with microorganisms.

Boiling Water or Steam Peeling

Boiling water or steam loosens the peel of very ripe peaches, especially freestone or melting-flesh types, in 10-30 sec. In general, early ripening cultivars peel much easier than those that ripen after the 'Elberta'. Middle and late season cultivars may be steam-peeled only after they are allowed to mellow for a couple of days. Since peaches for juice and freezing are usually riper than peaches for canning and pickling, steam peeling is much more adapted to use in juice and freezing plants.

When steam is used for peeling, the fully ripe peaches are pitted and placed on a solid belt, one layer deep with the peel side up. As the fruit passes through a steam box equipped with a series of spray heads, from which steam is sprayed directly on the peach halves, the peel is loosened. Usually a 30-sec treatment is sufficient to loosen the peel enough so that it can be easily removed in one piece with soft brushes and cold water. Although peach peels removed by steam have no value, this method of peeling saves water and does not contaminate sewage water with chemicals.

Lye Peeling

Lye dissolves the cell walls of peach peels at a rate dependent upon the lye concentration and temperature. The particular conditions selected for lye peeling depend on how the fruit is to be processed, the cultivar, the method of removing the peel, and the desire to reduce the water used in peeling and washing.

For peeling peaches for canning or pickling, a boiling solution of 1½% lye for 60 sec, followed by thorough washing and dipping in a solution of 0.5–3% citric acid gives good results. This method requires a considerable amount of water but is fast and cooks the surface of the fruit.

In peeling peaches for freezing or drying, the preferred method uses a higher concentration of lye (about 10%), a treatment time of 4 min, and a temperature no higher than 145°F (Table 3.1). In this manner, the surface of the peaches is not cooked. The last traces of lye are removed first by thorough washing, then by dipping in a citric acid solution.

Many variations have been introduced into the lye-peeling process. One of these is the use of wetting agents, which reduce the wetting time and cut the time of disintegrating the peel to about half. Other variations have been in the direction of increasing the lye concentration, and reducing the amount of water for removing the peel.

Lye peeling is by far the most common method of removing the peel of peaches and other fruits including pears, nectarines, and apricots. This method has the following advantages: (1) is efficient in removing not only the peel but areas of bruised or rotten tissue, thus reducing hand trimming; (2) is geared to quantity production, since a single lye scalder may handle as much as 30 bu/min; (3) is economical (the cost of lye peeling may be as low as 1¢/bu of peaches); (4) is suitable for peeling peaches of all sizes, shapes, and cultivars; (5) is suitable for peeling products other than peaches such as potatoes, sweet potatoes, carrots, and pears; and (6)

Table 3.1. Procedure for Peeling Peaches for Freezing

Alkali (89% caustic, 10% detergent, 1% wetting agent)	90 lb
Water	110 gal.
Time in contact with solution	1–3 min
Time of air exposure and drainage	20–30 sec
Rinsing in water with gentle brushing	30 sec
Dipping in 1% citric acid bath	30 sec

uses equipment that is easy to obtain and install, since all of it may be made of black iron.

In peeling with lye the temperature, concentration, and time must be carefully controlled. Since peach and other fruit tissue cooks at about 145°F, fruit intended for freezing should be held below this temperature at all times (the temperature of the fruit tissue seldom reaches that of the lye).

Dry-caustic peeling for canning is a modification of lye peeling in which little water is used and infrared rays are used as a source of heat. Other modifications in the formula are necessary depending on the cultivar, degree of ripeness, and uses to which the fruit is to be put. Generally, 3–4 lb of caustic are required to peel 1000 bu of peaches, about twice this amount for peeling pears, and about five times this much for peeling potatoes (Cyr 1971; Willard 1971).

Draper-Type Scalder. Several types of scalders or peelers are used for whole fruit, but the most common is the draper type. This is a vat with a width ranging up to 4 ft, a length of 30 ft, and a depth of 1½ ft. An endless, metal-slatted belt with flanges carries whole or half peaches in and out of the lye. With the most common design the peaches are carried above the belt, while with some designs the peaches float and are carried under the belt. In addition to the lye tank, there is a small prewetting bath at the entrance, and provision is made for air exposure and drainage of treated fruit on the other end. The capacity may be increased by extending either the length or width of the vat. A unit about 40 in. wide and 30 ft long should scald 20 bu of peaches per minute.

The scalder should be equipped with adjustable speed control, a thermostat, a lye concentration indicator, closed steam heating coils, and a V-bottom for drainage of sediment. The objection to this type of scalder is that it requires considerable floor space, heat, and water.

Lye-Spray Scalder. Spray scalders are used exclusively for mechanically pitted clingstone and freestone peaches. The peach halves are sprayed on a metal mesh belt with the peel side up, and passed under sprays of lye for the desired length of time. The lye is continuously pumped from a coil-heated vat onto the fruit.

The advantages of this method of scalding, compared with draper-type scalding, are that the peaches are pitted before peeling; they remain on the same belt during scalding and washing; the shape of the fruit is maintained better since the halves are not tumbled in the bath; and the lye does not react with the pit cavity of the peach. Any concentration of lye can be used in the sprays. The disadvantages are that at least four times the belt surface area is needed to scald a given quantity of peaches, as the halves

FIG. 3.2. Rotary scrubber machine for gently removing peel from lye-treated peaches. (Courtesy of Magnuson Engineering Co.)

cannot be piled on top of one another, whereas with the draper-type scalder, the halves can be piled more than four deep. Soft peaches survive a lye-spray scalder with much less loss than a draper-type scalder. There is more mechanical trouble from repumping the lye through spray nozzles, and from the pressure pump on the water line to get high pressure for washing.

The system can be modified to use any concentration of lye for as brief a time as desired, and to allow the wetted fruit to be exposed to air for any length of time. Also, the length of the belt may be extended to allow the disintegrated peel to be brushed off and discarded without its entering the wash water line.

Mill-Wheel Scalder. This is a scalder for immersing whole or halved peaches on a vertical wheel, the lower half of which turns in a bath of lye. It may be of any desired width and is equipped with flanges, placed at intervals, to carry the peaches into and out of the solution. The time of treatment is dependent upon the size and the speed of the wheel.

A mill-wheel scalder requires slightly less floor space, uses less power to operate, and holds less solution than a draper-type scalder. The disad-

Table 3.2. Alkali Required (lb) to Make Up Solutions from 1 to 10% in Peeling Vats of Six Sizes[a]

Desired lye concentration (%)	Capacity of peeling vats (gal)					
	360	384	408	432	456	480
1	30	32	34	36	38	40
2	61	65	69	74	78	82
3	93	99	105	112	118	124
4	125	134	142	150	159	167
5	158	169	179	190	201	211
6	192	204	218	230	244	256
7	226	241	257	272	287	302
8	262	279	296	314	331	349
9	298	317	337	357	376	396
10	334	356	378	401	423	445

[a]To prepare small quantities of lye (5%) for peeling peaches use 0.5 lb of alkali per gallon of water.

vantages are that it has practically no prewetting bath and provides less air exposure and drainage after the peaches are treated with the lye (Woodroof *et al.* 1947).

Preparation of Lye-Peeling Solution. Table 3.2 indicates the number of pounds of alkali required to make up lye solutions from 1 to 10% in peeling vats of 6 sizes. Calculations may easily be made for vats of any size:

Determine the capacity of the vat in cubic feet when filled to the desired level, then multiply by 7.09 to determine the capacity in gallons. In column one, find the lye concentration desired, then follow the horizontal line to the column indicating the capacity of the tank; this figure will be the pounds of lye needed. Hence, to make 360 gal of 5% lye solution 158 lb of alkali are required.

Vats should be filled to capacity with water and the alkali scattered over the surface with gently stirring. The alkali will cause the solution to heat rapidly, and if added in large quantities will form hard lumps (Woodroof *et al.* (1947).

Dry-Caustic Peeling

Dry-caustic peeling was developed in an attempt to reduce the waste generated in peeling potatoes, and became the basis of an industrial process known as the USDA–Magnuson Infrared Anti-Pollution Peeling Process (the Canner/Packer Gold Medal Award for 1970). The new process uses infrared energy at 1650°F to condition the surface of fruit treated with strong sodium hydroxide solution. The peel can then be removed mechanically by soft rubber scrubbing rolls rather than by water as is done in conventional caustic peeling. A final spray using low volumes of water removes residual peel fragments and excess sodium hydroxide. The effluent from the peeled fruit rinsing may be combined with the solid material generated to produce a thick, pumpable sludge.

With the dry-caustic peel system about 90% of the peel is removed as a thick, heavy "peanut butter-like" substance, which is trucked off as is. The remaining 10% is run through 20-mesh screens and pumped to a clarifier. With a conventional or wet-caustic system, the entire peel loss material goes down the flumes, passes through 4-mesh screens, and then is pumped to a clarifier.

Fruits are flumed into the processing plant from storage and held in hoppers for even flow to the peel lines. From the hoppers the fruit is weighed on conveyor-scales, and then passed over shuffles into infrared units, thence to scrubbers, and finally to washers. The scrubbers and washers are designed to use a minimum of water. Thus, the hydraulic BOD and suspended solids loads for primary treatment are markedly reduced when dry-caustic peeling is used (Cyr 1971).

The plant consists of a metering bin to regulate the flow of product, a caustic immersion unit, a holding conveyor, a gas-fired infrared treatment unit (which aligns the product for infrared treatment), a special scrubber to wipe off the softened peeling without the use of water, a special brush washer, which uses minimal water for final cleaning, and a neutralizing tank.

The process concentrates the peel waste into a thick, pumpable residue that can be converted into animal feed or returned to the soil. There is substantially less peel using this system than when using conventional methods of lye peeling or mechanical peeling (Anon. 1970).

Freeze Peeling

Freezing loosens the peel of very ripe, melting-flesh peaches. The procedure and equipment are very similar to those for steam peeling. Recently developed equipment for cryogenic freezing is causing a revival in the method; and, this, together with the clean environment resulting from freeze peeling, may render it more practical than was once thought.

In freeze peeling the peach is frozen very quickly to a depth slightly below the skin, then thawed rapidly. The flesh of the peach is not frozen and releases the skin easily and quickly. Since freezing stimulates enzymes that cause browning, the surface of freeze-peeled peaches should be treated with ascorbic acid or other anti-browning agents immediately (Woodroof *et al.* 1947).

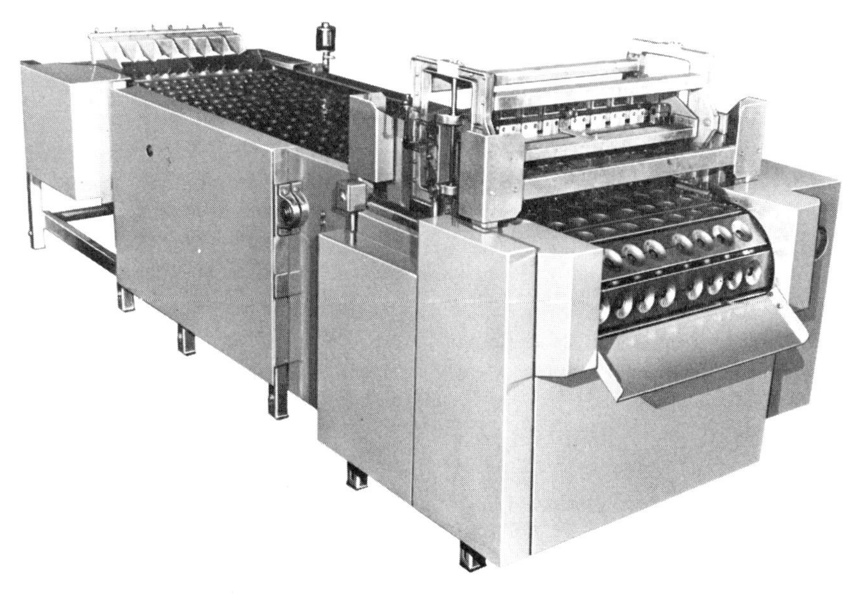

FIG. 3.3. Freestone peach pitter. Only two people are needed to pit 400 peaches per minute and the unit can replace eight hand-fed pitting machines. (Courtesy of Food Machinary Corp.)

Acid Peeling

Peaches have been successfully peeled by immersing in a hot solution of 0.1% hydrochloric acid, 0.05% oxalic acid, 0.1% citric acid, or 0.1% tartaric acid. Acids disintegrate the peel rather than loosening it, and removing the peel and chemical requires considerable water. However, there no browning or other forms of oxidation occur with this method of peeling. The big drawback to chemical peeling is the corrosive effect of the solutions on metal equipment (Woodroof *et al.* 1947).

High-Pressure Steam Peeling

Merkel (1970) developed a batch peeler with regulated pressure control that takes advantage of steam peeling but avoids the difficulties encountered with a conventional steam peeler. This system effectively separates the peel from the flesh for a variety of fruits.

The criteria for evaluating a peeling process are (1) the effectiveness of removing the peel, (2) the quality of the flesh remaining, and (3) the product loss due to peeling. The end result is primarily a function of the temperature-time relationship to which the fruit or vegetable has been exposed.

During high-pressure steam peeling an internal pressure is created within the product when it is exposed to a high-pressure temperature environment. This internal pressure is the result of vaporizing the liquids beneath the skin surface by raising them to above their boiling point. As long as the surrounding pressure is the same as the internal pressure, no reaction takes place. This condition exists while the fruit is exposed to high-pressure steam during the peeling time.

As soon as the chamber pressure is released, however, the pressure surrounding the product becomes lower than the internal pressure of the fruit. This pressure differential creates an internal force that acts against the skin, physically forcing it away from the flesh. Skin removal is facilitated by a softening of the cell tissue just beneath the skin layer. This action is most likely caused by several factors associated with the heat, such as the hydrolysis of carbohydrates and degradation of pectin substances within the cells. The other aspect that needs to be considered when processing produce is what effect the process has on the quality of the product. The quality of a product can be broken down into three main categories: namely, appearance, texture, and flavor.

Firmness of many products is considered to be affected by either a change in cellular organization, usually involving permeability changes, and/or activity of the pectic enzyme systems. Similarly, the polyphenolase system, present in many fruits, is responsible for discoloration or browning following peeling. Most fruits contain both the enzyme system (polyphenolase) and substrates (phenolic compounds required for browning and are susceptible to darkening when cellular organization is disrupted. Since there is very little that can be done with respect to the oxidizable substrate, attention will be focused on the behavior of the enzyme and what needs to be done to inhibit or inactivate its activity. When a product is heated to a temperature somewhere between 140° and 190°F, the normal respiratory processes become inactive while the polyphenolase system becomes more active. It is assumed that subjection to such temperatures permits reaction of the enzyme and substrate and thus completes the biochemical reaction resulting in browning. The phenolase enzyme can be inactivated if its temprature is raised to approximately 200°F. Thus a very precise time-temperature relationship in the area subject to browning must exist if discoloration is to be avoided.

In the case of apples or peaches, a brown ring is observed at some definite point below the surface of a steam-peeled fruit. The polyphenolase system in the portion within the ring and the outer surface has been inactivated. In the portion within the ring, the temperature rise has not disrupted the respiratory mechanism. But in the area between these portions, the temperature-time relationship requisite for browning has occurred, hence the development of the brown ring.

FIG. 3.4. Canning peaches,Sunnyvale, Calif. Peach halves in size grader. (Courtesy of USDA.)

According to the foregoing discussion, it is apparent that in order to prevent this enzymatic oxidative process from occuring, one of the following conditions must be met: (1) inactivation of the enzyme either by heat or by chemical inhibitors and/or antioxidants; (2) removal of oxygen from the product and its surroundings; (3) selection of a variety of fruit or vegetable having a low phenolase activity.

PEELING AND PREPARATION OF PEARS

Greatly increased yield of pear halves for canning has resulted from application of chemical and steam peeling, mechanical coring and slicing, and size grading by California canners and growers.

Traditionally, Bartlett pears to be used for canning are picked green and placed into bins in the orchards. The fruit is then trucked to the plant where it is then held in cold storage at about 30°F for a period of 5 days to 2–3 months. When removed from cold storage, the pears are size graded by a diverging rope sizer and conveyed to a bin filler that returns them, by grade sizes, to bins. Filled height of the bin is controlled by an electric eye which interrupts power to the bin filler when a predetermined height is reached.

In this size-grading operation, the fruit is graded into five sizes, three of which are for grade pack, while the other two are destined for cocktail production. Fruit used for grade pack ranges from 2⅜ to 3 in. diameter.

Bins of fruit are ripened in the canner's yard for a period of 5–7 days, depending on maturity. Ripeness is determined by measuring the mechanical pressure necessary to rupture the surface of the pear after peeling. An average of 2.5–3 lb is desirable. When ready for processing, bins of ripened pears, each containing up to 1100 lb, are dumped mechanically at speeds of 35–40 bins/hr, and are conveyed to the chemical application section of the peeling system.

Coming off the conveyor, the fruit enters a multilane, single-item shuffle-type feeder, the flight of which takes a predetermined number of fruit and deposits them into a bucket-type conveyor, which carries the fruit through the chemical applicator. This conveyor is driven by a variable-speed motor, which is manually regulated, thereby providing control of the length of time the fruit is subjected to the peeling solution.

Solution temperature is automatically regulated. Chemical concentration is manually regulated. The solution may be checked at any time by means of a permanent hydrometer mounted in the system. Concentration measurements are made by determining temperature and measuring specific gravity of the solution.

Today, chemical peeling systems use sodium hydroxide plus tergitol, or other wetting agent. Previous chemical peeling systems for pears left more dark spots on pears than are normally encountered. Current methods with a blast of steam both aid in peeling and tend to minimize spots.

Leaving the chemical bath, the fruit is subjected to pressurized steam, the intensity of which is predetermined, depending upon the maturity of the fruit. Regulation is automatic. The valve is built to ASME specifications and is capable of operating pressures to 120 psi. This steam–chemical method of peeling is similar to that used on sweet and white potato peeling lines.

Fruit is discharged into a washing section specifically designed for this system. It consists essentially of a rotating cylinder lined with a layer of rubber. Within the washer, a nylon brush rotates in the opposite direction of the drum's travel. Two bumpers, running the length of the cylinder, distribute and tumble the pears during the scrubbing action. The fruit is guided by an internal stainless steel spiral and is ejected upon completion of the scrubbing cycle into holding flumes. In normal operations these flumes merely constitute a surge area for the fruit, and use plain water.

Coring and halving complete the peeling system. Pears are aligned and the stem end is located. The fruit is transferred into the cups of the machine's coring section, stem end down. As the fruit enters each cup in this section, the cup centers the pear and holds it vertically for stemming, coring, and splitting operations. The stemming and coring machine consists of a series of flights, each containing 8 cups; each machine is capable of providing 50 flights per minute or 400 pears per minute.

Stems are removed with a stemming tube that bores a hole through the long axis of the pear. Next, a coring knife enters the pear through the hole made by the stemmer, expands inside the pear and cuts out the core. The coring knife is expanded in proportion to the height of the pear in the cup. With knife peelers, the coring knife must be changed with every change in pear size processed through the machine. Also with knife peelers, size grading ahead of the machine is required for good cored pear yields, otherwise excessive fruit is removed by using larger coring knives than necessary.

Inspection Table

After coring, pears are sliced into halves by a splitting knife and are flumed to an inspection table for cup-up inspection; then they are conveyed over two lines of inspection tables for cup-down examination with five or six inspectors on each line to examine pears for defects that were not removed during the operation. Spots are then hand-trimmed.

Pear halves next pass to a machine that turns the halves over, placing them cup down on six rubber conveyor belts.

Electronic Grading

Size grading is provided by a unique device consisting primarily of three beams of light plus their associated photo cells. Pear halves on the conveyor, cup down, break the beam of light between the first lamp and the photocell detector and then break the second pair. A third lamp stationed above the conveyor grades for proper width of the pear. Suitable guide rails channel the pear half through a light beam which has been adjusted for a predetermined width.

Neither nose-first nor nose-last position of the pear half is important since either will break the width light beam. As long as the initial length sensor light is held closed and the pear half has broken the width light, a solenoid is actuated that will eject the pear. If the pear is either too short or too narrow, it will continue on to the next station where a similar size evaluation is performed. The grader consists of six belts with four grading stations per belt and has an operating speed of approximately 300 ft/min. The six lane grader will handle 2400 pear halves/min.

Quality grading consists of three separate inspection tables for the assorted pear halves. Initially they are inspected in the cup-up position for any interior coring defects; then they are turned over for inspection, cup-down, for any surface blemishes. After the grading operation, a third inspection occurs; this is for any quality defects or mis-sizing prior to being packed by hand-packed fillers into all of the commonly used container sizes. They are then conveyed by cable to the syrupers and exhaust boxes. Filled cans are closed and cooked by conventional methods.

APPLES

Peeling Apples with Lye

Apples for processing are dewaxed by a 30-sec exposure to vapors of boiling isopropyl alcohol at 178°F, followed by a 2-min immersion in 10% lye at 140°F plus 0.5% detergent as a peeling aid, and rinsing in 5.0% citric acid to neutralize residual lye. The peel is removed by light brushing and rerinsing. Lye peeling is much faster with less waste than mechanical peeling.

Prevention of Browning

When apple slices are exposed to oxygen, the oxygen reacts with chemical compounds (phenols) in the apples causing apple pieces to turn brown. Now, treating apple slices in an alkaline medium changes the phenol compounds and inhibits the browning process. This is accomplished in a three-step process. Slices are dipped in a bisulfite solution immediately after they are cut to prevent color change. Then they are soaked in a weak solution of potassium phosphate to remove excess bisulfite and supply the necessary alkaline medium. Treated slices are then packaged.

The change from acidity to alkalinity allows the process known as methylation to occur. It does not occur naturally because of the natural acidity of the fruit. The chemically induced change from acidity to alkalinity is slight and does not cause alteration in the fruit's surface texture, flavor, or color. The treated apple slices can be held up to 6 weeks. Previously the natural flavor, texture, and color of apple slices could be held only by freezing. Untreated apple slices may be held for only a week under refrigeration.

Refrigerated Apple Slices

Refrigerated, unfrozen sliced apples have been used extensively for pies and other bakery products because of their convenience, year-round availability, and firm texture. Treatments to protect the color of these sliced apples have usually involved a sulfite dip, sometimes with an added treatment. Processing of apple slices for refrigerated storage differs from that for freezing, since the latter destroys cell organization and allows enzymes and substrates to mix. In frozen apple tissue there is a considerable amount of oxidation catalyzed by polyphenol oxidase which occurs rapidly and causes browning, especially during thawing.

In refrigerated apples, the cells remain intact and alive except on the surface. Thus, treatment of the surface is all that is necessary, in contrast to penetrating treatment necessary for freezing. For slices for refrigeration it is unnecessary and undesirable to use a penetrating sulfite treatment. Penetration of sulfites is influenced by pH; acid solutions soften apples and alkaline solutions harden them, except at high pH. The maximum firming effect is at about pH 9. The kind of acid used also influences not only the shear strength but the tendency to darken. Sulfurous acid is best for maintaining a light color but softens the fruit severely. Acetic acid both softens and darkens apple slices.

Softening of sulfited apple slices is more extensive at low than at high pH values. The degree of penetration of SO_2 is also greater at low pH values. An alkaline sulfite dip causes some softening, though it may approach zero. Since refrigerated apple slices are used mainly because their texture is crisp, the process that causes the least penetration of SO_2 but maintains color is best, provided the flavor is not adversely affected. This suggests an alkaline rather than an acid sulfite dip.

Either a sulfite dip followed by an alkaline buffer dip or a single alkaline sulfite dip effectively inhibits enzymic browning in the surface. The residual SO_2 on or in the slices rapidly drops to zero in storage. At this point browning may occur, especially if the surface is rough.

Calcium treatment of sulfited apples has been found very effective in firming apples in alkaline solution or suspension but not in acid solution. With unsulfited apples there is a firming effect in both acid and alkaline solution. Calcium sulfite solutions buffered with sodium carbonate-bicarbonate, in which the calcium should be practically insoluble, is effective. Actually, 0.2% calcium in such a buffer may make apples too firm and woody. Furthermore, SO_2 is considerably protected from oxidation in an alkaline calcium sulfite solution, so that SO_2 is retained during storage of apple slices and color is preserved longer.

Thus, by using suitable proportions of calcium and SO_2 in a dip of the proper pH, the qualities of firmness, color, and flavor can be balanced to give the most desirable product after storage. In this manner the storage life can be extended from a maximum of about 3 weeks to as long as 8 weeks at 34°F (Ponting *et al.* 1971).

Fortified Apple Juice

Fortification of apple juice with ascorbic acid tends to reduce cloudiness and sediment formation but increases coloring at higher storage temperatures (100°F) due to nonenzymic browning of ascorbic acid oxidized products. The apple catechins are similarly involved as proanthocyanidins in color and sediment formation but to a much lesser extent.

Apple proanthocyanidins play an important role in the processing of apple juice and the stability of the product during storage. The minor constituents of apples are extremely susceptible to enzymatic and nonenzymatic oxidative polymerization.

Enzymatic oxidation during milling and pressing apples greatly reduces the amount of proanthocyanidins (leucoanthocyanidins). The oxidation can result in complete disappearance of the lower molecular weight substances. The oxidized products remaining after the clarification process contribute to juice color. The polymeric material can further polymerize, particularly at higher temperatures, to produce an increase in color and sediment formation (Johnson *et al.* 1969).

CHERRIES

Bleaching Cherries

Fresh sweet cherries are harvested and delivered to processing plants in New York during late June. The cherries are immediately placed in 20,000-gal. wooded vats containing sulfur dioxide brine. After approximately 2 months the brine has bleached the fruit and removed the natural flavor. The brine also acts as a preservative, enabling the cherries to keep for as long as 3 years (see Chapter 9).

Cherries are removed from the brine and washed, sorted, and pitted. Electronic sorting machines scan each cherry for surface defects, rejecting those with blemishes. Passed cherries are conveyed to pitters rated at 700 lb/hr, thence to stainless steel processing vats holding 6500 lb of pitted cherries.

Brine is prepared by drawing softened water into 100-gal, cylinder mixing tanks. Spices, coloring, apple vinegar, and sweetener are added to the water and blended together. Treated water eliminates iron salts, which could affect the final color. Sweetened brine is pumped to the cherries in the processing tank; then they are heated as the brine is circulated once daily for 12 days. Additional sugar is added to the sweet brine to 45° Brix.

Spiced pickled cherries are filled into jars as they pass through at approximately 250 jars/min, using an automatic tumbling filler. A 49° Brix syrup is used as a packing medium, added just ahead of the screw cap closure machine. The jars are steam-pasteurized, rinsed with hot water, dried, cooled, and labeled.

Secondary Bleaching of Cherries. Cherries are brined by placing them in solutions containing 0.75–1.50% sulfur dioxide in addition to calcium ions, which firm and bleach the cherries to a pale yellow color and act as a preservative during storage. Sulfur dioxide does not completely remove natural coloration of highly pigmented cultivars such as 'Bing', 'Black Republication', or 'Windsor', nor does it effectively remove discoloration of the fruit caused by bruising, sunburn, limb rub, and wind whip. These discolorations appear as dark blemishes on the brined cherry and become more pronounced when the cherries are artificially colored. These blemishes give the colored product a mottled appearance.

To efficiently remove discolorations on the fruit after brining, several types of bleaching agents have been used as secondary bleaches following the normal brining process. Hypochlorites have been used as an effective means of color rcmoval. However, those agents frequently produce off-flavors, loss of firmness, and a brown color reversion in the bleached product. Hydrogen peroxide, 1.0% at pH 11.5, has been used as a secondary bleaching agent; but cherries treated with hydrogen peroxide still have a yellow color and soft texture.

Beavers and Payne (1969) successfully bleached brined cherries with sodium chlorite ($NaClO_2$). The bleached product is snowwhite in color, entirely free from off-flavors, and of firm texture. This is one of the chemicals "generally recognized as safe" (GRAS) by the FDA when used at a concentration not to exceed 0.75%, followed by leaching in cold water for removal of residues.They found reduction in bleaching time and chlorite consumption was most effective when the bleaching solution was below 110°F and in the pH range 4.0–6.0

Factors Affecting Cherry Scald

The threat of scald, a discoloration defect, forces severe restrictions on both growers and processors in the tart cherry industry. On hot days, growers may suspend harvesting operations because of excessive scald development. Processors often schedule plant production on the basis of scald development in cherries held in soak tanks. The coming of the mechanical harvester and its associated innovations in handling, sorting, and processing methods aggravate the scald problem.

Foremost among factors that affect scald is bruising. Unbruised cherries do not scald. Fortunately, cherries can withstand one bruise without scalding, provided they are maintained at a cool temperature. If, however, they are bruised a second time after a delay period, they rapidly develop scald even at a cool temperature.

In commercial practice, the rebruising that occurs during orchard sorting, handling at receiving stations, and unloading at processing plants may be particularly harmful. Many grower–processor teams now avoid secondary bruising by retaining cherries in their original orchard tanks until time for processing. This practice has reduced scald and increased pack-out yield.

Scald is sharply inhibited by cool temperatures. For instance, scald counts were 50% when cherries were soaked at 90°F, and only 4% when they were soaked at 40°F. The industry is making widespread use of temperature as a means of controlling scald.

Under some conditions, the bubbling of air through soak tanks inhibits scald. At cool temperatures (50° and 40°F), however, no beneficial effects from aeration were obtained. In fact, scald content was increased slightly. In semicommercial tests there were 13.9% of scald with nonaerated samples soaked for 11 hr at 60°F, and only 2.5% of scald with corresponding aerated samples. An additional 2.3% of the aerated cherries developed brown spots. Since aeration promotes oxidative browning of bruised tissues, aerated cherries may show increased numbers of dark brown defects in the heat-processed product.

PITTING DATES

One of the problems in pitting dates, cherries, apricots and other clingstone fruits is preventing adherence of juices, sugar, and other ingredients of fruits to the pit-removing apparatus or to other fruits.

Fehimann (1982) patented a complicated apparatus for pitting dates, cherries, and similar fruits. The machine positions the fruit in sockets on an endless apron, where it is deseeded and the belt washed for continuous refilling.

PESTICIDES IN HANDLING FOOD

The federal government is committed by law, policy, and tradition to assuring that the nation's food supply is safe, clean, and wholesome. Accordingly, the development and widespread adoption of DDT and other new chemical pest killers in food production following World War II prompted the government to strengthen and expand its regulatory authority over the entire range of economic poisons in the interest of both user and consumer protection. The result has been enactment by Congress of a series of statutes regulating pesticides and pesticide residues entering interstate commerce, with emphasis on safety.

Under the Federal Insecticide, Fungicide, and Rodenticide Act, all pesticide products must be registered with the U.S. Department of Agriculture before they can be marketed across state lines. Before registration is granted, the manufacturer must furnish scientific evidence that the product is effective as claimed and will not injure humans, crops, livestock, and wildlife when used as directed on the label.

Pesticide residue tolerances are established on raw agricultural commodities, and the residues in processed foods are a fraction of the amount permitted on the raw agricultural commodity.

The USDA has urged industry to develop more highly selective pesticides that will not persist beyond a single growing season so that no residue remains in marketed food crops. Much research has concentrated on the development of biological, cultural, and other alternative methods of control that will avoid residues and other pesticide problems while effectively controlling or even eradicating the target pests.

Contamination of the environment in which food crops are grown can be as important to the producer and processor as residues from direct pesticide applications (Campbell 1970).

IRRADIATION OF FRUITS

Beginning about 1945 extensive research was conducted by universities, government agencies, and commercial companies trying to use irradiation as a means of preserving fruit for the fresh market or for processing. Early optimistic reports were based on results with stationary equipment. Extensive studies in California (Maxie *et al.* 1971) employing actual and stimulated transit equipment, have yielded largely negative results. They concluded that apples, apricots, avocados, bananas, boysenberries, cantaloupes, lemons, limes, mushrooms, nectarines, oranges, peaches, pears, and other fruits offer no promise for commercial irradiation, in the light of alternative procedures that are cheaper, noninjurious, and more effective.

Strawberries are the only domestic fruit with even a remote potential for commercial irradiation. Conventional refrigeration is still the best means of reducing postharvest losses of fruits for processing.

Many fruits have not been studied in detail, but the diversity of species covered and the overwhelming negative results obtained indicate that irradiation holds little promise for fruits or most other perishable commodities.

REFERENCES

ANON. 1968. Bleaching process saves bruised sweet cherries. Canner/Packer *137*(11) 46.
ANON. 1970. Infrared peeling cuts wastes, ups yield. Food Eng. *42*(7) 70–71.

BEAVERS, D. V., and PAYNE, C. H. 1969. Secondary bleaching of brined cherries with sodium chlorite. Food Technol. *23,* 175–177.

CAMPBELL, J. P. 1970. A regulator's view of pesticides in foods. Canner/Packer *139*(2) 19–21.

COLOSIE, S. S., and BREEZE, J. E. 1973. An electronic apple redness meter. J. Food Sci. *38,* 965–967.

CYR, J. W. 1971. Dry caustic vs conventional caustic peeling and its effect on waste disposal. *In* Proc. 2nd Symp. Food Processing Waste. Natl. Canners Assoc., Washington, D.C.

FEHLMANN, VIKTOR. 1982. Apparatus for pitting cherries, dates and the like. U.S. Pat. 4,313,373. Feb. 2.

HELDMAN, D. R. 1974. Air as a substitute for water in food processing. Food Technol. *28,* 40–47.

JOHNSON, G., DONNELLEY, B. J., and JOHNSON, D. K. 1969. Proanthocyanidins as related to apple juice processing and storage. Food Technol. *23,* 1312–1316.

JULIEN, HENRI C. P. 1981. Sulfite treatment of cherries, U.S. Pat. 4,271,204. Jan. 2.

MAXIE, E. C., SOMMER, N. F., and MITCHELL, F. G. 1971. Infeasibility of irradiating fruits and vegetables. HortScience *6*(3) 202–204.

MERKEL, G. J. 1970. A high-pressure steam peeler for fruits and vegetables. Maryland Expt. Stn., Dept. Agr. Eng., Paper A-1660.

PAULSON, A. T., VANDERSTOEP, J., and PORRITT, S. W. 1980. Enzymatic browning of peaches: Effect of gibberellic acid and ethephon on phenolic compounds and polyphenoloxidase. J. Food Science *45,* 341–348.

PONTING, J. D., JACKSON, R. and WATTERS, G. 1971. Refrigerated apple slices: Effect of pH, sulfites and calcium on texture. J. Food Sci. *36,* 349–350.

U. S. PATENT 4,271,204, Jan. 2, 1981. Sulfite treatment of cherries, Henri C. P. Julien, Vaucluse, France. Assignee: Aptunion, Vaucluse, France.

U.S. PATENT 4,313,373, Feb. 2, 1982. Apparatus for pitting cherries, dates and the like. Viktor Fehlmann, Asignee: Ferrum AG, Rupperswil, Switzerland.

WILLARD, M. J. 1971. Infrared peeling. Food Technol. *25,* 27–32.

WOODROOF, J. G., SHELOR, E. and CECIL, S. R. 1947. Preparation of peaches for freezing. Ga. Expt. Sta. Spec. Rep. Bull. 251.

4

Seasonal Suitability of Fruits for Processing

J. G. Woodroof

Until about 1950, every major peach-, pear-, grape-, apple-, or cherry-producing area had a processing plant for canning, freezing, drying, or preserving the fruit as a primary or secondary product. Since that time there has been much merging and consolidation of processing operations resulting in fewer, but much larger, plants. While fruits have been hauled from one area to another for processing, many developments have led to more uniform products. These include growing the same cultivar for processing, mechanical harvesting and grading, following a succession of ripening dates, uniform system of purchasing from farmers, and broader sales programs.

At least one fruit is grown in practically every state for commercial processing, and in most cases there are several. When one fruit is grown, there are usually yearly, midseason, and late cultivars to extend the processing season. After a commercial fruit processing plant is established, efforts are made to extend the processing season by hauling in fruit from other areas and by developing additional fruits for processing.

For example, a peach canning plant in Fort Valley in middle Georgia operates for 3 weeks on locally grown fruit, 2 additional weeks on fruit from north-central Georgia, and 2 more weeks on fruit grown in South Carolina; thus, the peach canning season is extended for 7 weeks. This is

Commercial Fruit Processing, 2nd Edition

ISBN 0-87055-502-2

made possible by precooling and overnight hauling of the fruit, but without storage.

Some fruits—avocado, bananas, breadfruit, kumquat, crabapple, persimmon, pomegranate, and tucuma—are processed into just a few products; others, such as apples, blackberries, peaches, and raspberries, are processed into more than 20 products. (See Table 1.1, which shows 52 fruits and 36 products made from them.)

Some fruits have only one date and place of processing (dates and olives in California and pineapples in Hawaii), whereas apples are processed in 25 states, peaches in 24 states, and strawberries in 23 states.

PRIMARY CLIMATIC FACTORS

Many factors determine the suitability of a region for growth of fruit for processing. The most important factors are discussed in this section.

Latitude

The southern and northern boundaries for growing a fruit are determined by the length of day, mean temperature, dates of the first and last killing frosts, maximum and minimum temperatures, number of hours per year that the temperature is below 45°F, and other conditions. Within these limits will be found the optimum conditions for growing the fruit. Breeding more hardy or more tender cultivars, or introducing new cultivars or species from another land, are ways of extending the production area of fruits.

While fruits may be grown as a novelty out of their adapted area (bananas in Georgia, pineapples in Michigan or New Jersey, cranberries in Louisiana, for example), for commercial production they should be well adapted to the area. Attempts to grow large orchards of fruits outside of their optimum climatic growing conditions generally are unsuccessful.

There is a close similarity between fruits grown in the Northern Hemisphere (United States) and those grown—an equal distance from the equator—in the Southern Hemisphere (Argentina, Chile). Many cultivars of apples, peaches, grapes, nectarines, quinces, and others are grown in both areas in the opposite winter/summer seasons. This means that July peaches grown in the United States can be made available in Argentina in the middle of its winter; and fresh fruits ripening in Argentina and Chile in December can be made available in the United States during the coldest weather.

This continuous exchange of fresh fruits between these hemispheres greatly reduces the need for commercially canned, frozen, and dried fruits.

Effect of Location on Quality of Grapes. Comparative studies on 'Concord' grapes grown in South Carolina and New England show that ripening time, determined in large part by temperature conditions, contributes to variations in the biochemical composition of the grapes and, in turn, the final amount of processed product obtained. Although soil type, fertilization practices, and spray treatments all affected this biochemical makeup, temperature appeared to be the major factor contributing to the differences in the grapes. Time of harvest was cited as crucial in determining raw product quality.

Grapes should be harvested after final sugar builds up but before physiological breakdown. High temperature at harvest impedes sugar development and also accelerates decrease in acid content.

Time of ripening in the two growing areas under study differed considerably. In South Carolina, ripening occurs during peak heat in August. Under these conditions, ripening is very rapid, as is subsequent physiological breakdown. Researchers believe that respiration of fruit sugar and probably fruit acid also occur while fruit is still on the vine. In New England, final sugar buildup takes place late in September when temperatures are cool and often near freezing.

If harvest takes place when grapes contain below optimum sugar levels, color, aroma, and flavor also have lower values. Lower values for these properties result in fewer grape solids per acre and thus less final processed product.

Values for sugar as soluble solids, acid as tartartic, and methyl anthranilate, a major flavor component in 'Concord' grapes, were recorded from both growing areas after harvest. Grapes grown in Geneva, NY, had the following values: sugar, 18.4%; acid, 1.11%; methyl anthranilate, 1.67 ppm. Values for Concords grown in Clemson, SC, were sugar, 15.0%; acid, 0.95%; methyl anthranilate, 1.19 ppm.

Level of tartartic acid affects establishment of desirable sugar to acid ratios in pressed juices and is related to formation and removal of insoluble bitartrates during processing.

Altitude

There is a close relation between the optimum latitude and altitude for growing fruit—higher elevations are equivalent to higher latitudes. For example, apples and grapes can be successfully grown in Georgia only at elevations of 2000 ft or more; at lower elevations, the Georgia latitude is not suitable for growing apples. Furthermore, growing on the northern slopes of mountains is equivalent to growing at much higher latitudes than growing on the southern slopes.

At a certain latitude, some fruits exhibit a fairly specific optimum elevation for growth; at a higher or lower latitude, the optimum elevation would be lowered or raised, respectively. Only by experience can the optimum elevation on a mountainside for growth of various fruits be determined.

To provide air and water drainage and to avoid "dead air pockets" that favor frosts, fruits should be planted on "rolling land." To encourage early fruiting, the southern slopes should be used; fruits planted on northern slopes may ripen 10–15 days later. This difference in ripening time may be a means of extending the processing season.

Closeness to the Sea

Seashore orchards are usually on sandy soil that has been leached of many nutrients. The water table is high, and while air drainage is poor, sea breezes keep the air circulated. The high humidity is favorable to some fruits (blueberries, cranberries, citrus) and unfavorable to others (apples, dates, grapes, olives, pears). Seashore orchards are virtually free of extreme high or low temperatures, and are consequently less subject to cold injury to the trees or frost damage to the blossoms and fruit.

Closeness to Lakes

Fruit growing in the eastern United States owes much to the Great Lakes. If it were not for them, the fruit belt (apples, berries, cherries, grapes, peaches, pears) of Pennsylvania, Michigan, and New York would not be possible. The lake waters temper the northern winds for up to 40 miles on the southern and eastern shores of the Great Lakes and delay freezing in the fall. This permits growing and late processing of fruits that are impossible further inland.

Lakes in California, Oregon, Washington, Florida, and Tennessee are also favorable to fruit growing by tempering the air and reducing damage from frost, freezing, and ill effects of very low humidity. The high water table in these areas provides a source of water (by pumping) for processing and irrigation.

Soil Type

Good soil of suitable texture, structure, and composition is necessary for fruit production. The texture should be a loamy topsoil, 1–6 ft deep, to hold fertilizer and humus, with a semiclay subsoil to hold moisture and prevent leaching. Given this condition, fertilizing and irrigation are profitable; otherwise, both are without profit. Small pebbles in the topsoil and subsoil are beneficial, but large pebbles permit too rapid drainage and quick drying out.

The subsoil should be firm without an impervious hard pan and loose enough to be tilled. The surface should be pebbly/sand graduating into sandy/clay subsoil.

Given a suitable texture and structure, the composition of the soil (pH, nitrogen, phosphorus, potassium, calcium, magnesium, iron, humus, and minor elements) may be adjusted to the requirements of the particular fruit. Blueberries and cranberries require an acid soil of about pH 5; many others including apricots, grapes, olives, and quinces tolerate a fairly alkaline soil of about pH 7.5. Most other fruits including apples, berries, cherries, and peaches prefer a neutral soil (pH 7).

High soil alkalinity in arid fruit-growing regions of the world, especially in California, is a real problem. Alkalinity is increased by irrigation, because alkaline salts remain after the water evaporates from the soil or transpires from the leaves.

Throughout the United States and much of the world the most deficient element in fruit-growing soils is nitrogen and humus. Nitrogen can be added in the form of ammonium sulfate, which partially corrects alkalinity at the same time; nitrogen also may be added in the form of calcium nitrate, thus increasing the calcium content of the soil at the same time. Sodium nitrate has been found mildly toxic to peaches on alkaline soils, particularly in Argentina.

Other elements may be added to the soil as needed for the particular fruit, based on a soil analysis.

Rainfall and Available Irrigation Water

Water is deficient in more than half the fruit-growing areas of the United States and worldwide. Furthermore, additional areas could be brought into profitable production if water were available. Thus the conservation of rainfall and supplying additional water from streams and wells are major considerations in fruit growing. Water for the various steps in processing is an additional need. Most of the water for processing needs to be of a higher grade (chemically and biologically pure) than that for spraying, irrigation, and general cleaning.

Since water from streams or wells has been subjected to filtering through the soil, it contains many minerals that must be considered. The composition of well water varies depending on the depth of the well, the geological formation of the subsoil, distance from mountains, and other conditions. The quantity and quality of water for growing and processing fruit is extremely complex and variable, and can be controlled only by engineering skills based on chemical and biological analyses.

The accumulation of sulfuric acid and nitric acid from "acid rain" in fields and lakes is a problem in industrial areas. Sulfur dioxide and nitrous oxide gases from burning fossil fuels and volcanic ash accumulate wherever rainwater goes. The gases travel for thousands of miles in the air, and once in water are toxic to fish, animals and food products.

PRIMARY INDUSTRIAL FACTORS

In many cases, the market determines when, where, and how fruits are processed. For example, a few processors cater to supplying juices, jams, and canned fruits for military procurement in which Army Federal Specifications are followed as to type of containers, processing procedures, and labeling. The orders are delivered in a few large batches during the year. The location of the processing plant and the source of raw materials are of minor importance.

With fruits processed for institutional use or for reprocessing, the end users usually write the specifications and institutional-size containers are used. These products may or may not be exported.

By far the greatest quantity of processed fruits is for home use and in consumer-size containers. USDA Standards for Grades are used for selecting and grading the fruit, and FDA Standards of Identity specifications are followed for processing and labeling. These products are mostly canned or frozen concentrated juices, canned or frozen fruits, and jams and preserves; they are suitable for retail or wholesale sale anywhere in the world. Processing is in all major fruit-producing areas of the cou ntry.

Source of Water Supply

As fruit processing plants become larger, and operate more months of the year, more water is required; and as pollution and waste disposal become greater problems, there are efforts to use less water, or to recirculate some of the water. Thus, the quantity and quality of water needed for the operation of a processing plant is important. Due to the wide day-to-day, crop-to-crop, and season-to-season fluctuations in water demand, it is common practice for processing plants to use deep well water. From one to seven 7-in. wells may be in use at a single plant, and if the water is inadequate, another location may be necessary.

All of the water used is disposed of by (a) inclusion in the product, (b) reclaimation and use elsewhere in the plant, or (c) disposal in an acceptable manner. The water intake and output for each plant is custom-designed depending on the crops processed, time of operation, volume of each product run, and the efficiency of the operation.

Disposal of Waste

In addition to water, other liquids must be disposed of without contaminating the air or soil. These include fruit juices, rinsing acids, lye from peeling, detergents from washing, and other chemicals such as antioxidants, sugars, flavors, and coloring materials. There also are solid materials to be disposed of, such as peels, cores, seeds, trimmings, stems, trash, and soil. When run into a common sewer, some of these materials react with one another, some are digested, and some may precipitate.

As with the water supply, the disposal system for each plant is custom-designed to fit the kinds of fruits processed, the volume of each, the length of the operating season, the nature of the waste, and the amount of land available. In general, during waste disposal the solids are disintegrated, allowed to oxidize or react with air and then allowed to settle; the semiclear water is collected, treated, and reused in the plant, and the sediment is allowed to dry and applied to the soil. There are many ingenious ways of accomplishing each of these steps, requiring careful design and much space. The amount of area needed for waste disposal may be the determining factor in locating a processing plant. (See Chapter 15 for a longer discussion of waste disposal.)

Labor Supply

With the advent of more labor-saving devices, it is necessary for workers to be more skilled, and fewer are required. While most training of workers is done in the plant, there must be a reservoir of recruits to choose from. Thus, the available labor supply must be at least double the number actually needed. Due to the nature of the work, employees must be agriculturally oriented as well as industrially oriented, and a semi-agricultural area is a much better location for a plant than an urban area.

Since employment fluctuates from month to month, laborers need another job when they are not needed for fruit processing. In most cases, farming provides this. Therefore, a good location for a fruit processing plant is in a farming area.

CULTIVAR SELECTION

Every cultivar of fruit responds differently to processing, and to operate a successful processing plant there must be at least one cultivar in the area that gives an excellent product. For example, 'Concord' grapes are excellent for juice in the Ozark region of Arkansas and the Pennsylvania–New York grape-growing area; 'Bartlett' of pears for canning in California; 'Kieffer' pears for canning in Michigan; 'Phillips' peaches for canning in California; 'Blakemore' strawberries for freezing in Tennessee. In some cases, the processing industry in a particular area may depend on the performance of a single cultivar of fruit.

Once the canning, freezing, drying, preserving or juice industry has been established in an area, other less suitable cultivars are used to extend the operating season, to blend with the leading cultivars, or to develop variations in the original product. Many fruit processing operations have failed due to the lack of a single cultivar that is excellent for their processed products.

Breeding fruits more suitable for processing is revolutionizing fruit processing in the United States and throughout the world. There is an urgent demand for fruit cultivars that can be mechanically harvested, graded, and handled with a minimum of bruising. This is especially true of tuft fruits such as strawberries, raspberries, loganberries, grapes, and peaches.

There also is a demand for fruit cultivars that ripen at very nearly the same time to reduce the number of "pickings." This is especially true of tree fruits such as peaches, apples, oranges, plums, nectarines, and prunes. The ideal is for the entire crop to be mechanically harvested on a "one-way" trip.

New cultivars that produce of high-quality fruit are constantly being sought. The 'Elberta' has been replaced with cultivars that yield higher-quality peaches for canning and freezing in the southeastern United States. The 'Concord grape' is giving way to other cultivars of grapes for juice and wine in the Pennsylvania–New York area. More adapted cultivars of grapes for wine are being extensively planted in "the wine country" of California, including the Napa Valley, San Joaquin Valley, and southern California. Improved cultivars of bananas that are more suitable for shipping, drying, and puréeing are being grown in Ecuador, Brazil, and elsewhere. Coconut breeding in the Philippines and Central America is producing early-bearing, heavily loaded, dwarf palms that increase yields three to five times. Research is continuing around the world for fruit cultivars that give products with improved texture, flavor, color, aroma, and stability after being canned, frozen, dried, pickled, preserved, puréed or otherwise processed.

5

Factors Affecting Microflora in Processed Fruits

E. J. Hsu and L. R. Beuchat

Fruit processors are more concerned with microbial spoilage during the harvesting, transportation, and storage of fruits than during the growing of these products.

The relationship of microbiological spoilage of fresh fruits to loss of quantities of processed fruits that would otherwise be available for human consumption is simple enough to perceive, but estimates of the amounts of nutrients made unavailable because of spoilage on a worldwide basis are difficult to make. Certainly, large percentages of fresh fruits and vegetables expressly grown for human consumption are destroyed due to invasion and decay by microorganisms. Losses are greatest in developing countries where growing, harvesting, and handling practices are not adequate to fully control deterioration. Grains and legumes are less perishable than are fresh fruits and vegetables, yet several million tons of these crops are also lost annually because of microbial spoilage. The reduction of losses in fresh-market fruits and vegetables was recently reviewed by Eckert (1975) and Harvey (1978).

Fungi (yeasts and molds) that cause deterioriation of fresh fruits are strong pathogens—that is, they can invade healthy host substrates. Weak pathogens, on the other hand, generally infect crops that have been damaged in some way during cultivation, harvesting, storing, transporting, or marketing. Physical injury induced by mechanical means, natural

Commercial Fruit Processing, 2nd Edition

ISBN 0-87055-502-2

openings in the surface of fruits, or physiological injury resulting from exposure to adverse environments such as high or low temperatures or chemicals are preconditions for the development of weak pathogens. Both infection and detectable spoilage by microorganisms can occur at any point from preharvest through the time of consumption of fruits. Certain fungi are often more responsible for decay at specific stages of handling fruits than they are at others because of their capacity to invade and proliferate in host tissue.

The prevention and reduction of losses of fresh fruits due to microbial spoilage starts with the grower and ends with the consumer. The effects of changes in methods of handling fruits, as new technologies are developed, must be considered if decay is to be held at a minimum. Thus, for example, the influences of mechanical harvesting, new packaging materials, refrigeration systems, and atmospheric gas composition on the susceptibility of fruits to fungal invasion are important in the overall attempt to provide greater quantities of nutrients to consumers.

Spoilage of fresh fruits by microorganisms can be prevented or retarded in several ways. Selection of a control method depends largely on the fruit under consideration, handling procedures, and the elapsed time projected before consumption. Chemical inhibitors include chlorine, ozone, and various fungistats and fungicides. Hot-water washes may retard the decay of some fruits and vegetables, and refrigeration is effective for controlling the rate of growth of fungi. Eckert (1978) presented an excellent review of the strategy of postharvest spoilage and disease control, methods for applying treatments, and the properties and applications of specific fungicides.

Table 5.1 lists numerous spoilage diseases of fresh fruits that are caused by specific molds or groups of molds. This list is by no means comprehensive, but offers the reader some idea of the magnitude of the potential losses of nutrients that result from mold spoilage. Further discussions of spoilage of fresh fruits can be found in a monograph by Splittstoesser (1978).

The susceptibility of fresh and processed fruits to invasion and deterioration by various types of microorganisms is dictated to a great extent by the chemical nature of the fruit itself, both before and after processing. Fruits, including tomatoes, are highly acid, and about 90% of their organic matter is carbohydrate, chiefly sugar. The pH values are quite low, ranging from about 2–3 for lemons to around 5 for bananas and figs. Spoilage is therefore commonly due to molds and yeasts. Some fruits usually become moldy after a few days at room temperature or even in the refrigerator, and crushed fruits or fruit juices not only become moldy but may develop gas and an alcoholic flavor as a result of yeast activity (Lund 1971).

Table 5.1. Molds Commonly Associated with Spoilage of Raw Fruits

Fruit	Spoilage disease	Fungus
Apple	Black spot	*Alternaria, Venturia*
	Blue mold rot	*Penicillium expansum*
	Brown rot	*Phytophthora*
	Bull's eye rot	*Cryptosporiopsis malicorticis*
	Eye rot	*Nectria galligena*
	Gray mold rot	*Botrytis cinerea*
	Lenticel rot	*Cryptosporiopsis malicorticis*
	Soft rot	*Rhizopus* spp.
Apricot	Black to brown spots	*Alternaria* spp.
	Black rot	*Aspergillus* spp.
	Blue mold rot	*Penicillium* sp.
	Brown rot	*Monilinia fructicola*
	Gray mold rot	*Botrytis cinerea*
	Restricted rot with gray-black core	*Cladosporium* sp.
	Watery, soft rot	*Rhizopus stolonifer*
Avocado	Anthracnose	*Colletotrichum gloeosporioides*
	Stem-end rot	*Diplodia* sp.
	Watery, soft rot	*Rhizopus* sp.
Banana	Anthracnose	*Colletotrichum musae*
	Black rot, lesion rot	*Colletotrichum musae*
	Crown rot	*Ceratocystis paradoxa, Colletotrichum musae, Fusarium roseum, Verticillium theobromae*
Blueberry	Gray mold rot	*Botrytis cinerea*
	Mummification	*Monilinia fructicola*
	Wooly growth	*Alternaria* spp.
Cherry	Black to brown spots	*Alternaria* spp.
	Brown rot	*Monilinia fructicola*
	Gray mold rot	*Botrytis cinerea*
	Rhizopus rot	*Rhizopus stolonifer*
Cranberry	Black rot	*Centhospora lunata*
	Blotch rot	*Acanthorhynchus vaccinii*
	Bitter rot	*Glomerella cingulata*
	Early rot	*Guignardia vaccinii*
	End rot	*Godronia cassandrea*
	Ripe rot	*Sporonema oxycocci*
	Storage rot	*Diaporthe vaccinii*
Citrus fruits	Anthracnose	*Colletotrichum* spp.
	Black center rot	*Alternaria* spp.
	Black rot	*Aspergillus* spp.
	Black spot	*Guignardia* sp.
	Blue mold rot	*Penicillium italicum*
	Brown rot	*Fusarium, Phytophthora*
	Cocoa-brown rot	*Trichoderma* sp.

(*continued*)

Table 5.1. (*Continued*)

Fruit	Spoilage disease	Fungus
	Cottony rot	*Sclerotinia* sp.
	Gray mold rot	*Botrytis cinerea*
	Green mold rot	*Penicillium digitatum*
	Septoria spot	*Septoria* sp.
	Stem-end rot	*Alternaria citri, Diplodia natalensis, Phomopsis citri*
	Sooty blotch	*Stomiopeltis citri*
	Sour rot	*Geotrichum candidum*
Fig	Black to brown spots	*Alternaria* spp.
	Black rot	*Aspergillus* spp.
	Blue mold rot	*Penicillium* spp.
	Gray mold rot	*Botrytis cinerea*
	Soft rot	*Fusarium* spp.
	Spotting	*Cladosporium*
	Watery soft rot	*Rhizopus* spp.
Grape	Black rot	*Cladosporium*
	Blue mold rot	*Penicillium*
	Gray mold rot	*Botrytis cinerea*
Grapefruit	*See* Citrus fruits	
Guava	Anthracnose	*Colletotrichum psidii*
	Fruit canker	*Pestalotia psidii*
	Phoma rot	*Phoma psidii*
	Soft watery rot	*Rhizopus, Botryodiplodia theobromae*
Lemon	*See* Citrus fruits	
Lime	*See* Citrus fruits	
Mango	Anthracnose	*Colletotrichum mangiferae*
	Stem-end rot	*Diplodia* sp.
Melons[1]	Alternaria rot	*Alternaria tenuis*
	Blue mold rot	*Penicillium* spp.
	Cladosporium (green) rot	*Cladosporium cucumerinum*
	Downy mildew	*Phytophthora* sp.
	Fusarium (pink to white) rot	*Fusarium* sp.
	Rhizopus soft rot	*Rhizopus stolonifer*
	Stem-end rot	*Diplodia* sp.
Nectarine	Black to brown spots	*Alternaria* spp.
	Black rot	*Aspergillus* spp.
	Blue mold rot	*Penicillium* spp.
	Brown rot	*Monilinia fructicola*
	Gray mold rot	*Botrytis cinerea*
	Restricted rot with gray-black core	*Cladosporium* sp.
	Watery, soft rot	*Rhizopus stolonifer*
Orange	*See* Citrus fruits	
Papaya	Anthracnose	*Colletotrichum gloeosporioides*
	Stem-end rot	*Diplodia* sp.

Table 5.1. (*Continued*)

Fruit	Spoilage disease	Fungus
Paw paw	Anthracnose	*Colletotrichum* sp.
	Ripe rot	*Botryodiplodia* sp.
Peach	Black rot	*Aspergillus* spp.
	Blue rot	*Penicillium* spp.
	Brown rot	*Monilinia fructicola*
	Brown spots	*Alternaria* spp.
	Gray mold rot	*Botrytis cinerea*
	Restricted rot with gray-black core	*Cladosporium* sp.
	Rhizopus rot	*Rhizopus stolonifer*
	Sour rot	*Geotrichum candidum*
	Watery, tan rot	*Diplodia* sp.
Pear	Black spots	*Venturia* sp.
	Blue mold rot	*Penicillium expansum*
	Bull's eye rot	*Cryptosporiopsis malicorticis*
	Eye rot	*Nectria galligena*
	Gray mold rot	*Botrytis cinerea*
	Lenticel rot	*Cryptosporiopsis malicorticis, Phlyctaena vagabunda*
	Sooty blotch	*Gloeodes* sp.
	Watery, soft rot	*Rhizopus* spp.
Pineapple	Black rot	*Ceratocystis paradoxa*
	Black spot, brown rot, leathery pocket	*Penicillium, Fusarium*
Plum	Black to brown spots	*Alternaria* spp.
	Blue mold rot	*Penicillium* spp.
	Brown rot	*Monilinia fructicola*
	Gray mold rot	*Botrytis cinerea*
	Restricted rot with gray-black core	*Cladosporium* sp.
Quince	Bull's eye rot	*Cryptosporiopsis*
Raspberry	Gray mold rot	*Botrytis cinerea*
	Olive-green rot	*Cladosporium*
Rhubarb	Anthracnose	*Colletotrichum erumpens*
	Gray mold rot	*Botrytis* spp.
Strawberry	Gray mold rot	*Botrytis cinerea*
	Leathery rot	*Phytophthora* sp.
	Watery, soft rot	*Rhizopus stolonifer*
	Watery, white rot	*Sclerotinia* sp.
Tangerine	*See* Citrus fruits	
Watermelon	Anthracnose	*Colletotrichum lagenarium*
	Black rot	*Mycosphaerella citrullina*
	Phytophthora rot	*Phytophthora capsici*
	Stem-end rot	*Sclerotina sclerotiorum*
	Watery, soft rot	*Sclerotinia sclerotiorum*

[1]Includes cantaloupes, honeydews, Persians, Crenshaws, Catabas, and summer squash.

Many molds and some yeasts can tolerate salt concentrations greater than 15%, whereas bacteria are generally inhibited by 5–15% salt. Molds are inhibited by 65–70% sugar; 50% inhibits bacteria and most yeasts. Foods of high sugar or salt content are therefore most likely to be spoiled by molds; foods of low salt or sugar content may be spoiled by many kinds of organisms.

Bacteria do not play an important role in fresh fruit spoilage due to the inherent acidity associated with most fruits. Also, the presence of bactericidal substances in these products has a tendency to destroy many kinds of bacteria. Tressler and Joslyn (1971), however, found that bacteria, molds, and yeasts present on strawberries were viable after 3 years storage at 10°C.

Many types of fruit contain appreciable amounts of inherent sugars and acids, hence the juices are easily attacked by molds and yeasts. Mold spoilage of fruits during storage may be significant. Mold is common on citrus fruits. Moreover, mold induces spoilage very quickly when the fruit is injured. There are many kinds of spoilage associated with fruits, but the most characteristic one is the softening of the flesh followed by rotting, making the product inedible. In certain kinds of rot in fruits the organisms actually invade the fruit while it is attached to the plant.

The bacteria responsible for most soft rots of fruits and vegetables during transport or in storage are the soft rot coliform bacteria, *Erwinia carotovora,* and pseudomonads similar to *Pseudomonas marginalis* (Lelliott *et al.* 1966). Except for lactic acid bacteria, bacteria are of only minor importance in causing spoilage in most fruits, the reason probably lying partly in the acid pH (<4.5) found in most fruit juices, which may inhibit those bacteria capable of degrading plant tissue. Fruits that are particularly susceptible to bacterial spoilage include cucumber, tomato, and pepper (Ramsey and Smith 1961; McCulloch *et al.* 1968). Vegetable juices, however, generally have a pH of between 4.5 and 7, and bacteria are a significant cause of the market spoilage of vegetables.

Yeasts are responsible for much of the fermentation of fruit products and may eventually cause the food material to become distasteful. They may be killed easily by heat during preheating or processing, but a few resistant species have been found in canned orange juice, tomato products, and catsup. The acid tolerance of certain yeasts make them important in these products.

The microorganisms involved in the spoilage of fermented fruit juices were described by Splittstoesser (1982). Bottled wines, ciders, and perries containing sugar are susceptible to refermentation by yeasts and growth of lactic acid bacteria. The commonly found yeasts are strongly fermentative strains of *Saccharomyces,* whereas the lactic acid bacteria are acid- and ethanol-tolerant species that grow slowly and are fastidious in their nutrient requirements. Spoilage manifested by gas, haze, and off-flavor can be controlled by procedures such as filtration, pasteurization, and the use of sulfur dioxide (SO_2) and sorbic acid.

Probably the yeasts and the yeast-like fungi are typical of the most readily killed forms of microbial life. In contact with moist heat at 50°–60°C, vegetative yeast cells are usually killed in 5 min. However, in the spore stage these organisms may require a temperature of 70°–80°C for killing in the same period of time. Production of ascospores by yeasts in nature probably plays a part in adaptation and survival of yeast cells under changed environmental conditions. But Ingram (1955) has pointed out that unlike bacterial spores these are only slightly more resistant than vegetative cells to heat and other agents.

Yeasts are similar to bacteria in many ways as far as their growth processes are concerned. They show a geometric growth rate, ferment substances, and utilize many nutrients that bacteria use. They are unicellular and reproduce by budding or by spores. *Saccharomyces cerevisiae* has been used for most studies and for industrial wine-making processes. It produces asci which contain ascospores which germinate, conjugate, and begin vegetative growth by budding.

Yeasts generally grow at faster rates than molds and therefore usually precede them in the spoilage of fruit products. Spoilage of fresh fruits by yeasts usually results from their fermentative activity rather than degradation of the plant tissue by the action of cell wall degrading enzymes (Dennis and Buhagiar 1980). Pectinolytic yeasts do not usually infect fresh fruits; however, notable exceptions do occur. The cut ends of petioles of rubarb plants, for example, have been observed to undergo deterioration with *Trichosporon cutaneum,* often in combination with *Pseudomonas* and *Penicillium* spp. It is not yet clear whether molds are dependent upon the initial activity of yeasts in the process of food spoilage. Since other readily utilizable nutrients are present in fruits, molds should have no problem establishing themselves. Many yeasts are able to attack the sugars found in fruits and initiate fermentation of the organic material with the end result being the production of alcohol and CO_2; then, molds, which are able to utilize alcohols as sources of energy, move in and when the alcohols and other simple compounds have been depleted, the molds begin to destroy the remaining parts of the fruit such as the structural polysaccharides and rinds.

Molds will grow on many kinds of foods, especially where temperature, air, and humidity are favorable for their growth. Appreciable growth of mold may be seen only on the surface of food. It often changes the flavor and quality of the contents of the entire container. Molds are easily killed by moist heat.

Experimental results show that a temperature of 110°C for several minutes will destroy a large percentage of mold spores (Sussman 1966). In general, molds are not involved in canned food spoilage because of the unfavorable conditions for their growth in an airtight container. Moreover, most mold spores are unable to survive the temperature used in the processing of most foods.

The vegetative forms of molds or fungi are usually destroyed by a 30-min exposure to moist heat at 62°C, whereas certain spores may require a temperature of 80°C for killing in the same period of time. Most of the organisms classified as *Actinomycetes* are killed by moist heat at 60°C in 15 min (Austwick 1966). These organisms, including the spores, are killed by a time-temperature relationship of 30 min at 60°C to more than 1 hr at 72°C. In contact with dry heat, mold spores require a temperature of 100°–116°C for 90 min to ensure their destruction.

Horikoshi and Ando (1972) reported that *Aspergillus oryzae* was killed, under conditions of heat activation of a latent ribonuclease, by heating wet conidia at 55°C for 5 min and by heating dried conidia at 75°C for 5 min at pH 6.0. The thermoresistant molds, which are capable of withstanding the temperatures used in fruit juice pasteurization, are important. These include species of *Byssochlamys, Monascus,* and *Phialophora* (Jensen 1960), as well as *Talaromyces* and *Neosartorya.*

Splittstoesser *et al.* (1971) observed that 70% of the samples of fruit, vegetable, and soil obtained in surveys of New York orchards and vineyards were contaminated with heat-resistant molds. The mold counts generally were low, under one per gram. *Byssochlamys fulva* was the most common isolate. Other isolates were identified as *B. nivea, Paecilomyces varioti, Aspergillus fischeri, A. fischeri* var. *spinosus, A. fumigatus, Penicillium vermiculatum,* and *P. ochrochloron.*

DEHYDRATED FRUITS

Dehydration is, perhaps, the oldest and the most common form of food preservation. Since prehistoric time, fruits, vegetables, fish, and meats have been known to be preserved by drying. In tropical and other warm areas these foods were dried on trays in the sun. In cooler and humid regions fire and smoke were used. Each method, on a larger scale, is still used today. Modern procedures of lyophilizing (drying from the frozen state) are yielding more acceptable products than traditional methods.

Unfortunately, there have been few studies investigating the microbiology of dried preserved fruits in the last 30 years. Although standardized methods for the microbiological examination of these products are desirable, in view of their increased use throughout the processed foods industry, this area has received less attention than other segments of the food industry. Methods for the examination of dehydrated fruits and vegetables were developed principally during World War II, and research has been rather limited in recent years.

Almost every fruit that is preserved by drying has come into contact with and is probably carrying some microorganisms prior to dehydration.

The presence of water is mandatory in order for fruit or any food to undergo microbial spoilage. All organisms require water for carrying on their life processes. If the microorganism cannot acquire the water it needs, it either dies or its further growth is arrested. Potential spoilage of a dried fruit, then, depends upon how available water is to the spoilage microorganisms. It is, therefore, the thin demarcation line of water activity that establishes dehydration as a good preservative technique. The degree to which water is available is expressed by the term water activity, a_w, that is, the vapor pressure of the solution (fruit or fruit product) divided by the vapor pressure of the solvent (water).

Removing water from fruit lowers the availability of water to microorganisms. In a moist, solid substance, the water vapor pressure is lower than the vapor pressure of free water at the same temperature because, in a solid substance, water reacts with polar groups such as —CO—, —NH, and —OH. Still further, vapor pressure inside of capillaries (between plant cells) is lower than the vapor pressure of a plane surface of water. As the solutes present in the fruit are dissolved in water, the vapor pressure is depressed.

At any given temperature, the reduction of the a_w for molds causes a fall in the rate of spore germination. This decrease in the rate of germination is manifested in two ways. First, there is an increase in the latent period or time required for the first appearance of germ tubes. Second, there is a reduction in the rate of elongation of the germ tube.

The inhibition of fermentation by yeasts is increased in an almost linear relationship as the a_w is decreased. Yeasts need more water than molds, but less than bacteria. As the a_w is reduced below an optimum level for the organism, there is an increase in the lag or latent period, a decrease in the rate of growth, and thus a decrease in the amount of cell substance synthesized (Scott 1957).

Growth of the vast majority of bacteria, yeasts, and molds is prohibited at a_w below 0.90, 0.85, and 0.80, respectively. However, each organism has its own characteristic optimum a_w at which growth will occur. The range over which microbial growth has been demonstrated is from above 0.999 down to approximately 0.62. Inside these limits, each organism has its own characteristic range within which it can grow. The magnitude of the range varies with each organism, but shows a measure of constancy for each. The fundamental importance of a_w in the water requirement of microorganisms is strongly supported by experiments showing that the biological response to a particular a_w is largely independent of the type of solutes and total water content of a substrate (Scott 1957). Molds are the most troublesome group of microorganisms in dried fruits. *Aspergillus glaucus* will grow at a_w values below 0.70.

The control of a_w, then, is a very important means of stabilizing fruit against microbial deterioration. Whether a given microorganism can multiply in a particular fruit under particular conditions usually depends on factors related to the a_w of the fruit. It is important to note that the a_w of a fruit actually influences which microorganisms will have a chance to grow there. The lowering of the a_w of a fruit has an extremely microbiostatic effect. Depending on the duration of the drying operation and the actual a_w reached, the species and growth phase, the reduction of microbial population can be significant.

It has been shown that there are certain "osmophilic" (solute-requiring) yeasts and certain "xerophilic" (requiring little water) molds and fungi. Obviously, these microorganisms are able to live and proliferate at low a_w values. These are the microorganisms responsible for the spoilage of dried fruit.

The concentration of soluble solids is perhaps the most important factor affecting resistance to microbial deterioration after prunes are dried and processed (Miller and Tanaka 1963). The skins of dried prunes may be barriers against the diffusion of water molecules into the prune flesh, and diffusion in the prune itself may even be very slow. In stored dried fruit, these two factors are very important in preventing a rise in a_w value that would allow a microorganism to grow. The most important groups of osmophilic yeasts and xerophilic molds that attack prunes are the *Aspergillus glaucus, Saccharomyces rouxii, Xeromyces bisporus,* and *Chrysosporium* species (Pitt and Christian 1968; Tanaka and Miller 1963a,b). Dates and figs are attacked mainly by yeast species belonging to the genera *Zygosaccharomyces* and *Hanseniaspora* (Phaff 1946).

Concentrated juices, dried fruits, special products like "malt extract," honey, sugar preserves, and sugar confectionery, all possess two features in common—high proportions of sugar, together with low proportions of water, giving a_w values of the order of 0.75 or less. The overall a_w not the nature and amount of the carbohydrates present, determines the growth of yeasts under such conditions. This degree of physiological dryness completely prevents the growth of the common microorganisms, so that spoilage is effected by the osmophilic yeasts (Cook 1958).

The microorganisms responsible for the spoilage of dried fruit may sometimes be intrinsic to the fruit itself, so that spoilage does not depend on the nature of the infection to which the fruit is later exposed. An example of this can be seen in figs which have an essentially sterile internal tissue until visited by the pollinating fig wasp, *Blastophaga psenes*. This wasp introduces a specific microflora, *Candida guilliermondii* var. *carpophila* and *Serratia plymuthica*. However, these organisms do not cause spoilage, but serve to increase the attractiveness of figs for fruit flies, mainly, *Drosophila melanogaster,* which does carry spoilage yeasts on its exterior parts and introduces these organisms into the fruit cavity. These yeasts consist almost entirely of apiculate yeasts which can cause active fermentative spoilage. Thus, spoilage microorganisms can already be present in fruit before the actual drying process takes place (Miller and Phaff 1962). It is no wonder then, that sundried figs still have yeast counts of about 3.5×10^8. Also, the importance of keeping a low a_w can be seen directly here because as soon as the a_w increases, for some reason, the yeasts in question will proliferate and spoil the fig. In addition to certain insect vectors, spoilage microorgansisms may contaminate fruit through the air, dust, soil, mammals, birds, and by cross contamination with other spoiled fruit or by reusing previously contaminated containers.

The cuticle of intact fruit prevents the access of organisms, such as yeasts, that are able to proliferate in the acid environment of fruit flesh. When a fungus attacks an intact fruit, however, it damages the external structure and so permits the entry of yeasts. During the process the acids contained in the fruit tissue are dissimilated. This causes an increase in pH and enables nonacid-tolerant strains to take part in a spoilage association. The metabolism of the microorganisms generally results in the release of water also, and this results in local a_w increases. Thus, some of the less xerophilic organisms previously inhibited are allowed to grow. The development of xerophilic molds on dried fruit may ultimately permit less xerophilic species to take part in the spoilage association. Likewise, the growth of an osmophilic yeast in a fruit of highly concentrated sugar diminishes the total solids content and so makes growth of somewhat less osmophilic organisms possible (Mossel 1971). The ensuing spoilage is primarily the result of yeast and mold fermentation. Osmophilic yeasts produce numerous polyalcohols in the fermentive process (Onishi 1963).

The simplest and most commonly used method of removing contaminating organisms is by washing. However, certain necessary precautions and the limitations of this method should be pointed out. Washing can cause microorganisms to spread from an infected area to an uninfected one. Also, the water film left on the surface of a product may encourage the growth of microorganisms, and the water itself may be a source of contamination. Consequently, the use of a germicide in the water is advisable. Chlorinated water at a concentration of 50–125 ppm of available chlorine is quite effective in decontaminating surfaces and preventing the spread of microorganisms from one part of the product to another.

Dehydroacetic acid in water at concentrations between 0.5 and 1.5%

has been shown to be quite effective in lowering the incidence of rotting in peaches, strawberries, raspberries, cherries, and blackberries (Smith 1962; Von Schelhorn 1951). Other acids, such as sorbic and peracetic acid also have been shown to reduce spoilage of various products when applied in water solution in the form of postharvest washes, dips, and sprays.

Impregnating wrapping paper and box liners with diphenyl has proved highly effective against *Penicillium* and the organisms causing stem-end rots in citrus fruits (Smith 1962; Von Schelhorn 1951).

Originally fruit was dried on trays in the sun. This often resulted in microbial spoilage, especially when rain fell during drying seasons. Today, in the United States, dehydrators are used almost exclusively for fruit drying. In the dehydrator, heat can be applied according to a thermal death time measurement. In prunes, since contamination is practically all on the surface, drying is at 75°C for 18–24 hr. Prunes, when removed from the dehydrator, are free of viable yeasts and molds. Figs, on the other hand, seldom come from the dehydrator free from viable microorganisms because of their hollow structure (Natarajan *et al.* 1948).

Fruits are easier to dry than are vegetables because most fruits do not need to be blanched. However, certain chemical changes due to the activity of enzymes and air tend to discolor fruit during dehydration. For this reason most fruits to be dehydrated, except some cut fruits, are treated by exposure to the fumes of burning sulfur or to vaporized liquid sulfur dioxide. Solutions of sulfites are less suitable because they penetrate the fruit poorly and leach sugar. It has been estimated that the storage life of dried apricots is directly proportional to the initial SO_2 level. The gas disappears during storage at a rate approximately proportional to the logarithm of its concentration. Retention depends upon the time of exposure and temperature. The optimum temperature is 43°–49°C when SO_2 is used.

It is not possible to make a general recommendation as to the strength of the dip solution or the length of time that the product should be immersed, as there are many variables and experimentation is required to obtain satisfactory levels in the fruit. The recommended levels of SO_2 in some specific dried and dehydrated fruits at the beginning of the storage period are 2000 ppm for apricots, peaches, and nectarines; 1500 ppm for sulfur bleached raisins; 1000 ppm for pears; and 800 ppm for golden bleached raisins and apples.

It has been found, however, that a portion of from 0.3 gm up to 1 gm daily of sulfured fruit causes signs of harmful effect in some individuals, over a period of months. The symptoms included an increase in uric acid, a destruction of corpuscles in the blood, belching of sulfur dioxide gas, inflammation of the mucous membrane of the mouth, symptoms of malaise, headache, backache, sick appearance, nausea, albumin in the urine, sensation of cold, anemia, dull eyes, and a listless manner (Kallet and Schlink 1933). Even though these findings were presented, no action has been taken by FDA, and SO_2 is still being used.

Ethylene oxide was applied to packaged dried fruits to minimize microbial spoilage during adverse shipping and storage conditions encountered by the military during World War II. One of the techniques was to add an ethylene oxide solution or "snow" to the outer container of packaged fruit. Gradual permeation throughout the containers essentially sterilized the contents and the gas eventually escaped through the packaging material. Others have used the "snow" application (frozen, ground ethylene oxide in water) in packaged foods. Today, gaseous ethylene oxide is used in the sterilization of dehydrated vegetables, fruits, and spices that cannot be treated satisfactorily by other methods (Chichester and Tanner 1968).

Because ethylene glycol and diethylene glycol were encountered as residues and are suspected of toxicity, ethylene oxide decontamination has been limited to a few low-moisture food ingredients. On the other hand, propylene glycol is generally recognized as safe, and gaseous sterilization by propylene oxide is receiving more attention (Gunther 1969).

After coming from the dehydrator, most fruits and fruit packages are usually fumigated with ethylene oxide. Ethylene oxide has been used for the sterilization of spices for many years (Yesair and Williams 1942). This gas causes little damage to fruit material. It is very effective at room temperature and there is little residual effect, although it is relatively slow and requires special equipment to apply. Beta-propiolactone is relatively new and may prove very useful (Hoffman and Warshowsky 1958). Irradiation also reduces spoilage of fruits and vegetables due to surface contamination by microorganisms.

Consumers desire high moisture in dried fruits such as raisins, prunes, and figs, but this also increases their susceptibility to mold and yeast growth. Potassium sorbate, applied by dip or spray, has been shown to effectively protect raisins, prunes, and figs. It is suggested that initial trials be made with solutions containing 2–7% potassium sorbate. The deposit of the preservative on the fruit should be from 0.02 to 0.05%.

Commercially sold dates usually contain 12–25% moisture. The souring of dates by yeasts generally occurs when the moisture content exceeds 23–25%, and has been observed at all stages of processing from harvesting to packing and distributing. Yeast spoilage may be overlooked since the process is slow, even though undesirable odors may be present.

Yeasts isolated from this material were described as *Zygosaccharomyces cavarae*. Mostly *Hanseniaspora valbyensis* and *Candida guilliermondii* were isolated from soft dates. *Saccharomyces* species were absent probably due to a short incubation time or the low sugar content of the recovery medium.

Sugar-tolerant yeasts that grew well on dates containing 30% moisture or on date syrups containing 65% soluble solids (mainly invert sugar) and isolated in pure culture from soured dates produced souring in pasteurized sound dates containing more than 25% moisture. All of the yeasts grew abundantly at temperatures from 20° to 37.5°C, optimum about 30°C. Drying of dates to 23% moisture or pasteurizing at 71°C and 71% relative humidity (RH) controls date spoilage (Walker and Ayres 1970).

In addition to being highly subject to spoilage by osmophilic yeasts, dates are also often contaminated and sometimes deteriorated by filamentous fungi (Abu-Zinada and Ali 1982). These researchers reported that fungal counts of eight date palm varieties grown in Saudi Arabia differed greatly. *Aspergillus flavus, A. niger, Penicillium rubrum, P. oxalicum, Rhizopus stolonifer, Stemphylium verruculosum,* and *Fusarium* sp. were generally associated with date varieties. Apparent colonization of fungi was obtained by increasing the relative humidity to 90% at 30° and 40°C. Best growth of fungi isolated from dates was noted on an artificial medium containing 60% glucose. The yeasts that have been most frequently isolated from the gut and exterior of the dried fruit beetle are *Candida krusei* and *Hanseniaspora valbyensis*. Yeast populations increased greatly during sun-drying under unfavorable weather conditions or in cooler climates; they declined as the fruit dehydrated.

A white "sugar-like" coating often developed on dried figs and prunes during storage and gas pockets developed. This coating was a mixture of sugar and yeasts and formed more commonly on bin-stored fruit than on packaged fruit.

When sun-dried prunes were stored in sealed cans at 26.7°C, fermentation and gas formation occurred after several months, but freshly dehydrated prunes did not ferment under these conditions unless inoculated with a *Zygosaccharomyces* species isolated from prunes. This type of spoilage required 22% moisture in the prunes and was controlled by dehydration, rapid sun-drying to a low moisture content, or fumigation with ethylene oxide and proper storage.

Byssochlamys fulva, a heat-resistant mold, along with *B. nivea, Paecilomyces varioti, Aspergillus fischeri, A. fischeri* var. *spinosus, A. fumigatus, Penicillium vermiculatum, P. ochro-chloron,* and other similar molds were isolated in over 70% of the samples of fruit, vegetation, and soil in surveys of New York orchards and vineyards (Splittstoesser *et al.* 1971). The incidence of these microorganisms could have caused grave problems in the fresh fruit industry and in the dehydrated fruit industry except for the use of methyl bromide in low concentrations (Seeger and Lee 1972). Methyl bromide is also used as a fumigant on dates in the ratio of 1 lb/1000 $ft.^3$ Although the FDA claims that methyl bromide is not retained by the food, this assurance is based on outmoded tests. Methyl bromide is characterized as "only slightly soluble in water," and, therefore, dates may retain some fumigant even after being washed. Such chemical treatment usually causes some modification of the food product, as well as contamination.

In general it can be said that in dried fruit, spoilage microorganisms will be active whenever water is available to the degree needed for growth. If the fruit is not kept at low a_w and fumigated to kill any viable spores, spoilage will soon be brought about by various yeasts, mold, and fungi that are able to withstand conditions of low water and high sugar content.

REFRIGERATED AND FROZEN FRUITS

Growth of the vast majority of microorganisms is inhibited or severely retarded at refrigeration temperatures. Several species of microorganisms, however, can tolerate and in fact prefer to grow at refrigerated temperatures. Among these are yeasts and yeast-like organisms that may cause spoilage of fruits and fruit products. Davenport (1980) reviewed cold-tolerant yeasts and summarized their general features. Most (at least 75%) species are of the basidiomycetous type, which are more suited to colder habitats since their minimum temperatures for growth are lower than those for ascomycetous organisms. The latter yeasts also lack other characteristics such as absence of carotenoid pigmentation and intracellular lipid storage, high osmotolerance, certain vegetative features (e.g., capsules) and a complex slow-growing life cycle. Davenport (1982) also stated that it is highly probable that reproduction of yeasts in cold environments is restricted to those strains that have a minimum growth temperature of 0°–5°C. The ascomycetous yeasts that reproduce at 5°C and below all have rough thick-walled ascopores containing lipid droplets.

Most fruits can be kept several months in the frozen state without undergoing microbial spoilage. Various freezing operations, however, produce sterility on one hand or maintain microorganisms alive on the other. The rate of freezing, therefore, has a significant effect on both fruit texture and microbial quality.

Slow freezing, in which 3 hr to 3 days may be required for freezing food solid, preserves foods. However, when thawed, the foods are soft, juices exude, and the quality is impaired. This is due to the formation during freezing of large ice crystals that rupture fruit and cells of microorganisms.

On the other hand, quick freezing of food, usually in small containers at temperatures as low as −45°C, achieves a frozen state in 30 min or less. This rapid freezing forms only very small ice crystals that do not rupture the cells. Such foods keep for relatively long periods if stored at very low temperatures. Before quick freezing, some fruits may be blanched or scalded to reduce surface organisms and inactivate enzymes that might damage quality during storage. After thawing, microorganisms that survive in the frozen foods may begin to grow, impairing the quality or giving rise to food spoilage. With fruits that have not been blanched, controlling the numbers of microorganisms present when these products are frozen is more of a problem.

Many of the organisms associated with fruits are soil and air types and have little, if any, public health significance. Thorough washing and the blanching used with some fruits will mechanically remove most of the microflora present on the raw product before it enters the freezer. Intermittent freezing is more destructive to microorganisms than continuous freezing, but it is also more destructive to fruits, which may show a deterioration in quality due to defrosting and refreezing even though they may have a low microbial content. The microbiological population, therefore, should not be used as the sole criterion for judging the quality of frozen fruits.

Quick freezing reduces but does not entirely prevent tissue damage. Frozen food is therefore highly susceptible to microbial invasion after thawing. Thawed food should be used immediately, because the surviving microorganisms may begin to multiply as soon as they are exposed to their normal growth-temperature range.

Most, if not all, of the microbiological deterioration in relation to frozen fruits takes place during the stages of precooling or after defrosting. This is because microorganisms do not multiply or carry out normal enzymatic activities in fruits held at temperatures below −12° to −10°C. At temperatures between −6° and −7°C, psychrophilic bacteria can grow, but food-poisoning organisms cannot (Ingraham and Stokes 1959). This inhibition of food-poisioning organisms is not only due to low temperatures, but also to the typically low pH of fruits, which is below the level that generally favors bacterial growth. Low pH largely explains the absence of bacteria, including the food-poisioning organisms, in the spoilage of fruits; the wider pH growth range of molds and yeasts makes them the main spoilage agents of fruits. With the exception of pears, which sometimes undergo *Erwinia* rot, bacteria are of no known importance in the initiation of fruit spoilage (Nasuno and Starr 1966).

A number of the yeasts and molds that frequent fruits are psychrophiles. Included in the low-temperature-growing yeasts are species and strains of the genera *Candida, Debaryomyces, Rhodotorula, Saccharomyces,* and *Torulopsis*. Among the molds are species of the genera *Botrytis, Cladosporium, Geotrichum, Mucor,* and *Penicillium*. Species of *Botrytis* appear most frequently as one of the main spoilage organisms of fruits.

With respect to the minimal growth temperature of fruit spoilage organisms, it has been found that yeasts in general grow at lower temperatures than molds and bacteria. One pink yeast was reported to grow at −34°C and two others at −18°C. The molds grow at about −12°C. Some fruits and fruit juices can support the growth of molds and yeasts at subfreezing temperatures. A species of *Saccharomyces* has been found to be capable of growth at −2.2°C in apple cider, and spoilage of strawberries packed in 50% sucrose was attributed to a *Torula* species growing at −4°C (Wood and Rosenberg 1957; McCormack 1950).

Most workers have assumed that some unfrozen liquid is required for the growth of microorganisms in frozen foods. Citrus concentrates, for instance, are not completely frozen at −9°C.

In addition to the factors listed above, the freezing of fruits is also accompanied by changes in such properties as pH, a_w, titratable acidity, ionic strength, viscosity, osmotic pressure, freezing point, surface and interfacial tension, and redox potential—all of which can have a fruther effect on any yeast or mold present in the fruit material.

For example, a freezing temperature of −20°C is less injurious than a temperature of −10°C to yeasts and molds. It has been conclusively proven that more of these organisms are destroyed at −4°C than at −15°C or below. Temperatures below −24°C seem to have no additional effect. Gunderson (1962) and many others demonstrated that fluctuating temperatures in frozen foods caused migration of moisture to localized areas in the food. This resulted in sufficient available water for mold growth even during short periods in which the product did not thaw completely.

Ricci (1972) reported that the rotting of fresh figs during storage was caused mainly by *Botrytis cinerea,* while *Alternaria tenuis* and species of the genera *Penicillium, Mucor,* and *Rhizopus* were also harmful. Most of these fungi had been present on the figs at harvest. Freezing or cold storage resulted in a higher level of contamination, because of a slight increase in cell numbers, but rotting was delayed. *Alternaria citri* was isolated from Florida citrus fruits during extended frozen storage. Frozen storage for 2 months at 0°C was effective in controlling blue mold rot caused by *Penicillium expansum* in wounded and inoculated 'Golden' and 'Red Delicious' apples (Spalding and Hardenburg 1971). In blueberries held at various frozen storage times and temperatures, *Botrytis cinerea, Glomerella cingulata,* and *Alternaria tenuis* were isolated (Cappellini 1972).

With regard to yeasts, *Candida krusei* was isolated from frozen tomato pulp. *Candida krusei, C. chalmersii, Rhodotorula mucilaginosa, Saccharomyces rosei, Torulopsis albida, T. carpophila,* and *T. stellata* were all isolated from soured figs during storage at 0°C (Ricci 1972). *C. tropicalis, C. krusei, C. mycoderma,* and *T. inconspicua* were all isolated from soured dates at 0°C. *Saccharomyces rouxii* was isolated from frozen sugared cherries and pears, and *S. cerevisiae* was isolated from frozen strawberries and raspberries during frozen storage.

The a_w of fruits may be expected to decrease as the temperature falls below the freezing point. For water at 0°C, a_w is 1 but falls to about 0.8 at −20°C and to 0.62 at about −50°C. Yeasts and molds that grow at subfreezing temperatures, then, must also be able to grow at reduced a_w levels, unless a_w is favorably affected by the food constituents usually present in frozen fruit products. For example, yeasts have been found to survive better in tap water or 10% sucrose than in 2% sodium chloride or in cider (pH 3.2) during freezing and to survive better in storage at −20°C than −10°C.

Freezing results in the loss of cytoplasmic gases such as O_2 and CO_2. This is very important, as a loss of O_2 in aerobic cells suppresses respiratory reactions. The solubility of dissolved gases in the microbial cell decreases appreciably with decreasing temperatures. Gases become more concentrated during freezing and after a period of time may be forced out of the cell. Before these gases diffuse into the atmosphere, they may have a toxic effect on microorganisms and reduce their numbers. With the loss of oxygen to the surrounding environment, however, this would create a favorable environment for anaerobic bacteria to grow and multiply. Several experiments of this type using a controlled atmosphere have been reported. For example, decay caused by *Botrytis cinerea, Pullularia pullulalans,* and species of *Rhizopus* was severely retarded in an atmosphere at nitrogen at 0°C, during short-term storage (Lockhart *et al.* 1971). An atmosphere of 2.5 or 5.0% O_2 greatly reduced rot development caused by *P. digitatum, P. italicum, Diplodia natalensis,* and *Alternaria citri* on Shamouti oranges at 0°C (Aharoni and Latter 1972; Brown and McCormack 1972).

If freezing is not sufficiently rapid, microbial growth may occur. In large packages, the center may spoil as a result of microbial growth, particularly yeast growth, before it reaches a temperature low enough to cause the fruit to freeze solid. It is also possible for a number of microclimates to exist in stacked frozen products. A film of moisture may develop over the surface of the product and create an environment for slow multiplication of spoilage agents (Insalata and Roab 1969).

Fungal pectolytic enzymes are most active at a low pH and peak in activity at a pH of 3.5–4.5. Fungal enzymes probably can cause softening of low-acid refrigerated fruits and vegetables (Phaff 1959; Luh and Phaff 1951). Bacterial enzymes generally have a higher pH optima. This is one reason why molds or fungi are mainly responsible for spoilage in frozen foods (Vaughn 1962; Hsu and Vaughn 1969).

The inhibitory effect of the acidity of fruits helps to maintain good microbiological quality; on the other hand, the sugar present helps to protect microorganisms against the lethal effects of freezing. Although relatively few microorganisms can develop at −2° to −5°C, in cool storage tanks for grape juice, problems of yeast and mold contamination before freezing become serious at times.

Contamination in the handling and storage of grape juice may arise from yeasts harbored in the pores of wood or coating of tanks, the air in the room, foam on the surface of containers, intermediate holding tanks, valves, improper gaskets, and similar places where microorganisms may accumulate and grow (Pederson *et al.* 1958). Four types of yeasts capable of growing at the low storage temperature were isolated and described, one of which is a true psychorophile. In a later study, Pederson *et al.* (1960) have shown that several organic acids are effective fungistats even when used in very low concentrations in controlling this yeast growth at low temperature.

CANNED FRUITS

Since microorganisms are more easily killed in acid media, microbial spoilage of canned fruits were a pH below 3.7 seems unlikely. However, spoilage of canned tomatoes by yeasts and butyric acid bacteria has occurred and resulted in serious losses. Spoilage of canned pineapples has also been reported (Tressler and Joslyn 1971). Detailed discussion on the effects of thermal process and microbial spoilage of canned vegetables has been made (Hsu 1975).

Spoilage of canned and bottled fruits by *Byssochlamys,* a heat-resistant mold, was first recognized in Great Britain in the 1930s and has since been documented in several other countries. A review of the two recognized species of the mold, *B. fulva* and *B. nivea,* as related to their importance in processed fruits was made by Beuchat and Rice (1979). Its occurrence on various fruits, factors affecting growth, control of growth, sporulation, and enzyme production are covered. Of particular interest is the fact that *Byssochlamys* produces at least four by-products (patulin, byssotoxin A, asymmetrin, and variotin) that are toxic to animals. Patulin, probably the most potent of these mycotoxins, is produced at rather high levels (e.g., 6.11 mg/50 ml of blueberry juice) in a wide range of fruit juices (Rice *et al.* 1977).

The combination of low pH and light heat, therefore, assures sterility and the desirable fruit texture of most canned fruits. For example, adding an acidulant to adjust the pH to a standard value often permits a shortening of sterilization time. In some cases, lower sterilization temperatures are possible. This is important since prolonged heating of fruits, vegetables, and their juices not only destroys their natural structure but may also reduce nutritive value (Gardner 1966a).

Various fruits, including tomatoes, normally contain large amounts of organic acids, but additional amounts of acid may be required for proper sterilization during canning since drought, other weather conditions, and the variety and maturity of the fruit leads to a lowered acid content (Gardner 1966b). Addition of food acids, however, is only an aid to proper food processing. It is not a means for overcoming poor sanitary or processing conditions.

Unlike most fruits, figs are in the low acid range and thus represent a substrate for growth of *Clostridium botulinum*. The Standard of Identity (1977) for canned figs permits the addition of lemon juice, concentrated lemon juice, or organic acid to reduce the pH to 4.9 or below. Lemon juice is added to increase the acidity to a pH of 4.5, but is not necessarily noted on the label as an additive unless greater amounts are incorporated.

The outgrowth of spores of *C. botulinum* in figs packed in heavy syrup has been shown to be inhibited at pH 4.9 (Townsend *et al.* 1955). Subsequent studies by Ito *et al.* (1978) confirmed this observation. Anaerobic incubation of inoculated packs of fig purée mixed with water and light or heavy syrup and adjusted to pH 4.9 or below failed to show outgrowth of *C. botulinum* spores at 30°C for a period up to almost 1 year. Woodburn (1982) evaluated the acidification and pH of Oregon-grown figs before and after cooking. The proportion of lemon juice that must be added to lower the pH to 4.6 or below by the boiling waterbath method was determined. The pH of figs averaged 5.51 but was higher as figs ripened. The addition of 15 ml of lemon juice per pint of figs canned in syrup was sufficient to lower the pH of the processed fruit to 4.6 or below.

Acidulation is also mandatory in the canning of artichokes and is permitted for a number of fruits and vegetables.

Splittstoesser *et al.* (1977) made a study of viable counts versus the incidence of *Geotrichum* on processed fruits and vegetables. Of 51 samples of tart cherries collected from two factories and examined, none was positive for *Geotrichum*. Most of the microorganisms contaminating cherries were yeasts, and most samples gave counts in the range of 10^3–10^4/g. Some vegetable samples were positive for *Geotrichum,* but these samples did not yield higher viable counts than did those that were negative for this mold.

FRUIT JUICES

In contrast to other fruit products, bacteria are the most diversified microorganisms causing spoilage of processed fruit juices. The concentration of sugars in fruit juices is not high enough to inhibit bacterial growth, and the inhibitory factor for bacteria appears to be acidity rather than sugar content. Due to their resistance to acidity, lactic, acetic, and butyric acid bacteria are the most common spoilage microorganisms. The spore-forming bacteria of the genera *Bacillus* and *Clostridium* and bacteria significant to public health are also important.

The lactic acid bacteria comprise the most frequent spoilers of fruit juices and are commonly responsible for the spoilage of apple, pear, citrus, and grape juices (Vaughn 1955). They exist on the surface of plants and fruits, growing at the expense of secreted plant materials, and are, therefore, usually present in fruit products. They are capable of growing in a high concentration of CO_2 and at pH 3.5, but do not grow well at temperatures below 8°C. The majority of lactic acid bacteria isolated from fruit juices are heterofermentative, producing lactic acid, acetic acid, propionic acid, ethyl alcohol, and CO_2. They usually belong to the general *Lactobacillus* or *Leuconostoc,* although *Streptococcus* and *Pediococcus* species are found in lesser numbers. Some homofermentative strains have been found as well as some strains producing acetoin and diacetyl (Kitahara 1966).

Citrus juices are particularly subject to spoilage by yeasts and lactic acid bacteria at refrigeration temperatures. Murdock and Hatcher (1975) investigated the growth of lactic acid bacteria and yeasts in chilled orange juice. Yeasts grew at 1.7° to 10°C, their rates of growth increasing with the temperature. *Lactobacillus* grew at 10°C, but not at 7.2°C; *Leuconostoc* did not grow at 1.7°C, but grew slowly at 4.5°C and rapidly at 7.2° and 10°C. Spoilage due to growth of *Lactobacillus* was detected after 1–2 weeks of incubation at 10°C, whereas *Leuconostoc* required 13 days to 5 weeks at 4.5° to 10°C. Shelf life of chilled juice is dependent upon the initial microbial population at the time of packaging and the temperature maintained until the juice is consumed.

Owing to their ability to produce large amounts of acids, lactic acid bacteria often inhibit the development of other bacteria in juices and are capable of causing their own autolysis. The spoilage of fruit juices by lactic acid bacteria often leads to "ropiness" (the production of a gummy slime from the sugars utilized) and "buttery" off-flavors. While some members of the group, species of the genus *Streptococcus,* are capable of initiating disease states, the role of contaminated fruit juices in the spread of such disease is of very little practical importance.

The second major group, acetic acid bacteria, commonly reside on plant and fruit surfaces and are often found in contaminated fruit juices. However, they require oxygen to carry out the incomplete oxidation of sugars or ethyl alcohol to acetic acid. Most acetic acid bacteria can produce sugar acids and/or keto acids from polyalcohols and sugars. The two genera most often found in fruit juices are *Acetobacter* species. *Acetomonas* and *Acetobacter* species further convert ethanol to acetic acid, followed by oxidation to CO_2 (Vaughn 1957). Morphologically, both are motile, *Acetomonas* with polar flagellation and *Acetobacter* with peritrichous flagellation (Leifson 1954).

The third major group, butyric acid bacteria, ferment carbohydrates with the production of butyric acid. Most notable among this group are species of the genus *Clostridium* whose extraordinary diversity of mechanisms for anaerobic fermentation can well be appreciated by anyone happening to smell their end-products. *Clostridium butyricum, C. sporogenes, C. pasteurianum,* and *C. acetobutylicum* can obtain energy for growth by the fermentation of carbohydrates to butyric and acetic acids, CO_2, and H_2.

With the exception of *Enterobacter* species, members of the Enterobacteriaceae are not routinely found in raw fruit juices. In contrast, *Enterobacter* species are commonly found on virtually all types of unsound fruits, in wash waters, and in spoiled fruit juices (Dack 1955). However, faulty processing and preservation may facilitate spoilage and food poisoning by *Salmonella, Escherichia, Staphylococcus,* and *Clostridium* (Mossel and DeBruin 1960). Fortunately, toxins are not usually produced at $pH < 5$, and spoiled fruit juices are not considered to be a major source of food poisoning.

With the exception of a few molds, sporeforming *Bacillus* and *Clostridium* species are the only heat-resistant microorganisms found in fruit juices. However, owing to the low numbers found naturally on fruits and their inability to develop readily at low pH, their numbers in fruit juices are usually low unless contamination from an outside, processing source occurs or conditions become favorable for growth of existing bacteria in the juice (Beerens 1960; Csaba and Nikodemusz 1960).

Molds are generally considered to be the least important group of microorganisms causing spoilage in fruit juices because of their limited ability to grow in the absence of air (Lüthi 1959). However, molds are more capable of growing on dry substrates than either bacteria or yeasts; therefore, their importance becomes much greater when freeze-dried or powdered fruit juices are considered. As a result of their frequent rotting of fruit, the aerial-borne spores of molds can usually be found in most raw fruit juices and fruit products.

Also, because of the effectiveness of heat treatment and the storage of juices under a pressurized CO_2 atmosphere, mold development in fruit juices is rare, except in home situations where neither heat treatment nor CO_2 atmospheres are usually employed in the production and maintenance of fruit juices.

Fruits usually maintain low redox potentials as a result of the consumption of oxygen by tissue particles. However, juices are aerated during extraction causing the redox potential to greatly increase. This increase favors the growth of most microorganisms, especially yeasts which do not grow readily in the absence of oxygen. The enzymes responsible for oxygen consumption are often associated with the solid residue of the fruit. Therefore, if the juice contains fruit pulp, the redox potential falls, thereby favoring the growth of clostridia, yeasts, molds, and acetic acid bacteria (Mossel and Ingram 1955). The pH of pulp-containing fruit juices can be lowered by heating the juice to destroy the enzymes or raised by increasing the surface to volume ratio of storage.

The levels of certain fermentation end-products in fruit juices are commonly used as indicators of microbial alteration. The levels of ethyl alcohol, glycerol, acetic acid, succinic and dehydroshikimic acid, diacetyl, and acetoin can be used as relatively accurate indications of the extent of microbial alteration. Viable counts are not an accurate indication of the degree of microbial alteration.

Fruit juices are commonly heated immediately following pressing to destroy fruit-associated enzymes that might discolor or clarify the juice. This treatment, at 90°C for 30 sec, destroys most yeasts, bacteria, and molds, except the spores of *Bacillus, Clostridium,* and the heat-resistant species of *Byssochlamys, Monascus, Paecilomyces,* and *Phialophora* (Jensen 1960; Splittstoesser *et al.* 1971; Beuchat and Rice 1979). The organisms usually found in spoiled concentrated fruit juices are *Asperigillus glaucus, Saccharomyces rouxii,* other *Saccharomyces* species, *Schizosaccharomyces, Hansenula,* and *Debaryomyces* (Ingram and Lüthi 1961). The limiting concentration of fruit juices also depends on the aqueous activity of the juice as determined by the type of sugars present, with glucose or fructose usually being more effective than sucrose. Generally, molds and yeasts survive in frozen fruit juices better than bacteria (Ingram 1952). Even so, spore-producing bacteria and staphylococci are relatively unaffected. Low pHs and temperatures just below the freezing point accelerate the death rate, while greater concentrations of solutes (e.g., sugars) cause it to decrease.

Put *et al.* (1976) studied the heat resistance of yeasts most commonly found in spoiled canned soft drinks and fruit juices. Their results indicated that among the 35 asporogenous strains tested, a higher heat resistance was observed than in the 85 ascomycetous strains. The thermal death rate (*D* value) of yeasts depends upon initial ascospore number in the heating medium (Put *et al.* 1977). The *D* values of ascospores of two *Saccharomyces* species were about tenfold greater than those of vegetative cells. In a later study (Put and De Jong 1982), ascospores of *Saccharomyces* isolated from spoiled heat-processed soft drinks and fruit products proved to be 50–150 times more resistant to heat than were vegetative cells.

Potassium sorbate and sodium benzoate have a synergistic effect with heat to inactivate yeasts in fruit juices (Beuchat 1982). Supplementation of juices with sucrose (30 and 50%) resulted in protective against thermal death of yeasts. While neither sorbate nor benzoate exhibited a consistent superior lethal effect compared with the other, overall, sodium benzoate was generally more effective than was potassium sorbate.

Organisms vary in their abilities to withstand freezing temperatures in fruit juices. Whereas strains of *Escherichia coli* die in a few days in frozen, natural fruit juices and last longer in concentrates, some streptococci are capable of much longer survival in both natural and concentrated frozen fruit juices (Larkin *et al.* 1955).

Yeasts are usually responsible for the characteristic alcoholic fermentations of various fresh fruit juices. It has been observed that a variety of yeasts known as *Saccharomyces ellipsoideus* has been isolated from grapes during the vintage season. An extensive review of yeast flora of unpasteurized and fermented fruit juices has been made by Walker and Ayres (1970).

In some countries clarified grape juice is often preserved by filtering it through a sterilizing "germ-proofing" filter and holding it under carbon dioxide pressure of approximately 150 psi. The filtered grape juice is then bottled under aseptic conditions into sterile bottles. Tressler and Pederson (1936) have shown that clarified Concord juice is more easily sterilized by filtration through a Seitz filter than are juices from other fruits. However, the juice must be heated sufficiently to inactivate the enzymes if it is to be preserved in this way; otherwise undesirable changes in flavor occur.

The perservation of fruit juices, syrups, concentrates, and purées by SO_2 treatment is more common in countries with warm climates and those where frozen storage facilities are limited. This chemical method of preservation can be used effectively for bulk juices and purées that are eventually to be processed into consumer products. Suggested SO_2 levels range from 350 to 600 ppm. Higher levels may be necessary because of the considerable amount of sugar and other SO_2-binding materials. For optimum inhibitory action by SO_2, acidification of purées, concentrates, and similar products may be necessary.

If, in their final uses, sulfite-treated bulk fruit materials are to be canned, the sulfite must be reduced to less than 20 ppm. Otherwise, in contact with the metal of the can, sulfite may generate H_2S which will cause a black precipitate.

Various methods for removing SO_2 from juices, such as heating in vacuum pans, mechanical agitation in conjunction with bubbling of an inert gas, and spraying against baffles under vacuum, have been developed. It should be noted that sulfite in conjunction with heat processing permits the use of lower temperatures. In the case of jams or other highly sweetened products stored in glass, there is more latitude in the final SO_2 content as the sweetness covers the taste of the chemical.

Applications of diethyl procarbonate (DEPC) have been suggested by Harrington and Hills (1966). They found that 50 ppm DEPC rapidly destroyed most of the yeasts, molds, and bacteria in fresh apple cider. Viable counts were reduced by 99%, yet 200 ppm DEPC would not render cider sterile. Treated cider had a storage life of 4–5 days before regrowth. The use of 50 ppm DEPC and 350 ppm potassium sorbate inhibited regrowth so that the storage life of cider was extended to more than 80 days. There was no effect on flavor. Splittstoesser and Wilkinson (1973) found that 50 ppm DEPC reduced the viable count of *Saccharomyces cerevisiae* over 9 $\log_{10}$ cycles, whereas 200 ppm reduced the count of *Byssochlamys fulva* ascospores by only about 1 log. However, the use of DEPC as a food preservative in the United States is not permitted since it is known to react with ammonia to form urethan, a carcinogen.

Sodium benzoate at 0.05–0.10% or potassium sorbate at 0.025–0.10% can be used to preserve fruit juices. Both can also be used conjunctively, each at a lower level. Most of the conditions that apply to still beverages apply to fruit juices.

Fruit salads, juices, sauces, syrups, and fillings are among the foods reported to be preserved by parabens (para-hydrobenzoic acid). Combination with sodium benzoate is recommended. About 0.05% of a 2 to 1 mixture of methyl and propyl benzoate is suggested for experimental processing.

Bacterial spoilage is perhaps the most troublesome type of spoilage in tomato juice and results in an off-flavor. Overheating can bring about cooked flavors in the juice. *Bacillus thermoacidurans* may cause a high incidence of spoilage. This organism is very heat resistant and can withstand a processing temperature of 100°C for 20 min. It not only produces an off-flavor in the tomato juice, but may eventually produce a "flat-sour" type of spoilage.

Pederson and Becker (1949) suggest that if tomato juice of pH 4.15–4.25 could be subjected to moderate pasteurization conditions, the vegetative cells would be destroyed and the spores would be of no consequence, since they would be incapable of germinating and producing growth at these low pH values. However, Rice and Pederson (1954a,b) indicate that while a minimum heat treatment may kill the vegetative cells, if the spore concentration is reduced, the minimum pH of tomato juice at which growth can occur rises. A juice of pH 4.30 or even 4.25 under these conditions could not necessarily be inhibitory to the growth of *Bacillus thermoacidurans*. It appears that the organism is widely distributed in the soil so its entrance into tomato products is not unusual.

The effects of various germicides on microorganisms in can cooling waters was summarized by Ito and Seeger (1980). The relative effectiveness of chlorine, iodophors, and ozone water as influenced by manner of application, pH, and organic load is reviewed.

CATSUP, PICKLES, AND PRESERVES

Molds and yeasts find a favorable environment in catsup, and mold counts are routinely conducted following processing as a measure of quality control. Gas formation and yeasty flavor accompany the presence of yeasts. *Lactobacillus lycopersici, Lactobacillus plantarum,* and other species of lactobacilli cause gaseous spoilage and off-flavors. The lactobacilli can be destroyed, however, by a 2-min thermal processing at 75°C.

The usual sugar concentration of catsup is not sufficient to inhibit the growth of most microorganisms. However, when 15% sugar and 3% salt in the presence of acid were combined in catsup, microbial growth was reduced to a minimum.

For some pickled fruits, either sodium benzoate or sorbate can be used. The pH of fermented olives is more favorable to sorbate than to benzoates. Recommended levels are from 0.025 to 0.5%, the higher level being more appropriate to sweeter products. Tests showing effectiveness at or near these low levels have been reported (Vaughn *et al.* 1969). Vaughn (1962) showed that 0.1% of sorbate in olive fermentation controlled scum yeast and did not interfere with lactic acid production by bacteria. In practice even lower levels are effective.

A level of about 0.1% of parabens is recommended for olives and for pickled fruits. A lower level in combination with sodium benzoate is effective in pickled fruits.

Fruit salads, fruit cocktails and gelatin salads are generally kept under refrigeration which retards bacteria but may permit the growth of yeasts and molds. The addition of sorbate at 0.05–0.1% increases the shelf life of these and similar products. Potassium sorbate is most conveniently added to the liquid part of salads such as the cover syrup in fruit salads and cocktails; the dressing in cole slaw, and potato and macaroni salad; or the hot water in fruit gelatin. Sorbates have been evaluated in various other foods, usually at levels of from 0.025 to 0.1%. Extension of shelf life was noted in such products as tangerine sherbet base, maraschino cherries, strawberry purée, tomato juice, prepeeled carrots, and wax cucumbers (Vaughn *et al.* 1969).

From 0.2 to 0.4% sodium propionate has been used to retard the development of mold on syrup, blanched apple slices, figs, cherries, blackberries, peas, and lima beans.

Molin *et al.* (1963) reported that surface mold on fresh strawberries was checked by dipping berries in solutions of diethyl procarbonate (DEPC). Most effective levels were 0.01 and 0.1%. A minimum level of 0.1% was found essential for extending the shelf life of applesauce under various conditions of storage.

Jellies, jams, preserves, and marmalades depend upon a high content of soluble solids and high acidity to prevent spoilage. The possibility of spoilage in cooked jams and jellies by osmophilic yeasts is small because these organisms are killed during heating; any contaminants that are able to grow at all, grow slowly. Osmophilic yeasts in substrates with sufficiently high solids content to prevent growth do not necessarily die, but may remain viable for several weeks or months. Thus, the concentration of soluble solids in sugar-preserved products must be maintained at a level that prevents growth of yeasts and molds. The required legal-minimum concentration of soluble solids in jams and jellies is 65% (Oliver 1962; Desrosier 1970). It was found that *Zygosaccharomyces barkeri,* when cultivated at its optimal pH range of 4–5 and at 30°C, is inhibited only by a sugar concentration equivalent to an equilibrium relative humidity of 62% which corresponds to a pear concentrate containing approximately 82% solids. Conditions that favor the growth of yeasts, such as a favorable pH and temperature, also encourage spoilage; thus, slightly acid jams and jellies may ferment more readily than definitely acid products.

Subba Rao *et al.* (1965) have described spoilage of an Indian preserve called "murrabba" by osmophilic yeasts. However, osmophilic yeasts such as *Saccharomyces rouxii, S. mellis,* and *S. fermentati* have also caused fermentation of this food. Addition of acetic acid and sodium benzoate to the syrup was recommended to control growth. Acetic acid at a concentration of 0.2% with 50 ppm of sodium benzoate effectively inhibited growth of *S. rouxii* and apparently was lethal to *S. mellis*.

The addition of salt or sugar also controls a_w because availabe moisture is reduced. The available water can also be controlled by adding hydrophilic colloids, gels, and agar to tie up water. A 3–4% agar medium may prevent microbial growth by leaving too little moisture for microorganisms.

Fruit juice concentrates, soft drink concentrates, fruit purées, fruits in sugar syrup, and fruit jams and jellies are pasteurized and preserved by a combination of heat treatment and, in some instances, by adjusting the a_w below that required for growth of bacteria and most fungi. The heat resistance of vegetative cells and spores of microorganisms is known to be influenced by the a_w of the surrounding environment. Generally, in the a_w range of 0.70 to 0.99 common to fresh fruit and fruit products, increased thermal death rates are correlated with decreased a_w (Corry 1978; Beuchat 1981, 1983). Thus, knowledge of the behavior of microbial cells at reduced a_w as they are subjected to heat is valuable when designing time-temperature treatments for pasteurizing various fruit products.

Preservation of jellies, jams, maple syrup, and honey is mainly attributed to their high sugar content (65–80%), but poorly sealed or opened containers frequently allow the slow growth of molds. Osmophilic (''high osmotic pressure loving'') yeasts occasionally grow in honey and produce sufficient carbon dioxide to burst the jar.

Various chemicals, including formaldehyde, boric acid, benzoic acid, and sulfur dioxide, have been used in the past to prevent spoilage of certain food products, including fruits. Most of these chemicals are harmful, and their use is now strictly regulated (e.g., sodium benzoate). Sodium benzoate has very wide applicability as an anti-microbial agent in foods. It is used in carbonated and still beverages, syrups, fruit salads, icings, jams, jellies, preserves, salted margarine, mincemeat, pickles and relishes, pie and pastry fillings, prepared salads, fruit cocktails, and even some pharmaceuticals (Smart and Spooner 1972). Use levels range from 0.05 to 0.10%.

Under Federal Standards of Identity for jams, jellies, and preserves, sodium benzoate is permitted. Benzoate is also permitted in artificially sweetened jams, jellies, and preserves.

Organic acids are common preservatives that are particularly effective against putrefaction. They may be added directly (as in vinegar) or developed by fermentation of sugars in the food itself.

BIBLIOGRAPHY

ABU-ZINADA, A. H., and ALI, M. I. 1982. Fungi associated with dates in Saudi Arabia. J. Food Protection *45*, 842–844.

AHARONI, Y., and LATTER, F. S. 1972. The effect of various storage atmospheres on the occurrence of rots and blemishes on Shamouti oranges. Phytopathology *73*, 371–374.

ALEXOPOULOS, C. J. 1962. Introductory Mycology, 2nd Edition. John Wiley & Sons, New York and London.

ARCHER, S. A. 1979. Pectolytic enzymes and degradation of pectin associated with the breakdown of sulphited strawberries. J. Sci. Food Agric. *30*, 692–703

ARCHER, S. A., and FIELDING, A. H. 1979. Polygalacturonase isoenzymes of fungi involved in the breakdown of sulfited strawberries. J. Sci. Food Agric. *30*, 711–723.

AUSTWICK, P. K. C. 1966. The role of spores in the allergies and mycoses of man and animals. *In* The Fungus Spore. M. F. Madelin (Editor). Butterworths, London, England.

BEE, G. R., and HONTZ, L. R. 1980. Detection and prevention of postprocessing container handling damage. J. Food Protection *43*, 458–460.

BEERENS, H. 1960. Survival of anaerobic bacteria in apple and grape juices. Ann. Inst. Pasteur Lille *11*, 151–164. (French)

BEUCHAT, L. R. 1979. Comparison of acidified and antibiotic-supplemented potato dextrose agar from three manufacturers for its capacity to recover fungi from foods. J. Food Protection *42*, 427–428.

BEUCHAT, L. R. 1981. Microbial stability as affected by water activity. Cereal Foods World *26,* 345–349.
BEUCHAT, L. R. 1982. Thermal inactivation of yeasts in fruit juices supplemented with food preservatives and sucrose. J. Food Sci. *47,* 1679–1682.
BEUCHAT, L. R. 1983. Influence of water activity on growth, metabolic activities and survival of yeasts and molds. J. Food Protection *46.*
BEUCHAT, L. R. and RICE S. L. 1979. *Byssochylamys* spp. and their importance in processed fruits. Adv. Food Res. *25,* 237–288.
BEVERIDGE, E. G. 1969. Metabolism of *p*-hydroxybenzoate esters by bacteria. Pharm. J. *203,* 468.
BROWN, G. E., and McCORMACK, A. A. 1972. Decay caused by *Alternaria citri* in Florida citrus fruit. Plant Disease Reptr. *56,* 909–912.
BULLERMAN, L. B. 1979. Significance of mycotoxins to food safety and human health. J. Food Protection *42,* 65–86.
CAPPELLINI, R. A. 1972. Fungi associated with blueberries held at various storage times and temperatures. Phytopathology *62,* 10.
CHICHESTER, D. F., and TANNER, JR., F. W. 1968. Antimicrobial food additives. *In* Handbook of Food Additives. T. E. Furia (Editor). Chemical Rubber Co., Cleveland.
COOK, A. H. 1958. The Chemistry and Biology of Yeasts. Academic Press, New York.
CORRY, J. E. L. 1978. Relationships of water activity to fungal growth. *In* Food and Beverage Mycology. L. R. Beuchat (Editor). AVI Publishing Co., Westport, CT.
CSABA, K., and NIKODEMUSZ, I. 1960. Observations on the numbers of aerobic sporing bacilli in fruit juices and other drinks. Ann. Inst. Pasteur Lille *11,* 227–230. (French)
DACK, G. M. 1955. Significance of enteric bacilli in foods. Am. J. Public Health *45,* 1151–1156.
DAVENPORT, R. R. 1980. Cold-tolerant yeasts and yeast-like organisms. *In* Biology and Activities of Yeasts. F. A. Skinner, S. M. Passmore, and R. R. Davenport (Editors). Academic Press, London.
DENNIS, C. 1978. Post-harvest spoilage of strawberries. Agric. Res. Coun. Res. Rev. *4,* 38–42.
DENNIS, C., and BUHAGIAR, R. W. M. 1980. Yeast spoilage of fresh and processed fruits and vegetables. *In* Biology and Activities of Yeasts. F. A. Skinner, S. M. Passmore, and R. R. Davenport (Editors). Academic Press, London.
DENNIS, C., DAVIS, R. P., HARRIS, J. E., CALCUTT, L. W. and CROSS, D. 1979. The relative importance of fungi in the breakdown of commercial samples of sulphited strawberries. J. Sci. Food Agric. *30,* 959–973.
DENNIS, C., and HARRIS, J. E. 1979. The involvement of fungi in the breakdown of sulfited strawberries. J. Sci. Food Agric. *30,* 687–691.
DESROSIER, N. W. 1970. The Technology of Food Preservation, 3rd Edition. Avi Publishing Co., Westport, Connecticut.
ECKERT, J.W. 1975. Postharvest diseases of fresh fruits and vegetables—etiology and control. *In* Symposium: Postharvest Biology and Handling of Fruits and Vegetables. N. F. Haard and D. K. Salunke (Editors). AVI Publishing Co., Westport, CT.
ECKERT, J. W. 1978. Control of postharvest diseases. *In* Antifungal Compounds. M. R. Siegel and H. D. Sisler (Editors). Marcel Dekker, New York.
FOOD AND DRUG ADMIN. Canned fruits and juices. Code of Federal Regulations, Title 21, & 27.60, 27.70, 27.73. U.S. Government Printing Office, Washington, D.C.
FOOD AND DRUG ADMIN. Canned vegetables other than those specifically regulated. Code of Federal Regulations, Title 21, & 51.990 (a) (1), 51.990 (a) (3) (i). U.S. Government Printing Office, Washington, D.C.
FRANK, H. K. 1966. Aflatoxine in Lebensmittelen. Arch. Lebensmittelhyg. *17,* 237–242.

GARDNER, W. H. 1966a. Inhibiting undesirable chemical reactions. *In* Food Acidulants. Allied Chemical Corporation, New York.
GARDNER, W. H. 1966b. Role of acidulants in foods. *In* Food Acidulants. Allied Chemical Corporation, New York.
GAUMANN, E. A. 1964. The Fungi, 5th Edition. Birkhäuser-Verlag, Basel, Switzerland. (German)
GOVERD, K. A., BEECH, F. W., HOBBS, R. P., and SHANNON, R. 1979. The occurrence and survival of coliforms and salmonellas in apple juice and cider. J. Appl. Microbiol. *46,* 521–530.
GUNDERSON, M. F. 1962. Mold problem in frozen foods. Proc. Low Temperature Microbiol. Symp., Campbell Soup Co., Camden, N.J.
GUNTHER, D. A. 1969. Adsorption and desorption of ethylene oxide. Am. J. Hosp. Pharm. *26,* 45–49.
HARRINGTON, W. O., and HILLS, C. H. 1966. Preservative effect of diethyl pyrocarbonate and its combination with potassium sorbate on apple cider. Food Technol. *20,* 1360–1362.
HARRIS, J. E., and DENNIS, C. 1979. The stability of pectolytic enzymes in sulfite liquor in relation to breakdown of sulfited strawberries. J. Sci. Food Agric. *30,* 704, 710.
HARVEY, J. M. 1978. Reduction of losses in fresh market fruits and vegetables. Annu. Rev. Phytopathol. *16,* 321–341.
HOFFMAN, R. K., and WARSHOWSKY, B. 1958. Beta-propiolacetone vapor as a disinfectant. Appl. Microbiol. *6,* 358.
HORIKOSHI, K., and ANDO, T. 1972. Conidia of *Aspergillus oryzae*. XII. Ribonucleases in the Conidia of *Aspergillus oryzae*. *In* Spores V. H. O. Halvorson, R. Hanson, and L. L. Campbell (Editors). American Soc. for Microbiol., Washington, D.C.
HSU, E. J. 1975. Microorganisms in relation to vegetable processing. *In* Commercial Vegetable Processing. B. S. Luh and J. G. Woodroof (Editors). Avi Publishing Co., Westport, Conn.
HSU, E. J., and VAUGHN, R. H. 1969. Production and catabolite repression of the constitutive polygalacturonic acid *transeliminase* of *Aeromonas liquefaciens*. J. Bacteriol. *98,* 172–181.
INGRAHAM, J. L., and STOKES, M. L. 1959. Psychrophilic bacteria. Bacteriol. Rev. *23,* 97–108.
INGRAM, M. 1952. The action of cold on microorganisms, in relation to food. Proc. Soc. Appl. Bacteriol. *14,* 243–260.
INGRAM, M. 1955. Introduction to Yeast. Pitman Publishers, London.
INGRAM, M., and LUTHI, H. 1961. Microbiology of fruit juices. *In* Fruit and Vegetable Juice Processing Technology. D. K. Tressler and M. A. Joslyn (Editors). Avi Publishing Co., Westport, Conn.
INSALATA, N. F., and ROAB, I. 1969. Fecal streptococci in industrial processed foods. J. Milk Food Technol. *32,* 86–88.
ITO, K. A., CHEN, J. K., SEEGER, M. L., UNVERFERTH, J. A., and KIMBALL, R. N. 1978. Effect of pH on the growth of *Clostridium botulinum* in canned figs. J. Food Sci. *43,* 1634–1635.
ITO, K. A., and SEEGER, M. L. 1980. Effects of germicides on microorganisms in can cooling waters. J. Food Protection *43,* 484–487.
JARVIS, B. 1973. Comparison of an improved rose bengal-chlortetracycline agar with other media for the selective isolation and enumeration of moulds and yeasts in foods. J. Appl. Bacteriol. *36,* 723–726.
JARVIS, B. 1978. Methods for detecting fungi in foods and beverages. *In* Food and Beverage Mycology. L. R. Beuchat (Editor). AVI Publishing Co., Westport, CT.

JENSEN, M. 1960. Experiments on the inhibition of some thermo-resistant molds in fruit juices. Ann. Inst. Pasteur Lille *11*, 179–182. (French)

KALLET, A., and SCHLINK, F. J. 1933. 100,000 Guinea Pigs. *In* Dangers in Everyday Foods, Drugs and Cosmetics. Vanguard Press, New York.

KOBURGER, J. A. and FARHAT, B. Y. 1975. Fungi in foods. VI. Comparison of media to enumerate yeasts and molds. J. Food Protection *38*, 466–468.

KONOWALCHUK, J., and SPIERS, J. I. 1975. Survival of enteric viruses on fresh fruit. J. Milk Food Technol. *38*, 598–600.

LARKIN, E. P., LITSKY, W., and FULLER, J. E. 1955. Fecal streptococci in frozen foods. II. Effect of freezing on *Escherichia coli* and some fecal streptococci inoculated onto green beans. Appl. Microbiol. *3*, 102–104.

LEIFSON, E. 1954. The flagellation and taxonomy of species of *Acetobacter*. Antonie van Leeuwenhoek, J. Microbiol. Serol. *20*, 102.

LELLIOTT, R. A., BILLING, E., and HAYWARD, A. C. 1966. A determinative scheme for the fluorescent plant pathogenic pseudomonads. J. Appl. Bacteriol. *29*, 470.

LLEWELLYN, G. C., DUCKHARDT, N. E., FISHER, M. F., EADIE, T., and O'REAR, C. E. 1980. Growth, sporulation and aflatoxin production by *Aspergillus parasiticus* on strained baby foods. J. Food Protection *43*, 428–430.

LOCKHART, C. L., FORSYTH, F. R., STARK, R., and HALL, I. V. 1971. Nitrogen gas suppresses microorganisms in cranberries in short term storage. Phytopathology *63*, 335–336.

LUH, B. S., and PHAFF, H. J. 1951. Studies on polygalacturonase of certain yeast. Arch. Biochem. Biophys. *33*, 212.

LUND, BARBARA M. 1971. Bacterial spoilage of vegetables and certain fruits. J. Appl. Bacteriol. *34*, 9–20.

LUTHI, H. 1959. Microorganisms in non-citrus juices. Advan. Food Res. *9*, 221–284.

McCORMACK, G. 1950. Technical note No. 5. "Pink Yeast," isolated from oysters grows at temperatures below freezing. U.S. Fish Wildlife Serv. Com. Fisheries Rev. *12, 11a*, 28.

McCULLOCH, L. P., COOK, H. T., and WRIGHT, W. R. 1968. Market diseases of tomatoes, peppers and eggplants. U.S. Dept. Agr. Handbook 28.

MILLER, M. W., and PHAFF, H. J. 1962. Successive microbial populations of calimyrna figs. Appl. Microbiol. *10*, 394–400.

MILLER, M. W., and TANAKA, H. 1963. Relation of ERH to potential spoilage. Hilgardia *34*, 183–190.

MOLIN, N., SATMARK, L., and THORELL, M. 1963. Pyrocarbonic acid diethyl ester as a potential food preservative. Food Technol. *17*, 797–801.

MOREAU, C. 1974. Moulds, Toxins and Food. John Wiley and Sons, New York.

MOSSEL, D. A. A. 1971. Physiological and metabolic attributes of microbial groups associated with foods. J. Appl. Bacteriol. *34*, 95–118.

MOSSEL, D. A. A., and DeBRUIN, A. S. 1960. The survival of *Enterobacteriaceae* in acid liquid foods of pH below 5 when they are stored at different temperatures. Ann. Inst. Pasteur Lille *11*, 65–72. (French)

MOSSEL, D. A. A., and INGRAM, M. 1955. The physiology of the microbial spoilage of foods. J. Appl. Bacteriol. *18*, 233–268.

MURDOCK, D. I., and HATCHER, W. S. 1975. Growth of microorganisms in chilled orange juice. J. Milk Food Technol. *38*, 393–396.

NASUNO, S., and STARR, M. P. 1966. Polygalacturonase of *Erwinia carotovara*. J. Biol. Chem. *241*, 5298–5306.

NATARAJAN, C. P., CHARI, C. N., and MRAK, E. M. 1948. Yeast population in figs during drying. Fruit Prod. J. *27*, 242–243, 267.

OLIVER, M. 1962. Problems in preservation by the use of sugar. *In* Recent Advances in Food Science, Vol. 2. J. Hawthorn and J. M. Leitch (Editors). Butterworths, London.

ONISHI, H. 1963. Osmophilic yeasts. Advan. Food Res. *12,* 53–92.

PEDERSON, C. S., ALBURY, M. N., WILSON, D. C., and LAWRENCE, N. L. 1958. The growth of yeasts in grape juice stored at low temperatures. II. The types of yeast and their growth in pure culture. Appl. Microbiol. *7,* 7–11.

PEDERSON, C. S., ALBURY, M. N., WILSON, D. C., and LAWRENCE, N. L. 1960. The growth of yeasts in grape juice stored at low temperatures. IV. Fungistatic effects of organic acids. Appl. Microbiol. *9,* 162–167.

PEDERSON, C. S., and BECKER, M. E. 1949. Flat sour spoilage of tomato juice. N.Y. State Agr. Expt. Sta. Tech. Bull.

PHAFF, H. J. 1959. The production of certain extracellular enzymes by microorganisms. In *Hanbuch der phlanzenphsiologie* Band XI. *Heterotrophie.* W. Ruhland (Editor). Springer-Verlag, Berlin. (German)

PITT, J., and CHRISTIAN, H. 1968. Water relations in xerophilic fungi isolated from prunes. Appl. Microbiol. *16,* 1853–1858.

PUT, H. M. C., and DE JONG, J. 1982. The heat resistance of ascospores of four *Saccharomyces* spp. isolated from spoiled heat-processed soft drinks and fruit products. J. Appl. Bacteriol. *52,* 235–243.

PUT, H. M. C., DE JONG J., and SAND, F. E. M. J 1977. The heat resistance of ascospores of *Saccharomyces cerevisiae* strain 195 and *Saccharomyces chevalieri* strain 215, isolated from heat preserved fruit juice. *In* Spore Research II. A. N. Barker, L. J. Wolfd, D. J. Ellar, G. J. Dring, and G. W. Gould (Editors). Academic Press, London.

PUT, H. M. C., DE JONG, J., SAND, F. E. M. J., and VAN GRINSVEN, A. M. 1976. Heat resistance studies on yeasts causing spoilage of soft drinks. J. Appl. Bacteriol. *40,* 135–152.

RAMSEY, G. B., and SMITH, M. A. 1961. Market diseases of cabbage, cauliflower, turnips, cucumbers, melons and related crops. U.S. Dept. of Agr. Handbook *184.*

RICCI, P. 1972. Observations on the rotting of fresh figs after harvest. Ann. Phytopathol. Soc. Japan *4,* 109–117.

RICE, A. C., and PEDERSON, C. S. 1954a. Factors influencing growth of *Bacillus Coagulans* in canned tomato juice. II. Acidic constituents of tomato juice and specific organic acids. Food Res. *19,* 124–133.

RICE, A. C., and PEDERSON, C. S. 1954b. Factors influencing growth of *Bacillus Coagulans* in canned tomato juice. I. Size of inoculum and oxygen concentration. Food Res. *19,* 115–123.

RICE, S. L., BEUCHAT, L. R., and WORTHINGTON, R. E. 1977. Patulin production by *Byssochlamys* spp. in fruit juices. Appl. Environ. Microbiol. *34,* 791–796.

SCOTT, W. J. 1957. Water relations in food spoilage microorganisms. Advan. Food Res. *7,* 83–127.

SEEGER, M. L., and LEE, W. H. 1972. The destruction of *Byssochlamys fulva* asci by low concentrations of gaseous methyl bromide and by aqueous solutions of chlorine, an iodophor and peracetic acid. J. Appl. Microbiol. *35,* 479.

SMART, R., and SPOONER, D. E. 1972. Microbiological spoilage in pharmaceuticals and cosmetics, J. Soc. Cosmetic Chemists *23,* 721–734.

SMITH, W. L. 1962. Chemical treatment to reduce post-harvest spoilage of fruits and vegetables. Botan. Rev. *28,* 411.

SPALDING, D. H., and HARDENBURG, R. H. 1971. Postharvest chemical treatment for control of blue mold rot of apples in storage. Phytopathology *61,* 1308.

SPLITTSTOESSER, D. F. 1978. Fruit and fruit products. *In* Food and Beverage Mycology. L. R. Beuchat (Editor). AVI Publishing Co., Westport, CT.

SPLITTSTOESSER, D. F. 1982. Microorganisms involved in the spoilage of fermented fruit juices. J. Food Protection *45*, 874–877.

SPLITTSTOESSER, D. F., GROLL, M., DOWNING, D. L., and KAMINSKI, J. 1977 Viable counts versus the incidence of machinery mold (*Geotrichum*) on processed fruits and vegetables. J. Food Protection *40*, 402–405.

SPLITTSTOESSER, D. F., KUSS, F. R., HARRISON, W., and PREST, D. B. 1971. Incidence of heat-resistant molds in eastern orchards and vineyards. Appl. Microbiol. *21*, 335–337.

SPLITTSTOESSER, D. F., and WILKISON, M. 1973. Some factors affecting the activity of diethylpyrocarbonate as a sterilant. Appl. Microbiol. *25*, 853–857.

STANDARDS OF IDENTITY. 1977. Canned Figs. Identitiy; label statement of optional ingredients. 21 CFR 145.130. USGPO, Washington, D.C.

SUBBA RAO, M. S., *et al.* 1965. Spoilage of foods. *In* Food Science and Technology, Vol. 2, J. M. Leitch (Editor). Gordon and Breach Science Publishers, London, England.

SUSSMAN, A. S. 1966. Types of dormancy as represented by conidia and ascospores of Neurospora. *In* The Fungus Spore. M. F. Madelin (Editor). Butterworths, London, England.

TANAKA, H., and MILLER, M. 1963a. Yeast and molds associated with spoiled dried prunes. Hilgardia *34*, 167–170.

TANAKA, H., and MILLER, M. 1963b. Studies of the osmophilic nature of spoilage organisms. Hilgardia *34*, 171–181.

TOWNSEND, C. T., GETCHELL, R. N., PERKINS, W. E., and COLLIER, C. P. 1955. The pH necessary to inhibit the growth of *Clostridium botulinum* in canned figs. Rept. Natl. Canners Assoc. Western Branch Laboratories, Berkeley, CA.

TRESSLER, D. K., and JOSLYN, M. A. 1971. Fruit and Vegetable Juice Processing Technology, 2nd Edition, Avi Publishing Co., Westport, Conn.

TRESSLER, D. K., and PEDERSON, C. A. 1936. Preservation of grape juice. II. Factors controlling the rate of deterioration of bottled concord juice. Food Res. *1*, 87–97.

VAUGHN, R. H. 1955. Bacterial spoilage of wines with special reference to California conditions. Advan. Food Res. *6*, 67–108.

VAUGHN, R. H. 1957. *Acetobactor*. *In* Bergey's Manual of Determinative Bacteriology, 7th Edition. R. S. Breed, E. G. D. Murry, and N. R. Smith (Editors). Williams and Wilkins Co., Baltimore.

VAUGHN, R. H. 1962. Microbial spoilage problems of fresh and refrigerated foods. *In* Microbial Quality of Foods. L. W. Slanetz (Editor). Academic Press, New York.

VAUGHN, R. H. *et al.* 1969. Gram-negative bacteria associated with sloughing, a softening of California ripe olives. J. Food Sci. *34*, 224–227.

VON SCHELHORN, M. 1951. Control of microorganisms causing spoilage in fruit and vegetable products. Advan. Food Res. *3*, 431.

WALKER, H. W., and AYRES, J. C. 1970. Yeasts as spoilage organism. *In* The Yeasts, Vol. 3. A. M. Rose and J. S. Harrison (Editors). Academic Press, London and New York.

WILDMAN, J. D., STOLOFF, L., and JACOBS, R. 1967. Aflatoxin production by a potent *Aspergillus flavus* link isolate. Biotechnol. Bioeng. *9*, 429–437.

WOOD, T. H., and ROSENBERG, A. M. 1957. Freezing in yeast cells. Biochim. Biophys. Acta *23*, 78.

WOODBURN, M. 1982. pH of Oregon-grown figs and their acidification for home-canning. J. Food Protection *45*, 1245–1247.

YESAIR, J., and WILLIAMS, C. B. 1942. Spice contamination and its control. Food Res. *7*, 118–126.

6

Canning of Fruits

B. S. Luh, C. E. Kean, and J. G. Woodroof

Commercial canning of fruits and fruit juices is an important industry, particularly on the Pacific Coast of the United States. The total pack varies from year to year but has been about 163,000,000 standard cases since about 1972.

The top 10 canned fruit and fruit juice packs in 1980 were as follows: clingstone peaches—26,765,000; orange juice—19,146,000; grapefruit juice—19,303,000; applesauce—18,613,000; pineapples—17,057,000; fruit cocktail mix—14,426,000; pineapple juice—13,628,000; pears—10,926,000; cranberry sauce—7,023,000; and apples—1,791,000.

THE CANNERY

Plant Location

The factors that determine the suitability of any location for a cannery are availability of raw products; adequate supply of water of suitable quality; availability of labor during the canning season; transportation at reasonable rates between the cannery and markets; and adequate facilities for disposal of plant waste. Locating a factory near the raw material will permit the fruits to mature properly and will decrease injury from handling and deterioration from changes after harvesting.

The site should be easily accessible for the receipt of raw fruit and the

Commercial Fruit Processing, 2nd Edition

ISBN 0-87055-502-2

shipment of finished materials. It should have ample space and be in a clean locality.

Water

The water supply for a cannery must be adequate. If there is not reliable water source from city mains, then filtration and treatments are needed to remove impurities from the water.

It is advisable to install a continuous water-softening plant. Water for boilers should be softened to prevent the formation of scale. The water used for washing the plant and floors requires no treatment for hardness but must be clean.

When cooling of cans is done with circulated water, the water should be chlorinated. The chlorine content must, however, be controlled to avoid overdosing, as this will cause corrosion of cans. Automatic chlorine solution injectors are widely used. The amount of active chlorine in the water at the entrance will be determined by the quantity of organic matter present. It is not uncommon to have 1–2 ppm of active chlorine at the entrance end. If the cooling water is recirculated and used again it is extremely important to check the residual chlorine level at both the entrance and discharge end of the cooler. The continuous addition of chlorine solution to recirculated processing and can-cooling waters has made possible considerable reduction in water consumption without undesirable effects on the quality of products.

The remarkable germicidal efficiency of chlorine in water is attributed to its ability to attack and inactivate sulfhydryl enzymes essential for life of the microbial cell. Hypochlorous acid (HOCl) formed by chlorine compounds in solution is the germicidal agent. With a given chlorine solution, the germicidal efficiency increases as the pH becomes more acid. However, corrosion of cans also becomes more severe as the pH is lowered. Off-flavor may result from excessive use of chlorine in the processing of certain sensitive fruits. Canning syrups must be free of chlorine.

Lighting

In a fruit-canning plant, proper lighting is very important, particularly for sorting of prepared fruits, inspection, filling of cans, and the operation of automatic can-filling machines and can sealers. Proper lighting means adequate light of good quality, directed where it is needed, and with good diffusion.

The lighting system in a cannery requires careful planning by a specialist who will make sure that wiring is properly designed and of suitable current-carrying capacity. The switches, control panels, and outlets must be properly selected and installed so that they are safe for the workers in the plant.

The lighting should permit comfortable vision so as to avoid fatigue. It should be sufficiently intense to promote clean and sanitary conditions, efficiency, and pleasant working conditions. Low surface brightness is desirable in lamps used for general lighting. Fluorescent lamps usually have lower surface brightness and provide better diffusion of light than other light sources.

The Cannery Illuminating Committee recommends the following intensities of light illumination: 10 fc for receiving and dispatching, retort and exhaust-box areas, can unscramblers, dining room, stairways, and rest rooms; 20 fc for preliminary sorting, washing raw materials, cutting and pitting, canning, syruping, seaming, and labeling; 30 fc for the machine shop; and 30–50 fc for the laboratory.

Sorting and canning operations for peaches and apricots require critical inspection, and 85–125 fc of illumination are recommended. For color evaluation of foods the level of light illumination should be 200 fc. The MacBeth light is widely used for determination of USDA color score by visual grading.

Cannery Organization

Because cannery operations are always under pressure during the peak of the season, the work flow must be well organized. The plant operation should be studied carefully and the work planned well in advance of the season.

The Fieldman. Quality in canned fruits is at all times dependent on the raw material. The canner must employ a fieldman who understands fruit physiology and can deal with growers tactfully to ensure that only material of good quality is delivered to the cannery.

The cannery should have sufficient raw fruit to operate at full capacity. A contract that specifies the quality of fruit to be delivered should first be drawn up and inspection of each load made where possible.

The fieldman must keep in close touch with the cannery and report what raw material is due in advance. This gives the cannery the opportunity to adjust the labor supply in the case of shortage of raw material.

Receiving. The foreman of the receiving department should be thoroughly familiar with all varieties of fruits. He is the factory link with the fieldman and should keep him advised at all times as to the amount required and the condition of material as received. The foreman of the receiving department must be fair in his dealings with growers.

Preparation. One supervisor should be appointed for each 30 workers, and when piecework is employed an assistant may be required for checking the work done by this number. The work of the supervisors in each department should be coordinated by a foreman.

Details for washing, peeling, and preparation of fruit for processing are presented in Chapter 3.

Syrup and Brine. The preparation of syrup for fruit requires great care, and experienced personnel should be employed for this work. Automation of syrup mixing and delivery is now common practice in many canning plants. Since liquid sugar is becoming more popular, the plant should have facilities for holding liquid sugar.

Sterilization. Heat sterilization of the canned product is the most important part of the canning operation. The time and temperature required for heat processing will vary with the maturity of the fruit and other conditions. These operations must be done by experienced employees who understand the principles involved in heat sterilization of canned foods.

The cannery should employ competent technologists, and the heat sterilization foreman should be guided by the information received from the laboratory where testing of the product at various stages is carried out.

Labeling. The labeling and casing of the canned product should be directed by a person who is concerned with neatness and appearance. The appearance of an article is one of the main sale factors, and the labeling department must be made wholly responsible for this. The duties of the foreman are to supply the machine with labels, paste, resin, etc., and to see that all cans going into cases are correctly labeled and that the cases are fastened straightly and securely, with the proper markings outside.

Warehouse. The foreman of the warehouse controls all products from the cannery and arranges shipments as required. He should have expert knowledge of rail and road transport and should be capable of arranging loads from orders as received from the order department. Today, computers are being used by large canneries in order to have an accurate record of their inventory at all times.

Cans

Cans are made from tinplates, which are very thin sheets of steel lightly coated with tin. This tin coating serves two purposes: it covers the face of the steel sheet, preserving it from rust; and it acts as a medium by which parts of a sheet may be made to adhere to one another by soldering.

Sanitary cans can be cleaned readily. One end of the body is left completely open and is flanged at the mouth. The lid is made with a flange which hooks over the flange of the body and is finally rolled on without difficulty by means of two rollers, the first of which forms the seam and the second of which tightens it.

Cans are usually referred to by numbers that correspond to the size of the can. Table 6.1 lists the cans used principally in the United States. The first number of the symbol denotes the diameter, and the second, the height of the can. The first digit of either number represents the number of inches, and the last two digits, the number of sixteenths. Thus, a can that is 3 7/16 in. in diameter and 4½ in. high would be designated as 307 × 408. The corresponding dimensions in millimeters are also presented in Table 6.1.

Aluminum Cans. Although aluminum cans have captured a sizable portion of the beer and beverage market, they have not been popular for canned food uses except for the small-size drawn can. Aluminum cans become less attractive as can size increases because of the need for heavy gauges to prevent paneling. Even the higher strength aluminum alloys employed today still require an increase of about 35% in the gauge of aluminum to provide equivalent buckling resistance to conventional tinplate in 303 diameter cans.

Experimental and practical experience in food canning has shown that most canned fruits and vegetables packed in uncoated aluminum cans attack aluminum quite readily.

Lacquered Cans. The tin container by itself is perfectly satisfactory for preserving most foods. For foods that deteriorate in appearance when packed in a plain can (e.g., boysenberries, plums, olives), it is necessary to coat the insides of cans with a lacquer. Two kinds of lacquered cans are available: an acid-resisting lacquer, which is used for acid foods, mostly fruits; and a sulfur-resisting lacquer, which is used for products of the nonacid group, including vegetables, beans, and meats.

Fruits like apricots, grapefruit, peaches, pears, and pineapples are packed in plain tin containers. Cans for highly colored fruits such as boysenberries, raspberries, strawberries, and red plums must be coated inside with acid-resistant lacquer. Cans for jams made from fruits high in anthocyanin pigments should also be lacquered.

Corn, red kidney beans, and lima beans contain very little acid, but they evolve an appreciable quantity of sulfides when heat-processed. This, while quite harmless, causes an unsightly blackening of the inside surface of the can. To prevent this blackening a special sulfur-resisting lacquer has been developed. Its use is almost universal in the packing of vegetables where quality packs are desired. The lacquer contains zinc oxide, which reacts with the sulfide to form colorless ZnS, preventing the

Table 6.1. Common Consumer Can Sizes for Fruit Canning

Recommended product standards		
Can name	Dimensions[1]	Diam. × Hgt. (mm)
For applesauce, apricots, cherries, berries including blueberries, figs, fruit cocktail, grapes, peaches, pears, plums, prunes, cranberries, citrus salad, fruits for salad		
—	211 × 212[2]	68.3 × 70.0
8Z Tall	211 × 304	68.3 × 82.6
No. 300	300 × 407[2]	76.2 × 112.7
No. 303	303 × 406	81.0 × 111.1
No. 2 ½	401 × 411	103.2 × 119.1
For pineapple (sliced, crushed, tidbits, and chunks)		
—	307 × 201.25	87.3 × 52.8
No. 1 ¼	401 × 207.5	103.2 × 62.7
—	307 × 309	87.3 × 90.5
No. 2	307 × 409	87.3 × 115.9
For ripe olives		
—	211 × 200	68.3 × 50.8
8Z Tall	211 × 304	68.3 × 82.6
—	300 × 314	76.2 × 98.4
No. 300	300 × 407	76.2 × 112.7
Pt. Olive	211 × 600	68.3 × 152.4
For juices, drinks, or nectars (citrus juice, juice drinks, nectars, other fruit juices, pineapple juice, tomato and vegetable juices)		
—	202 × 214[3]	54.0 × 73.0
—	202 × 314	54.0 × 98.4
211 Cyl.	211 × 414	68.3 × 123.8
No. 2	307 × 409	87.3 × 115.9
No. 2	307 × 512	87.3 × 146.1
Cyl.	404 × 700	108.0 × 177.8
No. 3		
Cyl.		
For pie fruits and fillings: apples, cherries (RTP), pumpkin, squash, and prepared fillings		
No. 303	303 × 406	81.0 × 111.1
No. 2	307 × 409	87.3 × 115.9
No. 2 ½	401 × 411	103.2 × 119.1

Source: Canning Trade Almanac.

[1]Diameter × height. First figure in each series is inches, the two remaining figures are 1/16 in.

[2]Cranberries only.

[3]Baby food only.

formation of dark-colored iron or copper sulfide. Sulfur-resisting lacquered cans should not be used for acid foods because the lacquer may peel away from the tinplate. It is also unwise to use lacquered fruit cans for products where the processing temperature exceeds 100°C, as this will cause peeling.

Lacquer is usually applied to the tinplate before fabrication of the can. Sheets are fed through a machine with cleaning rollers, passing from there under a rubber roller coated with lacquer and adjusted to give the desired thickness of film to the sheet. The sheets are then carried through long ovens where the lacquer is baked on. The lacquered sheets are usually passed through the lacquering operation twice in order that any small defect in the first coat will be covered by the second.

The best lacquered cans are produced by a method know as "flush lacquering." After the can is made from plain tinplate, it passes through a machine where it is filled with lacquer or the lacquer is efficiently sprayed on the inside. The can is then inverted and drained, after which it is again filled or sprayed with lacquer. The can is again inverted and drained, after which it is baked in special ovens at a low temperature so as to prevent any softening of the solder in the side seam. Tinplate covered with lacquer may be fractured to some extent during fabrication of the can. The application of the lacquer on the finished open-top can is the safest guarantee of a complete film which will protect the fruit from any contact with tinplate or base metal. The ends or covers are also sprayed on a special machine and are baked at a low temperature.

Lithographed Cans. Lithographed cans are now widely used. If the canner packs individual lines in large quantities, it is advisable to use decorated cans, as they save labor and present a good appearance on the shelf. The best plan is to order 50–75% of the estimated requirements in lithographed cans, leaving the balance for plain cans with labels.

APPLES

Apples are canned primarily in the Pacific Northwest, New York, and Pennsylvania. Canned sliced apples are used for the preparation of pies in restaurants and hotels. Canning of apples is considered a by-product industry in most apple-growing areas and is a means of utilizing the best quality of culls. The fruit for canning purposes should be of fair size and reasonably free of blemishes. Apples unfit for canning may often be used for cider, vinegar, or applesauce.

Sliced Apples

Apples for canning should be firm and hold their shape in the can. They should be of good flavor, color, and texture. Acid cultivars with white flesh are preferred. On the Pacific Coast the 'Yellow Newton Pippin', 'Winesap', 'Jonathan', and 'Spitzenberg' are popular for canning. Other commonly used cultivars are 'Baldwin', 'Golden Delicious', 'Greening', 'Northern Spy', 'Rome Beauty', 'Roxbury Russet', 'Stayman', and 'York Imperial' (Jackson 1979). In Great Britain, the more suitable apples are 'Bramley Seedlings' and 'Newton Wonder'.

Ripeness of the fresh apples greatly influences the flavor, color, and texture of the finished product. In canning apple slices, underripe apples tend to produce an excessively firm green product of poor flavor; overripe apples, on the other hand, produce a soft or mushy pack having a bland flavor. Sometimes it is necessary to store the apples briefly between harvest and canning. For storage, it is preferable to use crates or 455-kg bins stacked so as to permit good air circulation. If bulk storage is necessary, small piles are more desirable, and some provision for air circulation is necessary. For prolonged storage, refrigerated storage or controlled atmosphere (CA) storage should be used.

The texture or firmness of canned apple slices can be modified by controlling pectinesterase activity and addition of calcium salts. Apples that are usually canned as slices are not covered by the FDA Standard of Identity. Use of the USDA quality grade standards is optional.

Washing and Inspection. Processors prewash the apples by fluming them in water from the receiving area to the washers. Mechanical washers equipped with paddles and high-pressure jets are used. A good water pressure must be maintained at all times to ensure that all dirt is removed. If lead arsenate sprays are used in the orchard, it is advisable to add 1–1.5% hydrochloric acid to the washing water in order to remove any spray residue. Washing with water containing a small amount of trisodium phosphate or food-grade detergent is a common practice.

After washing, apples pass over an inspection belt or conveyor, where the bruised or irregular fruits are removed. Irregularly shaped, off-size, and cull apples can be used for making applesauce.

Size Grading. The purpose of size grading is to improve the efficiency of the peeling operation and to divert small apples to the applesauce line. Divergent roller-type size graders can separate apples into two or three size groups for delivery to separate peeling lines. Through this operation, peeling loss is reduced and skips can be largely eliminated.

Peeling, Coring, Trimming, and Slicing. Mechanical peeling and coring are done on the same equipment. Use of 0.3% alkyl aryl sulfonate or 0.25% sodium octanoate as wetting agents improves both peeling rate and surface appearance of the peeled products.

Peeled apples are delivered to conveyor belts for inspection and trimming. To prevent enzymic browning, it is necessary to handle the apples rapidly after peeling, and to hold the pared, trimmed apples in fiberglass or plastic tanks containing 2% citric acid or 2% salt solution until they are sliced. Peels, cores, and trimmings may be used for vinegar production, jelly, or pectin manufacture.

The peeled, cored apples are fed to slicing machines, and the sliced apples are washed in a reel washer or shaker screen to remove small pieces.

Can Filling and Closure. In order to obtain sufficient fill weight, it is necessary to remove air from and blanch sliced apples. Vacuum treatment removes occluded air from the slices of apples and permits its replacement with water or a 2–3% salt solution either during vacuumizing or, subsequently, during blanching.

Apple slices are packed by semiautomatic or automatic fillers into plain or enameled cans. It is important to fill the cans hot, adding sufficient hot water or syrup to fill the spaces between the slices. A filling temperature of between 77° and 82°C is desirable.

Thermal exhausting may be used to obtain a proper closing temperature. An exhaust period of 3–10 min may be used, depending upon the style of pack and the can size. A vacuum in the cans is obtained by double seaming the filled cans at a temperature of 77°C employing atmospheric closure. Replacement of the headspace air with nitrogen gas is effective in reducing internal can corrosion.

Sterilization and Cooling. There should be no delay in placing the sealed cans in the cooker and starting heat sterilization. Cans should be processed in boiling water until a can center temperature of 87.8°C has been reached. The processing time is 20 min at 100°C for 307 × 409 and 401 × 411 cans (initial temperature of 71°C) and 35 min for 603 × 700 cans.

Immediately after heat processing, cans are water-cooled until the average temperature of the contents reaches 41°C. Casing canned apple slices at high temperatures may result in softening, loss of normal color, darkening, or pink discoloration of the product.

Baked and Glazed Apples

Firmer cultivars such as 'Baldwin', 'Rhode Island Greening', 'Gravenstein', 'Northern Spy', and 'Rome Beauty' are suitable for canning as baked or glazed apples.

Choice apples about 2¾ to 3 in. in diameter should be selected for this pack. The apples are washed, cored by making a small cut across the blossom end and removing the core without cutting through the stem end, submerged in a 2–3% salt brine, and rinsed with fresh water prior to baking. The fruits are not peeled.

Baked Apples. Apples are baked at 176.7°C for about 45 min, depending somewhat on the size and cultivar of fruit. The baked apples are placed in cans of suitable size, which are then filled with hot 40°–50° Brix syrup. If a closing temperature of 71.7°C cannot be maintained, a steam exhaust should be employed to attain this temperature. After closure, the cans are processed in boiling water for 20–30 min, depending on the size of the can, and then are water-cooled.

Glazed Apples. Glazing of apples may be more practical than baking as a method of preparation for canning. Instead of baking, the cored apples are cooked in a 40°–50° Brix sugar solution for 10–14 min and placed in cans, which are then filled by the addition of syrup in which the apples were cooked. Cans are then closed and processed in the same manner as with oven-baked apples. The glazed apples tend to sink in the syrup without exhausting. This method requires much less time than baking.

Corrosion and pinholing may occur in cans of baked or glazed apples. Corrosion is caused by the malic acid of the apples in the presence of air or oxygen. Corrosion is limited or reduced to a negligible degree if the air is thoroughly expelled from the fruit by blanching and from the can and contents by exhausting.

Apples are easily sterilized on account of their high acidity. A sterilization of 8–10 min at 100°C in a continuously agitating sterilizer has been considered sufficient if the cans have been filled and sealed above 71.1°C. After processing, cans should be cooled to an average can temperature of 40.6°C.

Apples for Cake Topping. The Colby process entails placing unblanched fruit pieces in a closed vessel and subjecting the occluded fruit to a vacuum of 38.1 cm mercury. The evacuated gas cavities are then infused with an edible liquid that may contain dissolved or suspended solids to enhance the flavor or to help prevent enzymatic browning. The excess liquid is completely drained and the infused fruit pieces are then placed in cans and sealed under a vacuum in excess of 38.1 cm (15 in.) mercury and at a temperature of less than about 43.3°C. Sealed cans are heated to sterilization temperatures (about 98.9°C) for 5–20 min, and then cooled rapidly in water to less than 37.8°C. The concentration of the additives in the water solution can be 0.01–3%, and the solution should be allowed to cover the fruit for several minutes in order to provide sufficient infusion of additives throughout the fruit tissues.

Apple Juice

Apples used for juice (sweet cider) processing should be fresh and sound. Immature apples tend to produce unsatisfactory juice due to a high percentage of starch in the fruit. On the other hand, overmature apples give a low yield and cause difficulties in pressing, clarification, and filtration.

Most apple cultivars do not make a satisfactory juice when used singly but are excellent when blended with other cultivars. Factors to be considered for blending include degrees Brix, the tannin content, the total acidity, and pH. The flavor of an apple juice is based on two factors: sugar–acid balance and aroma or bouquet characteristics.

Opalescent juice, made by retaining fine pulp particles suspended in the juice, is preferred by some to the traditional clear juice.

In preparation for extraction, apples are washed thoroughly to remove all adhering dirt and sorted to remove partially or wholly decayed fruit. Then the apples are subjected to grinding. Two types of equipment are used for grinding: one type grates the apples to a pulp, the other type is a hammer mill.

Juice Extraction. The most commonly used machines for juice extraction are the hydraulic cider press, pneumatic fruit juice press, continuous screw-type press, continuous plate press, horizontal basket press, and the screening centrifuge. After apples are pressed, the juice goes through a "cider" screen, which is a cylinder of monel or stainless steel screen of approximately 100–150 mesh. This procedure reduces the suspended solids content to around 2% (Moyer and Aitken 1980).

Before filtering, the juice is treated with several materials to make filtration easier. First, pectic enzyme is added to hydrolyze pectin. Then tannin and gelatin are added to form a coagulation and then precipitate, or the juice is heated which does the same thing as the tannin–gelatin reaction. Finally, there is a bentonite treatment in which juice passes from the heat exchanger to tanks where a suspension of equal parts of bentonite and filter aid is added with vigorous stirring. The treating materials are suspended in a small quantity of juice and the suspension added to the large volume. After treatment the juice is allowed to stand for at least 1 hr and then is filtered.

Asti (1970) developed a patented process to provide an apple juice suspension having the body, viscosity, and texture of fresh natural apples in liquid form. Fresh apples are washed, and then passed to one of two paths—A or B. In path A, the washed sound fruit is peeled, cored, sliced, and cooked. The cooked and prepared fruit is then crushed and forced through a screen. After the crushing step, it is usually convenient to provide a storage tank for holding the product of path A in order to coordinate the whole process. The fruit in path B is milled but not cooked. Instead, an antioxidant is introduced into the crushed fruit at a point where the natural oxidation has not proceeded beyond return. The crushed and disintegrated fruit is pressed to express the juice from the pulp. The juice is then put through a screen to remove any coarse pulp, seeds, or skins. The screened juice is deaerated and may then be stored, if necessary, for coordinating path A with path B.

The products of path A and path B are mixed. The crushed cooked fruit of path A, primarily apple pulp, contributes body, texture, and viscosity to the mixture. The product of path B contributes fresh uncooked, unfiltered juice that retains many of the characteristics of the fresh fruit. Mixing may conveniently take place in a proportioning pump. The juice and pulp are proportioned at any desired ratio; preferably, the juice should be at least 70% of the total weight of the mixture to keep the product a pourable liquid. If over 95% juice is mixed, the benefits derived from the pulp are largely lost. The mixture is pasteurized and homogenized and then transported to container-filling equipment.

Canning. Cans for apple juice are generally lined with a special enamel or lacquer that is resistant to the corrosive action of the juice. As the cans travel along the line to the filter, they should pass through a can washer where all dust particles or other debris are removed. The cans are filled on special filling machines and immediately sealed with covers in a can-closing machine. The closed cans should be inverted or rolled on their sides for approximately 3 min to bring the hot juice into contact with the cover to sterilize it. The usual temperatures for flash pasteurization are between 77° and 88°C for 25–30 sec. The cans are then cooled in a cooler to 38°C.

Bottling. Bottling apple juice requires specialized equipment and more supervision than canning because of the fragility of the containers. Conveyor lines for glass containers must be designed and operated so that the bottles are not "bruised" or broken by impact.

Before being filled with hot apple juice, bottles should be cleaned by passing them through a special cleaning unit to remove all dust particles. If new bottles are used, it is rarely necessary to wash them. This is especially so with bottles that are placed neck-down in cartons immediately after they leave the annealing furnace at the factory. The bottles should be preheated to within 6.6°C of the filling temperature in a special section of conveyor in which steam jets impinge on the containers. The most satisfactory fillers draw the juice into the container by evacuating the container.

Closures for bottles and jugs can be screw caps, crown caps, or vacuum caps. The latter are the most satisfactory because there is less chance of breakage during application and the headspace vacuum is slightly higher than with other types.

Hot-filled glass containers should be cooled gradually in a special cooler, usually of the spray type. Where bottles enter the cooler, the sprayed water is hot; as the bottles move through the cooler, the sprayed water is gradually reduced in temperature. The water spray should be very fine so that cooling can be done by evaporation as well as by conduction. Bottles emerging from the cooler should still be warm enough to dry completely but not above 37.7°C to avoid deleterious heat effects. Recently, bottles made from polyester or polyolefin polymers with air-barrier coatings have been tried for commercial processing of apple juice.

Bulk Storage. Apple juice has been stored in bulk by heating and then pumping it while hot into storage tanks. These tanks, with capacities up to 7600 liters, are vented with an air filter and allowed to cool. This method of bulk storage has been largely replaced by presterilizing the tanks with hot water, steam, hydrogen peroxide, or sulfite solution. The air in the tank is replaced by CO_2 at 0.10 kg/cm^2 to prevent fungal growth and then pasteurized and cooled juice is pumped into the tank. The success of this storage depends on the maintenance of aseptic conditions following pasteurization and during transfer of the cooled juice into the tank. The temperature of a tank room is usually 15.5°C or below. Bulk storage of pasteurized apple juice at -1.1°C would extend the supply and permit blending to achieve a desirable flavor.

To overcome the difficulties encountered in maintaining sterility in bulk storage, the Boehi process was developed in Switzerland for the "return" bottle trade. Clean tanks are completely filled with water, which is then forced out with 3.2 kg/cm^2 CO_2. When empty, juice at 4.4°C impregnated with 0.6–0.8% CO_2 is pumped in until the tank is 95% full. The headspace of the tank is further purged with CO_2 to eliminate oxygen before it is sealed through a safety valve. If the storage temperature is maintained below 4.4°C, growth of yeasts and lactic acid bacteria is inhibited.

Preservation with Chemicals. The principal preservatives used in commercial apple juice are salts of benzoic acid, sulfurous acid or its salts, and sorbic acid. Sodium benzoate is used chiefly to increase the shelf life of unpasteurized apple juice; it is frequently used in the United States for juice packed in 3.78- or 7.57-liter jugs that often are labeled "apple cider." Benzoate prevents spoilage when present in concentrations of 0.1–0.3%, the quantity necessary varying with the acidity of the juice. The salt is dissolved in water and added to the juice at the time of preparation. Sulfurous acid is used mainly for preserving juice in bulk for export or for manufacturing purposes. Concentrations necessary to prevent spoilage vary from 0.02 to 0.1% calculated as sulfur dioxide, depending on the juice preserved. It is added in the form of sulfites or sulfur dioxide, as gas from a cylinder, or as a solution in water.

In many countries, benzoic acid and SO_2 or their salts are the only preservatives permitted by law. When added they have to be declared on the label.

Pure sodium benzoate, when added in concentrations of 0.05–0.1%, generally does not impart any objectionable flavor to fruit juices, particularly when they are to be diluted before use. The concentration perceptible to taste depends also on the ratio of benzoate ion to benzoic acid. The amount of preservative required depends on the character of the juice, particularly its acidity. The concentration of metabisulfite or sulfite required to prevent growth in an acid juice is about 0.1% calculated as SO_2. When only used to inhibit oxidation, about 0.02% of SO_2 is sufficient under ordinary conditions. The preservative should be completely dissolved and thoroughly mixed with all the juice to be treated. The two preservatives may be used to advantage in combination with each other: sulfurous acid to retard oxidative changes and benzoic acid chiefly to check spoilage organisms. Recent trends are to avoid use of chemical preservatives.

Although SO_2 is frequently used in fruit juices, in the United States it is seldom used in apple juice. In some states its use is not permitted. Even in cases where it is not intentionally added, minor amounts may become included in the final products. This originates in the SO_2 used for disinfecting utensils and equipment, particularly in smaller juice factories.

Sorbic acid, a 2,4-hexadienic acid, is metabolized by humans to carbon dioxide and water, the only known preservative with this important characteristic. Sorbic acid or sodium sorbate is effective for the inhibition of yeast fermentation in unpasteurized apple juice. It is also effective against many common molds, but generally not against bacterial fermentation. Sorbic acid seems to exert its suppression of microbial growth by the blocking the normal functioning of certain sulfhydryl enzymes. Sorbic acid has the advantage of not affecting the taste of fruit juices to the same extent as benzoic acid. The sodium salt of sorbic acid is, however, not quite as effective as that of benzoic acid.

Applesauce

Approximately 10 million bushels, close to 25% of the apples used by industry, are converted into sauce. Manufacture of applesauce is concentrated mainly in Maryland, Pennsylvania, Virginia, West Virginia, New

York, and the three Pacific Coast states. In California, 'Gravenstein' apples are most popular for sauce production, while 'Golden Delicious', 'York Imperial', 'Jonathan', and 'Stayman' predominate in the East.

In a typical commercial plant, after the apples are sorted, trimmed, and sliced, they are discharged directly into a continuous stainless steel thermoscrew cooker where they are heated rapidly to 98.9°C and held there for 3 min. The cooked apples are discharged into a Langsenkamp pulper with a 1.52-mm (0.060-in.) screen and operating at 1000 rpm. The apple pulp is pumped into a holding tank where water and sweeteners are added to adjust the soluble solids to 19°–20° Brix at 20°C. The applesauce is then pumped through a stainless steel tubular heat exchanger to reach 90.6°C and then filled hot into No. 303 cans (303 × 406), sealed hot with steam injection, inverted, heated 5 min in a steam box, and then cooled in water to reach a can center temperature of 37.8°C in 25 min.

Then quality of canned applesauce is affected by varietal characteristics, maturity of the fresh apples, postharvest storage conditions, and storage temperatures of the canned product. The flavor can be improved by fortification with apple essence and citric acid. Higher storage temperatures have been reported to cause faster corrosion of the tin coating and the formation of hydrogen gas in the headspace. For best storage stability, temperatures of 20°C or lower are recommended.

Preparation of Applesauce Using a Pressure Cooker. With the usual methods of preparing applesauce, the fruit is peeled, cored, trimmed, chopped, and conveyed to a cooker in which the prepared fruit and the requisite quantity of sugar are cooked to the desired degree. The volume of fruit is so large that for economical and efficient operations the cooking must be accomplished in a very short time, seldom more than 4 min. To achieve thorough cooking in such a short period, the apples are cooked under pressure at temperatures of about 102°–107°C. Cooking temperatures and pressures are generally obtained by the injection of steam into a closed cooking chamber.

Sauce cookers operate continuously with cooked sauce being discharged from the cooker at the same rate that the raw fruit and sugar enter. Fruit and sugar enter and leave the cooker through enclosed worms or butterfly valves which, when full, serve as seals that prevent loss of steam pressure in the cooker.

After cooking is completed, the cooked mass is conveyed from the cooker to a finishing machine in which the coarse fibers, seeds, and peel particles are removed. When the cooked sauce leaves the cooker and enters the finisher, pressure drops from superatmospheric pressure to normal pressure. The drop in pressure is accompanied by a drop in temperature, which is effected by the evaporation of the flash-off vapor.

The sauce is filled into cans or jars, and the containers are sealed immediately. Containers are then inverted or turned upside down to sterilize the lids. The hot sauce serves as the sterilizing medium. Then the containers are cooled quickly in water. In normal practice the sauce comes from the finisher at 98.9°C and is filled into the containers at 96°–98°C.

It is necessary to equip the cooker with a suitable means of cooling. In a batch process, the cooking kettle is a double-walled vessel with means to admit either a heating fluid or a cooling fluid to the space enclosed between the walls. In a continuous-type cooker, the cooling means is a double-jacked discharge worm. The length and diameter of the cooling section are such as to prevent steam pressure losses in the body of the cooker and to allow sufficient cooling surface to enable the sauce to be cooled to the desired temperature.

APRICOTS

Apricots are grown and canned mostly in California. The average annual pack of canned apricots varies between 3 and 5 million cases. Fruit that is smaller than 31 to the kilogram (14 to the pound) is used chiefly for canning whole or for production of nectar.

Cultivars

The 'Blenheim' apricot is the most popular cultivar for canning. It is medium in size, deep yellow in color, and excellent in flavor. When properly ripened, it has uniform texture from the skin to the pit and retains its shape in the can during processing.

The 'Royal' apricot is grown in southern California and in the hot interior valleys. It is somewhat smaller in size than the 'Blenheim' and has a more intense orange color. Many pomologists claim, however, that the 'Royal' and 'Blenheim' are identical and that the differences in appearance noted in commercial culture are due to the effects of locality and climatic conditions. When grown in hot, dry regions, 'Royal' apricots often become soft near the pit, a condition that renders them less suitable for canning.

The 'Tilton' is an important cultivar grown in the hot interior valleys of California, in eastern Washington, and in British Columbia. It is large, but is rather pale yellow in color. Recently, because of occasional texture softening in canned apricots, and because of the rapid conversion of the apricot orchards in Santa Clara county to industrial and housing development projects, more 'Tilton' apricots are being planted in the hot interior valleys of California to replace some of the 'Blenheim' apricots.

Ripeness Level and Horticultural Factors

Apricots for canning are harvested at the optimum ripeness level. When harvested at the "canning-ripe" stage, the fruit is firm, of good color, and of pleasing flavor. Yet, it will not have reached the maximum flavor at this stage of ripeness. As the fruit firmness decreases, the volatile reducing substances increase. The physiological and biochemical changes in maturing stone fruits include an increase in soluble solids, decrease in firmness, loss of chlorophyll, increase in specific pigments, and decrease in acidity.

The optimum period for storage of apricots at 0°–4°C and 85% relative humidity is about 20 days. Under modified atmosphere conditions (0°C, 81% RH, 3% O_2, and 5% CO_2), the shelf life is about 30 days. Organoleptic properties of the apricots are maintained better in modified atmosphere storage at all periods.

The fungus *Rhizopus stolonifer,* when present on apricot fruit, may cause texture breakdown after canning. The mold growth may be arrested by dipping the fruit in dichloran (2,6-dichloro-4-nitroaniline), a fungistat useful against *Rhizopus* rots in stone fruits. Processing apricot halves with $CaCl_2$ improves canned product texture.

Canning Process

Receiving. Most canners examine each delivery of apricots to determine roughly the percentage of the different grades, and payment is made to the grower on the basis of the test and whether the sample shows texture breakdown after canning on a trial run. If the sample shows texture breakdown right after canning, the lots are diverted for processing into nectars, baby foods, jams, and preserves.

Pitting. Apricots are washed, halved, and pitted but are usually not peeled. Some apricots are processed as whole fruit after lye peeling. Fruit may be cut by hand around the pit suture, and the pits removed; now, more commonly, fruit is cut and pitted by a machine.

Grading. Screens with openings of 3.18, 3.81, 4.45, 5.10, or 5.40 cm are used for grading apricots for size before pitting. The average diameters of Fancy, Choice, and Standard grades are usually 4.45, 4.29, and 3.97 cm, respectively. The grades are based more upon color, texture, and absence of defects than upon size (grading for quality is done after pitting).

Filling and Syruping. The graded fruit is conveyed to mechanical or hand-pack devices for filling. Filled cans are fed to vacuum syruping machines where syrups of the concentrations recommended by the California League of Food Processors (55°, 40°, 25°, 10° Balling) and plain water are used, according to whether the grade is Fancy, Choice, Standard, Second, or Pie.

Exhausting and Double Seaming. Apricots contain some imprisoned gas, which will cause pinholing in the can unless the gas is driven out by exhaust. Cans are exhausted up to 10 min at 82°C in an exhaust box and then closed in a steam-flow machine. A more common practice now is to prevacuumize the canned product and close it in an atmosphere of steam. This prevacuumizing gives less syrup loss, uses less floor space, and requires less steam than the exhaust box method.

Heat Sterilization. After exhausting and double seaming, canned apricots are heat-procesed at 100°C sufficiently long for the center temperature of the product to reach 90.6°C.

Most canned apricots are heat-processed in continuous rotary cookers at 100°C for 17–19 min for No. 2½ cans, and for 20–30 min for No. 10 cans, depending on the initial temperature and the texture of the fruit. Whole fruit requires longer processing than halved. The heat-processed cans are water-cooled to 40.6°C and then transferred to the warehouse for storage.

Yields and Drained Weight. The yield of halved canned apricots per ton varies from 52 to 55 cases of 24 No. 2½ cans. Loss in canning of unpeeled halved fruit is about 10–15%; where the fruit is peeled, the loss may exceed 30%. Yields of canned whole apricots usually exceed 70 cases per ton.

One day after canning, the drained weight is about 84% of the fill weight because apricot fluid is lost to the cover syrup. On storage, drained weight increases rapidly for the first 5–10 days, and then slowly until an equilibrium drained weight (90–97% of fill weight) is reached in 45–60 days.

Texture of Canned Apricots

The ripeness level of the fresh fruit and processing time are the most important factors influencing the texture of apricots. Canned apricot halves soften during storage; there is an increase in water-soluble pectin and viscosity in the syrup, and a decrease in protopectin. Riper apricots contain less protopectin and the syrup contains more water-soluble pectin.

Besides ripeness level and the processing condition, softening of apricots may also be related to intrinsic and/or parasite-originated pectic and cellulytic enzymes. Perhaps control of mold contamination in the orchard, better sanitation, more rapid and careful postharvest handling, and more slective sorting on the grading table would help to alleviate the softening problem.

High acidity in apricots is related to the softening problem. Acidity in the fruit is influenced by the cultivar, climatic conditions, and the level of nitrogen application. It is thought that softening in high-acid apricots with pH ranging from 3.3 to 3.5 may result from acid hydrolysis of cell wall constituents. Some canners have successfully eliminated softening by avoiding lots of fruit with high acidity and those with possible mold contamination.

Pectin degradation in canned apricots can lead to textural problems due to lack of firmness. Processing apricot halves with $CaCl_2$ improves the texture of canned products.

Canned Apricot Pie Fruit

Considerable pie-grade fruit is now pitted mechanically by Elliott pitters, then steamed and canned as solid-pack pie fruit, without the addition of water or syrup. The No. 10 cans require heat processing at 100°C for 45–60 min in an agitating cooker because of slow heat penetration. If the product is canned boiling hot, a shorter process time can be used. Paneling of No. 10 cans may occur unless the cans are of reinforced type.

Sieved Apricots as Baby Foods

The operations and equipment for canning sieved apricots may vary somewhat in various baby food plants. The following description summarizes the more important steps.

Whole ripe apricots are washed in a water tank, followed by a spray washer. Green and defective fruits are removed on a sorting belt. The sorted apricots are thoroughly washed by sprays of water under fairly heavy pressure and then cooked by steam at 100°C for 6–8 min in a screw-type, steam-jacketed continuous heater. They are pulped in a stainless steel cyclone-type pulper with a coarse screen to remove pits and coarse fiber. The hot puree then passes through a fine finisher 0.508-mm (0.020-in.) screen to remove small pieces of fiber. Sugar is added to give the proper balance in flavor between the acidity of the fruit and the sweetness of naturally occurring and added sugar. Usually 7–10% by weight of sucrose is added. A small amount of farina (tapioca starch) or modified starch may be added and cooked a short time before canning. The success of this process depends on control of the consistency of the product. The total solids of the final product are about 21–22%. After passage through a homogenizer to impart a smooth consistency, the product is deaerated under high vacuum. The deaerated product is then flash-heated in a closed continuous heat exchanger to 115.6°C, cooled to 93°–96°C in a second heat exchanger or through a flash cooler, filled into 202 × 214 cans or 4.85-fl oz glass jars at that temperature, sealed, inverted or passed through a steam chamber for a few minutes to ensure sterilization, and cooled in water in the usual manner. The jars are sealed by a high-speed jar sealer. Recent models can seal 500–800 cans/min. The conveyors, fillers, and sealers must be highly synchronized or else delay and pile-ups will occur. The canned product is labeled by a high-speed automatic labeling machine and cased. The cases are sealed and warehoused.

Apricot Puree and Apricot Beverages

Apricots for nectar manufacture should be so ripe that they are soft. Puree or nectar prepared from firm fruit, such as is used for canning, will be of inferior flavor and color. Tree-ripened fruit possesses a better flavor than that permitted to ripen after picking. Since apricots ripen unevenly, it is necessary to harvest three or four times in order to obtain the best-flavored fruit.

Apricot Puree. Washed and pitted apricot halves are steamed until soft; then they are passed through an expeller screw extractor with a 0.838-mm (0.033-in.) screen. One part of sugar is added to three parts of pulp. The product is filled into No. 1 plain cans, exhausted 8–10 min, sealed, processed at 100°C for 20–25 min, and cooled in water. The undiluted product prepared in this way requires dilution with water or sugar syrup before use as a beverage.

A continuous sterilization process for sterilization of apricot, sour cherry, plum, and tomato juices in 3-liter bottles using HTST counter-current equipment has been shown to be suitable and superior to sterilization in autoclaves.

Apricot Nectar. To make apricot nectar, the fruit is steamed in a continuous steam cooker for approximately 5 min. The hot fruit is then run through a brush finisher equipped with a 0.635- to 0.838-mm (0.025- to 0.033-in.) screen. The resulting puree is then passed through a steam-heated tubular heat exchanger where it is brought to a temperature of 88°–93°C. The puree is sweetened with approximately 1.8 times its volume with 15°–16° Brix sugar syrup; citric acid is added so as to maintain a constant total solids–acid ratio throughout the season. The resulting nectar is filled into plain cans, exhausted for approximately 6 min, and sealed. No. 1 tall cans are processed for 15 min at 100°C; larger cans are given a longer processing.

Apricot Concentrate. Apricot purees are concentrated in vacuum pans for shipment to consuming centers where they may be combined with syrup to prepare nectars for distribution in cans.

Apricot puree can be concentrated to a 2.5 to 1 ratio by vacuum evaporation and then packaged in 208-liter (55-gal) drums by the aseptic canning method. The apricot concentrate may be used to make a pumpkin-type pie, or it can be whipped with other ingredients to produce a chiffon or cream pie. It blends readily with other fruit juices and nectars to give added body, nutrients, and flavor. Another possibility is to market apricot concentrate in 170-g (6-oz.) cans for use in cake mixes, muffins, and fruit cakes.

Volatile Components of Apricots

Ripe apricots have a strong characteristic aroma. Tang and Jennings (1968) subjected a charcoal adsorption essence of 'Blenheim' apricots to repetitive gas chromatographic separations. The isolated components were characterized by infrared spectroscopy as benzyl alcohol, caproic acid, epoxy-dihydrolinalool IV, γ-octalactone, *S*-octalactone, *S*-decalactone, γ-decalactone, and γ-dodecalactone.

BANANAS

Bananas belong to the family Musaceae, genus *Musa,* comprising 32 or more distinct species and at least 100 subspecies. The majority of edible bananas are from a subsection of *Musa* called Eumusa and originate specifically from two wild species, *M. acuminata* and *M. Balbisiana.* Most commercial bananas are from the triploid group of *M. acuminata.* In this group are the 'Cavendish' and 'Gros Michel' which are by far the principal banana cultivars in world commerce.

The largest production of bananas occurs in Brazil, Ecuador, Hawaii, Honduras, Philippines, Puerto Rico, China, and Venezuela. Latin America produces 62% of the world's banana crop.

Approximately one-half of the bananas of the world are eaten as fresh fruit and as salads. The more important canned banana products are puree, baby foods, and tropical fruit cocktail. Banana puree canned in No. 10 cans or 208-liter (55-gal) drums by the aseptic canning process is a new product for the baking and ice cream industry. Bananas are also canned as pastes, drinks, and slices.

Banana Puree

Bananas are ripened until the cut flesh has a translucent appearance and has developed the full flavor. Peeling yields depend upon a number of factors, including fruit size and maturity, and have been found to vary from 57 to 67%. The peeled and trimmed bananas together with 0.4% by weight of citric acid are placed in a vacuum tumble blancher (VTB) which is then closed and revolved at 6 rpm. The vessel is evacuated using a two-stage steam ejector to 71 cm (28 in.) Hg vacuum in 55–60 sec, and the chamber is isolated from the vacuum system. The vacuum is then broken by admitting steam until a positive steam pressure of 0.14 kg/cm^2 is reached. This takes about 30 sec and the steam pressure is maintained at this level for approximately 7 min to heat the banana puree to 93°C. The pulp is discharged from the VTB at 93°C into a preheated screw press with the screw running at 380 rpm and 1.75 kg/cm^2 air pressure on the solids discharge cone. Puree discharged from the screw press at 85°C is filled directly into plain cans without leaving a headspace; the cans are then closed and inverted.

Banana Drink. Puree as discharged from the screw press is diluted in the ratio of 1:3 with water, and the pH is adjusted to 4.2–4.3 by the addition of citric acid. The diluted puree is centrifuged and the opalescent liquid obtained is adjusted to 12°–15° Brix by the addition of sugar to produce banana drink. The drink is then filled into 301 × 411 plain or enameled cans leaving a headspace of 7.9 mm, vacuum closed, spin-cooked at 100°C and 150 rpm for 2 min, and spin-cooled under water sprays for the same period.

Analytical determinations made at various stages throughout the process are set out in Table 6.2. Banana drink has a soluble solids of 13.0° Brix and acidity of 0.21% as malic acid.

Canned banana drink in plain electrolytic tinplate cans has maintained quality for 18 months at ambient temperatures.

Banana Slices

For processing banana at boiling water temperature, the pH of the product should be reduced to 4.2–4.3. This may be achieved by canning

Table 6.2. Characteristics of Banana and Its Products at Various Stages

	°Brix	Acid[1] (%)	pH	Spc Gr	Brix/acid ratio	Total solids (%)
Raw peeled banana	20.5	0.37	5.4	1.027	55.4	24.24
Blanched pulp	21.5	0.61	4.3	1.099	35.2	23.52
Puree	21.5	0.63	4.3	1.097	34.3	22.71
Drinks	13.0	0.21	4.0	1.054	61.9	13.52

Source: Casimir and Jayaraman (1971).
[1]As anhydrous malic acid.

banana slices with acidic fruit such as passion fruit, pineapple, and grapefruit in tropical fruit salad or by acidification with citric acid. The United Fruit Company is canning acidulated banana slices in Honduras. The bananas are packed near the growing area in extra-heavy acidified syrup (28° Brix) to prevent color loss. Canned banana slices have a sugar content of 17–19% which can be carefully controlled. Heat processing for slightly over 2 min in the acid syrup deactivates enzymes and eliminates the possibility of bacterial contamination of other food products. Color was stable through at least 2 years of shelf life. The products can be stored in nonrefrigrerated areas.

BLACKBERRIES

Moderate quantities of blackberries are canned in the Pacific Northwest for use in the preparation of pies. However, frozen blackberries are supplanting the canned product for pies.

In Oregon and Washington, the 'Evergreen' cultivar is most popular. In California, the principal cultivar is the 'Boysenberry', which is a hybrid and similar in composition and flavor to the loganberry.

Harvesting

Blackberries should be harvested in shallow boxes and should be picked daily if possible so that the fruit may be at the optimum stage of maturity. It is desirable that the fruit be canned on the same day it is picked; otherwise, serious deterioration may take place (see Chapter 2).

Canning Process

The berries are sorted and then washed thoroughly. Since most of the berries are used for pie making, they are generally packed in water or in light syrups. Fruit for dessert purposes is packed in syrup of 40°–55° Balling. The berries are dumped from small baskets into water, transferred to a sorting belt, given a preliminary sorting, graded by machine into five size grades, again sorted from slowly moving belts, and then canned.

The smallest berries are canned in No. 10 lacquered cans in water for use in pie bakeries. Larger berries are canned with syrup in No. 303 lacquered cans for use as dessert. The cans are thoroughly exhausted at 87.8°C for 4–5 min for No. 2 cans and for 6–10 min for No. 10 cans, double-seamed, and processed in boiling water. A steam flow closing machine may be used, and the syrup filling temperature can be 82.2°C or higher. Processing takes 11–14 min at 100°C for No. 2 cans, and 23–27 min for No. 10 cans.

In plain tin cans the color of the syrup and of the fruit bleaches rapidly. Therefore, it is customary to can blackberries in enamel-lined cans. The preferred can is one made of Type L plate and coated inside with two coats of so-called "berry enamel."

The fruit may also be canned as a light preserve after boiling 3–4 min with an equal weight of sugar. In this case no syrup except that formed in cooking is added.

Flavor Components

Sixteen volatile compounds have been identified in the ethyl chloride extract of blackberries. The compounds, indentified by combined gas chromatography and mass spectrometry, include acetals, esters, alcohols, ketones, terpenes, and an aromatic. Karwowska and Ichas (1969) reported that blackberry press cake can be used as a valuable raw material for production of natural aroma essence. From 100 kg of blackberry products, flavor concentrates were obtained at yields of 0.7 liter from pulp, 0.80–1.18 liters from juice, and 0.75–0.80 liter from press cake. The condensate from press cake had the most intensive characteristic aroma.

BLUEBERRIES

Blueberries are grown in the Atlantic Coast states, in Michigan, Wisconsin, Minnesota, and the three Pacific Coast states.

The lowbush cultivars yield small berries and grow wild in Maine, New Brunswick, and upper Michigan. Highbush cultivars yield large berries and are cultivated in New Jersey, Maryland, southern Michigan, and elsewhere. A third type, the Rabbit Eye Blueberry, also has large berries and is cultivated extensively in Florida, Georgia, and elsewhere in the South. Highbush and Rabbit Eye cultivars are handpicked and therefore require less cleaning than lowbush cultivars, which are picked by raking. Cleaning the fruit after mechanical harvesting appears to be one of the chief problems encountered in processing blueberries.

Ballinger and Kushman (1970) studied the relationship of ripeness to composition and keeping quality of highbush blueberries. The acid content increased during early stages of development but decreased rapidly during later stages of development. The pH, soluble solids, sugars, anthocyanin content, and berry weight increased as the berries developed on the vine. Dekazos and Birth (1970) developed a maturity index for blueberries using light transmittance. Light transmittance curves of intact blueberries (cultivars 'Wolcott' and 'Blue Crop') were recorded in the visible and infrared region with the ASOC biospect. The index involves measurement of the optical density (OD) of intact fruit at two wavelengths and computation of the OD difference (760–800 nm) vs. anthocyanin content. A high correlation coefficient of 0.967 was found for 'Wolcott' blueberries in the IR region OD 760–800 nm.

Lowbush blueberry fruit contains 81–84% moisure, up to 13% sugar, a small amount of protein and phosphorous, a fair amount of calcium, and a relatively large amount of iron and manganese.

In the field, blueberries are cleaned mechanically through a fanning mill like that used for cleaning grain; this device removes the greater portion of the leaves, twigs and stems, and other light trash by air blast. In the cannery, berries are washed either in shaker washers or paddle washers. The washed berries are sorted for defects on white, slow-moving belts. They are then placed in enameled No. 10 cans and covered with water for the pie-baking trade, or with a sugar syrup (40% solids) for the retail trade and home use as a dessert fruit. The cans are exhausted in steam and sealed hot. The No. 2 cans are processed for 10–12 min in boiling water; No. 10 cans require 25–30 min.

Canned blueberries may vary from a highly attractive, free-flowing product to one that is clumped into a firm mass. Clumping can be greatly reduced by cooling with agitation, which apparently interferes with the binding together of berries by surface wax or cutin. Overcooking also contributes to clumping.

Blueberry Juice

Because blueberries contain more mucilaginous material than most other berries, the preparation of blueberry juice is rather difficult. Clear blueberry juice, like clarified tomato juice, possesses relatively little flavor. Therefore, the unclarified product is the type considered best. The washed blueberries are heated in a steam-jacketed stainless steel kettle with agitation. When the berries reach 82.2°C, they are put through a screw impeller-type juice extractor. The temperature of the extracted juice is raised to 82.2°C by passage through a heat exchanger. The hot juice is then run into carboys which are completely filled and then closed with a paraffined cork. After standing in a cool cellar for at least 2 months, the juice is siphoned from the heavy sludge on the bottom of the carboys. It is then flash-pasteurized at 82.2°C according to the usual procedure, filled into bottles or cans, sealed, and quickly cooled.

BLACK CURRANTS

Black currants have not been canned in large amounts owing to the high labor cost in harvesting and strigging the berries. The crop is used principally in the manufacture of jams and jellies, syrup, and beverages. The most suitable cultivars are 'Baldwin', 'Boskoops Giant', 'French', and 'Westeick Choice'.

Only firm, ripe fruit should be used for canning; smaller, underripe fruit is better suited for canning as purée or for syrup manufacture. Removal of stalks by hand is very costly. A new method has been tried out and very successfully developed. In this method the black currants are frozen and run over the strigging machine in a hard frozen state that allows the stem to be pulled from the fruit. After leaving the machine, the currants should pass over an inspection belt to remove split or broken ones. The fruit is filled into enameled cans either by hand or with a mechanical filler. After adding the syrup, filled cans are exhausted to reach a can center temperature of 82°C, sealed hot, heat-processed at 100°C in a rotary cooker for 8–12 min for No. 2 cans, and then cooled in water to 41°C.

Black Currant Puree for Reprocessing

Black currants are also processed as puree in enameled No. 10 cans, and reprocessed into jams, jellies, syrup, and beverages when needed. On arrival at the plant, the fruit is washed, sorted, and heated in stainless steel vats that have perforated steam coils situated at the bottom. The fruit is heated to a boil with a minimum amount of water. The batch is started with a small amount of water and one-fourth of the total amount of fruit; remaining parts are added in three lots as the pulp boils. The pulp should be stirred constantly while being cooked. Cooking time for a 182-kg batch should be between 15 and 20 min. On leaving the vats, the pulp is filled into enameled cans as quickly as possible with constant stirring to ensure a uniform mixture in all cans. The filled cans should be steam-exhausted to reach 88°C, sealed quickly, heat-processed at 100°C for 30–40 min, and water-cooled. This pack is largely used for jam manufacture.

Black Currant Syrup

Black currants are picked on the strig and brought to the plant in wooden boxes or trays, the depth of the fruit being 10 in. and 4 in., respectively. After inspection, the fruit is milled through a grater mill. The removal of pectin is achieved by adding Pectinol to the fruit as it enters the mill. The process of enzyme action is followed by checking the viscosity of the expressed juice. The pulp can be pressed in a matter of 24–36 hr using 0.3% of single-strength pectinase, or in 1–2 hr when the pulp is kept at 43.3°C in the presence of Pectinol. The entire quantity of pulp is pressed out in the normal type of cider press using cotton or nylon cloths and ash wood racks. When properly treated with enzyme, yields of 568–606 liters of juice can be obtained from 1 ton of fruit. This juice is centrifuged and converted into a 55° Brix syrup by the addition of solid cane sugar. The necessary quantity of food color and 0.035% by weight of sulfur dioxide is added, and the syrup is clarified through a diatomaceous earth filter. Syrup produced in this way has a shelf life of 18 months. Its acidity of approximately 1.3–1.6% citric acid by weight is an important feature in its stabilization.

Black currant syrup of this type can be used for milk shakes at the rate of 22 ml of syrup to 207 ml of milk, mixed with constant agitation at 4.4°C. It is important to add the syrup to the milk and not vice versa.

Syrup made by this process can be acidified with 1.5% of citric acid by weight and diluted with carbonated water in the proportion of 1 to 5 for bottling and distribution as a sparkling fruit juice.

Black Currant Beverage (Sussmost)

To make Sussmost, pure black currant juice is diluted to contain approximately 25–30% of juice and then sweetened. The beverage is filtered and heated in a scraped surface heat exchanger to 73.9°C and transferred in a continuous fashion to the reservoir of the bottle filler (Fig. 6.1). The hot beverage is filled into warm, clean bottles almost to the top. Caps are applied at once. The additional precaution of previously sterilizing the

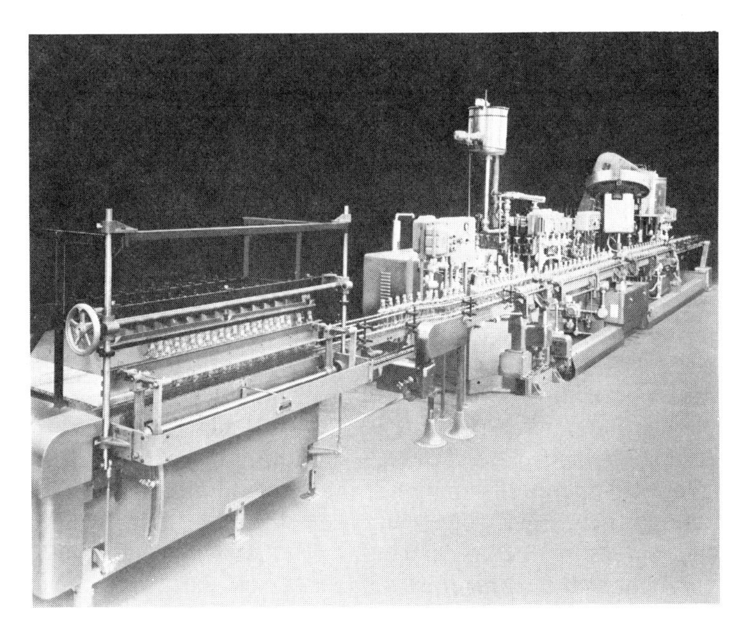

FIG. 6.1. Syncro bottling line including air cleaner, filler, and capper. (Courtesy Pneumatic Scale Corp.)

Table 6.3. Chemical Composition of Sweetened, Diluted Black Currant Beverages (Sussmont Type)

	From Boskoop giant fruit	Commercial black currant beverages		
		Min	Max	Mean
Specific gravity at 68°F	1.0590	1.048	1.0641	1.0573
Total acid (%)	1.03	0.97	1.27	1.04
Extract (%)	15.75	12.51	16.96	15.04
Sugar (%)	13.65	10.32	14.31	12.88
Sugar-free extract (%)	2.10	1.33	4.45	2.22
Ash (%)	0.234	0.176	0.482	0.204
Alkalinity of ash (ml *n*-NaOH/100 ml)	3.36	1.76	4.30	2.41
Alcohol (%)	0	0	0.48	0.12
Volatile acids (%)	0.01	0.008	0.035	0.016
Lactic acid (%)	0.005	0.002	0.017	0.004
Ascorbic acid (mg/100 ml)	37.1	15.0	54.2	25.0

caps with formaldehyde vapor is often taken, but this is not necessary with the hot-filled method if the bottles are inverted after filling.

The chemical composition of some black currant beverages is presented in Table 6.3. The beverage has a content of 1.04% as citric acid and 12.88% sugar.

Improved Production Methods

Technical improvements have streamlined the processing of fruit juices. Generally speaking, the methods of black currant juice extraction, clarification, pasteurization, and storage are similar to those used for other fruit juices. The following points in processing techniques have a special technical interest.

The pectic enzyme action is carried out on large batches of milled pulp held in stainless steel tanks with tapering sides leading to outlets to hydraulic presses.

Flash pasteurization of the centrifuged, filtered fruit juices for syrup production is at temperatures up to 93.3°C for 15 sec. Machines fitted with specially constructed plates can carry out this step with high efficiency, and in the same operation can reduce the temperature to 1.1°C. The lower temperature is necessary when juices are to be impregnated with CO_2 and filled under pressure (3.2 kg/cm^2) into tanks that will be maintained at 0°–1.1°C throughout the storage period.

The residue from the first pressing operation is repressed in a continuous fashion in expeller presses made of stainless steel.

For juices that are eventually destined to be diluted into consumable drinks with 25% of juice, it is necessary to extract the maximum amount of flavoring substances from the fruit. A satisfactory process is to apply steam to the fruit at 80°C. The apparatus often has a rotating central screw in the perforation along its length. The screw rotates at approximately 3 rpm. At this temperature undesirable astringent materials are not extracted, but the enzyme systems are inactivated. The pulp from the steamer can be pressed immediately after adding 2–3% of kieselguhr.

Large horizontal or vertical tanks are widely utilized for storage of black currant juice intermediates. Several new types of tanks are rapidly gaining in favor; resin-reinforced fiberglass tanks are prominent among the newcomers.

Chemical Composition of Black Currant Juices and Syrups

The chemical composition of black currants changes as the fruit approaches ripeness: sugar content increases, and ascorbic acid content, calculated on a weight basis, decreases. It is generally recognized that to produce a quality beverage, fruit should be at least 80% black before it is harvested. Black currant juice (or fruit) may vary greatly in acidity, sugar, tannin, and ascorbic acid contents.

Red Currant Juice

Although red currant juice is extensively used for making jelly and is used in a limited way for punches and other mixed fruit beverages, its use for beverage purposes is not of great importance.

The methods used for making grape juice may be employed in making red currant juice. However, since the juice does not contain potassium bitartrate, it is not necessary to hold currant juice to eliminate argols. The hot-pressed juice may be bottled immediately after filtration.

Red currant juice varies in acidity from 1.9 to about 2.9%. Juice of such high acidity is not suitable for use as a beverage without dilution with a sugar solution.

CRANBERRIES

Cranberries are canned commercially in Massachusetts, New Jersey, Oregon, Washington, and Wisconsin. The berries ripen in the fall and are picked by hand with a special rakelike device or by a machine that makes use of powerful air suction to strip the berries from the vines and convey them to a cleaner and hopper. At the cannery, berries are cleaned by screening and winnowing to remove leaves, trash, etc.

The ripe berries have an acidity of 2.90–3.17% as citric acid; total sugars, 3.5%; total solids, 13.0%; and ascorbic acid, 11–33 mg per 100 gm. The berries are canned in lacquered cans as strained sauce (jelly-like) or unstrained sauce (whole berry). Strained cranberry sauce represents a larger volume of the total pack.

The important cranberry cultivars for cranberry sauces are 'McFarlin' and 'Howes'. The 'Early Blacks' have a very deep color and may be used for blending with other cultivars that are lighter, as too dark a color is not desirable.

Unstrained Cranberry Sauce

Berries should be cleaned and washed, and then cooked in water in a stainless steel or enameled kettle for 8–10 min. The quantity of water used should be kept to a minimum to avoid the necessity of evaporation later. Sugar is added after the berries are well cooked, the quantity being determined by the product desired. Usually about 1 kg of sugar is added per kg of raw berries. The finish may be determined by cooking to 102.2°C, by the percentage of solids as determined by a refractometer, or by consistency. It has been observed that cooking equal weights of berries, water, and sugar to the temperature of 102.2°C gives a satisfactory product with about 43% solids content as determined by a refractometer. Whole cranberry sauce should not be cooked to a point where it sets to a solid gel, but should flow slowly when poured into a dish.

In another process, the cranberry pulp is utilized to manufacture whole cranberry sauce. The pressed cranberries, in the form of liquid crushed pulp, are placed in a kettle of water and brought to 85°–100°C. The mix attains a true cranberry color. Skins and seeds are removed by passing the mix through a 0.027- to 0.033-in. screen leaving a puree in the form of a strained aqueous suspension of cooked residue of crushed cranberries. Best-grade raw whole cranberries are then mixed with a sucrose–corn syrup solution and added to the cooked, pressed, strained cranberry puree. The mixture is brought to the boiling point as rapidly as possible. The resulting sauce is then filled into cans, sealed hot, and cooled in water to 38°C. The soluble solids content of the jellied product was 38%. The resultant product has a deeper red color and better flavor and consistency than a whole sauce prepared to the same soluble solids content by the same process, but lacking the added pressed cranberry puree. A whole cranberry sauce, made by boiling 10% whole cranberries with finely comminuted cranberries for 1–3 min, furnishes enough pectin to form a gel with water.

The cranberries that are to be introduced whole are processed in the usual manner, sorted, washed and stemmed. The process is equally well suited to the use of fresh or frozen cranberries. Likewise, the berries that are to be comminuted and mixed with the whole cranberries are sorted, washed, and cleaned, and finely comminuted by any suitable method such as by grinding or running through a blender. Once the whole cranberries are in the cranberry sauce, the soluble solids in the gel surrounding the whole berries gradually penetrate the whole berries since substantially all of them have been cracked open. The total soluble solids content of the sauce amounts to between 38 and 42% of the weight.

Strained Cranberry Sauce

Washed berries may be run directly into kettles containing water. The berries are then cooked 8–10 min and run through a cyclone, preferably with a nickel or monel metal screen with 0.03- to 0.04-in. openings to remove the skins and seeds. The pulp from the cyclone passes to another steam-jacketed kettle where sugar is added, and the evaporation may be determined by the usual jelly-sheeting test, by the boiling point (102.8°C), or with a refractometer when the solids reach 43%. The weight of sugar added is approximately equal to the initial weight of the berries. The amount of water added should be carefully controlled so that it will not be necessary to prolong the heating and evaporating process.

The sauce is filled at 88°–93°C into lacquered cans. For No. 10 cans, beaded cans are used to avoid paneling. The cans are sealed hot with steam injection and then pass directly to the cooler. The cans should be cooled to 38°C as they come from the cooler. It is desirable to stack them in a manner that will make thorough cooling possible and subsequent storage should be at a temperature below 20°C. Cans should not be disturbed after stacking until gel formation is complete.

Stainless steel of monel metal or other corrosion-resistant metal cookers should be used because of the high acidity and intense red color of cranberries. The No. 300 and No. 2 size cans are popular for home use; the No. 10 for institutional use.

Several constituents including pectin, pigments, flavor, and soluble solids that remain in the pressed cranberries are presently discarded after extracting the juice cocktail. This material, when utilized in the preparation of strained cranberry sauce, decidedly upgrades the resulting sauce with respect to color, flavor, and consistency, and lessens the weight of whole secondary-grade cranberries required per unit of production. In one study, highest-grade raw cranberries were crushed at room temperature in a Carver Press to extract 75–80% of the juice, which was subsequently diluted with water and sweetened with sugar to provide a deep crimson red cranberry cocktail. Light and mixed colored grade whole cranberries were placed in water to which was added the pressed pulp of the superior-grade cranberries. The aqueous suspension was heated to a temperature between 85° and 98.9°C. The mixture, while still hot, was strained through a finisher screen with 0.686- to 0.838-mm (0.027- to 0.033-in.) openings to reduce the average size of the contained particles and to remove seeds and skins. The resulting puree was blended with a sugar–corn syrup to obtain a 37°–40° Brix, and heated. The hot mix was then filled into consumer-size cans and cooled in accordance with usual practice. This procedure gave a jellied sauce that had a deeper color and firmer gel structure than would be present without the addition of the pressed cranberry pulp. Instead of screening, the seeds, skins, and other solids may be comminuted to fine particle size by passing the suspension through a suitable comminutor, disintegrator, or mill.

Cranberry–Orange Relish

Oranges have been used in combination with cranberries to form a fresh relish for some time. The relish contains the whole orange including peel, pulp, and juice. Such a relish tends to be watery due to the grinding of the cranberries and whole oranges, a process that extracts most of the natural juices as a liquid.

It is possible to make a cranberry–orange relish that can be processed and canned, thereby eliminating the necessity of immediate consumption in homemade relish or frozen relish after thawing. By using a small amount of orange peel, and by proper control of particle size, a true relish-type cranberry product that has a highly acceptable flavor and good textural qualities can be prepared. This process includes the steps of comminuting whole cranberries, reducing orange peel cuttings to form particles ranging in maximum size from about 3.18–12.7 mm, combining the comminuted cranberries and orange particles with sugar and water, and rapidly heating the mixture to between 93.3°–101.7°C to form a semi-jellied cranberry product with the orange peel distributed uniformly throughout; this product has a soluble solids content of 38–54% by weight.

In preparing cranberries for relish, the degree of comminution is important, for it is necessary to cut the cranberries so as to reduce the toughness of the skins without destroying the variation in texture resulting from discrete pieces of cranberries. The best source of orange is from freshly comminuted orange peel. This peel may be fresh or frozen and is conveniently the by-product peel remaining after the orange juice and pulp have been used in preparing frozen or canned products, such as forzen orange juice or canned orange sections.

In order to give the resulting relish a texture that exhibits a contrast between the size of the cranberry particles and orange particles, it is necessary to cut the orange peel to cubelike particles. After processing, there is a definite contrast between the size of the cranberry and orange particles present. Since the orange peel does not contribute additional liquid, the final processed relish possesses a certain homogeneity in the matrix surrounding the discrete pieces not exhibited by the fresh or frozen relishes.

Cranberry Sauce for TV Dinners

A quick-gelling cooked cranberry sauce may be mechanically handled in a pregelled state and will thereafter gel despite immediate subjection to commercial freezing environment, and remain gelled upon thawing. Such cooked cranberry sauce mix thus need not be held for any gelling period after cooking but may be rapidly metered hot directly from the cooking kettles onto individual TV dinner plates in the freeze-line of production without danger of losing gelation before or during the freezing operation or during thawing. Alternately, sauces of this type can be rapidly mechanically handled even after gelling without breaking the gel, provided appropriate delicate pumping mechanisms are used. Unlike previous gelled cranberry sauces, they may be metered cold into the TV package without losing the proper gel structure.

This sauce is composed of cooked whole or strained cranberry sauce mix prepared in accordance with conventional practice except for the addition of a small proportion of a gelling agent in the form of an acid- and freezing-tolerant gelling material before cooking. The preferred gelling material is one prepared from waxy maize starch, available on the market as a heat-soluble corn starch whose gelling capacity is not adversely affected by the acid content of cranberries, nor by freezing temperatures down to −40°C. Addition of the starch in an amount of between 1 and 2% by weight of the mix, whether a whole or a strained cranberry sauce, has been found effective in causing the mix to gel as it cools either before or during the freezing operation. In either case, it survives commercial quick-freeze processes and upon thawing, gives an attractive fully gelled cranberry sauce that does not liquefy or flow into the other ingredients of the TV dinner.

Jellied Cranberry Sauce

One of the more popular ways to market cranberries is in the form of a canned, jellied sauce. Previously, it had been thought necessary to remove the skins and seeds from cranberries via a screening process, add sugar and water to the resulting puree, and then cook the mixture to the desired end point. The removal of skins and seeds not only required extra steps and extra labor, but meant that there was a loss of approximately

10% by weight of the original cranberries. However, because skins and seeds contribute a distinct flavor to cranberries, they are desirable in a jellied cranberry sauce.

Nowadays, whole cranberries, either raw or cooked, either fresh or frozen and thawed, can be comminuted, heated with sugar and water, and canned. In preparing the puree from the entire cranberry, it is necessary to reduce the relatively tough skin and the seeds to a size such that the puree will pass through a screen having perforations no larger than 0.686 mm (0.027 in.) in diameter, without forming tightly rolled pieces of skin that could conceivably be forced through the perforations of the screen. A. Fitzpatrick comminutor or a Rietz disintegrator and their accessories are particularly well suited to attaining the desired degree of comminution of the whole cranberry.

Once the cranberries are properly comminuted, the resulting puree is then mixed with an amount of sugar and water sufficient to form a smooth

Table 6.4. Volatile Compounds Identified in Cranberry Juice

Compound	Peak no.	Gas chromatography			
		DEGS column	Mass spec	Infrared	Concentrate (%)
Aromatic					
Benzene	4	+	+		0.1
Benzaldehyde	18	+	+	+	9.6
Benzyl ethyl ether	19	+	+		1.0
Acetophenone	22	+	+		0.8
Methyl benzoate	23	+	+		1.0
Benzyl formate	24	+	+		0.7
Ethyl benzoate	28	+	+	+	1.0
Benzyl acetate	29	+	+		0.7
Benzyl alcohol	31	+	+	+	6.0
2-Phenyl ethanol	33		+		2.2
4-Methoxy	36	+	+		0.8
benzaldehyde	38	+	+		1.2
2-Hydroxy diphenyl	39	+	+	+	11.9
Benzyl benzoate	40	+	+		1.1
Dibutyl phthalate(s)					
Terpenes					
α-pinene	9		+		0.1
β-pinene	11		+		0.2
Myrcene	12	+	+		0.2
Limonene	14	+	+	+	1.1
Linalool	25		+		0.6
α-terpineol	30		+	+	13.0
Nerol	34		+		1.1

(*continued*)

gel. The water is adjusted to the amount of pectin contained in the cranberries. After the puree is mixed with sugar and water, it is heated to destroy all enzymatic and microbiological action and to cause the necessary interaction between the sugar, puree, and pectin to effect gel formation. The resulting hot liquid mixture is then sealed in cans or jars under vacuum.

Cranberry Juice Cocktail

The procedure for making cranberry juice cocktail is to put thawed cranberries through a tapered screw extractor. A yield of 66–70% of juice is obtained from each 100 kg of fruit. For cocktail, this product is diluted with twice its volume of water and sufficient sugar is added to bring the

Table 6.4. (*Continued*)

	Gas chromatography				
Compound	Peak no.	DEGS column	Mass spec	Infrared	Concentrate (%)
Aliphatic alcohols					
2-Methyl-3-buten-2-ol	6		+		0.9
2-Pentanol	7		+		0.8
Pentanol	10	+	+		0.9
Hexanol	15	+	+		0.7
1-Octen-3-ol	20		+		0.8
Octanol	26		+		2.3
Nonanol	32		+		0.8
Decanol	35		+		0.7
Octadecanol	37	+	+		0.8
Aliphatic aldehydes					
Acetaldehyde(s)	1	+	+		0.1
Pentanal	5		+		0.2
Hexanal	8		+		0.8
Octanal	17		+		0.9
Nonanal	21		+		1.0
Decanal	27		+		0.8
Other compounds					
Diacetyl	2		+		0.3
Ethyl acetate	3		+		0.7
2-Furaldehyde	13		+		0.8
Methyl hepanoate	16	+	+		0.6
Acids					
Benzoic acid	—	+	+	+	26.6
2-Methylbutyric acid	—	+	+		0.3
Total					95.2

Source: Croteau and Fagerson (1968).

specific gravity up to 15° Brix. The pomace may be used for making strained cranberry sauce.

Cranberry juice beverages are sometimes clarified by means of Pectinol or some pectic enzyme preparation. After treatment, the beverage is filtered using Hyflo Super-Gel as a filter aid. The filtrate should be heated to 85°C to inactivate the enzymes and thus prevent further action during storage. The hot product is filled into cans or bottles; these are then closed, turned on their sides, and cooled. Cranberry juice is very corrosive to tin plate; therefore cans lined with berry or fruit lacquer should be used.

The Standard of Identity states that cranberry juice cocktail (a juice drink under the Federal Food, Drug, and Cosmetic Act) is the beverage food prepared from cranberry juice or concentrated cranberry juice, or both, with water and nutritive sweetener (or nonnutritive sweetened product). It contains not less than 25% of the single-strength juice. The soluble solids are 14°–16° Brix as determined by refractometer. It may contain 30–60 mg vitamin C per 6 fl oz. The acid content, calculated as anhydrous citric acid, is not less than 0.55 gm/100 ml. It is sealed in containers and processed by heat so as to prevent spoilage (Anon. 1968). An aseptic HTST system has been applied to the canning of cranberry juice cocktail in brick packages made of Mylar-Al-Poly-Propylene laminated containers (Ito and Stevenson 1983).

Volatiles of Cranberry Juice

The major volatile components of the juice of the American cranberry are listed in Table 6.4. Forty-three compounds, comprising 87% of the concentration of the volatiles, have been identified. Twelve of these are aliphatic alcohols, 11 are aliphatic aldehydes and ketones, 5 are terpene derivatives, 8 are aromatic compounds, and 7 are other compounds. Terpineol (34%) is quantitatively dominant, while the amount of 2-methylbutyric acid is much smaller than in lingonberries (*V. vitis-idaea*) where it is the most important aromatic compound.

GOOSEBERRIES AND LOGANBERRIES

Gooseberries

Gooseberries are canned when they have reached their full size, but before they become soft or changed in color.

When brought to the plant they are first put through an Urshel snipper. The berries are then passed over a sorting belt to remove defective ones and foreign material, washed, drained, and filled into plain cans. Most gooseberries are packed in water in No. 10 cans for the bakery trade; some are packed in heavy syrup in No. 2 cans for the retail market.

After berries are filled into cans, boiling hot water or syrup is added and the cans are passed through a steam exhaust box at 100°C for 5–6 min. The cans are sealed under steam injection and then processed in boiling water. The amount of processing necessary depends on the temperature after exhausting. Assuming an initial temperature of 65.6°C, the processing time for No. 2 cans is 8–10 min at 100°C for water pack and 15–18 min at 100°C for syrup pack. Spin cooking in atmospheric steam for 3–4 min at 12 rpm improves the drained weight of the canned product. The cans should be cooled in water to 37.8°C after processing.

Loganberries

Oregon is a fairly large producer of loganberries, which are used for canning, frozen pack, jams, and juice. The berries are very large in size and deep red in color.

The canned fruit is used mostly for pie-making and therefore is canned in No. 10 enamel-lined cans, in water. The processes of harvesting, canning, and sterilizing are basically the same as for blackberries. "Double-enameled" Type L cans should be used to ensure the retention of color.

RASPBERRIES

Raspberries are canned in small commercial quantities in the northern and midwestern states, in New York, and on the Pacific Coast. Only 7000 cases of black raspberries and 26,000 cases of red raspberries were packed in 1971. The red raspberry is preferred to the black raspberry for canning, but is in even more demand for preserves of jam. The 'Willamette' cultivar accounts for more than half the acreage of red raspberries in Washington. Its popularity is based on its high yield of large, firm berries that are high in acid. Other cultivars include 'Puyallup', 'Summer', and 'Fairview'. In England, the best cultivars for canning are 'Cuthbert', 'Lloyd George', and 'Norfold Giant'.

Raspberries should be picked when they are ripe but firm. They are transported to the cannery in crates containing shallow 0.23- to 0.46-kg (0.5- to 1-lb.) baskets or perforated plastic containers. They should be canned promptly when received at the cannery and should not be held overnight except in a refrigerated room at −1.1° to 1.67°C.

Berries are washed and sorted to remove the deformed and overripe berries unfit for canning. Raspberries are size graded using slat riddles. The berries are canned in heavy syrup (50°–55° Brix) for dessert purposes, or in water for use in pies. Cans are exhausted until a center temperature of 18°C is reached. This requires approximately 5–6 min at 100°C. The choice of cooking method depends on the relative importance of the different quality characteristics. Still-cooking is preferred if wholeness, shape, and texture are of major importance; but spin-and-rotary cooking methods are preferred if drained weight is of greater significance.

Spin cooking of canned strawberries, raspberries, and gooseberries for 3–4 min in atmospheric steam at 12 rpm compares favorably with conventional cooking (15–16 min in boiling water). After 2 months, spin-cooked berries show higher drained weight, superior color, and firmer fruit than those cooked by conventional methods.

Mushiness and crumbliness occur commonly during processing of raspberries. Impacts during processing abrade and crush the fragile pulp tissues, freeing the pits and making the canned product mushy. Impact after cooling damages the berries only slightly. Crumbliness involves genetic and pathological problems that affect normal fruit structure. When cans are opened shortly after cooling, all the raspberries appear greatly shrunken, and many of the pits partly protrude. Those cooked in rotary cookers appear more shrunken than those still-cooked. Excessively rapid can rotation and, in particular, impacts during cooking and conveying abrade the shrunken berries and loosen the pits so they remain free after the berries have returned to normal size.

Increasing syrup concentration in canned raspberries does not affect their drained weight, but does tend to increase the proportion of broken fruit. The addition of 0.42% low-methoxyl pectin increases the drained weight of raspberries and improves their texture, but lowers acceptability for flavor and color. Syrups of 50°–55° Brix are preferred for flavor with raspberries and loganberries.

Tyramine Content of Raspberries

The tyramine content of fresh raspberries varies from 12.8 to 92.5 μg/g, and that of raspberry jams from 8.0 to 38.4 μg/g. The tyramine level in raspberries in comparison with that in other fruits indicates that tyramine is a useful indicator of the presence of raspberry in fruit products (Coffin 1970).

STRAWBERRIES

Strawberries are not a popular fruit for canning and are largely preserved by freezing in the United States. One reason for this is that addition of food coloring is not permitted by the FDA, and the color of canned strawberries is very unattractive compared with that of frozen products. The principal difficulty in canning strawberries is the softening of the fruit during heat processing, which results in the can containing only one-third its volume of berries. Canned strawberries are more common in England. The cultivars of strawberries used for canning in England are 'Huxley',

'Gautlet', and 'Talisman'. The 'Huxley' cultivar has a nonremovable plug but is the most reliable berry since it retains its shape in the can. The 'Huxley' has a dark red color and firm texture. It requires a longer processing than other cultivars to give complete sterilization.

The fruit is delivered to the plant in trays to prevent crushing during transport. When berries are held in cold storage, they should be removed only in small quantities as required. The canning must not be delayed, otherwise the strawberries will collapse as the temperature rises.

Berries are first delivered to the preparation belt where stemming is carried out. They are graded for quality and passed over a mechanical grader. Large berries are used for manufacture of jam or pulp; medium berries for canning. If strawberries are small, especially during the late season, the cost of stemming is high. Such strawberries should not be stemmed but pulped cold with SO_2 and used for jam manufacture.

It is advantageous to give the fruit a light spray washing. The washed fruit must be well drained before filling into enameled cans. The filled cans pass through an automatic drainer attached to the syruper, where 50° Brix sucrose syrup is used for fancy grade, and 30° Brix for the standard grade. Artificial color may be used in canning strawberries in England; 189 liters (50 gal) of syrup should contain 113.4 g of Ponceau 2 R, and 14.2 g of Erythrosine. Since there is a great deal of variation in strawberries, processors are advised to test a small batch before producing any quantity. At least 1 week should be allowed for dye penetration before judging the color of canned berries.

Canned strawberries should be clinched before exhausting because this fruit floats and the top berries will become soft during the exhausting and break up during processing. Exhausting time is usually 6–8 min at 82.2°C. Steam flow may be used here, but for better results the exhausting process is advised.

Attention should be paid to the exhausting of strawberries. The purpose of this process is to collapse berries slowly and to release the oxygen from their cell. If the exhaust is insufficient, berries collapse during cooking, with the result that the vacuum is not maintained and the berries will spoil quickly. The processing time for No. 303 cans is 7–8 min at 100°C.

The main aromatic substances in strawberries, as quantitatively determined by gas chromatography are methyl and ethyl butanoate, methyl and ethyl hexanoate, trans-2-hexenyl acetate, trans-2-hexenal, trans-2-hexen-1-ol, and 2,5-dimethyl-4-methoxy-3(2H)-furanone were.

CHERRIES

Two types of cherries—sweet and sour—are canned in this country and abroad. In New York, Michigan, and other eastern states the sour cultivars, 'Morello' and 'Montmorency' are commonly grown. On the Pacific Coast in Oregon and California the sweet cherry 'Royal Ann' predominates. Other sweet cultivars include Napoleon, Bigarreau, and Amber.

Since sweet cherries contain less acid, their keeping qualities are not as good as sour cultivars. On the other hand, sour cultivars require a heavier syrup to make them more palatable, and this tends to cause the fruit to shrivel. The main acid in cherries is malic, but citric and quinic acids are also present in reasonable quantities.

Whole Cherries

On arrival at the cannery the fruit is first stemmed by hand or mechanical stemmers. Mechanical stemmers have a series of rubber rollers on an incline, and the cherries roll down the incline from an automatic feed at the top of the machine. The rollers revolve toward each other, and as the cherry turns over, the stem is dropped between the rollers and pulled out. The cherries emerge uninjured with 95% or more of the stems removed.

After stemming, the cherries are thoroughly washed and graded for size. From here they go to the filler, syruper, and exhauster. Exhausting of cherries is most important because this fruit, containing stones, is very prone to pinholing and hydrogen swellls. An exhaust of at least 10 min at 73.9°–85.0°C is recommended for smaller cans. The center of the can should reach 82°C.

Most sweet cherries are canned without pitting, whereas most sour cherries are pitted. Pitting is accomplished by an automatic machine in which the cherries fall into small cups and the pits are removed by cross-shaped plungers. The loss in pitting is about 15% of the weight of the stemmed cherries. Considerable juice is expressed from the cherries in pitting and is usually recovered for canning as juice for use in syrups.

The standard pack of sour cherries in a No. 2½ can should be exhausted to give a can center temperature of 82.2°C and processed 16–20 min at 100°C. A No. 10 can of sour cherries requires 25–30 min at 100°C. After sterilization the cans should be quickly cooled before labeling and storage.

Red, tart 'Montmorency' cherries, when allowed to stand before being canned, either with or without having been previously bruised, are much firmer after canning than are similar cherries canned immediately after harvest. During the aging period a portion of the pectin is completely demethylated to form pectic acid, making the cell walls more rigid and less easily separated from each other.

A 1% increase in sugar content of raw fruit results in a 0.6% increase in the drained weight of the canned product. A copper spray results in smaller cherries, less juice loss, higher soluble solids, higher drained weights, and more red color than other sprays. The firmness of canned cherries increases as harvest is delayed.

As storage time and temperature of canned cherries is increased, sugars (total and free reducing), acidity, and hydroxymethyl furfural increases, while anthocyanins, carotenoids, pectins, volatile reducing substances, syrup viscosity, and organoleptic quality decline. To maintain high-quality canned cherries, storage temperatures of 4.4°C or lower are preferable.

Cherry Juice

Cherry juice has an attractive color and pleasing flavor; nevertheless, its manufacture and use is very limited compared with that of the more popular juices. Perhaps the reason for this is that the acidity and flavor of cherry juice are so strong it requires dilution to be pleasing to most persons.

Cherry juice is produced chiefly in Wisconsin although small amounts are processed in Colorado, New York, and Pennsylvania.

Ordinarily, no single common cultivar of cherry yields a juice of the proper acidity and sugar content for an ideal beverage. As a rule, the unsweetened juice of some sour cultivars (e.g., 'Montmorency', 'Early Richmond', and 'English Morello') is too acid and too low in sugar content to be entirely pleasing to the average palate. If 'Montmorency' cherries, however, are allowed to reach full maturity, their sugar content increases and the acidity becomes proportionally less. Such well-matured fruit produces a very desirable juice without blending. Sweet cherries, on the other hand, may be too low in acid to yield a juice of pleasing flavor. The flavor of juice made from 'Montmorency' and 'English Morello' cherries is excellent.

If a juice of excellent flavor and color is desired, the best-quality cherries must be used. Juice prepared from cull fruit generally possesses an off-flavor derived from spoiled or spotted fruit. The benzaldehyde-like flavor is probably derived from the enzymatic hydrolysis of cyanogenic glucosides similar to amygdalin. Juice made from underripe fruit is sour and of poor color.

Composition of Cherry Juice. Although there is little difference in the sugar content of different cherry cultivars, the total acidity varies widely. Sweet cultivars yield juice that is low in acid (0.47% as malic acid), whereas the common sour cultivars contain 1.3–1.8%. Malic is the principal acid present; cherries also contain small amounts of citric, succinic, and lactic acids. The principal sugars of cherry juice are dextrose and levulose, with only small amounts of sucrose. The reducing sugars of cherry juices range from 7.9–10.6%.

Sorting and Washing. Cherries for juice production may be harvested with or without stems. The freshly harvested cherries should be sorted to eliminate spoiled and damaged fruit. The cherries should then be washed in cold water (10°C), preferably with some time allowed for soaking. The soaking period should not be longer than 12 hr, or there will be a notable loss in soluble solids and some change in flavor. The pitting loss in a mechanical pitter amounts to about 7%.

Hot Pressing. The simplest method of making cherry juice is to heat the washed cherries to approximately 65.5°C in a steam-jacketed stainless steel kettle and then press the fruit through nylon cloths before it cools. The heating extracts a large proportion of the pigments of the cherries, and in the case of 'Montmorency' and 'Early Richmond' cherries produces a deep red juice. The 'English Morello' yields a very dark red juice.

A hydraulic press of the type often used for pressing grapes is suitable for the pressing of cherries.

The hot juice from the press is strained through a fine wire screen, made of corrosion-resistant metal, or a muslin bag. The strained juice is chilled to 10°C or lower and allowed to settle overnight. The clear juice is siphoned from the sludge, and then is mixed with a small amount of filter aid (e.g., Hyflo Super Cel) and filtered through canvas in a plate and rame filter press or some other filter. The yield obtained by hot pressing 'Montmorency' cherries varies from 62 to 68%.

Cold Pressing. Cold-pressed juice is not as brilliantly colored as the hot-pressed product, but its flavor closely resembles that of fresh cherries.

The washed fruit is drained and then cut to a coarse pulp in an ordinary apple grinder, such as the ones used for the making of cider. The knives are set so that the pits are not crushed during maceration. This comminution of the fruit results in a better extraction of pigments. The cold, macerated cherries are pressed in a rack and cloth hydraulic press. The yield obtained by cold pressing varies from 61 to 68%. The freshly pressed juice is rapidly heated to 87.7°–93.3°C, and then cooled. This operation inactivates enzymes, kills microorganisms, and coagulates colloidal matter.

It is usually necessary to give the juice a special clarification treatment before filtration or else the filter is soon clogged by the pulp. A simple method of preparing the juice for filtration is to treat it with Pectinol. The juice is cooled to 37.7°C, then 0.1% by weight of Pectinol M is added and held at this temperature for 3 hr. After this period, the juice is heated to 82.2°C, then cooled and filtered through a plate and frame filter press.

Cold Pressing Thawed Fruit. Deep red juice having a color nearly as dark as that obtained by hot pressing and yet possessing the fresh flavor of cold-pressed juice may be obtained by pressing frozen cherries. The cherries may be prepared for freezing either by packing pitted cherries, with or without added sugar, into enamel-lined tin cans or barrels, or by crushing the unpitted fruit to release only enough juice to cover the cherries when packed in enamel-lined tin cans or barrels. The cherries are frozen and stored at −17.7°C or lower. When needed for juice, they are thawed until the fruit reaches a temperature of 4.4°–10°C, and the thawed fruit is pressed in a hydraulic press. Juice obtained from thawed cherries should be treated with Pectinol and filtered as described for cold-pressed juice. Thawed 'Montmorency' cherries yield 70–76% and 'Early Richmond' 60–75% of juice; higher yields are obtained at 62.7°C.

Sweetening and Processing. Cherry juice from the sour cultivars—'Montmorency', 'Early Richmond', and 'English Morello'—is usually too sour to please the average palate. Therefore, unless the juice has been produced from especially sweet cherries, it is necessary to sweeten it by adding dry sugar or sugar syrup to bring the density of the juice to about 17° Brix. A more palatable beverage may be obtained by diluting the juice with half its volume of water and adding sufficient sugar to bring the percentage of total solids back to the original point. This procedure reduces the total acidity to 1% or less and maintains the solids content at approximately 10%.

If sugar, sugar syrup, or water is added to cherry juice, the addition should be clearly indicated on the label. It should be noted that a diluted product cannot be labeled "juice."

The juice of sweet cherries, such as the 'Bing', is somewhat lacking in acidity. Furthermore, most sweet cherries yield a juice that is not deeply colored. Because sweet cherries are ordinarily more valuable than sour cherries, they are seldom used for juice. If the juice of sweet cherries is available, it may be greatly improved by blending with an equal volume of the juice of a sour variety such as the 'Montmorency'.

Since hot-pressed cherry juice is of better color and cold-pressed juice of superior flavor, a blend of the two is more attractive than either alone. Equal parts of each or two parts of cold-pressed juice blended with one part of hot-pressed juice gives a product more desirable than either alone.

Cherry juice may be packed in either cans or bottles. If cans are used, they should be lined with a berry enamel.

Cherry juice and cherry beverages, containing one-half water and one-half cherry juice may be preserved by either holding or flash pasteurization methods. Flash pasteurization temperatures as low as 73.8°C may be used if care is taken to eliminate air in the headspace of the bottled product.

Pasteurized cherry juice should be held under refrigeration if it is to be stored, otherwise its flavor deteriorates markedly.

Factors Affecting Stability and Color of Cherry Juices. Blanching cherries for 1 min at 85°C before pressing yields a juice that is more intensely red than is obtained from unblanched fruit. The subsequent degradation of the color during storage is less in the blanched fruit juices. With diatomite filtration, anthocyanin pigments are absorbed, decreasing the color of the juices. No significant color loss occurs following treatment with a pectinolytic enzyme to clarify the juice. The concentrated juice is more stable at temperatures approaching 0°C. Exclusion of oxygen also has a beneficial effect on the color stability of cherry juices at all storage temperatures, but more markedly at the lower temperatures.

Carbonated Cherry Juice. Cherry juice carbonated with about 3 volumes of carbon dioxide is a very pleasing beverage. Cold-pressed juice does not yield a satisfactory carbonated beverage since the carbonation causes the deposition of a great amount of sediment. On the other hand, carbonated hot-pressed juice remains clear during carbonation, pasteurization, and subsequent storage. The method used for the carbonation of apple and other fruit juices may be used in preparing the product.

Maraschino Cherries

Maraschino cherries are prepared from 'Royal Ann' and other white cultivars. The cherries are picked after they have reached full size, but are not fully mature. They are stored in a preservative brine containing 0.75–1.5% sulfur dioxide and 0.4–0.9% unslaked lime. If too much lime is used, the sulfur dioxide will have no preservative effect and spoilage by yeast will ensue. If too little lime is used, the resulting low pH will cause splitting of the cherries and cracking of the skin.

After about 4 weeks storage in the calcium bisulfite brine, the cherries are usually ready for subsequent processing. The sulfur dioxide bleaches the cherries to a translucent white or cream-yellow color, while the calcium hardens the tissues by combining with pectin material (see Chapter 9).

Insufficient bleaching with sulfur dioxide is not uncommon and has led to schemes employing secondary oxidative bleaches. Beavers and Payne (1969) suggest leaching out most of the sulfur dioxide from the cherries (down to 100–200 ppm free SO_2) followed by immersing them in a 0.75% sodium chlorite solution at pH 4.5–6.0 and at a temperature not to exceed 43.3°C. Bleaching is completed in 1–7 days after which the residual sodium chlorite is removed by leaching in water for 24–36 hr. The cherries are then placed in the sodium bisulfite brine for firming and storage. The storage brine should be adjusted between pH 3.0 and 3.5 in order to maintain good texture. After 2 weeks in the storage brine, the cherries can be finished. Sodium chlorite-bleached cherries are free from off-flavors and possess a firm texture and excellent white color. Sodium chlorite does not hydrolyze or degrade cellulose to the same degree as the hypochlorite bleaches and is therefore preferred.

After removal from the calcium bisulfite brine, the cherries may or may not be stemmed or pitted. They are leached in running water until most of the sulfur dioxide has been removed. They are then boiled in several changes of water until tender and until the sulfur dioxide content is reduced to below 20 ppm.

The cherries are next boiled a few minutes in a 0.02–0.05% erythrosin dye solution or FDC Red No. 4. The dye is precipitated by the acid in the fruit and therefore penetrates only a short distance into the flesh unless the pH is 4.5 or higher. The fruit is allowed to stand for 24 hr in this dye solution after which it is boiled a second time. About 0.25% citric acid may be added to the water used for this boiling. This method of dyeing the fruit greatly reduces it tendency to "bleed" and stain other fruits with which it might be canned.

For use in cocktails, the cherries are put in a cherry-flavored syrup whose final cut-out density is about 40° Brix.

FIGS

Figs are canned extensively in California and Texas, and to a lesser extent in Louisiana and other southern states. In California, the 'Kadota' fig is the most important cultivars for canning. It is a white fig of moderate size, globular or oval in shape, with thin skin, firm flesh, and a small seed cavity. In Texas, the 'Magnolia' cultivar is most commonly used, and in Louisiana, the 'Celeste'. 'Magnolia' figs are light brown in color, of moderate size, and of excellent canning quality. 'Celeste' figs are very small and elongated, but have a very rich flavor; they are firm and retains their form and texture remarkably well in canning.

The Fresno–Merced district is the principal fig-growing area in California. Figs are picked frequently at the firm-ripe stage during the season. They should not be picked underripe or they will have an unsatisfactory flavor. If picked overripe, they will break up during canning and present a poor appearance.

Harvested figs should be taken promptly to the cannery for processing. They are usually shipped to the cannery in lug boxes with a 11- to 14-kg capacity. In the cannery, the first operation is to sort figs on a slow-moving belt and to remove fruit unsuitable for canning. Figs are then mechanically graded for size. Some packers use a greater variety of size grades for a given can size than others. For instance, the number of figs in each No. 10 can be 30–50, 51–70, 71–90, 91–110, or 111–140; and in each No. 2½ can 9–14, 12–20, 21–24, and 25–28.

Canning procedures for 'Kadota' figs vary considerably from one plant to another. Where figs are canned by hand, they are washed and then given a preliminary light blanch to remove the waxy coating. This blanch may be 2–8 min, in warm water sprays or steam, or in a mixture of both. They are then canned from a belt. The split and less mature figs are canned separately from the more mature unbroken figs. If semiautomatic fillers are used, the figs are given no preliminary blanch.

After filling, the general practice is to "dry exhaust" the container of figs for a considerable time through a steam exhaust box at 98.9°C before any liquid is added to the cans. The purpose of this operation is to expel the air present in the figs, destroy the latex, and soften the texture. During this procedure the cans become about half full of condensate. The blanching time varies with the condition of the figs, from 15–20 min for the smaller can sizes to 45–110 min for No. 10 cans. Cans "dry exhausted" are not drained before the syrup is added.

An alternative to the "dry exhaust" process is to fill the cans with water and pass them through a steam box at 99°C for 30–40 min, depending on the size of the container. The cans are then drained.

After exhausting, sugar syrup of desired strength (30°–48° Brix) is added at about 48.9°–65.6°C. The cut-out Brix measurements, as applicable, for the various designations are as follows:

Syrup	Testing
Extra heavy	26°–35° Brix
Heavy	21°–26° Brix
Light	16°–21° Brix

The Standards of Identity require acidification of canned figs to pH 4.9 or below. This inhibits the development of spoilage bacteria and improves the flavor. The pH of the syrup portion of canned figs is tested 15 days after canning. The quality control department can also macerate the figs and acidified syrup in the right proportion and then test for the pH of the blend. It is preferable to make the test before heat processing in order to know the exact degree of acidification and to have a sound pH control and adequate processing time. Some canners prefer to use a syrup that will yield a cut-out Brix reading at 21°–22°. Most canners add lemon juice concentrate or citric acid to the in-going syrup.

After addition of syrup, the cans are sealed under steam injection and heat-processed. They may also be exhausted at 99°C for 6 min for No. 2½ cans, and 12 min for No. 10 cans prior to double seaming.

The heat-processing time for canned figs varies greatly from one cannery to another. Canned figs are heat-processed in a continuous cooker at 100°C for 30–41 min for 8Z (211 × 304) cans; at 98.9°–101.1°C for 30–50 min for No. 303 × 406 cans; at 98.9°–101.1°C for 30–55 min for No. 2½ (401 × 411) cans, and at 100°C for 30–70 min for No. 10 (603 × 700) cans. When a still retort is used, figs in glass are heat-processed at 100°–102°C for 32–60 min for No. 303 (303 × 411) containers; and at 100°–102°C for 40–65 min for No. 2½ (401 × 414) containers. Since figs are low in acid, they should be heat-processed with the same caution as for canned mushchrooms, olives, etc., unless acidification with lemon juice is done properly to the right pH range. The glass packs are processed in still retorts, while canned figs are usually sterilized in continuous rotary cookers. The cans and jars are water-cooled immediately after heat processing.

When the addition of lemon juice (including concentrated lemon juice) or citric acid lowers the pH of the canned figs to less than 4.5, the label must bear the statement "with lemon juice" or "with citric acid." When two or more of the optional ingredients such as spice, flavoring (other than artificial flavoring), vinegar, unpeeled segments of citrus fruits, and salt are used, such words may be combined, as for example, "with added spices, orange spices, and lemon juice".

Canned Preserved Figs

In Texas, figs are peeled in a hot 2% lye solution for 10–15 sec. This is followed by washing the fruit in water. This treatment must be very thorough, and the mesh belt on which the fruit travels has heavy water sprays on both top and bottom, which also serves to remove the skins. The fruit is then put into a syrup of about 30° Brix and cooked in open kettles until a syrup density of 60°–65° Brix is reached. Cans are then filled hot, exhausted for 6 min at 82.2°C, and sealed at once. This product is called canned preserved figs. It must also be acidified with citric acid or lemon juice to a pH less than 4.9.

FRUIT COCKTAIL

Fruit cocktail is one of the more important canned dessert fruit items. The product consists of a mixture of diced yellow clingstone peaches, pears, seedless grapes, pineapple segments, and maraschino cherry halves, canned in medium syrup. The product is popular with consumers because of its attractive color, texture, and flavor. There are advantages of steriflame processing of canned fruit cocktail.

Diced Peaches

During the canning season, yellow clingstone peaches are harvested at canning ripeness based on a yellow skin color. The fruits are halved and pitted in a twist pitter or FMC knife pitter, peeled with 2% NaOH at 103.3°C for 20–38 sec in a cup-down spray-lye peeler, washed, and diced to give 1.27-cm cubes that will pass through a screen with 1.91-cm openings but will not pass through a screen with openings 0.953 cm sq (or size sections that conform with the regulations of the country concerned). The small fragments in the diced fruit must be removed in vibrating sieving machines.

Diced Pears

'Bartlett' pears are harvested in an average pressure test of 16–17 lb when tested with the Magness-Taylor pressure tester with 5/16-in. plunger. The pears are stored at 0°C under a relative humidity of 85% for 3–4 days to reach canning ripeness. For fruit cocktail, it is desirable to keep pears on the firm-ripe side so that they will not break down during dicing and canning. The pears are peeled either mechanically in an FMC peeler or lye-peeled in an FMC pear preparation system. The dicing operation is the same as that for clingstone peaches.

Because of the presence of polyphenoloxidase enzyme in pears, it is necessary to inhibit enzymic browning by handling and processing the fruit under controlled conditions. Enzymic browning in diced pears can be inhibited by rapid handling, spraying the diced pears with 2% citric acid or with 1% ascorbic acid, and inactivation of the browning enzymes by steam blanching at 90°C or higher for 2 min, followed by rapid water spray cooling or evaporative cooling.

Pineapple Segments

Canned pineapple segments or tidbits are usually used for fruit cocktail processing. The fresh pineapple should be cut in segments about 1.27 cm by 1.37 cm long, or in symmetrical segments or cubes the equivalent by weight of segments of the dimensions specified above, in keeping with the description of the peaches and pears used in fruit cocktail.

Other Fruit

Maraschino cherries are used in approximate halves, colored red with practically fast color approved by the USDA. Thompson seedless grapes at canning ripeness are stemmed by machine, sorted to remove unfit ones, and graded for size.

Each kind of fruit must be of a good color for that variety. A tolerance of 15% by weight of the pieces of pears and of 10% by weight of the pieces of each other variety of fruit that do not conform in size or shape to the above specifications is permitted by the USDA.

The proportion of ingredients for canned fruit cocktail is presented in Table 6.5.

Liquid Media and Brix Measurement

"Cut-out" requirements for liquid media in canned fruit cocktail are not incorporated in the grades of the finished product since syrup or any other liquid medium, as such, is not a quality factor for the purposes of these grades. The "cut-out" Brix measurement for the respective designations are as follows:

Designation	Brix measurement
Extra heavy syrup or extra heavy fruit juice syrup	22°–35°
Heavy syrup or heavy fruit juice syrup	18°–22°
Light syrup	14°–18°
Light fruit juice syrup	<14°
Slightly sweetened water	Not applicable
In fruit juice	Not applicable

Filling

Container fill for canned fruit cocktail is a fill such that the total weight of drained fruit is not less than 65% of the water capacity of the container. Canned fruit cocktail that does not meet this requirement is "Below Standard in Fill."

Diced pears and peaches are mixed continuously on their way to filling machines in some plants. More commonly, they are not mixed before addition to the can. Filling of the cans is usually done by automatic machines; syrup of desired concentration is added; the cans are exhausted and then processed for 20 min at 100°C in an atmospheric rotary cooker or for 15 min at 104.4°C in a continuous pressure cooker. Vacuum syruping and steam-flow seaming can be used. Fruit cocktail in No. 2½ glass jars is processed 18 min at 104.4°C in a continuous pressure cooker with added air pressure to give a total pressure of 1.41 kg/cm^2.

Quality Factors

The grade of canned fruit cocktail (Fancy, Choice, Substandard) is ascertained by considering, in conjunction with the requirements of the respective grade, the respective ratings (total of 100 points) for the clear-

Table 6.5. Proportions of Fruit Ingredients in Canned Fruit Cocktail

Fruit ingredient	Style	Proportions	
		Not less than	Not more than
Peaches (any yellow cultivar)	Diced	30% by weight of drained fruit	50% by weight of drained fruit
Pears (any cultivar)	Diced	25% by weight of drained fruit	45% by weight of drained fruit
Grapes (any seedless cultivar)	Whole	6% by weight of drained fruit	20% by weight of drained fruit
Pineapple (any cultivar)	Diced or sectors	6% by weight of drained fruit; but not less than 2 sectors or 3 dice for each 4.5 oz avdp of product and each fraction thereof greater than 2 oz	16% by weight of drained fruit
Cherries (any light, sweet cultivar; or artificially colored red; or artificially colored red and artificially flavored)	Approximate halves	2% by weight of drained fruit; but not less than 1 approximate half for each 4.5 oz avdp of product and each fraction thereof greater than 2 oz	6% by weight of drained fruit

Source: U.S. Dept of Agriculture (1971).

ness of liquid media (20), color (20), uniformity of size (20), absence of defects (20), and character (20). For the details of grading canned fruit cocktail refer to U.S. Dept. of Agr. (1971) and Chapter 13.

FRUITS FOR SALAD

Canning of mixed fruits for salad and dessert has become a fairly important part of the fruit-canning industry.

Pears and clingstone peaches are prepared by halving and peeling or coring as described earlier. Peaches are given a preliminary steaming to soften them sufficiently for canning with other fruits. Pear and peach halves are cut in half lengthwise or in thirds. Canned sliced pineapple in No. 10 cans cut to give 16 segments/slice is used. Apricot halves canned in No. 10 cans at thc firm-ripc stagc in Choicc gradc syrup with a light processing sufficient to prevent spoiling but not sufficient to soften them unduly are used. Maraschino cherries, dyed with erythrosin dye, are prepared in bulk during the season for canning fruits for salads or are purchased in bulk from cherry processors.

The syrup from the canned pineapple and apricots is recovered, mixed with a Hy-Flo infusorial earth or other filter aid, and filtered hot. Then it is mixed with sufficient water and sugar syrup to reach the desired Brix for canning the fruits for salad. In some cases it is strained to remove coarse particles of fruit.

Cans or jars are conveyed by straight-line conveyor. The required quantity of each variety of fruit is added. The sequence consists in adding peaches first, then pears, apricots, maraschino cherries, pineapple segments, and, last, one or more pieces of peach to top the filled can and give the desired fill weight. After the syrup is added in a syruping machine, the cans are exhausted, sealed, and processed in a rotary cooker. The process time is about 10 min at 98.9°–100°C for No. 1 tall cans, and slightly longer for No. 2½ cans. Exhausting should be thorough. However, prevacuumizing and steam-flow seaming can be used. Glass jars may be used and sterilized at 103.3°C in a continuous pressure cooker. They are cooled under air pressure in order to hold the lids in place.

Fancy Grades

The Canners' League of California has established specifications for Fancy grades of canned fruits for salads as follows. The fruit should be of good color, ripe yet not mushy, of uniform size, symmetrical, and free of serious blemishes. Maraschino cherries must not be smaller than seven to the ounce. Apricots must be in halves; pears and peaches in quarters, sixths, or eighths; pineapple in sectors; and the cherries whole. Apricots constitute 18–30%, pears 21–33%, peaches 24–40%, pineapple 9–16%, and cherries 4–8% of the total drained weight.

The syrup after canning and reaching equilibrium with the fruit should be at least 24° Brix, with a permissible tolerance for any single package of 10%, i.e., it may be as low as 21.6° Brix.

Choice Grades

The specifications for Choice grades are similar to those for the Fancy except that the fruit must be at least equal to the regular Standard canned fruit in quality and the pineapple equal to or better than Standard tidbits. The syrup after canning should test 20° Brix or above with a tolerance of 10%, i.e., not below 18° Brix for any package.

GRAPES

There are three broad classes of grapes that are grown in the three principal growing areas of the United States: the northeastern euvitis or bunch grape *Vitis lubrusca;* the grape, *Vitus vinifera;* and the southeastern Muscadine grape, *Vitis rotundifolia.*

In the United States, Thompson seedless grapes are used in canned fruit cocktail. There is a direct relationship between the soluble solids content of the berries and the percentage of berries that develop internal browning. Fruit susceptible to internal browning has high levels of polyphenoloxidase and low levels of dihydroxyphenolic substrate.

The unfermented grape products industry in the United States is developed largely around the Concord grape, which is grown throughout the cooler areas of the United States and Canada. The state of Washington now leads in the production of Concord grapes in the United States. The high cost of grape harvesting can be substantially reduced by using mechanical harvesters.

Concord Grape Juice

The majority of grape juices are made from Concord grapes. Approximately 96% of the juice made in New York state is prepared from Concord grapes. They are largely processed in 0.68- or 3.78-liter bottles for the retail market. Hot pressing of the grapes is important to extract the anthocyanin pigments, flavor components, amino acids, organic acids, sugars, minerals, tannins, and other ingredients. Maturity of the fresh grapes is an important factor affecting composition and quality of the grape juice. Usually, grapes at 16° Brix or higher are considered ripe for juice processing. If the juice contains more than 0.85% acid as tartaric acid, it may taste too tart to the consumer.

Hot Extraction. Grapes are washed with a spray of water and then fed through a conveyor into the hopper of a machine where they are forcibly moved into the receiving end of the drum by a rotating device. The grapes fall into the path of the rotating blades which not only beat the berries but also force them outward against beating bars and the surface of the drum. The beating and propulsion action breaks up the grapes, but not the seeds and stems. The centrifugal action produced by the rotating beating blades causes the juice and pulp to escape through slots and the opening in the drum. The stems are discharged from the outlet end by the propelling action of the spiral blades.

The juice and pulp are pumped into steam-jacketed stainless steel kettles equipped with rotating paddles to extract the pigments and other solubles. The grapes may be preheated first to 54.4°C and held there for 10 min. Then the product is heated to 62.8°C and held there for 10 min. more. It is necessary to adjust the time of heating to extract the desired color and soluble materials. The temperature and time of heating should be controlled and recorded by automatic devices.

Pressing the Juice. After removing about half of the free-run juice, more juice can be pressed from the heated grape mass in a hydraulic press through layers of nylon cloth over the racks. About 1–2% of diatomaceous earth can be added as a filtering aid. Hydraulic pressure is applied gradually until it reaches a pressure of 17–20 kg/cm^2. The yield of juice is 75%, or about 662–700 liters per ton of grapes. This method is time- and labor-consuming, and yeast fermentation may occur if the time of operation is delayed.

Continuous Screw Press. The hydraulic press method has been gradually replaced by continuous screw presses (Garolla Press) that require destruction of the pectin in grapes with pectic enzymes. The hot pulp is pumped into large stainless steel holding tanks equipped with slowly moving paddles. Pectic enzyme preparation is added at a rate of 0.09 kg/ton of grapes. The mixture is kept at 60°C for 30 min or longer to hydrolyze the pectin. Then 4.5–9.0 kg of wood fiber is added as a bulking agent. The digested pulp is drained through a 4-mesh screen or on sloping vibrating decks equipped with trapezoidal shaped rods. The free-run juice may contain 10–30% suspended solids as determined by centrifugal tests. The remainder of the pulp is then pumped to a continuous screw press. The residue from the press may have a moisture of 40–50%.

The free-run and expressed juice are combined, and 1–2% by weight of diatomaceous filter aid is added to aid filtration through a rotary vacuum filter or a pressure leaf filter. In the latter case, pressures of 5–6 kg/cm^2 are applied by filter pumps. Continuous desludging centrifuges may be used to remove suspended solids, followed by plate-and-frame filters using diatomaceous filter aids. The yield of juice may vary from 700–738 liters/ton of grapes. The cake from the screw press may be further extracted with water in a countercurrent system; this process can increase the juice yield by 5%.

Other types of presses also can be used for recovery of grape juice. The Willmes Press consists of a perforated rotating cylinder with an inner rubber sleeve into which compressed air is applied to inflate it and press the juice cake. It is desirable to apply the pectic enzyme treatment prior to pressing.

The serpentine fruit press (Fig. 6.2) is a very useful press for the grape juice industry. This continuous-flow fruit press can be constructed in a large number of configurations and sizes. Its light weight, low power requirements, and feeding flexibility make it possible to place this machine in any part of a processing line without disrupting the existing layout. Fruit can be fed to the bottom or top of the press by reversing the perforated belts. As the material rides on the belt into the press, guides and scrapers shape it to a height of 1.91 cm and leave a 2.54-cm margin on each side. As the material moves into the throat, the upper and lower belts meet and move at the same velocity and pressing begins. The juice flows by gravity into trays all along the operation and the finished pulp is released at the end of the operation. Between stages the pressure is lessened and the pulp has a chance to relax. Then the pulleys reverse the

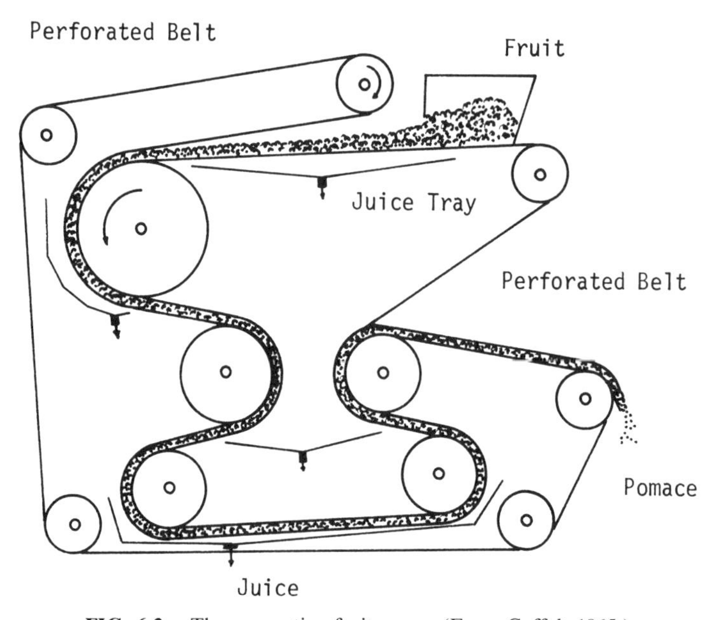

FIG. 6.2. The serpentine fruit press. (From Coffelt 1965.)

flex direction to open up new channels for the escape of liquid. The nylon belts are resilient and cup the crushed material, forming a tube while pressing, which keeps the material from leaking out the sides.

The prototype has 15.2-cm-wide belts, which move 100 meter/min, and operates the five pressing stages with a 1-hp motor. Commercial-size machines have been built with 30.5-, 61.0-, and 91.4-cm-wide perforated belts with from five to fifteen pressing stages. A press with a 91.4-cm-wide belt and nine pressing stages can handle 125 tons of crushed grapes per hour with only a 10-hp motor. The tension in the belts exerts a pressure of 0.5 kg/cm^2 on the material. Much greater pressures could be exerted through the pneumatic control system activators, but this has not been necessary for materials handled to date. The Serpentine fruit press is lower in operating cost than the basket or screw press by 25% or more.

Low-Temperature Storage. Extracted grape juice is flash-heated at 79.4°–85.8°C in a tubular, plate, or scraped-surface heat exchanger and then cooled in a second heat exchanger to 0°C. In many plants, the entire pasteurizing and cooling of grape juice is accomplished in single-unit heat exchangers. The cooled juice at 0°C is pumped into glass-lined metal tanks or concrete tanks with plastic coatings.

The juice is stored at 0°C for 1 month or longer to allow crystallization of tartrate salt (argols). The tanks usually are covered and sealed, although open tanks with ultraviolet lamps are also used to inhibit mold growth on the surface of the juice.

In an older method less used now, the grape juice is pasteurized at 77°–82°C and filled hot into steam-sterilized carboys or jugs through a tube leading to the bottom of the container. Sufficient juice is added to the container to cause the foam formed during filling to flow over the top and out of the container. A hot paraffined cork closes the opening, and melted paraffin is poured over it to obtain a tight seal. The carboys are cooled in a conveyor with a fine water spray, and then kept on racks in cool cellars.

The soluble solids content of ripe Washington-grown Concord grapes may vary from 16°–19° Brix at harvest and the total sugars from 14°–16°. The minimum soluble solids for U.S. Grade A Unsweetened Canned Grape Juice is 15° Brix, and the minimum acidity is 0.60% as tartaric acid.

The average solids content of cold-pressed grape juice is 16.36% sugars, 13.9% nonsugars, 2.43% acidity, and 0.78% as tartaric acid. After heating, the pressed juice has a solids content of 17.43% sugar, 14.03% nonsugar, and 3.40% total acid with 1.09% as tartaric acid.

Processing of Grape Juice in Bottles. The grape juice is reprocessed after the argols have formed and settled out. After removal of the clear juice, the thick juice and argols are partially filtered through a screen or through heavy cotton cloths on a frame. The juice flowing through is still muddy and thick and must either be resterilized and stored again to permit a second precipitation of argols, or filtered in a filter press using infusorial earth or other substances as a filter aid. Juice stored in large containers is more easily handled. The draw-off pipes in the sealed storage tanks are extended sufficiently above the bottom to prevent disturbance of the argol sediment. The contents of the tank are drawn off. The juice and argols remaining in the bottom of the tank are drained and treated as described for carboy juice, or are separated by means of a continuous bowl centrifuge. A continuous stream of centrifuged juice is obtained simultaneously with discharge of partially dry argols.

Improvements have been made by the industry in pasteurizing and bottling grape juice so that the deteriorative effects caused by oxygen and by prolonged heating are eliminated. The juice from which argols have been removed is conveyed to holding tanks. From there it flows by gravity through a tubular or plate that exchanger into the filler. The temperature employed is high enough to obtain a temperature in the bottled juice of 76.7°C or above. The hot juice from the heat exchanger is filled by means of an automatic filler into preheated bottles that have passed through an air cleaner and preheating hood. In the preheating hood, the bottles are subjected to steam heating.

After filling, bottles are capped under steam and are then discharged to a conveyor that carries the hot bottles to the pasteurizer. The pasteurizer serves to hold the bottles at pasteurizing temperature. The bottles are next subjected to mist sprays of cool and then cold water.

Muscadine Grape Juice

Well-ripened grapes are washed and crushed between rollers. One-third of the grapes are heated by simmering in a stainless steel steam-jacketed kettle with constant stirring. The one-third heated portion is pressed and the juice is blended with the juice obtained by cold pressing the other two-thirds of the grapes. The blended product is said to be more pleasing in color, flavor, and aroma than is the product obtained from either hot-pressed or cold-pressed juices alone. This is especially true if ascorbic acid is added to the crushed grapes immediately before pressing.

The yield of juice from Muscadine grapes is somewhat lower than that obtained from Concord. About 60–62% yields may be obtained by hot pressing and 50–55% by cold pressing. In juice cooled to 4.4°C, tartrates crystallize and settle within a few days. The cold storage and processing procedures are the same as those described for Concord grape juice.

Grape Juice Concentrate

Grape juice concentrate has assumed an important role in the grape processing industry. Concentration permits economies in storage and transportation. It also results in a more complete deposition of tartrates during low temperature storage at −5.6° to −2.2°C.

The unflashed juice from the separator may be concentrated in a forced circulation, falling-film or a single-pass evaporator having a relatively short retention time. Forced circulation evaporation may be carried out in one or several stages using temperatures ranging from 57.2° to 71.1°C and with a retention time of 1 hr. The retention time and heat exposure have been reduced to approximately 20 min by use of a falling-film evaporators at 54.4°–65.6°C under a vacuum of 68.6°–73.7°C Hg. More recently, in a single-pass evaporator of special design, the exposure time has been further reduced to approximately 2–3 min for complete concentration to 48° Brix in a multiple-stage unit using temperatures of 65.6°–71.7°C in the first effect and 46.1°C in the second effect. The same type of evaporator may be used to achieve a concentration of 72° Brix if the juice has been depectinized.

Grape juice concentrates are shipped in large tanks for remanufacturing purposes. They do not need refrigeration because the osmotic pressure of the grape sugars at 72° Brix can preserve the product provided that attention is paid to proper sanitation and aseptic filling operation.

Essence Recovery

The volatile components are removed by heating single-strength juice to 104.4°–110.0°C for a fraction of a minute in a heat exchanger, flashing a percentage of the liquid into vapor in a jacketed tube bundle, and then discharging the liquid and vapor through an orifice tangentially into a separator. The separator should be of sufficient size so that the vapor velocity is reduced to 3 m/sec or less. From 20–30% by weight of the original juice flashes off as a vapor that is led into the base of a fractionating column filled with ceramic saddles or rings. A reflux condenser on the base of the column is used to provide the necessary reflux ratio. The vent gases from the reflux condenser are then chilled in a heat exchanger and the condensate containing the essence is collected. The recovered essence can be added back to the grape concentrate.

Grape Drinks

The utilization of grape juice in grape drinks and in blends of juice has expanded considerably. Blends of apple and Concord grape juices have found a large market, and blends of apple with juices from several grape cultivars are also popular. Blends of grape with grapefruit and other juices are also very pleasing. It would be desirable to standardize the content of grape juice in grape drinks and juice blends.

GRAPEFRUITS

Grapefruit can be classified into two types: the common grapefruit and the pigmented grapefruit. The color of the pigmented fruit is due to the presence of lycopene, a carotenoid. Both seedy and seedless grapefruit cultivars exist. The most popular cultivar is 'Marsh Seedless', followed by 'Duncan', a seedy cultivar. The pigmented cultivars, such as 'Thompson' and 'Ruby,' however, are becoming more popular. Grapefruit is grown largely for the fresh fruit market except in Florida, where more than 60% of the crop is processed as single-strength juice, frozen concentrates, or canned sections.

Only tree-ripened grapefruits are used for canning. After size grading and sorting, the whole fruit is heated in boiling water for 4–7 min to loosen the pulp from the membrane covering the outer ends of the juice cells and to soften the rind. Then by quarter-scoring the rind through the pulp, the outer skin is peeled off, leaving a membrane-coated fruit. Highly developed mechanical equipment has been designed to perform almost the entire job of peeling, segmenting, and filling into cans. It is necessary to remove the membrane by passing the fruit through a boiling 2–3% sodium hydroxide (lye) solution for 15–20 sec or under a spray of boiling lye. This operation disintegrates the thin white membrane as well as the bitter element, naringin. The lye-treated fruits then enter a vigorous cold-water washer, followed by a water spray of sufficient pressure to wash off the particles of membrane and the residual alkali. If a lye-peeling facility is not available, knife peeling may be used.

The peeled segments are sorted to remove broken and defective pieces. They are then filled by hand into cans, often with the convex surface of the segment toward the can wall, in order to present a neat appearance. The fill weight for a No. 2 can is approximately 483 gm of fruit. It is preferable to weigh each can so as to assure a uniform fill weight. The usual syrup is 35°–40° Brix, depending on the sweetness desired. It may be added at 82°–90°C. Some canners add it cold, in which case the exhaust time has to be increased.

Cans are exhausted at 82.2°C, sealed hot in a double seamer (Fig. 6.3) or under steam flow, and processed at 82.2°C for 25–35 min or at 100°C for 10 min for No. 2 cans. For 404 × 707 cans, processing time is 35–50 min at 87.8°C. The can center temperature must reach 76.7°C. After leaving the process bath, cans are passed to a cooling bath by means of a conveyor that is adjusted so cans will enter the cooling bath after a lapse of 6 min. Cans remain in the cooling water bath until the average temperature has dropped to 40.6°C. Thorough exhausting to remove air and the use of Type L tinplate is desirable. Canned grapefruit should be stored in a cool place, preferably at 10°–15.6°C in order to retain the quality of the product.

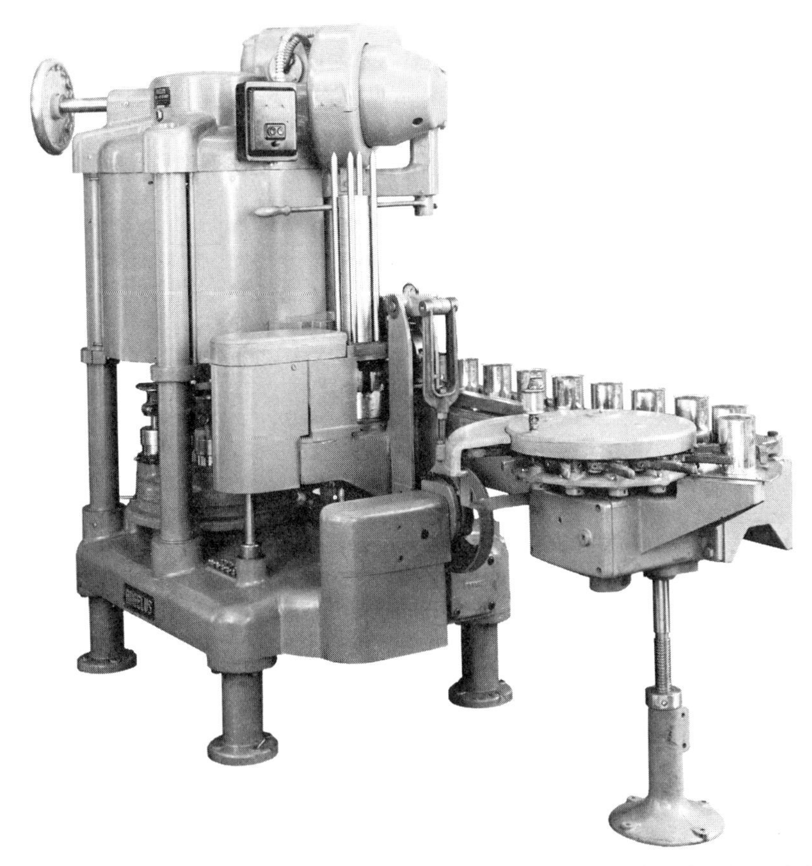

FIG. 6.3. Angelus model 40P MSLF double seamer with capacity of 275 cans/min. (Courtesy Angelus Sanitary Can Machine Co.)

Grapefruit By-Products

Citrus fruit processors have found it economically feasible to produce dry citrus peel, citrus molasses, and D-limonene (Lopez 1981). The peels and pulp residues are collected and ground in a hammermill. Sufficient amount of lime is mixed with the ground mixture to neutralize the acids present in the peels and pulps. The product is pressed to remove excess moisture. The pressed peel is conveyed to a direct-fired hot-air drier. The liquid from the press is screened to remove large solids, which are recycled back to the press. The press liquor is concentrated to molasses in an evaporator. The exhaust gases from the peel drier supply the heat energy for the concentration of the press liquor into molasses. D-limonene is recovered from the exhaust gases through condensation as they are released from the peel drier.

Grapefruit Juice

The canning of grapefruit juice is a well-established industry. Most of the commercial product is packed in citrus-enameled tin cans. Research is under way to use composite cans (Mylar–paper–aluminum–polypropylene) for aseptic canning of grapefruit juice. Results are very encouraging because of the improved quality of the canned juice.

Ordinarily, differential electrolytic tinplate lined with citrus enamel should be used in order to minimize development of a tinny flavor, although most grapefruit juices are packed in plain tin cans. In the early days of the industry considerable loss was encountered from hydrogen swelling by reaction of the acids of the juice with the tinplate. With improvement in tinplate quality and in processing and storage of the juice, such losses are now relatively slight.

Grapefruit juice can be canned in its natural state or slightly sweetened with sucrose or high-fructose corn syrup, depending on the composition of the grapefruit used.

In the recommended procedure for canning grapefruit juice, well-ripened grapefruits are washed and cut in half; then the juice is extracted by reaming, using a light squeeze and finishing process to avoid expressing essential oil from the rinds. The juice is strained through a finisher (Fig. 6.4), deaerated under a high vacuum (680–736 mm Hg) or in a continuous vacuumizer, flash-pasteurized at 190°–196°C for about 1–2 min, and cooled to 80°C. The hot juice is filled into citrus-enameled cans. The cans are sealed hot, inverted for 3–4 min to sterilize the lids. The canned juice is cooled thoroughly under sprays of chlorinated water to 38°C.

Soluble solids, % acid, Brix–acid ratio, and the levels of limonin and naringin are important factors influencing the acceptance of grapefruit juice. According to Nagy and Shaw (1980), the average soluble solids of mature California grapefruit was about 12° Brix, acidity about 2%, and Brix–acid ratio about 6. The average soluble solids of Texas grapefruit was 11° Brix and that of Florida 7°–10°. The acidity of Texas grapefruit was considerably lower than that from the other states. Texas grapefruit juice has a higher Brix–acid ratio and a sweeter taste. Both the FMC In-Line and AM Rotary juice extractors yield juice of a lower quality with higher bitterness when a harder squeeze is used.

Excessive bitterness results when immature grapefruit are processed and when excessive yields (45–48%) are obtained through hard squeezing and/or hard finishing. As a result of these findings, regulations were enacted in 1976 by the Florida Citrus Commission limiting the amount of limonin and naringen allowed in processed grapefruit juice during the early weeks of the season. The lemonin values in the grapefruit juice

FIG. 6.4. Brown model 202 pulper-finisher. (Courtesy Brown International Corp., Covina, CA.)

averaged 4–5 ppm. The early-season juice (Aug. to Sep.) contained 8–10 ppm limonin and late-season juice (Dec. to Jan.) only 2–3 pm. The average naringin value of the juice was 350–421 ppm.

The undesirable effect of prolonged storage on canned grapefruit juice at elevated temperature (above 21.1°C) can be measured by the increase in furfural content. Storage conditions that produce furfural levels approaching 1000 μg/liter cause flavor changes that can be detected by a taste panel.

To improve the acceptance of canned grapefruit juice, juice with low Brix–acid ratio should be blended with high-ratio juice to achieve a balance between tartness and sweetness. The ratio of the blend should be as high as practicable. Early-season or late-bloom fruit should be squeezed as gently as possible to minimize the extraction of naringin and limonin. Canned single-strength grapefruit juice should be stored at 21°C or lower.

Bitter Flavonoids in Grapefruit Juice. The bitterness in grapefruit is associated with the flavanone glycoside naringin and the triterpenoid lactone limonin. Two isomeric glycosides of naringenin have been isolated, both containing one mole each of the sugars rhamnose and glucose. One of these, naringin, is intensely bitter; the other, naringenin rutinoside, is not. It is the glycosidic linkage of the two sugars comprising the disaccharide unit that largely determines the bitterness. The neohesperidosyl group found in naringin consists of rhamnose and glucose linked C_1 to C_2, whereas in the rutinoside the linkage is the more usual C_1 to C_6. The neohesperidosyl group is associated with, although not essential for, bitterness. Thus, prunin (7-β-D-glucoside), although lacking the disaccharide linkage, still exhibits some bitterness.

Naringin—the main bitter component of grapefruit—is present in the albedo layer and in the carpellary membranes. The method of extraction of the juice or preparation of segments influences the amount of bitterness entering the product. If the fruit is hand-peeled, much of the naringin is removed with the albedo layer and outer membranes, but this method gives a lower yield of segments than if the segments are lye-peeled. Here the fruit is calded to soften the skin, and then the albedo layer is removed by an alkali dip and subsequent water spray to remove the loosened tissues. The efficiency of this process determines the bitterness of the product.

The level of naringin is highest in immature fruit. In one study, fruit sampled in November had only 38% of the level of bitter flavanone glycosides found in fruit harvested in July. However, it was shown that part of this decrease in bitter flavanones can be related to the increase in physical size of the fruit over this period. A solution to the problem of bitterness in canned grapefruit is the selection of mature fruit for processing.

Commercially canned Florida grapefruit juices contain more naringin than those from Texas. The juice from pink- and red-fleshed cultivars processed in Texas contains 0.026% naringin, while those from white cultivars contain 0.033–0.039% maringin. Under normal processing conditions, grapefruit processing residue (consisting principally of peel, membrane, and seeds) generally contains about 0.75% naringin (Nagy and Shaw 1980). The amount depends largely upon the maturity of the fruit, since naringin is more abundant in immature fruit.

The naringin content of the whole fruit ranges from 3.2% for fruit 5.9 cm in diameter to 0.4% in mature fruit. About 62% of the naringin is in the peel, 36% in the pulp and rag, and 2.4% in the juice. This illustrates how improper juice extraction and finishing can easily extract additional naringin from the pulp, rag, or even peel and increase bitterness.

Naringin is the bitter 7-β-neohesperidoside of naringenin; its tasteless isomer, the 7-β-rutinoside of naringenin is also present in grapefruit juice at one-third to one-half the level of naringin. Generally, flavanone glycosides with the neohesperidose linkage are bitter and those with the rutinose linkage are tasteless.

The Davis (1947) test has been widely used as a quality control index in grapefruit juice since it is proportional to naringin content. It measures total glycosides, including naringin and its tasteless isomer.

Naringin was formerly thought to occur only in grapefruit and the closely related shaddock, but it is now known to exist in several other plants including bitter orange and other citrus species. Naringin can occur in processed products as a fine crystalline suspension, which can impart a cloudy appearance to the syrup, or it can give rise to larger crystals forming "glassy" lumps in the fruit segments. It can be recovered as a by-product.

In order to obtain grapefruit juice of satisfactory appearance, the pulp should be blended and added to the juice to give 22–26% suspended solids. The use of the enzyme naringinase at 4°C partially hydrolyzes the naringin to give the less bitter prunin or, at 50°C, effects complete hydrolysis to the nonbitter aglycone, naringenin.

Other bitter flavanone glycosides containing the characteristic neohesperidosyl group have been isolated from citrus fruits. These include poncirin found in *Poncirus trifoliata* and neohesperidin isolated from Seville oranges. Recently, polyvinylpyrrolidone resins have been proposed to remove some of the undesirable flavonoids from grapefruit juice.

GUAVA

Guava is one of the more important pomiferous fruits of the myrtle family. It is used largely preserved as puree, juice, nectar, and beverages. The processing methods for guava products have been reviewed in Nelson and Tressler (1980).

LYCHEE

The lychee is a delicious fruit thought to have originated in southern China. The excellent flavor and aroma of this fruit is largely retained after canning.

Firm, ripe lychees are washed, peeled, and pitted by hand. To each No. 2 enamel can, 468 gm of fruit and 156 gm of 40° Brix sucrose syrup are added; 0.1–0.2% citric acid may be added to the syrup to improve the flavor. Cans are exhausted for several minutes to a can center temperature of 71.1°C prior to double seaming, or cans are sealed under vacuum at 38.1 cm Hg. The sealed cans are heat-processed in a high-speed spin cooker to 90.6°C for 2–3 min or in a boiling water bath for 12 min, followed by rapid water cooling. Excellent results are achieved when the fruit is processed in a spin cooker.

Excessive heat processing of lychees may cause phenols, leucoanthocyanins, and flavanols to leach into the syrup, causing pink discoloration in the canned product. It is believed that the level of leucoanthocyanins in lychees may be related to the pink discoloration problem. Excessive heat processing and delayed cooling of the canned product should be avoided.

ORANGES

The citrus fruits of commerce may be classified into four horticultural groups: oranges and mandarins; hybrid fruits; grapefruits and pummelo; and acid fruits such as citron, lemon, and lime. Citrus fruits also may be classified into 10 botanical groups: *Citrus sinensis* (sweet orange); *C. aurantium* (bitter or sour orange); *C. reticulata* (Mandarin orange); *C. paradisi* (grapefruit); *C. grandis* (pummelo or shaddock); *C. limon* (lemon); *C. aurantifolia* (lime); *C. medica* (citron); *Fortunella* sp. (kumquat); *Poncirus trifoliata* (trifoliate orange).

The USDA has reported that the combined orange and tangerine production in 22 selected countries was 36.2 MMT (million metric tons) for the 1980/1981 season. The fruit processed was 16.6 MMT, or 45.8%, of the orange and tangerine crops.

Single-Strength Orange Juice

Oranges intended for canning as juice should be of a quality comparable with that suitable for direct consumption. They should have a suitable balance of acidity and sugar, with fully developed aroma and flavor.

The major orange-producing areas have regulations concerning the maturity of the fruit that may be harvested. In order to ensure optimum quality, further selection is made. Most fruit used is well above the minimum values listed in the regulations. Attention is given to the blending of different lots to achieve a balance in solids, acidity, color, and flavor.

In California, the principal orange cultivar used for canning as juice is 'Valencia'. The 'Washington Navel' orange is used for certain types of beverage bases where the tendency for bitterness is not objectionable. In Florida, the principal cultivars oranges used for processing are 'Pineapple' (midseason) and 'Valencia' (late season). 'Parson Brown' and 'Hamlin' (early season cultivars) are also used but are considered less desirable because of somewhat lower soluble solids content, color, and flavor.

Harvesting

Sweet oranges for processing are harvested in the same way as are fruit for shipment as fresh fruit. A ladder is laid against the tree and the picker climbs the ladder carrying a canvas bag that holds 22–27 kg of fruit. Fruits are clipped and placed in the bag. The worker then descends and dumps the contents in a bin or other container. A number of mechanical tree shakers are in various stages of development and a number of abscission chemicals are being tried to loosen the fruit. Some devices have been developed for raking the fruit to the center of the aisle between tree rows and loading it into trucks (see Chapter 2).

Processing. Upon reaching the processing plant, the fruit goes through inspection lines for removal of bruised or broken fruit. The sorted fruits are conveyed to storage bins until enough fruit accumulates for continuous operation of the cannery. The fruit is never piled to a depth of more than 1.3 m.

As the fruit is conveyed to the bins, automatic devices divert a small portion to a laboratory where the titratable acidity, Brix, and juice yield are determined. These values are used to determine which bins are to be blended.

From the bins the fruit is conveyed to the washer, where it is soaked briefly in water containing a detergent, scrubbed by revolving brushes, rinsed with clean water, and inspected again to remove damaged fruit. The fruit is then separated into sizes automatically and conveyed to juice extractors. After finishing, the juice flows to large stainless steel tanks where it is checked for acidity and soluble solids. Sugar is added, if needed.

Control of Volatile Peel Oil Content. The standards for U.S. Grade A orange juice permit not more than 0.035% of peel oil by volume. Normally, it varies between 0.015 and 0.025%. One may control the peel oil content by adjusting the extractor or by softening the peel by immersing the fruit in hot water for 1–2 min. Excess peel oil in the juice can be removed by heating the juice at 51.7°–57.2°C in a vacuum evaporator until about 4–6% of the juice is evaporated. The vapors are condensed, the oil separated by centrifugation, and the water layer returned to the juice. This treatment can remove three-fourths of the peel oil present in the juice. The level of peel oil can be determined by distilling the juice with 2-propanol and oxidation of the recovered d-limonene with a standard potassium bromide–potassium bromate solution under acid conditions.

Pasteurization. Pasteurization of citrus juices is done to destroy microorganisms that would otherwise cause fermentation in the can and to inactivate enzymes that would otherwise cause cloud and other changes in the juice. Generally, higher temperatures are needed for enzyme inactivation than for destruction of microorganisms.

The juices are heated in tubular or plate-type heat exchangers for 30–60 sec, and then piped hot directly to the filling machine.

Filling and Storage. Juice is maintained at 85°C in the filler bowl and filled directly into cans. The juice is in the filler bowl from 1 to 2 min in most cases. The cans are closed in automatic steam injection sealers, inverted for about 20 sec, and rapidly cooled to 37.8°C by spraying with cold water while spinning in a conveyor. Plain tin cans are used for single-strength orange juice.

The temperature at which canned orange juice is held is a major factor influencing the flavor and vitamin content of the juice when it is consumed. At 21.1°C there is slight flavor change and approximately 85–90% of the ascorbic acid is retained for 1 yr. At higher temperature deterioration in flavor and loss of ascorbic acid progress more rapidly. At 0°–4.5°C the quality and ascorbic acid content of canned orange juice change very little during storage for 1–2 years.

Spectral Characteristics of Orange Juice. The visible and ultraviolet absorption curves of alcoholic solutions of orange juices from 'Hamlin', 'Pineapple', and 'Valencia' oranges are similar, except for their intensities. Absorption maxima for carotenoids were observed at 465, 443, and 425 nm, and for phenolic and flavonoid compounds at 325, 280, and 245 nm. A shift in wavelength or unusual change in intensity may be indicative of some type of additive. Maturity influenced the spectra of 'Hamlin' and 'Pineapple' orange juices, causing a general increase in the visible and ultraviolet absorption with increasing maturity. Extractor pressure also affected the spectra, the hard-squeeze juices having a higher ultraviolet absorption than the light squeezes.

Absorption characteristics such as spectral shapes, absorption wavelengths, intensities, and the ratios (443/325; 280/325 nm) may be helpful in analyzing and determining the quality of juice products. Spectral characteristics may be used to determine the orange juice content in citrus products.

Bitter Fractions in Orange Juice. Limonin, the main bitter fractions of 'Navel' oranges, is a triterpenoid derivative. It is a dicarbocyclic compound with two lactone rings, a cyclic ether ring, an epoxide group, a furan ring, and a ketone group. A nonbitter monolactone of limonic acid also has been isolated from citrus fruits. This compound is rapidly converted to the bitter-tasting dilactone limonin at pH 3.0, but slowly at pH 5.6.

Limonin is an intensely bitter limonoid of importance to citrus juice quality. The metabolically active form of limonin is the nonbitter limonoate A-ring lactone (LARL). LARL is synthesized in the leaves and transported to the fruit; it is gradually degraded in the fruit to nonbitter products by at least two pathways. LALR undergoes acid-catalyzed lactonization to limonin when fruit tissues are disrupted in juice preparation (Brewster *et al.* 1976).

Casas and Rodrigo (1981) studied changes in the limonin monolactone content during development of 'Washington Navel' oranges in Spain. From the start of development of the fruits, limonin monolactone accumulated to a maximum of 47 mg/fruit—around 130 days after anthesis. Throughout the development of the fruit, the total limonin monolactone content in the endocarp was greater than in the peel. The limonin monolactone content in the leaves decreased gradually from the start of the experiment until 130 days after anthesis.

Two theories are proposed to account for the delayed bitterness of citrus products. One hypothesis ascribes the delay to a physical phenomenon dependent on the solubility of the bitter fraction; the second hypothesis is that a chemical change occurs in which a nonbitter precursor is converted into a bitter compound by the process of extraction of the juice. This latter type of reaction is now accepted as the explanation of the delayed bitterness; the pH has an important influence on the reaction. In the growing fruit the bitter precursor is separated from the acids of the sap and no reaction occurs, but when juice is prepared this precursor comes into contact with the acidic juice and the reaction proceeds, liberating the bitter compound.

There are several methods by which the bitterness in citrus products can be reduced. During the preparation of the product, excessive extraction of bitter fractions from the rag and pulp should be avoided; also raising the pH to about 4 can prevent formation of limonin dilactone. The use of pectic enzymes to reduce bitterness has also been suggested. The dispersed colloids are coagulated by the enzymes, and in the subsequent precipitation they carry the bitter principles with them.

Polyamide resins have been used to remove limonin from 'Navel' orange juice. The bitterness was reduced by stirring juice with dry polyamide powders and subsequent centrifugation. Although the method is convenient for the removal of bitterness, a loss of up to 25% of ascorbic acid may also occur.

Mandarin Oranges

Mandarin oranges are canned in Taiwan and Japan on a commercial scale. The raw material of the canned product is a very popular fruit known as Mikan or Tunkan. In Japan 82.5% of the Mandarin oranges grown are used for the fresh market, 16% for processing, and 1.5% for export as fresh fruit.

Processing of Canned Mandarin Orange Segments in Syrup. The fruit is graded according to size ranging from 5.08–5.26 cm in diameter), blanched in hot water at 80°C for about 1 min, peeled by hand with bamboo knives, and dried slightly in the wind before being separated into segments. This segmenting is also done by hand, women being employed for this work as well as for the peeling operation. Then, acid and lye segment peeling is done automatically to dissolve the segments' outer skin. The divided segments are soaked in dilute acid solution (0.5% HCl) for 40 min at 30°C and then in dilute lye solution (0.15% NaOH) for about 20 min at 25°C. Peeled segments are then soaked in cold running water about 60 min. Segments are graded according to size by automatic grading machines and sorted for quality by hand while they are being conveyed by a wire net conveyer. Most mandarin oranges are placed in No. 5 cans, which are automatically filled with syrup by syruping machines. After sealing in a vacuum-sealing machine, the cans are heat-processed in a continuous-agitating cooker and cooler. The heating time for No. 5 cans at 82.2°C is about 13 min.

Prevention of Turbidity in the Product Syrup. Hesperdin contained in raw Mandarin oranges is found in the syrup of products in the form of needle-shaped crystals, which produce white turbidity and lower quality grades. Addition of gelatin, sodium carboxymethyl cellulose (CMC), or methyl cellulose (METHO-CEL) to syrup has proved to be very effective in preventing turbid syrup. Methyl cellulose is most effective and is added at a level of about 10 ppm. Segment peeling with a more dilute lye solution (0.1% NaOH) also is helpful in preventing turbidity.

Tangerine Juice

Because tangerine oranges are quite fragile and cannot be hauled at the usual depth in trucks or in bins, they are handled in boxes in trucks. Some orange conveyers and washers may crush tangerines and need remodeling in order to efficiently handle them. Regular citrus juice extractors, especially those fitted to handle small fruit, can be used. Special juice extracts (e.g., FMC Model 191 with a 60-mm cup size and AMC Model 1700) are often required to accommodate smaller tangerines (Nordby and Nagy 1980). Processed tangerine juice is highly susceptible to off-flavor development during storage. The extent of off-flavor development appears to be closely related to the amount of peel extractives and suspended materials that are processed into the juice. After finishing, the juice is centrifuged to remove excess pulp. The capacity of a given plant handling tangerines is only half that when handling oranges.

The procedures for canning are the same as those for canning of orange juice. The 'Dancy' tangerine is the predominant cultivar. More tangerines are used in the manufacture of frozen concentrate than single-strength juice because canned tangerine juice has a very limited shelf life, whereas frozen concentrate is quite stable. Tangerines are grown primarily for fresh market use and only the surplus is available for processing.

PEACHES

Peaches are usually classified either as clingstone or freestone types. In freestones, the fruit can be easily separated from the stone or pit; in clingstones, the flesh adheres tightly to the pit. Both types have yellow-fleshed and white-fleshed cultivars. Yellow-fleshed cultivars are most common and are preferred for both processing and fresh market use. Clingstone peaches are firmer in texture than freestones and hold their shapes well after canning. For this reason, processors prefer clingstones for canning, and only 17–25% of the U.S. peach pack are freestones. The share of the total peach crop used for canning is 55%, and that for fresh consumption is 39.0% (see Chapter 16).

California produces nearly all the U.S. clingstone peaches plus a substantial volume of freestone peaches. Most peaches produced in the East are freestone. However, Georgia and several eastern states have shown increasing interest in producing clingstone peaches for processing.

Total U.S. peach production for fresh use and processing averages 1.45 billion kg a year. California, South Carolina, and Georgia account for 60%, 10%, and 6% respectively, of total production. Other peach-producing states are New Jersey, Pennsylvania, Michigan, and Washington. Annual peach packs varied between 22 and 35 million standard cases (24 No. 2½ cans) during the years 1972–1981.

Cultivars

California, with its fertile soil, adequate irrigation water, and favorable climate, is ideally suited for growing canning peaches. This is particularly true in the great interior Sacramento and San Joaquin valleys where the climate is modified by coastal influences. Growers can expect regular crop production if they choose the proper location, select suitable cultivar, and manage their orchards properly.

Qualities that make a clingstone peach cultivar acceptable to growers and processors are good shape and symmetrical size; small pit with little or no red flesh color; clear, yellow-colored flesh that will withstand hauling, refrigeration, and handling; pleasing taste, aroma, and texture that will be retained and enhanced in the canned product; and regular production of good-sized fruit throughout the life of the tree. The important California clingstone cultivars include the following: extra early—'Fortuna', 'Carson', 'Loadel', 'Vivian', and 'Dixon'; early—'Cortez', 'Jungerman', 'Andross', 'Paloro', 'Johnson', 'Klamt', 'Peak', 'Andora'; late—'Gaume', 'Carolyn', 'Everts', 'Halford', and 'Stanford'; extra late—'Starn', 'Wiser', 'McKune', 'Sullivan No. 4', 'Stuart', 'Gomes', and 'Corona'.

Several freestone peach cultivars have been used for canning. In California, the 'Elberta' and 'Fay Elberta' are used commercially for canning. Other cultivars having good color and flavor after canning are 'Redhaven', 'Summergold', 'B-53624', 'B-53625', 'Redglobe', 'Southland', 'Keystone', 'FV-7-1226', and 'Loring'. Because of lower case yields and difficulties in canning, freestone peaches are less popular than clingstones in the canning industry. (Li *et al.* 1972.)

Peach Halves and Slices

The commercial canning of clingstone peaches in California is regulated by the provisions of the California Agricultural Marketing Act, which is administered by the State Director of Agriculture in cooperation with the Cling Peach Advisory Board. If needed, the size of the pack in a given year may be limited to a certain specified number of cases, each canner being assigned a quota based on his past pack. The fruit for passing the grade must be 6.03 cm or more in diameter, free of blemishes, of proper maturity, and of good canning quality in other respects.

If peaches need to be held several days or longer for canning, they should be placed under refrigeration at 0°–2.2°C. Peaches should be held no longer than 2 weeks to avoid development of off-flavor.

Receiving, Cutting, and Pitting. When delivered, the peaches may be graded according to maturity and size before pitting, since each machine is usually adjusted for peaches of approximately one size. The canning process for clingstone peaches starts with halving and pitting. All cling peaches nowadays are pitted and halved by machine because of the high labor costs associated with hand cutting and pitting. The fruit in 455-kg bins is dumped into flumes and then onto a sorting belt that delivers the fruit to the peeling machine. Several types of machines are used for pitting and peeling.

The FMC peach aligning and pitting system can feed, align, and pit clingstone peaches at rates up to 100 peaches/min (Fig. 6.5). The system consists of a horizontal aligner and a rotary pitter. The aligner incorpo-

FIG. 6.5. Clingstone peach aligning-pitting system. (Courtesy FMC Corp.)

rates 27 plastic carriages on an oval-shaped conveyor that delivers peaches to the pitter. Each carriage contains a finding device that rotates the fruit until it assumes a stem-down position and a positioning device that turns the fruit until the suture is aligned with the center line of the machine. An automatic detection and rejection device, located immediately in front of the pitter, rejects any improperly positioned or misaligned fruit. The rejected fruit is then recycled in the system.

Size-graded clingstone peaches are fed into the aligner section from the conveyor. Each fruit is fed in sequence and deposited onto a carriage. The pitter employs a series of rubber transfer cups positioned on a turret. The cups are indexed with each carriage, enabling gentle transfer of fruit from the carriers into the pitter. Upon being transferred, the fruit is carried to, and impaled by, two stationary blades, cutting the flesh of the peach around the pit's longitudinal axis. The peach is then engaged by a second turret with cups to carry it through a circular saw, cutting the pit in half. The two peach halves are then transferred to a pitting position in which the divided pit halves are cut from the fruit. The fruit and pits are subsequently discharged to a shaker-type separator.

The Filper torque pitter (Fig. 6.6) has been used for cutting and pitting

FIG. 6.6. Filper torque pitter for peaches. (Courtesy Filper Corp.)

clingstone peaches for canning. Automatic alignment systems are available to feed and align the peaches in the correct position prior to the cutting and pitting operation. The peach is first cut around the suture by the machine. Then the two halves in the tight grasp of the two cups are twisted in opposite directions to free them from the whole pits. The pits are separated from the peach halves by falling through a screen. The pitted peach halves pass through a mechanical shaking device that arranges them in single layers in the cup-down position. They are then conveyed on a moving stainless steel conveyor into the spray-type hot lye peeler.

Lye Peeling. Clingstone peaches are usually peeled cup down by a 1.5–2.5% hot lye spray for 15–60 sec. In the Dunkley peeler the hot lye is applied as a spray to the peach halves in the cup down position as they are carried through the peeling compartment on a perforated stainless steel conveyor. The lye is circulated by pump and is heated by steam in a heat exchanger.

The concentration of the lye is maintained by additions of 50% lye solution, controlled by a Honeywell electronic controller based on conductivity of the solution. Such a controller is necessary because the peeling process reduces the lye strength rapidly. If an automatic controller is not available, the operator can judge when lye addition is needed by observing whether the fruit is peeled satisfactorily or not. In many plants, the lye is titrated frequently with standard sulfuric acid or hydrochloric acid, using a 1% solution of phenolphthalein in 95% alcohol as indicator. A 50% solution of the lye (NaOH) is stored in a large metal tank from which it is pumped as needed.

Lye-treated peach halves are passed through a warm-atmosphere holding section for 30–60 sec to allow the lye to react with the peach skin; next the lye and skin are washed off with high-pressure sprays of cold water. Little difference is found in the flavor and composition of canned peaches peeled by the cup-down lye-peeling and the steam-peeling methods. (See Chapter 3 for additional discussion about peeling methods.)

Freestone peaches harvested at 5.9–7.7 kg (M2 maturity) pressure test ripen successfully at 20°C and 80% RH within 6–8 days; those harvested at 2.7–5.5 kg (M3) need 4 days for ripening.

With all pitters and peelers, large peaches have subsequently lower losses than smaller fruit. For example, in one study the average loss with the rotary knife pitter was 11.64% for smaller peaches (6.03–6.35 cm) and 9.90% for larger ones (7.30–7.62 cm). Likewise, loss with the torque pitter was 8.6% for smaller fruits and 7.76% for larger ones; loss with the cup-down lye peeler was 5.89% for smaller peaches and 5.15% for larger ones; and that with the immersion peeler was 13.48% for peaches of 6.35–6.99 cm, and 9.45% for peaches of 7.30–7.62 cm.

The amount of lye required to peel 1 ton of peaches may vary from 2.7 to 4.1 kg.

Dry-Caustic Peeling. Many canneries once located on the outskirts of cities are now surrounded by city growth. These plants usually have to dispose of their effluents through municipal sewage treatment plants, paying a charge for the volume of effluent and the biological oxygen demand (BOD). In many cases plants have to install systems to continually correct the pH of their effluent to avoid damaging sewage equipment or upsetting operation of sewage disposal systems. In addition, disposal of salt or caustic solution into the sewage contributes to water pollution problems.

Hart *et al.* (1970) developed a dry-caustic peeler for continuous removal of alkaline peach peel as a solid rather than the dilute slurry of common industrial practice. In this system, the peaches are first washed, halved, and pitted. The peach halves are placed face down on a stainless steel mesh belt and 3% boiling lye solution is poured over them for a specified time between 5 and 10 sec. After a short draining period, the peach halves are placed in a disk peeler. The peeler consists of rows of soft rubber disks rotated in such a manner that peel material is gently wiped from the surface of fruit and flung into collectors as a wet solid. Any remaining residue consists of peel material already wiped loose from the surface but not flung off by the disks, and is easily removed by a brief water rinse. Cling peaches have been successfully peeled by this method with peeling losses comparable to those normally obtained commercially by the cup-down hot lye spray method. A scrubber for peel removed from soft fruit is shown in Fig. 6.7. It is likely that this method will be applied more widely to minimize the water pollution problem facing the canning industry.

Steam Peeling. Freestone peaches, especially when thoroughly ripe, can be peeled with live steam at 100°C for 1–2 min and then chilled with sprays of cold water. The peels can then be slipped off the fruit easily with the fingers. For commercial operation, the lye-peeling method is more economical.

Washing. It is necessary to use water sprays to wash off the residue skins and excess lye, and to cool the fruit. The spray water should have sufficient pressure to remove lye-softened tissue from the outer surface. When peaches are delayed in heat processing after lye peeling, the enzyme polyphenoloxidase catalyzes the oxidation of natural phenolic compounds to form quinones, which subsequently polymerize to form brown pigments. The browning problem may be controlled by thorough washing of the peeled peach halves with a water spray, followed by a bath containing 1% citric or malic acids to inhibit polyphenoloxidase activity. The acidity of the bath should be carefully controlled so that the pH of the peach surface will always be below 4.0.

The phenolic compounds in cling peaches have been identified as chlorogenic acids, leucoanthocyanidins, catechin, epicatechin, isoflavone, *p*-coumaryl-quinic acids, and caffeic acid (Luh *et al.* 1967). Among these, chlorogenic acids, leucoanthocyanidins, and catechin are thought to be involved in enzymic browning.

The principal polyphenols separated from extracts of mature 'Elberta peaches' are, in order of prominence on paper chromatograms, leucoanthocyanins, chlorogenic acids, catechin, and flavonols. After enzymatic oxidation of blended peach tissue by endogenous phenolase, a major portion of the total phenols and leucoanthocyanins are no longer detachable, and presumably have been enzymatically oxidized.

Sorting and Grading. Peeled and washed peaches pass from the lye peeler on a slowly moving belt from which sorters remove blemished, broken, and partly peeled pieces. The prime fruit travels to a size-grading

FIG. 6.7. Magnuscrubber unit for removal of peels from peaches after lye treatment. Peaches emerge from discharge and with sharp corners intact. (Courtesy Magnuson Engineering Corp.)

machine, and the Pie Grade fruit is sent by separate conveyor to the pie-fruit department without size grading.

Peach halves are graded into five sizes in size graders. The smallest peach halves usually go to the Pie Grade line and the largest to the slicers.

The most nearly perfect fruit of large size is Fancy Grade; next are Choice, Standard, and Second, in that order. Fruit below Second in quality is canned as Pie Grade. The Pie Grade comprises blemished fruit, trimmed, overripe fruit, and green fruit. Second and Pie Grades are not graded for size.

Slicing. Sliced peaches are usually of Choice and Standard grades. Considerable Pie Grade peaches are also sliced in order to make them more uniform in appearance.

A peach-slicing machine consists of several circular revolving knives against which the halved, peeled peaches are carried by a rubber belt. A vibrating device automatically places the halves cup-side down before slicing.

Filling and Weighing. After size grading, sorting, and inspection, the halved or sliced peaches are conveyed to canning stations. The empty cans are carried to the canning stations by gravity conveyors. The cans are filled by semiautomatic hand-pack fillers, which greatly reduce the labor cost. The filler consists of a circular revolving stainless steel table or traveling stainless steel belt, in both of which are circular holes of the diameter of the cans to be filled. Workers scoop the fruit into the cans through the holes in the table or belt. The filled cans are delivered to a conveyor that carries them to the syruping machine. Most of the cling peaches are processed in No. 2½, 2, 303, and 1 Tall cans, with the exception of the Pie Grade, which is canned in No. 10 cans for use by bakeries, hotels, and other similar establishments. A limited quantity of the higher grades is also canned in No. 10 cans for the hotel trade. Sliced peaches are frequently canned in 303, No. 1 Tall, and 8-oz cans.

The weight of peaches placed in a No. 2½ can may vary from 511 to 596 g. Government regulations require that a No. 2½ can of peaches contain at least 524 g oz of drained fruit. Immediately after canning, the drained weight falls below the fill weight because peach fluid is lost to the cover syrup. Drained weight increases rapidly during the first week after canning and then slowly until a maximum is reached in 90 days as a result of penetration of syrup into the fruit.

Great care must be employed in filling the cans in order that they contain the required amount of fruit. Some factories fill the cans according to the number of pieces per can rather than by weight. If a semiautomatic filling machine is used, filling is by volume rather than by weight.

Syruping. Cans are automatically filled with syrup by a vacuum syruping machine. The peaches may receive a syrup of 40°, 25°, 10° Balling or water. Some canneries use syrups richer in sugar than these, but seldom use syrups more dilute. Peaches low in sugar should be given syrups of higher Brix degree than recommended by the Canners' League Standards. Fancy peaches should test at least 26° Brix on a cut-out test, Choice at least 21°, Standard at least 17°, and Second at least 11° Brix. The Standard of Identity for canned peaches specifies the Brix readings of the cut-out syrup that correspond to label names such as "heavy syrup" or "light syrup." Most of the canned Choice grade peaches have a cut-out Brix reading of 21°–22°. According to surveys, consumers prefer both the flavor and texture of canned clingstone peaches having a cut-out Brix of 22°–23°.

Sucrose sugar is bought in the form of a syrup of 67° Brix (liquid sugar), and corn sugar is delivered in a similar manner. Often a blend of the two—containing not more than 25% of corn syrup solids—is used in canning. High-fructose corn syrup is a new sweetner used in fruit canning.

Exhausting. Filled cans may be conveyed from the syrupers through a steam-filled exhaust box for 5–6 min at 93°–96°C. The exhaust box has been replaced to a large extent by prevacuumizing the can contents and sealing the cans in steam-flow double seamer which can save syrup loss, floor space, and steam. Cans sealed cold require about 5 min more processing than do cans exhausted in steam before sealing.

Marking and Sealing Cans. Coding of the cans is used to designate cultivar, grade, and date and hour of packing. Coding consists in embossing on each lid the letters and numbers of the code by means of metal dies as the lids pass to the double seamer. One may also allow the cans to roll through a device that marks the cans with the number of bands corresponding to the grade, the bands being in indelible ink. The label covers the bands completely, so that they are not visible to the purchaser.

From the vacuum syruper, the filled and syruped cans go to the double-seaming machine, which automatically places lids on the cans and seals them. Modern seaming machines can seal 275 cans or more per minute.

Double seaming is very important to the keeping quality of canned peaches. Can samples taken from the line after double seaming are cut and critically examined by experienced operators every 30–60 min so that faulty seaming can be detected.

If cans have been exhausted in steam, they are sent directly to the double-seaming machine, which automatically places lids on the cans to draw air out of the fruit and syrup. If they have not been heated in an exhaust box, the double seamer must be equipped for sealing the cans under a jet of steam, or a vacuum-sealing machine can be used in place of steam-flow.

Sterilization and Cooling. Heat sterilization of canned peaches should be thorough enough to cook the fruit sufficiently, but not to overcook it. The required processing time depends on the maturity and cultivar of the fruit. The usual time is 20–25 min at 100°C for No. 2½ cans in still cookers, and 14–20 min in rotary agitating cookers. The main factor is the initial temperature obtained just prior to processing, which depends on whether the cans are exhausted prior to double seaming or sealed with steam injection at room temperature. Cans sealed at room temperature with steam injection require a 5-min longer processing than those sealed at 71°C or higher. Other factors to be considered are the time and temperature of the exhaust, reel speed, can size, and the maturity and cultivar of peaches. A minimum can center temperature of 90.6°C prior to water cooling after processing is generally satisfactory for achieving commercial sterility.

Steriflamme heat processing of canned cling peach halves and fruit cocktail is a new method of thermal processing in which butane or natural gas flames at temperatures ranging from 1370° to 1650°C are utilized to sterilize canned fruits with rapid spinning of the container. The high-temperature short-time treatments provide good retention of the color, flavor, texture, and sensory appeal of the canned products. By steam blanching and virtually eliminating the covering liquid, and by implementing flame deaeration in place of steam heating, one can achieve a pack with high vacuum, retaining all the advantages of high-temperature, short-time flame sterilization.

Solid-pack pie fruit is largely trimmed, overripe, and low-quality peach slices or quarters. These are carried through a long steam box on a wire mesh belt, heated with steam at 100°C for 10–12 min, filled into No. 10 cans at 93°C with as little free liquid as possible, sealed hot, and heat-processed at 100°C in a rotary agitating cooker for 35–45 min or for 40–50 min in a nonagitating cooker at 100°C. In order to obtain the desired texture, solid-pack pie fruits are sometimes filled at 93.3°C, sealed hot, and cooled in air.

After sterilization canned peaches should be cooled in water to 38°C, at which temperature drying will take place satisfactorily and can rusting will not occur. Cooling is done in a rotary water spray cooler similar to a continuous sterilizer. The cooling water should be chlorinated with 2 ppm of available chlorine to preclude infection of the can contents by spoilage microorganisms.

Labeling, Packing, and Postcanning Storage. Canned peaches can be stored on a warehouse floor in stacks approximately 4–7 m high, with laths between each two tiers of cans to bind them, or on pallets. Cans may be labeled and packed into cardboard cases when needed, labeled and cased immediately after cooling and drying, or cased unlabeled for storage in the warehouse.

Canned peaches are labeled by portable automatic machines, except for those in lithographed cans. Cans may be delivered by conveyor directly from the can cooler after rolling and drying, then passed by gravity through the labeling machine. They pass first over small rollers that apply label paste, which may be rosin, glue, mucilage, casein preparation, or some chemical adhesive. They next roll across a stack of labels, one of which is picked up by the label paste on the can and is smoothed in place automatically by the machine. The adhesive is applied automatically to the end of the label, and the label end is sealed to the can. Labeled cans are packed by automatic casers into cardboard cases, 24 No. 2½, 48 small size, or 6 No. 10 cans/case.

The warehouse should be kept at 20°C with good ventilation. The cans must be protected against condensation of moisture caused by fluctuations of temperature and humidity. In cold climates, it is desirable to heat the warehousc in order to prevent freezing of the cans and condensation of moisture upon them with resulting rusting.

Postcanning storage conditions greatly affect the extent of brown discoloration of canned peaches. In one study, samples stored at 20°C for 4 months showed no brown discoloration, whereas those stored at 36.7°C showed 4% brown peach halves. A progressive increase in the percentage of brown peach halves was observed as the storage time lengthened. At 43.3°C, there was an abrupt increase to 26% brown peach halves during the 4-month storage period. Corrosion of the internal lining of canned peaches was related to the storage temperature. Higher storage temperature results in accelerated dissolution of the tin lining.

Steril-Vac Canning. The Steril-Vac flame sterilization process is a vacuum packing and sterilizing procedure. In the hot-fill process, sliced peaches are blanched to remove air from the fruit tissue. The blanching process permits higher fill weights and better control of the final headspace. After adding 426 g of blanched peach slices and 28 gm of 70° Brix syrup to each No. 1 can, the lids are clinched onto the cans. Clinching corresponds approximately to a loose first operation roll in the closing machine. The clinched can is deaerated by rotating it over a direct flame; during deaeration the can is tilted and the flame applied at the bottom. The flame vaporizes some of the liquid packed with the product. The steam generated in the can causes air to be vented past the loosely clinched lid. When the air is displaced, as indicated by a continuous discharge of steam, the can is sealed. The sealed can is flame-sterilized. The residual steam in the can acts as the heat transfer medium during processing.

When the steam condenses during cooling, high vacuum levels (greater than 63.5 cm Hg) are achieved. The high vacuum results from the removal of air from the fruit tissues during blanching, and from removing the air from the can. By eliminating the covering liquid, which may constitute approximately one-third of the weight of the pack, the tree-ripened fruit flavor is retained, undiluted. Alternatively, sugar or 70° Brix syrup may be added in sufficient amounts to enhance the flavor. This process works equally well in both three-piece and two-piece cans.

The time of flame application from the start of deaeration to the end of the holding section is approximately 3 min for the hot-fill process and approximately 7 min for the cold-fill process. The product is blanched, raising the temperature to 88°C. During filling, the temperature drops to 77°C. In flame deaeration, the vaporized water raises the temperature to 100°C, which is maintained through a steam-flow closing machine. The rising, holding, and cooling sections resemble conventional flame sterilization sequences in which the temperature is raised to 104°C for peaches.

The Steril-Vac flame sterilization process has the advantage of heating cans to sterilizing temperatures in a shorter time than the conventional process. Results indicate that the Steril-Vac process retains the quality of peaches better than the conventional continuous retort process. The process also is superior to the conventional one by eliminating the cover syrup, thus allowing more fruit in the container. A preliminary energy utilization study indicates that the elimination of covering liquid combined with hot-fill flame sterilization of peaches leads to a reduction in energy consumption of approximately 30% on an energy per mass of fruit basis compared with conventional packing in liquid media and steam retorting (Carroad *et al.* 1980).

Cut-Out Data and Case Yield. The quality factors for canned peaches are vacuum, headspace, drained weight, color, uniformity of size, absence of defects, character (texture), and flavor.

The case yield of canned peaches may vary from plant to plant, and with the season. On the average, the yield should be around 45–50 standard cases per ton.

Consumer Acceptance. Large-scale surveys of consumer response to the sweetness and acidity of canned clingstone peaches in Australia indicated that canned peaches with soluble solids to titratable acid (SS/TA) ratios of 54 and 40 and titratable acidities of about 0.35% were liked significantly better than sweeter products with SS/TA ratios of 73 and 80 and titratable acidities of about 0.26%. The optimum SS/TA value for canned peaches for the Australian market appears to be well below 70.

In another survey, samples of canned cling peaches prepared from these differently colored cultivars were viewed by 2808 individuals at the 1965 and 1966 Royal Agricultural shows in Melbourne. The medium-orange fruit was liked significantly better than the deep-orange and pale-yellow fruit.

Tests of consumer acceptance of canned freestone peaches varying in sucrose content from 18.46° to 31.40° cut-out Brix indicated that the optimum sweetness in all-sucrose packs was approximately 22.5° Brix.

Prepared Peach Baby Food

Clingstone peaches are more attractive for making prepared peach baby food, mainly because of the bright yellow color of the puree compared with that from freestone peaches. In genral, clingstone peaches store better than freestones and the flesh is nonmelting and thicker in consistency. In the southwestern United States emphasis has been placed on the 'Baby Gold' and 'Amber Gem' cultivars. They are used in blends with freestone cultivars such as 'Sullivan', 'Elberta', and 'Southland'. In California, many clingstone cultivars are used for preparation of baby foods.

On arrival at processing plants, peaches are weighed; tagged according to grower, cultivar, and arrival time; and placed in cold storage at 1°C. Fruit can be kept under cold storage for 2–3 weeks. At the time of actual manufacture, 20-bu pallet boxes of peaches are moved by lift truck to bulk hoppers from which the peaches start their movement through the processing line. Care is taken to see what the recommended blend is maintained. All equipment surfaces that actually come in contact with peaches are made of stainless steel or rubber belting.

The peaches are washed, halved, and pitted in a torque pitter, cup-down lye-peeled, and washed several times to remove spray residues. They are inspected and trimmed on a continuous belt. Finally, the prepared peaches are conveyed through a screw steamer at 100°C where they are rapidly and fully cooked in the flowing steam, inactivating enzymes. The time needed for blanching and preheating may vary from 10 to 16 min, depending on the size, maturity, and variety of the fruit. The emerging peaches are passed through a pulper with a 1.524-mm (0.06-in.) screen, and then a finisher with a 1.016-mm (0.040-in.) screen for junior baby foods or 0.686-mm (0.027-in.) screen for ordinary strained baby foods. The speed of the pulper and finisher may be kept at 800–1000 rpm. The distance between the paddle and the screen should be carefully adjusted to get the best yield and quality.

The hot puree is pumped to batch tanks where sugar syrup is added to maintain a balance with acid content. The batch is then heated to the proper temperature, adjusted to correct volume and consistency with water or syrup, and pumped to a holding tank where continuous blending takes place and surge condition is maintained for flash heating. The product is sterilized by heating in a Votator-type scraped surface heat exchanger to 110°C and held at this temperature under pressure in the product line. The sterilized product is discharged into a vacuum deaerator where the occluded air is removed and the product temperature is reduced to 93.3°–96.1°C.

At the filler, prewashed jars of 135- or 213-g capacity, or 202 × 214 cans, are filled within very narrow limits to ensure adequate fill with minimum waste due to overfills. The hot filled jars pass to the capper where jar headspace is steam evacuated immediately before twist caps are applied. For baby foods in cans, double seaming is done with steam injection at 0.21 kg/cm^2. The hot packages are conveyed to a sterilizer-cooler, a combination heating and cooling unit. The jars are carried on a belt through a steamheated section for 5–6 min at 100°C, and then gradually cooled by water spray at 82°C for 3–7 min, then at 66° and 49°C for 20 min each. When jars exit from the unit, they are cooled to 38°C.

The cool jars or cans arrive by conveyor to the labeling operation where they are washed, dried, labeled, cased, and taken by lift truck to the warehouse for storage. All case identification must be clearly and neatly printed.

The final peach baby food has a soluble solids content of 21–22%, a pH of 3.9–4.1, acidity of 0.35–0.45% as citric acid, and a Bostwick consistometer reading of 7.0–8.5 cm.

Freestone Peach Nectar

Yellow freestone peaches are preferred for the manufacture of peach nectar because of their delicate flavor and thin consistency after heat processing. Ripe 'Elberta' freestone peaches are generally used for nectar manufacture. The 'J. H. Hale', 'South Haven', and 'Golden Jubilee' cultivars are also suitable. 'Champion' is a suitable white cultivar.

Peaches are thoroughly washed, halved, pitted, and passed over an inspection belt to remove damaged fruit and foreign material. Peeling is also required for most cultivars, especially if the fruit is not fully ripe. Peaches are halved and pitted in a Filper torque pitter. The halves are peeled in a cup-down peeler by spraying with 1% sodium hydroxide solution at 100°C for 15 sec. They are held in the holding section for 60 sec and then spray-washed with cold tap water. Fully ripe freestone peach halves can be peeled with steam at 100°C for 1–2 min. Immediately after scalding, the fruit should be cooled either with water sprays or by immersion in cold water. After cooling, the skins are rubbed off. This method is more time-consuming than the lye-peeling method.

The peeled peach halves should be heated in a steam-jacketed kettle or a continuous steam cooker to 82°C and run through a fruit disintegrator. The resulting puree is then run through a finisher with a 0.508- to 0.838-mm screen. A ton of fruit yields approximately 492 liters of puree. To each 378.5 liters of puree 240 liters of 30° Brix sucrose syrup should be added. The syrup may be prepared from 3 parts sucrose and 1 part glucose.

It may be necessary to add a small quantity of citric acid to adjust the pH of the nectar to 3.7–3.9. It is advisable to pass the finished nectar through a vacuum deaerator prior to pasteurization. This procedure will eliminate the air that has been incorporated in the product during preparation. Excess air in the nectar will lead to deterioration of color and flavor.

The high-temperature short-time flash pasteurization process is the most convenient method of heat treatment for fruit nectars. If the pH of the product is below 4.5, the product may be flash-pasteurized for 30 sec at 110°C. Continuous flow, plate, or tubular heat exchangers are used for this purpose.

For peach nectars, cans made from differential electrolytic tinplate bodies and enameled ends are recommended. The empty cans should be protected from dampness and steam, and kept as clean as possible during storage.

Immediately after pasteurization, the nectar should be filled into the cans, closed at a temperature of 88°C, inverted, and given a holding period of approximately 3 min prior to cooling. If flash pasteurization methods are not available, the nectars can be filled hot, closed, and processed for 20–30 min at 100°C when packed in cans larger than No. 1 size. No. 1 cans and smaller ones should be processed at 100°C for 15–20 min. Some packers practice a method of filling at 88°C, closing, holding, and cooling, but this practice may not adequately ensure against spoilage caused by heat-resistant bacterial spores.

After holding or processing, the cans should be cooled immediately in water until the average temperature of the contents reaches 35°–41°C. The cooling water should contain 2 ppm available chlorine. If cooling is not thorough before the cans are cased, serious discoloration and a poor flavor may result.

Canned fruit nectars should be stored in a cool dry place. The warehouse must be dry at all times to prevent rusting of the cans, staining of the labels, and weakening of the fiber cases. Storage at 10°–16°C considerably increases the shelf life of canned nectars.

Peach Puree for Drinks. The steps in the preparation of the peach puree base, after washing and trimming of peaches, are as follows (Heaton *et al.* 1966):

1. Heat whole fruit in a continuous thermoscrew for 2 min at 93°C to aid in pulping, prevent oxidation, and stabilize cloud in puree. A jacket around the screw should maintain 1.41 kg/cm steam pressure.
2. Pulp by passing through a continuous rotary unit with 0.62-cm perforated screens to remove soft flesh from seed and unripe portions.

3. Finish the pulp by passing through a rotary unit with 0.838-mm or 0.610-mm perforated, stainless steel screen. This reduces pulp to liquid and removes fiber.
4. Accumulate in tank, add 0.14% ascorbic acid and mix, then feed uniformly to pasteurizer.
5. Pasteurize at 88°–93°C and cool quickly to 2°C.
6. Fill aseptically into sterile 208-liter (55-gal) drums or large cans for refrigerated storage. Filling hot in No. 10 or smaller size cans may also be used.
7. Close cans using vaccum and nitrogen, or vacuumize with steam jet.
8. Cool cans in canal or water spray.
9. Dry cans with warm air to remove water drops and avoid rusting or staining of labels.
10. Label containers, use code identification.
11. Store in cool dry place.

PEARS

The 'Bartlett' pear is the most commonly used cultivar for canning because of its uniform shape, fine texture, and excellent flavor. The 'Hardy' cultivar is sometimes used in canned fruit cocktail.

Harvest and Ripening

Pears for canning are usually harvested at full size while they are still hard and green, and are shipped in this condition direct to the cannery in bins for ripening. The bins should be stacked to insure good air circuila-tion. 'Bartlett' pears with average pressure test of 0.9–1.35 kg (2–3 lb) immediately before peeling are best suited for canning.

The use of ethylene gas is of no great benefit if ripening is satisfactory and uniform without the treatment. On the other hand, where several sortings are necessary because of nonuniform ripening, the use of ethylene is profitable and desirable. Ethylene not only hastens coloring and ripening of pears but also increases the rate of respiration as evidenced by carbon dioxide evolution.

Size Grading

'Bartlett pears' are graded for size mechanically in diverging cable or roller graders prior to peeling. They must have a diameter of 6.03 cm or larger for canning as halves. According to the Standards of the Canner's League of California, Fancy Grade canned pears must contain 8–10 pieces, Choice 10–12, and Standard 12–17 pieces per No. 2½ can.

Peeling and Coring

Peeling of 'Bartlett' pears is carried out either by mechanical peelers or lye peelers. A typical mechanical Atlas-Pacific pear peeler is shown in Fig. 6.8. This mechanical peeler is designed to achieve micro-thin peeling. As the machine is peeling, removing the stem, core, and seed cell, it is also being loaded for the next operation. The unit will handle up to 66 pears/min (six at a time) with one operator. The peeling heads are unique in that they peel from the inside out, thus lifting the majority of skin blemishes from the pear. Three sizes of quick change cups can handle practically all sizes of pears. The operator places the fruit stem end down into the self-centering cups. Removal of the stem end, coring, and peeling occur simultaneously. As the loading cups deliver and release the fruit, they return to their original positions for reloading and remain stationary while the fruit is being processed, thus allowing the operator ample time to reload.

In the FMC mechanical peeler, the pear is placed in a cuplike carrier or impaled on a forklike device by the operator. The fruit is then carried between safety-razor-like blades that remove the skin to a uniform depth. Other knives cut the pear in half and remove the core and stem. In the

FIG. 6.8. Mechanical pear peeler. (Courtesy Atlas-Pacific Engineering Co.)

FIG. 6.9. FMC trushape pear preparation system. (Courtesy FMC Corp.)

Edwald peeler, the pear is held in a clamshell-like cup the size of the pear. Knives peel the pear in uniform size and shape. Hand trimming may be required to remove blemishes and bits of skin.

The FMC Corporation of San Jose, California, has developed the C8 TRUSHAPE pear preparation system (Fig. 6.9). The system is designed to peel, stem, core, and optionally slice pears in a high-speed, continuous operation. The system can be used interchangeably for canned cocktail and sliced pears. It reduces operating costs substantially by eliminating the need to presize the pears. Self-adjusting flights and coring knives used in this system handle practically all sizes of pears, at random, with maximum yields. Only two attendants are required to check fruit alignment, thereby saving most of the feed labor required with hand-fed knife peelers. Maximum yield is achieved by peeling with boiling hot lye, which eliminates the peeling away of the flesh.

'Kieffer' pears are first blanched in steam or boiling water for about ½ min, followed by cooling and hand peeling with a special guarded knife. The direction of peeling is from the stem toward the calyx end, not around the pear. Because of the tough skin of 'Kieffer' pears, it is more difficult of peel them with a mechanical peeler.

The loss during peeling, coring, and stemming of pears is usually 30–35% of the original weight. The type of peeler and size of the fruit can influence the case yield.

Pears start to undergo browning very soon after peeling, due to polyphenoloxidase enzyme activity, in the presence of air. Because riper pears are more susceptible to browning, they should be handled promptly or dipped in a 1–2% salt solution or a 1% citric acid solution to inhibit the polyphenoloxidase activity before going to the canning line.

The cores and peels from pears can be used in the preparation of vinegar, brandy, or alcohol. However, in most plants, the waste material is discarded or used as a stock feed for livestock after drying.

Sorting and Filling

After peeling, coring, and washing, pear halves are sorted on a belt and filled into cans in semiautomatic or hand-pack fillers. Defective or off-color halves are sorted to the cocktail line or the sliced pear line. High speed in sorting, filling, and canning is essential in order to avoid enzymic browning.

Syruping

Pears are lower in acidity than peaches and apples and therefore require syrups of lower Brix readings for canning. The syrup may be 40°, 30°, 20°, or 10° Brix or water. Ingoing syrups of higher than 40° Brix will impart too sweet a taste. In many operations 25° Brix sucrose syrups are used for canning. A prevacuumizing syruper is shown in Fig. 6.10. This machine uses less syrup than other types, provides a constant fill of syrup, and eliminates the exhausting process when used in conjunction with a steam-flow sealer. Syruping may also be done with a piston filler with the syrup added at 71°–82°C. With a special filling valve, the headspace can be accurately controlled. If either hot syrup at 88°C or deaerated syrup is filled with this valve, the cans are double-seamed with steam-flow closure, eliminating the exhausting process. By varying the sweetness and acidity of the canning syrup according to the characteristics of the raw fruit, a more acceptable flavor in canned 'Bartlett' pears can be obtained. An addition of small amounts of citric or malic acid has significantly improved the flavor of the canned pears; however, the fruit is rated somewhat softer than untreated pears.

Exhasuting and Double Seaming

Cans of pears and syrup are exhausted by passing through a steam box until the temperature at the center of the cans is at 77°–82°C. The time of exhaustion may be 7 min for No. 303 or 2½ cans, and 9–12 min for No. 10 cans.

Following the exhausting period, cans are closed in a double seamer at

FIG. 6.10. FMC prevacuumizing syruper. (Courtesy FMC Corp.)

77°–82°C. If the minimum average closing temperature is below 71°C, then a steam-flow closure is needed to obtain a satisfactory vacuum. If a prevacuumizing syruper is used, cans must be sealed with a steam-flow closure or a mechanical vacuum seamer. The cans should have a headspace of 0.95 cm (⅜ in.).

Heat Sterilization

In an agitating cooker, 'Bartlett' pear halves are processed in No. 2½ cans for 17–20 min at 100°C or 14 min at 104.4°C. If non-agitating cookers are used, the processing time is 25 min at 100°C. The processing time for No. 10 cans is from 25 to 35 min at 100°C. Rapid cooling after heat sterilization is very important; otherwise, some of the canned products are liable to turn pink.

Pink Discoloration in Canned Pears

Pink discoloration in canned pears is due to growing the fruit on acid soil, excessive cooking; or delayed cooling.

The pink discoloration in 'Bartlett' pears results from conversion of colorless leucoanthocyanidins in pear tissue to a water-insoluble pink-colored leucocyanidin. The discoloration occurs in pears high in leucoanthocyanidin content, especially when excessive heating and delayed cooling are applied. The formation of pink pigment may be related to localization of leucoanthocyanins in certain parts of the fruit exposed to sunshine. A positive correlation exists between the intensity of skin blush on the fresh fruit and pink color formation in the canned product. The type of sweeteners used and acidification of the syrup with 0.2% citric acid also influences the extent of pink discoloration. In one study, pears sweetened with 27° Brix high-fructose corn syrup were distinctly pink when heat-processed at 110°C for 20 min in No. 2½ cans, whereas those in 27° Brix sucrose syrup processed under the same conditions did not turn pink. Addition of stannous ions to susceptible pear puree before processing partly or completely inhibits the discoloration.

Aseptic Canning of Pear Puree

Aseptic canning of pear puree involves a high-temperature short-time (HTST) sterilization process followed by cooling in a heat exchanger; sterilization of the containers and covers with steam; aseptic filling of the cold, sterile product into sterile containers; and sealing of the filled containers in an atmosphere of saturated or supersaturated steam. The four operations are done as a continuous process. The aseptically canned product is usually better in color and flavor acceptance and higher in vitamin content than products subjected to conventional heat sterilization. The aseptic canning process has been applied successfully on a commercial scale to fruit concentrates made from apricots, bananas, peaches, pears, and tomatoes.

The procedure for aseptic canning of pear puree with essence recovery is as follows: The pears are harvested at pressure tests of 7.3–8.6 kg on a

Magness-Taylor fruit tester (7.94-mm plunger). The fruits are stored at 1°C and then ripened at 20°C under 85% RH until an average pressure test of 1 kg is reached. The pears are washed, sorted, and fed into a Rietz thermoscrew at a rate of 9 kg/min. They are then heated rapidly to 100°C, disintegrated, and held at 100°C for 20 sec. The volatile compounds from the pears are condensed in three vertical condensers in series, held at 99°, 77°, and 11°C and trapped as aqueous essences.

The pear pulp is passed through a Langsenkamp paddle pulper with a 0.686-mm (0.027-in.) screen. The paddles are operated at 475 rpm. The puree is deaerated in a Dole centrifugal deaerator, and cooled in a creamery package heat exchanger to 21°C. The recovered essence is added to the puree at a rate of 5% (v/n) puree. The product is mixed, heated to 110°C in a creamery package heat exchanger, held there for 20 sec, cooled to 49°C in a second heat exchanger, and filled aseptically into containers. Filled containers are sealed aseptically in an atmosphere of saturated or supersaturated steam.

PINEAPPLES

Four cultivars account for virtually all pineapples grown for canning. 'Cayenne', the major canning cultivar, is grown exclusively in Hawaii, China, and the Philippines, and is increasing in popularity in other areas throughout the world. The 'Singapore' cultivar is grown in the Malaysian Peninsula; the 'Queen' in Australia and South Africa; and the 'Red Spanish' in the Caribbean. The chief sources of the world's canned pineapple and pineapple juice are Hawaii, the Philippines, China, Malaysia, South Africa, Australia, Thailand, Ryukyus, and the Ivory Coast.

In its first fruiting year, some 18–22 months following planting, a pineapple plant bears a single fruit on a central stalk or peduncle. Approximately 12 months later, one or more axillary suckers give rise to ratoon fruit, which are usually smaller than plant crop fruit. The pineapple is a composite fruit, that is, it is a collection of small fruits called fruitlets. As a result of evolutionary processes the individual fruitlets became fused, thus forming a composite fruit.

Quality Factors

Cultivar, nutrition, weather conditions, and ripeness are the more important factors affecting pineapple quality. Hawaii's warm days and cool nights are considered ideal growing conditions for pineapples and for the development of maximum fruit quality. Fruit size and weight are determined primarily by the adequacy of essential nutrients and mositure. In Hawaii 'Cayenne' pineapples reach peak fruit quality in June and July.

Every effort is made to provide adequate labor and canning capacity to maximize production during this period. During winter, the fruit tends to be paler, lower in Brix, higher in acid, and somewhat lacking in volatile flavor constituents.

Harvesting

Since the relative ripeness of the fruit is of primary importance in determining the ultimate quality of all products derived from pineapple, harvest control is a vital factor in the total production process. In general, pineapples do not improve after harvest. If harvested prior to prime ripeness, they remain inferior with respect to quality. Also, overripe fruit should be downgraded or rejected because of the deterioration of physical and chemical properties and, accordingly, of taste and appearance.

Processing

Harvested pineapples are unloaded from bulk bins at the cannery and mechanically graded for size. Each size goes to a Ginaca machine that has been adjusted for the proper fruit size. This machine is entirely automatic and removes the inedible portion of the fruit from the edible parts. It cuts a cylinder from the center portion of each fruit, removes the shell, and cuts off about 1.27 cm of shell portion at the bottom of the cylinder and approximately 1.91 cm at the top of the cylinder. The final operation of the Ginaca machine removes the core from the center of the fruit. The Ginaca machine handles some 90 pineapples/min. One part of the machine—called an "eradicator"—scrapes the edible flesh from the shell as completely as possible for use in crushed pineapple or for juice. Semiautomatic shelling or coring machines are used in pinapple canning plants in the Oriental countries. These machines can remove the shell and core from pineapples but at a slower rate of production than that of the Ginaca machine.

Pineapple cylinders from the Ginaca machine are conveyed to trimming tables where each cylinder is hand-trimmed to remove the last traces of shells and blemishes. The trimmed cylinders pass through a spray washer on their way to the slicing machine. This machine slices the cylinders transversely into rings about 1.27 cm (½ in.) thick for No. 2½ cans and 1 cm (25/64 in.) thick for No. 2 cans. Pineapple slices are visually graded and manually packed in cans. Slices that have been cut either too thick or too thin, and broken pieces not good enough for canning as slices, are used in crushed pineapple or juice.

Filled cans enter a chamber under vacuum (64 cm Hg) for 5–10 sec where air is removed from the fruit tissue, changing the appearance of the fruit from chalky white to semitranslucency. The cans are syruped in a conventional syruper, sealed under vacuum (38 cm Hg), and processed in a continuous pressure cooker at 102°–104°C until the center temperature of the can reaches 91°C. The process usually takes about 7–10 min for No. 2½ cans. The cans are then cooled, trayed, and stored until required for labeling.

Color

Carotenoids play an important part in the color of canned pineapple, and much of the carotenoids are lost during exhausting and processing. This color loss is due to the isomerization of highly colored 5,6-epoxides to the less intensely colored 5,8-furanoside form, rather than the actual destruction of the carotenoids. The isomerization takes place in an acid medium. Bruising of fruit during postharvest handling also will lead to pigment isomerization in the damaged areas. The isomerization causes a characteristic hypsochromic shift in the absorption maxima of the carotenoid pigment. The sharp absorption peak at 466 mlc is lost as isomerization progresses. Thus the ratio of absorbances at 466 and 425 μm can serve as a measure of the extent of isomerization of the pigment.

Standard of Identity

Canned pineapple, according to the U.S. Federal Food, Drug and Cosmetic Act is the food prepared from peeled, cored mature fruit of the pineapple plant. It can be sold as slices, half slices, broken slices, tidbits, chunks, diced cubes, spears, or crushed. It can be packed in either water, pineapple juice, clarified juice, light syrup (14°–18° Brix), heavy syrup (18°–22° Brix), or extra heavy syrup (23°–35° Brix). The syrup density measurements are those determined 15 or more days after the pineapple is canned. Optional ingredients include spices, vinegar, and flavoring other than artificial. Depending on the form of the canned pineapple no more than 5–15% defects (off-size, -shapes, etc.) based on drained weight are allowed. In all forms of canned pineapple, not more than 7% of core can be present in the drained fruit, and not more than 1.35 g of citric acid can be present in 100 ml of liquid drained from the product 15 days or more after canning.

Crushed Pineapple

To make crushed pineapple, shredded pineapple is pumped into steam-jacketed kettles and heated to 91°C. Some of the juice is drained away to give a product of optimum consistency. If it is to be sweetened, sufficient heavy sucrose syrup is added to give the desired sugar content. The hot mix is packed into cans automatically, sealed, given a short heat processing to ensure keeping quality, and cooled.

By-Products

Shells, trimmings, and other by-product material from the canning of pineapples are shredded and pressed in a continuous press to recover as much juice as possible. The juice is refined and is used in canning the pineapple after mixing with cane sugar syrup. The press cake is dried in rotary drum driers, and the dried product used for feeding livestock. Citric acid is recovered from the juice.

PLUMS

The plum is one of the important fruits canned in Great Britain. The principal canning cultivars are 'Pershore' (Golden), 'Purple Pershore', 'Victoria', and 'Early Laxton'. It was a favorite canned fruit at one time in California, but has been supplanted to a certain extent by canned apricots, cherries, peaches, and pears.

The large sweet cultivars of white plums, such as the 'Green Gage' and 'Yellow Egg', are preferable to the dark-colored 'Lombard' cultivar for canning. In Russia, Crimean plums and cherry plums are used in canning as juices, preserves, jams, and purées.

Plums are washed and stemmed simultaneously in a Herbort Strigger and washer. The machine has rubber-covered rollers running in opposite directions to grab and remove the stems. From this machine the plums pass over a sorting belt and then into a size grader with vibrating screens having circular openings of 2.54, 3.17, 3.81, and 4.45 cm (1, 1¼, 1½, and 1¾ in.) in diameter. A final inspection on the belt is made after size grading to remove imperfect fruits. Light-colored plums are then filled into plain tin cans made of Type L plate or glass jars. For dark-colored plums rich in anthocyanin pigments, lacquered cans may be advisable. The filling operation can be done either in a hand-pack filler or an automatic filling machine.

Since plums tend to soften badly during the heat sterilization process, it is important to pay attention to the filling and exhausting operations. Either hot syrup or water at 88°–93°C is added, depending on whether the plums are to be sweetened or unsweetened. The concentration of syrup for different grades of canned plums is as follows: Fancy, 55° Brix; Choice, 40°; Standard, 25°; Second, 10°; and Water or Pie, none. It is preferable to exhaust the cans thoroughly at 88°–93°C for 8–10 min to reach a can center temperature of not less than 82°C. The cans should be sealed with a headspace of 8 mm (5/16 in.), and the sealing temperature should be 82°C or higher. The average processing time for No. 2 cans is 12–15 min at 100°C and 28–35 min for No. 10 cans. For plums packed in water, the processing time is 10 min in boiling water for No. 2 cans and 25 min for No. 10 cans. The heat-processed cans are then cooled in a rotary cooler with water spray to 38°C.

PRUNES

'Italian' cultivars of prunes are canned in Oregon in the fresh state, sometimes labeled as purple plums. Prunes are a less important crop than peaches and pears.

Well-colored and properly ripened prunes are washed, sorted to remove imperfect ones, size-graded in a roller-type grader, and packed with 40° Brix syrup in enameled cans. The container should be made of Type L plate. The exhausting and processing procedures are the same as those described for canned plums.

French prunes are grown in California largely for dehydration. For canning fresh, they should be harvested firm-ripe. The fruits are washed, sorted, lye-peeled with 10% boiling hot sodium hydroxide for 1–2 min, washed thoroughly, sorted again, and then canned in plain Type L plate cans or glass jars with 30° Brix syrup. The cans are exhausted at 93°C for 4–6 min, sealed hot, heat-processed at 100°C in a rotary cooker for 20 min, and cooled in a rotary water cooler.

Canned Dried Prunes

Large (110–132 counts/kg; 50–60 count/lb) unprocessed dried prunes of the French cultivar 'Prune d'Agen' are used for canning. The dried fruits contain about 18% moisture before canning. They are sorted on a belt to remove defective ones, washed, blanched in boiling water for 4–5 min, rinsed, and packed in Type L cans or cans or similar plate. The "going-in" weights of the processed prunes are 397 g per No. 2½ can and 1.36 kg per No. 10 can. Allowance must be made for the dried prunes to take up a part of the syrup during processing and storage. The syrup for canning is usually 20° Brix. Cans are exhausted at 93°C for 12–15 min. The headspace should be 0.79 cm (5/16 in.). Cans are sealed at 88°C or higher, and processed 20 min at 100°C for No. 2½ cans and 35 min for No. 10 cans. Steam-flow closure is advisable to obtain a higher vacuum. The canned products are cooled in a rotary water cooler to 38°C.

The shelf life of canned, dried prunes is greatly prolonged if the syrup used in canning is acidified with 0.40% citric acid or an equivalent amount of concentrated lemon juice. It is important that the syrup used for canning be 20° Brix. If a syrup of higher Brix reading is used, the canned prunes will be shriveled.

"Dry-Pack" Prunes

To prepare "dry-pack" prunes, dry prunes of a French cultivar are heated in boiling water for 4–5 min, drained, and packed scalding hot in enameled cans. The lids are placed on the cans and given the first rolling operation. The cans are then exhausted 20 min in live steam, sealed, and allowed to cool in the air. The product is cooked later in water before serving.

A high-moisture prune that may be eaten satisfactorily without further treatment has been developed. The dried prunes are graded for size and are then cooked in boiling water until the flesh attains about 33% moisture. The scalding-hot prunes are packed into double-enameled Type L plate cans rather loosely in order to minimize hydrogen swelling. The cans are exhausted at 93°–96°C for 5–6 min, sealed, and processed at 100°C for about 30 min for 8 oz and No. 1 cans. No. 2 and No. 2½ cans should be processed 40–45 min and then cooled thoroughly. Such canned prunes are used as a between-meal snack.

Fresh Prune Juice

The fruits are sorted to remove unfit material, washed, and drained. They are steamed 8–10 min to soften the prunes and prevent browning by the fruit's enzymes. Then the heated fruit is passed through a pulper equipped with a very coarse screen to remove pits and obtain a coarse purée containing the skins. Next, the purée is cooled to below 49°C in a heat exchanger; 0.2% of Pectinol 0 or other pectic enzyme preparation of similar activity is added and mixed thoroughly. It is left to stand until juice can be obtained readily when tested by draining a sample on cheesecloth, normally about 6–12 hr. Then 2% of diatomaceous earth such as Hyflo Super Cel is added to aid in pressing. The purée is placed on light canvas or heavy white muslin which lies on heavy press cloth. This gives double press cloths and a cleaner juice.

The juice is filtered and its Brix adjusted to 22.5°–23°. Then it is flash-pasteurizied in a continuous heat-interchanger type pasteurizer to about 88°C and filled into steamed bottles or into reenameled (double enameled) Type L berry cans at 82°–85°C and sealed. The containers are placed on their sides to sterilize the tops and then cooled.

Alternatively, bottles are filled with cold juice, crown capped, and pasteurized by placing the bottles in cold water on their sides, and heating the water to 82°C for 30 min for quart and smaller bottles. They are cooled slowly with tempered water.

Inactivation of pectolytic enzymes in prunes improves the viscosity and consistency of prune juice. The optimum conditions are 2.5 min at 85°C.

Under the Federal Food, Drug and Cosmetic Act, vitamin C in the amount of 30–60 mg/180 g may be added to canned prune juice (Anon. 1966).

REFERENCES

ANON. 1966. Prune juice; order amending standard of identity. Federal Register *31,* 5957–5958, Apr. 19.

ANON. 1968. Artificially sweetened cranberry juice cocktail—a juice drink; order establishing identity standard. Federal Register *33,* 5617–5618, Apr. 11.

ANON. 1970. Fruit situation. USDA Econ. Res. Serv. TFS 175, July.

ASTI, A. R. 1970. Apple juice containing pulp. U.S. Pat. 3,518,093. June 30. *In* Food Process. Rev., Vol. 21. M. Gutterson (Editor). Noyes Data Corp., Park Ridge, N.J.

BALLINGER, W. E. and KUSHMAN, L. J. 1970. Relation of stages of ripeness to composition and keeping quality of highbrush blueberries. J. Am. Soc. Hort. Sci. *95*(2), 239–242.

BAZAROVA, V. I. 1964. Change of some sugars in the process of ripening of bananas. Sb. Tr. Leningr. Sov. Torgovli No. 23, 71–80 (Russian); cf CA 64, No. 1, 1966, No. 1269 b.

BEAVERS, D. V. and PAYNE, C. H. 1969. Secondary bleaching of brined cherries with sodium chlorite. Food Technol. *23* 175–177.

BEDFORD, C. L. and ROBERTSON, W. F. 1955. The effect of various factors on the drained weight of canned red cherries. Food Technol. *7,* 231.

BREWSTER, L. C., HASEGAWA, S., and MAIER, V. P. 1976. Bitterness prevention in citrus fruits. Comparative activities and stabilities of the limonoate dehydrogenases from Pseudomonas and Arthrobacter. J. Agric. Food Chem. *24,* 21–24.

CANSFIELD, P. E. and FRANCIS, F. J. 1970. Quantitative methods for anthocyanins. 5. Separation of cranberry phenolics by electrophoresis and chromatography. J. Food Sci. *35,* 309–312.

CARROAD, P. A., LEONARD, S. J., HEIL, J. R., WOLCOTT, T. K., and MERSON, R. L. 1980. High vacuum flame sterilization: Process concept and energy use analysis. J. Food Sci. *45,* 696–699.

CASAS, A. and RODRIGO, M. I. 1981. Changes in the limonin monalactone content during development of Washington Navel oranges. J. Sci. Food Agric. *32* (3) 252–256.

CASIMIR, D. J. and JAYARAMAN, K. S. 1971. Banana drink. A new canned product. Food Preserv. Quart. *31* (1/2) 24–27.

CHUNG, J. I. and LUH, B. S. 1971. Effect of ripening temperature on chemical composition and color of canned freestone peaches. Confructa *16* (5/6) 275–280.

CHUNG, J. I. and LUH, B. S. 1972. Brown discoloration in canned peaches in relation to polyphenol-oxidase activity and anthocyanin. Confructa *17* (1) 8–15.

COFFELT, R. J. 1965. A continuous-crush press for the grape industry. The Serpentine Fruit Press. Calif. Agr. *19* (6), 8–9. U.S. Pat. 3,130,667, Apr. 28, 1964.

COFFIN, D. E. 1970. Tyramine content of raspberries and other fruit. J. Assoc. Offic. Anal. Chemists *53,* 1071–1073.

COONEN, N. H., and MASON, S. I. 1970. Recent advances in rigid metal containers. *In* Third Intern. Cong. SOS/70 Food Science and Technology. Institute of Food Technologists, Chicago.

CREVELING, R. K. and JENNINGS, W. G. 1970. Volatile components of Bartlett pear high boiling esters. J. Agric. Food Chem. **18,** 19–24.

CROTEAU, R. J. and FAGERSON, I. S. 1968. Major volatile components of the juice of American cranberry. J. Food Sci. **33,** 386–389.

CZERKASKYJ, A. 1970. Pink discoloration in canned Williams' Bon Chretien pears. J. Food Sci. **35,** 608–611.

CZERKASKYJ, A. 1971. Consumer response to sweetness and acidity in canned cling peaches. Food Technol. (Australia) *23* (12) 606–611.

DAVIS, W. B. 1947. Determination of flavanones in citrus fruits. Anal. Chem. *19,* 476–478.

DEKAZOS, E. D. and BIRTH, G. S. 1970. A maturity index for blueberries using light transmittance. J. Am. Soc. Hort. Sci. *95* (5) 610–614.

FAO. 1971. FAO Production Yearbook. U. N. Food and Agriculture Organization, Rome.

FERNANDEZ DIEZ, M. J. 1971. The olive. *In* The Biochemistry of Fruits and Their Products, Vol. 2. A. C. Hulme (Editor). Academic Press, London and New York.

HART, M. R., GRAHAM, R. P., HUXSOLL, C. C., and WILLIAMS, G. S. 1970. An experimental dry caustic peeler for cling peaches and other fruits. J. Food Sci. *35,* 839–841.

HARTMANN, H. T. and OPITZ, K. W. 1966. Olive production in California. Calif. Agric. Exp. Stn. Circ. 540.

HEATON, E. K., BOGGESS, T. S., JR., WOODROOF, J. G., and LI, K. C. 1966. Peach purée as a base for drinks. Peach Products Conf. Rept. Georgia Exp. Stn., Experiment, Ga.

ITO, K. A. and STEVENSON, K. E. 1983. Sterilization of packaging materials by aseptic systems. Abst. #415. *In* Proc. 43rd Annual Meeting, Packaging Developments Related to Advances in Food Processing, Institute of Food Technologists, New Orleans.

JACKSON, J. M. 1979. Canning procedures for fruits. *In* Fundamentals of Food Canning Technology. J. M. Jackson and B. M. Shinn (Editors). AVI Publishing Co., Westport, Conn.

JACKSON, J. M. and SHINN, B. M. 1979. Fundamentals of Food Canning Technology, AVI Publishing Co., Westport, Conn.

KRAMER, A. and TWIGG, B. A. 1970. Quality Control for the Food Industry. Vol. 1, Fundamentals. 3rd ed. AVI Publishing Co., Westport, Conn.

KRAMER, A. and TWIGG, B. A. 1973. Quality Control for the Food Industry. Vol. 2, Applications. 3rd ed. AVI PUblishing Co., Westport, Conn.

LAWRENCE, L. N., WILSON, D. C., and PEDERSON, C. S. 1959. The growth of yeasts in grape juice stored at low temperatures II. The types of yeast and their growth in pure culture. Appl. Microbiol. *7,* 7–11.

LI, K. C., BOGGESS, T. S., and HEATON, E. K. 1972. Relationship of sensory ratings with tannin components of canned peaches. J. Food Sci. *37,* 177–179.

LOPEZ, A. 1981. A Complete Course in Canning, Book I and II. 11th ed., Canning Trade, Inc. Baltimore, Md.

LUH, B. S. 1980. Tropical fruit beverages. *In* Fruit and Vegetable Juice Processing Technology, 3rd ed. P. E. Nelson and D. K. Tressler (Editors). AVI Publishing Co., Westport, Conn.

LUH, B. S. and DIRDJOKUSUMO, S. 1967. Processing apricot concentrate in laminate and aluminum-film combination pouches. Fruchtsaft-Ind. *12,* 210–219.

LUH, B. S. and PHITHAKPOL, B. 1972. Characteristics of polyphenoloxidase related to browning in cling peaches. J. Food Sci. *37,* 264–268.

LUH, B. S., HSU, E. T., and STACHOWICZ, K. 1967. Polyphenolic compounds in canned cling peaches. J. Food Sci. *32,* 251–258.

MAIER, V. P. and DREYER, D. L. 1965. Citrus bitter principles. IV. Occurrence of limonin in grapefruit juice. J. Food Sci. *30,* 874–875.

MAIER, V. P. and GRANT, E. R. 1970. Specific thin-layer chromatography assay of limonin, a citrus bitter principle. J. Agric. Food Chem. *18,* 250–252.
McCARTHY, A. I., PALMER, J. K., SHAW, C. P., and ANDERSON, E. E. 1963. Corrleation of gas chromatographic data with flavor profiles of fresh banana fruit. J. Food Sci. *28,* 379–384.
MOHAMMADZADEH-KHAYAT, A. A. and LUH, B. S. 1968. Calcium and oxalate ions' effect on the texture of canned apricots. J. Food Sci. *33,* 493–498.
MOYER, J. C. and AITKEN, H. C. 1980. Apple juice. *In* Fruit and Vegetable Juice Processing Technology, 3rd ed. P. E. Nelson and D. K. Tressler (Editors). AVI Publishing Co., Westport, Conn.
NAGY, S. and SHAW, P. E. 1980. Processing of grapefruit. *In* Fruit and Vegetable Juice Processing Technology, 3rd ed. P. E. Nelson and D. K. Tressler (Editors). AVI Publishing Co., Westport, Conn.
NELSON, P. E. and TRESSLER, D. K. (Editors). 1980. Fruit and Vegetable Juice Processing Technology. 3rd ed. AVI Publishing Co., Westport, Conn.
NORDBY, H. E. and NAGY, S. 1980. Processing of oranges and tangerines. *In* Fruit and Vegetable Processing Technology, 3rd ed. P. E. Nelson and D. K. Tressler (Editors). AVI Publishing Co., Westport, Conn.
O'BRIEN, M., CARGILL, B. F., and FRIELEY, R. B. 1963. Causes of fruit bruising on transport trucks. Hilgardia *35* (6) 113–124.
O'BRIEN M., CARGILL, B. F., and FRIDLEY, R. B. 1983. Principles and Practices for Harvesting and Handling Fruits and Nuts. AVI PUblishing Co., Inc. Westport, Conn.
PALMER, J. K. 1963. Banana polyphenoloxidase preparation and properties. Plant Physiol. *38,* 508–513.
PEDERSON, C. S. 1980. Grape juice. *In* Fruit and Vegetable Juice Processing Technology, 3rd ed. P. E. Nelson and D. K. Tressler (Editors). AVI Publishing Co., Westport, Conn.
SILVERSTEIN, R. M. 1971. The pineapple: Flavour. *In* The Biochemistry of Fruits and Their Products, Vol. 2. A. C. Hulme (Editor). Academic Press, London and New York.
SIOUD, F. B. and LUH, B. S. 1966. Polyphenolic compounds in pear purée. Food Technol. *20* (4) 182–186.
STARK, R., FORSYTH, F. R., HALL, I. V., and LOCKHART, C. L. 1971. Improvement of processing quality of cranberries by storage in nitrogen. Can. Inst. Technol. J. *4* (3) 104–106.
STRAND, L. L., OGAWA, J. M., BOSE, E., and RUMSEY, J. W. 1981. Bimodal heat stability curves of fungal pectolytic enzymes and their implication for softening of canned apricots. J. Food Sci. *46,* 498–500, 505.
TANG, C. S. and JENNINGS, W. G. 1968. Lactonic compounds of apricot. J. Agric. Food Chem. *16,* 252–254.
TRESSLER, D. K. and JOSLYN, M. A. 1971. Fruit and Vegetable Juice Processing Technology. 2nd Ed. AVI Publishing Co., Westport, Conn.
U.S. DEPT. OF AGRICULTURE. 1971. U.S. Standards for Grades of Canned Fruit Cocktail. Effective Aug 5, 1971; see Federal Register Sect. 52, 1054. Amended Aug. 5, 1971, Federal Register *36,* 14377.
VAUGHN, R. H., MARTIN, M. H., STEVENSON, K. E., JOHNSON, M. G., and CRAMPTOM, V. M. 1969C. Salt-free storage of olives and other produce for future processing. Food Technol. *23* (6) 124–126.
VAUGHN, R. H., MARTIN, M. H., STEVENSON, K. E., JOHNSON, M. G., and CRAMPTOM, V. M. 1953. Lactobacillus plantarum, the cause of "Yeast Spots" on olives. Appl. Microbiol. *1* (2) 82–85.
VELDHUIS, M. K. 1971A. Orange and tangerine juices. *In* Fruit and Vegetable Juice

Processing Technology, 2nd ed. D. K. Tressler and M. A. Joslyn (Editors). AVI Publishing Co., Westport, Conn.
VELDHUIS, M. K. 1971B. Grapefruit juice. *In* Fruit and Vegetable Juice Processing Technology, 2nd ed. D. K. Tressler and M. A. Joslyn (Editors). AVI Publishing Co., Westport, Conn.
VON SYDOW, E. and KARLSSON, G. 1971. Aroma of black currants. V. Influence of heat measured by odor quality assessment techniques. Lebensm. Wiss. Technol. *4* (5) 152–157.
VYAS, K. K. and JOSHI, V. K. 1982. Canning of fruits in natural fruit juices I. Canning of peaches in apple juice. J. Food Sci. and Technology *19* (1) 39–40.
WYMAN, H. and PALMER, J. K. 1964. Organic acids in the ripening banana fruit. Plant Physiol. *39,* 630–633.

7

Freezing Fruits

B. S. Luh, B. Feinberg, J. I. Chung, and J. G. Woodroof

For centuries the freezing of fruits was to be avoided because the formation of ice in the tissues so altered the physical state of fruits that they were unpalatable and spoiled within a few days after thawing. This was true with fruit unintentionally frozen in transit, in warehouses, or in the home.

Research during the late 1920s showed that damage to fruit tissue was worse from large ice crystals, which resulted from slow freezing (within hours or days), than from small ice crystals, which resulted from fast freezing (within minutes), and that it was possible to freeze so rapidly that ice crystals were no larger than, and could be contained within, plant cells. Thus, there was a vast difference between "slow frozen" and "fast frozen" fruit. It also was found that the juice from slow frozen fruit (citrus fruit frozen in the orchard) could be salvaged if processed within about 48 hr. These early studies showed that the formation of ice (not the temperature per se) stimulates enzymes in fruit tissue that—unless inhibited by heat, exclusion of air, or chemicals—causes deleterious effects on the color, flavor, and aroma of the fruit. The "leakage" or "drip" from thawed fruit is due to rupture of plant cells by ice crystals; much less leakage occurs with fast freezing, with little or no cell breakage, than with slow freezing. However, refreezing fruit causes additional cell rupture and more tissue leakage. When it is desired to retain the natural cell

Commercial Fruit Processing, 2nd Edition

ISBN 0-87055-502-2

structure, texture, color, flavor, and aroma of fruit (such as strawberries, raspberries, gooseberries, peaches, and figs), it is important that they be frozen as rapidly as possible. The rate of freezing is much less important with fruits to be reprocessed into preserves, confections, dairy products, and bakery goods (strawberries, apples, peaches) or fruit to be made into juice (citrus fruits, grapes, apples, cranberries, brambles).

Freezing fruit commercially as a means of preservation began in a crude way about 1920. The pack was restricted almost entirely to strawberries in 400-lb wood barrels, for later use in preserves. As much as 25% sugar was sometimes added, which "preserved" the fruit until freezing was complete about a week later. Freezing occurred in commercial refrigerated warehouses held at about 0°F. Since the barrels were not turned, freezing began around the sides and slowly progressed to the center. The first ice to form was pure water; as freezing progressed, the syrup increased in concentration. There was a core in the center of each barrel that was so concentrated that it did not freeze at 0°F. The high sugar concentration plus the low temperature preserved the strawberries against spoilage for as long as 2 years. Some mold and considerable oxidation occurred near the surface. After the sugar drew a large part of the juice from the berries, the frozen mass was about half and half juice and berries. Most of the barreled frozen strawberries in the United States were packed in the Northwest (Washington and Oregon), with a few frozen in the Northeast (New York, Michigan, and New Jersey).

About 1925, other fruits—including cherries, blackberries, boysenberries, and blueberries—were frozen for reprocessing, and 30-lb tin cans replaced barrels. Many of these went to bakeries, dairies, lonfectioners, and preservers. Along with smaller containers came quicker freezing, less need for large quantities of sugar, better grading of fresh fruit, less mold and oxidation, and a wider acceptance of frozen fruit.

Several important developments in commercial fruit freezing began to be introduced about 1929: (1) consumer-size containers of about 1 lb; (2) emphasis on "quick freezing" at about −40°F; (3) formulas for freezing most fruits separately for direct consumption; (4) antioxidants to prevent development of off-colors and off-flavors due to oxidation; (5) small freezers for retail outlets and homes, enabling frozen fruits to be temporarily held without deterioration; (6) refrigerated trucks and railway cars for transporting frozen foods; (7) highly automated commercial plants for grading, preparation, packaging, freezing, and movement of frozen fruits; (8) new cultivars of fruits with superior freezing qualities; (9) listings of kinds and cultivars of fruits in order of suitability for freezing; (10) economic data on the most suitable geographical and climatic areas of the United States and the world for production of frozen fruits; (11) data on the shelf life of frozen fruits under commercial storage conditions; and (12) data on the nutritional qualities of frozen fruits in comparison with those of fresh, canned, and dried products.

About 1950, a major effort began to freeze combinations of two or more fruits, as well as to freeze prepared fruit products. The former included mixtures of fruits for cocktails, salads, desserts, and prepared dishes; the latter included frozen fruit pies, tarts, cakes, confections, dairy products, and specialty items (see Chapter 16). More recent trends in commercial fruit freezing have included (1) the rapid development of facilities for fruit harvesting, preparation for freezing, storing (including improvements in home freezers), displaying, and marketing; (2) an increase in frozen juices and concentrates (especially juices of citrus fruits, pineapple, apple, grape, cranberry, and many tropical fruits); (3) a decrease in freezing fruits for desserts and salads, where natural texture is expected (such as peaches, figs, most berries, and tropical fruits); (4) improvements in packages that make them more functional and standardized; and (5) reduction in weight by partial removal of water (e.g., concentration of juices and dehydro-freezing of fruits).

FACTORS AFFECTING QUALITY OF FROZEN FRUITS

Production and Harvest

The quality of the raw fruit is the most important factor in determining the quality of a frozen product. It is influenced by varietal characteristics, climate of the growing area, irrigation, cultural practices, and ripeness level at harvest.

New fruit cultivars are developed each year by various agricultural experiment stations and private seed companies. Plant breeders strive for higher yield, ability to withstand rough handling, and disease resistance. Uniformity of ripening and suitability for mechanical harvest also are important traits. Because mechanical harvesting is usually done only once in a growing season, fruit of a wide range of maturity is obtained. If the fruit does not ripen uniformly, the under- or overripe fruit must be sorted out. The labor cost and loss of fruit can make the whole operation uneconomical. Nevertheless, the trend is toward more mechanical harvesting. Intensive research effort currently is devoted to further improve fruit harvesters.

New transportation systems minimize deteriorative changes between harvesting and processing. For example, red tart cherries are collected and transported in tanks of cool water, a practice that decreases respiration rate and scald from bruising in addition to firming the texture. Bins holding about 1000 lb are widely used for peaches, pears, and other fruits. These bins require less transferring of fruit and may actually cause less bruising during trucking than smaller boxes. A logical extension of such procedures would be the moving of more prefreezing operations to the field, so as to minimize changes between harvesting and processing.

Refer to Chapter 2 for additional discussion of harvesting.

Climate

The temperature, humidity, and rainfall characteristic of the growing area may affect the quality of fruits grown there. For this reason, most fruits are grown in areas where they are particularly suited to the climate. There has been considerable shifting of major growing areas from one section of the country to another. For example, the southeastern states were once a major strawberry-producing area, but the high prevailing humidity caused a serious mold problem. The main production areas then shifted to Washington and Oregon, but production there was limited to one crop by a relatively short growing season. California, with a long harvest period from new cultivars and a long growing season, then became the dominant producing area. Raspberries are grown mostly in the Pacific Northwest.

Cultural Practices

Fertilization and irrigation practices have not only a quantitative but a qualitative effect on fruit crops. For example, apricots and peaches have been shown to be lower in acid and astringency and firmer in texture when highly fertilized with nitrogen. Apples with high nitrogen fertilization are less susceptible to core browning and softening in storage than those with low nitrogen fertilization.

Maturity Level

Both growers and processors need to know when a fruit has reached optimum maturity for harvest. With the trend toward mechanical harvesting this knowledge is even more important. Objective tests for maturity can help the fieldman to judge the optimum time for harvest. The pressure tester is frequently used to exclude fruit below a certain maturity level. Peaches testing more than 12-lb pressure on pared cheeks will not ripen to give a high-quality product. Ground color, weight, and size are not suitable for determining the time at which peaches should be harvested. Rather, a combination of pressure test and ground color has been found to be the best index for pickers. Apple maturity is best indicated by the time elapsed from full bloom; the color of apple seed also is an indication of maturity.

HANDLING FRESH FRUIT FOR FREEZING

Controlled-Atmosphere Storage

Storage life of apples can be extended by keeping them in an atmosphere relatively high in carbon dioxide and low in oxygen. Controlled-atmosphere storage (CA) is important for some apple cultivars that must be kept at or above 3.5°C (38.3°F).

Fidler (1970) studied storage of several cultivars of apples at 0°C (32°F), 3.6°C (38.5°F), 7.5°C (45.5°F), and 12°C (53.6°F) under air, nitrogen, 0% CO_2 plus 2.5% O_2, and 5% CO_2 plus 3% O_2 for as long as 105 days. The effect of temperature on sorbitol content of apples in air was slight at 12°C (53.6°F), but sorbitol accumulated as temperature was reduced. The effect was similar from 7.5°C down to 0°C for apples under CO_2–O_2 mixtures. Sorbitol accumulation accompanied injury by core flush and low temperature breakdown, but a causal connection was not established.

Fidler and North (1971) investigated storage of three apple cultivars in air or under CA conditions. They demonstrated that the initiation of ethylene production was not necessarily associated with the onset of the respiration climacteric, and therefore cast no doubt on the belief that ethylene is a ripening hormone. 'Gravenstein' apples can be held under refrigeration in good condition in air only about 2–3 months, but in CA storage they can be held about 6 months. Knee (1971) reported on CA storage of 'Golden Delicious' apples at 3.5°C (38.3°F) under 5% CO_2 plus 3% O_2. Loss of sugar and acid was retarded by CA storage.

Although CA storage has been used mainly for fresh apples, it is also used for holding processing apples from stored packinghouse culls. Some processors of frozen apple slices prefer to store their apples fresh, under a controlled atmosphere, rather than frozen. In this case, the quality differences are not significant but the processing season is extended. Controlled-atmosphere storage is highly effective in maintaining fruit acidity and in delaying scald and internal browning in apples.

The reason for using a controlled atmosphere high in carbon dioxide and low in oxygen is to slow down respiration; it should extend the life of any respiring fruit. Pears, peaches, nectarines, apricots, sweet cherries, strawberries, citrus fruits, and grapes have been stored successfully under a controlled atmosphere. A modification of CA storage combined with rapid cooling can be achieved by blowing cold nitrogen gas (from liquid nitrogen) over the fruit, resulting in both a low temperature and a low oxygen environment. This system was developed for shipment of fresh lettuce and may be applicable to fruits as well.

Ripening

Fruits for freezing are usually picked as near eating-ripe maturity as possible. Exceptions are apples that ripen slowly in storage, and Bartlett pears, which are picked at a hard-green but sweet stage. Pears are stored under refrigeration until a few days before use, when they are brought to a temperature of about 70°F (21°C) for ripening. Some tropical fruits are similar to pears in this respect. There is no satisfactory method for uniformly ripening the common temperate-zone fruits. If fruit is ripe on one side but green on the other, it may be used for jams, puree, or juice.

Mold Control

Mold growth is a serious problem in damaged fruits. Great care must be taken to control it. Thus, berry-picking equipment is steamed or treated with a fungicide such as orthophenylphenate after use. Since ripe fruits are subject to infection with molds and other microorganisms, a program of control by the processor is required. This includes chlorination of wash water, protection of fruit from bruising during handling, and sorting of damaged and moldy fruit, as well as frequent and thorough cleaning of equipment. Spraying or dipping the fruit to cover the surface with a mold inhibitor has been tried with varying success. One of the effective methods for preventing growth of microorganisms is rapid cooling to just above the freezing point of the fruit, but the procedure has not been used extensively. Often the only cooling is that obtained by exposing boxed fruit to the air overnight to dissipate field heat. A cooling system using liquid nitrogen might be effective in preventing mold growth in fruits, since it lowers oxygen concentration greatly.

Sorting

Sorting is necessary to separate green, overripe, and defective fruits with bird-pecks, bruises, etc. The amount of sort-out fruit can be crucial to the profitability of a freezing operation. Off-colored and overripe fruit may be used in making puree or juice. Mechanical harvesting makes the sorting operation particularly important because the whole crop may be harvested at one time. Labor cost is constantly increasing, and unless sorting is mechanized along with harvesting, the extra sorting cost can nullify the saving in labor from mechanical harvesting. There is a strong incentive to develop automatic sorting devices. Mechanical size graders have long been used. Sophisticated photoelectric devices have been developed to sort lemons, apples, cherries, and other fruits by color and flotation baths are used to separate fruit by density difference.

PROCESSING PROBLEMS

Color, Flavor, and Texture Changes

While the flavor and color of soft-ripe fruits are usually better than underripe ones for freezing, such fruit is easily bruised and crushed and does not slice well or retain its shape after freezing and thawing. Thus a compromise is usually made between the characteristics that give the best flavor and color and those that make processing easiest. Selection and development of cultivars for processing may be based on the intended product. For example, different cultivars of peaches are used for different frozen products. For sliced dessert peaches, a cultivar is used that has an attractive red pit cavity; for preserves, a cultivar is used that lacks red color around the pit, and for pie fruit a firm-textured cultivar is used that is more resistant to enzymic oxidation (Loeffler 1967).

Color is an important quality factor for frozen fruits. To maintain a bright color in the final product, chemical treatments or additives are often used in place of blanching by heat to inactivate enzymes. For example, ascorbic acid is added to the syrup in the freezing preservation of freestone peach slices. In some fruits, a combination of chemical treatment plus a mild heat treatment can be used. In others, the enzymes are only partially inactivated, especially on the surface of the fruit, and reliance is placed on quick consumption of the fruit after thawing.

Changes affecting color, flavor, and texture of fruit may occur at any stage of handling. Cultivar selection, method of harvesting, transport and storage, processing, packaging, frozen storage, shipping, and thawing may also affect the quality of frozen fruits. The best raw material can be quickly downgraded by poor processing and handling practices. See Chapter 11 for further discussion of these topics.

Freezing

Freezing preservation of fruits depends on the inhibition of postharvest physiological changes, along with inhibition of microbial action at low temperatufe. For prolonged storage, temperature must be well below the freezing point of water, preferably at −10°F. The deterioration of flavor and color of frozen strawberries, as well as of other frozen foods, is a straightline logarithmic function of the temperature. At temperatures near 0°F, a small decrease in temperature increases storage life markedly.

Water Removal

Water removed by freezing from a biochemical system is not entirely replaceable on thawing. Colloidal solutions become irreversibly dehydrated in cell membranes and this causes a change in their permeability and elasticity. The result is a loss of rigidity upon thawing, so that the fruit becomes soft and somewhat rubbery. At the same time there is excess fluid outside the fruit from the irreversible dehydration, and less juiciness inside. Less water will be frozen out at a given temperature when the concentration of soluble substances is high, since dissolved solids cause a lowering of the freezing point. This is one reason that freezing certain fruits in syrup or sugar has been found to improve quality. Fennema and Powrie (1964) stated that not all the water in foods is frozen above −67°F. In normally frozen foods a great deal of supercooling occurs before there is a change in state from liquid to solid.

Rate of Freezing

The effects of rate of freezing on quality characteristics of frozen fruit are controversial. The "quick freezing" method was developed because it speeds up production and produces frozen fruit of better quality. The enzyme activity that causes browning and off-flavors is inhibited by quick freezing. Quick cooling of the berries to a temperature below 40°F was considered more important to quality than quick freezing.

Rapid freezing in liquid nitrogen improves the texture of frozen strawberries. The decrease in amount of drip found after thawing may be related to the decrease in mechanical damage obtained by formation of fine ice crystals evenly distributed, and also to the decreased movement and interaction of nonaqueous constituents.

QUALITY CHANGES RESULTING FROM PROCESSING

Changes during Processing and Thawing

Biochemical and physiological changes during prefreezing, processing, freezing, frozen storage, and thawing can lower the quality of frozen fruits. Most changes proceed relatively slowly and decrease in rate as temperature is lowered; however, the changes occurring upon thawing can be drastic and rapid. The mechanical and chemical disruption of cells greatly increases permeability of cell membranes, permitting the mixing of cell contents that would otherwise be separated. As ice begins to melt, the cell contents, enzymes, and their substrates are mixed in a concentrated solution and at a temperature favorable to reaction. The result is often a rapid degradation in quality unless steps are taken in processing to prevent such reactions.

Adding ascorbic acid to the syrup can protect fruit from enzymic browning. Packing fruit in syrup has a beneficial effect on color, flavor, and texture. It not only excludes air but has an inhibiting effect on enzymic browning. A syrup of about 30% sugar concentration is beneficial to texture as well as to flavor of most fruits, but a higher concentration can cause dehydration by osmosis, with consequent shriveling and toughening. Furthermore, color may bleed from the fruit along with water, and this is not reversible. (Guadagni *et al.* 1957A,B, 1958).

Effect of Delayed Processing

Delay in processing raw fruit results in defects in the final product, either from microbial growth or enzymic action. Bruises in fruit will cause discoloration due to polyphenol oxidase activity. If the fruit is rapidly processed after harvest, the bruises may be hardly noticeable in the final product. If there is a period of several hours or longer for enzymic oxidation to proceed before the fruit is frozen, the bruises may become quite obvious. If delay is unavoidable, the fruit should be cooled as quickly as possible after harvest.

Effect of Mechanical Damage

Mechanical damage in handling results in broken and bruised fruit. Losses from such damage can occur during mixing with sugar as well as during conveying and filling. Mixing should be no more than sufficient to dissolve the sugar; even at best it is very hard on berries.

Subjective Quality Evaluation

Subjective quality evaluations of frozen fruit by a trained panel may be used to detect changes in color, flavor, and texture. Guadagni *et al.* (1957B) showed that laboratory panels could find significant flavor differences in about 90% of 55 lots of commercially frozen strawberries within 2 weeks of storage at 20°F, but quality control experts from some frozen food plants stated that it might take 4–6 weeks at 20°F to cause deteriorative changes serious enough to cause consumer complaints. The flavor quality of frozen strawberries was found to be related to retention of red color and ascorbic acid.

The anthocyanins in frozen raspberries and boysenberries migrate to the syrup at storage temperatures of 10°F or higher, but only slightly at 0°F or lower (Guadagni and Nimmo 1957; Guadagni *et al.* 1960).

The quality of frozen peaches is directly related to the extent of browning (Guadagni *et al.* 1957A). The most important factor in determining the extent of browning is the degree of container fill in the 10° to 25°F temperature range. Larger headspace will cause more enzymic browning. Peaches in hermetically sealed tin containers were found to retain color and ascorbic acid better than those in composite containers. Color retention was further improved if the fruit was packed under a vacuum of 15–25 in. Hg.

Red sour pitted cherries stored at 20°F showed texture and flavor differences to a trained panel in 3–4 weeks. Color differences in baked pie did not change appreciably for 6–8 weeks. In thawed unbaked cherries, browning was readily observed at 20°F in the fruit exposed to the headspace. Firmness was found to increase with storage time. As with peaches, sealed tin cans were superior to composite containers from the standpoint of preventing enzymatic discoloration at elevated temperatures. Storage under fluctuating temperatures and simulated distribution patterns indicated that changes that occurred under variable temperatures were similar to those that occurred at an equivalent steady temperature.

Orange juice quality depends to a large extent on retention of a stable "cloud," which is a colloid consisting mostly of pectin.

Most juice concentrates used in remanufacturing are not frozen. They are sold as high-density concentrates and kept at room temperature. The high concentration prevents spoilage. Although commercial storage of concentrates at room temperature is not uncommon, Ponting *et al.* (1960) found definite browning in a sixfold (approximately 70° Brix) boysenberry juice concentrate held for 6 months at 70°F; some browning occurred even at 40°F in 6 months. The color changes were attributed to a decrease in anthocyanin pigments accompanied by an increase in brown products of deteriorative reactions.

Objective Quality Evaluation

Evaluation of frozen fruit products by trained panels, based on statistical processes, is slow and requires many people. It is not suited to routine quality control or other quick applications. There is a demand for objective tests that will give equally good results rapidly and with fewer people.

Overall quality of frozen peaches stored in between 10° and 25°F is directly related to extent of browning. Both the extent and intensity of browning are directly related to ascorbic acid content.

In addition to browning, cherries lose their natural red anthocyanin color. This can be measured by a decrease in the "a" value of the reflectance meter. At the same time the syrup becomes increasingly redder, so that the ratio of red color in syrup to that in the fruit (color index) increases during storage at temperatures above freezing. This increase parallels loss of quality and can be used to measure it.

The ascorbic acid content of berries is a good indicator of adverse handling treatment. Guadagni *et al.* (1957B) found that little ascorbic acid was lost in frozen strawberries stored at 0°F, but in composite containers stored at temperatures between 10° and 30°F the loss in ascorbic acid per day varied exponentially with temperature. It was found that by measuring both reduced ascorbic acid and its oxidation products, dehydroascorbic acid and diketogulonic acid, an indication of the storage history could be obtained (Guadagni and Kelly 1958). Boysenberry quality changes can likewise be estimated from the total ascorbic acid analysis.

Fruits that undergo browning by enzyme-catalyzed oxidation of phenols, including apples, apricots, cherries, and peaches, lose ascorbic acid through secondary oxidation by the oxidized phenols. Furthermore, these fruits have a relatively low initial ascorbic acid content compared with berries. Therefore, ascorbic acid analysis is not as good a measure of quality in these fruits, although in peaches packed with added ascorbic acid, the latter has been shown to decrease with loss in quality.

FROZEN FRUIT JUICE CONCENTRATES

Fruit juices can be preserved by a number of methods. These include heat processing after canning, aseptic packing using high-temperature short-time sterilization and sterilized containers, concentration of juice to 70% solids or higher without further processing, and freezing. Of all these methods, freezing is regarded as best for retaining the delicate flavor and characteristic color of fruit juice or concentrate. But although concentrated frozen apple juice of high quality is commercially available, it is still a poor second to canned and bottled processed apple juice in volume marketed. The same is true for concentrated pineapple juice and grape juice.

Large quantities of apple, prune, and grape juice are concentrated to 70° Brix or higher and packed in 50-gal drums. Products at this concentration require neither heating nor storage at low temperatures, because the high osmotic pressure at these concentrations prevents growth of microorganisms. In recent years, aseptic canning processes (in which purees, such as apricot puree, are concentrated to 20°–25° Brix and heated at high temperatures for a short time, then rapidly cooled, and aseptically filled into sterile 50-gal drums) have given products of good quality which may be stored without freezing. For this reason there is little, if any, frozen concentrated apricot puree, since the product produced as described above is satisfactory for its end products such as nectars and jams.

One advantage of freezing juices and juice concentrates is that it permits the use of large containers. While slow thawing is a disadvantage of frozen juices or concentrates packed in 50-gal drums, special equipment has been designed to reduce bulk-packed frozen single-strength juice or pulp back to liquid form in 3 min, using a combination of chopping and heat. The juice or concentrate can be slush-frozen with a Votator or heat exchanger and filled into 30-lb tins, 55-gal drums, or a 10,000-gal refrigerated tank car. In Florida, slush-frozen single-strength orange juice of high quality has been shipped in refrigerated tank trucks and even in ships carrying large stainless steel refrigerated tanks. This is a simpler process than aseptic filling. The chilled or semifrozen juice is unloaded at the destination and repackaged at that time.

Preparation

Fruit juices may be classified as (a) clear juices, such as those prepared from cherries and apples; (b) cloudy juices containing a quantity of insoluble solids, such as pineapple juices; and (c) pulps containing fibrous and other insoluble matter, such as those obtained from apricots and peaches. Some products are available in both forms; for example, strawberries may be crushed and used directly as a pulp in jam, or the seeds and insoluble materials may be removed by various means and the resulting clear liquid used for strawberry jelly. The preparation of juices, pulps, and purées consists primarily of removing all, or a part, of the insoluble solids from the whole fruit. To accomplish this, ingenious equipment of many kinds is available. For a comprehensive review of the production of juices the reader is referred to Tressler and Joslyn (1971).

Apricot, peach, and pear concentrates are made commercially by washing the fruits, cooking them in a steam-jacketed hollow screw heat exchanger, and then pulping. Cooking the fruits has two functions: (1) enzymes that may cause discoloration are destroyed and, (2) the fruit is softened sufficiently so that it is possible to separate pits and fibrous materials from the pulp in pulpers and finishers. The pulpers and finishers have paddles revolving at 500-800 rpm to force the crushed, softened fruit through stainless steel screens. This is a two-step operation: the first screen may have relatively large (0.25-in.) perforations to remove pits, while the second or finisher screen with holes of 0.023–0.027 in. removes pit fragments, skin, and seeds. Where excessive pit fragments are a problem, paddles may be replaced by special nylon brushes that force the material through the screen without breaking the pits. In order to eliminate the gritty texture of stone cells in pear pulp, a disintegrator that grinds these materials into very fine particles has been successfully used. The stone cells may also be removed in a centrifugal machine.

Fresh and frozen berries are made into a coarse puree by a comminuting machine or paddle finisher having a screen with 0.25-in. holes. For clear juice, the puree is treated with 0.2–0.5% of a pectic enzyme preparation and held at 75°F for 3–8 hr. One to 2% by weight of press aid such as rice hull is added, and the resulting mixture is pressed in a hydraulic press. After addition of 0.25–0.5% of diatomaceous earth, the cloudy press juice is filtered in a pressure filter to obtain a clear juice. Gelation is sometimes a problem in apple juice and strawberry juice, as well as other fruit juices held in frozen storage. It is caused by activity of pectinesterase (PE) enzyme and can be prevented if the enzyme is inactivated by heat. This can be done by passing the product through a scrape-surface heat exchanger (Fig. 7.1) at 180°F for 2 min. In the production of some juice concentrates, pectic enzyme treatments eliminate the possibility of gelation, and heat inactivation of pectinesterase is not necessary.

Some processors heat berries before pressing because the heating reportedly solubilizes color, destroys enzymes, and increases yield. It has been observed that cold pressing yields a juice having both excellent color and flavor.

FIG. 7.1. Scrape-surface heat exchangers. (Courtesy Creamery Packaging Mfr. Co., Div. of St. Regis Paper Co.)

Frozen purees and pulps are used in flavoring ice creams. They are supplied either as a straight single-strength pack or a pack of four parts of fruits plus one of sugar. Apricot, blackberry, blueberry, boysenberry, cherry, grape, nectarine, peach, plum, raspberry, and strawberry are made into frozen puree as a flavoring for ripple-style ice cream.

Pectic Enzyme Treatments

Pectic enzymes have been used for many years in the preparation of fruit juices and fruit juice concentrates. A few examples will illustrate the many functions of pectic enzyme treatments. In the manufacture of grape juice, pectic enzymes break down the grape pectin and increase the capacity of the press as well as the yield of juice; they eliminate much of the burning-on or fouling of heat-transfer surfaces during concentration and permit higher concentrations to be reached; and they cause the elimination of gelation in juice concentrates during storage. For example, it is not possible to concentrate single-strength strawberry juice without occurrence of gelation in storage if the natural pectins are not destroyed. In addition, if pectin is present, nondispersible gels develop and cause a marked cloudiness in jelly and other products made from the juice. It is desirable in jelly manufacture to destroy the last trace of native pectin in a fruit juice or concentrate so that a standard pectin can be added to yield a product that will not vary in consistency.

Pectic enzymes are available as liquids of various enzymatic strengths or as a dry powder mixed with filter aid, dextrose, or gelatin. The amount of enzyme to be added, as well as the temperature and the length of holding time, varies with different fruits. Most enzyme suppliers have specific directions for the use of their preparations with each kind of fruit. For example, 6 lb of commonly used liquid enzyme per ton of grapes will cause complete depectinization in about 90 min at 135°F.

Concentration

It is expensive to package, store, and ship fruit solids as single-strength juice. Concentrates offer the advantages of lower shipping and storage costs. In addition to their use as beverages, frozen fruit juices and concentrates are important ingredients in ice creams, sherbets, sundae sauces, fruit syrups, and other soda fountain supplies. Fruit juice concentrates also are very useful in the production of jelly. By using concentrated fruit juice for making jelly, it is possible to obtain better color and flavor than by employing single-strength juice. The use of fruit juice concentrates not only eliminates much of the evaporation required, but permits the continuous manufacturing of jelly by combining metered streams of fruit juice concentrate, sugar syrup, and pectin solution. In this process the sugar syrup is heated separately to a high temperature. The flavor and color of the concentrate are protected from heat damage by confining heat treatment to a minimum; no evaporation is required. Frozen concentrate plus essence is particularly suitable for continous jelly manufacture and produces a jelly of excellent quality. The economics in packaging and storage make concentrates appealing to food processors and to the retail market.

Fruit juices are heat sensitive, and their color and flavor deteriorate rapidly at boiling temperatures. Thus, it is necessary to concentrate fruit juices in vacuum evaporators. Multiple-effect vacuum evaporator systems are sometimes used. There are many kinds of vacuum evaporators and these have been well described by Tressler and Joslyn (1971). In the early development of frozen orange juice concentrates, special low-temperature evaporators that concentrated juice at 70°F were designed. Evaporators operating at 120°F were felt to produce an inferior product. But evaporation at these low temperatures required strict observance of sanitation, since microbial growth was possible. A heating step was also required to destroy pectinesterase, otherwise gelation would take place in the concentrate during storage. Many of the older low-temperature evaporators operated at a low temperature for periods ranging up to several hours. More modern evaporators, on the other hand, employ a small amount of liquid passing rapidly over a large area of heating surface held at a temperature as high as 245°F. While the heating temperature is high, residence time of the liquid in the heating chamber is only a few seconds. This reduces holding time and results in better color and flavor. Such evaporators have higher capacity per square foot of heating surface than the older low-temperature evaporators and are rapidly becoming common in commercial production of orange juice and pineapple juice concentrate (Hass *et al.* 1966).

Methods of Concentrating Fruit Juices

Fruit concentrates are produced by (a) freeze concentration, (b) low-temperature vacuum evaporation, (c) high-speed, high-temperature evaporation, and (d) reverse osmosis.

Freeze Concentration. Water may be removed from juices by freezing out as crystals of solid ice. This process, called freeze concentration, has been commercially used for producing orange juice concentrate.

Freeze concentration consists of three fundamental elements: (a) a crystallizer or "freezer" that produces a slurry of ice crystals; (b) a separation device (a centrifuge, wash column, or filter press) for separating the ice crystals from the mother liquid; and (c) a refrigeration unit to cool the liquid and remove the heat of fusion and the frictional heat resulting from hydraulic flow, wall scraping, and agitation of the slurry.

With equipment available today, it is possible to freeze-concentrate most juices to approximately 50° Brix. Orange juice freeze-concentrated to 44.8° Brix need not be "cut back" with straight juice as is frequently done in conventional concentration by evaporation. Although freeze concentration can produce a frozen juice concentrate of superior quality, this method has not been used commercially since 1966. Juice processors believe that even the slight increase in cost required would not find a market among economy-minded consumers.

Freeze concentration is free of the drawbacks associated with heat evaporation. It is capable of concentrating fruit juices without appreciable loss in taste, aroma, color, or nutritive value. The principal disadvantage of freeze concentration of fruit juices is its high capital cost. The problems in freeze concentration are (a) inability to control ice crystal growth over a period of time and (b) excessive solids loss due to liquid entrapped in the ice crystals. The problems in concentration by freezing have been discussed by Tressler and Joslyn (1971).

Low-Temperature Vacuum Evaporation. Most vacuum evaporators are constructed in successive multistage units. The heating medium for each stage may be either boiler steam or water vapor coming from a previous stage. When water vapor from boiling juice is used as the heating medium to evaporate juice boiling at a lower temperature, the operation is known as an effect. Many of the modern orange juice concentrators are 7-stage, 4-effect, high-temperature and short-time (HTST) evaporators. The juice makes a single pass through each stage. Vacuum operation is sometimes used not only to remove water, but to reduce the temperature of the concentrate. For this purpose there is no heating stage and the liquid is flashed off to the final concentration by means of vacuum. The sensible heat of the concentrate is utilized as the latent heat required to boil off liquid and so reduce its temperature to a point corresponding to boiling at the pressure in this final step.

Rising-Film, Falling-Film, and RFC Evaporators. Evaporators are frequently classified either as natural circulation evaporators, in which the circulation of the product results from the reduction in density of the solution on heating and from pressure generated by vapor evolved at the heat exchange surface, or forced circulation evaporators, in which a circulating pump is used to ensure high velocity across the heating surfaces. Evaporators also are sometimes classified as rising-film or falling-film evaporators. In the former, the dilute liquid is fed into the chamber below a tube-sheet and rises in the tubes. As steam is admitted to the steam chest, the liquid reaches the boiling point and bubbles of vapor are formed in the column of liquid. As the bubbles rise, they expand and push the liquid forward with increasing velocity. The mixture of vapor and liquid exits from the tubes into a vapor separator. Evaporators of this kind are widely used in the food industry for concentration of clear solutions of low to moderate viscosity. In a falling-film evaporator the dilute liquid is introduced into the chamber above the upper tube-sheet and flows as a thin film down the inner surface of the tubes. Only a small amount of liquid is in the falling-film tubes at any one time, compared to the column of liquid in a rising-film evaporator.

A rising-falling concentrator (RFC) is a single-pass evaporator in which the feed juice is introduced at the bottom of one bundle of tubes at a temperature above its boiling point, and the discharged mixture of liquid and vapor is separated at the top and redistributed over the down-flow pass. This type of evaporator has been used for concentrating orange, grape, and apple juices.

Plate-Type Evaporators. The plate-type or AVP evaporator is a single-pass evaporator using specially modified stainless steel plates as heat exchange surfaces. There are more than 100 of these installations throughout the world concentrating orange juice, apple juice, grape juice, milk, and other liquids. In one Florida orange juice concentration plant, a 12,500 lb/hr double-effect plate evaporator and auxiliaries were installed in an 18 × 27-ft room with a 90-ft ceiling. This is considerably more compact than conventional vacuum evaporators. The advantage of this method of evaporation is the small amount of product in the system at any one time. In the concentrator described here, there is less than 5 gal. of product in each effect at any one time. The system may be described as a rising and falling film evaporator with liquid fed from the bottom and rising and descending through alternate plates. The Florida installation described used two effects to produce orange concentrate at 65° Brix, with a total retention time of about 1 min.

Mechanically Induced Film Evaporators. In another type of evaporator the liquid to be evaporated flows down the heated walls of a cylinder; a thin film is mechanically induced by rapidly rotating vanes, which both agitate and spread the film. Heat for evaporation is transferred through the jacketed cylinder wall. The vapor mixture produced during heating may be separated in an adjoining vapor separator or in an integral separator contained in the evaporator. The design is most effective for concentrates where high viscosities are encountered.

Rotary Steam-Coil Vacuum Evaporators. Rotating steam coils have been used for many years for evaporating liquid foodstuffs in open kettles, and such a coil has been adapted to vacuum operations. The equipment consists primarily of a rapidly whirling coil completely submerged in the boiling liquid under vacuum. Because of the movement of the coil, fouling is minimized and quality in the product is preserved. Evaporation rates of 55–65 lb/ft^2/hr of water have been obtained with such viscous materials as 50% solids cold-break tomato paste; by comparison, only 35 lb of water could be evaporated from similar feed material and the same temperature difference using a swept-film evaporator. This evaporator appears to be useful for small concentrators of fruit juices, since it reportedly combines high output with low capital cost.

High-Speed, High-Temperature Evaporation. The Thermally Accelerated Short-Time Evaporator (TASTE) is a high-temperature evaporator with very short holding times. The total residence time during evaporation may be less than 1 min and time in individual tube bundles only a fraction of a second. The short holding time permits high operating temperatures. In evaporation of citrus juices, the juice is heated in the first stage to approximately 195°F and passes through a pressure release valve where it flashes off under a pressure of 19 in. Hg. The heating and flashing-off at a low pressure is repeated through a series of successive tube bundles, each flashing-off at a lower pressure than the one preceding. The 4-effect, 7-stage evaporators of this type, having an evaporative capacity of 40,000 lb/hr, are now being used by many orange juice concentrating plants.

Evaporation produces a mixture of vapors and concentrated liquid which must be separated rapidly and thoroughly. The separated vapors are frequently used as a heating medium for an additional effect in another evaporator.

Reverse Osmosis. Osmosis occurs in many biological systems whenever a dilute liquid and a concentrated liquid are separated by a semipermeable material—one that selectively permits one kind of molecule to pass through but holds back other kinds. Under ordinary conditions water passes from a dilute liquid to concentrated liquid. By applying pressure on the more concentrated liquid, however, it is possible to reverse the flow and force water molecules through a cellulose acetate semipermeable membrane while other molecules (proteins, sugars, acids, etc.) are held back. The process may be thought of as molecular filtration, although the exact mechanism is not well understood.

Apple and orange concentrates up to 40° Brix have been made on a pilot plant scale from their respective juices, using a pressure of 2500 psig. Other tests on maple sap indicate that the reverse osmosis technique may be an economically feasible method for concentration of maple syrup. This process is in its infancy for concentration of fruit juices, but has been tried on recovery of proteins from whey with prospective industrial application.

Volatile Essence Recovery

Volatile components are important to the characteristic flavor of fruits. Various techniques have been used to recover these volatile components, and to add the essence back to the concentrated fruit juice. Volatile essence recovery is now a common procedure in the manufacture of Concord grape concentrate and of pineapple juice concentrate. Essences of strawberry, Concord grape, apple, and other fruits are available commercially.

These volatile components generally are distilled from a liquid food, concentrated by fractionation in a column, and recovered in concentrated form. The product is usually prepared as a stated concentrate by volume of the ingoing feed: 100-, 150-, or 200-fold being the usual degrees of concentration. Thus, 1 volume of essence (aroma solution) recovered from 100 volumes of juice will be labeled a 100-fold essence. This does not mean that all of the volatile flavor components in the original 100 volumes of juice are recovered in 1 volume of the essence. It depends largely on the efficiency of the column and the conditions under which the volatile flavor components are recovered.

An example of a Krenz essence recovery unit is shown in Fig. 7.2. The feed is heated very rapidly to boiling temperature, the desired percentage of water evaporated in a flash evaporative section, and vapor separated

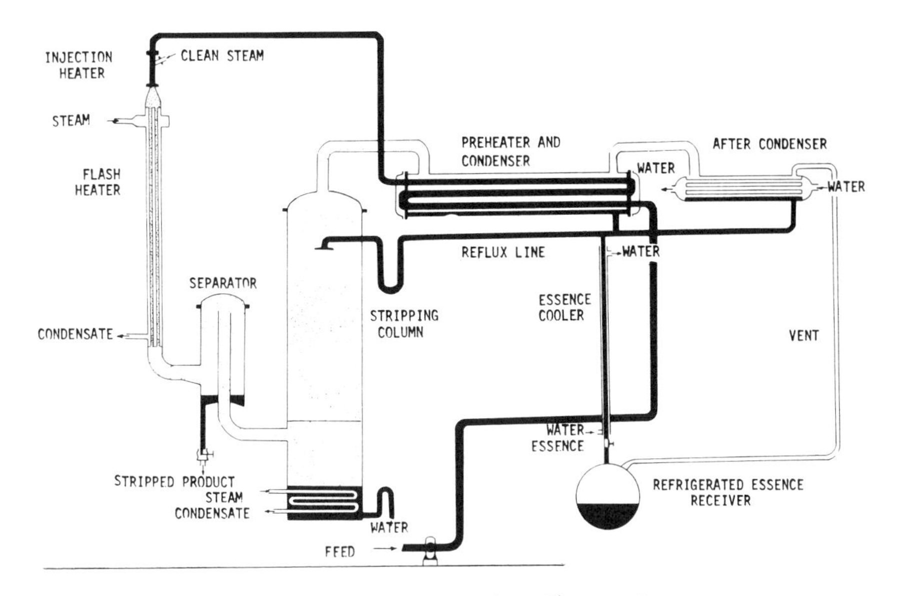

FIG. 7.2. Diagram of essence recovery plant. (Courtesy Oscar Krenz, Inc.)

from liquid in a vapor liquid separator. The partially concentrated liquid (stripped juice) is very rapidly cooled by some means to eliminate heat damage to it. The vapor from the separator, which contains both volatile flavors and water vapor, is led to a fractionating column provided with a reboiler and a condenser. In the column the volatile flavors tend to collect at the head of the column while the water tends to collect in the reboiler. When such a column is operated under partial reflux, it is possible to recover the volatile flavors in concentrated form (essence) by drawing from the refluxing system a set portion of the condensate flowing back from the reflux condenser to the head of the column. Water collecting in the reboiler is discarded. In some cases it is desirable to scub vent gases with either chilled water or chilled essence in order to recover the volatile flavors that these noncondensable gases contain.

Where volatile flavors are collected from all vapors given off during manufacture of a concentrate, the evaporator replaces these items of equipment together with means for cooling stripped juice. The heart of the process lies in feeding the water vapor and volatile flavor mixture to a column and in operating the column so that an enriched solution of volatile flavors in water may be recovered in the form of essence.

The greater part of the volatile components may be recovered from fruit juices by evaporating only a small fraction of the water present: for example, 8–10% of apple juice, 20% of strawberry juice, 30% of blackberry juice, and 40% of 'Montmorency' cherry juice.

A special method for recovering and concentrating heat-sensitive aromas from fruit products has been developed. The feed is boiled under vacuum at 100°F. The vapors are fed to a vacuum sievetray column where a noncondensable gas is used to strip the volatile aromas from the condensate, which is boiling at 100°F. A liquid-sealed vacuum pump compresses the noncondensable gas and absorbs the aroma in the sealant liquid. The process has successfully produced aroma solutions from orange juice, apple juice, peach puree, and apricot puree. It should be applicable to production of aroma solutions from passion fruit, guava, mango, papaya, and other tropical fruit juices.

Deaeration of Citrus Juices and Concentrates

Deaeration is the removal of dissolved and occluded oxygen from liquid food products. This operation has found particular application in the technology of fruit juices, carbonated beverages, tomato ketchup, and baby foods.

Removal of oxygen may be desired in order to avoid (1) oxidative reactions leading to deterioration in color or flavor; (2) oxidative reactions affecting nutritive value, e.g., loss of ascorbic acid; and (3) adverse effects on can lacquers and accelerated corrosion. Removal of other dissolved gases may be desired to avoid (1) frothing during filling of containers, (2) appearance of gas bubbles, and (3) flotation of pulp particles in glass-packed foods.

The general methods used for deaerating foods are vacuum deaeration (Fig. 7.3) and gas displacement. Both are based on the principle of reducing the partial pressure of a gas in contact with a liquid in order to cause the gas to come out of solution. The solubility of the gas may be further reduced by raising the temperature.

Vacuum Deaeration. Citrus juices at the time of extraction are supersaturated with air and have oxygen contents around 0.8 ml/100 ml. Deaeration to remove the oxygen is a desirable processing operation but the benefits derived will vary according to the nature of the end product. With

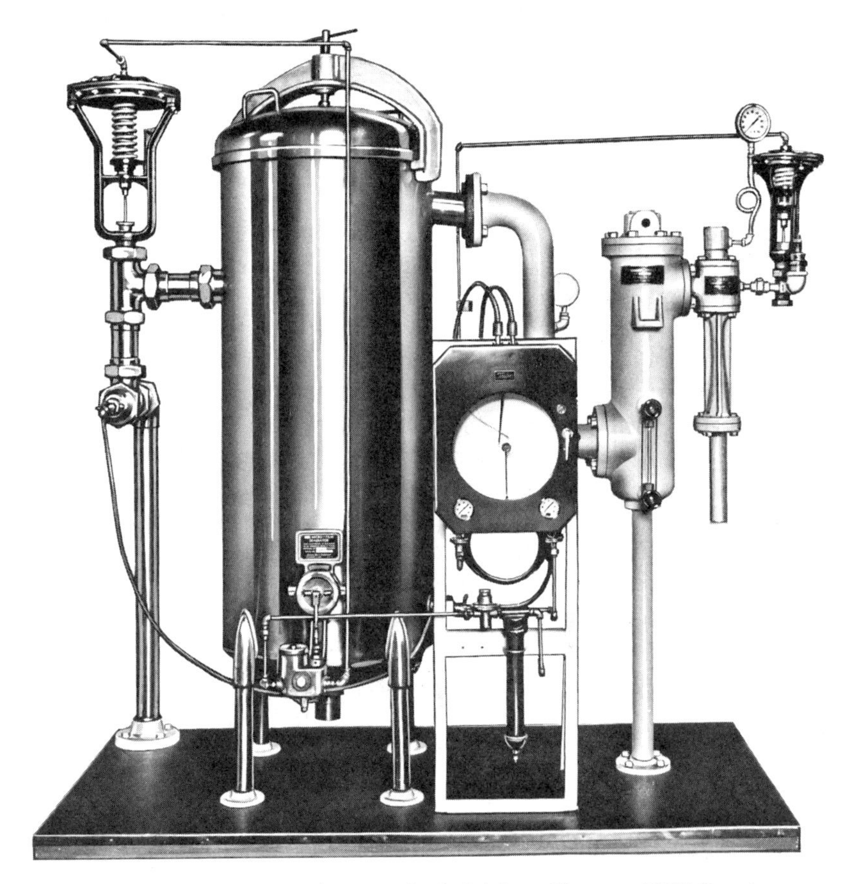

FIG. 7.3. Vacuum deaerator for fruit juices. (Courtesy FMC Corp.)

canned single-strength juice, deaeration will slightly improve quality immediately after processing and container performance throughout storage, but will not retard flavor deterioration during storage. With frozen citrus juice concentrates, deaeration is a refinement that may improve the retention of fresh juice flavor and nutritive value. In the manufacture of frozen orange juice concentrate, deaeration of the cutback juice or the blended final concentrate is recommended because oxidative reactions are the primary cause of flavor deterioration in storage. Kefford *et al.* (1959) found that the efficiency of air removal from orange juice in a vacuum-spray deaerator at 75°F under a pressure of 15 mm Hg was 90–92%. Centrifugal deaerators may also be used to accomplish the same objective.

Gas Displacement. Deaeration by gas displacement may be carried out in a gas saturation tower or simply by sparging with nitrogen in a pipeline. Inert gas displacement appears to be more efficient than vacuum deaeration in terms of removal of oxygen. It has the advantage that the loss of volatile constituents is negligible. Assessment of the efficiency of procedures for oxygen removal from liquid foods can be done easily with the Beckman oxygen electrode or the vibrating mercury-plated platinum electrode.

There is general appreciation of the need to minimize aeration by taking obvious precautions such as keeping juice lines full and filling tanks from the bottom. Some plants have deaeration tanks for cutback juice connected to the vacuum system of the evaporators. In the evaporators, of course, the bulk of the juice is deaerated during concentration and the only benefit from prior deaeration here would be to reduce the load of noncondensibles in the vacuum system.

Some plants practice nitrogen stripping and sparging in pipelines, and nitrogen blanketing of the cutback juice and blended concentrate up to the time of filling. These procedures are considered to be beneficial to flavor retention especially when the product is subjected to extreme temperatures during storage.

Freezing Methods

Fruit juices, purees, and concentrates have been frozen by rotating the filled containers in a refrigerated liquid such as glycol, packing the juice in small containers and placing them in contact with refrigerated metallic plates, freezing in an air-blast tunnel at temperatures of approximately −30°F, and slush-freezing the juice before packaging. The last method has become increasingly popular, because it permits freezing juices in containers of all sizes. While it is preferable to use small containers for blast freezing, it is not uncommon for small jelly processors or food processors to pack and freeze single-strength juices in 30-lb tins by this method.

The Votator slush-freezing unit is widely used for single-strength and concentrated juices. A thin stream of juice is forced at about 50 psi pressure into the space between the center shaft and the heat transfer tube. Just as rapid evaporation requires a large area of refrigerated heat transfer surface per unit of juice, the Votator freezes rapidly by passing a thin layer of juice over a large refrigerated surface. Floating scraper blades, affixed to the rapidly revolving center shaft, are forced outward against the heat transfer surface and automatically remove the frozen film which would otherwise accumulate. Ammonia is ordinarily used as the refrigerant in this equipment, although Freon or brine has also been used. For single-strength orange juice, an outlet temperature above 27.5°F is recommended since cooling below this point results in a stiff slush that cannot be handled by an automatic filler. By contrast, orange concentrate is commonly chilled at 20°F because it is still quite fluid at this temperature. After orange juice concentrate is slush-frozen, it is packed into cans and frozen hard by passage through a low-temperature blast tunnel at −20° to −30°F.

Cryogenic Railcar. According to the American Frozen Food Institute and the International Association of Refrigerated Warehousemen, final engineering is underway on a prototype refrigerated railcar that will preserve frozen food in transit through use of dry ice "snow" produced from liquid carbon dioxide contained on board.

Liquid carbon dioxide at 0° will be stored in a number of interconnected tanks beneath the floor of the car. The tanks will not only provide in-transit storage for the liquid carbon dioxide, but will also act as a barrier to heat entering the car and turn the floor into a giant "cold plate" to help keep the load at the desired temperature. An initial blanket of carbon dioxide snow will be deposited on loads prior to departure from loading points. While in transit, additional charges of snow will be triggered by the thermostatically controlled temperature sensors. Operation of the entire refrigeration system will require only the pneumatic power provided by the liquid carbon dioxide. No other power sources will be needed (Anon. 1982).

FREEZING PRESERVATION OF FRUIT PUREES AND CONCENTRATES

Apple Juice Concentrate

Apple juice is available in several different types including natural, a pulpy juice made by adding ascorbic acid during or soon after pressing; clarified, a centrifuged juice from which most of the larger apple particles have been removed; clear, a filtered, sparkling clear juice; and opalescent, a juice with various amounts of suspended insoluble material left, to produce a cloudy product. There are regional preferences for each type of juice. Most of the juice consumed in the United States is of the clear type, whereas in British Columbia, Canada, about one-third of the apple juice pack is of the opalescent type. There are also regional preferences for different Brix–acid ratios. Western consumers like a juice that is sweet; those in the eastern states like a rather acid juice. The blending of juices from different cultivars requires experience. In general, the blend should include enough acid cultivars to give the product an acidity between 0.40 and 0.50% as malic acid and a sugar content about 12.5° Brix, or a sugar–acid ratio between 25 and 31. It is desirable to include sufficient aromatic cultivars such as 'McIntosh' and 'Golden Delicious' to contribute bouquet. Apple cultivars also differ in pressing characteristics from mealy and soft to crisp and juicy.

Apple juice is made by washing and sorting apples, followed by grinding the fruit in a hammer mill or disintegrator (Fig. 7.4) and pressing out the juice in a hydraulic rack and frame press. The ground apple pulp is loaded onto a cloth, preferably nylon, and evenly spread; the pulp is then wrapped in the cloth by folding the corners; a wooden rack is placed on the filled cloth and the process is repeated until a sufficient number have been stacked for the capacity of the press. The assembly of racks and

FIG. 7.4. Rietz disintegrator. (Courtesy Rietz Manufacturing Co.)

cloth is then compressed in a hydraulic or mechanical screw-type press to press the juice from the pulp. One ton of raw apples pressed in such equipment will yield approximately 160 gal. of single-strength juice. This method leaves much to be desired from the standpoint of sanitation, and requires a considerable investment in labor, time, cloths, and racks.

More modern techniques for apple juice extraction include the continuous screw press; Willmes Press, which is essentially a large inflatable rubber tube inside a horizontal cylindrical screen lined with press cloth; several types of centrifuges; and vacuum filters. Some of these depend upon the use of some kind of press aid such as rice hulls or shredded cellulose. In a unique two-stage, thick-cake dejuicing process, apples are ground to produce a coarse pulp, shredded cellulose is added as a press aid, and the mixture is fed into a basket centrifuge where the free-run juice is extracted. The partially dejuiced pomace leaving the centrifuge is fed into a vertical screw press over a small vibrating screen that serves to remove coarse fibrous material. The term "thick-cake" refers to the formation of a press cake about 4½ in. thick on the centrifuge wall during the first extraction.

Extracted apple juice may be clarified with pectic enzymes (4 oz/100 gal.). The enzyme treatment may take 8–12 hr at 70°F or 3–4 hr at 120°F. The treated juice is then mixed with 1–2% diatomaceous earth and filtered through a filter press.

There are several commercial processes for producing frozen apple concentrate. Clear or opalescent apple juice is fed to an essence recovery unit; after removal of the volatile flavor components, the stripped juice is concentrated under vacuum to 45° Brix (4-fold). The concentrated essence is added to the cool concentrate which is packed in 6-oz cans, frozen, and stored at −10°F. Since the aromatic components of apple juice are more stable and easily recovered than those of citrus juices, this added essence technique is used for apples rather than the cutback process used for orange juice concentrate.

Berry Purees

The preparation of raspberry, boysenberry, youngberry, blackberry, and blueberry pectinized purees is a relatively simple matter. A tested recipe was presented by Tressler (1968) as follows:

Ingredients	Parts (lb)
Pulpy fruit juice or puree	100
Granulated sugar	50
Enzyme-converted corn syrup (43° Brix)	60
Pectin (100 grade)	1–2
Citric acid	1.5
Water	20

The pectin is stirred into 15 lb of the enzyme-converted corn syrup, which has previously been warmed to about 160°F. This suspension of pectin in corn syrup is then slowly stirred into the water that has been brought just to the boiling point. The solution is agitated rapidly and kept hot for 5–10 min, or until the pectin forms a smooth syrupy solution. Heating is then discontinued and, while agitation is continued, the remainder of the enzyme-converted corn syrup is added. The fruit purée, to which has been previously added the granulated sugar and powdered citric acid, is then stirred with rapid agitation into the solution of the pectin in the corn syrup. Slow agitation is continued while the pectinized puree is being cooled. When its temperature has been reduced to 60°F or lower, the product is run into 30- or 50-lb enamel-lined slipcover cans and frozen at 0°F or lower. The total soluble-solids content of the product will be approximately 50%.

The same formula and procedure can be used in making pectinized strawberry puree, except that berries which have been ground in a food chopper should be used instead of a seedless puree.

Orange Juice Concentrate

Frozen orange juice concentrate was the first among the citrus products to be processed in large commercial quantities. It continues to be the leading item among the frozen juice concentrates.

Normally, Florida orange groves produce over 75% of the total U.S. crop. The bearing acreage of oranges in Florida was estimated at 650,000 acres. California, Texas, and Arizona also produce citrus fruit for fresh market and for processing. Based on a 62% utilization of the total crop, approximately 125 million gal of frozen orange concentrate was produced in 1982. The total evaporative capacity of the 27 major plants in the Florida citrus area is in excess of 2 million lb of water/hr (Tressler and Joslyn 1971).

Converting oranges into a frozen concentrate is a highly mechanized, closely controlled process. The oranges are washed, sorted, spray-rinsed, and conveyed to the extracting floor. In Florida, an automatically proportioned sample of the fruit drops off the conveyor and goes to a Florida State Department of Agriculture laboratory, where it is analyzed to ensure that the state requirements are met. The only important cultivar of oranges for juice in California is 'Valencia'; the cultivars used in Florida for processing are 'Pineapple,' 'Valencia,' 'Parson Brown,' and 'Hamlin' (Veldhuis 1971A).

Extracting

FMC In-Line Extractor. A sketch showing the operation of the FMC in-line extractor, which is widely used in the citrus juice industry, is shown in Fig. 7.5. In the first position, the fruit has been deposited in the bottom cup and the upper half has begun to descend. As it does, the sharp upper end of the tube in the lower half of the machine cuts a hole in the bottom of the fruit and as the many fingers of the two halves mesh, the crushed, juice-laden segments pass into the tube as shown in the second position. There is a restrictor in the lower end of the tube to prevent loss of juice which is forced through the perforated tube and emerges at "J." When the upper cup is firmly seated, the central tube containing the restrictor rises to compress the contents of the tube, recover remaining juice, and eject the plug at "T," as shown in the third position. The machine illustrated is designed for simultaneous recovery of cold-pressed peel oil. As the fruit is squeezed, the peel oil runs down the outside of the fruit and water sprays wash it down the sloping plane until it drops into a conveyor at "O." To make recovery of oil easier, the pulp is discharged separately. An annular space between the center tube and the "fingers" in the upper cup facilitates separate discharge of peel upward at "P." The yield of juice and the type of juice obtained can be varied by changing the type of cup, by using perforated tubes with holes of different size, by changing restrictors to reduce clearance in the passage for pulp and seeds,

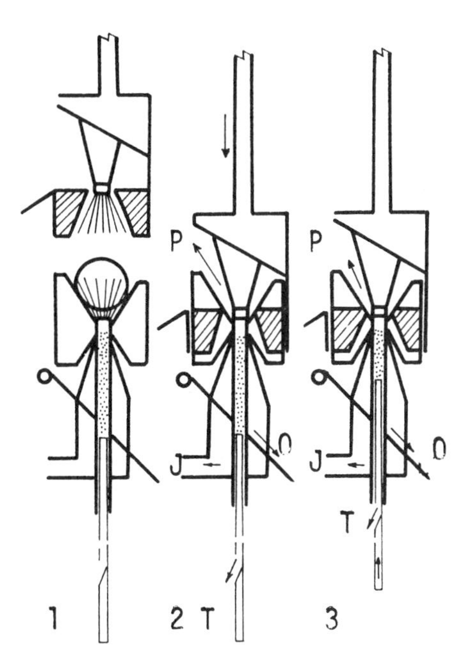

FIG. 7.5. Sketch showing principle of FMC in-line juice extractor. This model is arranged for simultaneous collection of cold pressed peel oil (O), shredded peel (P), and juice (J). (Courtesy FMC Corp.)

and finally by changing the height to which the central tube rises in the last operation. From the extractor the juice passes to a finisher where excess pulp is removed. A 5-head machine will handle from 200 to 400 fruit/min.

Brown Model 400. Used for many years in the citrus industry and a continuing favorite, the Brown Model 400 produces a high-quality juice very low in peel oil content. The machines are arranged in batteries adjusted to the size of the fruit being used. Each fruit is sliced in half, the two halves going to opposite sides of the machine. The fruit halves are carried in synthetic rubber cups revolving on a horizontal plane and reamed by revolving reamers mounted on a synchronized carrier revolving on an inclined plane. The juice collects in the machine and is conveyed to the finishers, while the peel is diverted to the waste conveyer. The maximum capacity is 350 fruit/min.

Brown Model 700. This machine operates on the same principle as the Brown Model 400 in that the fruit is halved and the juice removed with a rotating reamer. It produces a juice with the same high quality and low oil standards as the older model. In the older machine the cups and reamers are mounted on rotating disks but in the new machines the cups are mounted on tracks, which permit programming of the approach of the reamer into the fruit. The reamer can be made to penetrate rapidly at first and then more slowly as the point of maximum penetration is reached. This machine handles up to 700 fruit/min.

Brown Model 1100. Three parallel lines of single file fruit are fed into this extractor. Each orange drops to its proper position as it becomes wedged between pairs of revolving, tapered, circular, synthetic rubber disks. The smaller fruit falls nearer the center and the larger sizes are engaged nearer the periphery. The disks carry the fruit through a stainless steel knife, which slices it in half. Each half travels a converging path between a perforated stainless steel grid and the disk as the disk rotates. Spent peel is discharged out of the rear of the machine after having traveled approximately 320° around its circular path. Juice flows to the bottom of the collector and can be divided into two fractions, i.e., that which has been expressed lightly, containing low peel oil and pulp, and the juice recovered from the tighter extraction which contains higher levels of peel and oil pulp. The juice flows out from two outlets located at the bottom of the stainless juice collector. The extractor has a capacity of up to 12 tons of fruit/hr.

Finishing. The extracted juice is immediately strained free of pulp. Incidence of gelation in frozen concentrates is directly proportional to the time the pulp stays in contact with the fresh juice at temperatures above 35°F. Centrifuging removes essentially the remaining solid pulp.

The screw-type citrus juice finisher is used commercially with a capacity of 100–200 gal/min. In the center of the finisher there is a large screw that turns rapidly and carries the pulp and juice along. In operation, a reinforced screen is placed over the screw. As the material is carried along, juice passes through the screen and excess pulp, bits of peel and rag, and seeds pass out from the annular orifices. Even pressure is maintained by air against a flexible diaphragm pushing against the end of the shaft. The annular opening opens just enough to let the excess pulp and seeds pass.

An example of a commercial screw-type finisher is the Brown Model 2503 unit which is slightly different in design from the one just described. The unit has a capacity of 200 gal/min. Separation of the juice from the seeds and pulp is accomplished by means of a rotating screw within a cylindrical screen. The liquids and a certain amount of solids pass through the screen. The balance of the solids is discharged through a nonrotating air-loaded valve. Dryness of discharged pulp can generally be accurately controlled by varying the air pressure on the discharge valve. However, the amount of product going through the Brown Model 2503 also affects the dryness of discharged pulp when the air pressure is maintained at the same level. This model gives highly efficient liquid-solids separation with a minimum degradation of solids.

The FMC in-line citrus juice extractor does a preliminary removal of pulp in the perforated tube, so there is less pulp to be removed in the final finisher. This means that different settings must be made than for juice from Brown juice extractors where everything but the peel must pass through the finisher. Finishers with paddles rotating within a cylindrical screen (paddle finishers) have been used. In some cases the manufacturer may wish to provide some larger pieces of pulp in the juice to simulate the appearance of fresh juice. This can be done by increasing the size of the openings in the finisher and decreasing pressure on the finisher shaft. Also, the perforated tube in the FMC machine may be replaced by one with larger sized openings. This may be done for only part of the juice.

Yield is important to the grower who wants the highest return for his fruit and to the processor who is also responsible for the quality of the finished product. In Florida, yield is regulated by law. Test extractors are located in each processing plant and are used by state inspectors to determine a reasonable yield and the processor must comply. If fruit has been damaged by cold weather, permissible yield is reduced as a safeguard of quality.

Evaporation and Mixing with Fresh Juice. The TASTE evaporators utilize a flash concentration at a temperature above 190°F and require less than 8 min to convert orange juice from approximately 12 to 65% solids.

Evaporated juice is cooled to about 60°F and mixed with freshly squeezed pulpy juice. This step adds volatiles, which restore the fresh juice taste lost during concentration, and dilutes the concentrate to market specifications for soluble solids.

The cutback juice is sometimes prepared in separate extractors operating at lower pressure, and larger screen openings are used in the finisher so as to produce large particles of pulp. The pulp and juice are separated by centrifuging or screening. The pulp is pasteurized to inactivate the pectinesterase and then recombined with the juice for use as cutback to restore fresh flavor to the concentrate. Thus, the only unpasteurized part of the concentrate is the fluid part of the cutback juice.

Several orange juice processors have attempted to use recovered essence in lieu of fresh addback juice. Some of the methods used include freeze concentration, wherein the original aroma is retained and the juice is not subjected to heat; essence recovery under vacuum; and essence recovery from the condensate from the first stage of the evaporator. Most plants in Florida do not restore essence, because they do not believe the benefits warrant the cost. Orange essence recovered by refluxing the distillate from the concentrating operation is not stable, and the beneficial effect of such added essence disappears on storage. Orange concentrates with added essence recovered in a liquid-sealed vacuum pump system have retained excellent quality when stored for more than 6 months at 0°F. Most of the orange flavor in today's concentrate comes from the naturally accompanying peel oil plus the flavors in the fresh cutback juice.

Cooling, Filling and Freezing. The mixture of fresh unprocessed orange juice plus concentrate is cooled to 40°F in cold-wall tanks and is passed through refrigerated Votators to further lower the temperature to about 25°F. The concentrate at 25°F is filled into 6-oz cans at a rate of about 1000 cans/min/machine. Sealed cans are then passed through a freezing tunnel for approximately 1 hr at −50°F to ensure quick freezing before they go into storage at −10° to −15°F.

Quality Standards and By-Products. In 1966, the state of Florida made a strong effort to maintain high quality in frozen orange concentrate by prohibiting the washing of citrus pulp and by increasing the minimum fruit solids content from 42° to 45° Brix. For many years the yield of 4-fold frozen orange concentrate was in excess of 1.5 gal/90-lb box of fruit. Under the new quality measures, the yield is approximately 1.25 gal/box.

There are many by-products from the manufacture of orange juice concentrate. These include peels, pulp, and seeds, which are pressed and dried for cattle feed. The press liquid yields citrus molasses, which is also used for cattle feed, as well as terpene oils for paints and plastics. Orange oil goes to the food, beverage, pharmaceutical, perfume, and soap industries.

Synthetic Frozen Orange Concentrate. The shortage of frozen concentrate resulting from the Florida freeze of December 1962 led to the development of several substitute frozen concentrated "orange" drinks in 1963, among them a product named AWAKE (Black and Polopolus 1966). This synthetic concentrate is made of sugar, syrup, water, corn syrup, orange pulp and rind, citric acid, gum arabic, vegetable oil, cellulose gum, potassium citrate, calcium phosphate, vitamin C (ascorbic acid), natural and artificial flavors, vitamin A, artificial color, and vitamin B. In 1965, the product had obtained approximately 10% of the concentrated orange juice market. The capital requirements for plant and equipment required to produce a given volume of synthetic frozen concentrated orange juice are estimated to be approximately one-fifth of the amount that would be required to produce an equivalent volume of natural frozen orange concentrate. The cost of production for synthetic orange concentrate is lower than that for natural frozen concentrate. These cost advantages enabled the manufacturer of AWAKE to advertise and market the product aggressively. Severe freezing weather in Florida in 1980 and 1984 stimulated the production of synthetic orange juice from cold-damaged fruit, as well as from juice imported from Brazil.

Chilled Citrus Juice. Chilled citrus juice is made throughout the year from fresh citrus fruit and moves rapidly into consumption outlets after manufacture. In addition, substantial quantities of chilled orange and grapefruit juice are prepared by reprocessing single-strength bulk juice and reconstituting bulk-frozen concentrate.

Grapefruit Juice Concentrate

The volume of frozen grapefruit juice concentrate produced is much less than that of frozen orange concentrate. One reason for this may be that the flavor of the frozen grapefruit product is only slightly, if at all, superior to the cheaper single-strength canned juice. Processing of frozen grapefruit juice concentrate is similar to that of frozen orange juice concentrate, namely, washing, juice extraction, heat inactivation of pectic enzymes, vacuum evaporation, filling, and freezing. The details of the operation have been described by Veldhuis (1971B).

Frozen concentrated juices that have not been pasteurized or heat-treated to inactivate pectic enzymes are subject to gelation, that is, formation in the can of a jelly-like mass that is difficult to reconstitute. Grapefruit juice is more susceptible to gelation than orange juice because it contains more pectin and pectic enzymes. To inactivate these enzymes, it must be heat-treated prior to concentration for periods of a few seconds up to 1 min at temperatures ranging from 160° to 205°F. The juice is then concentrated fivefold under vacuum at temperatures below 80°F and diluted or "cut back" with fresh, unconcentrated, deaerated, heat-treated juice having a higher pulp and oil content to about fourfold. Often, cold-pressed grapefruit oil is added to replace that lost during concentration. The concentrated juice is filled into lithographed cans, sealed, and frozen. Often, it is frozen to a slush before filling to lower the freezing load of the final freezing system. Most frozen concentrated grapefruit juice is sweetened with added sugar.

Concentration and pasteurization are achieved in the same equipment used for frozen orange juice concentrate.

Both sweetened and unsweetened fruit concentrated grapefruit juices are covered by USDA Grade Standards. The soluble solids content of the final unsweetened product must contain at least 3.47 lb of soluble grapefruit solids per gallon exclusive of the added sweetening ingredients. In Grade A unsweetened concentrate the Brix–acid ratio may vary from 9-to-1 to 14-to-1 and in the sweetened product from 10-to-1 to 13-to-1. All the types are reconstituted by adding three parts of water to one part of concentrate. While either seedless or seeded cultivars can be used, seeded cultivars such as 'Duncan' are generally preferred.

Citrus Puree

Frozen purees made from whole citrus fruits are useful ingredients in the commercial preparation of baked foods, beverages, and frozen desserts. Orange, tangerine, lemon, and lime purees are processed in the same manner. Sound, mature fruit is thoroughly washed with detergent in water and rinsed well. After washing, the fruit is trimmed to remove stem ends and discolored spots. The trimmed fruit is either crushed or sliced and is then put through a rotary or tapered screw press fitted with stainless steel screens having 0.027- to 0.044-in. perforations, depending upon the intended use of the puree. Yield of puree is approximately 50–60% of the weight of the whole fruit; it should contain 0.40–0.75% of peel oil, depending on the kind of fruit being crushed. The oil content of puree may be adjusted by grating off the skin before crushing, or by adding sufficient single-strength juice to bring the peel oil content down. Purees are packed both with or without the addition of sugar. Sweet purees are usually mixed with one part sugar to five parts puree. The puree may be either directly filled into enameled tin containers or slush-frozen and then filled. Storage should be at 0° to −10°F.

Lemon Concentrate

Commercial production of lemon products in the United States is confined almost exclusively to California. The 'Eureka' cultivar makes up approximately 88% of total lemon production. The lemon fruit is picked for size only and held under controlled storage conditions for ripening from green to yellow color. About 40% of the lemon crop is processed into juice and frozen lemonade concentrate. The operations in processing lemons into juice are similar to those used for preparing orange juice. The extracted juice is pasteurized to inactivate pectic enzymes, and then screened to remove rag and seeds. It is held in brine-jacketed tanks to chill. The juice should be deaerated in these tanks by applying a vacuum of 25 in. Hg for 30 min. For frozen single-strength juice, the chilled juice is drawn from the cold holding tanks and further cooled to 30°F by passing through a heat exchanger, filled into enamel-lined cans, sealed, frozen, cased, and stored at 0° to -10°F.

Frozen lemon juice concentrate is made by concentrating single-strength lemon juice in a low-temperature vacuum evaporator. Evaporators of various types make use of falling-film heat exchangers, in which the juice runs in a thin film down the inside of a tube while the tube is gently heated from the outside. The temperature of the juice during evaporation depends on the design of the unit, but is kept in the range of 60° -80°F.

Stainless steel is the standard construction material for all parts in contact with the juice or concentrate. The lemon juice is concentrated to 43° Brix, chilled to around 30°F by passing through a heat exchanger, filled into drums lined with polyethylene bags, and frozen at -10°F. Sometimes the chilled concentrate is filled into cans, sealed, and frozen at -10°F for institutional use.

Lemonade and Limeade

Lemonade is a beverage prepared from lemon juice or concentrated lemon juice with water, sweetener, and optional ingredients (Anon. 1968). The optional ingredients are lemon oil, cold-pressed lemon oil, concentrated lemon oil, and lemon essence recovered during concentration of lemon juice; sodium benzoate and sorbic acid as preservatives; and buffering salts, emulsifying agents, and weighting oils. The proportion of lemon juice ingredients used is sufficient to yield not less than 0.70 g anhydrous citric acid/100 ml. It may be heat-treated and preserved by freezing or canning. Colored lemonade is lemonade colored with a color additive. Limeade is similar to lemonade except that fruit and flavoring ingredients are derived from mature limes of an acid variety.

Frozen Concentrate for Lemonade. Frozen concentrate for lemonade contains the proper amount of added sugar. It is reconstituted into lemonade when needed. The concentrate is principally a single-strength lemon juice with sugar added. Since most people prefer a tartness in lemonade, the acid content of the product is adjusted by the addition of approximately 10% of concentrated lemon juice to give the proper balance of sugar and citric acid.

According to the USDA (1962), a typical frozen concentrate for lemonade can be prepared as follows:

> Add sufficient concentrated lemon juice to 280 gal. of single-strength lemon juice and 2800 lb of granulated sugar so that each 100 gm of the final product will contain from 3 to 3.5 gm of citric acid. This will make approximately 500 gal. of 55° Brix concentrate. The frozen concentrate is reconstituted to lemonade by adding 4 volumes of water to each volume of concentrate.
>
> It is desirable that some juice cells be added to this concentrate to improve the appearance of the reconstituted lemonade. This is done by screening juice cells from juice after extraction and mixing them into the concentrate.

Frozen Concentrate for Limeade. Frozen concentrate for limeade is prepared by adding enough sucrose to single-strength lime juice to raise the Brix to about 48°. Some processors apply a mild heat treatment during preparation, and subsequently freeze and store the product at 0°F or lower. On reconstitution with 4–4½ parts of water, an excellent limeade is obtained. As limes vary somewhat in solids and acid characteristics, some processors concentrate batches of juice for blending to a uniform acid content in the finished product. Then by the addition of sugar to a 45° Brix level, a uniform composition is obtained.

When lime juice is concentrated, much of its characteristic aroma and flavor are lost. Efforts to regain the original flavor through the addition of fresh juice, as is the practice in the production of frozen orange concentrate, or through the use of emulsified lime oil have been unsatisfactory. The use of lime puree in amounts sufficient to give an oil content of 0.003% in the reconstituted limeade is a far superior method. The products prepared by this method include an eightfold sweetened concentrate, which requires the addition of seven parts of water to prepare the beverage, and a more concentrated product, which is diluted to 35 times its volume by the addition of sugar and water.

Blended Grapefruit and Orange Juice Concentrate

Frozen blended grapefruit juice and orange juice is of excellent quality and resembles the freshly prepared product. Although it is less popular than frozen concentrated orange juice, it has a place in any line of frozen food products.

The same procedures are used in producing this blend as are used in preparing the straight products. Blended products of both sweetened and unsweetened types are covered by USDA Standards for Grades. They should contain no less than 50% orange juice in the mixture. The USDA Grades call for 40°–44° Brix in unsweetened concentrates. In sweetened concentrates the Brix must be at least 38° before sweetening and 40°–48° after sweetening. For Grade A, the Brix–acid ratios in the packed concentrate may vary from 10-to-1 to 16-to-1 if unsweetened, and from 11-to-1 to 13-to-1 if sweetened.

Tangerine Juice Concentrate

Frozen tangerine juice concentrate, when properly prepared, is equal in quality to frozen orange juice concentrate.

While grapefruits and oranges will withstand considerable rough handling, tangerines have a loose, tender skin which is easily broken. If the skin is broken and the fruit bruised, bacteria and yeasts readily attack the fruit and undesirable enzyme actions occur. For these reasons tangerines cannot be handled in orange bins but must be handled in boxes, or loose in trucks to a depth of not over 2 ft. The 'Dancy' tangerine is the most common cultivar.

The processes and equipment used in manufacturing concentrated tangerine juice are practically the same as those used with oranges. Since tangerines are smaller than oranges, the yield of juice from a given number of extractors is smaller. About double the amount of extracting equipment is required to furnish juice to keep the evaporators operating at full capacity. To minimize off-flavors care should be taken during extraction to avoid incorporation of excessive oil from pulp and peel; tangerine juice should contain not more than 0.02% of recoverable oil and not more than 7% of free and suspended pulp. Unlike orange juice, frozen tangerine juice concentrate shows little tendency to gel and it is unnecessary to heat-inactivate enzymes prior to concentrating.

USDA Grade Standards have not been established for frozen concentrated tangerine juice. The values for Brix–acid ratio, peel oil content, and concentration have followed those prescribed for the orange product quite closely. Recently more of the pack has been sweetened. A Brix of 44° is common for a 3-plus-1 concentrate.

Grape Juice Concentrate

Most of the grape juice concentrate marketed today is prepared from Concord grapes (*Vitis labrusca*) grown in New York, Michigan, Washington, Pennsylvania, Ohio, and Arkansas. The grapes are harvested when the soluble solids have reached a concentration of 15–16%.

After washing in a twin-screw washer, the grapes are conveyed to a stemmer consisting of a perforated, slowly revolving (20 rpm) horizontal drum inside which several beaters revolving at a much faster speed (200 rpm) knock the berries off the cluster and partially crush them before they are discharged through the drum perforations; the cluster stems are expelled from the open end of the drum. The crushed fruit is then pumped through a tubular heat exchanger where it is heated to 140°–145°F for extraction of the pigments and juice. The hot pulp may then go to hydraulic presses where the juice is removed. The expressed juice may be clarified in a centrifuge or filter press. If a filter press is used, 1–2% of diatomaceous earth is used to maintain a high filtering rate and provide ample removal of suspended matter. About 185–195 gal. of juice are obtained from a ton of grapes. In some plants screw presses are used for all or part of the crushed grapes, but this increases the suspended matter that must be removed later.

The clarified juice is then pasteurized in tubular or plate heat exchangers to a temperature of 180°–190°F and quickly cooled to 30°F before storing in tanks kept in rooms refrigerated at 28°F. The cooling of the juice is usually accomplished in two or more steps. In some heat exchange systems, a regeneration cycle is used in the first step whereby the hot juice leaving the pasteurizer preheats entering juice.

The method of handling the cooled juice depends on the intended use of the concentrate. If it is to be used in later manufacture of jelly, the juice is stored at 28°F for 1–6 months to permit settling of the argols (potassium acid tartrate, tannins, and colored materials) that would give a gritty texture to the jelly or detract from its clarity. The clear juice is siphoned off the precipitate in the storage tanks and may be refiltered. If the concentrate is to be sold as a blend formed by mixing sugar and ascorbic acid before canning and freezing, the cold storage tank merely serves as a surge tank since the juice is pumped out within a few hours to the concentrator. Here a polishing filter is used before the concentrator to minimize fouling of the evaporator tubes. The concentrate for either jelly manufacture or blended juice may be stored in tanks at 27°F prior to subsequent processing. Whenever single-strength juice is bulk-stored at 27°F prior to concentration, spoilage may occur, so all pipelines and equipment from the pastuerizer to the cold room should be of a sanitary design for ready and frequent cleaning. The interior surfaces of storage tanks must be relatively smooth or free from crevices and the tank should be thoroughly cleaned before use.

Concentration is carried out in two steps. First, the volatile flavoring materials are stripped from the juice. The stripped juice is then concentrated to the desired density. The volatile components are removed by heating the single-strength juice to 220°–230°F for a fraction of a minute in a heat exchanger, flashing a percentage of the liquid into vapor in a jacketed tube bundle, and then discharging the liquid and vapor through an orifice tangentially into a separator. The separator should be of sufficient size so that the vapor velocity is reduced to 10 fps or less for minimal entrainment. From 20 to 30% by weight of the original juice flashes off as a vapor that is led into the base of a fractionating column filled with ceramic saddles or rings. A reflux condenser on the vapor line from the column and a reboiler section at the base of the column are used to provide the necessary reflux ratio. The vent gases from the reflux condenser are then chilled in a heat exchanger and the condensate containing the essence is collected at a rate equivalent to 1/150 of the volume of entering flavoring material.

One of the flavor components of Concord grape juice is methyl anthranilate, which is only slightly soluble in water. The high boiling point and low solubility in water has resulted in losses of methyl anthranilate when the efficiency of the stripping column is low. It has been suggested that these losses may be reduced by increasing the vaporizing temperature.

In a typical formulation, grape juice is concentrated to a little over 34° Brix, and essence or fresh cutback juice is added to reduce it to this concentration. Sucrose is added to 48° Brix and citric acid is added until the total acidity is 1.8% calculated as tartaric acid. When diluted with an equal quantity of water, the equivalent of sweetened single-strength juice is obtained; however, for a more palatable beverage, three parts of water are added. The product is labeled as concentrated, sweetened grape juice. The frozen grape juice concentrate sold on the market as a retail item is a sweetened product and is usually sold at about 48° Brix, the solids being composed of 39% from the fruit and 9% from sucrose. The concentrate may be cooled to 20°–30°F in a heat exchanger or cold-wall tank, filled into cans, sealed, cased, and allowed to freeze in subzero storage.

Frozen grape juice concentrate is often purchased according to its grade. The various types of concentrates, methods of scoring, and objective tests of quality are listed in the USDA Standards for Grades of Frozen Concentrated Grape Juice.

Buyers of bulk concentrated grape juice must decide whether they want a high-quality concentrate of 48° Brix packed in 50-gal. drums and frozen, or a lower-quality concentrate of 72° Brix, which can be preserved without either heat processing or freezing. The 72° Brix heavy-density concentrate is used in manufacturing jellies, while the frozen 48° Brix concentrate is used primarily for beverages.

Guava Puree and Concentrate

The guava is an important pomiferous fruit of the Myrtle family. The species (*Psidium guajava* L.) is grown commercially and has been introduced into the West Indies, Florida, India, West and South Africa, and several of the Pacific island groups (Boyle *et al.* 1957). The ripe fruit can be eaten fresh and has been popular for making nectars, juice drinks, and jams (Luh 1971).

The fruit of the common guava tree has a rough-textured, yellow skin and varies from round to pear-shaped. In the wild, individual fruits from 1 to 3 in. in diameter are found, but cultivated trees yield fruits up to 5 in. in diameter and weighing 1½ lb. The color of the inner flesh varies from white to deep pink to salmon-red. Desirable fruits for processing are those with a thick outer flesh and a small seed cavity since they yield more purée per unit weight than thin-fleshed types. The distinctive musky flavor of guava is more intense in some selections than in others and the acidity varies, too, from those with a pH of 3.0 to a few that have a pH of 4.0. The majority of the common guavas in Hawaii fall into the sour or subacid category with pH values ranging from 3.0 to 3.5.

Guava is a rich source of vitamin C and a fair source of vitamin A, calcium, and phosphorus. Guava pectins show a high methoxyl index and produce good gels at 65% soluble solids. The gel is stable at pH 2.1–2.4. The nonvolatile organic acids in guava have been identified as malic, citric, and galacturonic acids. Citric and malic acids are present in almost equal amount, and a small amount of lactic acid is also present in cultivated guavas.

Guavas with few seeds and small quantities of stone cells are desirable, provided they have the other essential qualities. The more acid fruits, pH 3.3–3.5, are better for processing than the sweeter fruits. Good color and flavor and high vitamin C content are also important factors in selecting fruit types for processing.

Transportation and Storage of Fresh Fruit. Because guavas picked at peak maturity do not keep well, there should be no delay in getting them to the plant for quick processing or refrigerated storage. It would be desirable to ship only fruits that are firm and slightly underripe, and to finish off the ripening under controlled conditions at the processing plant.

Fully ripe guavas should be processed without delay, but if necessary they can be held for about a week at 36°–45°F.

Puree Processing. When guavas are ready for processing, they are fed to an inspection belt where spoiled fruits are removed and green fruits set aside for ripening. Those fruits that have only small defects may be trimmed to make them acceptable for processing. A flow sheet for guava processing is shown in Fig. 7.6.

From the inspection belt the sound fruits drop into a washing tank or onto a washing belt. Mechanical or manual agitation and the addition of a detergent help to remove dirt, debris, and dried-on flower parts. An elevator belt removes the fruits from the washing bath and a clear water

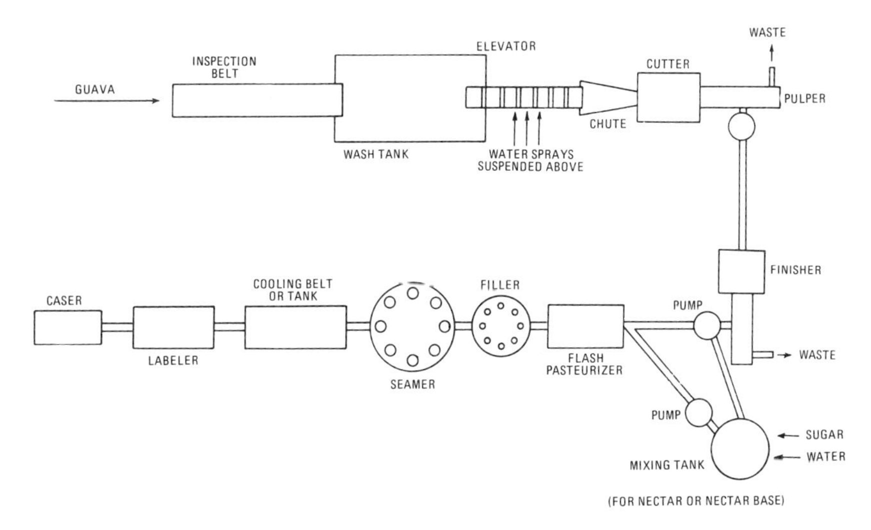

FIG. 7.6. Flow sheet for guava processing line. The pasteurizer and cooling belt are bypassed for frozen products. (From Boyle et al. 1957.)

spray rinses off the detergent. After washing, the whole fruit can be fed directly into a paddle pulper for maceration into a puree; if the fruits are rather firm, it may be necessary to attach a chopper or slicer to the hopper that feeds into the machine. To remove seeds and fibrous pieces of skin tissue, a 0.033- or 0.045-in. perforated screen should be used on the pulper. Chopping of all fruit before feeding into the pulper may be desirable since it allows for a more uniform feed rate.

Removal of Stone Cells. The outer flesh of some guava fruit has a considerable number of hard stone or grit cells. Removing most of these stone cells improves the texture of the final product. One method for getting rid of these unwanted cells is to pass the puree through a paddle finisher equipped with 0.020-in. screens. This machine is a pulper in which the steel paddles are replaced by neoprene rubber strips held in place with stainless steel or hardwood cleats. The rubber paddles are adjusted so that they almost touch the screen and the angle decreased from that at which the paddles are set for the pulping operation. For this operation, the speed of the machine is reduced to 600–800 rpm while the puree is fed into the hopper at a uniform rate. Adjustment of paddles, pitch, speed, and waste gate should be continued until the waste is slightly moist.

Another method of finishing is to run the puree through a mustard mill so that the stone cells are pulverized. This operation reduces the grittiness but does not improve the color. Both methods are used by guava processors.

Deaeration. After pulping and finishing to remove the seeds and stone cells, the puree is usually put through a deaerator. The removal of oxygen lessens deterioration in storage, because oxidation is the chief cause for loss in color, vitamins, and flavor. The removal of air makes for a more uniform and smoother-looking product with improved color. Also, the prevention of foaming caused by air allows correct and uniform fill of containers.

Frozen Guava Puree. After removal of stone cells, a slush freezer may be installed in the processing line to chill the product before it is filled into containers. The advantage of the prechilling is that the puree will freeze more rapidly, maintaining high quality and allowing labeling and casing to take place sooner. Because of the high acidity of this product, the containers should be enameled or protected with a plastic lining.

Frozen Guava Nectar Base. Frozen guava nectar base is a combination of puree and sugar in such proportions that it may be diluted with water by the consumer in the same manner that many other fruit juice concentrates are prepared. The optimum dilution is 2½–3 parts of water to 1 part of the base (Boyle *et al.* 1957).

The following formula is recommended for making guava nectar base: 100 lb guava puree (7% soluble solids), and 48 lb cane sugar. If the soluble solids of the puree vary from 7%, the amount of sugar should be adjusted in the formulation. The pH range of the finished product should be between 3.3 and 3.5. Most wild guava is in the sour or acid category, but occasionally sweet guavas may be brought to the processor. By adding citric acid, the pH of the final mixture may be kept in the desired range.

After blending the correct amounts of puree and sugar in a mixing kettle, the mixture should be pumped through a slush freezer before going to the filling machine. After filling and closing the containers (preferably lacquered), they should be placed immediately in the freezer and kept at 0°F or −10°F. Labeling and casing should take place after freezing without allowing the cans to thaw.

Mango Puree

The mango is a delicious tropical fruit. Certain processed mango products such as pickles, chutney, and canned slices are available in Jamaica, India, Mexico, and South Africa. Puerto Rico and the Philippines also produce mangoes (Ross 1960A). The cultivars of mango grown in Hawaii are 'Haden,' 'Pirie' (Yee 1963), 'Pope' (Hamilton 1960), 'Joe Welch,' 'Zill,' and several others (Orr and Miller 1955).

Mangoes are marketed chiefly as fresh fruit. Processing has not developed on a large scale, but occasionally small packs of frozen slices or of chutney have been marketed.

A procedure for freezing mango puree was described by Brekke *et al.* (1968) as follows: 'Haden' mangoes were sorted, washed in a rotary spray washer, and the seeds removed with knives by hand. A mechanical scraping device was used to remove the peel for some of the experimental lots. The fruit was put through a plate heat exchanger for rapid heating and cooling to inactivate the enzyme catalase. In the heat exchanger the puree was heated to 195°–200°F, held at that temperature for 1 min, and then cooled to 90°–100°F. It was then filled into 30-lb tins with polyethylene liners and frozen at −10°F.

Papaya Puree

Papaya or pawpaw is a widely used tropical fruit. The 'Solo' cultivar is grown in Hawaii largely for the fresh fruit market. In Australia, it is an important component in tropical fruit salads. Small quantities are also used in making nectars and mixed fruit drinks.

The Hawaii Agricultural Experiment Station has developed a procedure for making papaya puree that can be preserved by freezing. Fresh papayas are inspected and sorted to remove damaged and undesirable ones. The fruits are immersed in water at 120°F for 20 min. The warm water treatment is used to prevent undue spoilage losses during ripening. The fruits are ripened at room temperature for 5–6 days. It is advisable to cool the product to 35°F prior to processing to lessen the possibility of gel formation. The chilled fruits are washed, trimmed with a stainless steel knife or a hydraulic cutter to remove the ends, and then cut into four pieces in a slicer. Skins are separated in a skin separator. The mixture, containing the seeds, is passed to a paddle pulper with a 0.022-in screen. Seeds are removed in this operation. The puree is pumped through a scraped-surface heat exchanger at 210°F or higher, and held there for 1 min. It is necessary to heat-inactivate pectinesterase in the puree immediately after seed removal. The product is cooled to 85°F in a second scraped-surface heat exchanger, and then passed through a paddle finisher with a 0.020-in. screen to remove specks and fibers. The puree is then filled into polyethylene-lined 30-lb cans, sealed, and deep frozen to −10°F or below.

A pilot model machine for separating papaya skins from the seeds and flesh was used to lessen or eliminate the bitter flavor. Papaya puree made with this machine showed no sign of bitter off-flavor. The machine consists essentially of two reels rotating at different speeds with the papaya slices fed between them. The upper reel is made of wood covered with a thin sheet of corrugated rubber. The larger lower reel, which separates the flesh from the skin, is constructed of stainless steel rods that are approximately 5/16 in. apart. The machine is referred to as a skin separator.

Since the frozen puree might be stored for some time before reprocessing, it is necessary to heat-inactivate various enzymes that could cause development of off-flavor during storage. Gelation in frozen papaya puree can be prevented by heating the product at 200°–210°F for 1 min and cooling rapidly. Heating papaya puree to 210°F for 1 min immediately after seed removal consistently prevents gel formation. The treatment also inactivates catalase, peroxidase, and papain. Yamamoto and Inouye (1963) described the use of sucrose to inhibit pectinesterase, which can cause gelatin in reconstituted frozen papaya puree.

When papaya puree is heated to boiling, pectin esterase activity is completely lost. Demethylation of pectin in papaya caused by pectinesterase activity during processing may result in gelation in the final product.

Since the adoption of uniform standards through a marketing order granted by USDA, increasing amounts of off-grade papayas have become available for processing. The development of off-flavors has been a major obstacle to increasing the amount of papayas processed into puree (Luh 1971). Commercially frozen papaya puree, as now prepared, is not heat-treated. Hence, off-flavors develop by enzymatic reactions. The puree is frozen in bulk containers of 25 lb or more and is subsequently allowed to thaw at room temperature for about 24 hr before being used in the preparation of other food products. Conceivably, off-flavors could arise from microbial action in the center of the puree mass prior to freezing and at the periphery of the mass during thawing.

Brekke *et al.* (1972, 1973) developed an improved method for preparation of papaya puree. A flow chart of their process is shown in Fig. 7.7. Whole ripe papayas are steamed for 2 min to coagulate the milky latex in the peel, to inactivate enzymes in the peel, and to soften the outer portion of the fruit. The fruit is spray-cooled, sliced, and crushed in a deflesher or crusher-scraper. The peel is separated from the pulp and seeds in a centrifugal separator. The intact seeds are separated from the pulp by means of a paddle pulper fitted with rubber paddles to prevent rupturing the seeds and sarco testa. The pulp is then acidified with citric acid (in 50% solution) to pH 3.3–3.6, pumped through a heat exchanger where it is held at 205°F for 2 min, then cooled. It then goes through a paddle finisher fitted with a 0.02-in. screen to remove specks and some fiber. The puree is filled into 25-lb paperboard containers with plastic bag liners and frozen in a blast freezer at 0°F.

Off-flavor development in frozen papaya puree can be minimized by adding citric acid, which inhibits microbial growth and enzyme activity. Chan *et al.* (1973) reported that off-odor in frozen papaya puree resulted from both enzymatic and microbial activity.

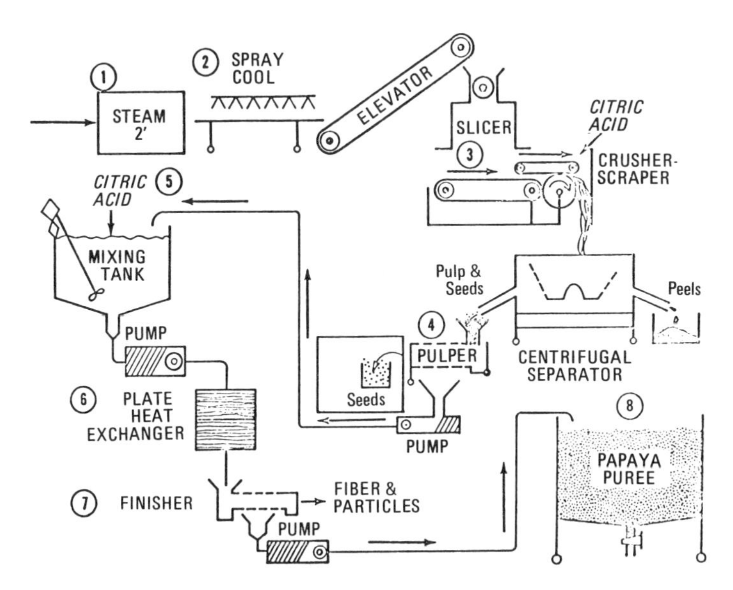

FIG. 7.7. Flow sheet for processing papaya puree by the improved method. (From Brekke et al. (1973); courtesy Hawaii Agriculture Experiment Station.)

Passion Fruit

The purple passion fruit (*Passiflora edulis*) has become a cultivated plant of commercial importance. The yellow passion fruit (*Passiflora edulis,* var. *flavicarpa*) was introduced into Hawaii from Australia and has become the favored commercial cultivars in Hawaii. The yellow cultivar grows more vigorously and yields more prolifically (average yield 10 tons/acre) than the purple cultivar (average yield 1–2 tons/acre).

The passion fruit is an oval or sometimes round fruit weighing approximately 1 oz and having either dark purple, leathery skin or bright yellow, waxy skin. Within the fruit are numerous embryo sacs containing black or dark brown seeds enclosed in orange-yellow gelatinous pulp. The pulp has an acid but highly attractive and distinctive flavor. There is a real though subtle difference in flavor between the purple and yellow cultivars, and the pulp of the yellow cultivars tends to be more acid.

Extraction of Pulp. Passion fruit pulp can be extracted, in a manner similar to citrus juice extraction, by reaming the halved fruit on a small burr or wire loop. A mechanical extractor used in Australia consists of feeding the fruit into the space between two revolving shallow-pitched cones. A centrifugal extractor has been developed in Hawaii and applied successfully in commercial practice. The fruit is sliced by means of a gang of rotating knives, and the slices (5/8 in.) drop directly into a perforated centrifuge bowl. The bowl has sloping sides and four baffles at right angles to the sides. When it is rotated at a centrifugal force of 175 g, the pulp and seeds from the slices are thrown out through the holes in the basket while the residual skins climb the walls of the basket and are thrown out over the edge. The pulp and skins are then collected from separate chutes. The extractor has a capacity of 4000 lb of passion fruit/hr, and an efficiency of extraction of 94% is claimed.

Extraction serves only to remove the pulp from the rind; separation of the juice is accomplished in a two-stage process that employs either a brush finisher, or a paddle finisher with the paddles faced with Neoprene. In the first stage, the pulp passes through a stainless steel screen with 0.033-in. holes; this is followed by a finishing operation with a screen of 60–80 mesh stainless steel to remove broken seed fragments. The average yield of juice from the fruit is 30–33%. A typical yield of screened juice is 70 gal/ton. A flow sheet for the processing of passion fruit products is shown in Fig. 7.8.

Frozen Passion Fruit Juice. After mechanical extraction and removal of seeds, passion fruit juice is deaerated, filled into lacquered cans, sealed under vacuum or steam flow, and frozen. Frozen juice retains satisfactory quality for at least 1 year at 0°F.

Several workers have investigated the concentration of passion fruit juice. Vacuum evaporators can be used to make passion fruit concentrate. It is usual to add back the first 10–15% of the distillate to the concentrate. The product is then hot-filled into cans and frozen. Seagrave-Smith (1952) prepared a 1:1 blend of passion fruit and pineapple juice with sugar added to bring the solids content to 38.4%. This blend was evaporated under vacuum, diluted with fresh blend or with fresh blend together with the first 10% of distillate, and then the final concentrate was frozen. It was claimed to have satisfactory aroma and flavor although not as good as the initial blend.

Composition of Passion Fruit Juice. Among the soluble carbohydrates of passion fruit juice, sucrose makes up 25% of the total sugars, and glucose and fructose are also present. The principal acid in passion fruit juice is citric acid, which contributes 93–96% of the total acidity, while malic acid contributes 4–7%. Passion fruit juice contains little pectin but significant amounts of starch, which may settle out as a white or grey precipitate during storage of the juice or beverages made from it.

The fruit pulp consists of juicy arillus tissue and black seeds, has 10–15% sugar, 2.3–3.5% acid as citric acid, and a pH value of 3.4. The amount of pulp and juice corresponding to 1000 kg of fruit is 580 and 426 kg, respectively.

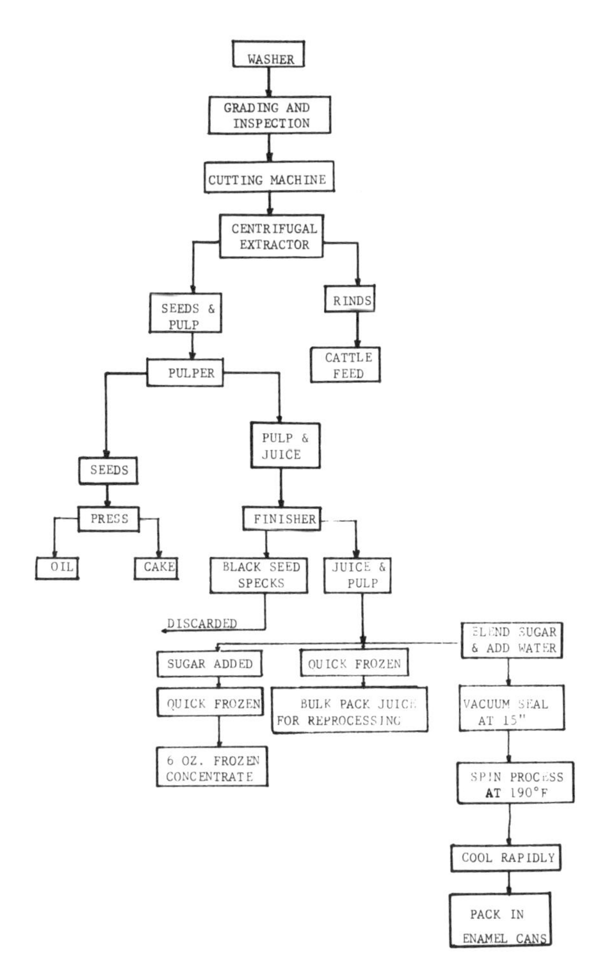

FIG. 7.8. Flow sheet for processing passion fruit. The same process may be used for blends. (From Luh 1971.)

Passion fruit juice is a good source of ascorbic acid. The purple cultivar appears to have higher ascorbic acid contents than the yellow cultivar.

The free amino acids in passion fruit juice are leucine, valine, tyrosine, proline, threonine, glycine, aspartic acid, arginine, and lysine.

The characteristic flavor of passion fruit is found in the water-insoluble oil, which constitutes about 36 ppm of passion fruit juice (Hiu and Scheuer 1961). Four components, *n*-hexyl caproate, *n*-hexyl butyrate, ethyl caproate, and ethyl butyrate, make up about 95% of the oil. Among these four, *n*-hexyl caproate is the principal component, accounting for about 70% of the volatile passion fruit essence.

The major components of passion fruit juice are listed in Table 7.1.

Table 7.1. Composition of Passion Fruit Juice

	Purple cultivar				Yellow cultivar			
	India					Hawaii		
Composition of juice	Max	Min	Mean	Australia	Queensland, Australia	Max	Min	Mean
Moisture (g/100 g)	82.5	76.9	80.4	76.0	71.1	84.2	79.6	82.0
Ether extract (g/100 g)	0.08	0.01	0.05	2.2	—	1.2	0.0	0.6
Soluble solids (g/100)	29.9	14.4	17.3	—	16.3	18.0	13.0	15.0
Acidity (g/100 g)	4.8	2.4	3.4	—	2.1	5.0	3.0	4.0
Brix/acid ratio	7.7	3.4	5.3	—	—	—	—	—
pH	3.2	2.6	2.8	—	3.3	3.3	2.8	3.0
Reducing sugars (g/100 g)	8.3	3.6	6.2	—	5.1	7.8	6.2	7.0
Nonreducing sugars (g/100 g)	7.9	2.3	4.6	—	4.2	—	—	—
Total sugars (as invert) (g/100 g)	13.3	7.4	10.0	18.4	—	11.6	9.3	10.0
Crude fiber (g/100 g)	—	—	—	—	14.2	—	—	0.2
Starch (g/100 g)	3.7	1.0	2.4	—	—	—	—	—
Protein (g/100 g)	1.2	0.6	0.8	2.4	2.4	1.2	0.6	0.8
Mineral matter (g/100 g)	0.52	0.36	0.46	—	0.70	—	—	—
Ca (mg/100 g)	18.4	9.7	12.1	11.0	—	—	—	5.0
P (mg/100 g)	60.4	21.4	30.1	—	—	—	—	18.0
Fe (mg/100 g)	4.0	2.3	3.1	1.2	—	—	—	0.3
Ascorbic acid (mg/100 g)	69.9	21.9	34.6	17.0	—	20.0	7.0	12.0
Thiamin (mg/100 g)	0.04	0.02	0.03	0	—	—	—	—
Riboflavin (mg/100 g)	0.19	0.12	0.17	0.10	—	—	—	—
Nicotinic acid (mg/100 g)	1.9	1.5	1.7	1.4	—	—	—	—
Vitamin A (IU/100 g)	1547.0	1073.0	1345.0	10.0	—	—	—	570.0

Source: Luh (1971); Boyle and Shaw (1955); Pruthi and Lal (1959).

Utilization of Passion Fruit Juice. Because of its unique intense flavor and high acidity, passion fruit juice has been described as a natural concentrate. When sweetened and diluted it makes a highly palatable beverage and the flavor blends well with other fruits and fruit juices in the preparation of fruit salads and punches.

For most general acceptance the sugar:juice ratio should not be more than 55:100 nor less than 45:100 and the rate of dilution should be 1 plus 3. In Australian experience, passion fruit juice sweetened to 50° Brix (sugar:juice ratio approximately 70:100) gives a very pleasing beverage when diluted 1 plus 4. Passion fruit juice is also highly suitable for flavoring ice cream, cake fillings, frostings, gelatin desserts, sherbets, and chiffon pies.

Blends of 5–10% of passion fruit juice with apple juice are very attractive in flavor. In South Africa, blends of passion fruit juice with pear, apple, peach, and orange juices have been prepared. A blended punch canned in Hawaii contains orange, pineapple, guava, papaya, and passion fruit juice.

Passion fruit cordials and squashes are popular beverage bases in Australia and South Africa. They consist of passion fruit juice with sugar syrup and acid added so that they normally contain 50–55% soluble solids and 1–2% acid. Cordials are diluted 1 plus 4 or 1 plus 5 with water for consumption. They are usually artificially colored and preserved with benzoic acid (600–770 ppm) or sulfur dioxide (220–350 ppm) since they are expected to be stable against spoilage after the bottle is opened. In Australia, the minimum juice content varies from 12.5 to 25% by weight in different states. Blends of passion fruit juice with grapefruit, grape, and plum juices are preferred to similar syrups made from passion fruit alone.

Pineapple Concentrate

The pineapple (*Ananas comosus* Merr.), sometimes called the King of Fruit, has its probable origin in Brazil or Paraguay. The major single area producing pineapples for commerce is the Hawaiian Islands. Substantial quantities of the fruit also are produced in Taiwan, Malaysia, Philippines, Thailand, Puerto Rico, Australia, and several countries in Africa. Although South America is its ancestral home, the pineapple industry in this area is relatively small compared with that in other areas. The principal cultivar of pineapple in Hawaii is 'Smooth Cayenne.' In other areas the cultivars of importance include 'Red Spanish,' 'Queen,' 'Singapore Spanish,' 'Selangor Green,' 'Sarawak,' and 'Mauritius.'

Frozen pineapple juice concentrate is used both as a beverage base in 6-oz cans and as an ingredient in the manufacture of such blended canned fruit drinks as pineapple–grape, pineapple-grapefruit, and other fruit drinks. Pineapple juice comes from several places on the processing line.

It is obtained by pressing the shell scrapings from the Ginaca machine, which peels the pineapple and forms the cylinder from which slices are made. The juice also comes from the pieces from the canning lines, from fruit too small for canning, and from drainage from the crushed pineapple operation. The solid material is shredded by various machines and filter aid such as infusorial earth is mixed with the finely ground material before it is fed to a hydraulic press. The liquid material is heated to coagulate some of the solids and the resulting thin slurry is passed through a continuous centrifuge which removes most of the suspended solids, including the fibers and other coarse small pieces. Pineapples too small for canning may be peeled and pressed for juice by machines similar to those used for oranges, or they may simply be crushed, heated, and pressed. Pineapple juice is sometimes homogenized to stabilize the slightly cloudy appearance. Detailed information on pineapple juice and concentrates has been presented by Mehrlich and Felton (1971).

For pineapple juice concentrate, the juice is concentrated in multiple-effect vacuum pans. The juice can also be concentrated in vacuum pans that operate at temperatures below 140°F. When short-time evaporators are used, temperatures as high as 180°F have been successfully used. The fresh pineapple juice at about 12° Brix is concentrated to 60°–65° Brix for remanufacturing purposes, or to 45° Brix for the retail market.

Essence recovered from the recovery unit is added and mixed with the pineapple concentrate, which is then slush-frozen and packed either into 6-oz metal containers at 45° Brix or in 75-lb polyethylene-lined fiberboard containers at 65° Brix. Freezing is completed by holding at −10°F.

Bulk pineapple concentrate is used as an ingredient for mixing with citrus concentrate in the production of frozen juice blends. Pineapple concentrate is also used as an ingredient in the production of many types of canned fruit drinks. A popular combination has been a pineapple-grapefruit drink that has reached a volume of production exceeding that of single-strength canned pineapple juice.

Pineapple Flavor Components. The volatile flavoring components of pineapple juice are frequently lost during commercial disintegration, extraction, heating, and centrifugation. Some 20–30% of the essence is lost if appropriate precautions are not taken to prevent losses. In general, processing should be done in closed systems at as low temperatures as are practical.

Pineapple does not have a single component that accounts for most of the characteristic flavor. Even the total volatile flavor fraction is not easily recognized as being dinstinctly pineapple although it does have fruity character. The pineapple residue from which the volatiles have been stripped is always identified as being pineapple. However, the addition of the distilled essence to the stripped base gives a superior product. The volatile flavor components in processed pineapple products cover up the cooked flavor and are important contributors to high-quality pineapple juice. The flavor components have been identified as alcohols, aldehydes, acids, esters, and lactones. A total of 58 compounds are known. Among these, 2,5-dimethyl-4-hydroxy-3 (2H)-furanone is present in a larger amount (1.2 mg/1 kg) than the others.

Prune Juice Concentrate

Prune juice is not a fruit juice in the usual meaning of the term, but rather a water extract of dried prunes. Because it is rich in mineral salts and acts as a mild laxative, prune juice has become a popular breakfast drink in the United States. Most prune juice and prune juice concentrate is now preserved by heat processing. However, frozen prune juice concentrate is superior in color, aroma, and flavor to most of the heat-pasteurized canned and bottled products.

Prune juice is made either by the diffusion method or the disintegration method. In the diffusion method the soluble substances are extracted from the dried prunes by repeated steeping in hot water, followed each time by draining off the liquid extract. After a thorough washing, the prunes are dumped into wooden or stainless steel tanks. About 25 gal of water/100 lb of fruit are added and the mixture is heated to 185°F by steam coils. The prunes are steeped in the hot water for 2–4 hr and the surrounding liquid (which now contains much of the soluble solids of the dried prunes) is drained and stored. The process is repeated again with fresh water except that about 15 gal of water are used for each 100 lb of fruit in the second extraction. The liquid from the second extraction is drained and combined with the liquid from the first extraction. The process is repeated a third time, with 10 gal. of fresh water being used for each of the original 100 lb of dried prunes. The three extracts obtained in this manner are mixed together and the vats are emptied of the residue of extracted prunes, which is now practically free of soluble substances and is discarded as waste. The concentration of single-strength commercial prune juice runs from 19° to 20° Brix. If the concentration of the blend of the three extracts is too low, it is increased either by evaporation or by extracting a fresh batch of dried prunes. Single-strength prune juice is usually preserved by heat processing in quart bottles at 190°F for 35 min. Cans and smaller bottles require less sterilization. It is too costly to pack single-strength prune juice as a frozen product.

A second commercially used method of making prune juice is extraction by disintegration. Several hundred gallons of water are added to 1200 lb of washed prunes. The water is brought to a boil and the fruit is cooked for 60–80 min under constant agitation until the prunes are thoroughly disintegrated. Pressure cookers may be used so that the cooking time can be reduced to as little as 10 min. The disintegrated prune mass is dropped from the cooker onto a cloth and the wrapped mass is placed in a hydraulic press. The liquid portion is then extracted under pressure. The extract obtained by this method usually has a density of about 9°–11° Brix. It can be clarified either by allowing the sediment to settle and siphoning off the clear juice or by filtering through a filter press, using 1% infusorial earth as a filter aid. The resulting extract is a clear liquid which may then be evaporated in open vats or a vacuum evaporator to bring the final concentration up to the desired 19°–21° Brix before final packing.

For making prune juice concentrate, the juice must be depectinized by addition of sufficient amount of pectinol or pectic enzyme (4 oz or more per 1000 gal). The enzyme is slowly mixed with the juice and the mixture is allowed to stand overnight. The juice is then filtered and concentrated under vacuum to approximately 60° Brix at a temperature not more than 120°F. The product is preserved by freezing in 6-oz cans and marketed as a frozen prune juice concentrate.

Strawberry Concentrate

A certain percentage of sound strawberries delivered to packinghouses are unsuitable for use in fresh and frozen packs because of size, shape, or blemishes. Such berries may be used for making frozen strawberry juice concentrate for jelly manufacture. Concentrates are more economical to store and ship than frozen single-strength juice, and they have certain advantages in jelly manufacture. FDA Standards require that formulation of the strawberry jelly be based on a juice of 8% solids. The concentrates permit accurate control for this requirement. Some lots of single-strength juice tend to form more or less permanent gels during storage, and these may be difficult to disperse in making jelly; in concentrates such gels are enzymatically prevented during processing. Good quality concentrate of density as high as 73° Brix (12 to 1 by volume) has been commercially prepared.

The juice is prepared by pulping cold, sound, whole berries through a ¼-in. screen in a hammer mill. Filter aid is added to the chopped berries in amounts of 3–10% and the slurry is pressed in a bag-type press. Depending upon cultivar and maturity, a yield of cloudy juice ranging from 70 to 80% of the original weight of the berries can be obtained.

The cloudy press juice is immediately treated with pectic enzymes to degrade the pectic substances. This permits production of a clear, stable juice and prevents gelling in the concentrated product. The treatment consists of adding pectic enzyme to the juice in a concentration of 0.5% and holding for 3 hr at 75°F. Filter aid is added to a concentration of about 0.25%, and the juice is clarified in a plate filter.

The clear depectinized juice should be concentrated as rapidly as possible to avoid deterioration of its color and flavor. When natural-circulation evaporators, using either calandria heaters or external heat exchangers, are used, with resulting long evaporation times, the maximum boiling temperature should not be more than 100°F, or a maximum boiling pressure of 1.9 in. Hg, to avoid flavor and aroma deterioration. Vacuum concentration of strawberry juice can result in removal of practically all volatile flavor constituents from the concentrate. For a high-quality product these essences may be recovered in essence recovery units as described earlier in this chapter. The recovered essence can be mixed into the concentrate to make a "full flavor" product, or can be packaged separately. The latter procedure may be advantageous, because a jelly manufacturer may wish to incorporate the essence into the completed jelly just prior to the filling operation, thereby eliminating volatilization of much of the essence during the heating of the ingredients. Both concentrated juice and concentrated essence should be packed in enameled plastic-lined bag containers, frozen, and stored at 0°F or lower.

Packaging of Fruit Juice Concentrate

Most concentrates are now packed in cans with a spirally wound fiber-foil body and metal ends. The ends are made of either tinplate or aluminum and are lacquer- or enamel-coated inside. The fiber body is usually a four-ply composite, composed of two plies of natural kraft liner-board sandwiched between two thin sheets of aluminum foil. Both metal and fiber-body cans come in a variety of sizes, 6-oz cans being most popular at the retail level, although there is an increasing proportion of 8-oz and 12-oz cans. For institutional use, 32 fl oz of concentrate is packaged.

The apparently simple act of opening a 6-oz can of frozen orange juice concentrate actually was followed by unexpected problems. When part of the industry switched to a spiral fiber-foil can with aluminum ends, consumers found that the magnet on can openers could not hold the aluminum top when it was removed, and they had to fish for the top when it fell into the concentrate. Can manufacturers tried to solve this problem by making one end of aluminum and the other end of steel with directions on the container to open the other end, i.e., the steel end. By 1965, various easy-open containers were being used. These consisted of fiber or metal cans with either a device on the lid such as a key-shaped or ring-shaped piece of metal enabling consumers to lift and pull, or a variety of tear strips, for which users pull a tape or string to remove the top portion of the can. By the end of 1966, approximately 20% of the orange juice concentrate pack was in easy-opening cans.

While the 6-oz container is still the most popular size for frozen orange juice concentrate, approximately 5–8% of the pack was packaged in 8-oz cans in 1966 and 30% in 12-oz cans. Some processors experimented with a small pack of 16-oz cans, which make a half gallon of single-strength juice. According to reports by the Florida Citrus Processors Association, the use of 4-oz containers for orange, grapefruit, and blended juices has declined since 1970 and has not been reported since 1974. At the same time, the use of 8-oz containers has continued to increase.

Several experimental containers other than the conventional metal or fiber-foil cans are being test marketed. These include 8-oz rectangular blocks of frozen orange juice concentrate encased in plastic material and packed in a carton; pouch freezing of orange concentrate; miniature milk cartons with aluminum foil liner and tear tab top; and containers with reclosable plastic lids.

PREPARATION AND FREEZING OF FRUITS

Fruits are harvested for freezing at the fully ripe stage and are soft in texture even before freezing and thawing. Conventional methods of freezing tend to disrupt the structure and destroy the turgidity of the fruit tissues. However, great advances have been made in the technique of freezing fruits. Individually quick frozen (IQF) and cryogenic frozen fruits are superior in quality and stand up better upon thawing than do fruits frozen slowly in packages, cartons, or bulk containers.

Apples

Apples are frozen commercially as slices in large containers and are widely used by the baking industry for pies. A small pack of frozen applesauce in both retail and institutional containers is put up each year.

For freezing, the apples should have good flavor and a texture that will not disintegrate or become mealy after thawing.

Cultivars. The leading cultivar for processing in the eastern states has been the 'Rhode Island Greening.' 'Baldwin,' 'Ben Davis,' 'Northern Spy,' ?6Cortland,' 'McIntosh,' 'Wealthy,' and 'Monroe' are also used in the frozen pack, but not all of them are ideal freezers. 'Cortland' and 'McIntosh' tend to disintegrate even when packed soon after harvest. Apple cultivars differ in susceptibility to fragmentation and disintegration even though of comparable firmness when prepared for freezing.

'Jonathan,' 'Yellow Newtown,' 'Golden Delicious,' and 'York Imperial' remain largely intact after cooking when frozen at all stages of ripeness. 'Golden Delicious' and 'York Imperial' are considered best for freezing.

The theory that the softening of fruits as they ripen is associated with changes in the pectic substances is widely accepted. According to theory, change in the texture of fruits from the firmness of the green to the softness of the ripe condition is the result of conversion of protopectin to water-dispersible forms through the action of pectin-degrading enzymes. Demethylation of the polymers by pectin methylesterase followed by hydrolysis of glycosidic linkages by polygalacturonase could account for diminished cell adhesion as fruit ripens. Apples were among the first fruits to be studied from the standpoint of textural changes associated with ripening. A higher cellulose content distinguishes firmer apples from soft ones, but the softening that accompanies ripening can not be accounted for on the basis of changes in the cellulose of the fruit.

In the Pacific Northwest, the best cultivars for freezing are 'Jonathan,' 'Stayman Winesap,' 'Yellow Delicious,' and 'Yellow Newtown.' In Colorado and Utah, the 'Jonathan,' 'Rome Beauty,' 'Stayman Winesap,' and 'Yellow Newtown' are important. In California, the 'Yellow Newtown' is suitable for pies; 'Gravenstein' can be processed for frozen sauce.

Processing. After sorting and size grading, apples are washed, mechanically peeled, cored, trimmed by hand, and sliced. To prevent browning, the peeled or sliced fruit can be held in a 1–3% salt solution.

Sulfite or sulfurous acid treatments have been used extensively on slices for frozen pie stock even though it has the disadvantage of giving an undesirable flavor. Surface browning on fresh peeled apples can be prevented by dipping apples in a 0.25% $NaHSO_3$ solution, followed by soaking in a 0.20% K_2 HPO_4 buffer solution. The same method may be beneficial in the freezing industry. A third method to prevent enzymatic browning consists of blanching apple slices in steam or hot water. The disadvantage of blanching is that leaching causes loss of solids and flavor compounds when the fresh apple slices are subjected to moist heat. It is also possible to prevent browning by treatment with ascorbic acid.

The texture of overmature apple slices may be improved by dipping them in solutions of calcium salts.

Apricots

The pack of frozen apricots has increased several fold in the last 10 years or so to over 16 million lb. Apricot halves are preferred to sliced fruit and can be packed with or without peeling. Practically all of the pack is frozen in 30-lb cans for the baking, jam, and preserve trade.

Apricots for freezing should have low browning tendency, even ripening, good color, good flavor, a tender and smooth skin, and firm texture. In California, 'Tilton' and 'Blenheim' are the most important cultivars, although 'Hymskirk' and 'Moorpark' are frozen in small amounts. In the Northwest, several cultivars have been tested on a small scale for freezing, such as 'Blenheim,' 'Tilton,' 'Earliril,' 'Blenril,' and 'Perfection.'

Apricots are inspected on a conveyer belt and then passed through a halving and pitting machine. After separation of the pits, halves are washed, either blanched in hot water or steam, dipped in a bisulfite solution, or treated with ascorbic acid. Blanching in a single layer on a mesh belt for 3–4 min in steam is satisfactory for firm fruit. It is probably best to treat softer fruit with SO_2 or ascorbic acid. The latter may be incorporated at a level of 0.05–0.1% into the syrup used for packing. The SO_2 treatments should be adjusted to leave a residue of not more than 75–100 ppm in the frozen apricots.

For baking or preserves, SO_2 treatment is satisfactory. The packing medium can vary from 15° Brix syrup, to dry sugar at three parts of apricots to one part of sugar or even a higher proportion of fruit, depending on the specifications of the buyer. When dry sugar is used, it may be sprinkled on the fruit as it is filled into the container or be distributed evenly by light mixing.

The treated apricot halves with sugar or syrup are packaged and frozen in air-blast, contact, or cryogenic freezers.

Avocados

California is the best avocado-producing state in the United States and has been a center of attempts to process and preserve surplus avocados. The 'Fuerte' and 'Haas' cultivars dominate the California production. Appreciable amounts of 'MacArthur' avocados are also grown in the state. Florida produces about 15,000 tons of avocados annually, almost all sold as fresh fruit. The principal avocado cultivars grown in Florida are 'Booth 7,' 'Booth 8,' 'Collinson,' 'Lula,' and 'Waldin.'

It is desirable to acidify avocado puree to pH 4.5 by adding lemon or lime juice and extra salt. This treatment permits retention of natural flavor and light green color for at least 1 year in frozen storage.

An avocado-peeling machine has been developed, consisting of two drums, both rotating downward toward the nip. One has a solid outside surface. Pitted avocado halves are placed one at a time with the seed cavity toward the perforated drum. As the drums rotate, the solid drum presses the meat of the avocado through the ¼-in. holes in the perforated drum. A doctor blade, mounted inside the perforated drum, cuts the meat from the peel, and a doctor blade on the outside of the perforated drum removes the peel.

Benson (1968) obtained a patent for a process of immersing avocado in liquid nitrogen and then subjecting the fruit to storage in a cold environment. The avocados are selected on the basis of maturity and size; they should be just on the immature side of the condition generally referred to as "eating ripe." At this stage of maturity the fruit is free of brown spots.

The avocado is cut in half along its longitudinal axis and the seed is removed. The open half is inspected for spots or areas of discoloration and for any soft or rotten places. The avocado half is then treated by coating it with an antioxidant such as 0.5% ascorbic acid, ½-strength lemon juice, or a solution of citric acid. The length of the antioxidant dip is not critical; a 1-min dip in 0.5% ascorbic acid or in lemon juice has been found satisfactory. After treatment with the antioxidant, the avocados are frozen quickly by immersion in liquid nitrogen or liquid nitrous oxide.

The time of immersion depends upon the initial temperature of the avocado, maturity, and cultivar, but most of all upon the size of the avocado half. An avocado half cannot be frozen all the way through by immersion in a liquefied gas at −100°F or lower without splitting or cracking. It is necessary, therefore, to freeze avocado halves part way through and then remove them from the liquefied gas. This does not ordinarily build up enough refrigeration in the frozen portion to freeze the inner unfrozen part by conduction. After withdrawal from the liquefied gas, the avocado half is placed in a cold environment below the freezing temperature of the avocado and preferably as low as 0°F.

The actual immersion time in the liquefied gas varies from about 15 to 50 sec to freeze at least 35–45% of the mass of the avocado half (the longer time being for a larger avocado) before the fruit will crack. The avocado half is kept in the cold environment until it freezes solid all the way through. The process is carried out more quickly by doing as much of the freezing as is safe from cracking and splitting in the liquefied gas. A substantial part of the freezing must be done in the liquefied gas otherwise the subsequent freezing becomes too slow to prevent cellular damage. For maximum keeping qualities, the frozen avocado half is glazed after freezing, to keep the fruit from having contact with oxygen that may be present in the package. The glaze is applied by dipping the frozen avocado half in the glaze material, which freezes on the surface of the avocado. The avocado must, of course, be colder than the freezing temperature of the glaze material.

Pure water can be used for the glaze, but the avocado half has better keeping qualities of glazed with an antioxidant, such as lemon juice. After glazing, the avocado half is packed in an atmosphere substantially devoid of oxygen. Vacuum packaging can be used, but packaging in a nitrogen atmosphere is preferred. The package is preferably stored under ordinary frozen food refrigeration and kept frozen until shortly before use.

Blackberries and Boysenberries

There are many species of berries varying in size, shape, flavor, color, and drupelet size. Blackberries, boysenberries, and raspberries are commercially the most important species. Cultivars suitable for freezing should ripen evenly, be resistant to bruising during transportation, and rich in flavor. Blackberries that tend to revert to red color when frozen are not desirable, as they may drop in grade by being judged as fruit of mixed maturity.

Blackberries. More than 80% of the U.S. production of blackberries in recent years has come from the Pacific Coast. The major blackberry-producing areas are Washington and Oregon. 'Evergreen' and 'Himâlaya' are the important cultivars of this region, and some small-seeded, high-flavored cultivars have been developed by George Waldo of USDA in cooperation with Oregon State Experiment Station in Corvallis. These include the 'Chehalem,' 'Cascade,' 'Pacific,' 'Olallie,' 'Marion,' and 'Aurora' cultivars.

Blackberries should be thoroughly ripe when picked for freezing; otherwise they may turn red on freezing. After aircleaning and washing, they go over a dewatering shaker and onto inspection belts. The fruits may be packed in 30-lb plastic-lined corrugated boxes, 30-lb slipcover enameled cans, or plastic-lined steel drums of 55-gal capacity. There is little production for the retail trade. If sugar is added, a ratio of one of sugar to three of fruit is common if the berries are to be used for preserves or jelly, but the ratio will vary according to the specifications of the buyer. At present, considerable quantities of blackberries are frozen without addition of sugar to be used in the production of wine in the Pacific Coast states. For this purpose they may be frozen in the trays in which they are picked or on a continuous-belt air-blast freezer. In the latter case they are transferred to barrels or slipcover cans for freezing storage or shipment to a winery or other purchaser. While the presence of sugar is believed to set the flavor better, many industrial users prefer to add sugar at their plants.

Boysenberries. In California, boysenberries dominate the brambleberry picture. Because the receptacle is small and there are few seeds, it provides a more suitable fruit for processing than most other berries. Boysenberries are red rather than black in color. They frequently have flowers and fruit on the plants simultaneously, and the berries tend to ripen randomly on the stems. It is thus possible to obtain fruit at different stages of maturity at a single picking. On the other hand, commercial fruit will not vary widely in maturity as a number of pickings are taken from the same bushes.

Accompanying the maturation of berries, there is a decrease in titratable acidity from 2.74% in underripe berries to 2.19% in medium-ripe and 1.54% in overripe berries (expressed as percentage of citric acid on the fresh basis). The soluble solids contents are 8.82° Brix in underripe, 10.32° Brix in medium-ripe and 11.42° Brix in overripe berries, respectively. The pH of the berries increases from 2.97 in underripe to 3.12 in medium-ripe and 3.39 in overripe berries. Correct harvest maturity is important for a proper balance of sweetness and acidity for frozen boysenberry pies.

Boysenberries are handled and frozen as IQF berries in the same manner as blackberries. In addition to their use in preserves, jellies, and jam, they are used extensively by commercial pie bakeries, as are frozen pack blackberries.

Blueberries

The blueberries of commerce originated in North America. The species grown commercially are *Vaccinium myrtillis,* the lowbush blueberry; *V. corymbosum,* the highbush berry; and *Vaccinium ovatum* or the "Evergreen" blueberry, grown in the Pacific Coast areas north of San Francisco Bay. Of these, the lowbush blueberry is the more important. It makes up much of the Eastern blueberry pack. The Evergreen blueberry has been packed in decreasing tonnage due to insect infestation problems.

The many available blueberry cultivars differ greatly in size, color, firmness, and flavor. Desired characteristics for freezing are tender skin, large size, and sufficient acidity and good flavor. The retention of the natural bloom on the berries is also a desirable characteristic.

Wild lowbush blueberries are largely grown in Maine; 'Bluecrop,' 'Jersey,' 'Rubel,' and 'Stanley' cultivars in the East and Pacific Northwest; 'Blue Crop,' 'Jersey,' and 'Rubel' cultivars in Michigan. The characteristics of highbush blueberries can be summarized as follows:

Bluecrop. Bush with average vigor, upright spreading, productive; fruit cluster loose; berry large, oblate, light blue, firm, resistant to cracking, slight aroma, above medium in dessert quality; small scar.

Earliblue. Vigorous, upright, productive bush; loose fruit cluster; large, light blue, firm berry, resistant to cracking, good dessert quality; good scar.

Jersey. Vigorous, erect, productive bush; fruit cluster very long and very loose; berry medium, round-oblate, good blue color, firm, lacking aroma, medium in dessert quality; scar good; season late.

Stanley. Erect, vigorous, productive bush; medium loose fruit cluster; medium size berry, oblate, of good blue color, very firm, very aromatic, of high dessert quality; scar above medium size; early midseason.

The prime factor to be considered in selecting blueberry cultivars is their adaptability to mechanical harvesting. Mechanical harvesting of highbush blueberries is increasing. Factors such as uniform ripening of berries, ease in shaking the fruits from the bush, and resistance of the bush to damage from shaking must be considered in mechanical harvesting of blueberries. The USDA Selection of 1613A is popular in the Pacific Northwest because it is one of the easiest cultivars to harvest by machine.

Fresh or preserved berries may be used in bakery products, juice blends, or dessert wines. Cleaning the berries after mechanical harvest and removing excess pigment from some products are problems in processing this fruit. Some processors find it advantageous to blast-freeze field-cleaned berries during the brief harvesting season, later removing them from frozen storage for juicing or repackaging as frozen fruit. Such berries keep satisfactorily in the frozen state and are easy to clean. Freezing before processing also facilitates the removal of fruit worms and maggots, since these insects leave the berries when chilled. Much of the frozen blueberry pack is packed in bulk for the preserve or bakery trade. The berries are cleaned by air-blast, washed, strained, surface-dried in a blast of air, sorted on a belt, and filled into polyethylene-lined fiberboard containers holding 30 lb of fruit, in 20-lb slip-covered enameled cans, or in 60-lb bags. For the retail trade 10-oz cartons are packed, either as straight dry berries, as a 4:1 dry sucrose pack, or with a 40–50% syrup.

Blueberries packed in either 50–60% syrup or 4:1 dry sucrose have good texture and appearance. Blueberries frozen in 60% sucrose syrup are too firm when baked into pies, but all others have good texture: the dry berry; water-packed; with 20°, 30°, 40°, or 50° Brix syrup; or as a 4:1 dry sucrose pack.

Skin toughness can be a problem with frozen blueberries. However, immersing berries in boiling water for 30–60 sec eliminates skin toughness. Different blueberry cultivars vary in skin toughness from year to year. Blanching does not prevent the toughening of skins during freezing and storage.

Ballinger and Kushman (1970) studied the relation of ripeness to composition and keeping quality of highbush 'Wolcott' blueberries. As the fruit developed, the percentage of acid decreased while pH, soluble solids, anthocyanin content, and berry weight increased. It was suggested that cultivars with high keeping qualities could be obtained through selection of clones high in acid. The increase in anthocyanin content as berries develop indicates the possibility of an electronic method of sorting the berries.

A large portion of the total vitamin C content of fresh blueberries consists of dehydroascorbic acid. No vitamin C losses have been observed immediately after deep-freezing, but losses on storage of unsealed deep-frozen berries at −20°C have been reported to be 20% after 6 months and 50% or more after 9 months. Stability of vitamin C was much higher in sugared deep-frozen berries.

Cranberries

Frozen cranberries maintain quality better than most fruits having a shelf-life of several years at 0°F or lower. Large quantities of whole berries are frozen in bulk for later processing into juice, sauce, or jelly.

Cranberries are harvested in much the same way as blueberries, by using a rake or a rake and a scooper. The main operation in preparing them for freezing consists of cleaning by blowing out leaves and chaff in a fanning mill. Soft or rotten fruit is eliminated by causing the sound berries to bounce over a barrier; they are then graded according to size. After washing in a tank of cold water, the berries are stemmed, drained, inspected, filled into containers, and frozen. Servadio and Francis (1963) studied the relation between color of cranberries and color stability of cranberry sauce. Five blends of red and pink cranberries were used to prepare jellied and whole cranberry sauce. The color was measured with a colormaster differential colorimeter. Pigment determinations were made by measuring the absorption at 535 nm of an extract in an HCl–ethanol mixture. The color and pigment content of the berries correlated highly with color and pigment content of the sauces. The ratio of absorption at 535–415 nm in acidic ethanol was a good index of color and pigment degradation.

Elderberries and Gooseberries

Several attempts have been made to commercialize frozen elderberries for making jelly and pie. In general, the preparation and freezing steps are similar to those given for blueberries, except for the stemming operation.

Gooseberries develop singly or in small clusters and the fruit may be picked at a single stage of maturity if desired. These berries freeze very well for subsequent use in jams and pies. The major commercial packs are made from 'Downing,' 'Poorman,' and 'Oregon' cultivars in the Northwest.

Strawberries

Frozen strawberries are used largely for remanufacturing into ice cream, jams, preserves, and bakery products. They are packed in 30-lb cartons, 30-lb tins, and 50-gal barrels. For the retail trade, 10-oz and 16-oz packages are used.

In Washington and Oregon, the 'Northwest' and 'Hood' cultivars are grown for freezing. In California, more than 60% of the strawberries are of the 'Tioga' cultivar; other cultivars are 'Shasta,' 'Lassen,' and 'Cupertino.' The criteria for selection of strawberry cultivars are uniformity of shape and size, texture of pulp, and favorable price. The 'Shuksan' strawberry is well adapted to freezing preservation and is very satisfactory as a preserving berry. Coupled with its winter hardiness and productivity, its processing characteristics foretell commercial potential for growing in areas where winter kill is a problem. The French cultivars 'Sengatigaiga,' 'Belrubi,' and 'Tioga' also are very suitable for freezing. The criteria for quality of frozen strawberries include percentage of drip after thawing, maintenance of shape, color, taste, and aroma.

Fresh strawberries to be frozen should have a deep color, firm texture, and rich flavor. They should be picked when fully colored and firm-ripe. For packing in barrels, the berries should be precooled for 24 hr in shallow pans or trays at about 32°F. Usually, they are picked in the field without hulls, thus eliminating hulling at the plant. Some strawberries are packed without sugar but they are not as good as the sugar-packed fruit. The net contents of a 50-gal barrel of berries and sugar vary from 400 to 450 lb for the 4-plus-1 pack.

The freezing plants should not be too far away from the fields. Freezing and storage of the product should be carried out at 0°F or lower. Efficient methods of freezing whole berries are the IQF process and cryogenic freezing (Fig. 7.9). A combination immersion and spray system utilizing liquid Freon has been reported for strawberry freezing (Anon. 1969). The advantages of this process are low freezing costs, virtually no dehydration during freezing, space saving as compared to other freezing units, and ability to handle whole strawberries.

Strawberries are picked several times each week so that the fruits do not become overripe. The berries are picked without stems or caps; thus they are ready for sorting and processing on arrival at the plant. Shallow trays or crates holding about 14 lb of berries are used. The alternative is to keep the harvested berries at 32°F for processing the next morning.

Although operations in strawberry freezing plants vary considerably, the following outline is fairly representative. Trays of berries are dumped gently into a small tank of water and conveyed to a vibrating, sloping riddle, or screen, on which they are washed with sprays of cold water and delivered to sorting belts that slope upward. Grades A and B berries are separated, and the culls and green berries are removed. In foggy weather a considerable number of the ripest berries are apt to be moldy and must be discarded.

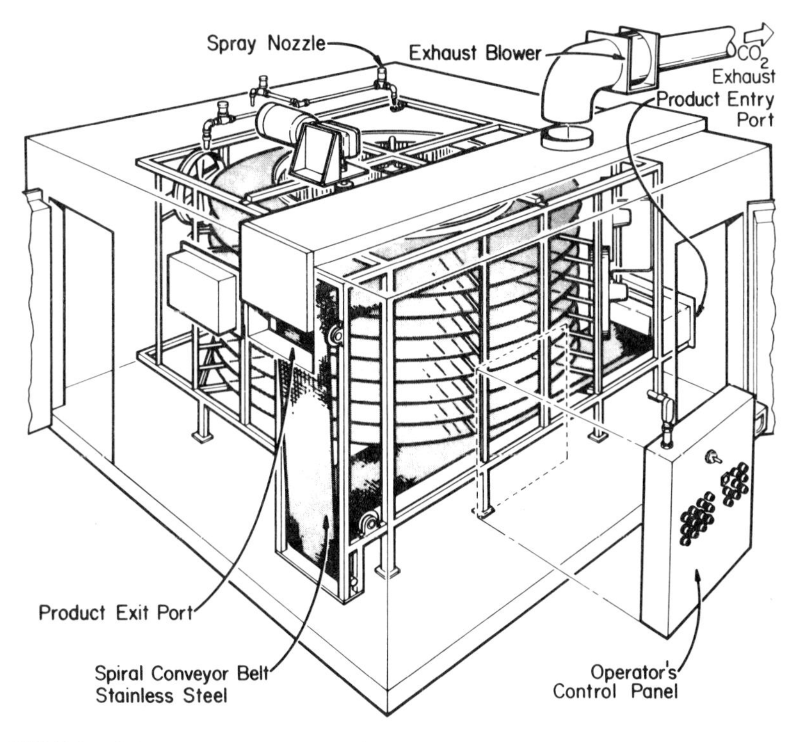

FIG. 7.9. Spiral cage cryogenic freezing unit using liquid carbon dioxide. (Courtesy Airco Cryogenics, Div. of Arco.)

The berries are then conveyed to a size grader consisting of diverging stainless steel rods or a stainless steel sheet with circular openings to two or more different diameters. Usually, three sizes are made by the grader. Most of the large and medium berries go to the slicer, while the small berries to to the bulk line. The slicer consists of a number of circular disk knives rotated by a common shaft. The slices are about ¼ in. in thickness. The sliced berries are next mixed with dry sugar in one of several types of mixing devices. A common arrangement consists of a stainless steel trough in which is a slowly turning augur. Sliced berries are filled into the trough to a predetermined level where they contact a vertically placed rod. This activates a sugar-measuring or -weighing mechanism operated by electric eye that adds the required weight or volume of dry sugar. The slowly turning augur mixes the berries and sugar and conveys them to the filling hopper at the end of the mixer. Considerable juice or syrup is formed by osmosis and the mechanical handling of the mix. A batch mixer, resembling a concrete mixer in general appearance and in operation, has also been used. Weighed amounts of fruit and sugar are mixed and delivered to the filling machine. A slowly revolving stainless steel cylinder has also been used for mixing.

From the filling hopper the mixed berries and sugar drop by gravity into an Elgin, FMC, or other filling machine that measures the product by volume and places it in the final retail package. The fiber can (Sefton or other) is a popular and widely used container. It is filled by machine, passed along by conveyor to the lidding and sealing machine where a lid is dropped on each can, and sealed by a machine resembling an ordinary can double seamer. Small cartons carrying an inner, heat-sealable plastic or parchment bag, or leakproof cartons without inner bag, are used. The plastic bags are sealed by electrically heated sealing devices as the cartons are carried along by conveyor.

Grade B sliced berries are usually mixed with sugar, as described above, and delivered to a hopper from which they are filled by a hand-operated valve into 30-lb tins for preservers' use. Others are packed, especially in the Pacific Northwest, in 50-gal barrels, the usual procedure being to add sugar intermittently to the berries as the barrel is being filled. The medium-size berries may be included in the barrel pack. Barrels and 30-lb tins of berries are usually slow-frozen in a large room which may be at −30°F or in a warehouse at 0°F. In order to hasten freezing the air is circulated by fan in some plants. Retail-size packages are quick-frozen in Birdseye or Amerio plate freezers or in an air-blast freezing tunnel, in either method at about −30° to −40°F. Storage is at 0° to −5°F in most warehouses.

Freezing with Liquid Nitrogen and Freezant-12. Wolford *et al.* (1971) used liquid nitrogen to freeze raspberries, strawberries, peaches, and cherries. The fruits were prepared for freezing after washing, slicing, and impregnation with syrup in some cases. Liquid nitrogen freezing was done by immersion of the fruit in the liquefied gas for the time required to freeze a shell 0.5–0.6 of the radius of the fruit pieces. Fruit was also immersed in Freon-12 precooled with a mixture of solid CO_2 and Freon-11. Half berries of uniform size froze well if presented to liquid nitrogen in a manner such that clumping did not occur. Sugared fruit could not be successfully frozen by this method. After thawing, berries had firmer texture and less drip than conventionally frozen berries. More rapidly frozen berries had a lower rate of drip loss. Perhaps the cost of freezing by this method needs to be critically evaluated for commercial application. For high-valued foods, liquid nitrogen or Freon 12 should be able to compete with the conventional methods of freezing.

Strawberry Syrup. In one plant in California, cull, green, and overripe berries of sound quality are pureed in a Rietz disintegrator, infusorial earth and a pectic enzyme are added, and the mixture allowed to stand several hours or overnight to permit the enzyme to hydrolyze most of the pectin, resulting in a product that can be pressed. It is pressed in a combination bag press and filter (Harris press). The juice is concentrated at low temperature under vacuum to a light syrup. The volatile flavor esters are condensed and concentrated to a highly flavored essence. About 1 gal of essence is obtained for each 100 gal of concentrate. The two are combined later at point of use, which may be for preparation of fountain syrups, jellies, etc. Berries packed with sucrose seem superior in flavor to those packed with other sweetening agents but drained weights are affected more by the ratio of sweetener to fruit than by the kind of sweetener.

IQF Strawberries. In processing individually quick frozen (IQF) strawberries, problems have arisen because the washed berries tend to stick together in large chunks or masses when put into the freezing tunnels. To break up these clusters, they are run through rubber roller "busters"; many of the berries break with a consequent lessening of the quality of the frozen product. Nakagawa (1962) eliminated the objectional freezing together of strawberries by providing a special tray that maintains berries in spaced relation to each other while they are being frozen. At the same time, better drainage from the berries and improved circulation of the freezing medium past the berries are obtained.

The special tray is a flat rectangular base plate made of a flexible plastic material such as polyethylene. Risers or nipples project upward from the base evenly in parallel rows to extend over the entire area. They leave straight and continuous intersecting "aisles" between all the different rows. The nipples prevent any berries seated in adjacent spaces from contact with each other.

The berries to be processed are graded and washed and after draining are distributed onto the trays. The loaded trays are conveyed by truck into the usual freezing tunnel. Any water that remains on the berries when they are placed on the tray tends to drain and collect on the base, instead of forming a bond between adjacent berries when they are frozen. When berries are frozen, the trays are removed from the tunnel and rack, inverted, and bent downward so that the base becomes convexly curved in a downward direction. The nipples are spread apart allowing the frozen and separated berries to drop cleanly. There is no fracturing of the berries or other damage to their appearance.

Tart Cherries

Tart cherries are preserved by freezing largely in Michigan, New York, and other eastern states. Mechanical harvesting of tart charries has progressed rapidly in recent years, especially in Michigan. Processors were at first reluctant to accept mechanically harvested cherries but now they find that high-grade cherries can be delivered to their plants at lower cost than hand-harvested fruit.

Tart cherries are harvested when bright red in color. Cherries are shaken from the trees and then passed over a short sorting belt right in the field. At the end of the belt the cherries fall into a tank of cold water and are taken immediately to the processing plant. Up to 92% of these cherries make U.S. No. 1 Grade or better.

One problem resulting from mechanical harvesting is the need to remove stems from the cherries at the plant. Development and testing of mechanical harvesting equipment have been discussed in Chapter 2. The mechanical harvester consists of two inertia shakers on two self-propelled units and is capable of harvesting 30–40 trees/hr. The conveyor is located at the bottom center and the elevator along the front. Tanks of 1000-lb capacity receive the fruit from the end of the conveyor. With this system three men can harvest as many cherries as 100 hand pickers would.

The 'Montomorency' cultivar dominates the red tart cherry industry. Some 'Morello' and 'Richmond' cherries are grown, but are of minor importance.

Transportation and holding of cherries in ice-cold water have greatly improved their quality. This technique also prevents crushing and bruising, and results in firmer fruit with less juice loss during pitting.

Scald or discoloration of the fruit has been a serious problem to which a great deal of research effort has been devoted (Whittenberger and Hills 1958, 1960). Scald is the localized loss or translocation of the red pigment from the skin of the cherry, the following oxidation being regarded as a separate but related process. The extent of damage depends on the initial handling of the fruit, bruising from picking and postharvest handling, and the product of time × temperature which accumulated between picking and chilling. Fruit bruised in a normal commercial picking will develop severe scald if held for 4 hr at a normal orchard temperature of 85°F.

Arnold and Mitchell (1970) reported that scald is a major grade-lowering defect resulting from mechanical harvesting of sour cherries. Histological sections of scald tissue showed no crushing or distortion of cells, but the epidermal cells appeared dense and the cell walls appeared to be thicker than those of nonscalded tissue. They stated that bruising apparently induced a chemical change as a result of membrane disruption, bringing about discoloration.

The cherries are drawn from holding tanks and are conveyed or flumed to the size grader, where fruit less than ⅝ in. in diameter drops through the belt. The larger fruit comes onto an inspection belt; the smaller cherries are graded for size again into those larger and smaller than ½ in. in diameter. The sorted fruit then goes to pitting machines. The pitted cherries are packed with sugar in 30- or 50-lb enamel-lined cans, according to the specifications of the buyer. Finally, the cans of cherries are frozen in an air-blast freezer.

Sweet Cherries

Most sweet cherries are preserved by canning or brining. The dominant cultivars of sweet cherries for freezing are 'Bing,' 'Black Republican,' 'Lambert,' 'Napoleon' (Royal Ann), and 'Windsor.'

The procedure for freezing sweet cherries is essentially the same as that for tart cherries. The maturity of dark sweet cherries can be determined by measuring the anthocyanin pigment content of the fruit.

Almost all whole frozen cherry pies are made with fruit that has previously been preserved through freezing. The frozen cherries are thawed, and the juices drained. The juice is usually mixed with starch and sugar. The mixture is then cooked for a few minutes until the desired viscosity and clarity are obtained. The cooked, starchy mass is then mixed back with the fruit from which the juices were drained, and used as the filling for the pies.

A novel method for making frozen cherry pies was developed. In this process, frozen cherries are thawed and the juice drained off. The cherries are then reimpregnated with a gel-forming solution. This solution may consist of previously drained juice in which a gel-forming hydrophilic colloid has been dissolved. It has been found that frozen and thawed fruit is readily impregnated with the colloid solutions. Thus, the operation is a relatively fast one, requiring no prolonged impregnating time. The gel-forming characteristics are not affected by freezing and thawing. Accordingly, when the frozen pies are cooked, the juice inside the fruit or berries becomes hard due to gel formation, and as moisture is evaporated during the cooking operations, the fruit or berry remains in firm, attractive form due to the concentration of the gel solution.

Coconuts

Fresh coconut can be preserved by freezing. Frozen products include shredded coconut, coconut chunks, coconut milk, and coconut cream.

Fully ripe coconuts are husked, shelled, peeled and sterilized with steam and/or sulfites, and then frozen in moisture-proof containers. The frozen and thawed products are used in confections, bakery goods, beverages, and in dozens of commercial and homemade salads, desserts, and other sweets (Woodroof 1979; Tulecke *et al.* 1961; Ross 1960B).

Currants

The seeds of currants are enclosed in a fleshy pericarp, the number and size of the seeds being characteristic of the fruit. They are arranged on short stems or "strings," with a single currant adjacent to the string on the main stem. The fruit ripens in order along this string, the single fruit ripening first and then the fruit nearest the branch with the terminal fruit last. As fruits mature, there is a gradual color change from green to the characteristic color of the respective fruit. During this ripening period an increase occurs in the pericarp area which pushes the seeds farther apart and there is a gradual loosening of the druplets from the core.

Under normal harvest conditions whole strings are picked and not individual currants, so that even at an optimum picking period, there will almost certainly be some underripe and some overripe fruit in any sample of fruit examined at a particular date; special further grading may be required. After harvesting, the berries are cleaned, washed, sorted, and stemmed. They may be packed whole, without sugar or syrup, and frozen at 0°F.

Red currants are preserved by freezing for remanufacture into jam and jelly. The 'Red Lake' and 'Wilder' cultivars are recommended for New York; the 'London Market,' 'Perfection,' and 'Red Cross' cultivars for Michigan. The 'Perfection' cultivar dominates the Northwest Production. Red currants are handled for freezing in the same manner as blacks.

Dates

Much of the date crop is allowed to ripen thoroughly on the palm and can then be packed without further ripening or drying in the packing house. Dates are preserved fairly easily so there has been little stimulus to develop methods of freezing. However, dates have been frozen experimentally both to ripen immature dates and to preserve dates.

Huxsoll and Reznik (1969) studied sorting and processing of mechanically harvested dates based on size, color, specific gravity, and resilience. The possibility of harvesting immature dates and artificially ripening them was examined and a freeze-heat ripening process was developed. Freezing at 0°F and heating at 100°F gave optimum results. Storage at 34°F for 1–2 days improved uniformity of ripened dates. Results show that when 75–80% of the bunch is ripe the entire lot can be processed to a high quality. Equipment for a vacuum hydration process for very dry dates was designed. Pilot models of a diverging roll sizer and resilience separator were developed for the industry. Those of normal moisture content and of fine quality are ripened as outlined above and then packed in fancy baskets, candy boxes, or wooden veneer baskets. A special carton with a cellophane window is also used extensively.

According to Nixon (1950), the 'Barhee,' 'Khadrawy,' 'Halawy,' and 'Deglet Noor' date cultivars would make excellent frozen products. The Barhee cultivar was tested as a frozen fruit and held for 2–3 years at −20°

to −30°F with good results. Frozen dates would be used as a fresh fruit rather than as a bakery product, since cooking would destroy the delicate fresh flavor which is preserved by freezing.

Figs

Most cultivars of figs are preserved by canning or drying. 'Calimyrna' is the principal cultivar dried in California, but its thick skin does not yield a good frozen product. The 'Kadota' fig, a canning cultivar is not very suitable for a frozen dessert fruit. 'Black Mission' figs make an exceptionally good frozen product. Woodroof and Shelor (1949) have described the use of ascorbic acid and citric acid for maintaining good color of figs frozen in Georgia. Hohl (1946) recommended dipping whole figs in a 2000-ppm solution of bisulfite for 2–3 min before packing the fruit in 35° Brix sugar solution to preserve flavor and color.

Tender-skin figs such as the 'Mission' have been recommended for freezing. They should be table-ripe, considerably riper than for canning or shipping. Airtight containers are recommended to prevent off-flavors induced by oxidation. The figs can be packed whole in a 35°–40° Balling cane- or beet-sugar syrup, or prepared by slicing and then packed with 1 part of sugar to 4–6 of sliced fruit. This product is excellent as a dessert with cream. A short dip in dilute sulfur dioxide solution helps to prevent changes in flavor of sliced figs. Addition of 0.10–0.15% ascorbic acid to the syrup can prevent oxidation and thus protect flavor.

Figs to be prepared for freezing must be carefully washed and sorted, and all that show internal rots or souring must be eliminated. Slicing or halving permits both easy detection of internal rots or souring, and good penetration of sugar or syrup.

No satisfactory method of peeling has been developed. The ripe fruit is too delicate to stand the temperatures and handling of lye peeling, and hand peeling is expensive, slow, and difficult. The fruit may be frozen whole and then peeled with abrasive or friction vegetable peelers, but the partial defrosting that occurs during this process almost completely destroys the structure of the figs, and allows oxidation to be accelerated. The skin of the 'Black Mission,' however, may be left on without impairing the quality of the frozen fruit.

Even though figs are less subject to oxidation than other fruits, they must be handled quickly during processing and freezing because of their high susceptibility to microbial spoilage.

Sliced figs covered with 35° Brix syrup make an attractive product for dessert use. Oxidation is so slow in the syrup pack that any benefit from the use of antioxidants is too small to justify the added expense involved. Blanching figs before freezing is effective in preventing oxidation, but imparts a cooked flavor and softens the tissues too much. Packaging and freezing are conducted in the same manner for this as for other fruits.

Grapes

Most of the grapes grown in California have been used for fresh market, raisins, and wines. Some grapes have been canned and some frozen, chiefly for fruit cocktail. According to Peynaud and Ribéreau-Gayon (1971), *Vitis vinifera* is the only species of European origin. *Vitis riparia, V. rupestris,* and *V. labrusca* are American species that generally bear black grapes. The crossing of American species with *V. vinifera* has resulted in the production of hybrids with good cultural qualitites. Chen and Luh (1967) reported the anthocyanin pigments in 'Royal' grapes, a hybrid of 'Alicante Ganzin' and 'Trousseau' grapes. The 'Royal' red is very rich in anthocyanin pigments and can be used to enrich the color of red grape juices and wines.

While grapes can be frozen fairly satisfactorily in syrup, they are rather unsatisfactory in texture, resembling small rubber bags filled with juice. After thawing, they rapidly oxidize. Thus, there is very little future for frozen grapes, except for the use of bakers, and for jam and jelly manufacture. The 'Thompson Seedless' cultivar may be packed in light syrup in 30-lb cartons for bakers. The desirable characteristics for freezing are Vinifera types of even maturity and tender skin; sweet, delicate flavor; relative freedom from seeds; and resistance to discoloration.

Native *Vitis labrusca* types are not usually suited to freezing except as juice. The 'Muscat of Alexandria' cultivar retains its flavor and appearance best after freezing. The 'Thompson Seedless' lacks flavor, except for sweetness. The other cultivars are not suitable. Woodroof (1945) described a process for freezing muscadine grapes.

The fruit should have attained its optimum fresh table qualities when it is harvested. 'Muscat,' 'Thompson Seedless,' and 'Ribier' grapes, which make the most acceptable frozen products, are best if harvested at about 20° Brix. At this maturity, they have their best fresh table qualities and the acidity is sufficient to prevent tartrate precipitation during freezing.

Grapes to be frozen for fruit cocktail processors are precooled and washed, stemmed and sorted, and packed into 30-lb tins or plastic-lined, fiberboard cartons, using 18 lb of grapes to 12 lb of 35° Brix syrup. If the grapes have been harvested at a maturity beyond the optimum, 0.5% citric acid added to the syrup will reduce tartrate crystallization.

A good-sized pack of "grapes and pulp" is frozen for remanufacture into jam and jelly. The principal cultivar is Concord. Muscadine-type grapes are frozen also for dessert products, consisting mainly of the 'Hunt' cultivar and smaller quantities of 'Scuppernong' and 'Thomas.'

Since frozen pulp retains flavor and color better than whole grapes, most of the product packed in bulk for later conversion into jam and jelly is heated to 140°–150°F, then the seeds are removed in a finisher and the pulp is frozen in large cans or barrels.

Lychee

The lychee is a popular fruit in the Orient. It has a white flesh and a very excellent flavor. The fruit finds a ready market as a canned product among Chinese-Americans. The fruit can be frozen in the hard shell, as a peeled fruit, or as a pitted product. 'Brewster,' 'Kwai Mi,' 'Hak Ip,' and 'Groff' are well-known lychee grown in Hawaii. 'Brewster' is the only cultivar of consequence in Florida, although it is sometimes sold as 'Royal Chen.'

Mangoes

Mango is a popular fruit in tropical and subtropical countries. Australia, Brazil, China, Mexico, India, Panama, Phillipines, and many other countries in South America and the Far East grow a large number of mango trees. Although it is a tropical fruit, there is some production in both Florida and Hawaii.

The Philippine Bureau of Science has shown that the mango behaves well in frozen pack. Freezing destroys the eggs and larvae of the fruit fly; hence freezing is a means of making this popular fruit available to Americans. The ripe fruit should be peeled, pitted, sliced, and frozen at once in syrup of 25°–40° Balling in proper containers.

The mango, one of the most delicious of all fruits, maintains its excellent qualities when frozen as slices or chunks in syrup. Cultivars such as 'Irwin,' 'Keitt,' 'Kent,' 'Sensation,' 'Smith,' and 'Zill' have been planted in Florida, but 'Haden' still remains the dominant cultivar. 'Pirie' and 'Haden' dominate production in Hawaii, but a number of seedlings, hybrids, selections, and sports have been described by Yee (1958). Orr and Miller (1955) evaluated the suitability of a number of cultivars for freezing in Hawaii. Boyle and Shaw (1955) summarized findings on changes in quality of frozen mangoes. For more detailed information about mango, readers may refer to Hulme (1971).

Gorgatti-Netto *et al.* (1973) studied the quality of frozen sliced mangoes in syrup. They added 0.1% ascorbic acid to the sucrose or glucose syrup (40° Brix) to inhibit polyphenoloxidase activity and 0.07% $CaCl_2$ to improve the texture. The mangoes were peeled, sliced, washed with water spray and held in a receiving tank with a solution containing 0.05% ascorbic acid to avoid browning. Slices were transferred to polyethylene plastic bags. The ratio of fruit and syrup was 3:1 on a weight basis. The plastic bags were vacuum-closed and frozen in a Dole Freeze Cell Model 2735-8 with double contact. The frozen product was kept in a cold room at −20°C up to 120 days. The $CaCl_2$ treatment results in a slightly firmer texture in the product.

Melons

There are striking differences among the cultivars of *Cucumis melo*—the netted, orange-fleshed melons—in growth pattern and ripening habit as well as in color, flavor, and shape. For example, cantaloupe, Honeydew, Persian, Crenshaw, and Casaba are all cultivars of *C. melo*.

Cantaloupes suitable for freezing should have salmon- or orange-colored flesh that is firm enough to permit the cutting of balls. Vine-ripened melons are of better quality than those that are picked green. However, the melons should not be too soft since this usually results in watery, mushy flesh. In judging maturity, the characteristic aroma of ripe cantaloupe, the condition of the stem attachment, and the appearance of the skin and netting are important factors to be considered.

Persian melons, Honeydew, cantaloupe, and Crenshaw are best frozen as balls or cubes when covered with a syrup of 30° Brix. All melons are more prone to develop an off-flavor after storage when frozen without sugar syrup. Frozen melon, if consumed in the partially defrosted stage while some ice still remains in the tissues, is less flacid and more pleasing in texture and flavor. There is no advantage to be gained by adding ascorbic acid to the syrup. Blanching results in mushy fruit of an unnatural flavor.

Melon cubes or balls add color to fruit cocktail mixes and are preferably frozen mixed with other fruits. This can be done with fresh fruits or by mixing frozen fruit after partial defrosting.

Muskmelon and Honeydew melon flesh cut into 1⅛-in. balls and mixed in about 50:50 proportions make a delicious frozen product; the mixture is now preferred to 100% Honeydews.

Melons are prepared for freezing by halving, removing seeds, and paring before cubing or "sphering." It is possible to do this mechanically with the smaller and firmer melons, such as cantaloupes, by paring them while they rotate on a large spindle. Then pieces can be separated from the cubes by passing them over vibrating shaker screens.

After inspection to remove imperfect pieces, the fruit is washed with water sprays and filled into either retail- or institutional-size containers, along with 28° Brix syrup. Racks of the packages are transported into an air-blast tunnel for freezing.

Nectarines

The nectarine is a smooth-skinned sport of the peach. The white-fleshed cultivars are 'Gower,' 'Quetta,' 'Rivers,' 'Rose,' and 'Stanwick.' 'Humboldt' is a yellow-fleshed cultivar grown on a commercial scale. Relatively little has been published concerning suitability of nectarine cultivars for freezing. 'Stanwick,' 'Gower,' and 'New Boy' are generally listed as the California-grown cultivars suitable for freezing. However, they are no longer of commercial importance. Fred Anderson of Merced, Calif., has developed 'Early Sun Grand,' 'Sun Grand,' and 'Late Le Grand' nectarines. These clingstone-type cultivars are largely for the fresh market. John Weinberger of USDA at Fresno, Calif., has developed 'Independence,' 'Fantasia,' and 'Flavor Top' nectarines. These freestone-type cultivars have potential for freezing preservation.

'Kim' and 'Bim' are yellow cultivars grown only to a limited extent. 'Kim' is good for freezing. Of the white-fleshed cultivars, 'Gower' has the best flavor and is good in texture. It has a large amount of anthocyanin pigment in both the skin and the pit cavity, which colors the syrup a bright red, but does not appreciably detract from the appearance, 'Stanwick' may be rated as medium good to fair, with 'New Boy' and a group of practically indistinguishable ('Ansenne,' 'Goldmine,' 'Diamond Jubilee,' 'Surecrop'), all known as "Australian Seedlings," closely following. This group of cultivars is of pleasing flavor, but very soft in texture.

Nectarines are essentially similar to peaches in properties. Proper maturity is even more critical for nectarines than for peaches. They must yield to thumb pressure with some firmness and not be mushy. When slightly overripe, they take on a characteristic disagreeable flavor which is very noticeable. On the other hand, fruit harvested before it has reached full "eating ripeness" never attains its maximum flavor quality. The fruit ripens rapidly at the usual temperatures prevailing at harvest, and hence should be frozen immediately after picking. If necessary, it may be held a few days at 32°F and 85% RH. Discoloration at the seed and loss of flavor occur if such storage is prolonged.

Nectarines may be frozen either in halves or slices. Generally, unpeeled halves are preferred for commercial operations. Peeled, frozen nectarines tend to be too mushy. Neither blanching nor sulfite treatment is satisfactory for natural flavor preservation in nectarines.

A typical commercial procedure for nectarine freezing is as follows: sort and grade, wash, cut and pit, inspect and trim, package. For 1-lb containers, use 10 oz of fruit and 6 oz of 30° Brix syrup containing 0.1% ascorbic acid or 0.03% ascorbic acid plus 0.5% citric acid. Seal, close, overwrap tray and freeze, case and store at 0°F. Of the large containers available, enameled slipcover tin cans or cellophane- or parchment-lined fiberboard cartons are most suitable for nectarine halves. Ascorbic acid, if added to the syrup, should be calculated to give 150–200 mg/lb of finished pack.

Olives

The freezing of cured, ripe olives has been successfully done experimentally at the University of California, but it has not been undertaken commercially. 'Mission' is the best cultivar for freezing. Other varieties, particularly 'Sevillano,' lose too much in texture unless they are heated before freezing; but this results in a flavor that is similar to that of canned olives instead of the desirable characteristic flavor of freshly cured olives. Curing of the fruit would be executed in the usual manner, and quick-freezing, which is important for texture retention, may be accomplished by CO_2, liquid N_2 spray, IQF, or Freon 12 methods. Since canned olives are quite attractive in flavor and texture, there is little justification for freezing storage of cured olives.

Papaya

Desirable processing characteristics of papaya are large size; fewer ridges on skin; smaller seed cavity; and thicker, firm flesh. A problem that must be solved is the development of an improved machine for peeling the fruit. Hand peeling is too slow and costly. Recently, the University of Hawaii and USDA have been testing devices for mechanical peeling.

By careful selection and proper ripening of the fruit, papaya chunks could be kept intact during the freezing and thawing process. If necessary, the pieces could be dipped in a dilute calcium chloride solution to increase firmness. Preparation and freezing procedures for papaya are similar to those for mangoes. Up to the present time, frozen papaya chunks have not been of commercial importance. The procedures for making frozen papaya puree are described in a previous section of this chapter.

The nonvolatile acids of papaya have been identified as α-ketoglutaric, citric, malic, tartaric, ascorbic, and galacturonic. Citric and malic acids are present in equal amounts, with α-ketoglutaric acid in much lesser amount; these three acids and ascorbic acid accounted for 85% of the total acid in papaya. Much remains to be investigated on the enzymic production of off-flavors in papayas at certain stages of maturity, and the methods for detecting and controlling the enzymic reactions in frozen papayas.

Peaches

Peaches have been more extensively studied than other fruits with respect to their adaptability to freezing. A peach is greatly affected by its growing conditions such as climate, soil, fertilizer, and cultivar characteristics. In general, clingstone cultivars are considered unsuitable for frozen dessert packs, and have only been frozen for special purposes such as reprocessing into fruit cocktails, baby foods, pies, and jams. The white-fleshed peach cultivars generally give less satisfactory frozen products than the yellow-fleshed. This is caused mainly by their usually soft texture and great susceptibility to oxidative browning.

Investigations on freezing preservation of freestone peaches have been reported by several investigators. Culpepper *et al.* (1955) stressed the importance of maturity and ripening procedures for canning and freezing preservation of peaches. A conference (Anon. 1967) on processing and utilization of peaches was held in Georgia, during which many of the problems involved in peach processing were discussed.

Desirable cultivar characteristics of the yellow-fleshed freestones include pronounced flavor, firm but tender texture, and relatively low tendency to darken oxidatively. The fruit should ripen uniformly, be easily peeled, and be able to withstand handling without bruising badly.

Nicotra *et al.* (1971) reported on suitability of some Italian peach and apricot cultivars for freezing. Peach cultivars tested were 'Coronado,' 'Coronet,' 'Goosen,' 'Loadel,' 'Redhaven,' and 'Shasta.' They concluded that the best peach cultivars for freezing are 'Coronado,' 'Loadel,' and 'Coronet.' Addition of calcium salts to the syrup to control damages to flesh structure caused by freezing gave negative results. Browning in frozen peaches may be controlled by adding 0.5% ascorbic acid to the syrup of 30° Brix.

In general, early-ripening cultivars, while less subject to darkening than others, lack characteristic flavor and are rather fibrous and mushy in texture. The best cultivars ripen in August during the main peach season. Later cultivars are less suitable, mainly because of their susceptibility to oxidation.

For retail packages of sliced peaches, cultivars with red around the pit cavity, such as 'Rio Oso Gem,' are desirable. Peaches frozen in bulk for remanufacture into preserves should have good flavor and should not have red color around the pit cavity since it would detract from the appearance of the jam or marmalade. 'Fay Elbertas' picked on the immature side to reduce the red center have proved to be satisfactory for preserves.

For the institutional market, principally pies, a highly flavored, firm-textured cultivar with resistance to oxidation is required. Pie and dessert markets in particular need cultivars of the Gem type.

Freestone peaches cannot be harvested at tree-ripe stage because of bruising and discoloration problems. It is important to harvest them at the firm stage and ripen them under controlled conditions to get the best fruit for freezing.

Processing.—Peaches are pitted, peeled, and sliced before freezing. The Filper peach pitter is currently used by the freezing industry. There is a cup on which the peach is placed which swings up into the cutting position. The fruit is held gently by spring-loaded arms while cutting blades from four directions sever it into halves, push the pit out, and move the halves on to the next operation. After halving, the peaches may be peeled by passing them, cup down, through a 10% lye bath at 140°F for about 2 min. A 2% sodium hydroxide lye spray to peel the cup-down peach halves at 212°F for 38 sec is recommended (Fig. 7.10). The loosened peel may then be removed by water sprays or by rubbing or brushing, and the remaining fragments are dislodged by hand.

After peeling, peach halves are rinsed, first in cool water and later in a 2% citric acid solution. This neutralizes excess alkali and retards browning. Peaches are sliced with rotary disks or fixed-blade knives that cut each half into 5–10 slices, depending on the size of the fruit. Another

FIG. 7.10. Continuous cup-down lye peeler for fruits and vegetables. (Courtesy FMC Corp.)

method of peeling is scalding the peaches in steam for 1–2 min, depending on the size, maturity, and cultivar of the peach. After scalding, the flesh is cooled by a water spray to loosen skins and prevent softening and darkening by the heat. This method needs more labor and is expensive for a large-scale operation. For small-scale production where lye-peeling facilities are not available, the steam-peeling method can be utilized where labor cost is not prohibitive. However, the uniformity of ripeness must be carefully controlled to get best results.

The sliced peaches are packed in 40° Brix syrup containing 0.1% ascorbic acid or isoascorbic acid. A ratio of two to four parts fruit to one part syrup generally is acceptable. For the bakery and ice cream trade, slices are packed in syrup in 30-lb enamel-lined slipcover cans. The cans are frozen in a tunnel air-blast freezer at −30°F.

The texture of fruit frozen in liquid nitrogen and dichlorodifluoromethane is more like that of fresh fruit than is that of fruit frozen in an air blast. With 10% drip loss as the cutoff point, liquid nitrogen-frozen and DCDFM-frozen fruits had 1 hr longer useful life after thawing than did air-blast frozen products.

Wolford *et al.* (1971) evaluated the quality of peaches, cherries, and berries frozen in liquid nitrogen and in Freon 12 (F-12). The fruits were prepared for freezing by washing, slicing, and by immersion in syrup in some cases. Fruit was also immersed in F-12 previously cooled with a mixture of dry ice and Freon 11. The texture of peaches frozen in liquid nitrogen and F-12 was firmer than that of conventionally frozen peaches. Recently, Lewis "Freon" flash immersion continuous freezers have been made available to the food industry. The unit is available from the Lewis Refrigeration Co., Box 167, Woodinville, Washington 98072. It has a mechanical refrigerating unit to recover the Freon 12, and is the latest step developed for individually quick frozen (IQF) foods. The Lewis Refrigeration Company also makes fluidized IQF freezing tunnels. More than 300 of the Lewis fluidized IQF freezing tunnels are in operation throughout the world.

The boiling point of Freon 12 is −22°F. The freezing of apricots, peach slices, strawberries, and other fruits takes only 1–2 min when the fruit is dipped in liquid F-12 followed by spraying with F-12 liquid onto fruits moving on a stainless steel conveyor belt.

Pears

In general, pears are not recommended for freezing, except as purées, because of losses of flavor and texture. Pears have been frozen commercially to a limited extent mainly for use in reprocessing operations. To prepare for freezing, peel as for canning, core, cut into halves or slices, fill into large containers, cover with 40°–50° Brix syrup, and freeze. Pears are extremely subject to browning, and should therefore be handled quickly after peeling. The brown discoloration is caused by the activity of the enzyme polyphenoloxidase after the tissue is damaged or when oxygen is available. Washing peeled pears in a 1.0% citric acid solution prior to freezing, or addition of 0.1% ascorbic acid to the syrup, is helpful in inhibiting browning.

One process for freezing peeled and cored pears involves a predip, prior to freezing, in a sucrose solution of slightly greater strength than the Brix reading (soluble solids) of the fruit itself, with added ascorbic acid and other fruit acids such as malic or citric acids.

Persimmons

The 'Hachiya' cultivar of persimmons makes a frozen puree with good color and flavor. It is important that the proper stage of maturity be selected for freezing, since underripe persimmons are very astringent. With added sugar this puree has been used as an ice cream flavor and for persimmon pudding.

The soft-ripe fruit is washed, decapped, and passed through a pulper with a stainless steel screen. It is important to avoid contamination with iron or copper which will cause discoloration. One part of sucrose is added to each five or six parts of puree and dissolved by thorough mixing; addition of 0.1% ascorbic or isoascorbic acid aids in retention of flavor and color. The product should be packed in hermetically sealed enameled cans or friction-top egg tins. If the latter are used, a layer of sucrose or powdered sugar is placed on top.

Pineapple

Frozen pineapple has appeared on the market for many years in the form of rectangular chunks. The sliced pineapple kept well in syrup of 25°–49° Brix for about 1 year, but eventually developed a "stale" taste. The addition of ascorbic acid to the syrup aids greatly in stabilizing the flavor.

The 'Smooth Cayenne' cultivar, grown in Hawaii and to a limited extent in Puerto Rico and other tropical regions, is the principal cultivar used for freezing. The 'Red Spanish' cultivar of Cuba develops off-flavors when it is frozen. Pineapple maintains a good texture after thawing due to its slightly fibrous structure. In recent years crushed and tidbits pineapple have been available as frozen products for the institutional market, principally the bakery trade.

Pineapple for freezing is prepared the same way as that for canning, except that the cylinders usually go through an additional coring operation to remove the last vestiges of the fibrous core. Bins of fruit arriving from the plantation are unloaded onto conveyor belts or into flumes of water. They are washed thoroughly, graded into three or four sizes, then peeled and cored on the Ginaca machine. The cored cylinders are inspected, trimmed, and diverted to a second coring machine followed by a fixed blade chunk cutter. The chunks are filled into 211 × 414 or No. 10 cans, syruped, seamed, and conveyed directly to a tunnel air-blast freezer at −30°F.

The bulk-frozen packages are filled directly from the product line in the form of tidbits, syruped, and frozen in a low-temperature blast freezer.

FREEZE-DRIED FRUIT

While not a major household or supermarket fruit item, freeze-dried sliced strawberries, diced fruits, and berries are important to military operations and other operations where weight is important.

Right Away Foods is an outgrowth of the first freeze-drying plant in the United States, which started in Edinburg, Texas, in the early 1950s. The company has been active in military rations since its beginning and in 1980 built the first facility in the United States, in McAllen, Texas, exclusively to produce new military ready-to-eat rations.

The core of the freeze-dry plant consists of 15 large batch vacuum freeze driers. A wide range of products, including fruits, are processed to produce lightweight versions that can be rapidly and easily rehydrated to their original composition by adding water.

Each drier consists of a number of hollow shelves, through which heating solution is circulated. The units are of different sizes. For example, one type of drier has 32 shelves and can handle 2000 lb of product/batchload; another has 56 shelves and can run about 3500 lb.

Products to bc freeze-dried are frozen prior to placing in the drier. The frozen product is cut or deposited in its final size and shaped onto trays that are placed on the shelves in the drier. After the doors are closed and sealed, the vacuum system is turned on and the chamber is pulled down to it final operating conditions. All products are processed in the narrow range of 1 to 4 mm Hg absolute pressure. At these conditions, ice in the product sublimes directly into vapor with no intermediate liquid phase.

To assist in driving off moisture, heating solution at 130° to 250°F is circulated through the hollow shelves. Even then the time seems very long; for example, as much as 30 hr is required for a single load of 4-oz fruit slices to dry.

Individual pieces leaving the freeze driers are immediately wrapped in heavy laminated film on an intermittent motion horizontal machine. In this operation, after the bottom web of film is formed into cavities as it enters the machine, operators place the dried product into the cavities as the film moves forward at about 10 cycles/min. The top layer of film is then heat-sealed over the product, and the four-wide packages are cut apart (Mans 1982).

FREEZING PREPARED AND PRECOOKED FRUIT PRODUCTS

A great variety of prepared and precooked fruit products are frozen. They include baby foods, baked apples, cobblers, fritters, spreads, pectinized purees, pies, pie fillings, relishes, sherbets, shortcakes, toppings, and sundae sauces. Freezing retains the color and flavor of many prepared and precooked fruits.

Apples, blueberries, blackberries, cherries, and raspberries are excellent when cooked by immersion in boiling 50% sugar syrup for 15 sec. When these fruits are steamed in an open vessel, considerable flavor is lost and they become watery. The loss of flavor by steaming in an open vessel is caused both by leaching and the distilling off of flavors.

Plums, pears, and freestone peaches are less desirable as preheated products except when prepared as a puree or a solid pack. Apples, blueberries, raspberries, blackberries, and plums are superior when preheated before freezing provided proper methods of heating and suitable packages are used. Preheated fruits should be solid packed without bubbles or air spaces, in rigid, leakproof, moisture-proof containers.

Applesauce

At one time, applesauce was packed and frozen commercially, but the demand for the product has not held up. The frozen product is superior in flavor but darker in color than canned sauce. This may be explained by the bleaching of canned sauce by the action of the malic acid of the apple on the tin of the can, and also due to enzyme browning in the frozen sauce.

Apples are washed, mechanically peeled, and cored. The product is conveyed to a rotary spindle-type washer, where powerful jets of water knock off the loosened skins and force them through the bottom openings in the revolving cylinder to a waste conveyor below. The apples then pass over a trimming table where adhering bits of skin and defective parts are removed. The fruit is sliced and the slices cooked by a screw-type preheater where the proper quantities of liquid sugar are added. The cooked slices are pureed in a stainless steel pulper. The applesauce is inspected for defects then pumped to a stainless steel tank equipped with agitator and heater. The hot puree is pumped through a scraped-surface heat exchanger where it is rapidly cooled. The cooked sauce passes to an automatic filler which fills measured amounts into cartons. Filled cartons are put in trays which are placed on racks and the racks are moved into air-blast freezing tunnels. When frozen, the cartons are put into fiberboard shipping containers for storage or shipment.

Concentrated Apple Segments

Powers and Miller (1971) made pilot trials on frozen concentrated apple segments. The apples were peeled, cored, radially sliced, sulfited, and dried in a tray-type laboratory drier. The samples were packaged in polythene bags and then frozen at 14°F. Commercial trials were also carried out and it was concluded that 3–3.5 times frozen concentrated apple segments represent an optimum product. Apple segments so produced rehydrate readily for bakery use.

Cranberry–Orange Relish

Frozen cranberry–orange relish is packed either in 11-oz cans for the retail trade or in 40-oz cans for the wholesale trade. Fresh washed cranberries are ground in a food chopper. The oranges are washed, quartered, and the seeds removed. The unpeeled orange quarters are put through a food chopper. The ground cranberries, ground oranges, and sugar are mixed together in equal proportions by weight. The product is packaged in liquid-tight, moisture-vapor-proof cartons or fruit-enamel-lined cans, then frozen. For 11-oz cans, a contact freezer (Fig. 7.11) and air-blast tunnel may be used.

Cranberry Sauce

The first step in making jellied cranberry sauce is to prepare a cranberry pureAe. This may be done by boiling cranberries with an equal weight of water in a steam-jacketed kettle for 3 min, partially cooling, and pureeing in a juice extractor. The berries may also be steamed on stainless steel trays for 2 min, cooled to 125°F, pureed, and then diluted with water (118 lb of water to 100 lb of puree). For each 100 lb of diluted puree, 0.42–0.58 lb of rapid-set, 150-grade citrus pectin mixed with 13 lb of granulated sugar is added. The sugar–pectin mixture is stirred into the puree; stirring is continued for 15 min to dissolve the pectin. Then 53.7 lb of sugar are added to the puree–pectin mixture, which is stirred until all of the sugar dissolves. After the sugar has dissolved, the liquid mix is packaged in liquid-tight, moisture-proof cartons or fruit-enamel-lined cans which are closed and allowed to stand at 70°–80°F for approximately 24 hr. The product is then frozen and stored at 0°F or lower. The yield from 100 lb of cranberries (85 lb puree), 100 lb water, 123.3 lb sugar and 0.8 lb pectin is 309 lb of frozen jellied cranberry sauce. The jellied sauce prepared from

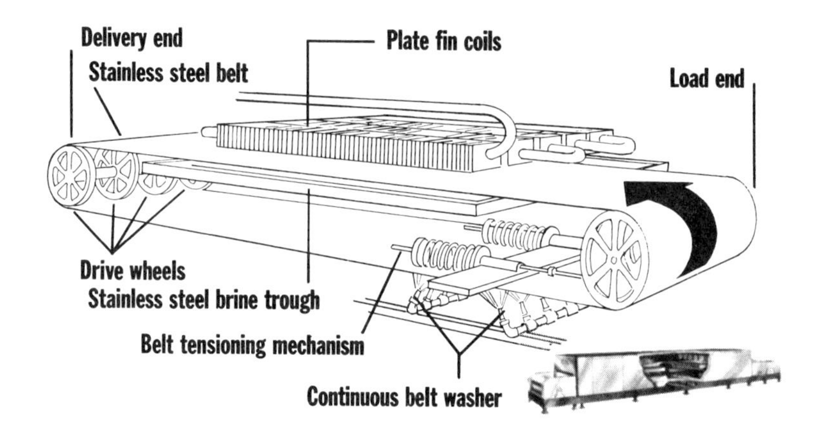

FIG. 7.11. Frick "Kontack" freezer. (Courtesy Frick Co.)

the 'Howes', 'Early Black', and 'McFarlin'' berries does not differ significantly in gel strength.

Fruit Cocktail

Fruit cocktail is a popular dessert item. Frozen fruits are sometimes used by various institutions and restaurants as components in making so-called fresh fruit cocktails or fresh fruit salads. Often these cocktails or salads are a mixture of fresh and frozen fruits. Sometimes canned fruits are also used as components.

In one study, a comparison was made of two methods of preparing frozen cocktail mix consisting of diced peaches, diced pears, pineapple pieces, whole Thompson seedless grapes, and dyed cherries, frozen with 40° Brix syrup. Half of the lot was packed and frozen in 40° Brix syrup without heating the syrup; the other half was heated in the syrup to boiling for 5 min, cooled, packed, and frozen at 0°F. It was observed that the unheated frozen cocktail developed a noticeable hay-like flavor and odor after storage for 1 year at 0°F. That which had been heated was unchanged from its original flavor and was much superior in flavor to that which had been preserved by heat sterilization in cans.

Molded Fruit Salad

Molded fruit salads can be prepared with either frozen or fresh fruits. If frozen fruits are used, the fruits are thawed and drained. Gelatin, which has been dissolved in warm water, and a little ascorbic acid are added to the drained syrup. The drained fruits are mixed in the desired proportions. The fruit mixture is placed in cartons or other containers. The syrup containing the gelatin is allowed to cool until it forms a soft jelly, and is then poured over the fruit mixture in the cartons, which are then closed, overwrapped, and placed in a freezer.

If fresh fruit is used, the procedure is the same except for the preliminary preparation of the fruit and the addition of corn or sugar syrup. Soft-ripe fruit is selected, washed, peeled, cored or pitted, and cut into pieces of the proper size. The fruits are mixed together. Gelatin and a small amount of ascorbic acid are dissolved in warm water. The solution of gelatin and ascorbic acid is mixed with sugar syrup and then allowed to cool until it forms a soft jelly. The fruit mixture is placed in cartons or other containers, covered with the soft jelly, and containers are closed and placed in the freezer.

Fruit Salad

Another method of packaging, freezing, and defrosting fruit salad and prepared fruits involves the packing of the fruit salad to which has been added a little sugar and ascorbic acid, or a small amount of sugar syrup containing 0.1% ascorbic acid, in a heat-sealable polyethylene pouch. A high vacuum is pulled while the pouch is being sealed after which it is packaged in a rectangular carton and quick frozen. When used, the pouch is immersed in a pan of running cold water or lukewarm water until the fruit is thawed but is still cold.

Packaging the fruit under a vacuum maintains the color and flavor of the product. Rapid freezing and rapid thawing aids materially in retaining color and flavor, and results in a product of better texture than is obtained if the fruit is thawed slowly while exposed to air.

Pie Fillings

Fruits are prepared for freezing either whole, halved, or sliced, or as fruit puree. Consumers can use these various styles in making pie fillings. If the frozen fruit packer prepared and froze the complete pie fillings, one step would be eliminated and pies of superior flavor should result. Some blueberry, elderberry, and pumpkin pie fillings are frozen, but the business is relatively small.

Pie Filling Mixes. Johnson (1950) studied the preparation of frozen fruit pie mixes and methods of preparation for apple, apricot, blueberry, peach, and cherry mixes. He recommends the use of either an Irish moss extractive or low methoxyl pectin. These thickening agents have a specific gelling temperature and will not gel until the pie is cooled to this temperature. If low methoxyl pectin is used, a small amount of monocalcium phosphate should be added to make the calcium pectate gel. Johnson prefers fillings made with Irish moss extractives, since the latter make a very desirable clear, soft gel. He describes the following steps in preparing the dry mix:

1. A mixture containing the dry ingredients is first prepared. The sugar, thickening agent, antioxidant, and salt are thoroughly mixed. The composition of the dry mix may vary slightly with the particular fruit used but, in general, the composition will fall within the following limits: 2–3 lb of Irish moss extractive (or 2.5–3.5 lb of low methoxyl pectin containing 0.25–0.3 lb of monocalcium phosphate), 0.2 lb ascorbic acid, and 0.5 lb of sugar.
2. The prepared fresh fruit is mixed with the correct amount of the dry mixture. The mixture of the dry ingredients should be added in such a manner that uniform distribution is obtained for better protection of the fruit. This uniform distribution is more important for peaches than for some of the other fruits. When using cans, the dry mix can be added as the fruit is packed. Care should be taken to add the last part of the dry mix on top.
3. After the fruit and dry mix are packaged, it is frozen immediately and stored at 0°F or lower.

Apple Filling. Apples are peeled and cored. Large apples are sliced into 12 pieces and smaller ones into 8 pieces. The slices are dipped 1–2 min in a bath containing 500 ppm sulfur dioxide as sodium bisulfite and 0.25% citric acid. After draining, slices are mixed with the dry mix in the proportion of 17 oz of slices to 4.5 oz of mix. This quantity is sufficient for a 9-in. pie.

Apricot Filling. Fully ripe apricots are washed, halved, pitted, and then mixed with the dry mix in the proportion of 6 oz of dry mix for each 20 oz of halved apricots, which is a sufficient quantity for a 9-in. pie.

Blueberry Filling. After cleaning, washing, and inspecting the blueberries, they are mixed with the dry mix in the proportion of 18 oz of berries to 4½ oz dry mix (enough for a 9-in. pic).

Red Tart Cherry Filling. Fully ripe cherries are chilled in ice water, washed, inspected, and pitted. The pitted cherries are mixed with the dry mix in the proportion of 1⅓ lb of fruit to 6½ oz of dry mix (enough for a 9-in. pie).

Peach Filling. Soft-ripe freestone peaches are peeled and cut into 8 slices and are then mixed with the dry mix in the proportion of 1⅓ lb of slices to 5 oz of dry mix (enough for a 9-in. pie).

Deep Dish Pie Fillings. Sayles and MacLennan (1965) give the following recipes and procedures for preparing freezing and fillings for deep dish pies.

Apple (McIntosh)

Ingredients for One Pie	Parts (oz)
Apples (peeled, cored, sliced)	6.0
Lemon juice	0.2
Water	3.5
Sugar	0.4
Apple cider	0.4
Modified starch	0.3
Cinnamon	0.05
Orange "Tang"	0.2
Vanilla extract	0.1

Procedure: Sprinkle lemon juice over apples. Cook slowly in covered sacepan 2-3 min. Combine water and sugar and cook 10-15 min to reduce volume ¼. Slowly add cider to the modified starch and cinnamon. Stir until smooth. Add Tang and vanilla. Add to syrup. Heat to 180°F, stirring constantly. Combine apples and syrup. Package in vacuum pouch. Blast freeze.

Blueberry

Ingredients for One Pie	Parts (oz)
Blueberries, fresh	8.0
Water	3.5
Sugar	1.2
Orange "Tang"	0.2
Vanilla extract	0.1
Modified starch (e.g., Purity 69)	0.3

Procedure: Combine ½ the water and sugar and cook 10-15 min to reduce volume ⅓. Slowly add remaining water to modified starch. Add to syrup. Heat to 180°F, stirring constantly. Combine blueberries and syrup. Package in vacuum pouch, blast freeze.

Cherry

Ingredients for One Pie	Parts (oz)
Frozen cherries (thawed, well-drained)	8.00
Cherry juice syrup	2.75
Vanilla extract	0.10
Orange "Tang"	0.20
Modified starch (e.g., Purity 69)	0.30

Procedure: Slowly add cherry juice syrup to the modified starch; add vanilla and Tang. Heat to 180°F, stirring constantly. Combine cherries and syrup. Package in vacuum pouch, blast freeze.

Pectinized Puree for Ice Cream

Pectinized fruit puree can be used directly in fruit ice creams. The product should have pronounced flavor and color, and yield a frozen product of smooth texture. It must have a total solids content 10% higher than the ice cream in which it is used, so that it will not change the texture of the ice cream.

If sucrose is used as the only sweetening agent, the puree will be too sweet and, when frozen, will be somewhat sandy because of the crystallization of the sugar. Replacement of one-half of the sucrose with enzyme-converted corn syrup will decrease the sweetness.

The addition of pectin to the puree is advantageous because of the increased body, causing the puree to melt to a very soft jelly instead of to a liquid.

Most ice cream makers prefer a tart puree, which may be obtained by the addition of citric acid. Further, increasing the acidity increases the viscosity of the puree.

Formulas for preparation of pectinized strawberry and peach puree are as follows (Tressler 1968):

Ingredients	Parts	
	Berry	Peach
Fruit (lb)	100	100
Sugar (lb)	80	76
Slow-set 150 grade citrus pectin (oz)	17.3	17
Monocalcium phosphate (oz)	1.7	1.7
Citric acid (oz)	0	4

The amount of sugar used is such that the final soluble solids content is about 50%.

(1) The puree is divided into 2 portions in the ratio of 60 to 40.

(2) Slow-set 150 grade citrus pectin, approximately 0.6% of the weight of final product (puree plus sugar), is dispersed and dissolved as follows: The pectin is mixed with about ten times its weight of sugar. An amount of puree equal in weight to this sugar is taken from the 60% portion and added to the pectin-sugar mixture. The mixture is stirred until all of the sugar is dissolved. This step helps dispersion by allowing the particles to become wetted without much swelling. The concentrated pectin mixture is then added to the remainder of the 60% portion of puree to form mixture A, and is mixed for about 20 min with a mechanical mixer of a type that does not beat air into the product. The soluble solids content at this stage should not exceed 25%. A higher sugar concentration will interfere with proper dissolving of the pectin. As an alternative and somewhat simpler procedure, the dry pectin can be mixed with approximately 2.5 times its weight of glycerin or 10 times its weight of invert sugar, and then added slowly to the 60% portion with constant stirring for about 20 min.

(3) The amount of sugar required to increase the soluble solids of the puree to 50% is mixed into the 40% portion. This will be referred to as mixture B. The greater part of

the sugar will dissolve. Amount of sugar required to increase soluble solids to 50% can be calculated from the following equation:

$$x = 0.50y - \{[(s.s/100)\ (y)]\}/0.50$$

where x = weight of total amount of added sugar, y = weight of fruit puree, and $s.s$ = percent soluble solids determined by refractometer using the sucrose scale.

The amount of sugar to be added to the 40% portion then becomes x, less the amount of sugar, glycerin, or invert sugar added to the pectin.

(4) The mixture B is added to the 60% pectinized portions (mixture A) and stirring is continued until all of the sugar is dissolved and a uniform mixture is obtained (approximately 5 min).

(5) Monocalcium phosphate is made into a slurry by the addition of a small amount of water. This slurry is then mixed with the pectinized puree with slow agitation. When desired, citric acid is also added at this stage.

(6) Two parts of the pectinized puree are then packed with 1-1.5 parts of sliced peaches or strawberries or whole berries.

(7) After packaging, the product is placed into freezing storage.

The product has excellent fresh fruit flavor and good consistency for toppings for ice cream and shortcake. The soda fountain is a potential outlet for this product.

Punch Blends

Seale and Sherman (1960) developed a Polynesian Punch with an enticing flavor; the formula is as follows:

Ingredients	Parts
Passion fruit juice	18
Pineapple concentrate	5
Acerola purée	4
Pineapple juice	3
Orange concentrate	4
Lemon juice	2
Water	8
Sugar	58
Citric acid, enough to give an acidity of 1.7%	

The ingredients are mixed in a blending tank; then citric acid is added to bring the acidity to 1.7%. The product is filled into enamel-lined cans and frozen. When used, each can of blend is diluted with 4½ cans of water.

REFERENCES

ANDRUS, M. 1966. Differential in Quality Between Handpicked and Mechanically Harvested Red Tart Cherries. Mich. Dept. of Agriculture, East Lansing, MI.

ANON. 1946. Velva-fruit, a new frozen fruit dessert. U.S. Dept. Agric. Bur. Agric. Ind. Chem., West. Reg. Res. Lab. Mimeo Circ. *AIC-40,* Revised Mar. 1946.
ANON. 1967. Conference on peach processing and utilization, Experiment, GA. U.S. Dep. Agric. ARS 72–67.
ANON. 1968. Diluted fruit juice beverages: Order establishing identity standards for lemonade, colored lemonade, and limeade. Fed. Regist. *33,* 6864–6865.
ANON. 1969. Freon freezer to compete directly with IQF vegetable, fruit systems. Quick Froezn Foods *31* (8) 109–111, 153.
ANON. 1972. Frozen Food Pack Statistics. American Frozen Food Institute, Washington, DC.
ANON. 1982. Cryogenic railcar under development. Prepared Foods *151,* 8, 22.
ARMERDING, G. D. 1966. Evaporation methods as applied to the food industry. Adv. Food Res. *15,* 303–357.
ARNOLD, C. E. and MITCHELL, A. E. 1970. Histology of blemishes of cherry fruits (*Prunus cerasus* L., cv. Montmorency), resulting from mechanical harvesting. J. Am. Soc. Hort. Sci. *95* (6) 723–725.
AUNG, T. and ROSS, E. 1965. Heat sensitivity of pectinesterase activity in papaya purée and of catalase-like activity in passion-fruit juice. J. Food Sci. *30*(1) 144–147.
BALLINGER, W. E. and KUSHMAN, L. J. 1970. Relation of stage of ripeness to composition and keeping quality of highbush blueberries. J. Am. Soc. Hort. Sci. *95* (2) 239–242.
BAUCKMANN, M. 1970. Suitability of strawberry varieties for wet preservation and freezing. Obst. Garten. *89* (5) 172.
BEATTIE, B. B., HALL, E. G., COOTE, G. G. and BAXTER, R. I. 1971. Effects of temperature and time in cool storage on the ripening and storage life of pears. Aust. J. Exp. Agric. Anim. Husb. *11* (52) 576–581.
BENSON, E. J. 1968. Freezing of avocados by two stage process. U.S. Pat. 3,398,001. Aug. 20.
BLACK, W. F. and POLOPOLUS, L. 1966. Synthetics and substitutes and the Florida citrus industry. *In* Synthetics and Substitutes for Agricultural Commodities. Univ. Fl. Inst. Food Agric. Sci. Publ. *1,* 2–11.
BOYLE, F. P. and SHAW, T. M. 1955. Freezing and canning of mangos. Univ. Hawaii Mango Packet.
BOYLE, F. P., SEAGRAVE-SMITH, H., SAKATA, S. and SHERMAN, G. D. 1957. Commercial guava processing in Hawaii. Univ. Hawaii Agric. Exp. Stn. Bull. *111,* 5–30.
BREKKE, J., CAVALETTO, C. and STAFFORD, A. E. 1968. Mango purée processing. Univ. Hawaii Agric. Exp. Stn., Tech. Prog. Rep. *167.*
BREKKE, J. E., CHAN, H. T. and CAVALETTO, C. G. 1972. Papaya purée: A tropical flavor ingredient. Food Prod. Dev. *6* (6) 36–37.
BREKKE, J. E., CHAN, H. T. and CAVALETTO, C. G. 1973. Papaya purée and nectar. Univ. Hawaii Agric. Exp. Stn. Res. Bull. 170.
CHAN, H. T., FLATH, R. A., FORREY, R. R., CAVALETTO, C. G., NAKAYAMA, T. O. M., and BREKKE, J.E. 1973. Development of off-odors and off-flavors in papaya purée. J. Agric. Food Chem. *21,* 566–570.
CHEN, L. F. and LUH, B. S. 1967. Anthocyanins in Royal grapes. J. Food Sci. *32,* 66–74.
COBIN, M. 1961. The lychee in Florida. Univ. Fl. Agric. Ext. Serv. Bull. *176.*
CULPEPPER, C. W., HALLER, M. H., DEMAREE, K. D. and KOEH, E. J. 1955. Effect of picking maturity and ripening temperature on the quality of canned and frozen Eastern-grown peaches. U.S. Dept. Agric. Tech. Bull. *1114.*
EAVES, C. A., FORSYTH, F. R., LEEFE, J. S. and LOCKHART, C. L. 1964. Effect of varying concentrations of oxygen with and without CO_2 on senescent changes in stored McIntosh apples grown under two levels of nitrogen fertilization. Can. J. Plant Sci. *44,* 458–465.

ENOCHIAN, R. V. 1968. The rise, present importance, and future of frozen fresh food. *In* The Freezing Preservation of Foods, 4th ed., Vol. 3. D. K. Tressler, W. B. Van Arsdel and M. J. Copley (Editors). AVI Publishing Co., Westport, CT.

FENNEMA, O. and POWRIE, W. D. 1964. Fundamentals of low-temperature food preservation. Adv. Food Res. *13,* 219–347.

FIDLER, J. C. 1970. Sorbitol in stored apples. J. Hort. Sci. *45* (2) 197–204.

FIDLER, J. C. and NORTH, C. J. 1971. Effect of conditions of storage on the respiration of apples. J. Hort. Sci. *46* (2) 237–243.

GASTON, H. P., LEVIN, J. H. and WHITTENBERGER, R. T. 1966. How to use cherry harvesting machines. Mich. State Univ. Ext. Bull. *532.*

GORGATTI NETTO, A., BLEINROTH, E. W., and LAZZARINI, L. C. 1973. Quality evaluation of frozen sliced mangoes in syrup. *In* Proc. of the XIII Intern. Cong. of Refrigeration, Washington, DC, Vol. 3, 1971. AVI Publishing Co., Westport, CT.

GUADAGNI, D. G. and KELLY, S. H. 1958. Time-temperature tolerance of frozen foods. XIV. Ascorbic acid and its oxidation products as a measure of temperature history in frozen strawberries. Food Technol. *12,* 645–647.

GUADAGNI, D. G. and NIMMO, C. C. 1953. Effect of growing area on tannin and its relation to astringency in frozen Elberta peaches. Food Technol. *7,* 59–61.

GUADAGNI, D. G., and NIMMO, C. C. 1957. Time–temperature tolerance of frozen foods. XI. Effect of time and temperature on color distribution in retail packs of frozen strawberries. Food Technol. *11,* 604–608.

GUADAGNI, D. G., NIMMO, C. C. and JANSEN, E. F. 1957A. Time-temperature tolerance of frozen foods. II. Retail packages of frozen peaches. Food Technol. *11,* 33–42.

GUADAGNI, D. G., NIMMO, C. C. and JANSEN, E. F. 1957B. Time-temperature tolerance of frozen foods. VI. Retail packages of frozen strawberries. Food Technol. *11,* 389–397.

GUADAGNI, D. G., NIMMO, C. C. and JANSEN, E. F. 1958. Time-temperature tolerance of frozen foods. XI. Retail packs of frozen red sour pitted cherries. Food Technol. *12,* 36–40.

GUADAGNI, D. G., EREMIA, K. M., KELLY, S. H. and HARRIS, J. 1960. Time-temperature tolerance of frozen foods. XX. Boysenberries. Food Technol. *14,* 148–150.

GUEST, R., MARKWARDT, E. D., LABELLE, R. L. and FRENCH, O. C. 1960. Mechanical harvesting of red tart cherries. N.Y. State Agric. Exp. Stn. Prog. Rep.

HAMILTON, R. A. 1960. Pope mango. Hawaii Agric. Exp. Stn. Circ. *60.*

HASS, L. V., KRENZ, W. and ROBE, K. 1966. Makes better concentrate in semi-vacuum HTST evaporator. Food Process. Mark. *27* (10) 159–161.

HIU, D. N. and SCHEUER, P. J. 1961. Volatile constituents of passion fruit juice. J. Food. Sci. *26* (6) 557–563.

HOHL, L. A. 1946. Freezing California fruits. Figs. Food Packer *27* (11) 66, 68, 70.

HOLGATE, K. and KERTESZ, Z. I. 1948. The comparative usefulness of various calcium salts in the firming of canned and frozen sliced apples. Fruit Prod. J. *28,* 37–38, 42.

HULME, A. C. 1971. The mango. *In* The Biochemistry of Fruits and Their Products, Vol. 2. A. C. Hulme (Editor). Academic Press, London and New York.

HUXSOLL, C. C. and REZNIK, D. 1969. Sorting and processing mechanically harvested dates. Date Grow. Inst. Rep. *46,* 8–10.

JOHNSON, G. 1950. New frozen fresh fruit pie mixes. Quick Frozen Foods *13* (1) 50–51, 100.

KAUFMAN, C. W. and CAMPBELL, H. A. 1949. Some fundamental considerations in the processing of frozen orange juice concentrate. Food Technol. *3,* 395–404.

KEFFORD, J. F., McKENZIE, H. A. and THOMPSON, P. C. O. 1959. Effects of oxygen on quality and ascorbic acid retention in canned and frozen orange juices. J. Sci. Food Agric. *10,* 51–63.

KNEE, M. 1971. Ripening of apples during storage. III. Changes in chemical composition of Golden Delicious apples during the climacteric and under conditions simulating commercial storage practice. J. Sci. Food Agric. *22* 371–377.

KULP, K. and BECHTEL, W. G. 1962. Frozen fruit pies. Food Technol. *16,* 104–106.

LEVIN, J. H., GASTON, H. P., HEDDEN, S. L. and WHITTENBERGER, R. T. 1960 Mechanizing the harvest of red tart cherries. Mich. State Univ. Agric. Exp. Stn. Q. Bull. *42*(4) 133–141.

LOEFFLER, H. J. 1967. Peaches for freezing. Paper presented at 13th Annu. Meet. Calif. Freestone Peach Assoc.

LOWE, E., DURKEE, E. K., HAMILTON, W. E. and MORGAN, A. I., JR. 1964. Better apple dejuicing through thick-cake extraction. Food Eng. *36* (12) 48–50.

LUH, B. S. 1971. Tropical fruit beverages. *In* Fruit and Vegetable Juice Processing Technology, 2nd ed. D. K. Tressler and M. A. Joslyn (Editors). AVI Publishing Co., Westport, CT.

MANS, J. 1982. Right Way Foods—15 vacuum freezer dryers process everything from ham slices to fruits and vegetables. Prepared Foods *151* (8) 48–50.

MARKWARDT, E. D., GUEST, R. W., CAIN, J. C. and LA BELLE, R. L. 1964. Mechanical cherry harvesting. Trans. Am. Soc. Agric. Eng. *7* (1) 70–74, 82.

MEHRLICH, F. P. and FELTON, G. E. 1971. Pineapple juice. *In* Fruit and Vegetable Juice Processing Technology, 2nd ed. D. K. Tressler and M. A. Joslyn (Editors). AVI Publishing Co., Westport, CT.

MENZIES, D. J. and KEFFORD, J. F. 1949. Apple juice blends. Food Preserv. Q. *9* (2) 31–32.

NAKAGAWA, F. 1962. Supporting tray for quick freezing berries. U.S. Pat. 3,031,311. Apr. 24. *In* Food Process. Rev. *21,* 151–152 (1971).

NICOTRA, A., FIDEGHELLI, C., and CRIVELLI, G. 1973. Considerations of the suitability for freezing of some peach and apricot varieties. *In* Proc. XIII Inter. Cong. of Refrigeration, Washington, DC, Vol. 3, 1971. AVI Publishing Co., Westport, CT.

NIXON, R. W. 1950. Imported varieties of dates in the United States U.S. Dept. Agric. Circ. *834.*

OLSON, R. L. and MORGAN, A. I., JR. 1970. Factors affecting quality preservation in frozen foods. *In* Proc. 3rd Intern. Cong. Food Sci. Technol., Washington, DC. Institute of Food Technologists, Chicago.

ORR, K. J. and MILLER, C. 1955. Description and quality of some mango varieties grown in Hawaii and their suitability for freezing. Hawaii Agric. Exp. Stn. Tech. Bull. *26.*

PEYNAUD, E. and RIBÉREAU-GAYON, P. 1971. The grape. *In* The Biochemistry of Fruits and Their Products, Vol. 2. A. C. Hulme (Editor). Academic Press, London and New York.

PONTING, J. D., SANGSHUCK, D. W., and BREKKE, J. E. 1960. Color measurement and deterioration in grape and berry juices and concentrates. Food Res. *25,* 471–478.

POWERS, M. J. and MILLER, W. J. 1971. Frozen concentrated apple segments. Food Res. *31* (4) 76–78.

PRATT, H. K. 1971. Melons. *In* The Biochemistry of Fruits and Their Products, Vol. 2. A. C. Hulme (Editor). Academic Press, London and New York.

PRUTHI, J. S. and LAL, G. 1955. Technical aspects of manufacture of passion-fruit juice and squash. Chem. Age India *6* (2) 39–48.

PRUTHI, J. S., and LAL, G. 1959. Chemical composition of passion fruit (*P. edulic*). J. Food Sci. Agr. *10,* 188–92.

ROSS, E. 1960A. Present and future of mango processing. Hawaii Farm Sci. *9* (2) 3.

ROSS, E. 1960B. Trends in coconut processing research. Hawaii Farm Sci. *9* (1) 3, 8.

ROUSE, A. H. and ATKINS, C. D. 1952. Heat inactivation of pectinesterase in citrus juices. Food Technol. *6,* 291–294.

SALUNKHE, D. K., COOPER, G. M., DHALIWAL, A. S., BOE, A. A., and RIVERS, A. L. 1962. On storage of fruits: Effects of preharvest and postharvest treatments. Food Technol. *16* (11) 119–123.

SAYLES, C. L., and MacLENNAN, H. A. 1965. Ready foods. Cornell Univ. School of Hotel Admin. Res. Rep. *10*.

SCOTT, F. S. 1956. Consumer preferences for frozen passion-fruit juice. Hawaii Agric. Exp. Stn. Agric. Econ. Rep. *29*.

SEAGRAVE-SMITH, H. 1952. Passion fruit is projected for frozen concentrates. Food Eng. *24* (7) 94.

SERVADIO, G. J. and FRANCIS, F. J. 1963. Relation between color of cranberries and color and stability of sauce. Food Technol. *17*, 632–636.

TENNES, B. R., DIENER, R. G., LEVIN, J. H. and WHITTENBERGER, R. T. 1967. Firmness and pitter loss studies on tart cherries. Am. Soc. Agric. Eng., 60th Annu. Meet. Trans. *67–333*.

TRESSLER, D. K. 1968. Prepared and precooked fruit products. In The Freezing Preservation of Foods, 4th ed., Vol. 4. D. K. Tressler, W. B. Van Arsdel and M. J. Copley (Editors). AVI Publishing Co., Westport, CT.

TRESSLER, D. K. and JOSLYN, M. A. 1971. Fruit and Vegetable Juice Processing Technology, 2nd ed. AVI Publishing Co., Westport, CT.

TULECKE, W., WEINSTEIN, L. H., RUTNER, A. and LAURENCOT, H. J. 1961. Biochemical composition of coconut water (coconut milk) as related to its use in plant tissue culture. Contrib. Boyce Thompson Inst. *21*, 115–128.

USDA. 1962. Chemistry and Technology of Citrus, Citrus Products, and Byproducts. USDA Agric. Hdbk. *98*.

VELDHUIS, M. K. 1971A. Orange and tangerine juices. *In* Fruit and Vegetable Juice Processing Technology, 2nd ed. D. K. Tressler and M. A. Joslyn (Editors). AVI Publishing Co., Westport, CT.

VELDHUIS, M. K. 1971B. Grapefruit juice. *In* Fruit and Vegetable Juice Processing Technology, 2nd ed. D. K. Tressler and M. A. Joslyn (Editors). AVI Publishing Co., Westport, CT.

WHITTENBERGER, R. T. and HILLS, C. H. 1958. Bruising of red cherries. Canner Freezer *126* (10) 33.

WHITTENBERGER, R. T. and HILLS, C. H. 1960. Bruising of red cherries in relation to the method of harvest. U.S. Dept. Agric., Agric. Res. Serv. *ARS 73–27*.

WHITTENBERGER, R. T., GASTON, H. P. and LEVIN, J. H. 1964A. Effect of recurrent bruising on the processing of red tart cherries. Mich. State Univ. Res. Rep. *4*.

WHITTENBERGER, R. T., McADOO, D. J., GASTON, H. P. and LEVIN, J. H. 1964B. Electric sorting machines for tart cherries. Performance studies in 1963. U.S. Dept. Agric., Agric. Res. Serv. *ARS 73–45*.

WOLFORD, E. R., JACKSON, R. and BOYLE, F. P. 1971. Quality evaluation of stone fruits and berries frozen in liquid nitrogen and in Freezant 12. Chem. Eng. Progr. Symp. Ser. *67* (108) 131–136.

WOODROOF, J. G. 1945. Freezing muscadine grapes. Food Packer *26* (12) 48.

WOODROOF, J. G. 1979. Coconuts: Production, Processing, Products, 2nd ed. AVI Publishing Co., Westport, CT.

WOODROOF, J. G. and SHELOR, E. 1949. Preparation of frozen foods. G. Exp. Stn. Bull. *261*.

YAMAMOTO, H. Y. and INOUYE, W. 1963. Sucrose as a gelatin inhibitor of commercially frozen papaya purée. Hawaii Agric. Exp. Stn. Tech. Progr. Rep. *137*.

YEE, W. 1958. The mango in Hawaii. Univ. Hawaii Agric. Ext. Serv. Circ. *388*.

YEE, W. 1961. The lychee in Hawaii. Univ. Hawaii Coop. Ext. Serv. Circ. *366*.

YEE, W. 1963. The mango in Hawaii. Univ. Hawaii Coop. Ext. Serv. Circ. *338*.

8

Dehydration of Fruits

L. P. Somogyi and B. S. Luh

Drying of fruit is one of the oldest techniques of food preservation known to man—and one of the newest. Its essential feature is that the moisture content of food is reduced to a level below that at which microorganisms can grow.

Early American colonists used sun-dried fruit as a regular part of their diet in winter months, and in rural areas of the United States drying of fruit was a part of the family farming enterprise.

During the past century the volume of dehydrated fruit produced has fluctuated widely in response to needs of large-scale military conflicts. During the Franco-Prussian War, Boer War, and World War I, there was a tremendous increase in demand for dried food products. World War II focused even more attention on the industry, as transportation and deployment of men and supplies involved much greater areas than ever before. But during the intervening years of peaceful commerce little progress was achieved in gaining domestic consumer acceptance of dehydrated fruits. The value of dehydrated foods under wartime conditions was undeniable, but the technology was not sufficiently advanced to make any impact on the public in general. Sun drying still accounts for the major part of the dried fruits consumed in the world today; mechanically dehydrated fruits are produced in relatively small amounts but the quantities of these are rising rapidly. This may be partially attributed to the high degree of sophistication attained by dehydration technology during the past 25 years, resulting in products that are winning consumer acceptance in retail markets.

Commercial Fruit Processing, 2nd Edition

ISBN 0-87055-502-2

Today the potential for dehydration is greater than ever. Much has been done to improve the quality of dehydrated foods. Among the things that have led to improvement are the use of raw materials better adapted to the requirements of dehydration, new processing technology, more careful application of known processing procedures as well as sophisticated quality control procedures, improved equipment, lower moisture content in finished products, better control of sulfur application, and improved packaging (Van Arsdel *et al.* 1973; Potter 1978; McBean 1971).

Although preservation is usually the principal reason for dehydration, other considerations often are important. Significant reductions in the weight and bulk of foods are particularly attractive to backpackers and campers. Production of convenience items, such as dehydrated fruit snack and fruit drinks, instant applesauce, pie filling and mixes, and "natural" foods (i.e., fruit products processed without the use of chemical additives), and their retail marketing are rapidly expanding (Kitson and Britton 1978; Salunkhe *et al.* 1974). The accent today is increasingly on convenience foods in almost every field and the word "instant" has a new connotation in our vocabulary. Development of new forms of dried fruit and successful preparation of intermediate moisture food has led to products that are stable to storage and pleasant to eat directly (Silge 1981).

Fruits that are properly dehydrated, particularly to a moisture level below 5%, have the following advantages:

- They have an almost unlimited shelf life under proper storage conditions, because a high degree of inhibition of bacteria, enzymatic, and mold actions is achieved.
- They have substantially lower transportation, handling, and storage costs and do not require costly refrigeration during transport and storage. Their average weight is 1/7 to 1/9 of the raw, canned, or frozen counterparts. Shipping and handling weight is therefore reduced by approximately 90%.
- Dehydration hardly affects the main calorie-providing constituents of fruits. It leaves the mineral content virtually unchanged. Therefore, the process is helpful in preserving the nutritive content of the final product. Vitamin losses are no greater with dehydration than with other preservation methods, and low-moisture fruits can be conveniently fortified with vitamins.
- They provide a consistent product, an important modern marketing requirement. Seasonal variation in product quality is either absent or at a minimum with low-moisture fruits.
- They provide opportunities for maximum convenience, flexibility, and economics as industrial or foodservice ingredients, because they can be sized, shaped, formed, etc., to fit almost any requirement. With

Table 8.1. U.S. Dried Fruit Production in Tons (Dry Basis), 1970–1980

Year	Apples	Apricots	Dates	Figs	Peaches	Pears	Prunes	Raisins	Total
1970	11,862	5,600	18,200	12,280	2,275	585	158,360	193,450	402,612
1971	6,012	4,000	19,500	10,370	1,880	750	91,850	194,830	329,172
1972	9,288	3,000	15,700	8,950	1,500	980	41,750	105,350	186,496
1973	15,481	3,280	23,100	11,220	1,500	790	161,760	224,550	441,651
1974	12,325	2,640	22,700	11,430	1,750	700	106,221	242,150	386,916
1975	14,344	4,500	24,800	8,840	2,375	980	102,695	283,650	442,184
1976	14,331	4,650	22,400	7,000	2,500	1,270	106,322	218,400	376,873
1977	14,094	5,100	25,400	11,000	2,500	1,130	116,414	248,900	424,538
1978	13,813	3,600	21,550	6,930	2,000	855	94,966	172,500	316,214
1979	15,982	5,300	22,350	10,240	1,800	1,530	95,451	303,400	456,063
1980	12,357	3,600	22,500	9,900	3,100	1,310	131,626	311,200	496,553

Source: Clampet (1981).

Table 8.2. Commercial Production of Dried Fruits in Specified Countries, in Metric Tons, 1978–1980

Commodity and country	1978	1979	1980
Apples			
South Africa	247	118	251
Total	247	118	251
Apricots			
Australia	2,051	2,380	1,500
Iran	5.000	4,000	3,500
South Africa	1,139	1,505	1,704
Spain	680	900	700
Turkey	8,000	12,000	7,000
Total	16,870	20,785	14,404
Currants			
Australia	4,375	6,124	6,450
Greece	66,800	57,000	64,500
South Africa	894	691	973
Total	72,069	63,815	71,923
Figs			
Greece	18,350	18,230	17,000
Portugal	4,250	5,000	3,500
Spain	4,250	6,000	5,000
Turkey	50,000	52,000	57,000
Total	76,850	81,230	82,500
Peaches			
Australia	233	180	200
Chile	1,300	1,300	1,300
South Africa	1,931	2,354	2,707
Total	3,464	3,834	4,207
Pears			
Australia	291	153	200
South Africa	570	636	888
Total	861	789	1,088
Prunes			
Argentina	9,500	10,000	10,000
Autralia	2,214	4,000	2,400
Chile	5,200	5,400	5,500
France	22,190	24,308	16,500
South Africa	2,081	1,583	1,673
Yugoslavia	14,350	10,798	15,000
Total	55,535	56,089	51,073
Raisins			
Australia	64,518	54,077	90,307
Greece	81,000	79,000	61,350
Iran	70,000	60,000	60,000
South Africa	12,881	17,520	25,396
Spain	2,500	3,500	3,300
Turkey	82,000	83,000	100,000
Total	312,899	297,097	340,353

Source: Clampet (1981).

low-moisture fruits, the purchaser uses all that he buys thus eliminating waste disposal and pollution problems. Further, they go a long way towards attaining price stability throughout the year.

- They utilize the most economical and disposable form of packaging. The two major considerations in packaging dried fruits are the exclusion of moisture and oxygen. Metal cans, plastic bags, and laminated bags and boxes effectively limit the passage of moisture and oxygen.
- They offer many distinctive conveniences as snack products.

INDUSTRY LOCATION AND PRODUCTION STATISTICS

There were 184 fruit and vegetable dehydrating plants in 1977. Of these plants, 63% were located in the western region, and about 90% (446,000 tons) of the dried fruit output was produced in California in 1980. Apples, apricots, figs, peaches, pears, prunes, and raisins are the important dried fruits, with raisins accounting for the greatest volume. Dried fruits, particularly apples, are also produced in Washington and Oregon, and some fruit drying is practiced in Arizona, Idaho, New York, and Virginia.

As shown in Table 8.1, total production of dried fruits increased by approximately 10% in the United States during the 1970's (Clampet 1981). In recent years, however, increasing interest has arisen in the use of dehydrated products, particularly in convenience foods, foodservice systems, and cereal and bakery products.

The following countries are important dried fruit producers: Australia, Argentina, Algeria, Chile, France, Greece, Iran, Portugal, Spain, South Africa, Turkey, and Yugoslavia. Dried fruit production of the more important fruits in these countries is shown in Table 8.2. The data shown represent mainly sun-dried fruits; only a few mechanical driers are used in these countries to produce low-moisture fruits.

DEFINITION OF DEHYDRATION TERMS

Dehydration terms are often misused or applied indiscriminately.

Dried is the term applied to all dried products, regardless of the method of drying.

Evaporation refers to the use of sun and forced-air driers to evaporate moisture away from fruit to a fairly stable product. Usually, drying conditions, such as humidity, temperature, and air flow, are now carefully controlled during the processing of evaporated fruit. The moisture level of evaporated fruit is approximately 25%. In general, sun drying will not lower the moisture content of fruit below 15%; therefore, the shelf life of such fruit products does not exceed 1 year, unless they are held in cold storage.

Dehydration refers to the use of mechanical equipment and artificial heating methods under carefully controlled conditions of temperature, humidity, and air flow. Although the term "dehydrated" does not refer to any specific moisture in the finished product, it is usually considered to imply virtually complete water removal to a range of 1–5% moisture. Products with such low water content can be stored at room temperature for periods well in excess of 2 years with no detectable change in quality, and indications are that most low-moisture fruits will remain acceptable for periods of 5 years or more when stored under temperate conditions (about 65°F) in moisture-proof containers.

Freeze drying is a method of drying in which the fruit is frozen and than dried under high vacuum to around 2% moisture. The ice sublimes off as water vapor without melting. Controlled heat is usually applied to the process without melting the frozen material; this is called *accelerated freeze drying* (Hanson 1961).

Intermediate-moisture or *semimoisture, fruits* (IMFs) are defined on the basis of water activity level rather than on the percentage of moisture. Restraint of the water molecules to a degree that prohibits spoilage by microorganisms occurs at different moisture contents depending on the amount and nature of the dissolved material present and to some degree on the insoluble components. Food may be classified as an IMF if it has an a_w greater than that of common low-moisture food (0.2) and less than that of most fresh food (0.85). In practice most IMFs have an a_w in the range of 0.65–0.85 and contain 15–30% moisture.

"Water activity" (a_w) is defined as the ratio of the water vapor pressure in equilibrium with a food to the vapor pressure of water at the same temperature. Bacteria will not grow if a_w is below 0.9, and yeasts and molds are inhibited below a_w of 0.7 (Heidelboug and Karel 1975).

Vacuum drying is a method of drying in a vacuum chamber under reduced atmospheric pressure to remove water from the food at less than the boiling point under ambient conditions.

PREDRYING TREATMENTS

Preparation of raw fruit for drying is similar to that for canning and freezing except sulfur is applied to preserve the color of dried fruits. The steps involved in the common predrying treatments applied to the important fruits are (1) selection and sorting for size, maturity, and soundness; (2) washing; (3) peeling by hand, lye solution, or abrasion; (4) cutting into halves, wedges, slices, cubes, nuggets, etc.; (5) alkali dipping, used for raisins, grapes, and prunes; (6) blanching, used for some fruits; and (7) sulfuring.

The general procedures and equipment used for fruit preparation prior to various types of processing are discussed in detail in Chapter 3 of this volume. Only preparation practices used specifically for dried fruits are discussed here.

Sulfuring

For many years sulfur dioxide (SO_2) has been used to help preserve the color of dried fruits. It is the only chemical additive widely added to dried fruits for its antioxidant and preservative effects. Several sulfite salts and SO_2 gas are generally recognized as safe (GRAS) for use in foods by the FDA. Recently the confirmation of sulfites as a GRAS ingredient has been opposed by various consumer groups because of allergic reactions by certain individuals resulting from consumption of sulfite-treated food (Pintuaro *et al.* 1983). The FDA has ordered that consumers in restaurants or in retail stores must be informed if sulfites are applied to preserve fruits or vegetables.

The presence of SO_2 very effectively retards the browning of fruits in which the enzymes have not been inactivated by sufficiently high heat normally used for drying. Sun-dried fruits, (e.g., apricots, peaches, and pears) are usually exposed to the fumes of burning elemental sulfur before being put out to dry in the sun. Apples often are treated with solutions of sulfite before dehydration. Solutions used range from 0.2 to 0.5% (as SO_2) made up of sodium sulfite and sodium bisulfite in about equal proportion. Sulfite solutions are less suitable than burning sulfur because the solutions penetrate the fruit poorly and leach its natural sugar, acid, and flavor components. In addition to preventing enzymatic browning, SO_2 treatment reduces destruction of carotene and ascorbic acid, which are the important nutrients of fruits.

Sulfuring dried fruits to preserve their natural color must be closely controlled so that enough sulfur is present to maintain the physical and nutritional properties of the product throughout its expected shelf life, but the amount should not be so large that it adversely affects flavor. Control of the level of SO_2, which is usually set in the finished product specification, often presents some problems. In a typical product, such as low-moisture apples, the rate and the amount of sulfite absorption by the fruit depend on piece size, type and maturity of fruit, drying method and conditions used, and the method of sulfite application. In particular, the gas concentration in the sulfuring chamber greatly affects the content of SO_2 in the finished product. Bolin and Boyle (1972) studied SO_2 absorption during drying for several cultivars of apples at various maturities.

They found the final SO_2 content increased directly with the soluble solids level of the fresh fruit. The average increase was found to be 200 ppm SO_2 per degree Brix.

The usual levels of SO_2 that are desirable in dried fruit products are shown in Table 8.3. Fruits high in carotene, such as apricots and peaches, require higher SO_2 levels to retain natural color.

Ahlborg *et al.* (1977) reported a survey of SO_2 and sulfite levels in various commercial dried fruits. They found the highest concentration in dried apricots (5.4 g/kg) and in dried pears (7.0 g/kg). Bolin *et al.* (1976) demonstrated that packaging material and packaging atmosphere are important to control SO_2 loss from dried peaches during extended storage and in preserving the light color of the fruit. Nitrogen packing reduced the loss of SO_2 from fruit.

To control the SO_2 content of dehydrated apples under commercial practice, usually three independent applications of sulfite during the process are required. Peeled and cored apple pieces are exposed to 2–3% bisulfite solution for a few minutes while in the flume water, after the cutting operation. This is followed by exposure to SO_2 fumes in the "kiln" during the first 3–5 hr of drying, which is designed to yield a product containing approximately 24% moisture. In order to control the content of SO_2 in the finished product, it is common practice today to dry apples in the kiln below the 24% moisture level to about 16–18% moisture, then determine the SO_2 content and remoisturize the apple to 24% moisture with a solution containing a sufficient amount of sulfite to achieve the desired SO_2 content in the finished product.

During processing of low-moisture fruit from SO_2-treated, evaporated material, substantial loss of SO_2 is expected. The volatilization of SO_2 can be as high as 50% during the vacuum drying process.

Marked reduction of SO_2 in fruit before consumption can be induced by immersing the fruits in hot water (Bolin and Boyle 1972). Sulfured apricots, peaches, and pears held in boiling water lost SO_2 rapidly and continuously while being hydrated (Table 8.4). When the boiling water

Table 8.3. Usual Levels of Sulfur Dioxide in Processed Dried Fruits

Fruit	SO_2 (ppm)
Apples	1000–2000
Apricots	2000–4000
Peaches	2000–4000
Pears	1000–2000
Raisins (sulfur-bleached)	1000–1500

Table 8.4. Sulfur Dioxide Loss (MFB[1]) and Moisture Gain During Boiling in Water

Boiling time (min)	Apricots		Peaches		Pears	
	SO_2 loss (ppm)	Moisture gain (%)	SO_2 loss (ppm)	Moisture gain (%)	SO_2 loss (ppm)	Moisture gain (%)
0	6430	28.6	970	35.3	830	38.0
10	5440	58.6	560	58.6	650	47.5
20	4900	63.7	570	61.3	620	54.6
30	3750	68.0	523	62.7	480	56.2
90	3350	76.5	240	70.1	380	63.8

Source: Bolin and Boyle (1972).
[1]Moisture-free basis.

was replaced by 20° Brix sugar syrup, the SO_2 loss was increased by about 15%. These treatments produced the most rapid reduction of SO_2 from dried fruit, with a 50% decrease observed in about ½ hr. Although a moderate increase in storage temperature over a long time can induce accelerated SO_2 loss, dry heat at high temperature for up to 1 hr does not induce rapid loss of SO_2. On the contrary, due to a decrease in moisture content, a slight increase in the percentage of SO_2 was observed then evaporated apricots containing 3645 ppm SO_2 were exposed to dry heat for 1 hr.

Stafford and Bolin (1972) showed that a 30-sec dip treatment of dried apricots, peaches, and pears in a 7% solution of potassium metabisulfite substituted for the time-consuming and difficult control process of resulfuring dried apricots by the conventional method involving burning sulfur in a sulfur house for 8–12 hr over the fruit spread on trays.

To achieve comparable levels of retained SO_2 in fruits such as mangos, nectarines, and peaches prepared for solar drying, higher levels of sodium bisulfite were required for solar drying, higher levels of sodium bisulfite were required for predrying treatments than for fruit prepared for hot-air drying (Wagner *et al.* 1978). For solar-dried nectarines and peaches 1.25% SO_2 concentration in the soaking solution was sufficient to attain a concentration of at least 2000 ppm SO_2 in the dried product. When hot-air drying was used, 2.00 to 2.25% SO_2 was required in the soak solution to achieve the same SO_2 level in the finished product. Retention of SO_2 was higher in peeled nectarines than in unpeeled fruit.

A combination of ascorbic acid with SO_2 has also been recommended (Voirol 1972). Such an approach has the advantage of replacing some of the SO_2 with a natural constituent of fruits, while still retaining the enzyme-inhibiting property of SO_2.

Replacement of Sulfur

Although SO_2 is the most widely used compound to prevent browning of dehydrated fruit, it (1) causes corrosion of equipment, (2) induces off-flavors, (3) destroys some important nutrients such as vitamin B_1, and (4) is not approved for use in some countries.

Important marketing countries such as Germany and Japan have regulations that substantially limit the use of SO_2 in low-moisture fruits, and within the United States there are increasing demands for SO_2-free dried fruits. Therefore, alternative treatments and the use of more acceptable food additives that retard enzymatic browning of fruits have been considered (Roberts and McWeeny 1972).

Several additives or special treatments to retard enzymatic browning and other oxidative reactions during drying have been investigated. These include lowering pH (using citric or other organic acids), rapid dehydration to very low water contents, use of other antioxidants (ascorbic acid, tocopherols, cysteine, glutathione, etc.), heat inactivation (Individual Quick Blanching), and reduction of the water activity (osmotic treatment).

Miller and Winter (1972) reported that when peaches were dehydrated or sun-dried after a 3-min dip in a 1.0% ascorbic acid and 0.25% malic acid solution, the dehydrated fruits had very good color and were far superior to nontreated fruits. Panelists judged the flavor of the dipped peaches superior to that of commercially dried fruit treated with SO_2.

Ponting *et al.* (1972) found that when 'Golden Delicious' and 'Newton Pippin' apple slices were dipped into solutions of ascorbic acid, calcium sulphate, and sulfur dioxide, the color of the fruit was protected better when both ascorbic acid and a low concentration of SO_2 were combined with calcium, but combinations of ascorbic acid with SO_2 did not improve the results more than an increase of one ingredient by itself. Likewise, three-way combinations of ascorbic acid, SO_2, and calcium were no more effective than two-way combinations of either ascorbic acid or SO_2 with calcium.

Dipping apples and pears into solutions containing 200 g NaCl in 100 liters of water was recommended by Voirol (1972) to preserve color. Even better results were obtained when the fruits were dipped into solutions of 200 g ascorbic acid in 100 liters of water, or in a solution of both substances (100 g NaCl and 10 g ascorbic acid in 100 liters of water). Good results were obtained with cherries, peaches, and plums by treating them at room temperature in solutions containing either 5% ascorbic acid, or 0.1% ascorbic acid and 2% citric acid, or 0.5% ascorbic acid and 0.5% NaCl.

The revival of home dehydration for fruit preservation stimulated development of oven dehydrators operating at low temperature (between

120°–150°F or 50–65°C) with air velocities of about 750 fpm (228 mpm). Peaches, apricots, bananas, mangos, pineapples, nectarines, apples, pears, papayas, and plums cut into ¼–⅜ in. (6–9.5 mm) were placed one layer deep into a forced-draft oven with cross-flow air and dried at 120°F (50°C) without sulfite or blanching pretreatment (Gee *et al.* 1977). Drying required 16–24 hr to reach 0.5 a_w. The resulting dried fruit pieces had a bright natural color and were stable for many months when stored at room temperature in the absence of light.

A new blanching process, IQB or Individual Quick Blanch (Lazar *et al.* 1976), may be used prior to dehydration to inactivate enzymes and at the same time improve the nutritional value and texture of the processed product. In the IQB system, pieces of fruits or vegetables are spread in a single layer at a density of about 1 lb/ft^2 (5 kg/m^2) or on a mesh belt moving rapidly through a steam chest where maximum heating rates result from complete exposure of each piece to live steam. Before the interior of the pieces becomes very hot, the product is discharged as a deep bed onto another belt moving slowly through an insulated chamber, where the heat is already added, and the holding time is sufficient to equilibrate the product temperature at a mass average temperature high enough to stop enzyme activity.

Osmovac Drying

Ponting (1973) has described osmovac drying for dehydration of fruit pieces, slices, or chunks. Fresh fruit is exposed to concentrated sugar syrup (dry sugar) or to salt to remove water from the fruit by osmosis. Over 50% of the initial weight of the fruit can be removed as water by this means (Farkas and Lazar 1969). The partially dehydrated fruit piece is then further dried by other conventional dehydration techniques—most commonly, but not necessarily in a vacuum shelf drier–to a low moisture content. This two-stage combination of osmotic and vacuum drying is referred to as "osmovac" process.

The high concentration of sugar surrounding the fruit pieces prevents enzymatic browning of the fruit, making it possible to produce a dry product of good color with little or no SO_2 or other reducing agents. It is reported that fruits such as apples (Ponting 1973), peaches, bananas (Bongirwar and Sreenivasan 1977), mangoes and plantains (Hope and Vitale 1972; Jackson and Mohamed 1971) have been dehydrated with significantly improved flavor and excellent color; also, interesting textures were produced with the osmovac technique and the application of little or no SO_2. The following additional advantages are claimed for this technique: (1) reduced time during which the product is exposed to high temperature, (2) minimized heat damage to color and flavor, (3) the use of sugar syrup as the osmotic agent reduces the loss of fresh fruit flavor, and

(4) some fruit acid is removed by osmosis which, combined with the residual sugar from the osmotic treatment, produces a blander and sweeter product than conventionally dried fruit. However, the decrease of acidity and the addition of sugar may be disadvantageous in certain products.

Dixon and Jen (1977) identified some of the chemical changes that occur in osmovac-dried apple slices. They found that the sugar-to-acid ratio of the final product increases by as much as threefold compared with that of the initial apples. However, the osmovac-dried apples had a sweet pleasing taste, appropriate for a snack item or breakfast cereal component.

Ponting (1973) summarized other problems with the osmovac process: (1) storage stability may be changed with certain products due to rancidity development in products treated with sugar and dried to low moisture; (2) cost of the process including the unresolved problem of utilization of the excess sugar solution; and (3) sticking together of sugar-treated fruit pieces into large clusters that are difficult to separate without inducing a large amount of fines.

Farkas and Lazar (1969) studies the effects of various factors related to the osmotic dehydration of nonsulfured 'Golden Delicious' apple pieces. They correlated data for time, temperature, sugar concentration, and weight reduction. Their results are presented in Fig. 8.1. They found that under the most favorable condition of 70° Brix sugar solution at 50°C (122°F), about 8 hr were required to reduce ½-in.-thick (12.5-mm) apple slices to 50% of their untreated weight.

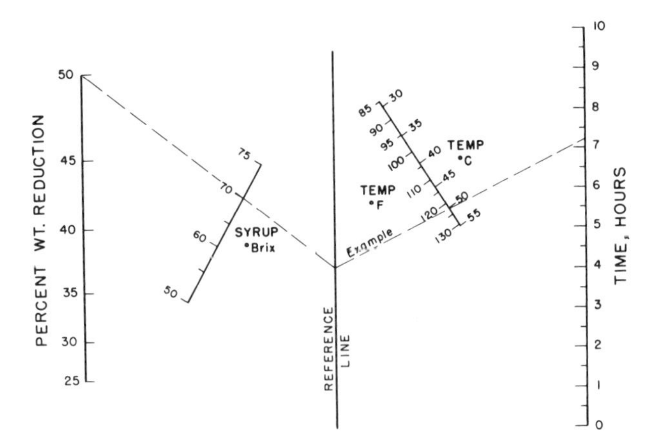

FIG. 8.1. Nomograph for determining percentage of weight reduction for ½-in. (1.27-cm) thick half-rings of 'Golden Delicious' apples during osmotic drying. [Copyright by Institute of Food Technologists and reprinted with permission from Food Technology *23* (5) 91 (1969).]

Centrifugal Fluidized Bed (CFB)

The CFB process can achieve blanching plus about 50% water reduction in less than 6 min. The treatment can be followed further by any conventional dehydration process. Brown *et al.* (1972) reported that peeled and cored 'Newton Pippin' apples diced to ⅜ × ⅜ × 5/16 in. (9.5 × 9.5 × 8 mm) were prepared successfully for air or vacuum drying without the use of SO_2 in a CFB system operated at 2400 ft/min (730 m/min) air velocity at 240°F (115°C) process temperature. This process eliminates the disadvantages associated with the introduction of sugar or salt to the product during the osmotic method. The CFB method reportedly offers the advantages of simplicity of design and an intimate gas-to-particle conduction that provides uniform particle exposure without mechanical agitation. The equipment can be designed to operate as a continuous process. However, it is limited to small particles, about ½-in. (1.25-cm) cubes or smaller pieces.

DRYING METHODS

Several drying methods are commercially used, each better suited for a particular situation. Sun drying of fruit crops is still practiced for certain fruits, such as prunes, grapes, and dates. Atmospheric dehydration processes utilizing kiln, tower, and cabinet driers are used for apples, prunes, etc. Continuous processes (e.g., tunnel, belt trough, fluidized bed and foam-mat drying) are mainly used for vegetables. Spray drying is suitable for fruit juice concentrates, and vacuum dehydration processes are useful for low-moisture fruits with high sugar content such as peaches, pears, and apricots.

The selection of drying methods depends on the following factors:

- Form of raw material—liquid; paste, slurry, pulp, thick liquid; large aggregates; small aggregates.
- Properties of raw material—very sensitive to oxidation; sensitive to temperature damage; thermoplastic residues; none of these.
- Desired product characteristics—powder; instant solubility; excellent rehydration; retention of shape (complete or partial).
- Cost—low; medium; high; very high.

There are three basic types of drying process:

- Sundrying and solar drying.
- Atmospheric dehydration including (1) stationary or batch processes (kiln, tower, and cabinet driers) and (2) continuous processes (tunnel, continuous belt, belt-trough, fluidized-bed, explosion puffing, foam-mat, spray, drum, and microwave-heated driers).

- Subatmospheric dehydration (vacuum shelf, vacuum belt, vacuum drum, and freeze driers).

The principles of these basic processes are described in this section. Various modifications are discussed in the next section under descriptions of individual types of equipment.

Sun Drying

Sun drying of fruit crops as a method of food preservation is still practiced largely unchanged from ancient times in many parts of the world including the United States. It is limited to climates with hot sun and dry atmosphere, and to certain fruits such as prunes, grapes, dates, figs, apricots, and pears. These crops are processed in substantial quantities by this primitive method without much technical aid, by simply spreading fruits on the ground, on racks, trays, or roofs, and exposing them to the sun until dry.

Only small capital investment is required with this simple procedure. Since sun drying depends on uncontrolled factors, production of uniform and high-quality products is not expected. Some overdrying and contamination by dust, dirt, and insects of the finished product are usually tolerated. The most obvious disadvantage of sun drying is its complete dependence upon the elements. It is a slow process, unsuitable for producing high-quality products. Since sun-dried products generally have moisture levels no lower than 15–20%, they have a limited shelf life.

Solar Drying

In recent years considerable interest has been focused on the use of solar energy for hot-air dehydration because of the rapid increase of fuel costs (Flink 1977). In commercial application, solar energy is used alone or may be supplemented by an auxiliary energy source, including geothermal energy, wastes, and biomass (Anon. 1982).

A simple method of accelerating the sun-drying rate of fruit on trays is to paint the trays black; this causes a greater portion of the incident solar radiation to be absorbed and transmitted to the drying fruit (Bolin *et al.* 1982). Halved apricots that were sun-dried in black trays lost 16% more moisture in 1 day of drying than fruit dried on unpainted wooden trays.

A solar trough was designed by Bolin *et al.* (1982) in which fruit is heated by direct incident radiation and indirect reflected radiation. This system requires 40% less time to dry apricots in 24% moisture than the conventional tray drying method. A solar trough could be used to reduce fruit moisture to 50% after which the fruit could be air-dried in bins to the desired moisture level.

Another type of direct solar drier employs mirrors to increase solar energy (Wagner *et al.* 1977; Bryan *et al.* 1978). In two-stage drying procedure, the first stage depends on direct and reflected solar radiation, followed by heated air blown over the fruit when solar radiation is insufficient for drying.

In indirect solar driers (Bolin and Salunkhe 1982), solar energy is collected by a solar collector that, in turn, heats the air as it blows over it before being channeled into the dehydration chamber.

Among the various collector designs, greatest interest is in flat-plate collectors. One of the largest dehydrators of this type is currently used in California for raisin drying (Anon. 1978). This solar drier unit consists of 22,000 ft^2 (2044 m^2) of single-glazed solar flat-plate collectors, a 700-ton rock heat storage system, and a heat recovery wheel. All of these are connected on one dehydration tunnel. Natural gas is used to supply supplemental heat when necessary. The heat recovery wheel scavenges heat from the exhaust, providing a heating efficiency of over 80%.

Dehydration

Dehydration includes the application of artificial heat to vaporize water and some means of removing water vapor after its separation from the fruit tissues. The removal of water involves mass transfer, and the application of heat in some manner also involves heat transfer. Energy must be supplied to vaporize the water and to remove the resultant water vapor from the drying surface. The quantity of heat energy required to vaporize water depends upon the temperature at which vaporization occurs. In practice, the efficiencies are usually between 20 and 50%.

The general equation for heat transfer is:

$$q = hsA(t_a - t_s)$$

where q is the heat transfer rate in Btu/hr, hs is the heat transfer coefficient, A is the area, t_a is the air temperature, and t_s is the temperature at the drying surface.

Typical heat transfer coefficients (hs) for certain dehydration equipment were given by Williams-Gardner (1971) as follows:

Type of drier	Heat transfer coefficient (Btu/hr/ft² °F)
Vacuum shelf	1
Agitated tray	5–60
Rotary vacuum	5–50
Indirect rotary	2–10
Jacketed through	2–15
Drum	200–300

Heat may be applied to the drying material by conduction, radiation, and convection. While all three modes of heat transfer can occur during drying, depending on the method used, one mode usually dominates to such an extent that its influence is predominant. A current of air is the most common medium for transferring heat to a drying fruit, and convection is the main principle involved. Conduction and radiation are usually associated with vacuum drying. Once heat is supplied to the drying material's surface, it is distributed throughout the material by conduction.

Two important aspects of mass transfer in dehydration are (1) the transfer of water to the surface of the material being dried and (2) the removal of water vapor from the surface. The drying curve (Fig. 8.2), which relates the amount of moisture with time, usually consists of two phases: a constant rate period and a falling rate period. During the constant rate period, water is readily available at the surface of drying foods, and therefore drying rate is determined by the temperature, the relative humidity, and the flow rate of the air. This is a rather short time period during the initial stage of the drying process, and during this period the drying rate is high. When the product has lost most of its surface water, the remaining moisture must diffuse from inside to the surface before evaporation can take place. This results in the "falling rate" period of drying, which in the later stages of the process becomes extremely slow. During this stage the relative humidity of the air will fall below 100%, and

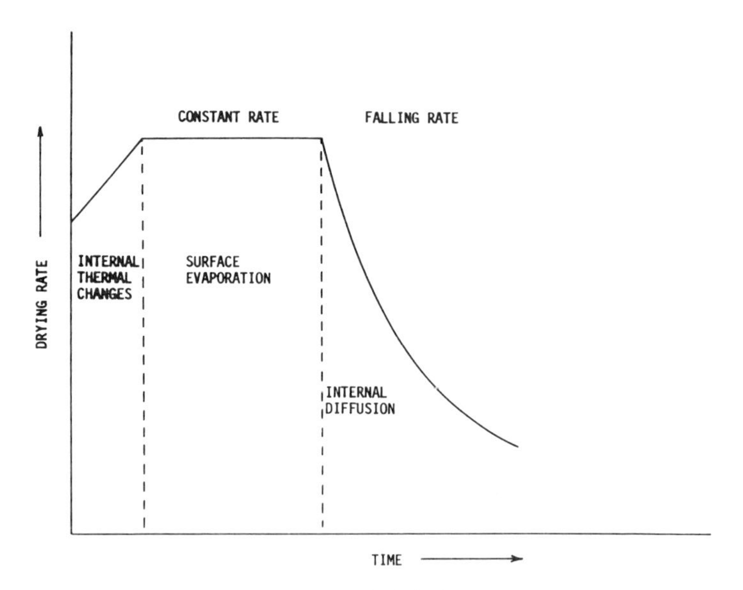

FIG. 8.2. A typical drying curve.

the dominant factor limiting drying rate is no longer heat supply, but the availability of water at the evaporation site.

To achieve dehydrated products of high quality at a reasonable cost, dehydration must occur fairly rapidly. Four main factors affect the rate and total time of drying:

1. Properties of the Food Product.—The biochemical and biophysical characteristics of individual foods define what type of dehydration procedure may be applied without causing unacceptable changes in flavor, color, texture, and nutritional qualities. The most important characteristics are the structure and composition of raw materials, which affect the migration of water toward the surface during drying; shrinkage, which is caused by the stress during dehydration and results in a retardation and limit of the rehydration of the dried product; browning reactions (both enzymatic and nonenzymatic browning reactions); and rehydration characteristics of the finished product.

2. Particle Size and Its Geometry.—The geometry of product in relation to heat transfer, surface, and medium play an important part in determining the overall drying time: the thicker the product, the longer time is required to remove moisture. Generally, fruit is cut into small pieces prior to dehydration. This provides a large surface area that can be exposed to the heating medium and from which moisture can escape. Smaller particles arranged in thinner layers also reduce the distance heat must travel to the center of the piece and, at the same time, reduce the distance moisture must move to reach the surface and evaporate.

The depth of the product in a layer (tray loading) has a great effect on drying time. The lower the tray loading and the greater the distance between the particles, the shorter the drying time and more uniform the finished product.

3. Physical Properties of the Drying Environment.—The temperature, humidity, and velocity of air and the atmospheric pressure greatly influence the rate of drying. The greater the temperature difference between the heating medium and the food, the greater will be the rate of heat transfer, which provides the force for moisture removal. The hotter the air, the more moisture it will hold; therefore, in the vicinity of the dehydrating food it will take up more moisture being given out from the food. In the later stages of drying, heat damage is more likely to occur because the temperature of the product will rise gradually as the drying rate falls and evaporative cooling decreases. Thus, in drying processes it is customary to commence with high temperature, followed by gradually falling temperatures to levels at which deterioration due to heating is reduced.

Beside air temperature, other important factors are air velocity (air in motion more effectively removes water) and the humidity of air (dry air holds more moisture). The combined effect of humidity and temperature of air is determined by the psychrometric relationship as measured by the wet bulb temperature. The drying rate has been shown to be proportional to the wet bulb depression.

Finally, the effect of atmospheric pressure vs vacuum should be considered at this point. At an atmospheric pressure of 760 mm Hg, water boils at 212°F (100° C); at pressures below 760 mm Hg, boiling of water occurs at a lower temperature. Therefore, if fruit is dehydrated in a heated vacuum chamber, its moisture can be removed at a lower temperature than if it is dehydrated at atmospheric pressure. This provides a system to dehydrate products at a lower temperature thus reducing the degradation in color, flavor, and texture of the product.

4. Characteristics of the Drying Equipment.—It is necessary to use the kind of equipment most suitable to a particular operation. A guide to drying equipment and design/selection factors that can contribute to improved operations was published by Levine (1977). In the following section the most important dehydration equipment used commercially is described.

DEHYDRATION EQUIPMENT

Atmospheric Forced-Air Driers

Several methods of artificial drying involve the passage of heated air with controlled relative humidity over the food to be dried, or the passage of food through heated air. Various devices are used to control air circulation and to recirculate air.

Kiln Drier. The kiln drier, in which the natural draft from rising heated air brings about the drying of the food is the simplest and oldest type of dehydration equipment still in commercial use. Kiln driers generally have two levels: gas burners on the lower floor provide heat, and the warm air rises through a slotted floor to the upper level. Food material such as apple slices are spread out on the slotted floor in a layer about 10 in (25 cm) deep and turned over periodically. Kiln driers are still widely used in producing evaporated apples slices. After being dipped in sulfite solution, peeled and cored apple rings or slices are dried to about 14–44% moisture in about 6–8 hr. The sulfite dip may be replaced by use of burning sulfur during the kiln-drying process. This type of drier is inefficient in the use of heat, results in slow drying, and does not permit accurate control of the process.

A number of modifications have been developed to increase the efficiency of kilns and to speed up the drying process. Fans may be set in the wall of the furnace room or in the roof vents to force the heated air more rapidly through the fruit, thereby shortening the drying time. During the initial drying period, when drying is rapid, all of the hot gases are allowed to escape through the roof vents. As drying progresses, much of the air may be recirculated by fans, thus increasing the efficiency of heat consumption.

Tower ("Stack") Drier. A tower drier consists of a furnace room containing a furnace and heating pipes and cabinets in which trays of fruit are dried. In a typical design each "stack" or cabinet holds about 12 trays, usually 3 ft^2 (0.33 m^2) in size; and a furnace room accommodates about six stacks of trays (Cruess 1958). Heated air from the furnace rises through the trays holding the fruit. As the fruits on the bottom trays of the stack become dry, they are removed and are replaced with freshly loaded trays on the top of the stack. This necessitates that each time a fresh tray enters the stack, the entire set of trays are shifted downward.

Cabinet Drier. A cabinet drier is similar in operation to a stack drier, except that the heat for drying is supplied in steam coils that are located between the trays. This type of equipment provides some control and uniformity of the temperature; thus it represents a substantial improvement over the "stack drying" system. It is suitable, however, only for small-scale operations. The equipment is inexpensive and very convenient for drying fruit and vegetable pieces. The duration of a drying cycle is 10–12 hr. A cabinet drier is particularly useful for establishing the drying characteristics of a new product, prior to a large-scale commercial run. Because of its small capacity and high operating cost, a cabinet drier is commercially economical only for high-valued raw materials.

Tunnel Drier. The most flexible and efficient dehydration system is the tunnel drier, which is widely used in drying fruits. The equipment is essentially similar to a cabinet drier, except that it allows a continuous operation along a rectangular tunnel through which move tray-loaded trucks. The tunnel is supplied with a current of heated air that is introduced at one end. A tunnel drier provides rapid drying without injury and permits a uniform drying process.

The size of tunnel driers varies greatly. A typical system for fruit (Holdsworth 1971) consists of two or three single-stage tunnels about 30 ft (9.1 m) long, each 6¼ ft (2.0 m) wide and 7 ft (2.1 m) high. The air is usually directed in a counterflow direction and reaches about 180°–200°F (82–93°C) and 90°F (32°C) wet bulb temperature at the end. The air velocity is between 600 and 1200 ft/min (183 and 366 m/min). Such a unit could dry apple slices that are 5⁄16–¼ in. (8–6 mm) thick from 23–24%

moisture to 2.5% moisture in 2–3 hr, at a rate of about 1000 lb/hr (454 kg/hr) of finished product.

Tunnel driers are classified by the direction in which the air traverses the product. In a "parallel-flow" unit, the fresh material encounters the direst and warmest air initially, and leaves the drier at the coolest end; in a "countercurrent-flow" unit, the air direction is opposite to the movement of the product, so the dry product leaving the drier encounters hot dry air as it enters the system.

A parallel-flow system has high initial rates of evaporation and presents little danger of overheating the product, since the surface temperature of the fruit is below the dry bulb temperature as heat is removed by the evaporation process. But since the product is in contact with progressively cooler air, which results in a decrease in drying rate, very low moisture contents cannot be achieved with this system. Parallel-flow equipment is generally used only for drying grapes, or it is used in combination with counterflow equipment.

The countercurrent-flow process is more economical and is often used in fruit dehydration equipment. Most tunnel driers, however, operate as a double-stage plant, and the two stages are often arranged as separate tunnels. A countercurrent/parallel system is so arranged that the product first encounters air flowing in a countercurrent direction, followed by air flowing in a parallel direction; this is achieved by feeding the air in at the central point. The parallel/countercurrent flow system is the most widely-used arrangement in commercial two-stage operations and takes advantage of the high rate of initial evaporation provided by parallel-flow systems. This system has been found to result in more uniform drying, increased output, and good overall quality. Often, the first stage is shorter in length than the second to compensate for the low drying rate in the second stage. A double-stage system enables independent adjustment of air temperature. In a parallel-flow system a higher air temperature is followed by a lower temperature in the finishing countercurrent position, which is very advantageous.

Multistage driers consisting of three, four, or five drier stages are also used. Such systems are very flexible and can achieve close to optimum drying conditions for a wide variety of products.

Continuous Belt (Conveyor) Drier. A continuous belt (conveyor) drier is similar in principal to a tunnel drier, except the food material is conveyed through a hot air system on a continuous moving belt without the use of trays. Therefore, the system has the obvious advantage of eliminating the costly handling of products on trays before and after drying. It also allows continuous operation and automatic feeding and collection of the dried material. A commonly used continuous belt drier is equipped with a belt about 75 ft (23 m) long and 8 ft (2.4 m) wide, which takes 2½ hr to travel through the system. The raw fruit is loaded uniformly 4–6 in. (10–15 cm) deep onto the belt, which is made out of woven metal mesh or interlocking plates. The speed of the conveyor is variable to suit both the product and the heat conditions. Furthermore, process conditions are usually controlled by designing the system in sections, thus allowing different flow rates, humidities, and temperatures to be set in each section, and by rotating the product when it moves from one section belt to the next.

Lower initial inlet temperatures are normally used in the first heat zone of a belt drier than in a tunnel drier because the effect of passing the hot air stream through the product, rather than over it, produces a higher rate of evaporation. Temperature, therefore, must be controlled carefully to avoid scorching, case hardening, and protein denaturation because of the high rate of evaporation. The temperature in the second zone is usually kept at 10°–15°F (5°–8°C) lower than in the first, and in the third zone about 10°F (5°C) below the second zone. Some fruits with high sugar content tend to adhere to the belt at the discharge end; these require a rotating brush or other scraping device to remove them from the plate surface. Adhesion may be minimized by applying a coat of "dehydrator's wax" or food-grade mineral oil spray to the belt.

Belt-Trough Drier. In a belt-trough drier a continuous stainless steel wire mesh belt forms a trough about 10 ft (3.3 m) in length and 4 ft (1.2 m) in width. The raw material is fed onto one end of the trough and is dehydrated by forcing heated air upward across the belt and the product. The belt moves continuously, keeping the food pieces in the trough in constant motion and continuously exposing new surfaces. The movement of the belt and lateral inclination of the drier away from the input end, plus continuous feeding of fresh product into the input end, forces the product across the trough surface toward the lower discharge end. Trough driers are used in two-stage series for dehydration of product to 10–12% moisture; single-stage units are suitable for processing partially dried products such as dehydro-frozen products with about 50% weight reduction. Belt-trough driers have been successfully used for dehydrating vegetable pieces; however, they are not suitable for drying fruits because fruit pieces that exude sugar on drying tend to stick together and clump with the tumbling motion.

Fluidized-Bed Drier. In a fluidized-bed drier, a modification of the belt-trough drier, airflow from beneath is sufficient to lift particles of food and at the same time convey them toward the outlet. The moist air is exhausted at the top of the equipment. The process is continuous, and the length of time particles remain in the drier can be regulated by the depth of the bed and by other means. Fluidized-bed driers offer the advantages of simplicity of design, intimate gas-to-particle contact, and uniform particle exposure without mechanical agitation. Their use, however, is limited because if air velocity becomes too great, channeling will occur and most of the air will escape without performing its function; at even higher air velocities, particles may be ejected from the bed (Brown *et al.* 1972). Thus, the use of conventional fluidized beds is limited to the preparation of food powders. They are often installed as secondary driers to finish the drying process initiated in other types of driers.

The minimum air velocity to produce fluidization was found to be 375 ft/min (114 m/min) for ⅜-in. (9.5-mm) diced apples (Holdsworth 1971). The initial hot air treatments were 30 min at 212°, 194°, and 176°F (100°, 90°, and 80°C, respectively) and finally 3–3½ hr at 140°F (60°C). The initial temperature had a negligible effect on drying rate or overall length of drying time. The apple dice dried rapidly to 10% moisture in 1½ hr, followed by very slow drying.

Lazar and Farkas (1971) have extended the fluidized-bed technique by developing the centrifugal fluidized bed (CFB), which achieves high drying rates of 7.5–25 ft/sec (2.3–7.6 m/sec). They designed a pilot unit that employed a centrifugal force greater than the gravitational force, which has the effect of increasing the apparent density of the particles and allows smooth, homogeneous fluidization at much greater air velocities. Increased air velocity provides improved heat transfer so that moderate temperatures can be used, thus eliminating problems of scorching or surface heat damage associated with high-temperature drying. Brown *et al.* (1972) reported that ⅜ × ⅜ × 5⁄16 in. (9.5 × 9.5 × 8.0 mm) 'Newton Pippin' apple dice was reduced 50% in weight in less than 6 min in the CFB operated at an air velocity of 2400 ft/min (132 m/min) at 240°F (115°C).

Explosion Puffing. In explosion puffing, fruit pieces are partially dehydrated in a conventional manner and then heated in a closed vessel, known as a "gun," having a quick-opening lid. When the water contained within the pieces is heated above its atmospheric boiling point, and pressure at a predetermined value has thereby developed in the chamber, the pieces are instantly discharged to atmospheric pressure. The flashing water vapor from within each piece creates a porous structure that permits much faster dehydration and much more rapid rehydration of the dried product. The fruit particles are then dried to 4–5% moisture by conventional drying methods.

The moisture content of the fruit pieces entering the gun is critical to achieve successful puffing. For apples the acceptable range of moisture content for puffing is between 20 and 30%, for blueberries 19–30%. Below this range little puffing is achieved, and the product will become scorched. At higher moisture content the pieces tend to collapse after puffing (Eisenhardt *et al.* 1968). The process is estimated to cost more than conventional dehydration processes, but the rehydration time of the finished puffed product is much shorter. It is claimed that this is the first low-cost process that produces relatively large pieces of dehydrated fruits that will reconstitute rapidly. Explosion puffing has been particularly successful with apples and blueberries.

Eskew and Cording (1968) developed a process for greatly reducing the bulk of explosive puffed fruits without in any way impairing their rapid rehydration characteristic, appearance, flavor, or nutritional value on reconstitution. With this process, the bulk of the compressed dry product may be reduced below that of conventionally air-dried material. The process involves compressing the explosive-puffed pieces of fruit in their slightly moist plastic state after puffing, but before final processing to a stabilized form. The compression can be done in one or more stages between closely set rolls, which may or may not be heated, and the fruit pieces may be compressed while still warm from the gun or after cooling. The compression in rolls resulted in a reduction in one dimension of about $\frac{1}{7}$–$\frac{1}{15}$ of the dimension prior to compression of puffed apple pieces.

Foam-Mat Driers. Foam-mat drying involves drying liquid or puréed materials as thin layers of stabilized foam by heated air at atmospheric pressure. The foam is prepared in a continuous mixer by the addition of gas; when required, a small amount of edible foam stabilizer is added. The prepared foam is spread on perforated trays and dired by hot air, followed by crushing into powder.

Foam-mat-dried foods are characterized by a very porous structure, which makes them capable of nearly instant rehydration even in cold water. Many liquid or puréed fruits have been successfully foamed and dehydrated. Low-moisture powders of orange, grapefruit, lemon, lime, pineapple, apple, and grape juices have been produced.

The principal advantage of this process is that products can be dried at relatively low temperatures at atmospheric pressure because the foamed material high initial rates of water removal. Typical drying requirements for $\frac{1}{8}$-in. (3-mm) thick fruit juice concentrate is 15 min in 160°F (71°C) air to achieve 2% moisture content. Compared to ordinary air-dried fruits, foam-mat-dried powders have superior flavor and color and are nearly instantly soluble in water. The processing cost of foam-mat-dried products is expected to be less than that of vacuum- or freeze-dried products (Hertzendorf *et al.* 1970).

Foam-mat drying is limited to specific products, such as fruit powders for preparation of instant drinks. Because of their open, porous structures, these products have limited storage stability. Their structure is conducive to the adsorption of oxygen and moisture, which may promote reactions deleterious to quality. Also, because of the hygroscopicity of foam-dried fruit powders and the mild temperatures employed in the foam-mat processes in some cases, it is difficult to obtain sufficiently low moisture in single-stage drying. Several finish-drying procedures have been suggested, such as supplemental drying in a vacuum chamber or the use of an in-package desiccant.

Spray Driers. Spray drying involves the dispersion of liquid or slurry in a stream of heated air. Followed by collection of the dried particles after their separation from the air. The process, widely used to dehydrate fruit juices, has several features that favor high-quality products. The fine dispersion of the particles provides a large surface area, resulting in short drying times with high heat transfer rates. The temperature of the droplets remains below the wet bulb temperature of the drying gas until almost all the water has been removed due to the high evaporation rate. The final dried product is delivered as a free-flowing powder. Spray drying also is a continuous and simple to operate system, and because of its large throughput, its cost of operation is relatively low.

The essential components of the numerous types of spray-drying equipment available are (1) an atomizing system operating at 250–500 psi (17.6–35.2 kg/cm^2); (2) a hot-gas-producing unit; (3) a chamber for the sprayed particles to meet with hot gas; and (4) a recovery system. The specific design of these components may vary with the product being dried since each product needs its own set of drying conditions (e.g., size of atomized particles, type of air flow in the drying chamber, air temperature, separation and collection method). Three types of atomizers are used commonly: the pressure nozzle, or jet, type; the two-fluid nozzle type, in which the fluid is broken up into a spray by means of a jet of air or steam introduced into a slowly moving fluid; and the centrifugal or disk type. Spray driers are designed with three types of air flow: horizontal cocurrent driers; vertical cocurrent driers, which can be upward or downward, and simple or complex; and vertical countercurrent driers. Product collection is normally achieved by means of a cyclone and scrubber system or by a filter with the dry product being discharged from the base of the collection unit.

A recently developed spray drier is the BIRS drier (Hussman 1963), in which droplets of liquid fall from the top of a 200-ft (61-m) tower through countercurrent cool (about 86°F or 30°C), dry, and dehumidified air. Droplets descend in the tower in about 90 sec. Juice products, such as lemon and orange, that are difficult to dry in hot air can be successfully dried in this drier. Although the resulting dry products have more dense particles than conventional spray-dried products, they retain much more natural flavor because of the low-temperature process. Due to the high capital and operating costs of the BIRS drier, it has not yet been widely accepted in commercial operations.

Drum Driers. In drum drying, which is suitable for a wide range of liquid, slurried, and pureed products, a thin layer of product is applied to the surface of a slowly revolving heated drum; and in the course of about 300° of one revolution, the moisture in the product is flashed off, and the dried material is scraped off the drum by a stationary or reciprocating blade at some point on the periphery. The drum is generally heated from within by steam; the outer surface of the drum offers a drying interface with a good heat transfer. Drum driers are capable of increasing the solids content of puree from about 9–30% to 90–98% solids. The residence time of the product in the drier is on the order of 2 sec to a few minutes.

Drum driers can be divided into two broad classifications: single- and double-drum driers. The drums of a double-drier unit turn in opposite directions, and the feed material is applied to both drums. The equipment may be used either under atmospheric or vacuum conditions. Double-drum machines employ a "nip" feed with the space between the drums capable of adjustment, thus providing a means of controlling the film thickness. With a single-drum drier the feed is usually at the top, where the drum passes a shallow trough for the feed material.

Although drum drying is an inexpensive drying method, its commercial application is limited to less heat-sensitive products. Its usefulness for fruit dehydration is quite limited because the high temperatures required, usually above 250°F (121°C), impart a cooked flavor and off-color to the fruit product. Also, the high sugar content of most fruit juices makes them difficult to remove from drum driers because of the high thermoplasticity of such products. Application of chilled air directed to a narrow strip just before the removal blade cools the thin sheet of fruit and makes its removal easier.

Microwave Driers. Microwave heating has been tried experimentally for dehydration of fruits. Since microwaves selectively heat water with little direct heating of most solids, rapid and uniform drying can be achieved throughout the product at relatively low temperatures. In a process developed by Tobby (1966), fruit pieces having a dimension of ⅝ × ⅝ × ⅝ in. (16 × 16 × 16 mm) and an initial moisture content of about 85% were quickly dried to about 14% moisture; the dried pieces were sweet, unwrinkled, and stable. The process was completed in a specially designed drying cabinet using electromagnetic radiation at 100 megacycles.

The use of microwave energy for freeze drying was investigated by Gould and Kenyon (1970). Since microwave energy has the ability to selectively heat ice crystals, thus eliminating the problem of heat conduction across the dried food layer during the middle and end of the freeze-drying cycle, the final stages of the freeze-drying cycle were accelerated. The use of microwave energy reduced the drying time required in conventional freeze drying by 1/10 to 1/20. The main problems to be overcome are nonuniform heating, impedance matching, and ionization causing slow discharge.

Microwave equipment also is complicated and expensive, and no commercial installations are using it at the present for fruit dehydration. Microwave driers are more likely to be used for finish-drying than for the complete dehydration process. Synergistic effects between hot air and microwaves were demonstrated during the finish-drying of fruits by Salunkhe *et al.* (1974). Microwave energy also can be used to equilibrate food pieces that have a low moisture content on the surface layer and a high moisture content at the center. The use of microwave energy would make it unnecessary to draw down the average moislure content as much as is now being practiced prior to compression (Anon. 1982).

Vacuum Driers

The main purpose of vacuum drying is to enable the removal of moisture at less than the boiling point under ambient conditions. Vacuum drying provides important advantages for certain products in terms of final quality. Because of the high installation and operating cost of vacuum driers, they are used only for high-value raw materials or products requiring reduction to extremely low levels of moisture without damage. An important feature of vacuum drying is the virtual absence of air during dehydration; this makes the process attractive for drying material that may deteriorate as a result of oxidation or may be modified chemically as a result of exposure to air at elevated temperatures. Products that may decompose or change in structure, texture, appearance, and flavor as a result of the application of high temperature can be dried under vacuum with minimum damage.

All vacuum-drying systems have the following essential components: vacuum chamber, heat supply, vacuum-producing unit; and device to collect water vapor as it evaporates from the food. All vacuum driers also must have an efficient means of heat transfer to the product in order to provide the necessary latent heat of evaporation and a means for removal of vapor evolved from the product during drying. Such driers must be designed to establish and maintain a vacuum, and they must be vacuum tight to keep pumping requirements to a minimum. A vacuum drier and associated vessels must be of adequate strength to withstand the differential pressure of the atmosphere on the outside and the vacuum maintained inside. The outside pressure may exceed internal pressure by as much as 2000 lb/ft^2 (90 kg/m^2).

Vacuum Shelf Drier. A vacuum shelf drier, the simplest type of vacuum drier, consists of a vacuum chamber containing a number of shelves arranged to supply heat to the product and to support the trays on which the product is loaded into the chamber. The shelves may be heated electrically, or more often by circulating a heated fluid through them. The heated shelves are called platens, and they convey heat to the food in contact with them by conduction; where several platens are placed one above another they also radiate heat to the food on the platen below. The rate of heat transfer is slow in this type of equipment compared with that in driers in which the drying material is moved or agitated by some means. The vacuum chamber is connected to suitable vacuum-producing equipment, located outside the vacuum chamber, which may be a vacuum pump or a steam ejector. Another essential part of a vacuum drier that has a vacuum pump is a cold-wall condenser, which collects water vapor. This may be located inside or outside the vacuum chamber, but must be ahead of the vacuum pump to prevent water vapor from entering the pump. A steam ejector also may be used to create the vacuum. It is a kind of aspirator in which high-velocity steam jetting past an opening draws air and water vapor from the vacuum chamber. In units with a steam ejector, a condensor is not necessary because the steam ejector can condense water vapor as it is drawn along with the air from the vacuum chamber.

A shelf drier is suitable for batch-type operation only. The equipment is easy to maintain and is very suitable for high-vacuum operation. A wide range of fruit products—liquids, pastes, powders, discrete particles, chunks, slices, and wedges—can be processed in this type of drier.

Conical Rotating Vacuum Drier. The conical rotating vacuum drier is a batch-type drier. The rotation of the vessel provides a very gentle sliding action of the product over the internal walls of the vessel, which is jacketed for the circulation of hot water, steam, or other heating medium. The sliding movement results in close contact of the product with the heat-transfer surface. The movement of the product ensures an even temperature throughout its mass. This type of drier is suitable for powders or discrete particles, providing that they do not tend to form lumps or to adhere to the walls of the vessel, thus impeding heat transfer and drying.

Rotary Vacuum Drier. The rotary vacuum drier, a very efficient dryer, has a horizontal stationary cylindrical vessel with a jacket for the heat-transfer medium. The unit is capable of batch-type operation only. It will handle a wide range of products and is suitable for high-vacuum operation.

Vacuum Belt Drier. Continuous vacuum operation can be achieved in a vacuum belt drier. This type of drier consists of a horizontal tanklike chamber in which there are one or more conveyor belts. The chamber is connected to vacuum-producing and moisture-condensing systems. Appropriate isolation locking arrangements at the charging and discharging ends permit a continuous flow of material through the drier. A series of infrared heater panels or heated platens are located above and sometimes below the conveyor to supply heat. A tumbling effect is produced at the end of each conveyor band when a multiconveyor belt system is used to ensure exposure to the heat sources on each side of the particles as they progress through the drier. The continuous vacuum belt drier is particularly suitable for the drying of fruit pieces, granules, and discrete particles at a relatively high vacuum. The capital cost of this equipment is much greater than for a batch unit of similar capacity.

Freeze Driers. In conventional vacuum drying, moisture in the foods is evaporated from the liquid to the vapor phase. In freeze drying, the moisture is removed from the product by sublimation, i.e., converting ice directly into water vapor. Therefore, no transfer of liquid occurs from the center of the mass to the surface. As drying proceeds, the ice layer gradually recedes toward the center, leaving vacant spaces formerly occupied by ice crystals. The advantages of freeze drying are high flavor retention; maximum retention of nutritional value; minimal damage to product structure and texture; little change in product shape, color, and appearance; and finished products with an open structure that permits fast and complete rehydration. The disadvantages of the process include high capital investment; high processing costs; and the need for special packaging to avoid oxidation and moisture pickup in finished products.

The freeze-drying process involves two basic steps. The raw fruit is first frozen in the conventional manner and then dried to around 2% moisture in a vacuum chamber while still frozen.

The freezing rate may affect the reconstitution property of freeze-dried foods because the porous nature of the product is controlled by this factor. The process of vapor removal is influenced by the size, shape, and tortuosity of the pores. In general, the faster the freezing rate, the smaller the voids, and the slower the freeze-drying rate (Karel 1963).

The most common type of freeze-drying equipment is a batch chamber system similar to a vacuum shelf drier but with special features to meet the needs of the freeze-drying process. The material to be dried is placed on trays arranged between the heated plates. The plates are either electrically heated or internally heated with steam, pressurized hot water, or oil. Before heat is applied, a vacuum must be drawn in the chamber. The vacuum is produced either with a mechanical pump or with steam ejectors. Rapid evacuation of the chamber is essential once the product is loaded to ensure that there is no thawing before freeze drying is started. The vacuum system must maintain a pressure under 1 Torr so that the product remains frozen as long as water is present duirng the drying cycle. Figure 8.3 shows the relationship between pressure and temperature that makes freeze drying possible.

Refrigerated condensers are usually applied to condense the water vapor that is removed from the product. The surface of the refrigerated condenser must be maintained at a lower temperature than the ice within the product to ensure mass transfer. To improve storage stability of freeze-dried fruits, it is a common practice at the completion of the drying process to break the vacuum in the chamber with nitrogen gas, thereby preventing the instantaneous absorption of oxygen by the open pores of the dried product. The nitrogen-impregnated product is then packed under nitrogen in airtight, moisture-proof containers.

A great deal of research has focused on methods to improve heat transfer during freeze drying. The early technique of supplying by conduction from plates to food placed between them restricted vapor flow and also provided uneven contact. To overcome this, expanded metal inserts are used between the plates and the metals. This process, referred to as "accelerated freeze drying," is described in detail by Hanson (1961).

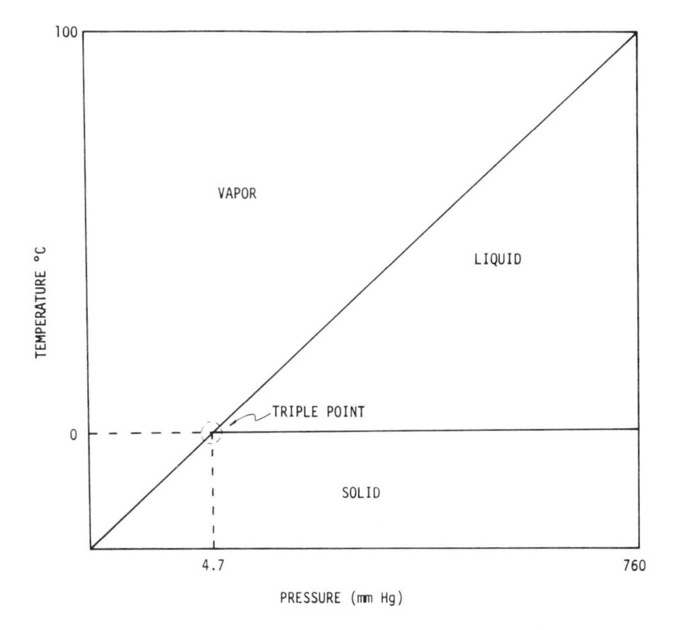

FIG. 8.3. Phase diagram of water shows relationship between temperature and pressure that makes freeze drying possible.

To improve the efficiency of the freeze-drying process, multiple batch chamber operations are employed by some plants. Continuous systems have also been designed (Togashi and Mercer 1966). Considerable attention is being given to hybrid schemes that take advantage of the positive effect of freeze drying on the cellular structure of food but reduce the cost of the process by removing some of the moisture before freezing or freeze drying (Anon. 1982).

Freeze drying is applicable to a wide range of fruit products. The major problem is its very high cost compared with the costs of canning, freezing, or other dehydration methods. Freeze drying has proved to be the superior method of dehydration for many common fruits, including blueberries, cherries, and strawberries (Anon. 1982).

In freeze drying, as in other dehydration techniques, it is difficult to dehydrate large pieces of product. As the size of product pieces increases, so does the cost of freeze drying, quality problems, and reconstitution problems.

POSTDEHYDRATION TREATMENTS

Treatments of finished dehydrated fruits vary with the kind of fruit and intended utilization of the product. Common postdehydration treatments are described in this section.

Sweating

During the sweating process dehydrated fruits are held in bins or boxes in order to equalize moisture. In the case of sun- and kiln-dried evaporated fruit, sweating involves readdition of moisture to a desired level. Bins are used for secondary drying to reduce moisture levels of particulate fruits from 10–15% to 3–5%, a range at which drying rates are limited by slow diffusion of water. Temperatures of 100°–120°F (38–49°C) and air flow, provided by a small blower fan, of about 100 ft min (33 m/min suit the nearly dry product. These conditions minimize the risk of heat damage at a stage when fruit products are most susceptible to degradation. During bin drying, which may take up to 36 hr, water contents are equalized as well as reduced. Air for the process is frequently dehumidified by condensation of water through refrigeration or by a desiccant such as silica gel. Bins used for secondary drying may vary in size and capacity. The usual size is about 4 × 4 × 4 ft (122 × 122 × 122 cm) with a perforated mesh or plate bottom, which is hinged for dropping down to a suitable angle to allow the dry material to discharge through a sliding door at the front of the bin; or a false bottom can be set permanently at an angle of about 45° to assist forward flow of the dry product.

Most continuous dehydration equipment, or bin dryers when used for final drying, usually discharge the dried product into a conveying system that brings the material to a sieving and aspirating plant to remove fines. In the course of this transfer, the product becomes well mixed. In this way, any unevenness in moisture is corrected. At this point in the process, when the low-moisture material leaves the bins or dehydrators, the area of the plant where final sieving, grading, selection, and packaging take place should be air conditioned and dehumidified to below 30% RH.

Screening

Most dehydrated fruit products have a specification for acceptable screen size distribution. During production of dehydrated fruit, fines are formed in the cutting operation and in the normal movement of product through the processing line. Screening is therefore required to remove the unwanted size portion of the dried product which can be utilized in other products. Sometimes the fines represent a loss due to operation. The removal of unwanted size pieces of fines is usually accomplished by passing the dry product over a vibrating wire cloth or perforated metal screens and collecting the fractions separately. The acceptable fraction passes onto the final inspection operation.

Inspection

The dried product is inspected to remove foreign materials, discolored pieces, or other imperfections such as skin, carpel, or stem particles. Manual and visual selection of most dehydrated fruit products is necessary and is carried out by inspectors who remove undesirable particles while the product is moving along on a continuous PVC belt at a speed of about 15–20 min (4.6–6.0 m/min). In addition to inspectors, magnetic devices are usually installed over the belt to remove metal contaminants.

Instantization Treatments

Various treaments are often used to improve the rehydration rate of low-moisture fruit products. A "flaking" treatment developed by Roberts and Faulkner (1965) involves warming fruit particles of less than 5% moisture and passing them between rollers spaced 0.001 in. (0.025 mm) apart. This process results in a shaped product having a thickness of approximately 0.01 in. (0.25 mm) because the product is resilient and partially tends to assume its initial thickness. The application of steel rollers 15 in. (40.6 cm) in diameter and 36 in. (76.2 cm) long rotating at about 300 rpm is recommended. The flakes are only compressed; because their cellular structure is unruptured, they rapidly resume their original particle size and shape during rehydration. Flaked fruit products rehydrate much more rapidly than regular products, and the rehydration rate of the flakes may be controlled by adjusting the thickness of the final product.

Another flaking treatment (Puccinelli 1968) recommended for low-moisture fruit differs from the Roberts and Faulkner method in that it tries to rupture the cells. This is achieved by passing the fruit pieces (dehydrated to 12–30% moisture) between counter-rotating rolls rotating at different speeds. Then the products having broken cellular structure are dehydrated to a moisture content of 2–10%. Products with unruptured cells will reconstitute as a piece, while those with ruptured cells tend to become mushy after rehydration.

Another instantization process involves perforating partially dried apple segments at about 16–30% moisture content, then dehydrating the perforated segments to an ultimate moisture content of less than 5% (Dorsey and Strashun 1962). The perforation treatment shortens the time required for dehydration to an ultimate moisture content. The dehydrated apple segments will rehydrate much more readily than apple segments that have not been perforated (Fig. 8.4). The punctures caused by this treatment largely disappear after rehydration. Perforation is accomplished by using a pair of rollers carried on spindles adjustably spaced;

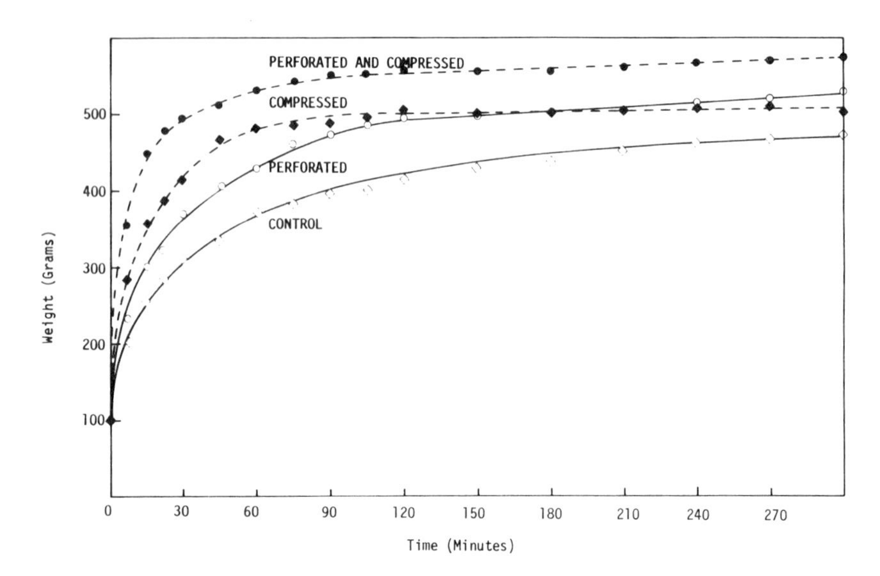

FIG. 8.4. Rehydration curves for instantized versus noninstantized low-moisture apple slices in hot (150°F or 65°C) water. (Courtesy of Vacu-Dry Co.)

one of the rollers is formed with a plurality of perforators or studs. The rollers operate in counter directions and the partially dried apple segments are passed through the rollers to perforate the segments.

For larger fruit pieces such as apple slices or wedges the improved rehydration effect resulting from compression and perforation treatments are additive; often both treatments are applied simultaneously to such products.

Packaging

The shelf life of a dehydrated fruit product is influenced to a large extent by its packaging, which must conform to certain special criteria: (a) protection of the dehydrated product against moisture, light, air, dust, microflora, foreign odor, insects and rodents; (b) strength and stability to maintain original container properties through storage, handling, and marketing; (c) size, shape, and appearance to promote salability of the product; (d) composition approved for use in contact with foods; and (e) acceptable cost.

Evaporated apples are packed tightly in fiberboard boxes lined with 2- to 4-mil polyethylene bags from which most of the air between the individual pieces has been expelled. Fruit dried to 24% moisture may be expected to retain its moisture content during considerable periods under ordinary conditions because it is in approximate equilibrium with the atmosphere at a relative humidity of about 75%.

The hygroscopic nature of low-moisture fruit tissue makes it imperative that special precautions be taken against moisture absorption. Low-moisture products must be packaged as soon as possible after removal from the dehydrator; and hermetically sealed containers are required to prevent absorption of moisture with subsequent caking and loss of quality during extended storage.

Dehydrated fruits are packed for institutional and remanufacturing use in large units such as bags, drums, bins, cartons, and cans. Heat-sealed polyethylene liners are usually required for bulk packs. For smaller retail market or catering packs, metal cans of foil-laminated, flexible pouches often are used. In case of high-value, freeze-dried products, small flexible containers prepared from three-ply laminates such as polyolefin–foil–Mylar are recommended.

Some dehydrated commodities, particularly freeze-dried products, must be packed in inert gas to ensure storage stability. Nitrogen gas is most commonly used to extend the storage stability of oxygen-sensitive products (Villota *et al.* 1980). In inert gas packing, oxygen levels of 1–2% can be routinely attained. Tin cans are used for nitrogen packing. Gas packing of cans is a well-established process. The simplest method of insert gas packing consists of piercing a hole in a filled and sealed can and placing it in a cabinet that is then evacuated. This operation removes the air from the can. When a sufficiently low pressure has been reached, nitrogen gas is admitted until atmospheric pressure is attained again. The cabinet is then opened and the holes are sealed by soldering. The more common method of gas packing is to run the filled cans through a sealing machine that applies only the first clinch. The partially sealed cans are placed in a vacuum chamber, evacuated, flushed with nitrogen, and sealed completely.

Vacuum Packing. Vacuum packing has been used to some extent with powdered or small grain and compressed dehydrated products. The container used must withstand the pressure differential without leaking. The vacuum needed to effectively extend shelf life is difficult to attain in commercial practice.

In-Package Desiccation. In-package desiccation has been used successfully for many dried fruit products, particularly for powders. The desiccant compound is placed in the container inside a small envelope made out of a moisture-permeable material that prevents contamination of the product with the desiccant. Calcium oxide (silica gel), a high-capacity adsorbent and desiccant, is usually applied. It is a granular, amorphous form of silica that can adsorb approximately 40% of its weight in moisture at 100% RH, and even when saturated remains dry and free-flowing. Package desiccation is effective if a storage period at reasonably low temperature (around 70°F or 21°C) is allowed for reduction of moisture to a suitable low level before any high temperatures are encountered.

To ensure the free-flowing property of fruit powders, particularly those high in sugar content such as prunes, figs, dates, and apple powders, an anticaking agent is mixed with the low-moisture product, usually during the milling operation. Calcium stearate is the most commonly used anticaking agent in dehydrated products. It is mixed with the fruit powder at a rate of approximately 0.25–0.50%. Silica gels and hydrated sodium silica aluminate have also been recommended as anticaking agents in dehydrated fruit powders. Fumed silicas at a rate of 1–2% have also been recommended to prevent caking of powdered fruits such as orange juice crystals (Salunkhe *et al.* 1974).

Compression

Freeze-dried fruits retain their original size, thus the space-saving advantage of conventional dehydration is lost. However, freeze-dried products can be compressed to reduce their bulkiness and packaging costs. Research at the U.S. Army Natick Laboratories resulted in significant

FIG. 8.5. Compressed freeze-dried cherry disk (center bottom). After rehydration, each disk (3 × ⅝ or 7.6 × 1.6 cm) provides sufficient filling for a 9-in (23-cm) pie. (Courtesy of U.S. Army Natick Laboratories.)

FIG. 8.6. Compressed freeze-dried blueberries. The compression process reduces the volume of freeze-dried berries 7 to 1. (Courtesy of U.S. Army Natick Laboratories.)

volume reductions of blueberries and cherries without impairing their rehydratability, appearance, flavor, or texture on reconstitution (Rahman *et al.* 1970; Do *et al.* 1975). The process involves freeze drying of sulfite-dipped IQF blueberries or red tart cherries to a moisture of less than 2%, then subjecting them to dry heat in an oven at 200°F (93°C) for approximately 10 min. The fruit becomes thermoplastic after heating and is compressed in a Carver Press with forces between 100 and 1500 lb/in.^{2} (7 and 20 kg/cm^2) and a dwell time of approximately 5 sec. The compressed, dehydrated fruits (Figs. 8.5 and 8.6) were either in the form of bars (3 × 1 × ½ in.; 7.6 × 2.5 × 1.3 cm) or disks (3 in diameter; 7.6 cm) which can fit into a No. 2½ can. Cherry disks were ½ in. thick and blueberry disks were ⅝ in. thick. The resulting reduction in volume was 1:7 for blueberries and 1:8 for cherries. This represents a reduction of 12- and 13-fold when the volume of dehydrated and compressed fruit is compared to those of loose frozen products.

Dehydrofreezing

Dehydrofreezing aims to combine the best features of both drying and freezing. The process consists of drying fruit—after peeling, coring, or pitting, and sulfiting treatments—to about 50% of its original weight and volume. Drying is usually accomplished in a tunnel drier, and then the product is frozen for preservation. The quality of dehydrofrozen fruit is equal to that of frozen products, as the drying process is discontinued at a stage where quality impairment usually does not occur. The advantages of the process are a 50% reduction in storage and freight charges and even greater savings in packaging costs in comparison with frozen products.

This process has been applied commercially to apples and apricots, and experimental work has been conducted with cherries, blueberries, and peaches with promising results (Lazar *et al.* 1961; Kitson 1970). Commercial production of dehydrofrozen apples, which are suitable for use in pies, exceeds that of any other dehydrofrozen commodity (Salunkhe *et al.* 1974). The apples are cored, trimmed, sliced, sulfured by immersion, dried to 50% weight reduction and frozen rapidly, preferably by air blast at −20 to −30°F (−29° to −34°C) and stored at 0°F (−18°C), or lower.

QUALITY CONTROL

The objective of quality control procedures is to ensure that the finished product shipped from a plant is within the specifications that have been established for that product. To ensure the acceptability of the product, the raw material must be inspected, important processing data (drying temperature, vacuum, process time, etc.) must be monitored and recorded, and the finished product must be tested. The development of meaningful sampling procedures and enforcement of proper plant sanitation are also important functions of the quality control personnel. However, these aspects are basically the same for all food processing operations and will not be specifically discussed here.

Quality control measures that are unique for dried fruit products are moisture content, sulfur dioxide content, screen analysis, physical characteristics of dried fruit, reconstitution ratio, bacterial count, and oxygen content of gas-packed products.

Moisture Content

Several methods are available for determining moisture in fruits. The vacuum oven method involves the measurement of the weight loss due to the evaporation of water. The procedure is to granulate the sample and dry it in a vacuum oven at 70°C (158°F) for about 5 hr.

A more rapid calculation of moisture in dry products can be made by using an infrared moisture meter. The sample first must be finely ground and passed through a 10-mesh screen. A given weight is dried under an infrared lamp, and the loss in weight is recorded by a sensitive scale on the tester. This test provides results in about 7–10 min. However, this method is not as accurate as the vacuum oven method.

The Karl–Fisher titration method for moisture determination is very convenient for low-moisture fruit products. This sensitive test is based on the non-stoichiometric reaction of water with iodine and sulfur dioxide in pyridine-methanol solution. For the details of moisture determination methods, the reader should consult Joslyn (1970).

Sulfur Dioxide Content

Sulfur dioxide determinations in fruit are usually carried out by the Monier-Williams method in which SO_2 is removed from the fruit by distillation in the presence of a strong acid (HCl) and subsequently measured in the distillate gravimetrically or by titration (Joslyn 1970).

Screen Analysis.

Screen analysis and the proportion of fines are often important quality characteristics of dehydrated fruits. The number and size of the screens to be used depends on the range of particle sizes expected or defined in the product specification. If desirable, a preliminary screening will establish these parameters. The analytical procedure involves weighing the sample and transferring it quantitatively to dry sieves. The sieves are then placed in a shaker and allowed to shake for a specified time, usually 3–5 min. The fraction left on each sieve is weighed after carefully transferring the residue in each sieve to a weighing pan. The results of screen analysis are presented as the percentage of residue on each sieve:

$$\% \text{ residue on sieve} = \frac{\text{weight of residue} \times 100}{\text{weight of sample}}$$

Physical Characteristics

Determination of the physical characteristics of a product involves a system of visual and organoleptic quality tests for such attributes as color, uniformity of size, absence of defects, texture, flavor, and odor. The U.S. Standards for Grades of Dried Fruits use a score system in which the relative importance of each of these factors (except flavor and odor) is scored and expressed numerically on a scale of 100. The maximum number of points that may be given to each factor are as follows: color, 20; uniformity of size, 20; absence of defects, 40; and texture, 20.

Reconstitution Ratio

Reconstitution ratio should be checked periodically since it is an important quality attribute in dried products. The reconstitution curve provides a good indication as to whether various processing conditions have been correctly applied. The recommended procedure is to soak 100 g dried fruit in excess water for a specified and closely controlled time and temperature. At various time intervals the soaking water is drained off, and the drained weight is compared with the original dry weight to give a reconstitution ratio (Fig. 8.7).

Microbial Spoilage

Microbial spoilage does not occur in dried fruit that contains less than 18–25% moisture because the osmotic pressure of the sugars in the product suppresses microbial growth. It is not necessary to do bacterial counts on every batch; periodic counts, on the other hand, are useful. They are a good indication of the quality of the raw material and, more importantly, of the effectiveness of sanitation practices used in a dehydration plant.

Bacterial growth is possible once dehydrated fruit is immersed in water for reconstitution; thus, major industrial users of dehydrated fruits have established limits on the microbial count of products they will accept. With normal handling of raw material, and after the usual drying processes, finished products can easily meet these requirements because of the relatively high concentration of SO_2 and low pH of most fruits used for dehydration (except for figs and dates, the pH values of fruits are below 4.5). The extended heat treatment given to fruit during drying also

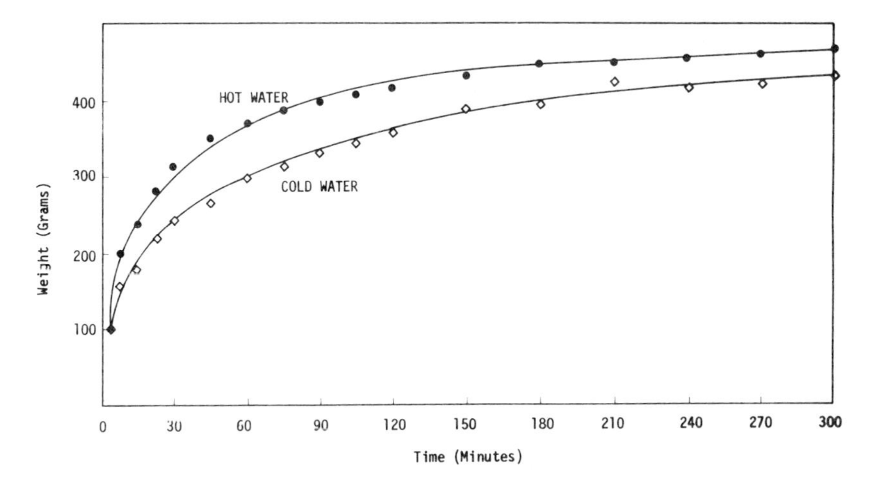

FIG. 8.7. Reconstitution curve for low-moisture apple dice in cold and hot water. (Courtesy of Vacu-Dry Co.)

reduces the number of microorganisms. Information on microbial contamination of dried fruit is very limited. King *et al.* (1968) conducted a study on the microbial quality of processed dried fruits after collecting them at various processing plants in California. Their results on aerobic bacteria, yeast, and mold counts are summarized in Table 8.5.

Headspace Analysis

When products are gas-packed, the residual oxygen content of the headspace can be determined by several standard techniques. Specially

Table 8.5. Aerobic Bacteria, Yeasts, and Mold Counts per Gram of Processed Dry Fruit

	Apples	Other cut fruits[1]	Figs	Dates	Prunes	Raisins
Bacteria						
Average	274	835	6,487	662	3,718	2,542
90% of counts less than	730	757	21,000	93	4,600	8,200
Range	0-2,600	0-11,000	0-50,000	0-7,700	0-50,000	0-60,000
Yeasts and molds						
Average	261	23	2,478	17	851	3,934
90% of counts less than	730	50	720	110	6,000	20,000
Range	0-1,500	0-210	0-30,000	0-120	0-7,100	0-30,000

Source: After King *et al.* (1968).
[1]Apricots, nectarines, peaches, and pears.

designed apparatus such as the Orsat type is widely used. More recently, gas chromatography has been applied to the determination of headspace gases in canned foods (Joslyn 1970). The accepted standard for residual oxygen content in nitrogen-packed dehydrated fruit products is 2.0% or less.

PRACTICAL DRYING PROCESSES

Fruit drying processes that are commonly employed at the present time to produce evaporated or dehydrated fruits are described in the following sections. Important characteristics of the raw material and the fruit cultivars specifically grown for drying purposes are listed, along with expected yield figures. The commonly marketed prices, size, shape, and important uses of various dried fruits are briefly discussed also.

Pome Fruits

Apples. Dried apple products are graded for texture and appearance rather than flavor. Cultivars that are firm in texture and capable of yielding a dried product of white color are preferred. In California, the 'Gravenstein' and 'Yellow Newton Pippin' cultivars have excellent processing qualities and are the most commonly used apples for drying. In the Northwest, 'Winesap,' 'Jonathan,' 'Rome Beauty,' and 'Golden Delicious' cultivars are acceptable for drying. The 'Delicious' cultivar is used frequently although its slices break badly, and peeling and coring losses are heavy. The largest quantities of apples are dried in the United States; significant but smaller quantities are also dried in Canada, Australia, the Union of South Africa, and Italy.

Apples are either dried immediately after harvest or after being held in cold and/or controlled-atmosphere storage until a convenient processing time occurs. Only artificial driers are used by commercial apple-drying plants. Water is removed relatively easily from apples to the level usually attained in dried fruit products (10–25% moisture), and simple driers, such as kiln or tunnel driers, are commonly used to produce dried (evaporated) apples. Secondary processing of the dried fruit to attain products with less than 3% moisture requires the use of a forced-air drier such as the continuous belt dryer. Air drying is sufficient for products intended for bakery utilization; if faster rehydration, greater bulk density, or better texture in the dried stage is required (i.e., dehydrated fruit snack), the use of vacuum equipment is necessary.

The processing of dried apples starts with sorting, peeling, and trimming the fruit. Sizing of the fruit is very important for mechanical peeling. Usually apples less than 2½ in. (6.35 cm) in diameter are removed by a screen. Peeling and coring losses average 30–35% of the raw material.

After peeling and trimming, apples are dipped into a sodium sulfite solution and then sliced to pieces ⅜–½ in. (9.5–12.5 mm) in thickness. These pieces are dried to the evaporated state, most frequently in a kiln or tunnel drier. Prior to or during drying, the fruit slices are exposed to the fumes of burning sulfur. Drying in a kiln requires 14–18 hr, depending on the rate of air circulation. Air temperatures of about 150°–165°F (65°–74°C) are maintained, and the fruit is turned several times during drying.

If a tunnel drier is used, the fruit slices are loaded onto trays at a rate of about 2 lb/ft^2 (10 kg/m^2). The air entering the tunnel is at 165°F (74°C) with a relative humidity of 25%. At the outlet, the air is at about 130°F (54°C) with a relative humidity of 35%.

Approximately 7 tons of fresh unpeeled apples will yield 1 ton of evaporated rings. If the product is marketed as evaporated fruit, it may be moisturized with a solution containing a sufficient concentration of sodium bisulfite to bring the SO_2 content of the product to 2500–3000 ppm. The product is then compressed and packaged.

If the product is processed to a low-moisture state, the evaporated slices are cut to the desired size, such as ¼-in. (6 mm) and ⅜-in. (9.5 mm) dice, slices, etc. Frequently, the first piece is instantized by compression and/or perforation treatment and dried to the final moisture content in a continuous belt air drier or in a vacuum drier. In this process, 100 kg of evaporated apples (24% moisture) will yield approximately 77 kg of low-moisture apples. Thus, 100 kg of fresh unpeeled fruit will yield about 10 kg of dehydrated low-moisture apples.

Other methods for making low-moisture apple products, such as explosive puffing (Eisenhardt *et al.* 1968), freeze drying of pieces (Hanson 1961) or product (Rahman and Schmidt 1970), foam-mat drying (Hertzendorf *et al.* 1970), drum drying (Escher and Neukom 1971) of applesauce, and tray drying of apple purée to give a dried apple leather (Moyls 1981) have been tried, but apparently there has been no large-scale commercialization of these methods to date.

Both dried (evaporated) and dehydrated (low-moisture) apple products are marketed on a large scale, and their utilization—particularly of low-moisture products—has been expanding in recent years.

Federal standards define dried (evaporated) apples as "apples that have been peeled, cored, and cut into segments and from which the greater portion of moisture has been removed. The fruit is sulfured sufficiently to retain a characteristic color. The finished product contains not more than 24% moisture by weight." The fruit may be prepared in the form of rings obtained by slicing cored apples at right angles to the core, or cut longitudinally into three, four, six, or eight approximately equal units. When apples are cut into more than eight approximately equal units, the product is referred to as sliced.

Dehydrated (low-moisture) apples are defined as "apples prepared from fresh, or previously dried (evaporated) apple segments. The dried (or evaporated) apple segments may be cut further into smaller segments in preparation for dehydration, whereby practically all of the moisture is removed to produce a very dry texture." Several forms of dehydrated apples are available: (1) pie slices, which consist predominantly of parallel-cut, irregular-shaped pieces, approximately 3⁄16 in. (5 mm) or less in thickness and ¾ in. (19 mm) or larger in their largest dimension; (2) flakes, consisting predominantly of parallel-cut, irregularly shaped pieces, approximately 3⁄16 in. (5 mm) or less in thickness and less than ¾ in. (19 mm) at their longest dimension; (3) wedges, which are fairly thick sectors, approximately no more than ⅝ in. (16 mm) at their greatest thickness; and (4) sauce pieces, which are small, popcorn-like units of varying shapes and sizes. Commonly used commercial sauce piece sizes are ⅜ in. (9.5 mm), ¼ in. (6 mm), nugget (⅛ in. or 32 mm), and granule (less than ⅛ in.).

Although the ultimate uses of evaporated and low-moisture apple products are similar, the secondary dehydration process results in some important advantages. The reduced bulk and weight of low-moisture products leads to lower shipping and storage costs. Evaporated apples have limited storage stability and require refrigeration for extended storage, whereas the low-moisture products can be stored for several years in common storage in proper containers. Low-moisture products have a lower SO_2 content because about 50% of the sulfite applied during preparation of the evaporated product is volatilized during the dehydration process. Finally, low-moisture apples are free-flowing, an important feature for certain uses.

Several other forms of dried apples are available. Apple chops are prepared by slicing and drying washed whole apples, frequently without treatment to prevent darkening. Small fruit unsuitable for peeling are used to prepare apple chops. It is a very inexpensive source of apple solids and may be used in preparing apple butter and other products where color and blemishes are not important quality factors. Dried applesauce is defined as a dried product made from sound, ripe apples after suitable preparation, pulping, and processing; dry sugar, acidulant, and flavor may be added as optional ingredients (Rahman and Schmidt 1970). Apple powder flour consists of dehydrated apples that have been reduced to a powder form by grinding. The low moisture content necessary to permit grinding to powder form results in a highly hygrosopic product that must be held in sealed containers to prevent caking.

Pears. Dried pears are utilized only to a limited extent, and are largely sun-dried. The 'Bartlett' cultivar is used almost exclusively for drying. Ripe fruit is peeled, cored, and then sulfured for 6–24 hr by exposing the fruit to the fumes of burning sulfur in a sulfur house. The pears are then spread on trays and kept exposed to the sun for ½–2 days, after which the trays are stacked beneath long sheds open at the sides to allow air circulation freely. It takes 3–6 weeks to dry the product to 24% moisture. The dried fruit may be cubed, sliced, or cut in Julienne strips and is expected to be translucent or nearly transparent.

Drupe Fruits

Apricots. Evaporated apricots are usually sun-dried. Iran, the United States (California), Australia, Spain, and the Republic of South Africa are the major dry apricot-producing countries. The 'Royal' and 'Blenheim' cultivars are the principal ones used for drying.

Apricots for drying should be allowed to fully ripen on the trees. Harvested fruits are pitted and cut into halves along the suture but not peeled. The halves are placed on wooden trays and exposed to sulfur fumes for at least 3 hr. Approximately 2.3–3.6 kg of sulfur is applied per metric ton of fruit to achieve an SO_2 concentration between 2000 and 5000 ppm in the dried product. The fruits are exposed to the sun for about 4 days; the trays are then stacked so that they are facing the prevailing wind. The dry product is sorted and stored in wooden boxes to undergo sweating for several weeks, and resulfured in the sulfur house if necessary before final packaging. The normal yield of evaporated fruit is about 20% of the raw fruit input.

Dehydrated (low-moisture) apricots are produced from evaporated fruit to a limited extent. The vacuum dehydration process is necessary to dehydrate apricots to less than 5% moisture. Vacuum shelf driers are commonly used for the process. Evaporated fruit halves are sliced or diced before loading on the drying trays. To preserve the bright orange color of the fruit, it is extremely important to keep the drier temperature at 160°F (71°C) or lower at all times. Apricot nuggets and granules are processed in vacuum chambers from extruded apricot paste, which is dried as a sheet on the tray, and the low-moisture product is cut to the desired piece size.

Low-moisture apricot products are utilized in dry turnover pies, tart filling mixes, and in fruit cocktail-type mixes in combination with other dehydrated fruits to be consumed as stewed fruit or compote. Flakes can be formed according to the process of Roberts and Faulkner (1965), and low-moisture apricot flakes may be used as a cereal additive.

Peaches. Freestone peaches are used almost exclusively for drying. The important cultivars used for drying in California include 'Elberta,' 'Muir,' 'Lovell,' and 'Rio Oso Gem.' In Australia, Chile, and the Republic of South Africa, limited quantities of freestone peaches are also preserved by drying.

The preparation and drying of peaches is identical with that described for apricots. Since peach halves are larger than apricots, the sulfuring process takes somewhat longer, about 4–6 hr. The drying ratio of peaches varies from 3.5:1 to 7:1, depending on cultivars and maturity.

The dehydration process for low-moisture peach products is the same as that described for apricots. Peaches can be dried to under 4% moisture only under vacuum. Low-moisture sliced or diced peaches are utilized as pie, tart, and turnover fillings. Low-moisture peach powder, which is usually a by-product of the process, provides excellent purées, spreads, or glazes after proper rehydration and preparation.

Cherries. Sweet or sour cherries are dehydrated or sun-dried in very limited amounts. The 'Bing' cultivar is regarded as the best of the sweet cherries for dehydration. Other cultivars that have been dehydrated successfully in commercial quantities are 'Napoleon,' 'Lambert,' and 'Black Republican.' The only red tart cherry that has been dehydrated on a commercial scale is 'Montmorency.'

Cherries can be dried as whole fruits with or without stems, or as pitted fruit. If whole cherries are dehydrated, it is necessary to dip the fruit in a dilute boiling solution of 0.5% NaOH to crack the skin, followed by rinsing in water. White and pink cherries should be sulfured for about 30 min; black cherries require a shorter sulfur treatment. Dehydration is more desirable than sun drying for cherries. The fruit may be dried in about 8–12 hr in a countercurrent tunnel, with the hot-end air temperature less than 170°F (77°C), wet bulb temperature at 105°–115°F (40°–46°C), and wet bulb depression of about 15°F (8°C) at the cool end. Drying ratios for pitted sweet cherries average about 4.2:1, while those for whole fruit average about 3:1 based on fresh fruit weight. Dehydrated pitted cherries are excellent ingredients for pie fillings.

Maraschino cherries in a sugar solution can be dehydrated under vacuum to less than 5% moisture. Such a product is decorative, but without flavor, and is used in dry fruit cocktail mixes.

Prunes. About 75% of the world's supply of dried prunes are produced in California and in the Pacific Northwest. Sun drying was very common earlier, but today nearly all prunes are dehydrated because of the high risk of early rains. The 'French' prune cultivar is most commonly selected for dehydration because of its high sugar content and full-bodied flavor. Other California-grown cultivars recommended for drying are the 'Imperial' and 'Sugar.'

Prunes are prepared for drying by cleaning with air blast and water sprays, followed by dipping in cold or hot water. The fruits are then spread in a single layer on trays and dehydrated to about 18% moisture in a forced-draft tunnel dehydrator. The drying process usually requires 24–36 hr, depending on the size and solids content of the prunes. The tunnel is operated under 165°F (74°C) dry bulb temperature, with the wet bulb 15°F (8°C) lower than the dry bulb at the cool end.

Conventional dried prunes are sometimes dehydrated to low moisture in vacuum shelf driers. The finished low-moisture prunes contain less than 4.0% moisture. Because they are free-flowing and highly hygroscopic, dehydrated prunes require very careful handling and packaging. Whole pitted prunes are dehydrated after perforation treatment to achieve better dehydration and reconstitution character. Diced prunes, prune flakes, nuggets, and granules are also commercially available. Such products are used for prune fillings and other bakery spreads. Low-moisture prune powder is well utilized as a sweetening and flavoring agent for whole wheat or rye breads, pumpernickel, etc., or for the preparation of instant plum spread. One pound of low-moisture prunes is equivalent to 1½ lb of evaporated product.

Small Fruits

Grapes. Dried grapes, commonly known as raisins, represent the greatest quantity of dried fruits produced today. Over 400,000 tons of raisins are produced annually in the world, of which approximately 50% is produced in California. Other important raisin-producing countries are Turkey, Australia, Greece, Iran, the Republic of South Africa, and Spain. The most important raisin grape is the 'Thompson Seedless.' 'Muscate of Alexandria' is also dried to some extent, particularly in Spain. A soluble solids content of 23% at harvest is considered the minimum for raisin grapes. Since raisins are produced primarily by sun drying, harvest time is an important factor in selecting a raisin grape cultivar. In the San Joaquin Valley of California, where most of the raisin grapes are grown, mid-September is the deadline for harvesting.

Sun drying of raisins is done in the vineyard on wooden trays or strips of plastic. Clusters are turned periodically to expose all grapes to the sun for uniform drying. Drying is continued until 16% moisture content is achieved, then the berries are stripped from the stems, cleaned, fumigated, and packed.

Before drying, grapes often are dipped in alkali to remove the waxy layer and thus hasten drying. The recommended dipping treatments are as follows: (a) soda dip—0.2–0.3% NaOH at 200°F (93°C) for 2 sec, followed by water rinse; (b) Australian mix dip—0.3% NaOH + 0.5% KCO_3 + 0.4% olive oil at 180°F (82°C) for 2–3 sec, without rinsing; (c) Australian cold dip—5% K_2CO_3 emulsified with 0.4% olive oil for 1–4 min at 95°–100°F (35°–38°C); (d) California soda-oil dip—0.4% solution of Wyandotte powder (soda ash and Na_2CO_3) for 30–60 sec at 95°–100°F (35°–38°C).

Stafford and Guardagni (1977) showed in extended storage tests at 70°F

(22°C) that the color and flavor of on-the-vine dried and tunnel-dehydrated raisins treated with a spray or dip of safflower oil ethyl ester were equal or superior to that of traditional sun-dried raisins. These treatments accelerate moisture loss by causing wax pellets on the grape skin to dissociate, thus facilitating water diffusion.

In California, dehydrated grapes are also produced to a limited extent. Mostly the "golden bleached" type grape is used for dehydration. The process involves dipping fruits in 0.25% hot lye bath, washing, spreading them on wooden trays, and sulfuring in a sulfur house for about 4 hr. During that time 1.8–2.3 kg of sulfur is burned per metric ton of grapes. After sulfuring the fruits are dehydrated in a tunnel drier to about 12–13% moisture. Dehydration in a tunnel drier requires 1–2 days. The SO_2 content of the finished product ranges between 1500 and 2000 ppm.

Low-moisture, whole "puffed" grapes are produced in a vacuum shelf drier, usually mainly golden-bleached 'Thompson Seedless' raisins as the raw material. The vacuum-puffed grapes contain less than 4% moisture. They are free-flowing, rehydrate in 2–5 min, and are used for instant breakfast cereals, prepared cakes, desserts, and pudding mixes.

Berries. Drying of strawberries, raspberries, blueberries, cranberries, etc., cannot be achieved successfully with the common sun drying or fruit dehydration techniques. However, freeze drying of these fruits has given excellent results, but the high cost of the process makes it largely prohibitive for large-scale application. Hanson (1961) recommended that strawberries be cut longitudinally prior to freeze drying to overcome the water vapor barrier effect of the impermeable skin. On the other hand, he found that whole raspberries, gooseberries, and currants can be easily freeze-dried.

Vollink *et al.* (1971) discovered that low-moisture strawberries can be produced with a modified freeze-drying process that yields fruit capable of rehydration in milk within 30–90 sec. The process involves cooling the fruit to just above the freeze point of the water present in the fruit, then slowly freezing this water to develop a growth of ice crystals that expand the cellular structure of the fruit, followed by freeze drying the slowly frozen fruit to under 3% moisture content.

Scharshmidt and Kenyon (1971) recommended perforating blueberries before freeze drying in such a way that not only are the outer skin portions punctured but also a substantial amount of the interior core. Before puncturing, the blueberries should be frozen, ideally at a slow rate for 10–20 hr, to a temperature of 0°F (18°C) or lower. The punctured freeze-dried product does not shrivel or otherwise change in physical appearance. If dried substantially in an unwrinkled bulbous spherical shape, it rehydrates in an aqueous medium in 30 sec to 3 min.

The explosion-puffing process has also been successfully applied to blueberries. The fresh berries require little preparation except size grading and can undergo the preliminary drying process without sulfiting or blanching. Unlike the freeze-dried product in which puncturing or scarifying causes problems, intact berries can be successfully dried. After preliminary drying to a moisture range of 11–18%, the berries are puffed in a gun, which has a surface temperature of 340°–350°F (171–177°C) and a pressure adjusted to 20 psi. The puffed product is finish-dried to about 6–9% moisture in about 2½ hr in a cabinet drier, operated at 150°F (60°C) dry bulb and 90°F (32°C) wet bulb temperatures.

Tropical and Subtropical Fruits

Figs. Figs are mostly sun-dried, but dehydration is also practiced to a limited extent to produce low-moisture figs. Turkey, Greece, Italy, and Portugal are the important fig-producing countries, and California is rapidly becoming a major supplier of figs. The 'Smyrna' fig, a white cultivar, is the principal cultivar grown in the Mediterranean countries for drying. 'Calimyrna,' a strain of the 'Smyrna' cultivar is successfully grown in California. The 'Adriatic' fig, a white fig with pink flesh, and the 'Kadota,' another white fig grown in California for canning and preserving, are also used sometimes for drying. The 'Black Mission' fig has been grown in California since the day of the Spanish missions. It yields a dry product of black color, tender texture, and excellent flavor. This fig is not harvested for drying but is allowed to ripen and partially dry on the tree, and is collected from the ground after falling.

'Calimyrna' and 'Adriatic' figs are dipped in a solution of 10 lb salt and hydrate lime/100 gal. of water (1.2 kg salt and hydrated lime in 1 kliter of water) to remove the hairs from the surface, to improve the color, and to soften the skin. After careful sorting, figs are spread on trays. 'Adriatic' and 'Kadota' figs are usually exposed to the fumes of burning sulfur for 4 hr or longer, but 'Calimyrna' and 'Mission' figs are never treated with sulfur. Since figs arrive from the dry yard partially dried, they are usually finish-dried in the shade, not exposed to the direct sun. Exposure to the sun toughens the skin of 'Calimyrna' figs, and a dry product of better quality is obtained by drying the fruit in the stack.

Some dried figs are dehydrated to less than 3.5% moisture in vacuum driers. Both white 'Kadota' figs and black 'Mission' figs are processed to low moisture as slices of approximately ⅝ × 5⁄16 in. (16 × 8 mm) thickness or as a powder (20-mesh). These products can withstand moderately high processing temperatures without detrimental changes in color, texture, or flavor.

Dates. Dates are grown commercially in hot, dry regions that have an abundance of water supplied naturally or by irrigation. Such areas are in North Africa, (Egypt, Tunisia, and Algeria) and the Sahara and Arabian deserts. California and Arizona produce all the dates grown in the United States. Date cultivars are classified into three groups (Cruess 1958): (1) dry dates, in which sucrose is the predominating sugar; (2) semidry dates, in which sucrose and invert sugar are both present in about equal amounts; and (3) invert sugar (soft) dates, in which invert sugars predominate. 'Sakhoti,' or bread, dates, which are grown in Egypt, belong to the dry date class. They become hard and very dry on the palm before harvest. The principal U.S. sun-dried date cultivar is the 'Deglet Noor' which belongs to the semidry class. Another important semidry date cultivar is 'Amri of Egypt.' The 'Khadrawi' and 'Halawi' cultivars are invert sugar (soft) dates.

Dates harvested from the palm in the fall may be astringent. To eliminate astringency, the fruits are incubated at 90°–100°F (32°–38°C) in ripening rooms for a few days. The sucrose in the date is converted by enzymes to invert sugar during the ripening process. Most dates are sufficiently low in moisture content at harvest time that they can be packed without further drying. If the moisture content exceeds 24%, they are treated with preservatives that inhibit growth of molds and yeasts.

In recent years, low-moisture dates have been used primarily in bakery goods. These products are dehydrated to a moisture content of less than 4% in vacuum shelf driers using sliced fruits that were sun-dried on the tree. Free-flowing, low-moisture dates are cut to uniform size as nuggets or granules, or milled to a powder, before packing in moisture-proof containers with a desiccant or anticaking agent included.

Citrus Fruits. Citrus products are mostly dehydrated as juice products. Almost every type of dehydration process has been evaluated for citrus juices. Experimental studies have shown the feasibility of preparing orange, lemon, grapefruit, and lime juice instant products by freeze drying, foam-mat drying, vacuum-puff drying, and some types of drum or spray drying (Berry 1979).

Low-moisture orange granules are processed in limited quantities in California, using fully matured 'Valencia' oranges. The whole seedless fruit is first dried in a kiln drier to 12–16% moisture, followed by secondary drying in bins to less than 3% moisture. The low-moisture product is then milled to about ⅛-in. (3-mm) granules. These orange granules are extremely fragrant and pungent. To protect the aroma, the product must be treated with an antioxidant to prevent oxidation of the flavonoid compounds. Spraying the granules with an alcoholic solution of leutylated hydroxyanisole (BHA) provides for a long shelf life of the product in moisture-proof containers. Low-moisture orange granules are used as a bakery spread ingredient and in spreads, jams, fruit fillings, and candies. One lb of low-moisture orange granule is equivalent to 6 lb of fresh ground orange.

Finisher pulp from citrus juice processing operations consists primarily of juice vesicles or "sacks." These juice sacks may be drum-dried to a moisture content of less than 10% (Braddock and Kesterson 1974). The dried sacks have potential uses as natural ingredients in the food and beverage industries (e.g., to absorb water and fat in comminuted meat products, breading mixes, pie filling, dehydrated beverage mixes, etc.).

Pineapples. Peeled and cored pineapples can be dehydrated in trays to about 15–20% moisture. Sulfuring is required prior to dehydration. Further drying to low-moisture levels can be achieved under vacuum.

Banana. Peeled whole ripe bananas are dried in Central America (Cruess 1958); although the resulting products are unattractive and dark in color, they have a pleasant flavor. A more attractive product is produced from ripe bananas when they are sliced lengthwise and sulfured for 20 min before drying.

Bowrey *et al.* (1980) used solar drier cabinets to dehydrate bananas. The fruit was placed in a single layer on a plastic mesh supported by perforated steel trays. Air was circulated by a centrifugal fan. When the air in the cabinet reached 35% RH, the air was exhausted and fresh air was drawn in by an automated system. If the temperature of the air inside the cabinet was less than 129°F (50°C), an electric heater was turned on. This process operated continuously until the fruit reached 14–15% moisture. The process usually required about 72 h.

Brekke and Ponting (1970) recommend an osmovac process for bananas. Peeled ripe bananas are cut into ¼-in. (6-mm) slices, which are transferred to a sugar syrup at about 70° Brix; water is withdrawn from the fruit by osmo-drying while the fruit is submerged in the syrup for 8–10 hr. After the surface syrup is strained from the fruit, the slices are dried in a vacuum shelf drier at 150°–160°F (66°–71°C) and 10 mm Hg. Vacuum drying is completed when the moisture content of the fruit is reduced to 2.5% or less. Bongiwar and Sreenivasan (1977) recommend an osmotic treatment at 50°C (122°F) with agitation of the syrup to reduce the time required for osmosis to occur. The finished vacuum-dried bananas retained crisp texture, original color, and flavor for 1 year in aluminum laminate pouches.

Mango. Peeled and diced green mangos can be air-dried to low-moisture levels after sulfuring and "osmotic" treatments in a sugar or salt brine solution (Jackson and Mohamed 1971). The production of dehydrated, brined green mangos was demonstrated in a remote area of Sudan where there were no market outlets for fresh fruit. The dehydrated mangos were utilized commercially as a substitute for the mangos used by chutney manufacturers.

Dabhade and Khedhar (1980) reported the successful drying of mango slices of 'Totapuri' and 'Seeding' cultivars in India. Drying required 10 hr with 1.3 lb/ft^2 (6.5 kg/m^2) tray load, while sun drying was completed in 15 hr. Blanched slices dried quicker than unblanched ones. Soaking the mango slices in 1% potassium metabisulfite solution for 30 min improved the retention of ascorbic acid, and sulfur-treated slices were lighter in color and rated higher by an organoleptic panel after 6 months of storage. Pulverized mango powder has been used in the preparation of curry.

Slices of Florida-grown mangos, pretreated with a 5-min dip in 2% sodium bisulfite solution, were successfully dried in solar driers (Coleman *et al.* 1980).

Kiwifruit. Slices of kiwifruits too small for fresh market have been dried successfully without sulfite treatment (Simons 1978). The best color retention was obtained with 4-mm slices dried below 122°F (50°C). Dried pieces may be consumed without further treatment or after rehydration to nearly full shape or size, or they may be candied for use as a confection.

Persimmon. Testoni and Maltini (1978) reported that 5/16-in. (8-mm) thick slices of Japanese persimmon grown in Italy were dehydrated to 20% moisture in a shelf oven with hot-air circulation at 40°–50°C (104°–106°F).

REFERENCES

AHLBORG, U.G., DICK, J. and SJOGAND, D. 1977. Utilization of sulphites in foods. Var Foda *29,* 331–353.

ANON. 1978. Raisin-grapes are solar dried at 12 tons per 24 hours. Food Eng. *50* (6) 202.

ANON. 1982. Dehydration and Compression of Foods. Report No. 116. Advisory Board on Military Personnel Supplies, Washington, DC.

BERRY, E. R. 1979. Citrus juice products. *In* Tropical Foods: Chemistry and Nutrition, Vol. 1. G. E. Inglett and G. Charlalamboy (Editors). Academic Press, New York.

BOLIN, H. R. and BOYLE, F. P. 1972. Effect of storage and processing on sulfur dioxide in preserved fruit. Food Prod. Dev. 6 (7) 82–86.

BOLIN, H. R. and SALUNKHE, D. K. 1982. Food dehydration by solar energy. CRC Critical Rev. Food Sci. and Nutrition *13,* 327–354.

BOLIN, H. R., STAFFORD, A. E. and YANASE, K. 1976. Color stability of dried peaches. Food Prod. Dev. *10* (8) 74, 76.

BOLIN, H. R., HUXSOLL, C. C. and SALUNKHE, D. K. 1983. Fruit drying by solar energy. Confructs *25,* 147–160.

BONGIRWAR, D. R. and SREENIVASAN, A. 1977. Studies on osmotic dehydration of banana. J. Food Sci. and Tech. (India) *14,* 104–112.

BOWREY, R. G. 1980. Use of solar energy for banana drying. Food Tech. (Australia) *32,* 290–291.

BREKKE, J. E., and PONTING, J. D. 1970. Osmo-vac dried bananas. Hawaii Agric. Exp. Sta. Res. Rep. *182.*

BRADDOCK, R. J. and KESTERSON, J. W. 1974. Dried citrus juice sacs for the food and beverage industries. Proc. Fla. State Hort. Soc. *86,* 261–263.

BROWN, G. E., FARKAS, D. F. and De MARCHENA, E. A. 1972. Centrifugal fluidized bed. Food Technol. *26,* 23–29.

BRYAN, W. L., WAGNER, C. J., JR., and BERRY, R. E. 1978. Food drying with direct solar augmented by fossil energy. Trans. ASAE *21,* 1232–1236, 1241.

CLAMPET, G. L. (Editor). 1981. Agricultural Statistics. USDA, Washington, DC.

COLEMAN, R. L., WAGNER, C. J., JR., and BERRY, R. E. 1980. Solar drying mango slices and mechanically deseeded muscadine grapes. Proc. Fla. State Hort. Soc. *93,* 334–336.

CRUESS, W. V. 1958. Commercial Fruit and Vegetable Products. 4th ed. McGraw-Hill Book Co., New York.

DABHADE, R. S. and KHEDAR, D. M. 1980. Studies on drying and dehydration of mango powder (Amchur). Parts I–VII. Indian Food Packer *34*(3) 1–59.

DIXON, G. M. and JEN, J. J. 1977. Changes in sugars and acids of osmovac dried apple slices. J. Food Sci. *42,* 1126–1127.

DO, J. Y., SRISANGNAM, S., SALUNKHE, D. K., and RAHMAN, A. R. 1975. Freeze-dehydrated and compressed sour cherries. J. Food Technol. *10,* 191–201.

DORSEY, W. R. and STRASHUN, S. I. 1962. Method for making dehydrated apples. U.S. Pat. 3,049,426. Aug. 14.

EISENHARDT, N. H., CORDING, J., JR., ESKEW, R. K. and HEILAND, W. K. 1968. Dehydrated explosion puffed apples. USDA Agric. Res. Serv. *ARS-73-57.*

ESCHER, F. and NEUKOM, H. 1971. Non-enzymatic browning and optimization of the drying conditions for the drum drying of applesauce. Lebensm-Wissen. Techno. *4* (5) 145–151.

ESKEW, R. E. and CORDING, J., JR. 1968. Process for manufacturing rapidly dehydratable dehydrated fruits and vegetable of high density. U.S. Pat. 3,408,209. Oct. 29.

FARKAS, D. F. and LAZAR, M. E. 1969. Osmotic dehydration of apple pieces: Effect of temperature and syrup concentration on rates. Food Technol. *23,* 688–690.

FLINK, J. M. 1977. Energy analysis in dehydration processes. Food Technol. *31* (3) 77–84.

GEE, M. FARKAS, D. F. and RAHMAN, A. R. 1977. Some concepts for the development of intermediate moisture foods. Food Technol. *31* (4) 58–84.

GOULD, J. W. and KENYON, E. M. 1970. Microwave applications to freeze dehydration. U.S. Army Natick Lab. Tech. Rpt. *71-15FI.*

HANSON, S. W. F. (Editor). 1961. The Accelerated Freeze-Drying Method (AFD) of Food Preservation. M. J. Stationary Office, London.

HEIDELBOUGH, N. D. and KAREL, M. 1975. Intermediate moisture food technology. *In* Freeze Drying and Advanced Food Technology. S. A. Goldblith (Editor). Academic Press, New York.

HERTZENDORF, M. S., MOSHY, R. J. and SELTZER, E. 1970. Foam drying in the food industry. CRC Critical Rev. Food Technol. *1* (1) 25–70.

HOLDSWORTH, S. D. 1971. Developments in preservation. Food Process. Ind. *40,* 27–31, 44.

HOPE, G. W. and VITALE, D. G. 1972. Osmotic dehydration. Canada Dept. Agric. Food Res. Inst. *IDRC-004e.*

HUSSMANN, P. 1963. Food dehydration by the BIRS process. *In* Lebensmitteltechnik, Dechema Monograph 46, pp. 761–780. Verlag Chemie GmbH, Weinheim-Bergstrasse. (German)

JACKSON, T. H. and MOHAMED, B. B. 1971. The shambat process. New development arising from the osmotic dehydration of fruits and vegetables. Sudan J. Food Sci. Technol. *3,* 18–23.

JOSLYN, M. A. (Editor). 1970. Methods in Food Analysis. 2nd ed. Academic Press, New York.

KAREL, M. 1963. Exploration in future food processing techniques. S. A. Goldblith (Editor). MIT Press, Cambridge, Mass.

KING, A. D., JR., FIELDS, R. K. and BOYLE, F. P. 1968. Dried fruits have low microbial counts. Food Eng. *40* (12) 82–83.

KITSON, J. A. 1970. A continuous process for dehydrofreezing apples. J. Can. Inst. Food Technol. *3,* 136–138.

KITSON, J. A. and BRITTON, D. 1978. Metamorphosis of fruit trimmings creates sweet alternative to dried fruit. Food Prod. Dev. *12* (2) 29, 42.

LAZAR, M. E. and FARKAS, D. F. 1971. The centrifugal fluidized bed. 2. Drying studies on piece-form foods. J. Food Sci. *36,* 315–319.

LAZAR, M. E., CHAPIN, E. O. and SMITH, G. S. 1961. Dehydrofrozen apples. Recent developments in processing methods. Food Technol. *15,* 32–35.

LEVINE, K. F. 1977. Drying equipment. Food Process. *38* (9) 122–140.

McBEAN, D. McG. 1971. Recent advances in dehydration process. CSIRO Food Res. Quart. *31* (3) 41–46.

MILLER, M. W. and WINTER, F. H. 1972. Dried fruit without sulphur. Food Process. *33* (5) 25.

MOYLES, A. L. 1981. Drying of apple purées. J. Food Sci. *46,* 939–942.

NURY, F. S., BREKKE, J. E. and BOLIN, H. R. 1973. Fruits. *In* Food Dehydration, 2nd ed. Vol. 2. W. B. Van Arsdel, M. J. Copley and A. I. Morgan, Jr. (Editors). Avi Publishing Co., Westport, CT.

PINTUARO, S. J., TAYLOR, S. L. and CHICHESTER, C. O. 1983. A Review of the Safety of Sulfites as Food Additives. The Nutrition Foundation, Inc., Washington, DC.

PONTING, J. D. 1973. Osmotic dehydration of fruits: Recent modifications and applications. Process Biochem. *8,* 18–32.

PONTING, J. D., JACKSON, R. and WATTERS, G. 1972. Refrigerated apple slices: Preservative effects of ascorbic acid, calcium and sulfites. J. Food Sci. *37,* 434–436.

POTTER, N. N. 1978. Food Science. 3rd Ed. AVI Publishing Co., Westport, CT.

PUCCINELLI, W. 1968. An edible product consisting of fruits and vegetables in form of flakes and process and apparatus for preparing the same. Brit. Pat. 1,129,972. Oct. 9.

RAHMAN, A. R. and SCHMIDT, T. R. 1970. Process of making dehydrated applesauce. U.S. Pat. 3,535,127. Oct. 20.

RAHMAN, A. R., TAYLOR, G. R., SCHAFFER, G. and WESTCOTT, D. E. 1970. Studies of reversible compression of freeze dried RTP cherries and blueberries. U.S. Army Natick Lab. Tech. Rpt. *70-52F1.*

RAO, M. R. R. 1970. Food technology recent development. Assoc. Food Drug. Officials U.S. Quart. Bull. *34* (3) 138–146.

ROBERTS, A. C. and MCWEENY, D. J. 1972. The uses of sulfur dioxide in the food industry. J. Food Technol. *7,* 221–238.

ROBERTS, R. L. and FAULKNER, R. E. 1965. Flaked comestibles and process for preparing same. U.S. Pat. 3,174,869. Mar. 23.

SALUNKHE, D. K., DO, J. Y. and BOLIN, H. R. 1974. Development in technology and nutrition value of dehydrated fruits, vegetables and their products. *In* Storage, Processing and Nutritional Quality of Fruits and Vegetables. K. K. Salunkhe (Editor). CRC Press, Cleveland.

SCHARSCHMIDT, R. K. and KENYON, R. E. 1971. Freeze drying of blueberries. U.S. Pat. 3,467,530. Sept. 16.

SIMMONS, I. D. 1978. Drying and candying of Chinese gooseberries. Food Tech. (Australia) *30,* 236–239.

STAFFORD, A. E. and BOLIN, H. R. 1972. Improves fruit resulfuring. Food Eng. *44* (11) 128–130.

STAFFORD, A. E. and GUARDAGNI, D. G. 1977. Storage stability of raisins dried by different procedures. J. Food Sci. *42,* 547–548.

TESTONI, A. and MALTINI, E. 1978. Preliminary experience in drying of persimmons. Annali dell Istituto Sperimentale per Iol Valorizzazione Tecnologica dei Prodotti Agricoli *9,* 111–117.

TOBBY, G. 1966. Method for drying foods. U.S. Pat. 3,249-446. May 3.

TOGASHI, H. J. and MERCER, J. L. 1966. Freeze-dried product and method. U.S. Pat. 3,293,766. Dec. 27.

VANARSDEL, W. B., COPLEY, M. J. and MORGAN, A. I., JR. 1973. Food Dehydration, 2nd ed. Vol. 1—Drying Methods and Phenomena and Vol. 2—Practices and Applications. AVI Publishing Co., Westport, CT.

VILLOTA, R., SAGUY, I. and KAREL, M. 1980. Storage stability of dehydrated food. Evaluation of literature data. J. Food Quality *21,* 123–212.

VOIROL, F. 1972. The blanching of vegetables and fruits. Food Process. Ind. *41,* 27–33.

VOLLINK, W. I., KENYON, R. F., BARNETT, S. and BOWDEN, H. 1971. Freeze dried strawberries for incorporation into breakfast cereal. U.S. Pat. 3,395,022. July 30.

WAGNER, C. J., JR., BRYAN, W. L. and BERRY, R. E. 1977. Preliminary solar drying. Studies of some Florida fruits and vegetables. Proc. Fla. State Hort. Soc. *90,* 158–161.

WAGNER, C. J., JR., COLEMAN, R. L., BRYAN, W. L. and BERRY, R. E. 1978. Preliminary studies on SO_2 absorption in mangos, nectarines and peaches prepared for drying. Proc. Fla. State Hort. Soc. *91,* 117–119.

WILLIAMS-GARDNER, A. 1971. Industrial Drying. Leonard, London.

9

Brining Cherries and Other Fruits[1]

G. G. Watters and J. G. Woodroof

Sweet cherries are held in calcium bisulfite brine for remanufacture into brightly colored, tasty products. The object of preserving cherries with brine is to bleach the fruit to an even, light straw color so that the subsequent dyeing will give the desired shade. It also firms the tissue so that the cherry will retain its original shape throughout the remanufacturing processes and subsequent life in commercial channels.

The use of sulfur dioxide to preserve fruit is an ancient process. Published reports on the use of solutions of sulfur dioxide for cherry preservation with and without hardening agents appeared in the late 1920s and early 1930s.

Sulfur dioxide as a gas or in a water solution occupies a unique position in the field of food preservatives (DeEds 1961). It serves a greater variety of purposes than any other food additive. Its antioxidant properties retard loss of ascorbic acid and carotene. Sulfur dioxide reduces or prevents microbiological spoilage. It inhibits enzyme-catalyzed oxidative discoloration and nonenzymatic browning during the processing, storage, and distribution of foods.

[1]Reference to a company and/or product named by the U.S. Department of Agriculture is only for purposes of information and does not imply approval or recommendation of the product to the exclusion of others which may also be suitable.

Commercial Fruit Processing, 2nd Edition

ISBN 0-87055-502-2

Atkinson and Strachan (1962) pointed out other advantages of sulfur dioxide as a preservative. It is inexpensive to use and provides a quick method of handling large quantities of fruit. The product may be placed in bulk containers for shipment, and the preservative is effective for a sufficient period of time to allow for storage and remanufacture. In addition, most of the sulfur dioxide can be removed easily and inexpensively.

PRODUCTION AND HARVEST

'Bing', 'Napoleon' ('Royal Ann'), and 'Lambert' cherries are the chief sweet cultivars used for brining in the western United States. The 'Napoleon', 'Windsor', 'Schmidt', and 'Emperor Francis' cultivars are brined in Michigan and in the cherry-growing areas of the eastern states. In the past, surplus quantities of red tart cherries, such as the 'Montmorency', were routinely brined. However, the increase in production of sweet cherries due to the increase in plantings usually ensures a surplus of the larger, more desirable sweet cherry.

Between 40 and 45% of the total sweet cherry production in the United States is brined. Michigan leads the individual states with 75% of her production going into brine processing. Production data for 1973–1981 are listed in Table 9.1.

Before the advent of mechanical harvesters, sweet cherries were handpicked into field lugs and boxes for transportation to a packing or processing plant. Fruit not meeting fresh market standards, called packing house culls, made up the principal source of fruit for brining. Now, whole orchards are grown and harvested solely for brining.

In Michigan and Oregon, a substantial quantity of cherries for brining are mechanically harvested. Much of the fruit is bruised by the violent shaking required to separate the cherry from the branch. However, Whittenberger *et al.* (1968) and Stebbens *et al.* (1967) found little adverse effect of bruising on quality if the fruit was brined immediately after harvesting. To minimize the effect of bruising, some of the mechanical harvesters are constructed to hold barrels half full of brine. In a matter of seconds, the cherries are conveyed into the brine. When the barrel becomes full, it is taken from the harvester by forklift truck to the end of the row from where it is transported to the processing plant. Over the years, processors have developed a preference for a slightly immature cherry for brining. Levin *et al.* (1969) found the more mature the cherry the better the recovery during mechanical harvesting. The yields increased from below 50 to more than 90% during a 20-day harvest. It is customary now to brine all different maturities together with only different cultivars kept separate.

Table 9.1. U.S. Production of Brined Sweet Cherries (in Tons), 1973–1981

	1973	1974	1975	1976	1977	1978	1979	1980	1981
Great Lakes states	16,045	23,300	30,900	9,000	19,600	25,200	22,000	19,750	12,000
Washington	4,150	3,800	3,700	2,700	3,400	12,900	12,800	5,300	7,200
Oregon	20,600	17,950	19,200	22,400	22,400	18,850	21.300	18,850	24,800
California	12,600	6,600	9,300	14,000	5,900	—	15,500	13,600	—
United States	54,483	53,600	63,890	51,390	54,110	62,050	75,100	59,500	55,017

Source: *The Almanac of the Canning, Freezing, Preserving Industries.*

PROCESSING

Commercial Brines

Sulfite Brine Formulations. Brine composition was more or less a company secret until the late 1950s when several collaborative studies were undertaken to determine optimum quantities of each ingredient and its effect on dyeing characteristics and texture. However, experience often dictates changes in brine formulation and ingredient proportions. Some briners change the proportions of ingredients as the season progresses and as maturity and quality of fruit change. Others modify formulations for fruit from different areas and different cultivars in attempts to get the desired texture.

Commercial brines now in use can be divided into two general classes: (1) the older form, in which dissolved SO_2 is the source of the bleaching agent and (2) brine in which $NaHSO_3$ (sodium bisulfite) is the bleaching agent. Several methods of formulating calcium bisulfite brines are used: (1) bubbling sulfur dioxide gas into a suspension of calcium hydroxide; (2) bubbling sulfur dioxide gas into a suspension of calcium carbonate; and (3) dissolving calcium chloride and sodium bisulfite in water and adjusting the pH with commercial hydrochloric acid. The following three formulations have been used extensively in the western states and can be expected to give satisfactory brined cherries:

1. Add commercial hydrated lime [$Ca(OH)_2$] to water at the rate of 6 lb/100 gal. Stir well to form a suspension. Introduce 10.5 lb of sulfur dioxide gas into the lime slurry by means of a perforated tube or other submerged bubbling device. The brine will turn nearly clear when the proper amount of sulfur dioxide has been dissolved. The pH of this brine will be about 2.7 ± 0.2.
2. Add commercial whiting ($CaCO_3$) to water at the rate of 8 lb/100 gal. Stir well to form a suspension. Introduce sulfur dioxide gas into the slurry of whiting by means of a perforated stainless steel tube or other submerged bubbling device. Dissolve 10.5 lb sulfur dioxide/100 gal. to give a solution containing about 1.25% of sulfur dioxide. During the addition of sulfur dioxide gas, considerable bubbling will usually take place due to the formation of carbonic acid and evolution of carbon dioxide. Loss of sulfur dioxide may be minimized by keeping a floating lid on the surface of the tank in which the brine is made. The brine will turn nearly clear when the proper amount of sulfur dioxide is dissolved. The pH of this brine will be 2.0 ± 0.2.
3. Add 14 lb of anhydrous sodium bisulfite ($NaHSO_3$) or 12.5 lb of anhydrous sodium metabisulfite ($Na_2S_2O_5$) to 100 gal. of water with stirring. Add 5 fl oz of commercial hydrochloric acid. Add 7 lb of commercial

anhydrous calcium chloride and adjust the acidity to about pH 3.5 by adding more acid. This brine will contain about 1% sulfur dioxide and 0.85% calcium chloride (or 0.3% calcium ion).

The solution may become cloudy and much insoluble material may be precipitated if the ingredients are mixed in a different order than that indicated here.

Most cherries are brined considerable distances from the location of the plant where they are syruped and dyed. The trend is away from shipping in barrels to using plastic-lined tote bins and tank trucks.

Secondary Bleaching. Highly pigmented cherries such as 'Bing', 'Black Republican', 'Lambert', and 'Windsor' cultivars often are not completely or evenly bleached by sulfur dioxide brines. In addition, dark blemishes resulting from wind-whip, limb-rub, sunburn, and bruising are only partially bleached, providing an uneven background for the artificial coloring operation. These discolorations become more pronounced when the cherry is artificially colored and yield a final product with a lower market value.

Early attempts to remove blemishes by a secondary bleach were only partially successful. Wagenkneckt and Van Buren (1965) effectively removed the discolorations with hypochlorites as a secondary bleach; however, these bleaches often produce off-flavors and loss of texture. They also found that hydrogen peroxide bleached cherries to a uniform light yellow but the texture was unacceptable. Beavers and Payne (1968, 1969) successfully bleached brined cherries with sodium chlorite ($NaClO_2$), yielding a snow white product of firm texture and free of off-flavors.

In general, the process is to leach pitted brined cherries in cold water to a residual SO_2 content of 100-200 ppm. The secondary bleach is best carried out in an acidified solution of sodium chlorite (pH 6-4) and below 110°F. When the secondary bleaching is complete, the cherries are leached free of of sodium chlorite and returned to a calcium bisulfite brine. Beavers *et al.* (1970A) have outlined a slow and a fast method for commercial application of secondary bleaching.

The use of sodium chlorite for secondary bleaching of brined cherries as well as other fruits and vegetables is protected by U.S. Patent Application SN 700,389 assigned to Oregon State System of Higher Education.

Reclaiming Used Brines

Brine disposal is a major problem facing the brining industry today. The nature of the effluent-high in dissolved salts, soluble solids, and sulfur dioxide precludes putting it through a sewage plant without extensive pretreatment. The cost of this treatment is passed on to the processor and, in recent years, has amounted to several thousand dollars a month for large food manufacturers.

The most obvious answer to this problem would be to reuse the brine year after year. Brekke *et al.* (1966) found that brines, after adjustment of the sulfur dioxide content to the original level, could be used at least 3 years without adversely affecting the cherry texture. Beavers *et al.* (1970B), in a more extensive study, were able to dye cherries successfully if the dissolved natural pigments were removed from the brine with activated carbon. Cherries packed in reclaimed brine were of higher quality than the control. They also found fruit processed in reclaimed brine had fewer solution pockets, brine cracks, torn pitter holes, or soft fruit. Reconditioned brines should be tested for pectin-degrading enzymes; if present, these should be inactivated by heat.

Pectinolytic enzyme activity can be detected and evaluated by the method of Dingle *et al.* (1953). The brine may be inactivated by passing through a suitable heat exchanger.

Quality Control

The most important function of the quality control laboratory is to determine the sulfur dioxide content of brine and fruit. It is necessary to be able to analyze accurately for sulfur dioxide while formulating the brine and also to be able to follow changes in the sulfur dioxide content of the brine and cherries during the brine-curing process. Whether the brine is in a barrel, bin, or large tank, it is important to get a representative sample. Sampling large tanks is usually done after pumping the brine from the bottom to the top of the tank long enough to ensure complete mixing. Samples of brine are obtained from bins and barrels by a glass tube or a small bore stainless steel pipe. Cherry samples can be withdrawn from a barrel through the bung by means of a stout wire bent at an angle 2–3 in. from the end. The stems of the cherries are caught in the sharp bend and they can be lifted out through the bung. Samples from bins and tanks may be removed by a pronged device called a cherry rake. The end of the leaching cycle is determined by the sulfur dioxide content of the fruit.

It is also important to be able to follow the firming of the cherry. Any change from the firming curve shown in Fig. 9.1 alerts the processor to possible problems in the texture of the final product. Brekke and Sandomire (1961) measured texture with a Chatillon spring gauge; Beavers *et al.* (1970B) modified a Hunter Texture Instrument for measuring cherry firmness.

The quality control laboratory for dyeing and syruping operations must be able to determine pH, total and reducing sugar, sucrose content, and total acid and sulfur dioxide. Analytical procedures for these are in the

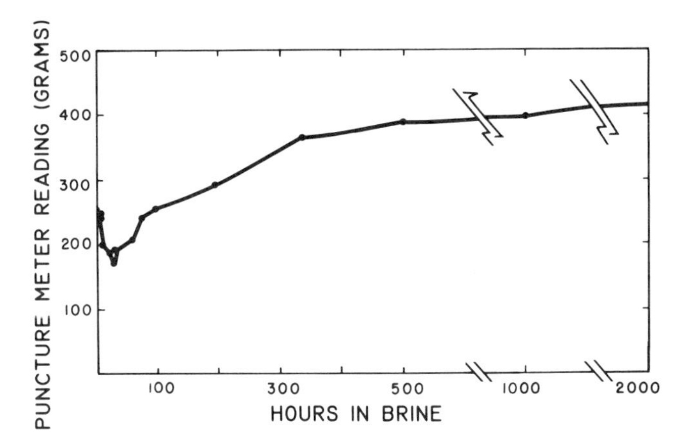

FIG 9.1. Changes in cherry texture during brine curing process.

Official Methods of Analysis (AOAC 1970). Atkinson and Strachan (1958) have published a bulletin on laboratory control for the candying process.

CHANGES DURING BRINE CURING

Normal Changes

In the first several days in brine, texture changes take place as shown in Fig. 9.1. The cherries soften rapidly for 2–3 days and then gradually

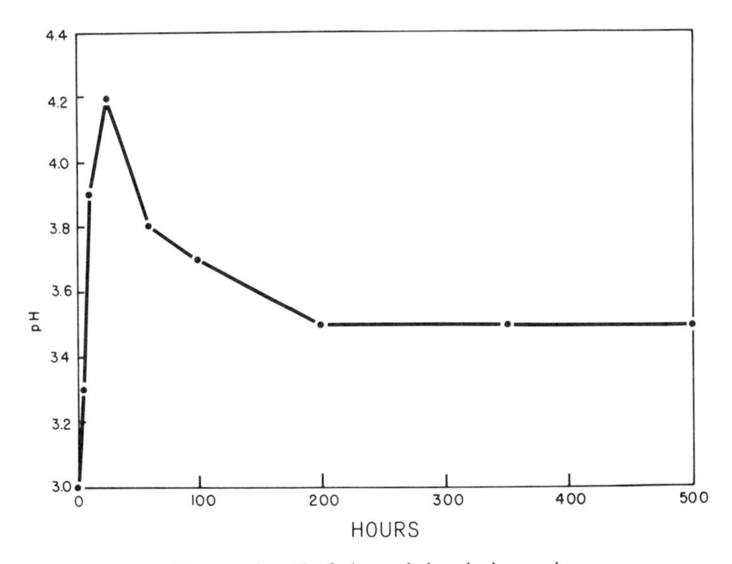

FIG. 9.2. Changes in pH of cherry brine during curing process.

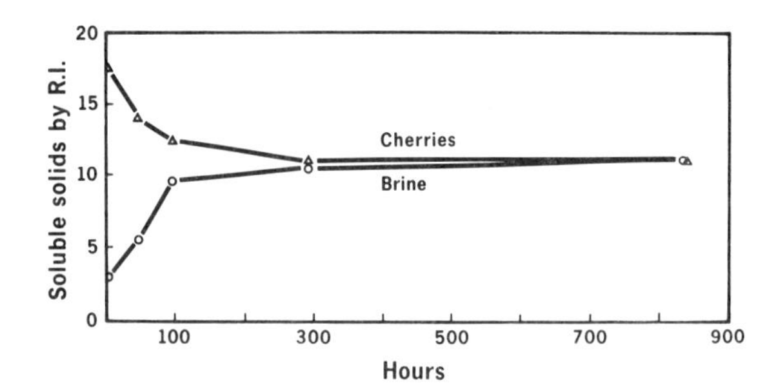

FIG. 9.3. Changes in soluble solids content of cherries and brine.

became firmer (Brekke *et al.* 1966). Concurrently, rather extreme changes occur in the pH of the brine (Fig. 9.2). The pH increases sharply for the first 20–40 hr and then decreases slowly until equilibrium is reached, usually in 300 hr. The soluble solids equilibrates between cherry and brine in 500–600 hr (Fig. 9.3). The sulfur dioxide reaches equilibrium between cherry and brine in 800 hr (Fig. 9.4) and continues to decrease slowly in the brine at ambient temperatures.

Normally, after 4–6 weeks in brine, the cherries are bleached to a light straw color and have become firm, ready for remanufacture into the final product.

Deteriorative Changes

Occasionally, cherries do not become firm in calcium bisulfite brine but become softer instead, which adversely affects the quality of the final

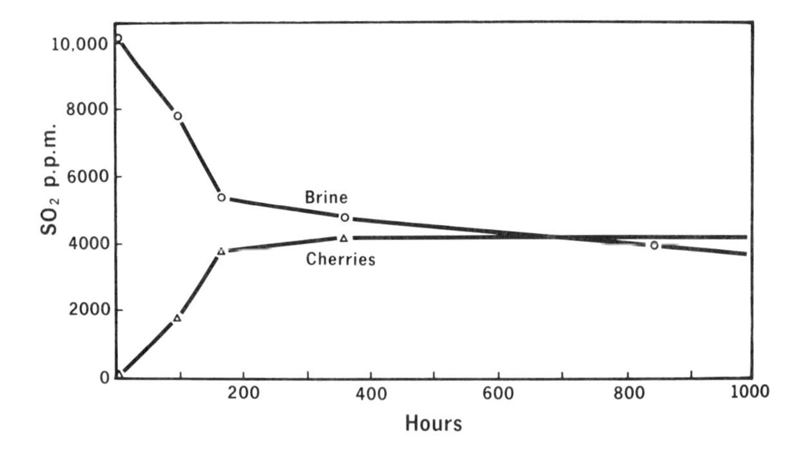

FIG. 9.4. Changes in sulfur dioxide content of brine and cherries.

product. Serious losses also occur from cracked skins and "solution pockets." As early as 1931, gaseous pockets were observed in brined cherries, followed by complete destruction of the cherry flesh. In 1935, Cruess (1935) reported on the splitting of brined cherries. Sporadic occurrences of cherry softening due to pectinolytic enzymes have been reported by McCready and McComb (1954). In 1957, softening occurred on a much larger scale, mostly in the western states. Additional studies further implicated pectic enzymes elaborated by certain fungi. Pectinolytic enzyme action in model systems was inhibited by alkylaryl sulfonate (Steel and Yang 1960), by adjusting the pH between 1.0 and 1.6 (Yang *et al.* 1960), or by adding extra calcium chloride to the brine (Brekke *et al.* 1966). Positive identification of the causative agents or their source has not been made.

Solution pockets result from internal rupture of the flesh, sometimes extending from the skin to the pit, clearly visible through the waxy skin.

Skin crack can be caused by improper brine formulation resulting in a too-acid brine. Internal pressure buildup due to osmosis can also cause rupture of the skin.

Factors Affecting Quality and Yields

Solution pockets occur more frequently as maturity increases and the harvest season advances (Beavers *et al.* 1971). Cherries held up to 12 hr before brining had fewer solution pockets than those brined immediately after harvest. A prebrine treatment with a solution containing 3% sodium chloride and 3% calcium chloride decreased solution pocket formation without adversely affecting texture and color of the final product.

Stebbins *et al.* (1967) found stemless cherries had neither solution pockets nor solution cracks when brined, probably because the soluble solids leach more quickly, reducing or preventing osmotic buildup. There is considerable evidence that these disorders are closely related.

Weight loss in brine, popularly called brine shrink, is decreased with additions of extra calcium chloride. Whittenberger *et al.* (1969) found that cherries brined immediately after harvest lose approximately 1% more weight than cherries brined 8 hr after harvest. Bruised cherries lose about 2% more weight in brine than unbruised cherries. Cherries with stems attached lose about 0.8% less weight than cherries without stems. Pitting loss and total loss decreased with the addition of extra calcium chloride to the brine and with increasing maturity. Van Buren (1965) reported that the commercial value of brined cherries is closely related to their color and firmness. As cherries mature, they yield increasingly darker and softer brined products.

Cherries processed in the presence of heavy metal ions, such as iron, copper, and lead, yield an undesirable, flat color in the final product. Kitson and Strachan (1955) found that both the artificial dye and the natural pigment were involved.

REMANUFACTURE

Maraschino, candied, and glacé cherries are prepared from sulfited cherries by leaching out the sulfur dioxide and then coloring with a certified dye. Leaching with water also removes other water-soluble components such as sugars, natural coloring matter, and flavor, leaving behind a firm, bleached ball consisting mostly of cellulose and pectin. The final process is a matter of coloring with a stable dye and adding flavor and sugar. The finished product is used chiefly for decorative purposes.

After the firming process is complete, the cherries are removed from the brine, then passed over a cluster breaker and through a mechanical stemmer (Fig. 9.5). Cocktail cherries are not stemmed. Figure 9.6 shows stemmers and inspection tables where cherries are inspected visually for blemishes prior to the pitting operations (see Chapter 2 for information on pitting cherries). In more modern, larger plants, electric sorters scan for cherries with discolorations, blemishes, and insufficient bleach and remove these from the processing line (Fig. 9.7). Lightly blemished fruit is diverted to a secondary bleaching line for further treatment. Heavily blemished fruit is dyed, syruped, and finely chopped for ice cream mixes.

FIG. 9.5. Brined cherry stemmer. (Courtesy Atlas-Pacific Engineering Co.)

FIG. 9.6. Modern cherry-brining plant with stemmers and inspection tables. (Courtesy Atlas-Pacific Engineering Co.)

Cherries for maraschino manufacture are graded into sizes of 2-mm increments from 14 to 24 mm. The pitted, graded fruit is now ready for remanufacture. If the fruit is not to be used immediately, it can be returned to the bisulfite brine. Total shrinkage from stemming, pitting, culling, and leaching may range from 22 to almost 35%. Losses vary according to the size and maturity of the cherry.

There are many variatios in the methods of leaching, dyeing, and syruping cherries. A processor may vary the procedure depending upon maturity, cultivar, and equipment available. Even the hardness of the water can affect the amount of acidulants necessary in some of the dyeing steps. Previously, by choosing a combination of dyes, a processor could obtain the particular shade of red that he considered best for the final product. Now, however, these red dyes have been disallowed by the FDA with the exception of Red No. 4 (Ponceau SX) which is limited to 150 ppm in the final product. Erythrosine dye is still on the GRAS list although it is considered by some manufacturers to be less desirable in that it does not impart a clear, deep red color to the cherry.

The following are general methods for bleaching, dyeing, and syruping.

FIG. 9.7. Electric sorters fed by pipe flumes. (Courtesy Atlas-Pacific Engineering Co.)

Atkinson and Strachan (1958) have published variations of these processes.

Candied Cherries

Leaching. Stemmed and pitted cherries are soaked for 24–48 hr in running water to remove most of the sulfur dioxide. The fruit is then boiled in several changes of water until the desired tenderness is obtained and the sulfur dioxide content is below 100 ppm. The pH of the fruit should be around 3.5–4.0. Cherries for fruit salad must be leached to a sulfur dioxide content of 20 ppm or below to prevent blackening of the tinplate of the can interior. Fruit properly leached for erythrosine dyeing has a pH of 4.1–4.4. Other dyes are absorbed more evenly and maintain a bright color if the pH is more in the acid range, around 3.5–4.0. Citric acid is used to adjust to a more acid pH.

Dyeing. When the desired level of sulfur dioxide is reached, the cherries are covered with the first syrup containing dye. Usually an initial syrup of 30% sugar is satisfactory. With fruit low in soluble solids, higher sugar concentrations may cause shrivel, which reduces the quality of the final product. All the dyes may be added to the first syrup, but with erythrosine it is necessary to use new neutral syrup to keep fruit in the pH range of 4.1–4.4. When the dye has penetrated sufficiently, the syrup is acidified with citric acid to pH 3.6–3.9. Cherries are dyed with erythrosine when they are used for fruit salad or in any application where the dye must be set.

Dye quantities vary depending upon individual requirements and variations in the degree of bleaching of the natural pigments.

Syruping. After the first 24 hr, the syrup strength drops to 17–20%. The syrup concentration on the cherries may be gradually increased at the rate of 8% per day. If the syrup temperature is maintained at 140°F, it will penetrate the tissue more easily and favor the formation of invert sugar, which helps prevent crystallization at high concentrations. Increasing the syrup concentration on the cherries is usually accomplished by evaporating some of the water in a vacuum pan evaporator or by simply pumping the syrup from the bottom of the tank through a heat exchanger and then back to the top of the tank through a fine spray. Candied cherries are brought to a final strength of 72–74% sugar. At this concentration it is necessary to maintain the proper ratio of sucrose to invert sugar to prevent crystallization. Table 9.2 gives these ratios for several temperatures.

The fruit must be drained sufficiently dry so that only a minimum of free syrup remains on the fruit. In large operations, syruping tanks are unloaded mechanically into draining trays which are placed in an air-blast dehydrator tunnel at 140°F. The tunnel is equipped with a tank or large pan to collect the syrup. The elevated temperatures and air blast aid in shortening the time necessary to remove the syrup from the fruit surface.

Table 9.2. Syrup Saturation Concentrations at Different Temperatures

Temp		Composition of sucrose and invert sugar solutions of maximum solubility (solutions saturated with both sucrose and dextrose)		
(°F)	(°C)	Total sugars (%)	Sucrose (%)	Invert sugar (%)
32	0	70.9	43.7	27.2
50	10	72.7	40.9	31.8
59	15	73.9	39.1	34.8
73.67	23.15	76.2	36.3	39.9
86	30	79.0	33.6	45.4
104	40	81.8	31.1	50.7
122	50	85.7	27.7	58.8

Source: Bates (1942).

The product is ready for retail packaging or storage after the dehydration step.

Where storage is necessary, the candied fruit is held in heavy syrup at room temperature or more preferably at 32°F. Sodium benzoate or sorbic acid is used at the legal limit where there is a possibility of the presence of sugar-tolerant yeasts.

Maraschino-Style Cherries

Cherries for the maraschino pack are leached, dyed, and syruped as for candied cherries. The syruping process is stopped when 48% sugar is reached in the fruit. The drained cherries are sorted, packed into desired containers, and covered with new 45% sucrose solution containing maraschino flavoring and sufficient citric acid to lower the pH to 3.6. If a preservative is not used, the containers are closed with a vacuum-closing machine and sterilized by heating until center temperatures reach 185°F.

Cherries for Fruit Salad

Cherries for canned fruit salad must contain no sulfur dioxide and have the dye sufficiently set to prevent coloring the adjoining fruit or the syrup. Sulfur dioxide in the leached fruit must be sufficiently low so that it disappears during subsequent processing. The pH of the cherry should be between 4.5 and 5.0. The fruit is added to a water solution of erythrosine (5–7 g/100 lb of pitted fruit) and remains there until the desired penetration is achieved. Application of heat accelerates the dye penetration. The dye solution is drained off, the fruit rinsed thoroughly, and then adjusted to a final pH of 3.0–3.4 by boiling in water acidified with citric acid.

RECENT DEVELOPMENTS

Some improvements have recently been made in brining cherries by a process that was patented in France and the United States. According to Julien (1981), after the fruit has been removed from the salt solution, the salt solution rather than being discarded is recovered. The salt solution containing sulfite ions is then treated with a reagent that causes sulfite to precipitate. The sulfite precipitate is then separated from the remaining salt solution, usually by filtration. Preferred reagents for precipitating sulfitions include calcium compounds (e.g., calcium oxide).

After the sulfite precipitate has been removed, the salt solution, which retains many of the natural sugars leached from the fruit, is subjected to a "debittering" step in order to decompose the amygdalosides such as amygdalin or Laetrile present. Amygdalin, present initially in the fruit stone and during the initial treatment of the fruit with the sulfite-containing solution, passes from the fruit into the salt solution and must be decomposed because it imparts an extremely bitter, unpleasant taste to the liquor. The debittering process is carried out in acid solution which decomposes amygdalin. Preferably, this acid hydrolysis is effected at a pH less than 3.8. Any conventional mineral acid or organic acid (e.g., hydrochloric acid, sulfuric acid, nitric acid, oxalic acid can be used for debittering.

After undergoing acid hydrolysis, the salt solution is then decolorized. The color of the solution before this step is generally brownish, due primarily to phenols present in the salt solution. Passing the salt solution over a bed of activated charcoal will remove phenols from the salt solution. The solution is then demineralized by passing it through a cationic exchange resin, fortified with acid, to remove mineral impurities.

The product obtained by this process is a purified sugar solution that is easily reusable and has a sugar content of 5°–6° Brix.

BRINING OF OTHER FRUITS

Cherries make up the largest tonnage of fruit preserved by brining in solutions of sulfur dioxide. However, during the war years of 1940–1945, large quantities of fruit, mostly berries, were preserved with sulfite brines for the Lend-Lease program (Woodroof and Cecil 1943A,C). Today, few or no berries are preserved in this manner in Canada or the United States. Sulfite preservation of berries is still important in some areas of Europe, however.

Appreciable quantities of citrus and pineapple are still processed in calcium bisulfite brine. The fruit is held in brine consisting of 1% sulfur dioxide and sufficient calcium ion concentration to give the desired texture in the final product. Leaching, candying, and dyeing are similar to the process described for cherries (Atkinson and Strachan 1958). Some manufacturers modify the number of leaching boils to remove the sulfur dioxide in order to obtain the desired tenderness. The syruping is more rapid than for cherries since there is no danger of shrivel.

REFERENCES

AOAC. 1970. Official Methods of Analysis. 11th ed. Assoc. of Official Agricultural Chemists, Washington, D.C.

ATKINSON, F. E. and STRACHAN, C. C. 1935A. Cherry processing. Part I. Sulphur dioxide treatment with special reference to a useful acid generator. Fruit Prod. J. *14*, 136–137.

ATKINSON, F. E. and STRACHAN, C. C. 1935B. Cherry processing. Part II. Leaching of sulphured stock. Fruit Prod. J.*14,* 174–175.
ATKINSON, F. E. and STRACHAN, C. C. 1936A. Cherry processing. Part IV. The effect of metals on color of glacéd cherries. Fruit Prod. J. *15,* 199.
ATKINSON, F. E. and STRACHAN, C. C. 1936B. Cherry processing. Part V. Glacéing of fruit by dehydration. Fruit Prod. J. *15,* 232–234, 249.
ATKINSON, F. E. and STRACHAN, C. C. 1936C. Invert sugar for candying fruit. Fruit Prod. J. *16,* 41–42, 53.
ATKINSON, F. E. and STRACHAN, C. C. 1940. Preservation of strawberry pulp with sulphur dioxide in British Columbia. Summerland Exp. Stn. Leaf., *F.P. 188,* Rev. 1940.
ATKINSON, F. E. and STRACHAN, C. C. 1941A. Candying of fruit in British Columbia with special reference to cherries. Fruit Prod. J. *20,* 132–135, 166–169, 185, 199–201, 217, 219, 229–232, 262–264, 289, 291, 310–312, 323, 324.
ATKINSON, F. E. and STRACHAN, C. C. 1941–1942. Preservation of fruits with sulphur dioxide in British Columbia. Fruit Prod. J.*21,* 5, 43, 72, 110, 141.
Atkinson, F. E. and STRACHAN, C. C. 1958. Preparation of candied fruits and related products. Part I. Can. Dep. Agric. Summerland Exp. Stn. Leafl.
ATKINSON, F. E. and STRACHAN, C. C. 1963. Sulphur dioxide preservation of fruits. Can. Dep. Agric. Summerland Exp. Stn. Leafl.
ATKINSON, F. E. and STRACHAN, C. C. 1962. Sulphur dioxide preservation of fruits. Can. Dep. Agric. Publ. *1176.*
BATES, F. S. 1942. Polarimetry, saccharimetry and the sugars. U.S. Dep. Commer. Natl. Bur. Stand. Circ. *C440.*
BEAVERS, D. V. and PAYNE, C. H. 1968. Upgrades brined cherries. Food Eng. *40,*(7) 84–85.
BEAVERS, D. V. and PAYNE, C. H. 1969. Secondary bleaching of brined cherries with sodium chlorite. Food Technol. *23,* 175–177.
BEAVERS, D. V. *et al.* 1970A. Reclaiming used cherry brines. Oreg. State Univ. Agric. Exp. Stn. Tech. Bull. *111.*
BEAVERS, D. V., PAYNE, C. H. and MILLEVILLE, H. P. 1970B. Procedure for secondary bleaching brined cherries with sodium chlorite. Oreg. State Univ. Agric. Exp. Stn. Inf. Circ. *632.*
BEAVERS, D. V., PAYNE, C. H. and CAIN, R. F. 1971. Quality and yield of brined cherries. Oreg. State Univ. Agric. Exp. Stn. Tech. Bull. *118.*
BREKKE, J. E. and SANDOMIRE, M. M. 1961. A simple, objective method of determining firmness of brined cherries. Food Technol. *15,* 335.
BREKKE, J. E., TAYLOR, D. H. and STANLEY, W. L. 1963. Determination of calcium in cherry brine by versenate titration. Elimination of anthocyanin interference by means of carbonyl reagents. Agr. Food Chem. *11,* 260.
BREKKE, J. E., WATTERS, G. G., JACKSON, R. and POWERS, M. J. 1966. Texture of brined cherries. USDA, Agric. Res. Serv. *74–34.*
BUTLAND, P. 1952. The darkening of maraschino and glacé cherries. Food Technol. *6,* 209.
CAMERON, H. R. 1966. Solution pockets in brined sweet cherries. Annu. Rept. Oregon Hort. Soc. *58,* 74.
CAMERON, H. R. and WESTWOOD, M. N. 1968. Solution pocket: A physiological problem of brined sweet cherries. HortScience *3,* 2.
CRUESS, W. V. 1935. Splitting of cherries in brine. Fruit Prod. J. *14,* 271.
CRUESS, W. V. 1937. Maraschino cherries. Fruit Prod. J. *16,* 263–265, 277, 279, 281, 283.
CRUESS, W. V., and NICHOLS-ROY, S. 1947. Some observations on barreled fruits. Fruit Prod. J. *26,* 292.

DEEDS, F. 1961. Summary of toxicity data on sulphur dioxide. Food Technol. *15,* 28.

DINGLE, J., REID, W. W. and SOLOMONS, G. L. 1953. The enzymatic degradation of pectin and other polysaccharides. II. Application of the "cup-plate" assay to the estimation of enzymes, J. Sci. Food. Agr. *4,* 149.

JACOBS, E. E. 1954. Tank trucks replace barrels. Food Eng. *26*(4)125, 189–190.

JACOBS, E. E. and TRUXELL, J. H. 1956. Mass production revolutionizes classic fruit process. Food Eng. *28,* 69.

JULIEN, H. C. P. 1981. Sulfite treatment of cherries. U.S. Pat. 4,271,204. Jan. 2.

KITSON, J. A. and STRACHAN, C. C. 1955. Metallic discoloration of candied fruits. Food Technol. *9,* 582–584.

LEWIS, J. C., PIERSON, C. A. and POWERS, M. J. 1963. Fungi associated with softening of bisulfite brined cherries. Appl. Microbiol. *11,* 93.

LEVIN, J. H., WHITTENBERGER, R. T. and GASTON, H. P. 1969. When to harvest sweet cherries mechanically. Mich. State Univ. Exp. Stn. Res. Rep. *87.*

MACGREGOR, D. R., KITSON, J. A. and RUCK, J. A. 1964. Sugar absorption in the fruit candying process. Food Technol. *18,* 109–111.

MCCREADY, M. M. and MCCOMB, E. A. 1954. Texture changes in brined cherries. Western Canner Packer *46,* 17.

MRAK, E. M., CAMPBELL, S. V. and ROYER, H. P. 1934. Barreling fruits in sulfur dioxide solution for export. Food. Ind. *6,* 507–512.

PAYNE, C. H., BEAVERS, D. V. and CAIN, R. F. 1969A. The chemical and preservative properties of sulfur dioxide solution for brining fruit. Oreg. State Univ. Agric. Exp. Stn. Inf. Circ. *629.*

PAYNE, C. H., BEAVERS, D. V. and CAIN, R. F. 1969B. Stratification in brined cherry tanks and its effect on quality. Oregon State Univ., Dept. Food Sci. (Unpublished.)

ROSS, E. 1949. Effect of temperature on brined cherries. Western Canner Packer *41,* 40,42.

ROSS, N. D. 1968. Relationships of cellulose and pectic substances to the texture of secondarily bleached cherries. M.S. Thesis. Oregon State Univ.

STEBBINS, R. L., CAIN, R. F. and WATTERS, G. G. 1967. Mechanical harvesting of sweet cherries in Oregon. Annu. Rept. Oregon Hort. Soc. *59,* 84.

STEEL, W. F. and YANG, H. Y. 1960. The softening of brined cherries by polygalacturonase, and inhibition of polygalacturonase in model systems by alkyl aryl sulfonates. Food Technol. *14,* 121.

STRACHAN, C. C. 1953. Can corrosion and discoloration tests with salad pack cherries. Can. Dep. Agric. Summerland Annu. Rept. *1953.*

STRACHAN, C. C. and ATKINSON, F. E. 1956. How to stop dye bleeding in cherries. West. Canner Packer *48*(9) 20–24.

TURNER, E. L. A better way to glacé fruit. Canner *85,* No. 15, 22–23, 26.

VANBUREN, J. P. 1965. The effect of Windsor cherry maturity on the quality and yield of brined cherries. Food Technol. *19,* 98.

WAGENKNECHT, A. C. and VANBUREN, J. P. 1965. Preliminary observations on secondary oxidative sulfited cherries. Food Technol. *18,* 658.

WEAST, C. A. 1940. Preparation of solution to be used in brining cherries. Western Canner Packer *32,* 26–27.

WEAST, C. A. 1941. Dyeing maraschino cherries with erythrosine. Fruit Prod. J. *20,* 332–333.

WHITTENBERGER, R. T., LEVIN, J. H. and CARGILL, B. F. 1969. Weight to volume relationship of sweet cherries in brine. Mich. State Univ. Agric. Exp. Stn. Res. Rept. *89.*

WHITTENBERGER, R. T. LEVIN, J. H. and GASTON, H. P. 1968. Maintaining quality by brining sweet cherries after harvest. Mich. State Univ. Agric. Exp. Stn. Coop. Ext. Serv. Res. Rept. *73.*

WIEGAND, E. H. 1946. The brined cherry industry. Western Canner Packer *38,* 60–63.

WILSON, L. A., BEAVERS, D. V. and CAIN, R. F. 1972. Structure of the Royal Ann cherry cuticle and its significance to cuticular penetration. Amer. J. Botan. *59,* 722.

WOODROOF, J. G. and CECIL, S. R. 1943A. Preserving fruits with sulfur dioxide to aid Lend-Lease Program. Food Ind. *15*(5) 59–61.

WOODROOF, J. G. and CECIL, S. R. 1943B. Good jams can be made from sulfur dioxide treated fruits. Food Ind. *15*(6) 67–68, 124–125.

WOODROOF, J. G. and CECIL, S. R. 1943C. Preserving fruits with sulphur dioxide solution. Fruit Prod. J. *22,* 132–135. 155, 166–169, 187, 202–205, 219, 221, 237–241, 253.

WOODROOF, J. G., CECIL, S. R. and THOMPSON, H. H. 1943. Improving the quality of preserves made from sulphited peaches. Fruit Prod. J. *22,* 269–272, 283.

YANG, H. Y., ROSS, E. and BREKKE, J. E. 1959. Suggestions for cherry brining. Oregon State Univ. Agric. Exp. Stn. Inform. Circ. *597.*

YANG, H. Y., ROSS, E. and BREKKE, J. E. 1966. Cherry brining and finishing. Oregon State Univ. Agric. Exp. Stn. Inform. Circ. *624.*

YANG, H. Y., STEELE, W. F. and GRAHAM, D. J. 1960. Inhibition of polygalacturonase in brined cherries. Food Technol. *14,* 644.

10

Other Products and Processes

J. G. Woodroof

In addition to canning, dehydration, and freezing there are scores of other methods of processing fruits commercially. Some of these involve the manufacture of primary products such as "Concord grape juice," "Catawba wine," "muscadine sauce," cranberry juice, orange juice, apple cider and other juices, pickles, spiced fruit, juice concentrates, preserves and cocktail fruit.

The majority of minor fruit products are secondary or by-products. These include jellies, marmalades, fruit butters, essences, flavor extracts, pectins, citric acid, tartaric acid, syrups, vinegar, peel thickeners, and fruit powders. In many cases several commercial by-products come from a single primary operation. Most such products are made from mixtures of cultivars and identified only as "apple jelly," "citrus pectin," "cherry essence," "blackberry syrup," "apple peel thickener," "orange powders," or the like.

Also, minor fruit products are often mixtures such as strawberry–apple jelly, cranberry–apple juice, refrigerated sliced fruit, citrus cooler (mixture of juice of four citrus fruits), cranberry–prune drink, and fruit punch (mixture of juice of 12 fruits).

Furthermore, fruit products are commonly mixed with synthetic or imitation colors, flavors, vitamins, proteins, minerals, thickeners, sweeteners, or texturizers. Examples are imitation jelly; apple drink containing caramel color, sodium citrate, fumaric acid, corn syrup, natural flavor, and water; orange breakfast drink containing water, orange juice pulp,

Commercial Fruit Processing, 2nd Edition

ISBN 0-87055-502-2

citric acid, sodium citrate, gum acacia, cellulose gum, orange oil, propylene glycol, and artificial color; isotonic drink containing orange, lime and artificial flavor, glucose corn syrup, citric acid, hydrogenated vegetable oil, cellulose gum, salt, sodium citrate, sodium orthophosphate, potassium citrate, calcium saccharin, and artificial color; milk–orange drink containing orange juice, milk protein and minerals and vitamins. Of the hundreds of fruit products that are already on the market, those discussed in this chapter are representative and typical. New products and processes are being developed constantly.

ASEPTICALLY PROCESSED FRUIT PUREE

Purees of peaches, pears, grapes, apples, bananas, oranges, apricots, plums, and other fruits have been aseptically packaged in metal and fiber barrels and drums for many years. The aseptic bulk storage system for acid fluid products was developed by utilizing a relatively new and advanced technology for aseptic canning (see Chapter 6). This bulk process involves pulping, concentration, preheating, deaeration, heating to achieve microbiological and enzymatic stabilization, and cooling prior to storage. Storage of the sterilized product can be done in drums or in large silos or tanks that have been previously sterilized by steam and/or chemicals.

The bulk storage system reduces the problems associated with seasonal processing, and with warehouses and rehandling costs. Transferring the product to portable units, such as railcars or truck tankers that are capable of maintaining aseptic conditions during shipment, reduces the costs of packaging and transportation. Due to rising energy costs and the development of bag-in-box packaging, there has been a renewed interest in this process, especially for tropical fruits.

Kanner *et al.* (1982) found that nonenzymatic browning, the main form of deterioration in these tropical fruits, was satisfactorily retarded at 12°C or lower. The rate of ascorbic acid destruction was dependent on temperature between 5° and 25°C and was affected by juice concentration. Orange juice concentrate of 58° Brix did not show flavor changes after storage at 5° or 12°C for 17 or 10 months.

Transoceanic shipments of tropical fruits from lesser developed areas to industrial markets will likely increase, since refrigerated transport and storage are essential. Presently purees of avocado, papaya, guava, and banana are in this class.

Chan and Cavaletto (1982) reported that during aseptic processing of guava puree there was virtually no loss of ascorbic acid (AA) and flavor, but significant loss of color. After 6-month ambient storage the AA loss was about 30% and further color changes and flavor losses occurred. Storage at 38°C for 3 months resulted in an AA loss of about 47% and losses in color and flavor.

For papaya puree, AA losses of about 6 and 56% occurred during aseptic processing and 6-month ambient storage, respectively. Color changes during aseptic processing and the first month of storage were characterized by a hypsochromic shift of the carotenoids' absorption spectra. After the first month of storage, further color changes were attributed to the products of nonenzymatic browning. Papaya flavor was stable during both aseptic processing and 6-month ambient storage. The flavor of papaya puree stored at 38°C for 3 months changed significantly, and AA retention was 39%.

To aseptically process bag-in-box papaya puree, the puree was acidified to pH 3.9 with citric acid and had a soluble solids content of 13.5%. It was then heated in a Cherry–Burrell Thermutator at 93°C for 60 sec before cooling in a scraped-surface heat exchanger to 24°C. A Scholle aseptic filler (Model Auto-fill X-1) was used to fill 1-gal multi-ply, metallized, polyester/evapolyethylene bags that were previously sterilized using gamma irradiation. The filler was first steam-sterilized at 120°C for 30 min after which sterile-heated air was introduced into the filling-head chamber. The heated air activated chlorine, which was constantly misted around the filling-head area. The chlorine mist provided a sterile wash for the outside of the bag while it was in the chamber. The bag cap was removed inside the chamber and the filling-head was then put into the bag spout. After the bag was filled, the filling-head was removed from the spout, and the cap replaced. The machine then ejected the bag from the filling chamber for placing in cartons. The same procedure was used for aseptic packaging of guava puree except the thermal process was 26 sec at 93°C.

Early in 1981, the FDA approved hydrogen peroxide (HP) for sterilizing materials used in aseptic systems. Since that time flexible, multilayer cartons have been rapidly replacing can and glass containers. It is projected that by 1990 more than 600 new HP aseptic fillers will be installed in the United States for packaging self-stable, single-serving fruit juices. Packages of all sizes from a few ounces to a half gallon will be used.

The quality of HP-packaged juices equals that of fresh or frozen juices and is superior to that of canned or bottled self-stable juices. The new HP-processing and -packaging technology gives the fruit juice industry an unparallel opportunity for both product and process innovation. Along with HP processing of single-strength juice will come new methods of sterilizing without heat (e.g., reverse osmosis). Also improvements will be made in concentrating juice without heat (e.g., freeze concentration).

The introduction of HP-sterilizing processes for flexible packages is one of the most meaningful breakthroughs in the fruit juice industry in five decades.

JELLY

Fruit jelly is one of the oldest and most popular fruit by-products. It may be made from underripe, undersize, and off-grade fruit, or even from peels, cores, and wind-fall fruit. The primary goal in making jelly is to obtain a product of uniform and desirable color, flavor, firmness, texture, and clearness.

Three essentials are pectin, sugar, and acids in the proper proportions. With some fruits (tart apples, black quinces, red and black raspberries, and blackberries), pectin and acid may be obtained from the fruit. In other cases (ripe apples and plums), the fruit contains sufficient pectin, but lacks sufficient acid. With most fruits, pectin or acid will need to be added. Sugar always is a necessary additive.

To achieve the proper pectin/sugar/acid ratio, the juices of underripe or overripe fruit may be blended, the juice of different fruits may be mixed, or artificial colors or flavors may be added.

For jelly making, the juice of crabapples, grapes, apples, currants, sour blackberries, lemons, sour oranges, and grapefruit are dependable sources of either pectin or acid. Strawberries, rhubarb, and apricots are usually sufficiently acid but may lack pectin; sweet cherries and quinces may lack acid, yet have abundant pectin. If no pectin-rich fruit is available to combine with one lacking in pectin, the latter may be added in the form of an extract or a commercial pectin. The viscosity of a fruit juice is an index of its jelling power.

Fruits give each jelly its characteristic flavor and color and furnish a part of the pectin and acid required for successful gels. They also provide mineral salts. Flavorful cultivars of fruits are best for jellied products because the fruit flavor is diluted by the large proportion of sugar necessary for proper consistency and good keeping quality.

Making Jelly

The fruit juices necessary for jelly making are obtained by boiling the fruit with water and later straining or filtering the solid pulp from the juice. The amount of water required depends on the type of fruit. Solid fruits such as quinces and apples require more water and longer heat treatment, even though they are cut up before heating; in comparison, berries need only slight heating and hardly any added water. Some fruits contain sufficient juice capable of extraction to justify two extractions. The duration of heating must be no longer than is absolutely necessary because flavors may be dissipated or the pectin may be hydrolyzed and converted to pectic acid.

Heating is usually carried out in steam-jacketed metal kettles, which are sometimes glass- or enamel-lined. On completion the hot mass is run into cloth strainers; more solid raw materials such as apples may be discharged into press cloths and pressed in a manner similar to that used for cider. The pomace may be moistened, broken up, and boiled, or pressed again for higher yields of juice.

Subsequent clearing of the juice is accomplished by numerous devices, including settling tanks, flannel jelly bags, mechanical pressure filters using pulp or kieselguhr as a filter medium, or mechanical centrifuges. After the hot juice has been filtered, it may go directly to the jelly kettles for immediate jelly making or it may be canned, bottled in large bottles; placed in kegs, barrels, or other large containers, and stored at low temperatures until needed.

When the juice goes to the kettles, sugar is added in amounts predetermined by experience or preferably by laboratory tests, in accordance with the acidity and pectin content of the juice. If the sugar content is too low, the resulting jelly is likely to be tough; excessive sugar results in a soft, easily broken final product. The pH of the mixture is of great importance and for satisfactory results should be approximately 3.2–3.5. Increased acidity lowers the amount of sugar necessary, but if the juice is too acid, "weeping" or syneresis of the resulting jelly is likely to occur. Fruit acids or acid fruit juices may be added if the acid content is below the optimum amount. Deficiencies in pectin may be encountered, in which case either a juice richer in pectin or a commercial pectin must be added.

The contents of the kettle are boiled to concentrate the mixture of juice, pectin, and sugar until the mixture will, on cooling, form a jelly of desired characteristics. During the boiling it is necessary to skim off the film or foam that forms on the top, as otherwise the appearance of the finished jelly will be impaired. Long heating at boiling temperatures may cause changes in color or flavor and lower the pectin content. Such changes may be minimized by use of vacuum pans and lower temperatures, in which case it may be desirable to add fruit acids such as citric acid or malic acid to facilitate the inversion of sucrose.

The point at which boiling should be stopped is of the utmost importance and may be determined in several ways. The use of a refractometer enables determination of the index of refraction, which is an accurate means of judging the end point. Thermometers or thermocouples may be used to indicate the temperature, which should be at approximately the boiling point of a 65% sugar solution (219°–221°F) when the process is complete. This determination is of value but should be checked by a sheeting test of the jelly with a spoon. Hydrometers may also be used to determine the specific gravity of the material for the same purpose. The maximum jelly strength is usually attained when the sugar concentration of the jelly is approximately 65–69%.

When boiling has been completed, jellies are run into the final containers while still hot. Glass jars are used for most jellies destined for the retail trade, while cans and pails may be the containers for institutions and the bakery trade. If the small containers are cleaned before use, the heat of the boiling jelly is usually sufficient to eliminate any spoilage microorganisms present. The high sugar content of jelly inhibits growth of most microorganisms encountered subsequently with the exception of molds. After the containers are filled, they should be handled as little as possible until the jelly has set because shaking injures the formation of the physical structure.

Standards of Identity for jellies may be obtained from the U.S. Food and Drug Administration, Washington, DC.

Jelly Failures

Crystals in Jelly. Sugar crystallization may result from (1) too much sugar, (2) too little acid, (3) overcooking of jelly, or (4) too long a delay in sealing the container of jelly.

Crystals are sometimes found in jelly because, in the boiling, syrup spatters on the side of the pan and dries; then, in pouring the finished product these crystals are carried into the glasses of jelly and the jelly becomes seeded with crystals.

Cream of tartar crystals in grape jelly may be greatly reduced, if not prevented entirely, by letting the juice stand for several hours in a cold place before making into jelly.

Cloudy Jelly. Cloudiness may result from imperfect straining and usually occurs with the red juices. Restraining of juice without pressure brings a lower yield but ensures a clear product.

Failure to Jell. An improper balance of pectin, acid, sugar, and mineral salts may prevent jelling. An improper balance may come about in several ways: (a) the fruits used may lack sufficient pectin or acid or both of these essentials; (b) overcooking may destroy so much pectin that a gummy mass is formed instead of jelly; (c) undercooking may result in failure to get a jelly formation because of insufficient concentration; (d) and too much water may have been used for the extraction of juice, so that the proportion of sugar is too great for the pectin and the long time required for evaporation may destroy some of the pectin.

Tough Jelly. Jelly becomes tough or stringy when too little sugar is used for the quantity of fruit juice used, or when boiling is continued after the jellying point has been reached.

Making Jelly from Concentrates

Jelly made from fruit concentrates rather than single-strength juices began about 1958 as a new use for concentrates, an economical procedure, and a food product with good color and flavor.

Continuous processing is used in making the jelly, with hot sugar syrup, cold concentrate, and cold pectin solution being pumped together in measured amounts, heated sufficiently for pasteurizing, and filled into glasses. Flavor essence (apple, grape, cherry, citrus, and other fruits) and citrus and other fruit acids are added as desired. The low temperature avoids an overcooked flavor that sometimes develops in long cooking of single-strength juices to reduce the water content.

The concentrates themselves are produced by low-temperature concentration under vacuum. They can be frozen and shipped to processors with lower storage and shipping charges than are associated with handling single-strength juices.

PRESERVES AND JAMS

Preserves and jams are similar to jelly except whole or large pieces of fruit are used for preserves and small particles of fruit are used in jams.

Jams

The following definition of fruit jams was issued by the FDA in 1936: "Preserve, fruit preserve, jam, fruit jam, is the product made by cooking to a suitable consistency properly prepared fresh fruit, cold-pack fruit, canned fruit, or a mixture of two or all of these, with sugar or with sugar and dextrose, with or without water. In its preparation not less than 45 lb of fruit are used to each 55 lb of sugar or of sugar and dextrose. A product in which the fruit is whole or in relatively large pieces is customarily designated a Preserve rather than a Jam."

This description covers jam made from all fruits, although small fruits such as raspberries and strawberries are used in larger quantities in this country. At one time, it was common practice to use sugar "pound for pound" in respect to fruit, and only cane sugar could be used without other declaration on the label, but this is no longer the case. When fresh fruit is available, it is used for jam production, but many jam manufacturers are too far from small fruit growing areas. In this instance, frozen cold-pack fruit, with some sugar added before the mixture is barrel-frozen, often serves as the raw material. Canned fruit may also be used if preferred.

When fresh fruit is used, the fruit should be of good quality and must be thoroughly washed and sorted to remove extraneous materials, stems, hulls, etc. Some of the larger fruits are peeled, cored, or stoned; or they may be cooked and when soft, run through pulping machines. When cold-pack fruit is used, it may be heated with water in the kettle in which the jam is to be made without additional preparation. Pectin is sometimes added to give body to the product; if cold-pack fruit is being used, the pectin should be added to the water and dissolved before adding the fruit. The acidity of the product is sometimes increased by the addition of fruit acids.

The exact amount of sugar required depends upon a number of factors, including the acidity of the fruit, the sugar content of the fruit, its maturity, and the type of product being made. If less sugar is used than necessary to bring the final product up to the standards cited above, the label must be modified to indicate properly the nature of the product.

After the preliminary crushing or other preparation of the fruit, the required amount of sugar is added and the mixture is heated in steam-jacketed kettles or preferably in vacuum pans if the fruits have delicate flavors. Heating is continued until the proper boiling temperature is reached, which will depend on the sugar concentration desired in the finished product. In general, the finishing temperatures range from 7° to 12°F above the builing point of water; at higher altitudes the boiling point of water is higher than at sea level (212°F) and therefore finishing temperatures are correspondingly higher. Care must be taken to prevent heating to the extent that the color or flavor of the fruit is affected, which will occur if heating is prolonged. The smaller the batch, the less likelihood there is of such changes.

Jams for domestic use are commonly packed in glass containers and usually have sufficient sugar content to preserve them against spoilage due to microorganisms. However, they may be heated in the container after sealing of it is necessary. Heating in containers may cause undesirable changes in color and flavor just as readily as these occur in the kettles, so the containers should be cooled as soon as possible. Lower-grade jams used for bakery products are frequently packed in large wooden pails. In this case the problem of avoiding losses in color is likely to be of greater concern unless cooling is carried out rapidly.

Preserves

Fruit for preserving should be in a firm-ripe rather than a soft-ripe stage. It should be uniform in size or in uniform pieces so as to cook evenly.

Preserves are made from practically all fruits including peaches, pears, plums, strawberries, grapes, muscadines, quinces, and tomatoes. By using up to 25% of firm-ripe fruit the tartness is increased and less pectin is required in the formula (DuPree *et al.* 1953).

Following are formulas for making two of the more unusual types of preserves, from figs and watermelon rinds.

Fig Preserves

Select 6 qt firm, sound fruit, discaring all overripe or broken figs. Sprinkle 1 cup baking soda over the selected figs and cover with the 6 qt of boiling water. Allow to stand for 15 min, drain off soda solution and rinse figs thoroughly in clear, cold water. Let figs drain while the syrup is prepared. Make syrup using 8 cups sugar and 2 qt water; boil 5 min and skim. Then add well-drained figs gradually so as not to cool syrup. Cook rapidly until figs are clear, about 2 hr. Lift figs out carefully and place in shallow pans.

If syrup is too thin, continue boiling until it reaches desired thickness, then pour it over the figs, being careful that figs are entirely covered. Let stand overnight. Next morning pack the figs cold in sterilized jars, placing figs so that all stems will be upward. Fill jars with syrup. Process at simmering temperature for 15 min.

Watermelon Rind Preserves

Prepared rinds	10	lb
Sugar	17½	lb
Citric acid and imitation lemon	1	oz
(or sliced lemons)	(1	lb)
Water	2	gal

Firming watermelon rinds is necessary to produce the desired crispness, otherwise the preserves will be spongy, soggy, and sometimes leathery. Calcium hydroxide is preferred to other calcium salts (carbonate, oxide, chloride, sulfate, or monophosphate) for firming the peeled, sliced rinds, because of its rapid firming action, low cost, and lack of imparted flavor when properly used.

Treating the rinds in 0.4% calcium hydroxide solution for 4 hr at 75°F gives excellent results. More rapid firming occurs by agitating a saturated solution (about 0.15%) for 30 min. Freshening the rinds consists of thoroughly removing the excess calcium salt from the rinds by washing and soaking in a continuous flow of water for 45 min. Next the rinds are parboiled to produce the desired texture, color, and translucent appearance for the rinds to "take the sugar" in the next operation.

Boil rinds, sugar, and water moderately for 1 hr. Add citric acid or sliced lemons and continue simmering for 30 min, or until the rinds are light in color, and the syrup reaches 224°F, or a soluble solids of about 65%. Transfer to sterile jars, seal while hot then cool quickly.

CANDIED FRUIT

Candying is one of the oldest methods of preserving fruits and antedates the manufacture of refined sugars. Doubtless, the first candied (glaze) fruit was soaked and heated in honey. Preservation is by dehydration of the fruit by osmotic pressure of the sugar solution. Candied fruit differs from fruit candy only in the relative amount of fruit to sugar; and it differs from canned fruits only in the relative amounts of fruit to water.

Ingredients

Candied fruit, peels, and mixes are unique in that much of the finished product is tailor-made as to size, color, flavor, special ingredients, and methods of packaging. Processing methods are unique and often call for custom-made equipment. Thus, most of the products are specialities made according to the customers' needs.

About 70% of the total production is for the institutional trade, particularly bakers and confectioners. Supermarkets are outlets for products such as glacé cherries, pineapple, citron, citrus peels, and mixes in clear plastic containers with white polyethylene lids. Economy, 1-lb plastic bags of individual fruits or mixes are available. While these are in demand in many areas, most retail items are packaged in reusable, rigid containers.

Many of the raw materials for candied fruits come from around the world. Typical examples of supplies and sources are ginger from Australia; cherries from France, Oregon, or Michigan; pineapple from Mexico, Hawaii, Malaya, Puerto Rico, Bahamas, and Venezuela; citron from the West Indies; cinnamon and other spices from Pacific Islands; pears from Oregon; apples from Washington state; cranberries from Massachusetts; and lemon and orange peel from Italy, Florida, and California. Since much of the processing is in Florida or California, the largest volume of most of the products is from these states.

Sugar is the second most important ingredient. (The Paradise Fruit Company, Plant City, Florida, uses more than 10 million lb/year.) Some of it is in liquid and some in granular form, some is cane and some is corn sugar. The Paradise Fruit Company has vats for 300,000 lb of fruit and various liquid formulae used in processing.

Preparing the Fruit

The initial steps in the preparation of candied fruit are the same as those for canning the respective fruits. Cultivar and maturity are important, as well as freedom from defects and bruises. To avoid excessive softening and to preserve the fresh aroma and flavor, heating and freezing as a means of preservation from the point of production to the candying plant are avoided. Sulfur dioxide and other chemical preservatives are used instead. Cherries are pitted and brined, as described in Chapter 9. Brining with sulfur dioxide and firming with calcium is also followed with strawberries, raspberries, citron peel, and other berries. The same general procedure is followed with sliced apples, pears, pineapple, and other fruits to be held for more than a few days from harvesting to final processing.

Firming Cherries. Preheating of freshly pitted cherries at about 140°F for 5–20 min promotes firming sufficiently to permit pasteurizing the fruit in bulk without excessive tearing. Improved texture from this pretreatment is shown in terms of increased extrusion force, drained weight, and bulk volume. Addition of calcium up to 0.04% of final product leads to 50% greater firmness. Addition of calcium up to 0.04% of final product leads to 50% greater firmness. Hot-fill temperatures of 180°F or higher are required to protect color and flavor from enzymatic degradation. While firmness or chewiness of the fruit is notably greater in the preheated, hot-filled product, the conventional cold-fill, exhaust and retort process provides superior drained weight and bulk with about equal color (LaBelle 1971).

Although the details of the preparation of each fruit product for candying vary, thc following procedure is followed in Florida for citrus peels. The fresh peels are received from canners and processors in the immediate vicinity of the production area. Upon arrival the first step is to feed the peel onto a conveyor that leads to a washer. As peels emerge from the washer, they are hand-checked for color and defects.

Peels are then diced and sifted into polyethylene-lined barrels and cured with sulfur dioxide. During this process, known as brining, the barrels are stored in a temperature-controlled warehouse. Samples are removed regularly and checked by laboratory technicians for sulfur dioxide content, pH, and general condition of cell structure. When the proper stage of brining has been reached, the peels are heated and agitated to thoroughly remove the brining solution. Barrels of brined fruit are moved from the storage area to washing tanks with motorized dump lift vehicles.

Color is introduced into cherries and pineapples immediately after debrining and washing instead of with syrup. Additional color may be added in the syrup if desired. The precise amount of color for each fruit depends upon customer specifications. The acidity of the particularly fruit greatly influences the color. The fruit is automatically pumped into the glazing vats for processing. The following general procedure applies to many candied fruits:

Fruit	10 lb
Sugar	10 lb
Water	5 lb
Light corn syrup	1 lb
Cinnamon, cloves, ginger	Trace
Color	As desired

Boil all ingredients except the fruit together to 234°F, or until the syrup spins a thread when dropped from the paddle. Add whole or sliced fruit or peel, and simmer without crowding until transparent. Remove from syrup, drain, roll in granular sugar, allow to dry until pieces no longer stick together, seal in moisture-proof containers, and store in cool place.

Prepare Tender Fruits: Prepare apples, berries, cherries, peaches, pears, and plums by washing, peeling, coring or slicing, and grading, and add directly to boiling syrup.

Prepare Less Tender Fruits: Prepare cranberries, figs, pineapple, and quinces by washing, paring or slicing, and simmer in clear water until tender; then transfer to boiling syrup until transparent.

Prepare Fruit Peels: Cut citron, grapefruit, lemon, and orange into quarters or large pieces; let stand 12 hr in 2% salt solution; drain and wash thoroughly; simmer until tender in clear water, changing water several times until there is no bitter taste; drain; cut into strips or cubes; transfer to boiling syrup; and boil until tender.

After glazing, fruit that is not shipped immediately is stored in polyethylene-lined drums where it may be held for many months. All candied fruit for retail market is hand-packed because no satisfactory filling equipment has been developed.

Candied fruits are very sensitive to high humidity and high temperatures. They should be held at 50% RH and below 50°F.

The time required for candying fruits such as cherries, apricots, plums, prunes, and jujubes may be shortened by preliminary immersion of the fruit in an aqueous solution of the enzymatic composition produced by cultivation of *Aspergillis niger,* and then carrying out the regular candying procedure. This patented process (Mochizuki 1971) reduces processing time from 21 to 5 days.

PICKLED FRUIT

Sweet pickled fruit includes dozens of products—peach pickles, pear pickles, spiced pears, spiced grapes, spiced muscadines, spiced plums, spiced crab apples, spiced currants, pear–tomato chutney, apple–tomato chutney, and many others.

Peach Pickles

One of the most popular fruit pickles is made from peaches, and the demand is far greater than the supply. While formerly peach pickles were made from undermature, small, freestone fruit from the grading tables of fresh or canned fruit, since about 1960 they have been made only from nonmelting, medium-size, small-seeded, clingstone peaches. Typical peach pickle cultivars set a heavy crop, are seldom thinned, and bear fruit too small for canning or shipping fresh. For this reason, growers are reluctant to grow large quantities of peaches for pickling. Peach pickles are made in California and Georgia.

Peaches for picking are picked in the firm-ripe stage and graded for freedom from defects and uniformity of size. They are peeled by dipping in boiling 2% lye, rinsed with water, and dipped in 1% citric acid solution. Pickling operations should be begun immediately.

Peach Pickles

Peeled peaches	10	lb
Sugar	4	lb
Vinegar	1	qt
Cloves, whole	1½	Tbsp
Stick cinnamon	4	2-in. pieces

Combine sugar, vinegar, and spices (in bag or tea ball so as to remove after cooking). Bring to a boil and let simmer for 30 min. Drop peeled peaches in boiling syrup for about 5 min, or until they are translucent and tender. Pack in jars or cans, cover with hot syrup, exhaust for 6 min, seal and process in boiling water for 20 min, cool in water bath. Yield 6 qt.

Sometimes apple cider vinegar or brown sugar is used to darken the color of the pickles. Regular sweet peach pickles are made with 50° Brix syrup, with a cutout of about 38° Brix; real sweet pickles are made with 65°–67° Brix syrup with a cutout of about 40° Brix. With the heavier syrup peaches shrink about 25%. Shrinkage is less with lighter syrups (Thrash 1968).

Peach pickles tend to float in the syrup; but this decreases as they equalize and is entirely overcome if the seal of the containers is broken and the product placed in a refrigerator overnight.

Peach pickles are excellent served with a variety of entrees. Later the syrup may be utilized in many ways; in barbecue sauce, basting ham or poultry, potato salad dressings, cole slaw, cheese dip, meat and vegetable sauces, etc.

Pear Pickles

The preparation and uses of pear pickles are similar to those for peach pickles. Sickel, Kieffer and other "sand pear" types are used whole and usually with stem attached. The firm-ripe pears are graded for defects and uniformity of size, peeled by hand or with boiling lye, washed and rinsed in weak citric acid solution, and the pickle-making process is begun.

Pear Pickles

Peeled pears	10	lb
Sugar	3	lb
Vinegar, white	1	qt
Cloves, whole	2	Tbsp
Stick cinnamon	10	2-in. pieces
Allspice, whole	2½	Tbsp
Water	1	qt

Combine sugar, vinegar, water, and stick cinnamon; add cloves and allspice in a bag or tea ball; bring to boil and simmer 30 min.

Add pears to boiling syrup and continue to simmer for 20 min; pack hot in glass or tin containers; seal and process 20 min in boiling water (Thrash 1968).

Spiced Fruit

Spices and ginger are frequently added to preserves and jams for special uses. The flavor of spiced grapes blends well with the flavor of white meat such as pork, chicken, and turkey.

Spiced pears are a delicacy having characteristics of both pear preserves and pickled pears. They are popular for buffets, to be eaten with meats, vegetables, salads, or alone.

Spiced Pears

Peeled pears, quartered	10	lb
Sugar	5	lb
Vinegar	1	qt
Stick cinnamon	10	2-in. pieces
Allspice, whole	2½	Tbsp
Ginger root	2	oz
Lemon, sliced	1	

Peel pears by hand (or in 2% boiling lye and rinse), dip in weak citric acid solution, quarter and core. Mix sugar, vinegar, lemon, and spices (in a tea ball) and bring to simmer, add pears and bring to boil. Boil slowly until fruit is clear and tender. Pack gently while hot into sterile jars, cover with hot syrup that has been boiled down to half volume, seal, and cool with water spray or bath.

A crisper texture results if the cooked pears are allowed to cool overnight in the syrup and reheated the following day. The texture is further perfected if this procedure is repeated 2–3 times (Thrash 1968).

Spiced Muscadines

Spiced Muscadines are a southern specialty that are used: on hot biscuits, rolls or bread, as marmalade; in sandwiches as jelly; with fowl, or pork, as cranberry sauce; and with vegetables on a pickle tray. They are made according to the following formula:

Deseeded grapes	50	lb
Sugar	45	lb
Vinegar, 40-grain	4½	lb
Powdered cinnamon	2½	oz
Powdered mace	1½	oz
Clove oil (4.9 ml)	1	tsp

Boil grapes for 15 min, or until the hulls are tender. Add sugar and cook until thick. Add spices and vinegar and cook until product gives a light jell test (219°F). Pack hot in sterile jars, seal and cool.

MARMALADES

Marmalades have the characteristics of jellies and preserves combined. They contain the fruit pulp and may contain the skins, suspended in jellied juice. They are made from underripe fruit, rich in pectin and acid, chiefly citrus fruits, alone or in combination with other fruits. Popular marmalades are combination citrus, orange–peach, orange–pear, ginger–pear, pear–pineapple, and grape.

CONSERVES

Conserves are similar to jams with chopped nuts (pecans, walnuts, or others) added for texture and flavor. Conserves are mixtures of two or more fruits with the chief ingredient in specific conserves being figs, peaches, pears, plums, oranges, or carrots. Conserves contain higher proportions of fruit than preserves or marmalades. Although formulas vary to suit individual tastes, the following formula (in pounds) for grape conserve is typical:

Grape pulp and hulls, ground	36
Sugar	12
Raisins, ground	6
Oranges, chopped	9
Pecan meats, chopped	3

Boil the grape pulp and hulls about 15 min, or until tender. Add sugar, raisins, and finely chopped oranges. Cool slowly until moderately thick, add chopped nuts, and boil for 5 min. Pack hot in sterile jars, seal, and cool quickly.

FRUIT BUTTER

Fruit butters are made by cooking the pulp of any fruit to a smooth consistency, thick enough to hold its shape but soft enough to spread easily. Fruit butters differ from jams in that the product is pressed through a coarse strainer and is more concentrated. Since butters are very thick, special care is needed to avoid scorching.

Butters may be made from grapes, peaches, plums, mayhaws, and other fruits, but apple butter is most popular. It is made from tart apples that have been washed, cored, and usually peeled. The apples are placed in a kettle with just enough water to cover and cooked until tender. After being put through a sieve, sugar is added at the rate of ⅔ unit of sugar to 1 unit of fruit. For more flavor, cider, cinnamon, and mace are added as desired. The mixture is cooked to a heavy consistency, packed in hot sterile jars, sealed, and cooled quickly.

FRUIT LEATHER

Fruit leathers (known commercially as fruit rolls) are manufactured by the dehydration of fruit purees into leathery sheets. They are eaten as a confection or cooked as a sauce.

Fruit leather can be prepared from most fruits (e.g., papaya, guava, peach, plum, and grape) that have high sugar and flavor and low fiber. It is a fruit by-product made from culls and overripe fruit, which must be available in abundance. Since fruit leather is usually poorly packaged and unrefrigerated, it loses color and flavor gradually and quality standards are difficult to maintain.

Chan and Cavaletto (1978) have described a procedure for making papaya leather, as follows. Papaya puree was prepared by steaming the whole fruit 1 min; slicing; separating flesh, skin and seed; pulping; acidifying to pH 3.5; finishing and heat-inactivating enzymes; adding 10% sugar and sodium bisulfide ($NaHSO_3$) to 552-1105 ppm SO_2. The puree was spread in teflon-coated pans; dried in a forced-draft oven at an air velocity of 110 fpm at 84°C to 12% moisture; and wrapped in plastic film and stored at −18°C. Drying rate was decreased by high SO_2 levels. The color of the finished leather depended upon the drying and storage temperature and presence of SO_2. Residual SO_2 declined at high or long storage temperatures. Sensory color, flavor, and aroma were correlated with the alcohol-soluble color index.

FRUIT EXTRACTS

The flavor of natural products are due not to specific chemical entities but result from a very complicated blend of aromatic chemicals produced in the ripening process. The identity of all constituents of fruit flavors has only partially been established, but we know that these natural flavor constituents cover the complete range of aliphatic, terpene, and aromatic chemicals and can be classified as acids, alcohols, aldehydes, hydrocarbons, lactones, and ketones.

The principle of volatile flavor recovery was first applied in the manufacture of Atlas Genuine Fruit Extracts. Alcohol is used as an aroma carrier which permits distillation and recovery of flavor constituents at low temperatures. The fruit residue containing pulp and soluble fruit solids is pressed for removal of insoluble solids. Juice obtained from the pressing operation is depectinized, clarified, and concentrated under high vacuum to preserve color. This concentrated juice is combined with the alcoholic distillate to yield finished genuine fruit extracts.

Concentrated fruit extracts represent the strongest extracts on the market today. They are used principally by extract manufacturers and industrial users who require the highest possible concentrations. They have become the standard of quality for the food and flavor industry due to their outstanding flavor characteristics. This has been achieved by the selection of the most highly flavored fruits, which are picked at the optimum stage of maturity. Cultivars chosen are selected for top flavor.

Extra concentrated fruit extracts can be reduced 50% to equal the flavor strength of commercial true fruit extracts and can also be used as bases to produce fruit extracts with other natural flavors or imitation flavors containing a true base. Extra concentrated fruit extracts are made from black raspberry, red raspberry, cherry, strawberry, blackberry, peach, apricot, pineapple, grape, and loganberry.

ESSENCES

Apple Essence

One of the first fruit essences to be trapped, separated, and used commercially was that of the apple. It is now recovered in large quantities, filtered to remove about 98% of the yeast cells and mold organisms, and used in candy jellies, household jelly, preserves, beverages, and bakery products.

The fresh juice is pasteurized, filtered, treated with ascorbic acid to preserve color and flavor, treated with sorbate to further reduce organisms, and distributed in bulk to dairies and beverage companies for packaging in plastic or paperborad containers. Apple essence is added to enhance apple flavor. The finished juice is bottled semiasceptically and refrigerated; it has a shelf life of a month or more.

The essence from different apple cultivars varies widely. The flavor potency of some lots of apple juice is as much as 40 times that of other lots.

Cherry Essence

Commercial-grade 150-fold cherry essence, a flavor concentrate, is produced by stripping and rectification of the juice from 'Montmorency' cherries. The degree of vaporization of the juice is 25%. The essence has a strong, pleasant, "cooked cherry" aroma typical of pasteurized cherry juice. It is shipped in 1-gal polyethylene bottles and stored at 34°F until used.

Ethanol and methanol are the most abundant low-boiling substances, estimated to comprise 9 and 0.5% of the essence, respectively. The next most abundant compound is acetaldehyde, estimated at 0.029% (Stinson *et al.* 1969).

Orange Essence

To recover orange essence, peel is finely ground and diluted with enough water to make a slurry. About 20% of the slurry is evaporated and the vapors conducted to an essence recovery unit. The resulting condensate is part citrus oil and part a solution of essence in water.

CITRUS PEEL OIL

Citrus peel oil, along with the essential oils of many fruits, is widely used to add flavor, aroma, and character to beverages.

Citrus peel oil is the essential oils of oranges, lemons, limes, tengerines, and grapefruit. It is the first product recovered from cannery or juice manufacture waste. The oils are contained in oval, balloon-shaped oil sacs or vesicles located in the outer rind (flavedo) of the fruit adjacent to the chromoplasts. The oil sacs are located irregularly at different depths in the flavedo of fruits. These glands or saclike intercellular receptacles have no walls, but are bounded by the debris of degraded tissues. The cells surrounding the oil glands contain an aqueous solution of sugars, salts, and colloids and exert some pressure on the glands.

To secure the oil from the peel of citrus fruits, oil sacs must be punctured by either pressure or rasping. There are several methods of expressing citrus peel oil, but all of them give an emulsion of oil and water. The oil is separated centrifugally from the aqueous phase by passing the emulsion through a sludger and then through a polisher. Following separation, the oil is stored for approximately 1 week at 32°–40°F; during this winterizing treatment undesirable waxy materials separate from the oil and settle. The clear oil is decanted into stainless steel storage tanks, which are maintained at a temperature of about 40°F. Air is usually excluded from the container to prevent deterioration. This is accomplished either by filling the tanks full or by displacing the air with an inert gas.

Some of the citrus peel oil becomes mixed with the juice as it is extracted by various types of juice extractors. Excessive amounts of peel oil in the juice are harmful to the quality of the juice; therefore, in most canning plants the oil content of the juice is reduced to a desirable level by passing the juice through a deoiler. The juice is usually flashed in the deoiler, which is operated under a vacuum of 11 in. Hg (190°F) to 25.5 in. Hg (130°F), and a vapor mixture of oil and water is removed. Then the mixture of oil and water vapors is condensed and the oil is separated from the condensate by decanting or centrifuging. Vacuum steam distilled oils manufactured in this manner have properties slightly different from oils obtained by steam distillation at atmospheric pressure.

Essential oils are composed of mixtures of hydrocarbons, oxygenated compounds, and nonvolatile residue. More than 14 compounds have been identified. The hydrocarbons are primarily terpenes; the oxygenated compounds are made up of a variety of compounds—aldehydes, esters, acids, alcohols, ketones, ethers, and phenols. Nonvolatile residues consist of resins and waxes. The oxygenated compounds, the principal flavoring agents of citrus oils, may constitute from 1.5 to 6.5% of the original oil.

Cold-pressed orange oils subjected to atmospheres of either nitrogen or hydrogen at 105°F undergo a chemical reaction evolving gas, but apparently this is not harmful to the aroma or color of the oil. Autoxidation is the primary cause of the "terpeney" character that develops in orange oil. The stability of the oil is also related to the quality of oil extracted from the peel.

It has been shown that antioxidants such as BHA or BHT effectively stabilized D-limonene and orange oil by retarding oxidative deterioration of the oil. BHA showed slightly better results.

Chemical and physical properties of citrus peel oil vary with the fruit cultivar, degree of maturity, seasonal variation, storage of fruit, and method of extraction. The aldehyde content of the oil is largely determined by the quantity of aqueous phase that comes in contact with the oil during processing.

AROMATICS FROM CITRUS FRUITS

Demand is increasing for essence-like aromatic materials for use in citrus products to improve fresh juice-like aroma and flavor. During the manufacture of frozen concentrated orange juice, many commercial processing plants remove fresh juice essence and reincorporate it into their final product. Future demand is expected to increase for citrus juice products such as blends, juice drinks, and dehydrated formulations. Some products will require higher levels of aromatic flavoring components than would normally be obtained from their equivalent content of fresh juice. In addition to fresh juice essence, other sources of flavoring components will be needed for manufactured and formulated products. Various fruit parts, unusual cultivars not suitable for juice processing, and certain processing waste streams are potential sources for such additional components.

Another important consideration of high current interest is the disposal of liquid waste from citrus processing plants. One source presenting particular problems is the peel-oil mill, where 1–3 gal of liquid waste are generated for each 90-lb box of fruit processed. Although only about one-third of the citrus peel is processed for oil recovery in this manner, this waste poses difficult disposal problems because it is about 2° Brix and may contain 0.5–2.5% oil. This oil adversely affects fermentation and bacterial decomposition.

Flash evaporation of apple juice in a tubular heat exchanger under turbulent conditions followed by fractionating in a packed column at atmospheric pressure has produced volatile flavor concentrates 100 times as strong as those in the original juice. Morgan *et al.* (1953) investigated the recovery of essence from orange juice under vacuum at 110°–115°F. Such essence enhanced the floral character of reconstituted juice, but peel oil was also required for full flavor. Other methods have included the recovery of essence from the first stage of a low-temperature evaporator and the use of a pump with a liquid seal to recover vapors. The most common method currently used in the citrus industry is the recovery of essence from the first stage of a citrus juice evaporator, operating at 210°F and at atmospheric conditions. This method could be extended to waste liquid streams (Veldhuis *et al.* 1972).

EXTRACTING RAW GRAPE JUICE

California's wine and grape juice industry, interested in extracting the most juice at minimum operating costs, has turned to the interrupted reducing-pitch horizontal screw S-press. This press has been used successfully with citrus peels and other slippery materials, and those that tend to pack hard.

Typical yields with several cultivars range from 190 gal/ton of 7.2% solids juice from 'Palomino' and 'Chenin blanc' grapes to 180 gal /ton of 8.0% solids juice from 'Thompson Seedless' grapes. Operating costs to achieve these yields are reduced because of lower horsepower requirements for comparable capacity machines.

In operation, free liquid is drained immediately upon the material's introduction to press through the inlet hopper's 180-degree drain screen. Material is then moved through the unit by rotating action of the graduated-pitch screw whose flight pitch gets shorter with each turn to increase pressure on product and allow for smaller volume as juice is pressed out. Moving material is additionally compressed by interaction of screw resister bars that exert additional pressure and inhibit premature formation of cake. Bars also turn the mateiral to expose all sides to the 360-degree drain screen.

With most liquid removed, the material is forced into contact with the screen cushion cone. Here, the opposing pressures of screw and suction cone (which both rotates and moves longitudinally) accomplish final removal of liquid. Pressed material, with a 62–67% moisture content, then passes around cone into cake outlet. The cone, rotating at a different speed from the screw, assures maximum liquid removal by making it possible for press cake to pack hard in discharge area.

Press cake from the three stages of pressing can be controlled separately. Capacities range from 3 tons/hr of equivalent fresh grapes (crushed and free-run juice removed) for the 6-in. diameter screw, to 20 tons/hr for the 12-in.-diameter screw, to 80 tons/hr for the 16-in.-diameter screw.

The press is available in all stainless steel, or with product contact parts only in stainless. "Easy-clean" models have counterbalanced swing-open screen frames. There is a wide selection of perforated drain screens and various cone options. Variable-speed drives are optional.

It is necessary to add a press aid to grapes prior to grinding, at the rate of about 3% (Anon. 1972B).

CONTINUOUS PROCESS FOR MAKING PRUNE JUICE

Prune juice, essentially a water extract of dried prunes, is produced mainly by two time-consuming batch processes. In the diffusion method, soluble solids of the prunes are leached out in a succession of hot water extractions with each extraction lasting 2–4 hr. This process yields about 500 gal of juice per ton of prunes. In the disintegration process, prunes are cooked for about 2–3 hr, then put through a disintegrator. After filter aid is added, the slurry is filtered through cloth or a filter pad to obtain the desired juice. This process yields about 600 gal of juice per ton of prunes, is faster than the diffusion process, and yields product with more body.

A new process based on the ability of pectic enzymes to rapidly hydrolyze the pectins in a prune slurry so the juice can be easily filtered was developed by Bolin and Salunkhe (1972). The juice is prepared by making a slurry of whole ground unprocessed French prunes, water, and Pectinol R-10. Or the pits can first be taken out after steaming the prunes. This slurry is heated to 60°C and held for 10 min; a longer heating time does not accelerate filtration. Temperature of the product is then raised to 90°C to inactivate enzymes. Pits, skins, and sediment may be removed by filtration or centrifugation, or a combination of these methods. Filter aid is stirred into the blend and it is filtered through a filter-aid pad. Recovery by this filtration procedure is calculated at 450 gal of juice per ton of prunes.

The resultant juice is about 19° Brix and is lighter and slightly clearer than regular prune juice. It has a consistency (88 min, Ostwald viscosimeter) between that of commercial juices prepared by diffusion (45 min) and disintegration (125 min); and no sediment forms during storage. The juice does not have any caramelized taste of the regular water-extract juice, but a mild fruit flavor.

The juice process can be modified to give a darker product having more caramelized taste by submitting the whole prunes to a baking step before grinding. The extent of the change depends on time and temperature of baking.

Instead of filtration, centrifugation can also be used to remove fiver from the juice. No filter aid is required but enzyme treatment and addition of about 3% of a bulk material, such as rice hulls, is necessary. A definite density gradient is noticed upon centrifuging the slurry. The denser material, which settles on the bottom, is the endocarp of the seed. The next layer consists of the fruit exocarp or skin, followed by the fine pulpy material and then the juice. The top layer contains the seed nucleus or kernel.

COUNTERFLOW COOLER

Very often it is necessary to cool products coming from preheating, pasteurizing, or sterilizing equipment. Counterflow coolers reduce the temperature of a product quickly and efficiently.

A counterflow cooler consists of a series of tubes of varying lengths depending on the product to be cooled and the volume to be handled. The tubes are fabricated of Type 316 stainless steel, each length of tubing being encased by an outer shell of mild steel. The product tubes are connected by 180° return bends, secured by tri-clamp fittings, allowing for easy access for brush cleaning and sanitary inspection. The outer casings are connected by close nipples welded in position to form a continuous flow of coolant to the outer surface of the product tubes. The counterflow cooler may be provided in 10-ft, 15-ft, and 20-ft lengths. Cooling tubes are supplied in multiples of four or six passes; capacity can be increased by use of additional multiples of four or six passes. Units are supplied without frames and may be mounted or suspended as desired. Figure 10.1 shows a diagram of a typical design.

In operation, the hot product to be cooled is pumped through the series of jacketed tubes, where the outer surface of the product tubes is exposed to the coolant, flowing in an opposite direction to the flow of the product. The cooled product passes on to succeeding equipment, while the coolant may be recirculated.

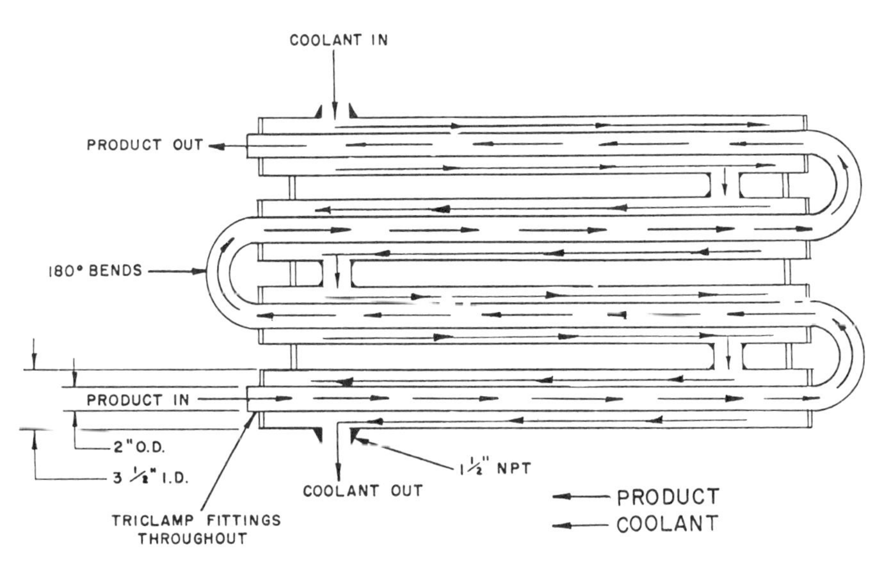

FIG. 10.1. Schematic illustration of counterflow cooling principle. (Courtesy Chisolm-Ryder Co.)

WINE

The traditional manner of making wine commercially is to first crush the grapes in a roller crusher equipped with a steam separator. Sulfur dioxide (100 ppm) and pectin enzyme (120 ppm) are added immediately; 4 hr later an active culture of yeast is added. Fermentation is at 68°–78°F with twice daily punching of the cap (skins and pulp). Wine used for making port is screened from the pomace at about 3% alcohol and fortified to about 20% alcohol. The dry wines are held on the skins until the Brix degrees fall below two.

In the production of red wines, the extraction of color from the grapes has always presented a problem. With the exception of a few cultivars with colored pulp, the pigments are located in the epidermal cells of the skins. To effect color release, the membranes must be permeable to the passage of the pigments. Alcohol and heat are the two agents commonly used to release color.

The traditional method for extracting color has been intermittent pumping of the fermenting juice over the cap. This does least damage to quality; but, where fermentation is limited, as in the production of port, color extraction is often inadequate. For this reason heat has long been used to effect color release. In the most common method of heating, the juice is pumped through heat exchangers and back over the cap for several hours, followed by cooling to fermentation temperatures. In another method crushed grapes are pumped through heat exchangers that heat the must to about 180°F and then cool it to fermentation temperatures. Though heating methods presently in use usually effect color release, they frequently result in quality impairment and are fairly costly in terms of labor and steam. They also require handling of pomace in the winery and do not easily adapt to continuous operation.

Coffelt (1969) reported that steam was the best means of heating grapes; also that heating should be of whole grapes in the clusters, as no pretreatment would then be necessary. Heating whole grapes permits high skin temperatures without high interior temperature. A color-release machine capable of heating grapes in a steam atmosphere as great as 50 psig for 5–50 sec was patented by Coffelt and Giannini (1965). This equipment makes possible heating grapes in a pressure chamber in which the steam temperature and heating time can be varied at will.

The following procedure for handling heat-treated, color-release grapes was adopted by Coffelt (1969). The box of the heat treater was filled with grapes in clusters. The grapes were heated for 10–20 sec at 30–40 lb of steam pressure, depending upon the cultivar and region grown. Immediately after heating they were crushed in a roller crusher with stem separator. This mass of heated, crushed grapes was stirred for 5 min, pressed, and the juice immediately cooled to about 70°F. From this point, the juice is treated in the usual manner—addition of sulfur dioxide, pectic enzyme, and yeast—except punching and screening are not required during fermentation.

The amount of color extracted by heating compared favorably with that extracted by fermentation; the same was true of the amount of color remaining after 3 months of aging of dry red wines. Increased fermentation time before pressing resulted in increased color extraction.

With port wines, 17 sec of heating generally resulted in deeper colored wines than did pressing in 3% alcohol. Heat treatment of grapes was less suitable for dry red wines than color extraction by fermentation.

Following fermentation or fortification, the wines are racked at intervals until clear and then stored in glass at about 55°F.

New California Winery

Sales of California wines doubled from 1965 to 1972. To accommodate the trend in consumer demand, wineries accelerated activities to improve plants and processes. Improvements were centered largely around modernizing equipment to achieve greater production while maintaining high quality. Each plant utilized engineering innovations and equipment unique to the plant.

Mondavi Winery, Oakville, Calif. (Mondavi and Havighorst 1971), for example, spent 3½ million dollars in replacing most metals throughout the plant with stainless steel, because of its sanitation and resistance to corrosion. The dump-trough, crusher-stemmer, and pipeline leading to the heat exchanger are all stainless steel. This winery introduced many other improvements.

New design of the crusher-stemmer eliminated inside horizontal support bars, which interfered with sanitation of this unit. Axles at the end of the cylinder provide support. The new design also permits operation of impellers at a speed of 300 rpm—a reduction from the normal 500 rpm. Reduced impeller rotation is sufficient to separate grapes from stems without macerating stems. Cut stems impart undesirable tannin flavors to the wine.

The plant also has an advanced heat excahnger—a dual, shell/tube unit—used for both heating and cooling incoming must or wines pumped to it from inside tanks. One shell/tube is the chiller for the 50% propylene glycol refrigerant, which is pumped to either the second shell/tube unit or to the jacketed tanks. The refrigerant in the chiller is Freon 22, which is also supplied to fin-coil units that refrigerate storage and other rooms. The second shell/tube unit is stainless steel, sanitary construction with 3-in. diameter stainless tubes. Refrigeration is supplied by two 125-hp compressors. The Freon is condensed in a 300-ton, air-cooled condenser. Safety controls on this system shut off refrigeration to all but essential units if demand exceeds capacity.

Hot glycol is supplied by a package boiler that has two separate copper tube bundles: one for heating glycol, and the other for heating process and wash-down water. Two pipe systems, each with its own pump, circulate heated glycol to the heat exchanger and to jacketed tanks.

Manual labor during fermentation is eliminated through use of tanks that slowly and gently rotate at regular intervals at a speed of 35 rpm. A helical screw inside the tank breaks up the pomace and distributes it throughout the fermenting wine.

Before entering the tanks, the must of red grapes is adjusted to fermentation temperature by passing it through the heat exchanger. Yeast culture and sterilant are added, and the must is left to ferment in the Roto-Tanks, for 4–6 days. Besides distributing pomace throughout the wine, the rotating tank also forces residue into an open port leading to a hopper, screw-conveyor, and hydraulic press. Juice from the press is pumped to fermentation tanks. Pomace is removed from the press and sluiced by water to open field disposal.

For white grapes, the must remains in a rotating tank only long enough to acquire sufficient flavor and body from the skins and seed. Time of residence is determined by the wine maker, and may be as long as 7–8 hr. Then free-run juice flows by gravity to a fermentation tank where it is inoculated with a pure yeast culture. A sterilant is added to control yeast and bacteria.

During seasonal peaks, volume of red grapes may exceed available capacity of the rotating tanks. When this happens, the must is pumped directly to thermostatically controlled, jacketed tanks for fermentation. Tanks are purged with nitrogen before filling, and after filling the headspace is filled with nitrogen to prevent oxidation of wine.

The tanks have individual temperature controls with ±2°F. Time and temperature of fermentation are critical, and they vary with nearly every type of wine. To help assure precise control over temperature, the tanks are baffled to permit better circulation of glycol refrigerant.

Following fermentation, wines are clarified and moved via nitrogen pressure, or pumps, to temporary storage and aging. Although low-speed centrifugal or positive displacement pumps are used, wine is often moved by applying inert gas at 8 psi to force the flow between equipment. This minimizes the chance of undesirable reactions taking place in the wine.

Wines are removed from storage at a time determined by the wine maker and filled into barrels and/or casks. Red wines are aged in 58-gal barrels (French white oak) for up to 24 months. White wines are aged in 300-gal casks (German white oak) up to 12 months, and subsequently in stainless steel storage or aging tanks. Fin-coil units in the oak storage room have aluminum housing, fins, and tubes to resist attack by sulphur fumes.

Before bottling, the temperature of wine is brought to 55°–60°F. Then it is filtered and moved to a bottling tank under nitrogen pressure. The wine is fed through a sparger and a polishing or finishing filter, then to the bottle filler. For sanitary reasons and to protect the wine, bottling takes place inside closed doors. The bottle washer is on the outside and feeds clean bottles to the filler. Filled bottles move on a U-shaped conveyor to the corker, capper, and labeler. Visitors to the winery may view the bottling operation from the outside through large picture windows.

The aim of many wineries is to provide flexibility that will permit separate development for each variety of wine throughout the entire winemaking process. Each wine variety is aged in the particular oak that will best enhance its character and depth. Such an operation contributes to higher costs—but it is the superior technique for making premium varietal wines.

Mondavi has 10 tanks of 7500-gal capacity; 18 tanks from 2800- to 7500-gal capacity; 48 from 1000–2000 gal; and 3000 58-gal barrels.

APPLE PEEL BY-PRODUCTS

Apple pomace is the primary by-product of the apple processing industry. The peels, cores, and trimmings from apple slices and sauce processing operations are usually macerated and their juice expressed, leaving a potentially useful but not well used by-product. The resulting press cake is frequently used as an ingredient in animal feed, often being trucked directly from the processing plant to local farms. Being low in protein, it is a poor feed. However, it is relatively high in pectin which renders it useful in certain food products.

Apple-Flavored Thickener

A peel powder was prepared from apple waste by Bomben *et al.* (1971). Apple peel and cores were obtained in a sanitary manner, using conventional peeling and paring machines. The peels were separated from the cores, which passed through a finisher to remove stems and seeds and to be macerated. Peels were held for 24 hr at 60°–65°F in shallow stainless steel tanks to increase the aroma up to threefold. They were then macerated through a 0.03-in. screen with a disintegrator and were mixed with the macerated core material. The mixture was pressed in a hydraulic press at 500 psig and the essence was recovered.

The press cake was dried, using a 12-in.-diameter × 18-in.-long double drum dryer operating at 300°F and 1.5 rpm. The dried press cake was cooled in a low-humidity room to remove additional moisture and to allow equilibration. It was then ground to pass through a 100-mesh screen, and stored in a cool, dry place. The powder contained about 50% of the aroma components originally present in the peel, which gave it a strong apple aroma. Drum drying and air drying resulted in low aroma retention; freeze drying gave the most significant aroma retention. The best means of obtaining a highly flavored peel powder, without the expense of freeze drying, was by the addition of aroma solution (essence) from the peel juice.

Essence from peel juice of some apple cultivars improved the aroma of apple pies, while essence from other cultivars did not. It appears that this method of salvaging apple peels is profitable for certain cultivars.

MISCELLANEOUS FRUIT PRODUCTS

Fresh Banana Puree without Heat Sterilization

A process that eliminates heat sterilization when preserving fresh banana puree has been developed by the USDA and the University of Hawaii. Bananas are pureed and pasteurized at 200°F and then preserved with potassium sorbate.

The fresh-tasting puree, devoid of any cooked taste, has possible wide application in the Hawaiian ice cream and bakery industries. Surplus puree can be exported as the process makes 6 months of storage possible at temperatures slightly above freezing (40°–45°F).

Hawaiian bakers and ice cream makers have been working with test samples of the puree. Microbial and taste-test data on stored samples have also been obtained. The tests indicate a promising outlook for the puree as a food ingredient.

Ripe bananas to be made into puree are peeled and immersed for 3 min in a 1.5% solution of sodium bisulfite ($NaHSO_3$). Bananas are then milled in a cutting mill and reduced to a puree. The small amount of bisulfite picked up in the previous step prevents browning and gelation. The puree is next pumped through a plate heat exchanger where it is held for 1 min at 200°F, and then cooled to 100°F within 1 min. This heat treatment inactivates enzymes and thus prevents discoloration, gelation, and loss of flavor. Enough citric acid is added to adjust the pH to 4.1; 200 ppm (0.02%) of potassium sorbate is added to prevent microbial spoilage.

Dehydrated Protein-Fortified Fruit Juices

The protein content of most deciduous fruits is low and its quality in terms of essential amino acids is poor. Therefore, an attempt has been made to fortify fruit juices with protein and then dehydrate them to nutritious powders. Apple juice, peach nectar, and cherry juice have been produced. The respective juices were fortified with protein that had been enzymatically modified commercially to increase its water solubility and foaming properties. This product (Gunther D-100 from Gunther Products Div., A. E. Staley Mfg.) contained 62% protein and was added as a 20% solution.

Progressive expansion of the fruit juice industry since 1970 is due to increased demand brought on by the improvement in the quality of dehydrated fruit products, including liquid concentrates and powders. Detailed analyses of the color, flavor, and body of reconstituted juice powders indicate that they are similar to those of canned juices.

Dehydrated juice products offer savings in weight and space. Hence, they are economical in transportation, storage, packaging, and handling. Juice powder is stable at elevated storage temperatures providing extended shelf life. Ease in handling and convenience, and rapid reconstitution, make it suitable for domestic and military requirements; fortified juice powders provide high quality and nutritive value at the time of consumption.

The foam-mat drying method has been used to make juice powder. The process consists of whipping together various juice–protein–methyl cellulose combinations for 5 min. The resultant foam is spread 1 mm thick on trays and dehydrated at 160°F for 25 min. After the trays are cooled, the friable, puffed material is removed easily. The powder is ground to pass through a six-mesh screen. Detraying, crushing, screening, and packaging are carried out in a room with the humidity below 10% to prevent the hygroscopic powders from picking up moisture and becoming sticky. All lots of powder for each fruit cultivar are mixed together for uniformity, placed in appropriate size cans, and sealed under ½ atmosphere air pressure. The moisture content of the powder is usually about 2%.

The viscosity of the reconstituted products is similar to that of malted and shake products—rather thick. The color of reconstituted apple juice is creamy; peach nectar has an orange-yellow color; and the tart cherry juice is pink. When the products are served after refrigeration, they have an excellent flavor and consistency.

Apricot Concentrate

Preparing a homogeneous pulp of the fruit and then concentrating it to about a 2.5 to 1 ratio by removing some of the water is the most economical method for preserving apricots, because it requires a minimum of hand labor. This concentrate is packaged in tins or 55-gal drums. It is bright orange in color, has a good apricot flavor, and is a good starting material for use in new product development because of its year-round availability, good tart apricot flavor, and economical method of preparation.

Apricot concentrate may be used to make a pumpkin-type pie, or can be whipped with other ingredients to produce a chiffon or cream pie. It blends readily with other fruit juices and nectars to give added body, nutrients, and flavor. Frozen products may be produced from concentrate, such as 100% natural fruit bars or popsicles. Another possibility is to market apricot concentrate in small cans for use in cake mixes, muffins, breads, fruit cakes, etc., in the home. A manufacturer of pourable salad dressings also might find it an interesting ingredient.

Besides being used directly, apricot concentrate can be processed further into products with more varied uses. One such product is a thin, flexible apricot "cloth" or sheet. This versatile material is produced easily and quickly using a double-drum drier operating at 280°F with a clearance of 0.015 in. (0.4 mm), rotating at 3–4 rpm. Sodium bisulfite (0.5%) added prior to drying helps protect the natural vitamin A in the concentrate and to ensure the product's light color during protracted storage. The resultant sheets of "cloth," both with and without bisulfite, retain about 12% moisture, have a bright orange color, and good apricot flavor.

This "cloth" can be cut into small strips, packaged, and marketed as is; it also lends itself to shaping and pressing. Multiple layers of the product can be pressed and molded into a number of apricot materials, including an apricot bar that can be cut into cubes or disks and coated with granular or powdered sugar, spices, and other materials. Also, layers of the product can be pressed easily into a mold (could be shaped like an apricot) to produce an artificial, dried apricot that has texture and taste similar to regular sun-dried fruit. One advantage to such molded products is that protein powders, vitamins, or other nutrients could be added to the apricot concentrate before drying or sprayed on the sheets of cloth before lamination and pressing. A uniform fortified "apricot half" with a wide variety of nutrients can be produced also.

From a nutritional standpoint, the thick sheets of air-dried product contain approximately one-third less β-carotene than dried apricot halves. This lower amount is to be expected since the product does not contain any sulfur dioxide. The nutritional composition varies with the means of production. If sulfur dioxide is added initially, the vitamin A content of the final product is high because sulfur dioxide retards the loss of vitamin A.

Since it is possible to mechanize production and thus use a minimal amount of hand labor, apricot concentrate products are economical to make (Bolin *et al.* 1973).

Pineapple Granules

These are 88% soluble solids, primarily sugars, 10% insoluble solids, and 2% moisture; they are yellow in color and pass through a No. 10-mesh screen. Titratable acidity of reconstituted granules is 0.6; pH, 3.7; and Brix, 15°. Granules are packed in controlled humidity with a polyethylene bag inside a fiberboard box. They are used as a fruit extender, flavor and color carrier, decorative material, and as fruit solids in sauce preparations in bakery goods, dry mixes, cereals, candies, ice cream, gelatin, and sauces.

The rather weak pineapple flavor of the granules is enhanced by the use of cloudless pineapple concentrate (72° Brix) and reconstituted by mixing 1 part concentrate to 5½ parts water (Anon. 1972A).

Fruit Flakes

Dry flaked fruit is made from orchard run fruits that are washed, graded, sliced, diverted to a continuous cooker, and then pumped into 1000-gal holding tanks where the product is sweetened with up to 10% sugar.

The cooked sauce is screened and piped hot to double-drum driers where it is spread thin and dried to 2% moisture in less than 17 sec. The sheet of dry material is sheared off and broken by an auger into flakes, or broken into a powder in a hammer mill, and sealed in drums with pry-off lids. The flakes or powder are stored in the same way as other dried fruit products, and may be reconstituted with either hot or cold water.

With tissue-thin flakes at less than 2% moisture, instant sauce was made of apples, pears, purple plums, and various types of berries. This process was developed by the U.S. Department of Agriculture, Washington State University, and Valley Evaporating Company (Livingston 1967).

Fruit Tidbits

Fruit tidbits are a new product, developed by the USDA, in which diced apples, bananas, peaches, raspberries, and strawberries are dried by a osmosis vacuum process. The color and flavor of the fruits is said to be retained without refrigeration or chemical preservatives. The fruits can easily be reconstituted for use in baking, salads, or desserts, and they absorb water to make a sauce. The tidbits can also be coated with chocolate or hard cream, as a confection.

Orange Juice Pellets

Orange juice pellets are a sugarless, candy-like product that was developed by the USDA at Winter Haven, Florida. They are made by compressing dehydrated orange juice into tablets. They require no refrigeration, have all the nutrients of liquid orange juice, and are low in calorie count. They should be ideal for campers and others who need lightweight, low-volume, nutritious foods.

Banana Chips, Flakes, and Slices

The conventional technique for making (dried) banana chips is to use unripe fruit having a solid texture and total sugar content of about 2%. The slices are immersed in a solution of SO_2 or other browning inhibitor (aluminum chloride, citric acid or ascorbic acid) and then blanched in hot water or steam. The resulting product lacks the sweetness, unique color, and flavor inherent in ripe banana products. To enhance these qualitites, the slices are treated with sugar solution, a gelatinous substance, or artificial flavors and dyes before being fried as banana chips.

On the other hand, banana powder or banana flakes are made from ripe fruit. To facilitate drying, the bananas are mashed into a paste and then dried by foam-mat drying, drum drying, or low-temperature vacuum oven drying. The final products take the shape and texture of fresh bananas.

Numata and Sugano (1980) patented a process for producing fried banana slices having superior flavor, crisp texture, and good storage stability from fully ripe bananas. The product is made by frying slices of fully ripe raw banana in an edible oil as rapidly as possible after slicing while maintaining a pressure of about 30–60 Torr and frying temperature of 75°–85°C, to dehydrate the slices. The slices are removed from the frying oil and excess frying oil adhering to the banana slices is removed while still maintaining a pressure of 30–60 Torr; thereafter the pressure is returned to atmospheric conditions. This process is so rapid that loss of flavor, color, and texture does not occur, and all of the constituents of ripe bananas, except water, are maintained.

Fully ripe raw banana flesh, by weight, comprises about 75.5% water, 1.3% protein, 0.4% lipid, and 21.4% carbohydrate. About 90% of the carbohydrate comprises sucrose and reducing sugars, and 2% starch. In addition, ripe raw banana flesh contains some crude fiber, ash, organic acid, inorganic acids, vitamins, pigments, tannins, and enzymes.

Explosion-Dried Blueberries

Explosion drying may be one of the best ways of preserving blueberries, and saves 40% of the energy required by conventional food-drying systems.

Inside the heated chamber (which blanches and prevents woodiness or grittiness in the dried blueberries), heavy pressure is exerted on small batches of blueberries. As the pressure is suddenly released, water in the blueberries literally explodes and evaporates from the fruit. Each explosion-dried batch, popping from the machine at regular beats, looks like soft blue gravel. The berries are porous, slightly puffed, and crunchy, and make a good snack food. After reconstituting in boiling water, they look very much like fresh blueberries.

Chemical engineer John Sullivan, who helped design and build USDA's explosion drier, believes that this energy-saving method of preserving blueberries is cheaper and preferred to either canning or freezing. Research on exploding blueberries is sponsored by the Georgia Blueberry Association, USDA, and Coastal Plain Experment Station, Tifton, Georgia. It is expected that a goodly portion of the 1983 crop of blueberries will be processed by this method.

Mango Puree

Mango is the second largest tropical crop after banana in terms of production and acreage. It is relished for its succulence, exotic flavor, and delicious taste. In spite of its excellence, mango has not been developed as a commercial and export processed crop because of its poor stability during storage and irregular productivity. It still lacks recognition in the market place and tends to lose its flavor during thermal processing.

Preservation of mango fruit remains at the "home canning" level in most countries because of difficulty in removing the skin and seed. Industrial utilization of mangos has been limited due to lack of adequate procedures for extracting the pulp. Peeling of the fruit and removal of the pulp by hand is not commercial feasible because of the high labor requirements, resulting in high production costs. Ripened mangos may be peeled by use of lye and/or steam and by a combination of mechanical devices, such as brush pulpers, special centrifuges, and paddle pulpers.

Although there are many ways to convert mango into nonperishable products, mango puree offers definite advantages because it requires less labor and can be used for manufacture of other products. The use of sucrose and high-conversion corn syrup are acceptable sweeteners.

Mango puree can be used for manufacture into jellies, jams, beverages, dairy and bakery products. A processing time of 5 min at 100°C is recommended for canning mango pulp with an initial temperature of 76°–79°C. For frozen mango puree, it is necessary to blanch the product by raising the temperature to 93°C for 1 min, then rapidly lowering it to −20°C.

Passion Fruit Juice

Passion fruit is a relative newcomer in the list of fruits used for juice. So little is known about this exotic fruit that processors in Hawaii had to pioneer techniques for growing, harvesting, and processing it. The Hawaiian Fruit Growers Exchange, with more than 300 acres, may be the largest passion fruit grower in the world. After detergent spraying, chlorinating, and fresh-water spraying, passion fruits go through a slicer and into a centrifuge juicer. From the centrifuge the fruit pulp goes through a finishing pulper to tanks, from which it is drawn through sterilizers, filling, closing, and casing units. The waste pulp is protein rich and palatable to cattle.

The procedure for pulping and juicing guava and papaya are similar to that for passion fruit. All of these juices are blended with each other and with other juices in a variety of punches and drinks.

Papaya Fruit Juice

The papaya is one of nature's oddities, a melon that grows on trees, varying in size from small cantaloupes to medium-size watermelons. The flesh has a consistency of very ripe cantaloupes, has a deep, rich orange color, and contains papain, a soluble enzyme similar to pepsin. It is also high in vitamin A and C and is considered a health specialty. The tree is extremely susceptible to cold and hurricane damage.

Special methods and equipment are used to extract papain from the skins and incorporate it into nectar and other products. Because the papain content and flavor of the melons vary considerably according to season and rainfall, fresh fruit and processed juice are blended to maintain consistent uniformity of finished product. Each batch is checked for papain and pH content.

The production of papain and papayas is limited to tropical areas, particularly Central America, south Florida, and the Philippines. Papaya nectar is a ready-to-drink beverage sold primarily in 1-qt decanters at dairy counters and supermarkets. The pasteurized, vacuum-packed beverage contains no preservatives and does not require refrigeration, though it is recommended at the retail level.

A 3-to-1 papaya concentrate is also packed in quarts, 12-oz bottles, and 1-gal jugs. It has been sold to hospitals, health food stores, and juice bars since about 1960.

Processed Kiwifruit

Kiwifruits (Chinese gooseberries) have a bright emerald green flesh covered with a thin furry skin. When ripe, they have an attractive characteristic sweet flavor and many very small seeds. The fruit is cultured in New Zealand and exported to most of the countries of the world. In recent years, kiwifruit has been cultivated in Australia, California, Japan, and the Mediterranean. It grows on vines that begin bearing 5 years after planting and continue to bear for 40 years. The fruit is used fresh, in salads, cake decorations, desserts, beverages, and jellies; it can be frozen, canned, pureed, or concentrated.

Kiwifruit is more than 90% pulp/seed. When firm ripe, the fruit contains 4.6% starch and 2.3% amylose, but when soft ripe, these are 2.4 and 1.25%, respectively.

Total solids content is 15.8–26.2% throughout ripening. Soluble solids increase during ripening, and are markedly accelerated by the presence of ethylene gas. Fructose, glucose, and sucrose are present in almost equal levels (33 to 18%) during ripening; there is no maltose present at any stage of ripening. Ripening is accompanied by a decrease in pressure test, acidity, starch and amylose content, and an increase in soluble solids, sugars, and pH. Changes in fructose and sucrose are related to the decrease in starch. Chlorophyll content is not affected by the ripening process. The ascorbic acid content of kiwifruit decreases from about 210mg/100 g to 190 mg upon ripening.

The production of kiwifruit is expanding rapidly in New Zealand, and other parts of the world including California, France, Chile, Italy, and Israel. Products include canned slices, frozen pulp, wine, and leather. The freshly harvested fruit is stored at 0°C for up to 90 days, moved to 25°C for ripening to 13° Brix, and sliced for canning in syrup or freezing with sugar.

The total chlorophyll remains constant during storage and ripening, but about 90% of it is degraded during canning. The total protein content increases by more than 100% during storage and ripening, but decreases significantly during canning. There is an inverse relationship between firmness of unpeeled kiwifruit and the water-soluble, high-methylozyl pectin content. The canned product is yellow-brown in color.

LARGE-VOLUME FRUIT PRESS

After proving itself in Europe, a large-volume, Swiss press for apples, pears, grapes, and berries is being used in several installations in the United States. Daily throughput of raw material of 100 tons allows 4 hr for plant clean-up. The new process replaces conventional vertical, hydraulic, rack-cloth equipment. Capital investment, installed, for each press is about $100,000; best economic results are obtained in plants with 3000 tons/year throughput. A similar model is available for plants with 1000 tons/year.

Sanitation and high productivity are aided by use of stainless steel and chrome nickel in the press wherever mash or juice contacts the equipment. Side-cylinders, drawbars, and the main piston are chromium plated.

The hydraulic unit is separated from the press. An electric switchboard serves the dual purpose of central control panel and switchboard for the operator. All operations of the line may be controlled from the panel. It is batch-type operation, with 10 tons of macerated fruit (mash) in each batch, which requires 2 hr for pressing, and gives 10 charges/day.

The fruit is fed from a hopper to a washer-destoner-elevator unit, which delivers it to a disintegrator (25 hp, 20 tons/hr). The mash is pumped to a large surge tank. A pump fills the horizontal cylinder press completely with the mash under slight pressure. When actuated, the main piston of the press compresses the mash and juice is forced through perforated rubber hoses, covered with filter cloth. This series of hoses is located in the front of the cylinder. Press, or dwell, time is determined by tests made of the raw product maturity and by pomace moisture tests.

When the press is completed, the press cycle is reversed. The piston is first retracted, then the product basket is slowly retracted to break up the pomace. This section of the cylinder is slowly rotated and the crumbled pomace drops through the filter elements into the helical-screw discharge conveyor. The juice, which is quite clear direct from the press, is treated in holding tanks, then filtered to give a sparkling brilliance. It is ready to can or bottle and pasteurize. The waste from the filter is combined with other solid waste from the plant and trucked to a sanitary fill, or spread on soil for improvement (Weisser and Havighorst 1972).

SALT-FREE OLIVE CURING

California producers are developing methods for curing and storing olives that upgrade quality and attack an ecological problem.

Saline solutions, as fresh fruit storage agents, are being replaced with food-grade chemicals. This substitution has no effect on the taste of the fruit. And open-top redwood curing tanks have been replaced with totally enclosed fiberglass-reinforced polyester tanks that permit anaerobic conditions.

The salt-free system for curing olives uses either FDA-approved sodium benzoate or potassium sorbate as preserving agent, the former being preferred because of lower cost. Both are biodegradable by bacteria, which break them down into solutions that present no particular disposal problems. Saline solutions are not biodegradable.

These solutions must be used under anaerobic conditions to keep aerobic microorganisms from growing and attacking the fruit. The tanks accomplish this and, at the same time, keep out any form of airborne contamination. To packers adding new capacity or replacing existing tanks, the new tanks offer efficient, contamination-free storage for presently used saline solutions and permit immediate changeover to the salt-free anaerobic system.

Tanks are 8 ft high and 10 ft in diameter. They hold some 1.6 million olives of giant size. Inside the tank is an FDA-approved gel coat, providing a nonporous and easily cleaned surface. Tops of the tanks have a 5-ft gasket-sealed manhole, a fiberglass-reinforced manhole cover with anaerobic seal, and 20-gal supply reservoir with removable top.

After a tank is filled with olives, liquid is brought up to 1 in. below the manhole. The cover, with its seal and reservoir, is then bolted in place. Final filling of the tank is through the reservoir. In the process, entrapped air is vented through brine sampling and product inspection ports, which are a part of the cover. Liquid level in the reservoir is then raised to about 9 in.

As gases are generated during fermentation, they pass through a screened hole at the apex of the manhole cover, then into the reservoir. Finally, they leave through screened vents on the sides near the top of the anaerobic reservoir. The reservoir provides a partial or complete anaerobic closure to the tank, allows liquid replacement of intercellular gases and expulsion of gases formed in fermentation, and compensates for expansion and contraction of the liquid in the tank. In general use, these tanks are semianaerobic but they become completely anaerobic when food-grade mineral oil is floated atop liquid in the reservoir.

Storing benzoate or sorbate solutions from one season to another also requires anaerobic conditions, and may also be used for storing cucumber pickles and possibly other food products.

USE OF VITAMIN C (ASCORBIC ACID)

Long before vitamin C was known, man found certain fruits and vegetables helpful in preventing and treating scurvy. These had been a remedy for scurvy for about 300 years when, in the first quarter of the twentieth century, researchers began to examine foods to discover the factor that made them antiscorbutic.

Between 1918 and 1925, Zilva concentrated this antiscorbutic substance from lemons. Zilva's work established the important properties of the vitamin, its molecular construction, and its activity as a "reducing factor"—its ability to absorb oxygen. In 1928, Szent-Gyorgyi, after 8 years of investigation, isolated a strong reducing compound from adrenal glands, oranges, and cabbage that he called "hexuronic acid." In 1931, C. G. King established the chemical structure of vitamin C, thus opening the way to new and important findings. The following year various investigators identified the "reducing factor," "hexuronic acid," and vitamin C as the same compound.

In the United States today, the average weekly production of man-made vitamin C now approaches 50 tons—the amount contained in approximately ¾ of a billion average oranges. Synthesis on such a gigantic scale has led to economic costs. Because of this and its proven value to health, vitamin C has an increasingly widespread use in food products. There is no substitute for vitamin C!

One kilogram (2.2 lb) of pure vitamin C will supply 13,333 people for 1 day with the amount recommended (75 mg) for daily intake by the National Research Council. Or it will supply 92 people for 1 year with the adult Minimum Daily Requirement (MDR) (30 mg) set by the U.S. Food and Drug Administration.

The designation, vitamin C, is used interchangeably with ascorbic acid. "Ascorbic" is a relatively new word in our language. It was devised by Szent-Gyorgyi and Haworth in 1933 from antiscorbutic. Among chemists and nutritionists, vitamin C is known as L-ascorbic acid. The formula is $C_6H_8O_6$, which means that it is composed of carbon, hydrogen, and oxygen.

L-Ascorbic acid is a white crystalline compound with a slightly acidic taste. It is readily soluble in water, alcohol, glycerin, and propylene glycol; and is insoluble in oils, fats, ether, benzene, or chloroform. Since it is able to "attach" itself to oxygen it is one of our most important antioxidants for certain foods that have a tendency to discolor and lose their natural flavor.

It is desirable to standardize the vitamin C content of processed fruit and vegetable juices at levels that will guarantee a significant amount of the vitamin per each serving of juice. Diet experts believe that such standardization is in the public interest. It may be achieved by the simple addition of pure, crystalline vitamin C. The Canadian government has recognized the value of this idea and has set a standard that all "vitaminized" apple juice must meet.

Vitamin C has been used since 1938 as the preferred antioxidant in frozen peaches, apricots, apples, berries, bananas, pears, pineapple, fruit purées, and fruit cocktail. It has been used also in canned fruit and fruit juices.

Bunnell (1968) found the overall use of juices and juice drinks to be the highest in homes where children were present, particularly teenagers. The relative consumption of the various varieties of fruit juices and drinks depends on the type of pack. Among frozen juice concentrates, orange juice claims about 75% of the market, with the remainder consisting primarily of grapefruit, orange–grapefruit, and pineapple juice. The consumption of single-strength fruit juices is more diverse with the pattern changing considerably in the case of the ready-to-serve fruit drinks.

Fruit juices and drinks are the primary fruit products that are enriched today. By far the most common nutrient for fruit product enrichment is vitamin C; vitamin A, usually in the form of β-carotene, which also supplies color, is also used quite often. A few juices such as grape juice and drinks, frozen fruit drink concentrates, and fruit beverage powders are enriched with B vitamins. The basic philosophy behind fruit enrichment is to standardize vitamin levels so that the various types of fruit juices, drinks, etc., are nutritionally interchangeable.

The ascorbic acid enrichment of fruit juices and drinks has a sound nutritional basis and therefore has become accepted as common practice. Most juices and drinks, unless Standards of Identity do not permit, are enriched to provide 30 mg ascorbic acid/4- or 6-oz serving, the minimum daily requirement for adults in 1 serving.

Modern processing and canning techniques of both single-strength and frozen concentrate fruit juices is effective in preserving the natural ascorbic acid content; for juices that have an adequate ascorbic acid content, enrichment is usually not necessary. There can be considerable cultivar and seasonal variation, in ascorbic acid levels, even in fruits that are considered good sources of ascorbic acid, so it is not always possible to guarantee that a standard serving will provide the MDR for ascorbic acid. Since the average consumer tends to use fruit juices interchangeably to provide variety in the diet, ascorbic acid intake would be extremely variable. The average amount of ascorbic acid contained in a 4-oz serving of various fruit juices is shown in Table 10.1. Of the fruit juices listed, only grapefruit, tangerine, and orange juice could potentially supply 30 mg of ascorbic acid in a 4-oz serving.

All data clearly indicate the sound nutritional basis for standardizing the ascorbic acid content of fruit juices and drinks, so that regardless of consumer preference, an adequate intake of ascorbic acid is provided. On the basis of experience, it is usually anticipated that juices to which 40–50 mg of ascorbic acid are added per 4 fl oz will contain 30–40 mg after processing and market storage.

In general, ascorbic acid is most stable in low pH frozen concentrates and juices such as orange, grapefruit, and orange–grapefruit blends. Other fruit beverages can be variable, so that experience in manufacture with a particular juice or drink dictates the enrichment necessary to meet label claims after processing and storage. Proper processing, minimum inclusion of air, and low iron and copper content all contribute to good retention of ascorbic acid as well as flavor. This can be achieved in practice by the use of stainless steel or glass-lined equipment and mixing that does not incorporate air during processing. Deaeration of the juice before pasteurizing is recommended. The enrichment of fruit juices and drinks does not affect the appearance, taste, or odor of the product. Ascorbic acid, however, does retard the development of oxidized flavor resulting from the introduction of air during processing. Although ascorbic acid improves nutritional value, it will not upgrade color, odor, or flavor.

In addition to conventional fruit juices and drinks, dry fruit-flavored beverage powders are easily fortified with ascorbic acid by dry blending with the other ingredients. The storage life of ascorbic acid in such dry blends is excellent as long as the product is protected from humidity.

Table 10.1. Ascorbic Acid Supplied by 4 oz of 8 Canned Fruit Juices

Fruit	Ascorbic acid (mg)
Apple	1.5
Apricot nectar	1.5
Pineapple	10
Tomato	20
Tangerine	32
Grapefruit	34
Orange and grapefruit	48
Orange	53

Many carbonated beverages are also enriched with ascorbic acid; sufficient excess of ascorbic acid is added to react with the headspace oxygen remaining in the bottle. Theoretically, 3.3 mg of ascorbic acid will react to 1 ml of air. The addition of sufficient excess of ascorbic acid to combine with the dissolved and headspace oxygen is an economical means of prolonging the shelf life of beverages. The enrichment of fruit juices, drinks, and beverages has, therefore, become a standard practice.

FORTIFYING APPLE JUICE

Apple cider, or "sweet cider," is made from a blend of many cultivars—'Jonathan,' 'Rome Beauty,' 'Delicious,' and others—to produce a palatable sugar–acid balance. Many of the apples are rejects from packing houses due to injury or poor shape. Only firm-ripe fruits are used, since immature or overripe ones lower the quality (Gerber and Salunkhe 1960).

The fruits are washed, sorted for decay, and pulverized in a greater or hammermill. Measured quantities of pulp are enclosed in heavy cloths—known as "cheeses"—and five or six are interspersed with slatted boards and pressed at a time. A yield of approximately 5 gal. juice/bu of apples is obtained. The juice is placed in a storage tank at ambient temperature and pectinase enzyme (Pectinol A) is stirred into it at the rate of 0.1–2.0%. The mixture is allowed to stand overnight—usually 12–15 hr. At the end of this time, most of the small particles have settled to the bottom and the clear juice can be drawn from the top of the tank. Diatomaceous earth is added to the residue in the bottom of the tank to aid in filtering so that the rest of the juice can be salvaged.

The juice is then run through a flash-pasteurizer in which the temperature of the liquid is raised from 60° to 180°F in 30 sec. Warm, sterile bottles are filled with the hot juice and they are capped with sterile, rubber-ringed metal caps. The bottles are placed on their sides for 5 min and then stored in a vat of water at 150°F until 20 are accumulated. Cold water is then introduced into the vat and the juice is cooled in a period of 30 min.

Sweet cider produced in this manner has a composition approximately as follows: water 87.1%, protein 0.1%, ash 0.25%, total carbohydrate 12.5%, sugar 10.5%, malic acid 0.52%, and 50 cal/100 g. Sweet cider is compared nutritionally with orange juice in Table 10.2.

The greatest difference between orange juice and apple juice is in the amount of vitamins A and C. Orange juice contains six times more vitamin A and an average of 21 times more vitamin C than does sweet apple cider. Experiments were started in Canada to determine the feasibility of fortifying sweet cider with vitamin C. Success was attained and a local drink was produced that contained as much vitamin C as orange juice. For

Table 10.2. Comparison of the Nutritional Value of Apple Cider and Orange Juice (in Units per 100 ml)

						Vitamins				
	Calories	Protein (g)	Total carbohydrates (g)	Calcium (mg)	Iron (mg)	Thiamin (mg)	Riboflavin (μg)	Niacin (mg)	A (IU)	C (mg)
Sweet apple cider	53	Trace	14	6	0.5	0.02	0.03	Trace	38	0.2–3.6
Orange juice	47	0.9	12	11	0.4	0.07	0.02	0.3	212	9.7–70.0

several years, this was the only kind of sweet cider sold commercially in Canada.

In 1947, experiments in the United States were carried out with ascorbic acid in the production of stable, unclarified sweet cider. A small amount of ascorbic acid could be added for the antioxidant effect only; or, a large amount (6–12 g/bu fruit) of ascorbic acid could be added for both antioxidant and nutritional effects. With a 50% efficiency, this would leave 25–50 mg vitamin C/100 ml of juice. Several successful methods for adding ascorbic acid were developed: (1) as a spray to the fruit before crushing, (2) as a spray to the pulp in the cheeses, or (3) with a metering pump in the juice line while it was being transferred from the tank below the press to the holding tank. Some juice manufacturers tried selling this fortified apple juice for a year or two but discontinued it because the public would not pay the slightly higher price necessary for a product with greater nutritional value.

In 1959 at Utah State University, experiments were begun with ascorbic acid (vitamin C) and sweet cider that have resulted in the production of two new (for Utah) apple juice drinks. The first drink looks much like standard sweet cider; has a vitamin C content equal to that of canned, unsweetened orange juice; and tastes much like standard sweet cider. The second drink has a vitamin C content equal to that of canned, unsweetened orange juice but has the flavor of a fresh apple and is much lighter in color than standard sweet cider. This light-colored apple juice is produced by adding ascorbic acid to the juice as it leaves the press. It is then clarified with Pectinol A, filtered, and pasteurized in a manner similar to that used with standard sweet cider. Sufficient ascorbic acid is added (about 70 mg/100 ml) so that the finished product will have 40 mg/100 ml.

The juice within an apple is light colored—almost white—but will darken within 1 hr after being pressed from the fruit. This is caused chiefly by the action of the enzyme polyphenol oxidase on the catechol and pyrogallol in the tannin of the juice in the presence of air. The addition of ascorbic acid to the light-colored juice immediately upon pressing inhibits the darkening process by combining with the oxygen present. Stirring, shaking, or pumping the juice in the presence of air "uses up" the ascorbic acid; when that occurs, the juice will darken.

Analyses have been made of fortified apple juice produced by the Utah State process for ascorbic acid content. The preliminary results may be summarized as follows:

1. Each time the juice is stirred, poured, or pumped before pasteurization, there is a decrease in ascorbic acid.
2. Under the conditions of manufacture, it was necessary to add ascor-

bic acid at the rate of 70 mg/100 ml so as to have 40 mg/100 ml in the light-colored juice.

3. Pasteurized, fortified juice from which air is excluded stored either cool (35°F) or warm (90°F) for a 2-month period loses little ascorbic acid.
4. As soon as the fortified apple juice is exposed to air it begins to lose ascorbic acid at the rate of 1–4 mg/100 ml per day depending on the amount of air mixed with the juice.

Chemical Preservatives for Apple Juice

Benzoate of soda has been used commercially for many years as a preservative for apple juice. However, the objectionable flavor of this additive can easily be detected.

Studies have been conducted at Utah State University using sorbic acid and sodium benzoate as preservatives. It was found that sorbic acid is more effective against yeast growths and subsequent alcoholic fermentation in apple juice. A concentration of 0.05% of sorbic acid or potassium sorbate in combination with mild pasteurization was satisfactory. The process consisted of heating the raw juice to 170°F, cooling immediately (with a milk cooler) to 100°F, adding Pectinol A for clarification, allowing product to stand 12–15 hr at ambient temperatures for settling, decanting the clear juice into 1-gal. bottles, and storing at 35°F with no further preservative treatment.

The advantage of the use of potassium sorbate is that this chemical is completely soluble in the apple juice whereas sorbic acid is not. (See Chapter 11, Flavors of Processed Fruit.)

FORTIFYING ORANGE JUICE

There is considerable variation in the vitamin C content of oranges and orange juice. Values ranging from 27 to 67 mg ascorbic acid/100 cc orange juice have been reported. California oranges were found to have a significantly higher mean ascorbic acid content than Florida oranges. The vitamin content varied with the season of picking, cultivar, root stock differences, and geographical location; oranges on the outside of trees, where there was more sunlight, had more ascorbic acid than oranges within the trees.

It is generally accepted that a supplementary source of ascorbic acid should be added to the diets of bottle-fed infants and others in need of it; 25–30 mg/day is recommended for this purpose. The supplementation is usually given in the form of orange juice, either freshly squeezed, canned, or reconstituted frozen concentrate, or as a multivitamin product.

Since 1971, the USDA has required that fruit juices be fortified with vitamin C equal to the level of orange juice.

PREPARING FRUIT JUICES FOR BEVERAGES

Fruit juice processing is one of the nation's largest food industries, requiring some of the most advanced developments in food technology. Every fruit is different in size; nature of peel or skin; nature of the meat or pulp; sugar and acid levels; enzymatic systems; and ease of bruising. The most satisfactory method of removing the juice from each type varies. In some cases, notably apple and cranberry juice, only the clear juice is wanted; in others, such as tomato and peach, not only the juice, but the pulp is desired; while with most citrus, berry, and grape juices, a small amount of pulp is desired.

The detailed methods of juice recovery are as numerous as the fruits themselves. Not a year passes but that one or more improvements are made in methods of extracting juice of some fruit. The fruits range from oranges, apples, grapes, pears, tomatoes, lemons, loganberries, peaches, papayas, and pineapple to passion fruit and pomegranates. Following are the general steps in extracting fruit juice for beverage purposes:

1. Harvest fruit in fully ripe but slightly firm stage, using containers that will not permit the fruit to bruise. Mechanically harvested fruit may be used only if processed immediately, before enzymatic changes resulting from bruising take place.
2. Grade to remove overripe, underripe, diseased, and otherwise defective fruit. This is done by a roller grader–sizer followed by hand inspection.
3. Weigh to ascertain pay rate, yield, and product rate and cost.
4. Allow toe fruit to mellow and become slightly soft, to develop full flavor, color, and aroma and to facilitate peeling, coring, and slicing. This period may be reduced to as short as 12 hr by raising the temperature, or extended to several days by refrigerating at 38°F with 90% RH.
5. Rinse gently, but thoroughly, to remove dirt or anything that would impart color or flavor to the juice. A detergent may be used provided it is thoroughly rinsed off.
6. Prepare fruit for processing; this includes peeling, coring, deseeding, slicing, dicing, or pureeing, depending on the type of fruit.
7. Blanch or heat to about 190°F to inactivate enzymes that would cause discoloration and changes in flavor, aroma, and texture; to destroy yeast, mold, and bacteria; and to coagulate the cell contents. Fruits that are heated before pureeing or peeling need not be reheated. One of the reasons for heating is to lower the viscosity of the juice to render pressing out easier.

Most juices and purees, including apple, boysenberry, cranberry, Concord grape, Muscadine grape, and citrus juices, may be bulk-stored for year-round processing. The most recent technique is to pack the juices aseptically in 50-gal drums or in 1000- to 5000-gal tanks. All quality attributes are retained better if the juices are held under refrigeration.

8. Press out juice while pulp is hot. There are many types of presses including rack presses, expeller type, Willmes press, and others. Some apply direct pressure up to 2000 psi; some exert pressure in a sliding motion; and some apple a moderate pressure, rework the pulp, and apply pressure again. This may be repeated several times, resulting in a slightly different quality of juice with each pressing. Sometimes a small amount of hot water or steam is added after the first pressing.

9. Clarify juice with pectinase enzyme by adding a corrected amount to the juice and allowing it to stand at room temperature for a given number of hours. This is applicable only to those juices (e.g., peach, nectarine and apple) that are undesirably "heavy." The purpose is to dissolve the pectin so that the juice can be filtered.

If it is desired to maintain, or build up, the "body" of the juice as is the case with grape, tangerine, or tomato juice, pectinase enzyme is not added; instead a natural gum or starch emulsifier is added.

10. Remove suspended material by filtering and/or centrifuging. This should follow immediately after depectinization of the clarified juices or addition of gum to heavier beverages. Following this, the beverage will have assumed its final character.

11. After the juice has been identified as to its natural color, flavor, aroma, acidity, sugar content, pH, viscosity, vitamins, minerals, protein, starch, and minor constituents, it is time to determine (a) if any one of these needs to be modified for an acceptable and competitive beverage, or (b) if a particular juice is to be blended with one or more juices for a marketable beverage. To answer these questions, every beverage manufacturer needs active departments for research and development and for quality control.

12. The beverage is now ready to preserve by an appropriate method: (1) bottled, canned, or pasteurized; (2) concentrated, canned, or frozen; (3) spray-dried, drum-dried, freeze-dried or foam-mat dried; (4) treated with a chemical, such as sodium benzoate, sorbates, or other preservatives; or (5) preserved by refrigeration as a fresh fruit drink.

Before processing it is necessary to determine the "image" of the final beverage product. The sugar, pectin, and acid content, as well as the color, flavor, and consistency should be decided upon; and the container and method of preservation should be chosen. For example, one should decide if the beverage will be clear, semiclear, or opaque; whether it will be concentrated, drunk undiluted, or served with ice; and whether it will be a complete drink or blended with another liquid to make a cocktail or punch.

Questions in reference to the container are whether the beverage will be marketed as individual drinks, split-size, king-size or family-size. And finally, it should be determined if the beverage will be carbonated, vacuum-packed, or in reclosable containers.

REFRIGERATED FRUIT SLICES

The growth of glass-packed, refrigerated, sliced, ready-to-eat fruit has been phenomenal since about 1960. Hardly a fruit has escaped this method of short-time preservation. The advantages are that the fruit is thoroughly ripe; there is no heating, freezing, or dehydration; the fruits are sweetened to taste, and packed alone or in combination with other fruits in consumer-size or institutional-size containers; and the refrigeration plus a preservative extends the "season" of a particular fresh fruit by 3 or more months.

The trend toward marketing fresh fruits in prepared, ready-to-serve forms has increased with the demand for convenience foods. Examples are citrus sections, peach slices, cantaloupe wedges, pineapple chunks, melon balls, and combination fresh fruit salads. The process of preserving these and other fruits by the use of chemical agents and low temperature (short of freezing) is well known. In addition to refrigeration, acidulants, antioxidants, microbial inhibitors, and sweeteners are used.

The cultivars used are usually those that are preferred for eating in the fresh state. Harvesting and handling are similar to that for fresh fruit or frozen fruit. The critical factors in processing are selection of the fruit, fast preparation, applying the preservatives immediately, and chilling quickly. The time between beginning of peeling and covering with a preservative is in minutes.

Apples

Refrigerated, unfrozen sliced apples have been used extensively for pies and other bakery products because of their convenience, year-round availability, and firm texture. Treatment to preserve the color of these sliced apples has usually involved a sulfite dip, sometimes with an added treatment. Processing of apple slices for refrigeration differs from freezing, since the latter destroys cell organization and allows enzymes and substrates to mix. In frozen apple tissue there is a considerable amount of oxygen, so that oxidation catalyzed by polyphenol oxidase occurs, resulting in rapid browning, especially during thawing. In refrigerated apples, the cells remain intact and alive except on the surface. Thus treatment of the surface is all that is necessary, in contrast to the penetrating treatment necessary for freezing. For slices to be refrigerated, it is both unnecessary and undesirable to use a penetrating sulfite treatment.

There is a definite effect on shear strength from variations in pH. Acid solutions soften apples and alkaline solutions harden them, except at high pH. The maximum firming effect is at about pH 9. The kind of acid used also influences not only the shear strength but the tendency to turn dark. Sulfurous acid is best for maintaining a light color, but it severely softens fruit. Acetic acid both softens and darkens apple slices. The degree of penetration of SO_2 is also greater at low pH values.

Since refrigerated apple slices are used mainly because their texture is crisp, a process that causes the least penetration of SO_2 but maintains the color is best, provided that flavor is not adversely affected. Calcium treatment of sulfited apples has been found to be very effective in firming apples in alkaline solution or suspension, but not in acid solution. With unsulfited apples there is a firming effect in both acid and alkaline solutions. Thus, by using suitable proportions of calcium and SO_2 in a dip of the proper pH, the qualities of firmness, color, and flavor can be balanced to give the most desirable product after storage. The storage life can be extended by this means from a maximum of about 3 weeks to as long as 8 weeks at 34°F.

Ponting *et al.* (1971) found texture and flavor parallel color retention in apple slices treated with ascorbic acid or SO_2 combined with calcium. Sulfur dioxide is, like ascorbic acid, synergistic with calcium. A treatment with 0.03% SO_2 and 0.1% calcium is sufficient to protect the color of apple slices for several weeks at pH 3.5–7.

Peaches

Interest in processing refrigerated fresh peach slices began about 1967 when two plants went into production and studies were made on suitable cultivars, pretreatments, antioxidants, microbial inhibitors, sweeteners, and containers.

Studies showed that ascorbic acid protected natural peach color more effectively than a peroxidase inhibitor, oxygen acceptor enzyme, or water emulsion of butylated hydroxyanisole. Sodium benzoate was found to be equal or superior to potassium sorbate, sodium bisulfite, diethylpyrocarbonate, or mixtures of two or more of these for preventing microbial spoilage. The primary reasons for the limited shelf life of refrigerated peach slices were the loss of fresh natural flavor, followed by development of a "seedy" almond flavor, and gradual softening of the texture. The stability of flavor varied with cultivars; some remained stable for 24 weeks and others only for 12 weeks.

Heaton *et al.* (1969) have outlined the steps in preparing refrigerated peach slices:

1. Harvest melting-flesh, freestone peaches in firm-ripe stage; deliver to processing plant and allow to mellow 1–2 days.
2. Grade fruit for maturity and uniformity.
3. Peel with boiling 2–5% lye spray or dip.
4. Remove peel and lye with soft brushes and enough water to clean brushes.
5. Neutralize residual lye in 1% citric acid spray or dip.
6. Slice in convenient sizes for packing.
7. Grade slices for defects and uniformity.
8. Coat 8 parts peaches with 1 part sugar containing 0.15% ascorbic acid and 0.06% sodium benzoate.
9. Mix gently and fill into sterile glass or plastic jars.
10. Close jars with vacuum or screw caps.
11. Cool jars quickly to 40°F in center.
12. Pack cooled jars in cases and store at 30°–32°F for up to 20 weeks.

Pasteurized Cantaloupe

Highly acceptable products may be made from firm-ripe cantaloupes by acidifying to pH 3.8 and processing in glass jars. The fresh product is cut into 1-in. cubes, dipped in 0.5% sodium metabisulfite solution for 3 min, steam-blanched for 30 sec, packed into jars and covered with 72° Brix syrup, vacuum closed at 120°F, and processed to a jar temperature of 160°F. The pasteurized product has a fresh flavor, but must be held at 32°F.

BENZOIC ACID AS A PRESERVATIVE

Benzoic acid and benzoate salts are found in many natural products, including cranberries, green gage plums, prunes, huckleberries, raspberries, currants, and others. The excellent keeping qualitites of these natural foods and their juices is largely due to the presence of benzoates.

The acidity of the particular product is important. As the acidity decreases, so does the preservative power of sodium benzoate. Higher concentrations of benzoates are required for preservation when the acidity is less than pH 4.5. Because of this, sodium benzoate is only recommended for preserving acid products. The most effective preservation is done with foods that naturally have a pH of 4.5 or lower, or products where acidity can be adjusted. Most fruits and fruit juices do not require any pH adjustment since their natural acidity is usually pH 3.5. The choice between the use of the acid or sodium salt is largely dictated by conditions of use.

Directions for the use of benzoic acid:

1. Add sodium benzoate promptly as the juice or pulp comes from the press. Even a few hours' delay may permit fermentation to begin.
2. Do not add sodium benzoate in powder form. The best method is to dissolve 1 lb in sufficient juice or water to make 1 gal. of solution. Each pint will then contain 2 oz of sodium benzoate.
3. When preserving very thick, viscous products, dissolve the preservative in water and add the viscous ingredient last. Then mix thoroughly to assure distribution of the benzoate.
4. An amount of 0.1% of sodium benzoate is usually sufficient to preserve a product that has been properly prepared and adjusted to pH 4.5 or lower.
5. Clean all receptacles such as bottles, carboys, and barrels thoroughly with steam, hot water, or a 0.02% solution of sodium hypochlorite.

PREVENTION OF FRUIT BROWNING

Oxidative browning, which is caused by the action of oxidase with catechol tannins, is a very important problem in handling most fruits, especially peaches, nectarines, apples, and cherries. Discoloration is encountered in all methods of processing, but mainly with methods that do not involve cooking, such as freezing, drying, and making juices. Changes in flavor accompany changes in color. Browning may take place before peeling due to bruising, but is accelerated once the skin is broken or the tissue cells are ruptured. There are several means of controlling browning in particular fruits for special purposes.

Sulfurous Acid

A sulfurous acid dip for 2–5 min in a solution of 2000- to 4000-ppm SO_2 effectively prevents peeled peaches, apples, apricots, nectarines, and cherries from browning. A 1-min dip in a sodium bisulfite solution of 2000–3000 ppm SO_2 concentration, followed by a holding period determined by the temperature and the solution, may be used. Shorter times or weaker concentrations will often prevent browning. Some fruits and cultivars have a much greater tendency to turn brown than others. When the residual concentration of SO_2 in the fruit is greater than 75 ppm an objectionable sulfur taste occurs in peaches.

One way to use SO_2 to prevent browning is as a gas dissolved in water;

if the solution is weak enough, the fruit can be packed in the acid solution. Penetration of the solution can be greatly facilitated by the use of vacuum.

High Acid

High acidity prevents browning. The juice of a wide variety of peaches ranges from pH 3.48 to 3.82 with an average of 3.64. These values vary slightly from year to year, from different locations in the orchard, and rise as the fruit ripens. Enough residual lye from peeling operations to raise the pH to 4 will markedly affect browning of peaches and apricots. It is very necessary to remove traces of lye from the surface of lye-peeled fruits.

In making purees, juices, or cocktail fruit, the acidity of some fruits can be raised by blending with fruits of higher acidity such as lemons, limes, pineapples, or oranges. Acids used to raise acidity include citric, fumaric, tartaric, acetic, phosphoric, and ascorbic.

Ascorbic Acid

Ascorbic acid, or vitamin C, came into general use as an antibrowning agent for frozen peaches and apricots about 1940. The small amount used does not affect the taste of fruits and juices, and its use soon became common in most fruit juices and canned fruits. It may be used alone, in dry sugar/citric acid/ascorbic acid mixes, or in syrups according to the following formulas (in pounds):

Formula I for Very Ripe Peaches Packed in Dry Sugar

Sliced peaches	1000
Sugar	100
Citric acid	1
Ascorbic acid	½

Mix ascorbic acid, citric acid, and sugar; then add 10 lb with each 1000 lb of fruit.

Formula II for Ripe Peaches Packed in Syrup

Sliced peaches	1000
Sugar	100
Citric acid	1
Ascorbic acid	½

Mix dry ingredients in water and add 10 lb to each 100 lb fruit.

Since high temperatures cause ascorbic acid solutions to lose strength, these should be made up with cold water as required. The powders are stable for more than 1 year, provided they are kept dry and cool. Citric acid is a stabilizer for ascorbic acid and is needed only in medium-to-low acid fruits and juices.

As most of the browning of frozen and dried fruits occurs on the surface in large containers, a "cap" of the dry mix is often as effective as mixing it with the entire batch, and requires much less ascorbic acid.

Sugars

The addition of sugar, either dry or syrup, is an excellent means of preventing browning in peeled and sliced fruit—especially peaches, apricots, apples, pears, cherries, or berries—by partially excluding air. Very ripe fruit will release enough juice to dissolve dry sugar and form a saturated syrup. The juice of peaches averages about 10.5% soluble solids, most of which is sugars. Sugar, when added to fruit for freezing, helps to hold the flavor and texture, as well as the color. Of course, the chief purpose in adding sugar is to sweeten the product.

The amount of sugar added to fruit depends mainly on the finished product—frozen, glacéd, pickled, canned, dried, concentrated. In frozen fruits, it is added in the ratio of 1 part sugar to 2, 3, 4, or 5 parts of fruit and is referred to as 1:2, 1:3, 1:4, 1:5. The amount of sugar used also depends upon the kind and cultivar of fruit, and personal preferences. The increased interest in low-calorie foods and artificial sweeteners has altered the sweetening method of many kinds of fruit products.

Dry sugar, or any concentration of syrup above that of the juice of the fruit, will cause the fruit to shrink and the amount of liquid in the container to increase. A syrup-pack has several advantages over a dry-sugar pack of frozen fruit: (1) there is little or no change in the volume of the fruit and the volume of the liquid in the container; (2) discoloration due to air pockets in the container is reduced; (3) it is easier to get an even distribution and coverage with syrup than with dry sugar; (4) syrup is better aid in preserving the color and texture on freezing and defrosting; and (5) the use of chilled syrup assures a more rapid rate of cooling of the fruit.

To be most effective in preserving the color, flavor, and texture of the fruit, sugar might well be used in conjunction with ascorbic acid and citric acid.

Vacuum Packaging

Vacuum packaging may reduce browning. While vacuum packing frozen or dried fruit, preserves, jams, or glacéd fruit is not adequate for prevention of browning, it is a valuble adjunct in prolonging the period before browning becomes noticeable. Vacuum packing is also valuable because it removes air from the fruit tissue, and increases the speed and amount of penetration of syrups.

USE OF ASPARTAME

The FDA has issued a final regulation allowing the use of aspartame as a sweetener in carbonated beverages, thus paving the way for the marketing of low-calorie soft drinks that do not contain saccharin. Aspartame, made from L-aspartic acid and phenylalanine, is about 180 times as sweet as sucrose and, unlike saccharin, has no aftertaste. The manufacturer, F. D. Searle & Co., initially petititioned FDA for use of aspartame in dry products in 1973. After several delays and an extensive review of all safety data, FDA issued a final regulation in July 1981 allowing use of aspartame in dry products. Searle petitioned for use of aspartame in carbonated beverages in the United States in August 1982. Aspartame has been accepted for use in carbonated beverages in four other countries and has been used in soft drinks in Canada since 1981 (Anon. 1983B).

IRRADIATION

Research of the use of irradiation as a means of preserving foods in the United States was begun about 1945 by the U.S. Food Container Institute for the Armed Forces. In 1955, the FDA took the program over; and in 1980 it was transferred to the USDA.

According to the *Federal Register,* February 14, 1984, a proposed regulation would permit food to be irradiated to inhibit the growth and maturation of fresh fruits and vegetables, to disinfect food of insects at doses not to exceed 100 kilorads (krad), and to disinfect spices of microbes at doses not to exceed 3 Mrad (Anon. 1983A).

Fruit irradiation, in the United States, is regarded by some as an additive (Roe 1967), by some as a "cold processing method" (Anon. 1983A), and by others as a "method of fruit processing" (Takeguchi 1983), and as such it is subject to the food and drug regulations.

USDA scientists are considering the use of irradiation of fruits with caution because of phytotoxin damage to the fruit even at the very low dose rates required for insect quarantine purposes (Thayer and Harland 1983). Burditt *et al.* (1981) reported that doses of 25 krad prevented the development of adult Caribbean fruit flies in irradiated grapefruit; but the irradiated fruit had increased thin pitting, scald, decay, and changes in the taste.

Irradiation is used as a preprocessing treatment against certain orchard and storage pests. Fruit flies of the Tephritidae family are among the most important pests to be considered in developing quarantine treatments. Three species are important: the Oriental fruit fly, *Dacus dorsalis;* the melon fly, *D. cucurbitae;* and the Mediterranean fruit fly, *Ceratitis capitata.* In addition to fruit flies, there are many other pest species of insects, not found in the United States, that are of quarantine importance; these may result in total exclusion of certain fruits and vegetables from being brought into the United States. These include a number of moths, beetles, flies, mites, and other pests.

In order for a proposed irradiation quarantine treatment to be accepted the following must be demonstrated: (1) the treatment must be effective against the pest species, under the conditions specified; (2) the treatment should not have any adverse effect on quality, storage, or composition of the fruit treated; (3) the treatment should not have a residual effect on the fruit; (4) the treatment must be feasible to implement.

Since about 1950 there have been many studies on the effect of gamma radiation on fresh fruits and vegetables. For the purpose of quarantine treatment against fruit flies and other pests, concern has been primarily on dosages below 100 krad, applied to papayas, mangoes, citrus, and other fruits.

Dennison *et al.* (1966) conducted extensive studies on the effects of irradiation on various types of citrus. They found an increase in peel injury when fruit were irradiated at 100, 200, or 300 krad and stored for 7, 14, or 21 days at 40° or 70°F. The severity of injury increased with larger dosage, longer storage time, and increase in temperature. Peel injury was reduced by storage at 2°C either alone or in combination with a 5-min hot-water dip at 53°C (Dennison and Ahmed 1975).

Recently, research was conducted on the effects of radiation from different sources on grapefruit in Florida (Burditt *et al.* 1981). In these studies, fruit were exposed to 25 or 50 krad of X-rays, 25 or 50 krad from a cobalt-60 source or, to 30 or 60 rads from a cesium-137 source. Treated fruit stored for 4 weeks at either 10° or 15°C had a higher incidence of scald and decay than nonirradiated fruit, and this became more pronounced after the fruit was held for an additional week at 21°C. The button of treated fruit had a black necrotic injury. Taste panels found statistically significant differences in the taste or odor of pasteurized juice samples and fresh fruit sections of some treatments when compared with other treatments as well as with controls. Some irradiation treatments reduced the vitamin C content of fresh juice 3–7.5% but did not significantly affect the oil and acid content. Analyses of the peel of irradiated and untreated grapefruit did not show significant quantitiative or qualitative differences in major volatile constituents.

Papayas have been studied intensively in Hawaii as part of a project to develop background information on tolerance and shelf-life extension, as related to potential use of γ-irradiation for disinfestation of papayas. Dollar*et al.* (1971) reported that irradiation caused no significant differences in total soluble solids, dry matter, ascorbic acid, color, or texture of papayas. They found that the shelf life of irradiated papayas was equal to or better than that of nonirradiated fruit. Softening was delayed by irradiation treatment.

Butditt *et al.* (1981) showed that irradiation caused dark sunken areas on mangoes in Florida and that the degree of such injury increased with dose from 25 to 150 krad. This injury was similar to that reported by Sreenivasen *et al.* (1971), who found that coating fruit with a 6% emulsion of an acetylated monoglyceride or treating fruit in nitrogen reduced skin discoloration. Akamine and Goo (1979) reported that surface scale occurred in irradiated mature green 'Haden' mangoes and in more mature fruit that was refrigerated and irradiated at 75 or 100 krad.

Studies have shown that harmful effects may result from application of gamma irradiation at doses useful as a quarantine treatment. These effects may vary with the type and maturity of fruit treated, as well as the dosage and method of application.

REFERENCES

AKAMINE, E. K. and GOO, T. 1979. Effects of ionizing irradiation on 'Hagden' mangoes. Univ. Hawaii Agric. Exp. Stn. Res. Rep. *205*.

ANON. 1969. Fresh banana purée keeps 6 months without heat pasteurization. Food Process. *30* (1) 38.

ANON. 1970. The use of citrates in electrolyte drinks. Pfizer Chem. Div., New York, Inf. Sheet *2016*.

ANON. 1971. Salt-free olive curing. Food Eng. *43* (4) 134–135.

ANON. 1972A. Pineapple granules have wide usage possibilities. Food Process. *33* (5) 28–29.

ANON. 1972B. Extracts 195 gal. juice per ton of raw grapes. Food Process. *33* (6) 36.

ANON. 1973A. Standards, proposals. Food Prod. Devel. *6* (8) 10.

ANON. 1983A. Radiation preservation of foods. Food Technol. *37,* (4) 55–60.

ANON. 1983B. Aspartame for use in soft drinks. Food Technol. *37,* (8) 48.

AVENA, R. J. and LUH, B. S. 1983. Sweetened mango purées preserved by canning and freezing. J. Food Sci. *48,* 406–410.

BEROLZHEIMER, R. 1940. American Woman's Cook Book. Butterick Publishing Co., London.

BOLIN, H. R. and SALUNKHE, D. K. 1972. Develops continuous process for prune juice. Food Eng. *44* (4) 129–130, 133.

BOLIN, H. R., FULLER, G. and POWERS, J. 1973. Product development and application for a dry apricot concentrate. Food Prod. Devel. *7* (2) 30, 32.

BOMBEN, J. L., GUADAGNI, D. C. and HARRIS, J. G. 1971. Apple-flavored thickener from apple peel. Food Technol. *25,* 1108–1117.

BUNNELL, R. H. 1968. Enrichment of fruit products and fruit juices. J. Agr. Food Chem. *16,* 177–183.

BURDITT, A. K., JR., MOSHONAS, M. G., HATTON, T. T., SPALDING, D. H., von WINDEGUTH, D. L. and SHAW, P. E. 1981. Low-dose irradiation as a treatment for grapefruit and mangoes infested with Caribbean fruit fly larvae. Agric. Research Results, *ARR-S-10*. U.S. Dept. Agric., Agric. Res. Serv.

CHAN, H. T., JR. and CAVALETTO, C. G. 1978. Dehydration and storage of papaya leather. J. Food Sci. *43*, 1723–1725.

CHAN, H. T., and CAVALETTO, C. G. 1982. Asceptically packaged papaya and guava puree: changes in chemical and sensory quality during processing and storage. J. Food Sci. *47*, 1164–1169.

COFFELT, R. J. 1969. A system for heating wine grapes for color release. Trans. Amer. Soc. Agric. Eng. *12* (6) 873–875.

COFFELT, R. J. and GIANINI, G. R. 1965. Color releasing machine. U.S. Pat. 3,203,339. Aug. 31.

DUAM, R. J. 1981. Private communication. U.S. Dept. Agriculture, APHIS, Hyattsville, MD.

DENNISON, R. A., GRIERSON, W. and AHMED, E. M. 1966. Irradiation of Duncan grapefruit, pineapple, and Valencia oranges and Temples. Proc. Fla. State Hort Soc. *79*, 285.

DEHNISON, R. A. and AHMED, E. M. 1975. Irradiation treatment of fruits and vegetables. *In* Postharvest Biology and Handling of Fruits and Vegetables. N. F. Haard and D. K. Salunkhe (Editors). AVI Publishing Co., Westport, CT.

DOLLAR, A. M., McCLISH, G. A. and MOY, J. H. 1971. Semi-commercial-scale studies on irradiated papaya. *In* Disinfestation of Fruit by Irradiation. IAEA, Vienna.

DuPREE, W. E., WOODROOF, J. G. and SIEWERT, S. 1953. Watermelon rinds in food products. Ga. Exp. Stn. Bull. *85*.

GERBER, R. K. and SALUNKHE, D. K. 1960. Vitamin enriched apple juice. Univ. Utah Farm Home Sci. *21*, 48–49, 63–64.

HEATON, E. K., BOGGESS, T. S., JR. and LI, K. C. 1969. Processing refrigerated fresh peach slices. Food Technol. *23*, 96–100.

KANNER, J., FISHBEIN, J., SHALOM, P., HAREL, S. and BEN-GERA, I. 1982. Storage stability of orange juice concentrate packed aseptically. J. Food Sci. *47*, 429.

LaBELLE, R. L. 1971. Heat and calcium treatments for firming red tart cherries in a hot-fill process. J. Food Sci. *36*, 323–326.

LIVINGSTON, L. 1967. Develop new low-moisture process for fruit flakes. Canner/Packer 136 (3) 36–37.

MATSUMOTO, S., OBARA, T., AND LUH, B. S. 1983. Changes in chemical constituents of kiwifruit during postharvesting ripening. J. Food Sci. *48*, 607–611.

MOCHIZAKI, K. 1971. Method for producing candied fruit. U.S. Pat. 3,615,687. Oct. 26.

MONDAVI, R. M. and HAVIGHORST, C. R. 1971. Winemaking surges ahead. Food Eng. 43 (2) 59–62.

MORGAN, D. A., VELDHUIS, M. K., ESKEW, R. K. and PHILLIPS, G. W. M. 1953. Studies on the recovery of essence from orange juice. Food Technol. *7*, 332.

NUMATA, M. and SUGANO, K. 1980. Process for producing fried banana slices. U.S. Pat. 4,242, 365. Dec. 30.

PONTING, J. D., JACKSON, R. and WATTER, G. 1971. Refrigerated apple slices. Effects of pH, sulfites and calcium on texture. J. Food Sci. *36*, 349–350.

PRESCOTT, S. C. and PROCTOR, B. E. 1937. Food Technology. McGraw-Hill Book Co., New York.

ROBERTSON, G. L. and SWINBURNE, D. 1981. Changes in chlorophyll and pectin after storage and canning kiwi fruit. J. Food Sci. *46*, 1557–1559.

ROE, R. S. 1967. Radiation of food. FDA food additive requirements. FDA Papers *1*, 25.

SREENIVASAN, A., THOMAS, P., and DHARKER, S. D. 1971. Physiological effects of γ-radiation on some tropical fruits. Bhabha At. Res. Center, Bombay. Disinfestation Fruit Irradiation, Proc. Panel 1970 SAEA, Vienna, 65–91.

STINSON, E. E., DOOLEY, C. J., FILIPIC, V. J. and HILLS, C. H. 1969. Composition of Montmorency cherry essence. J. Food Sci. *34,* 246–248.

TAKEGUCHI, C. A. 1983. Regulatory aspects of food irradiation. Food Technol. *37* (4), 44–45.

TAYLOR, P. V. 1966. Candied fruit firm expands. Canner/Packer *135* (11) 34–35.

THAYER, D. W. and HARLAND, J. W. 1983. Statue of USDA food irradiation programs. Food Technol. *37* (4), 46–47.

THRASH, N. 1968. Canning for your family. Univ. Georgia Coop. Ext. Serv. Bull. *602.*

VELDHUIS, M. K., BERRY, R. E., WAGNER, C. J., LUND, E. D., and BRYAN, W. L. 1972. Oil- and water-soluble aromatics distilled from citrus fruit and processing waste. J. Food Sci. *37,* 108–112.

WEISSER, B. AND HAVIGHORST, C. R. 1972. Fruit press handles 100 tons per day. Food Eng. 44 (1), 74–75.

WILSON, E. L. and BURNS, D. J. W. 1983. Kiwifruit juice processing J. Food Sci. *48,* 1101–1105.

WOODROOF, J. G., CECIL, S. R. and DuPREE, W. E. 1956. Processing muscadine grapes. G. Exp. Stn. Bull. N.S. *17.*

11

Flavor and Color of Fruits as Affected by Processing[1]

R. L. Shewfelt

Flavor and color are critical factors in the evaluation of fruit products. Color and other appearance defects may cause a consumer to reject a product before tasting it, or more subtly, the appearance of a product may change flavor perception. If appearance is acceptable, flavor is the primary sensory characteristic used in judging the quality of fruit products. Thus, any fruit processor must be aware of the role of color and flavor in product quality and the controlling factors in maintaining these characteristics.

Although processors may have little or no control over a fruit crop before it is harvested, several factors contribute to product yield and quality. Cultivar, weather conditions, spraying schedules, and cultural techniques are all important. The time between picking and the first step in the processing plant is known as the postharvest period. Since most postharvest changes in quality of fruit for processing are undesirable, this is an area of concern. As a general rule, once quality is lost during a process it can never be recaptured. Prudent postharvest management of the raw product can pay off in final product yield and quality.

Postharvest quality changes in fruits are the result of normal metabolic processes and physiological responses to stress. When viewed from a

[1]This chapter prepared for first edition by the late A. Lorne Shewfelt

Commercial Fruit Processing, 2nd Edition

ISBN 0-87055-502-2

biological perspective, the time associated with optimum quality is merely an arbitrary point on a continuum of a series of metabolic reactions. Manipulation of environmental conditions (e.g., temperature, relative humidity, and atmosphere) can slow metabolism and help to maintain quality. Physiological stress such as microbial invasion, chilling of chill-sensitive species, and mechanical damage can lead to speeding up of normal physiological changes or result in abnormal compounds known as phytoalexins or stress metabolites.

Fruit ripening is a complex process that results in the flavor and color attributes desired by consumers. Certain fruits such as grapes cease ripening when picked, whereas other fruits such as apples and peaches continue to ripen; certain pears will not ripen until harvested. The optimum time of harvest should be based not only on the quality at picking but also on quality changes during postharvest handling. Postharvest changes in the BAR (Brix-to-acid ratio) or in the delicate balance of flavor components may have deleterious effects on final product quality. Browning is usually the most adverse color change in fruit products.

Flavor can be subdivided into taste and aroma components. The basic taste sensations (sweet, bitter, sour, salty, and astringent) combined with the myriad of volatile flavor components result in characteristic fruit flavors. The BAR, a measure of the sweet-to-sour composition, is the standard quality control test for processed fruit products. Further flavor analysis may be performed using instrumental techniques and sensory panel evaluation. Instruments produce more reproducible results but cannot predict consumer acceptance.

The three major groups of plant pigments in fruits are chlorophyll, carotenoids, and anthocyanins. The chlorophylls (green) normally fade during ripening to give way to the carotenoids (yellow, orange, and red) or the anthocyanins (red and purple). Browning reactions, which can be either enzymic or nonenzymic, usually result in lowered quality. Fruit color can be analyzed by measuring pigment content, but is frequently analyzed by colorimeters, which define the color in a three-dimensional color space.

Fruit processing involves sacrificing flavor and color characteristics to provide longer shelf life and greater customer convenience. Processed products have found their niche in our society, and in some cases the flavor or color of the finished product is preferred to the fresh fruit. Processing is considered to be detrimental to fruit quality and the processor's goal is to minimize flavor and color changes while producing a safe, wholesome product.

Heat preservation is the most destructive method of fruit processing. Fortunately, because most fruits are high in organic acids, they require a less rigorous heat treatment than low-acid foods. Use of innovative techniques such as aseptic packaging or retortable pouches also reduce process times, result in an improvement in flavor and color retention. Although the BAR is not readily affected by heat, delicate flavor components are destroyed. Plant pigments are also degraded by heat, and anthocyanins can react with metal cans.

A much milder form of fruit preservation is freezing. Frozen fruits are closer in flavor and color to their fresh counterparts than are products from other processes. Fruits are not normally blanched prior to freezing. Although blanching will prevent enzymic browning of fruit tissue, it also results in destruction of flavor components. Enzymic browning is usually inhibited by adding ascorbic acid or sulfur dioxide. Proper packaging of frozen foods is critical. Exposed areas can become dehydrated resulting in "freezer burn" which is characterized by off-odors and off-colors.

The quality changes that occur during fruit dehydration are primarily dependent on the method of drying. Sun drying produces the most severe flavor and color changes, while freeze-drying produces the least changes. Fruit juice concentration results in a loss of volatile flavor components, which evaporate with the water. Recovery of flavor essences and reincorporation into the concentrate is a way to circumvent these flavor losses. Nonenzymic (Maillard) browning reactions between reducing sugars and proteinaceous materials is a major quality problem in dried and concentrated fruit products.

Thus, when considering flavor and color changes during processing, processors cannot restrict their view to the processing plant. Growing conditions, raw product quality, and postharvest handling play a major role in determining the quality potential of the final product. Careful monitoring of process conditions can allow processors to achieve this quality potential. In this chapter, a discussion of the effects of processing on flavor changes in specific fruits and fruit products is followed by a similar discussion of color changes.

FLAVOR

Although saltiness is foreign to fruits, the other taste sensations—sweet, sour, bitter, and astringent—are all present. Sweetness results from the presence of sugars and the sugar alcohol sorbitol. Both the total and relative concentrations of sugars are important in the perception of sweetness. The three major sugars in fresh fruits are glucose, sucrose, and fructose (listed in increasing order of sweetness). Organic acids in fruits provide the taste attribute of sourness. In most fruits, acids of the citric acid cycle predominate (e.g., citric acid in citrus fruits and malic acid in pome fruits and stone fruits). Although sourness is considered detrimental in most fruits, it is considered an attribute in grapefruit, lemons, and limes.

Bitterness is present in many fruits and is ordinarily considered detrimental. Several different compounds contribute to bitter flavor sensations. Astringency is usually attributed to tannins and is generally most pronounced in immature fruits. The persimmon is an example of an astringent fruit that normally turns sweet on ripening.

The aroma component of fruit flavors is due to many delicate volatile compounds that are easily lost during processing. The volatile flavor components for most fruits have been isolated, but the precise contribution of these compounds to flavor is not known. Character impact compounds, which have been identified in many fruits, are those compounds responsible for the major aroma sensation. The subtleties of fruit flavor, however, are due to other compounds that have been more difficult to determine. Although great strides have been made to elucidate the compounds responsible for fruit flavor, it is still not possible to duplicate fresh fruit flavors in the laboratory.

Fruit flavor develops during the ripening process. Volatile constitutents are diverse in their chemical structures and functional roles in fruit tissue. Many of the components highly valued for their contribution to aroma and flavor are waste products of metabolic pathways and accumulate in fruit until processing or consumption. Many factors occur during ripening, but an excess accumulation of some compounds may contribute to an "overripe" flavor. Among the volatile components contributing to fruit flavor are alcohols, aldehydes, esters, ketones, lactones, organic acids, phenolics, sulfur compounds, and terpenoids. Metabolic pathways that result in these compounds include amino acid degradation, β-oxidation of fatty acids, fatty acid synthesis, and carotenoid degradation. For specific examples, see Nursten (1970).

Careful attention to postharvest handling techniques is necessary to maintain fruit quality. Mechanical bruising will break down the delicate barriers inside cells that separate enzymes and substrates and result in reactions that can lead to the development of off-flavors and other undesirable characteristics. Elevated temperatures increase reaction rates, hasten the conversion of ripe to overripe fruit, and worsen the damage due to bruising. Ethylene generation by certain fruits and gas-powered equipment stimulates ripening and senescence leading to changes in flavors. The fruit processor must plan harvest date to produce the highest quality at the time of processing, which is not necessarily the highest quality at the time of harvest. Minimizing bruising and cooling the fruit as soon as practicable are essential in good postharvest management.

Freezing, the most gentle processing technique, generally results in the least flavor loss. Loss of volatiles in fruit dehydration or concentration is dependent on the method used and is directly related to the amount of heat applied. Freeze-drying and vacuum concentration of juices results in little volatile loss. Essence recovery is a process by which certain volatiles are collected separately from the water vapor for possible reincorporation into the product after concentration. Any heat preservation technique destroys or converts some volatile compounds.

The primary analytical tool for identifying volatile compounds has been gas chromatography, although procedures using high-performance liquid chromatography are being developed. The major problem in flavor identification is not analytical but correlation of the sensory perception of flavor with the instrumental data.

The remainder of this section is devoted to a discussion of the effects of handling and processing on specific groups of fruit. Relevant information on taste characteristics, volatile flavor components and effects important in specific products are described.

Citrus Fruits

Over 80% of the fruit juices consumed in the United States are citrus juices. Orange juice, primarily as a canned concentrate, accounts for 85% of the citrus juice consumed (Woodroof 1980). Although orange juice is predominant, grapefruit, lemon, and lime juices are also important products. Three of the basic taste sensations play a role in citrus products. Sweetness is an important characteristic in orange juice. Sourness or tartness gives lemons and limes their "bite." A certain degree of bitterness is a desirable quality in grapefruit, although it is undesirable in orange juice. Added to these basic taste sensations are complex mixtures of volatile aromatic compounds.

The flavor of citrus products is affected by cultural techniques, postharvest handling, and processing method. The effect of preharvest factors can be summarized as follows (Kefford and Chandler 1970):

1. Rootstock affects juice yield, soluble solids, acidity, and bitterness.
2. Fruit maturity is important for optimal BAR.
3. Fruit size is inversely related to sweetness in oranges and grapefruit.
4. Nutrient status of the tree affects juice yield and BAR.
5. Pesticides can lower BAR, and plant growth regulators affect fruit maturity.
6. Soil moisture and solids content are inversely related; heat affects BAR; freezing results in higher BAR and increased bitterness.

The effect of environmental and cultural factors on citrus products has also been reviewed by Ramana *et al.* (1981).

Factors involved in the harvesting and subsequent handling of fruits have been described in Chapter 2. The harvest date is determined on the basis of color break, minimum content of juice, acid, and soluble solids as well as the BAR. Most citrus fruit harvesting is done by hand, although most ladders have been replaced with hydraulic fruit pickers. Mechanical harvesting used in conjunction with abscission compounds reduces harvesting expenses but also leads to increased bruising. The practice of dropping fruit to the ground from fruit pickers leads to hidden bruising, which is particularly harmful to grapefruit (Smoot and Melvin 1975). In addition, abscission compounds change the composition of peel oil (Moshonas and Shaw 1978) and can adversely affect juice flavor (Moshonas *et al.* 1976). Citrus fruits are usually hauled in semi-trailers from the orchard to the processing plant. Usually little deterioration in product quality is observed at this point, as citrus fruits have low rates of respiration, the distance from the orchard to the plant is relatively short, and ambient temperatures during citrus seasons are generally 70°F or below.

When the fruit arrives at the plant, it is dumped onto conveyor systems and graded. As the fruit passes through grading, extraction, and finishing, the processor must decide whether yield or quality is most important; increasing yield frequently results in decreasing flavor quality. Quality control sampling at this point is necessary to determine the BAR. Later, juice with a low BAR will be blended with juice of a higher BAR to produce a product of consistent quality (Berry and Veldhuis 1977).

Of all methods of juice preservation, canning is the harshest on volatile flavor components. Freeze concentration provides a much better quality product, particularly when the juice is concentrated to 55°–60° Brix and then diluted with fresh chilled juice back to 45° Brix. Citrus essence can be recovered and returned to the juice for a fuller flavored product (Berry and Veldhuis 1977).

Heikal *et al.* (1977) found that freeze concentration provided higher-quality lemon and orange juice products than vacuum evaporation. Freeze-drying produces an excellent product, but this method is also very expensive. Use of an encapsulated citrus aroma powder in soft drinks has been suggested by Kopelman *et al.* (1977B). Aseptically packaged citrus juices preserved by UHT provide a shelf-stable, nonrefrigerated alternative to frozen concentrates (Johnson and Toledo 1975).

Oranges. Orange flavor is a combination of volatile and nonvolatile fractions. The structure of an orange is shown in Fig. 11.1. The *flavedo* or outer peel contains the oil sacs. The white pulpy inner peel is known as the *albedo,* which may be the site of bitterness formation. The *juice sacs* are located in the inner portion of the fruit. Flavor components in processed orange products such as juice include not only compounds present

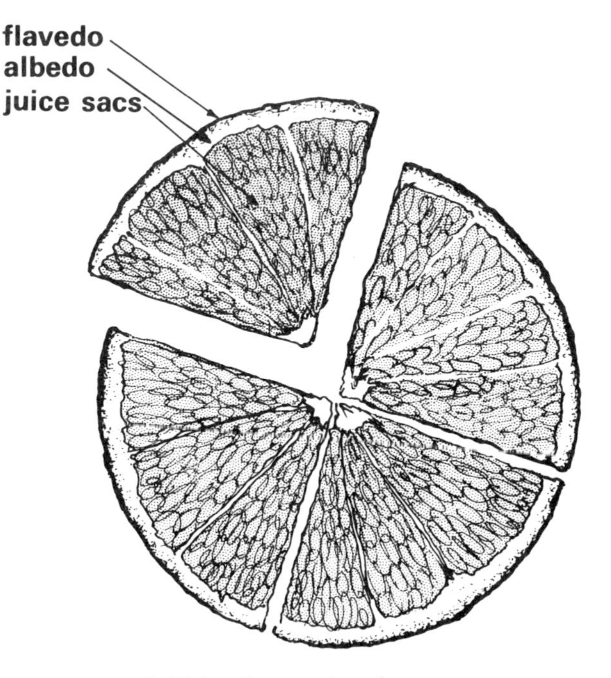

FIG. 11.1. Cross-section of an orange.

in the juice sacs but also those formed during processing and/or storage, as well as those recovered from the peel oil and recovered essence. The voluminous scientific literature on the subject has been extensively reviewed (Kealy and Kinsella 1979).

Like other fruit juices, orange juice taste is primarily the result of the BAR. Sucrose accounts for more than 50% of the total sugar present, which can range from 7.5 to 11% of the juice, depending on maturity and variety (Ting and Attaway 1971). Citric acid is the major organic acid present in the orange and the juice, with malic and succinic acids also present (Vandercook 1977).

Bitterness is an objectionable taste characteristic in orange juice. Although high acidities are associated with bitter oranges (Ting and Attaway 1971), the primary bitterness agent in orange juice is the terpene limonin. Intensive scientific investigation has led to our current understanding of limonin and its role in the bittering of orange juice. The problem of bitterness in orange juice was noted as early as 1932 and associated primarily with Navel oranges, which are more susceptible to bittering than other cultivars. Through a long, slow process limonin was isolated and identified as the bittering agent. Structural determination (see Fig. 11.2) and

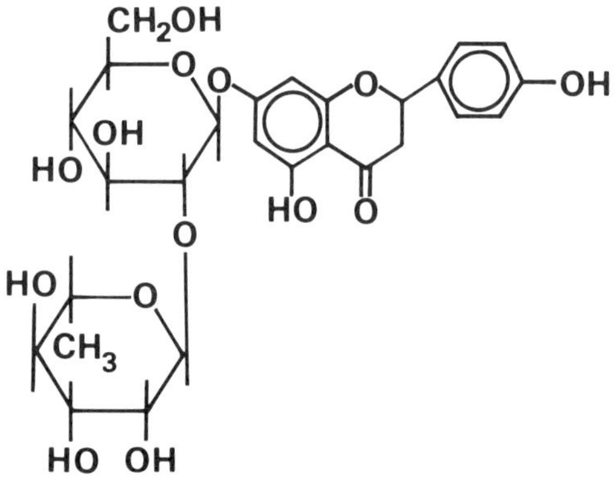

Naringen

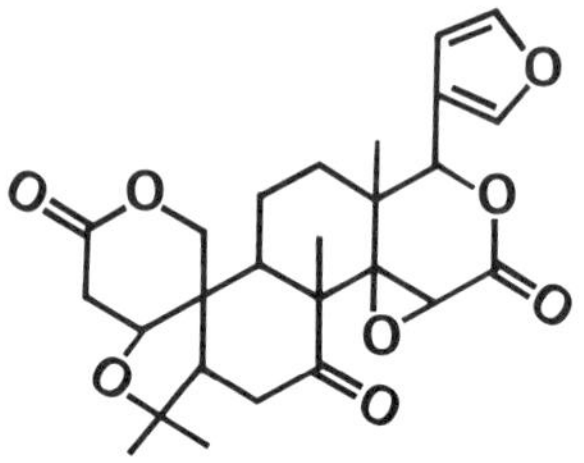

Limonin

FIG. 11.2. Chemical structure of bitter components found in citrus fruits.

methods of chemical assay led to further study on its biogenesis and metabolism. An understanding of the biochemical factors responsible for its formation and degradation led to methods to suppress or control bitterness in orange juice.

Limonin is not present in sufficient quantity in oranges or fresh orange juice to cause bitterness. Bitterness is the result of the conversion of the limonate A-ring lactone (nonbitter) to limonin (bitter) during storage, a reaction that is aided by pasteurization. This and other nonbitter precursors apparently enter the juice from the albedo during extraction and are converted to bitter compounds in the acid conditions of the juice. The bitterness associated with limonin is least likely to be detected at pH 3.8 (Guadagni *et al.* 1973). Kefford and Chandler (1970) provided evidence that the site of limonoid biosynthesis is in the albedo, but this contention has not yet been clearly established (Maier *et al.* 1977).

Orange juice has been debittered by (1) selection of cultivars that are low in limonoids or by using only late-season fruit that contain little or no limonoids; (2) using gentler methods of extraction and finishing to reduce limonoid contamination from the albedo; (3) precipitation of limonin by enzymatically removing pectin and thus reducing limonin solubility; (4) absorbing limonin in inert support systems; (5) enzymatically degrading limonoids to nonbitter components; and (6) adding compounds, such as dihydrochalcone, that counteract the bitterness perception of the limonoids (Maier *et al.* 1977). Each of these methods have their drawbacks and none are used appreciably in the industry. Absorption of limonoids on various support systems, such as cellulose acetate (Chandler and Johnson 1977) or its gel-bead form (Johnson and Chandler 1981) shows some promise. Earlier polyamide supports were unacceptable because they removed ascorbic acid as well as limonin (Kealy and Kinsella 1979).

Orange juice flavor is much more complex than the sweet, sour, and bitter taste sensations. Some 141 volatile components have been identified in aqueous orange essence (Shaw 1977B), and 56 components in distilled essential oils (Shaw 1977A). Extensive lists of these compounds can be found elsewhere (Kefford and Chandler 1970; Kealy and Kinsella 1979). The thresholds of volatiles contributing to orange juice flavor and odor have been determined (Ahmed *et al.* 1978A). Compounds found to be most important in orange flavor were acetaldehyde, citral, ethyl butyrate D-limonene, and octanol (Ahmed *et al.* 1978B). Acetaldehyde appears to be the most important flavor constituent in aqueous essence, and D-limonene shows the greatest importance in essence oil (Lund and Bryan 1977).

Finished product flavor of orange juice is dependent on preharvest, postharvest, and processing factors. One of the most important preharvest factors is rootstock. Oranges produced from lemon and lime rootstock tend to be much more bitter than those produced from orange rootstock (Kefford and Chandler 1970). Planofix applied as a preharvest spray decreased bitterness in the finished juices; lesser effects were observed using etherel and lead arsenate (Chakrawar 1978). When applied as postharvest sprays, etherel decreases bitterness, but the holding time required is usually longer than practical under current industrial practices (Maier *et al.* 1977). Preharvest application of Acti Aid, Release, and Pik-Off to aid abscission for mechanical harvesting led to off-flavors in orange juice (Moshonas *et al.* 1976). Bruising is the biggest postharvest handling problem with oranges. A sound orange is relatively stable, but bruised fruits deteriorate rapidly. Enzymes are released upon bruising that lead to the production of off-flavors.

Preparation of juice from oranges is a two-step process. First the juice is extracted by squeezing to release it from the juice sac; then it is finished to remove excess pulp, peel, rag, and seeds (see Chapter 7). Fellers and Buslig (1975) found that a "soft" extraction resulted in more acceptable flavor in canned orange juice. No differences in flavor were detected between "soft" and "hard" finishing of the juice.

The heat treatment required to can orange juice destroys many of the delicate aromatic compounds contributing to its characteristic flavor. Canned orange juice tends to be more sour and thus less sweet than frozen juice (Ennis *et al.* 1979). Only a small portion of orange juice sold in the United States is canned single-strength (Woodroof 1980). An alternate heat process is aseptic packaging. In this method the juice is heated outside the container and then filled under aseptic conditions. This milder heat treatment results in less destruction of aromatic compounds and thus a more desirably flavored beverage. Concentrates packaged aseptically show excellent stability at refrigerated temperatures (Johnson and Toledo 1975). Deaeration of concentrates before aseptic packaging increases shelf life so that good quality is retained for 8 months at 17°C (Kanner *et al.* 1982).

Some of the off-flavors observed in canned orange juice are the result of lipid oxidation. Oxidation is thought to proceed through a series of steps from the formation of free fatty acids by hydrolysis of the constituent phospholipids to oxidation of the fatty acids and their subsequent degradation to water-soluble compounds responsible for the off-odors and flavors. Although lipid oxidation is confined to the components of the oil such as D-limonene, the resultant off-odors and flavors are primarily carbonyls, which are present in the essence (Kealy and Kinsella 1979).

Juice concentration is achieved by evaporation. The two types of evaporators are the low temperature/high vacuum evaporators and the Temperature Accelerated Short Time Evaporator, known as TASTE. The industry is gradually replacing the former with the latter (Berry and Veldhuis 1977). Freeze concentration provides a beverage with a more acceptable flavor than that obtained by evaporation (Heikal *et al.* 1977), but the length of time required to remove water in this way has prevented its widespread use (Berry and Veldhuis 1977). Regardless of the method of concentration, frozen orange juice should be stored at temperatures of −18°C (0°F) or below to prevent off-flavor development. Flavor of concentrates can be enhanced by any of several methods, including incorporation of cutback juice or reincorporation of aqueous essence and essence oil.

Dehydration of juices also results in the loss of volatiles. Excellent products have been obtained by puff drying, foam-mat drying, and freeze-drying, but the expense of these processes, the availability of good-quality frozen concentrates, and high solids content have limited their production (Berry and Veldhuis 1977). Off-flavors can develop in orange juice products upon storage. In addition to bitterness, the degradation products of fatty acids and peel oil constituents can cause off-flavors. Furfural development is generally used as a chemical index of flavor quality (Kealy and Kinsella 1979). Compounds identified as extremely important in the off-flavor of single-strength orange juice are α-terpineol, 2,5-dimethyl-4-hydroxy-3(2H)-furanone, and 4-vinyl guaiacol (Tatum *et al.* 1975). Diacetyl and furfural are the major objectionable flavor compounds in moisture (2–15%) powders, while α-terpineol and 3-methyl-2-buten-1-ol cause off-flavors in intermediate-moisture (20–40%) powders (Papanicolaou *et al.* 1978). The type of container also has an effect on off-flavor development. Because fluorescent light catalyzes degradative reactions, orange juice flavor is maintained longer in fiberboard than in glass or plastic containers (Ahmed *et al.* 1976).

Grapefruit. The main grapefruit product is juice. Most grapefruit juice is canned, and it is the top-selling canned juice in the United States (Woodroof 1980). An additional manufactured product is canned grapefruit sections.

Grapefruit contains more acid and less sugar than oranges and thus produces a much more tart juice. Although this tartness is considered by some to be a quality attribute, most grapefruit juices are either sweetened to raise the BAR or blended with other juices.

Bitterness in grapefruit juice is attributed to the presence of limonin and naringin. Limonin concentration was found to correlate better than naringin content with bitterness in canned grapefruit juices (Ting and McAlister 1977). Limonin occurs at the highest concentrations in the cotyledon. It also is present in the seed coat, pith, segment membranes, albedo, flavedo, and juice vesicles in order of decreasing concentration (McIntosh *et al.* 1982).

Naringin (Fig. 11.2), the most common and most bitter of several flavanones that occur in grapefruit, is found primarily in the albedo. Flavanones do not occur as free aglycones but are glycosylated with either neohesperidose or rutinose. The former glycosides are bitter and the latter are tasteless. Thus, it is the naringin 7-β-neohesperidoside that is partially responsible for grapefruit bitterness. It is one of nature's little paradoxes that the derivatives naringin chalcone and naringin dihydrochalcone are very sweet (Horowitz and Gentili 1977). Debittering of grapefruit juice by enzyme treatment to convert naringin to nonbitter components has been achieved (Roe and Bruemmer 1977; Versteeg *et al.* 1977), but this process has not been adapted commercially.

Less information is available on the volatile flavor components of grapefruit than on those of orange. The composition of grapefruit oil (Wilson and Shaw 1978) and essence (Shaw 1977B) has been reported and in general is similar to that of other citrus fruits. Nootkatone, a bicyclic conjugated sesquiterpene ketone, is the most important flavor constituent of grapefruit (MacLeod and Buigues 1964). This compound, which is responsible for the uniqueness of grapefruit flavor, can impart bitterness at concentrations higher than normally found in intact fruit. Another important flavor constituent is 1-*p*-menthene-8-thiol (Demole *et al.* 1982). Although this thiol is only present in very small amounts, it provides a characteristic grapefruit juice aroma at proper dilutions, making it one of the most potent flavor components discovered.

Grapefruits, like other citrus fruits, are not climacteric and do not require picking during a short peak period like many other fruits. Citrus fruits can be essentially "stored" on the tree until just before processing. As the season progresses, fruit yield increases (Ramirez and Krezdorn 1975) and bitterness potential decreases (Maier *et al.* 1977). If harvest is delayed too long, however, tree yields the following year will be reduced.

Canned grapefruit juice develops off-flavors during storage resembling those of canned orange juice. The extraction and finishing steps are critical to the quality of grapefruit juice. "Soft" extraction and finishing lowers limonin and naringin concentrations in the juice, resulting in higher flavor scores (Dougherty *et al.* 1977). "Softer" juicing procedures sacrifice yield for improved quality, but too "soft" an extraction will result in a thin or weak flavor. Deaeration and deoiling of the product can reduce the susceptibility of the product to lipid oxidation. Vacuum distillation is used to remove excess oil. The high acidity of grapefruit juice lowers heating requirements for pastuerization, resulting in less volatile destruction. A fuller flavored beverage can be produced by adding back grapefruit essence or producing a frozen concentrate, but the difference between these and canned grapefruit juice is much less than that between the orange juice counterparts (Berry and Veldhuis 1977).

Lemons and Limes. Unlike juices made from other citrus fruits, lemon and lime juices are not normally consumed as such. Lemons and limes have much lower sugar concentrations and much higher acid (primarily citric) concentrations than oranges and grapefruit. Both lemonade and limeade contain large amounts of added sugar. Both products are used as flavor enhancers in many applications. Lemon oil also is an important citrus product.

Citric acid is the major taste component in both lemons and limes. In these fruits the acidity is much more important in assessing quality than the BAR. Bitterness is a problem in lemon and lime products; limonin is the primary bitter principle present. Although the volatile fractions of lemons and limes are not as well characterized as those from oranges, some composition data are available for essential oils (Shaw 1977A) and essences (Shaw 1977). Citral is the major volatile flavor component of lemons (Lund and Bryan 1976) and contributes to the flavor of limes. The other major component of lime flavor is 1,8-cineole (Shaw 1977A,B).

Much of the effect of processing on the flavor of lemon and lime products is the same as that for oranges and grapefruit. However, lemons and limes are picked on the basis of size not color. As they arrive at the processing plant, bruised and overripe fruit go immediately to the juicing operation. The remainder of the crop is "cured" to increase juice yield and citric acid. A green-fruit flavor is observed in juice from uncured fruit and a stale flavor is observed if curing is too long.

Pome Fruits

In contrast with the strong character of citrus flavors, apples and pears are noted for their subtle, delicate flavors. Perceived sweetness as measured by percentage of soluble solids and the BAR is very important. Cultivar differences are pronounced, and it is of critical importance to match the cultivar with the processing application. Pome fruits are unique in that fructose is the predominant sugar present.

Pome fruits provide the processor with some unique possibilities with respect to product availability. Unlike most fruits which are highly perishable and must be processed as soon as practically possible, apples and pears are quite stable to storage, allowing the processor greater flexibility in scheduling. Cold storage and controlled-atmosphere (CA) storage further extend raw product shelf life, but at a cost that might be prohibitive for a given process.

Apples. The most important flavor consideration in most apple products is the BAR. Great variations in the BAR are observed as the result of cultivar, preharvest conditions, harvest date, and postharvest handling conditions. Unfortunately, there are no consistent guidelines for proper BAR values for specific products. Regional preferences vary greatly, and products valued in one part of the country may be rejected in another area (Moyer and Aitken 1980). Thus, a product aimed at a state or region should excel in characteristics prized locally, while one marketed nationally must strive for general acceptability. Fructose accounts for 50% or more of the total sugars present in fresh and processed apples. Malic acid is the predominant organic acid of apples, although citric and numerous other acids are present in measurable quantities. The presence of tannins, particularly in immature fruit, are responsible for an astringent taste.

Volatile flavor components in apples vary with cultivar, preharvest conditions, fruit maturity, and storage conditions (Dimick and Hoskin 1983). An example of the effect of cultivar and processing on apple volatiles is shown in Fig. 11.3. The character impact compounds of "Red Delicious' essence are ethyl-2-methylbutyrate (Fig. 11.4), hexanol, and *trans*-2-hexenal (Flath *et al.* 1967). Hexanol and *trans*-2-hexenal are also

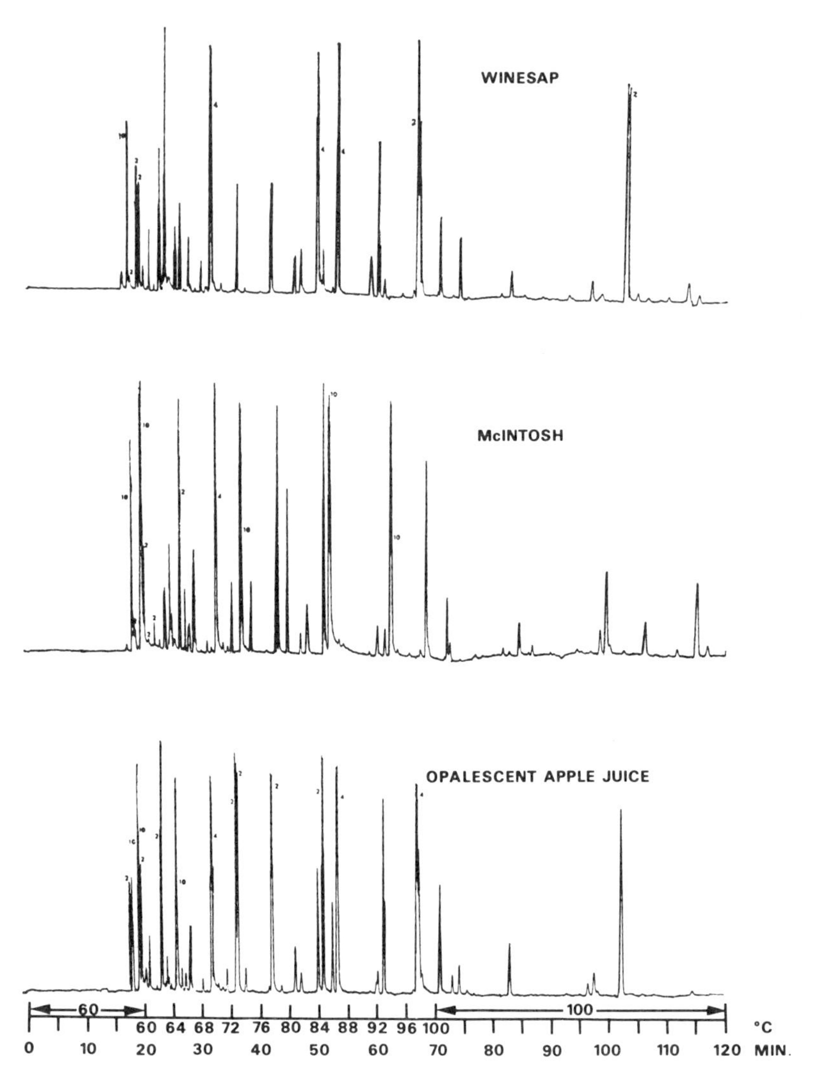

FIG. 11.3. Gas chromatograms showing flavor components of fresh apple cultivars and an apple product. (From Flath et al. 1969.)

important volatile constituents of 'McIntosh' apples (Panasuik *et al.* 1980); another important constituent observed in several cultivars is 4-methoxyallylbenzene (Williams *et al.* 1977). The apple peel is a source of much of the apple flavor and an important contributor to apple essence (Guadagni *et al.* 1971A).

An increase in the amount of acetic acid, butanol, and other objection-

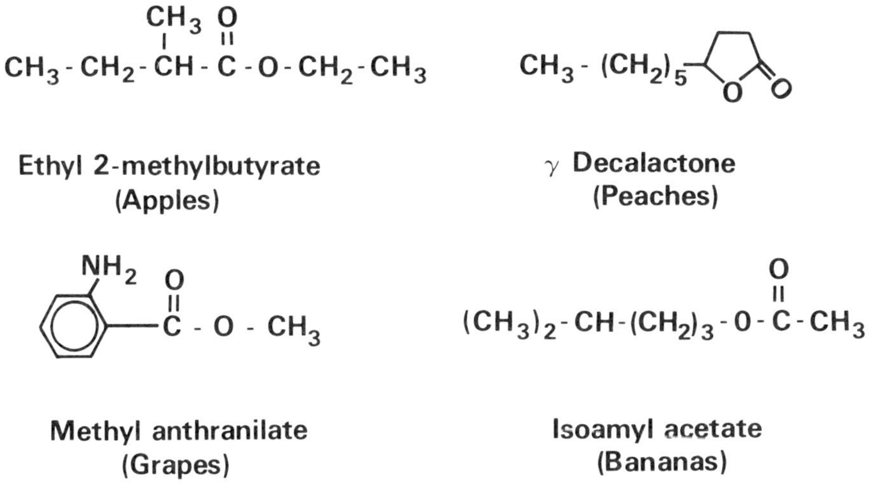

$CH_3-CH_2-CH(CH_3)-C(=O)-O-CH_2-CH_3$

Ethyl 2-methylbutyrate
(Apples)

$CH_3-(CH_2)_5-$ (lactone ring, O, O)

γ Decalactone
(Peaches)

NH_2 (benzene ring) $-C(=O)-O-CH_3$

Methyl anthranilate
(Grapes)

$(CH_3)_2-CH-(CH_2)_3-O-C(=O)-CH_3$

Isoamyl acetate
(Bananas)

FIG. 11.4. Chemical structure of character impact compounds of apples, peaches, grapes, and bananas.

able volatiles was observed in apples from trees fertilized with nitrogen fertilizers. This effect was reversed by using phosphorous and nitrogen plus phosphorous fertilizers (Wills and Scott 1976). Harvest date and the level of ethylene in the apple when harvested also appear to be important in flavor development (Blanpied and Blak 1976). Fruit maturity also greatly influences apple juice flavor: juice from immature apples tends to be harsh, starchy, sour, and astringent, whereas juice from overmature apples lacks flavor and is low in yield.

Controlled- or modified-atmosphere storage helps to extend the apple processing season. The time of harvest determines how long apples can be stored with retention of their quality. The earlier in the maturation process apples are harvested, the longer storage time is available. Early harvesting generally leads to less flavor development in the fruit; however, harvesting too early can lead to a lack of development of the enzyme precursors required for flavor generation. Care must also be taken to prevent the atmosphere from going anaerobic because this will result in ethanol formation (Dimick and Hoskin 1983). When apples are removed from CA storage, they must be ripened before processing. Modified-atmosphere storage can lead to improper flavor development even after removal to air storage (Lidster *et al.* 1983). Best flavor development in 'Red Delicious' apples was achieved at 22°C (Guadagni *et al.* 1971B).

Enzyme systems in the peel are very active upon cutting. Guadagni *et al.* (1971A) observed a marked increase in the volatiles characteristic of apple flavor in peels held for 24 hr after separation from the apple. Careful attention to storage conditions of these peels should be observed by the processor who wishes to take advantage of the flavor-generating properties of cut peels (Dimick and Hoskin 1983). Freeze concentration results in better retention of apple volatiles for juices than reverse osmosis, membrane diffusion, or foam-mat drying (Bolin and Salunkhe 1971), although improvements in the membrane diffusion process show some promise (Matsuura *et al.* 1975).

Careful attention to sanitary conditions must be followed during processing of apple juice. Undesirable musty or stale flavors develop from microbial action in the juice. Prompt pasteurization and filling following pressing will maintain the fresh apple juice flavor. Fresh apple juice should be refrigerated, and its shelf life can be extended by addition of sodium benzoate, sulfurous acid, or carbon dioxide in accordance with food regulations. Clarification of juices can lead to some flavor loss although addition of 6 g ascorbic acid/bu apples during the crushing operation helps maintain flavor retention. Although deaeration of apple juice reduces off-flavor development due to lipid oxidation, it also results in the removal of volatiles.

In canned juices, high-temperature short-time heating greatly reduces the conversion of volatile flavor components to secondary reaction products (Schreier *et al.* 1978). Frozen apple juice concentrate flavor is lost during concentration and is usually enhanced by addition of recovered essence. Loss of carbonyl compounds is a limiting factor in apple dehydration, since volatile carbonyls disappear even in the gentlest method, freeze-drying. Jezek and Smyrl (1980) observed an increase in carbonyl concentration during osmotic drying, with little or no loss of the volatiles in subsequent dehydration. For a more detailed discussion of the effect of processing on apple flavor see Dimick and Hoskin (1983).

Pears. A unique characteristic of pears is their low acid concentration and thus high BAR. Malic and citric acids, the two major organic acids in pears, are normally found in equivalent amounts. Ripe pears are characterized by high levels of fructose and low levels of sucrose. Sorbitol is present in concentrations near that of sucrose and glucose.

The limited research on volatile flavor components of pears and pear essence has been reviewed recently by Luh (1980). Hexyl acetate and methyl *trans*-2,*cis*-4-decadienoate have been identified as the volatile compounds most affecting pear flavor (Jennings and Sevenants 1964). Volatile constituents of pears are particularly influenced by the degree of ripeness of the raw products.

Little information is available on the effect of canning on pear halves, but, as with most heat preservation techniques, heating leads to destruction of volatile flavor components. Pear puree and nectar are also canned and are susceptible to similar losses in flavor quality. Adding half of the essence back to pear puree resulted in better flavor than adding all or none of it. Addition of sucrose decreased flavor acceptance of the puree. No significant differences were detected in purees packed aseptically from those filled conventionally (Leonard *et al.* 1976).

Stone Fruits

The four major stone fruits of processing significance are the peach, apricot, plum, and cherry. Malic acid is the predominant acid in these four fruits. Each has a desirable fleshy interior and is frequently preserved by canning. Proper care of the fruit during postharvest handling is critical to final product quality. For peaches and apricots, the firmer the fruit, the lower the temperature, and the earlier in the day they are harvested the less mechanical damage occurs (O'Brien *et al.* 1978).

Peaches. Peaches are canned whole, in halves or slices, and pickled. They may also be frozen or dehydrated, and are an important ingredient in pies, ice cream, cakes, and preserves. An excellent peach juice drink has been developed (Heaton *et al.* 1973) but has not been successfully marketed.

Ripe peaches are characterized by a high BAR. Citric and malic acids are present in about equal concentrations. Lactones present as flavor components contribute to the titratable acidity of peaches. Sucrose, the major sugar present, accounts for about 70% of the total sugars. Sweetness is a major quality characteristic of canned peaches. A heavy sucrose syrup was commonly added to canned peaches produced a few decades ago, but this practice is no longer as common because of greater awareness of sugar consumption by consumers. Romani and Jennings (1971) have reviewed the studies on peach volatile aroma compounds. A series of lactones including γ-penta-, γ-hexa-, γ-octa-, γ-nona-, γ- and δ-deca-, and γ- and δ-dodeca- are primarily responsible for the characteristic peach flavor (see Fig. 11.4 for structure of γ-decalactone).

Cultivar selection and preharvest care are important determinants of peach product quality. Texture and size, rather than flavor, are the most important factors in selection of cultivars for processing. Cultural techniques do influence final product quality and have been described by Woodroof *et al.* (1965). Peach product quality is also influenced by postharvest handling methods. The peach is a delicate fruit that bruises easily. Mechanical harvesting leads to a reduction in the number of top-quality fruits. In addition to damage to texture and appearance, bruising leads to the release of enzymes that can result in off-flavor production. Peaches deteriorate rapidly in warm weather. Hydrocooling shortly after harvest can extend the storage life of the fruit. Boggess and Heaton (1973) reported that mellow-ripe peaches stored at 80°F were unacceptable for processing within 24 hr, whereas those held at 33°F could be held for 14 days prior to processing with little loss of sensory quality. Certain cultivars are susceptible to chilling injury and would not withstand these low temperatures. Controlled-atmosphere storage is a possible method of extending storage time of peaches before processing and producing more evenly ripened fruit (Brecht *et al.* 1982). However this treatment leads to development of off-flavors in some cultivars.

Canning leads to a loss of volatile flavor components in peaches. Since the flavor of a peach is mild and delicate, fruits should be allowed to reach the mellow-ripe stage, when flavor is most pronounced, before processing. Lactone components change very little during canning, but hotrienol and α-terpineol are formed and are thought to be responsible for the uniqueness of canned peach flavor (Souty and Reich 1978). The flavor of canned peaches is also dependent on the type of peach. Freestone cultivars, grown mainly in the southeastern United States, are known for their fuller flavor but less desirable appearance. Clingstone cultivars, common in California, are more bland in flavor with better appearance and textural characteristics. A list of the acceptability for canning of several freestone cultivars has been developed by Boggess and Heaton (1973). In the preparation of purees and juices, better flavor was obtained from peeled than from unpeeled fruit (Heaton *et al.* 1973).

Freezing, although milder than canning, results in flavor changes as well. Linalool, an oxidation product of terpene alcohols, produces off-flavors in frozen fruit (Souty and Reich 1978). Cultivar selection plays an important role in freezing suitability, and a list of some acceptable cultivars has been developed by Sistrunk and Rom (1976). Marked differences in the volatile constituents of fresh peaches, puree, refrigerated fresh slices, clear juice concentrates, and canned, dehydrated, frozen and irradiated peaches have been observed.

Apricots. Citric and malic acids are found in apricots in near equal amounts, and sucrose is the major sugar present. The major flavor components are lactones; terpenes also contribute to flavor impact. Apricots are the only stone fruit that contain appreciable amounts of terpenes. Controlled-atmosphere storage of apricots before processing tends to decrease off-flavor development (Brecht *et al.* 1982). Canning is the primary method of preserving apricots.

Plums. The major organic acids found in plums are malic and quinic acids. Glucose is the predominant sugar in culinary and damson plums, while sucrose and glucose are found to be nearly equivalent in dessert plums. The volatile flavor components have been characterized by Ismail *et al.* (1980). The major plum products are canned plums and preserves. Most of the sucrose present in plums is converted to glucose and fructose during canning.

Cherries. In addition to malic acid, citric, quinic and shikimic acids are present in appreciable quantities in ripe cherries. Cherries are high in fructose and glucose and contain little or no sucrose. Major cherry products are maraschino cherries and ingredients for jellies, pies, and ice cream. Some canned and frozen products appear on the grocery shelves, but most processing is for use as ingredients. Cherries canned under pressure for shorter times have a better flavor than those canned in boiling water. Likewise, the faster cherries are frozen, the better the flavor retention.

Berry Fruits

Berry fruits tend to be highly perishable and generally require quick cooling upon harvest to maintain product quality. Most berry fruits develop harsh, stale off-flavors when frozen if not protected from oxygen.

Blueberries. Three types of blueberries are grown in the United States: highbush, lowbush (found in the northern states), and rabbiteye (grown in the southeastern states). Ripe blueberries are characterized by a very high BAR. Reducing sugars are reported to be present in a much higher concentration than sucrose, and the acid concentration is very low (Woodruff *et al.* 1960). The volatile flavor components that characterize blueberry flavor are linalool, *cis*-3-hexanol, and *trans*-2-hexenol (Parliment and Scarpellino 1977). Processed blueberries are usually machine-harvested, which can result in mechanical damage to fruit. Since this fruit is highly perishable, it is essential to transport the berries to the processing plant as quickly as possible and cool them as soon as they arrive. The introduction of explosion puffing to preserve blueberries (Sullivan *et al.* 1982) shows promise in providing a shelf-stable product year-round.

Strawberries. The sugar concentration of strawberries is generally lower than that of blueberries, while the acid concentration is higher. Sucrose is found in lesser quantities than glucose and fructose. Citric acid is the predominant acid; malic, quinic, and succinic acids are present in lesser concentrations.

The numerous studies of volatile aroma components of strawberries have been summarized by Pyysalo *et al.* (1979), but no clear agreement has been reached on the compound(s) primarily responsible for strawberry aroma. It appears that while many compounds are important contributors to strawberry aroma, none are considered major character impact compounds. Character impact compounds of other fruits—for example, ethyl-2-methylbutyrate (apples), hex-2-3-enol (tomatoes), and linalool (blueberries)—are thought to be contributing factors. Volatile sulfur compounds have also been suggested as important flavor components (Schamp and Dirinck 1982).

A machine-harvesting system for strawberries intended for further processing has been developed (Nelson *et al.* 1978). The machine-harvested strawberries produced an acceptable puree which was lower in Brix and acidity than one produced from hand-picked fruit (Morris *et al.* 1980). Although the sugar concentration decreases rapidly at room temperature, little loss of sugar was observed at 1.7°C for 7 days (Morris *et al.* 1976).

Strawberries have particularly active enzyme systems for volatile flavor component production. In one study, cut strawberries were incubated anaerobically in the dark at 30°C with the corresponding aliphatic acid. Greater production of the acetate esters, particularly isoamyl acetate, was observed when strawberries were cut into two to four pieces than when homogenized (Yamashita *et al.* 1975). This reemphasizes the stimulation of enzyme systems by mechanical damage.

Strawberries are used as ingredients in baked goods and dairy products. The major use of processed strawberries is in jams and preserves. The frozen product is also popular. Flavor compounds are concentrated by freezing but 2,5-dimethyl-4-methoxy-3(2H)-furanone is found in frozen berries in higher levels than accounted for by concentration alone and is thought to be important in frozen strawberry flavor (Schreier 1980).

Cranberries. The unique characteristics of cranberries is a very low BAR. The major acids in cranberries are citric and benzoic. Glucose is the predominant sugar present. Unlike most fruits, the major quality attribute of a cranberry is its tartness.

Grapes. The two major species of grapes are *Vitis vinifera* (European) and *Vitis labrusca* (American). In addition, *Vitis rotundifolia* (muscadine) is grown in the southeastern United States. All species are used in wine production, but most of the juice and jelly is produced from the 'Concord' cultivar of the *V. labrusca* species.

Grapes have a high concentration of tartaric acid. Malic acid is present in nearly equivalent amounts, and citric acid is present at much lower concentrations. Glucose and fructose are present in nearly equivalent amounts, with sucrose present at lower concentrations.

A total of 225 aroma compounds have been identified in *V. vinifera* grapes (Schreier *et al.* 1976), and 49 aroma compounds have been confirmed in muscadine grapes (Welch *et al.* 1982). Methyl anthranilate (Fig. 11.4) is believed to be the character impact flavor of *V. labrusca* cultivars, particularly 'Concord.' The most important volatile constituents of muscadine grapes are apparently isoamyl alcohol, hexanol, benzaldehyde, and 2-phenyl ethanol (Welch *et al.* 1982).

Flavor of grape juice is affected by juice extraction temperature. Sistrunk (1976) observed that 185°F was the optimal extraction temperature for final product flavor as determined by a sensory panel. Addition of ascorbic acid just before crushing helps to retain grape juice flavor during clarification. Essence recovery is an important step in the preparation of grape juice for juice concentrate or jelly. Volatile flavor components are stripped from the juice using a series of high distillation plates. Ethyl butyrate and *n*-hexanol are easily stripped from the juice due to their high vapor–liquid equilibrium constant. Because methyl anthranilate has a low vapor–liquid equilibrium constant, several plates are necessary to remove it (Saravacos *et al.* 1969). The essence is then reincorporated into the juice before filling.

Cane Fruits. Raspberries, blackberries, cloudberries, and brambles are higher in reducing sugars than sucrose. Citric acid is the primary organic acid in cane fruits; malic and phosphoric acids are present in lesser amounts.

The most important volatile flavor component of raspberries is 1-(*p*-hydroxyphenyl)-3-butanone (Buttery 1981). Arctic bramble flavor is most closely associated with 2,5-dimethyl-4-methoxy-2,3-dihydro-3-furanone (Kallio 1976; Pyysalo *et al.* 1977). Important aroma constituents of the cloudberry include ethyl hexanoate, 4-vinylphenol, 2-methoxy-5-vinylphenol, 2-methylbutanoic acid, hexanal, benzyl alcohol, and 2-phenylethanol (Pyssalo *et al.* 1977).

Subtropical and Tropical Fruits

Subtropical and tropical fruits that are subject to further processing include bananas, dates, mangoes, and pineapples. These fruits are known for their high concentration of sugars, particularly sucrose, when ripe. Subtropical fruits are susceptible to chilling injury. When they are near ripe or ripe, their flavor components change rapidly and storage temperatures can be reduced to the optimum temperature—14°C for bananas (Marriott 1980), 7°C for mangoes (Hulme 1971), and 8° for pineapples (Dull 1971).

Bananas. A climacteric fruit, bananas are characterized by a rapid conversion of starch to sugars. Although this reaction is responsible for most of the development of sweetness in ripe bananas, the predominant sugar is sucrose and not glucose, which is the usual metabolite of starch hydrolysis. Malic acid is the predominant organic acid present, followed by citric and oxalic acids.

The numerous studies of banana flavor volatiles have been reviewed by Marriott (1980). The characteristic flavor and aroma of bananas have been attributed to several esters, including the acetates and butyrates of 3-

methyl butanol, 2-methyl propanol, 1-butanol, and ethanol. Isoamyl acetate (3-methyl butyl acetate) is believed to be the most important ester (Fig. 11.4). Chilling injury results in an inability of the fruit to develop flavor components, differing ratios of the major volatile esters, and development of off-flavors due to α-farnescene, acetaldehyde, and ethanol.

Green bananas or plantains are more likely to be processed than ripe bananas. Green bananas have up to 10 times the amount of starch and about 5% of the sugar found in ripe bananas. Harvesting of the bananas at the proper level of maturity is very important in the production of frozen slices of green bananas. Sanchez Nieva *et al.* (1980) observed that a pulp: peel ratio of 1.5 resulted in the best flavor scores. Production of intermediate-moisture foods from ripe bananas can result in an astringent product. This astringency has been attributed to tannins located in the latex cells of the fruit which may be ruptured during dehydration but not ruptured during mastication of the fresh fruit (Ramirez-Martinez *et al.* 1977).

Mangoes. Sucrose is the major sugar in most varieties of ripe mangoes. Asian mangoes tend to be sweeter than their U.S. counterparts. Citric is the major acid present, but many other acids have been reported (Hulme 1971). The volatile components of fresh mangoes have been identified by MacLeod and de Troconis (1982), who report car-3-ene and an unidentified dimethylstyrene as the major flavor impact compounds. Hunter *et al.* (1974) identified furfural and 5-methylfurfural as important flavor components of canned mango purée.

Pineapples. As in other fruits of this group, sucrose is the major sugar present in pineapples. Citric acid is the predominant acid with malic and oxalic acids also present (Dull 1971). Studies on the isolation and identification of pineapple volatiles have been reviewed by Silverstein (1971). A volatile flavor component unique to pineapple is 2,5-dimethyl-4-hydroxy-2,3-dihydro-3-furanone. Other important pineapple volatiles include methyl- and ethyl-3-methylthiopropionate, ethyl butyrate, and ethyl hexanoate (Buttery 1981).

Certain cultivars of pineapple (e.g., 'Red Spanish' and 'Smooth Cayenne') develop off-flavors upon frozen storage. Blanching of the product before freezing improves product quality, but the frozen slices do not have the flavor quality of canned slices (Sanchez Nieva and Hernandez 1977). Acetic acid, 5-hydroxymethylfurfural, furfural, formaldehyde, acetaldehyde, and acetone were the major volatiles obtained from canned pineapple juice (Gawler 1962). Acetaldehyde, acetone, ethanol, and the ethyl, isobutyl, methyl, and propyl esters of acetic and formic acids as well as five other esters were the major compounds isolated from canned pineapple (Howard and Hoffman 1967).

COLOR

The importance of color as a quality attribute in fruit products must not be underrated. Prior to the first bite or sip, the consumer has already made a quality judgment based on product appearance. Johnson and Clydesdale (1982) noted that dark red beverages with 1% less sucrose tasted sweeter than light red control samples. The art of preserving color in fruit products involves the selection of fruit with optimal pigmentation and the prevention of pigment degradation during handling, processing, and storage. The major pigments of fresh fruit are chlorophylls, carotenoids, and anthocyanins. Undesirable pigments such as melanoidins and melanins are formed in browning reactions. For a more detailed description of plant pigments see Francis and Clydesdale (1975).

Chlorophylls and carotenoids are lipid-soluble pigments. Chlorophylls are complex tetrapyrrole ring structures with a phytol chain and chelated with magnesium. They are the green photosynthetic compounds and normally disappear during fruit ripening. Substitution of hydrogen for magnesium in the tetrapyrrole ring results in a loss of green color to form pheophytins, which are brown in color. The phytol chain is responsible for the lipid solubility of chlorophylls; its removal results in the chlorophyllides, which are water-soluble green pigments. Chlorophylls *a* and *b* differ in an alkyl group on the ring structure. Chlorophyll *a* normally makes up about 70–80% of the total chlorophyll concentration of green plant tissue.

Carotenoids are long hydrocarbon chains built from eight isoprenoid units, which may terminate with cyclic structures. This class of pigments may be further subdivided into oxygenated (xanthophylls) and nonoxygenated (carotenes) forms. The xanthophylls in fruits are primarily yellow and are responsible for the color of ‘Golden Delicious’ apples, bananas, and various other fruits. Some carotenes are red, for example, lycopene, the major pigment of watermelon and pink grapefruit flesh; others, such as β-carotene found in apricots and oranges, are orange. For an extensive review of carotenoid distribution in fruits, see Goodwin and Goad (1970).

The third major group of fruit pigments is the anthocyanins. Anthocyanins are water-soluble pigments responsible for the reds, blues, and purples of many fruits, flowers, and vegetables. Anthocyanin structure is based on a basic flavylium cation ring system. Six major anthocyanidins (cyanidin, delphinidin, pelargonidin, peonidin, petunidin, and malvidin) differ from each other by the functional groups present on the 3′, 4′, and 5′ positions of the single ring. Anthocyanins are formed by sugar substitutions at the 3, 5, and 7 positions of the double ring system. Anthocyanin structure may be further complicated by acylation of the sugar substituent at the number 3 position. For a more detailed description of anthocyanin structures, see Shrikhande (1976).

Browning reactions in fruits and fruit products are generally undesirable. The two major browning processes in fruits are (1) nonenzymic, or Maillard, browning and (2) enzymic browning. Maillard browning is started by condensation of a free amino group with a reducing sugar. A subsequent series of reactions results ultimately in brown compounds of undetermined structure known as melanoidins. Maillard browning is enhanced by increased concentrations of reducing sugars, increased temperatures during processing or storage, and a pH above the isoelectric points of the amino groups. A more detailed discussion of Maillard browning is presented by Hodge and Osman (1976).

Enzymic browning consists of a series of reactions involving both hydroxylation and oxidation steps. Phenolic compounds (e.g., catechol, caffeic acid, and chlorogenic acid) serve as substrates for these reactions and melanins are the resultant brown pigments. The potential for enzymic browning is within the cell. Cutting of the first surface provides oxygen needed for catalysis. Bruising of fruit tissue disrupts cell membranes, allowing enzymes and substrates to interact. Enzymic browning can be inhibited by careful postharvest handling, application of sulfites or ascorbic acid, or heating to inactivate the enzymes. Richardson (1976) provides detailed explanations of enzymic browning.

A rapid disappearance of chlorophyll during the ripening of most fruits leads to a change from green to other colors. The loss of chlorophyll may be accompanied by biosynthesis of carotenoids or anthocyanins, but the resultant color may be due to a simple unmasking of pigments already present. Chlorophyll is not lost upon ripening in limes and certain cultivars of apples that are considered green when mature. Some fruits, such as bananas, pears, and apples, will continue to ripen after harvest, showing an improvement in color development upon holding. Carotenoids tend to be more stable during postharvest handling than anthocyanins. Enzymic browning, resulting from mechanical bruising, is probably the most common cause of fruit discoloration during the postharvest period.

Heat preservation is destructive of each of the pigment groups. Chlorophylls are degraded to brown pheophytins; carotenoids are converted to epoxides; and anthocyanins are rapidly degraded. Two other types of discoloration are associated with anthocyanins in canned products. Leucoanthocyanins can be converted to anthocyanins during canning; this causes a pink discoloration of pears or excess red coloration of peaches. Anthocyanins react readily with the metal in cans and require a special enamel to protect the can and the product. Enzymic browning is usually prevented by blanching and the addition of ascorbic acid.

Very little pigment degradation is observed in frozen fruit products, as enzymic browning is the major factor in the discoloration of frozen fruits.

Because fruit dehydration results in the concentration of plant pigments, it intensifies the coloration. Brown pigments may result from pigment degradation, enzymic browning, or Maillard browning. In products such as dried apricots and peaches, browning is avoidable and unacceptable. Other dried fruit products (e.g., raisins and prunes) are expected to be brown. More detailed discussions are available on the effects of processing (Chichester and McFeeter 1971) on plant pigments as well as on the stability of carotenoids (Gordon and Bauerenfeind 1982) and anthocyanins (Markakis 1982).

Color Measurement

There are several ways to measure fruit color. One obvious method is the determination of pigment concentration. Spectrophotometers and colorimeters can be used to measure the color of the raw and/or finished product. Electronic color-sorting is replacing human graders in some operations. Finally visual evaluation by human judges is still a widely used method. Measurement of pigment concentrations is not usually the best way to determine product color. Rarely is one pigment predominant in a fruit product, and the detailed isolation techniques are too complex for routine quality control analysis. Crude pigment analyses—for example, absorbance of acetone extracts at 663 and 645 nm for chlorophyll (Lebermann *et al.* 1968) or the total anthocyanin and degradation index (Fuleki and Francis 1968)—are useful in some products. In many products, however, the initial degradation compounds are similar in color to the original pigment. For example, in red wines polymeric degradation compounds are also red, so their formation causes little change in the color of the beverage (Somers 1971).

The subject of food colorimetry is described in detail in two excellent books (Francis and Clydesdale 1975; Hunter 1975). Many instruments are available using different optical principles and color scales. Three of the more versatile instruments are Hunterlab's Labscan spectrocolorimeter (Fig. 11.5), the Gardner XL-845 Tristimulus colorimeter (Fig. 11.6), and the Agtron Model E-5W colorimeter (Fig. 11.7). The first two instruments use the Hunter color solid (*L, a, b*) system based on the color-opponent theory (Fig. 11.8). The *L* value provides a measure of lightness and darkness; the *a* value measures the red-green character of the sample and the *b* value measures the yellow-blue character. Using these basic functions, equations have been developed to calculate values for hue, saturation, and chroma. Options are available to read in other color scales or to develop specific indices for a special product. Agtron instruments, which are less complex and less versatile, provide a number that describes a specific color ratio. Several models are available to fit specific applica-

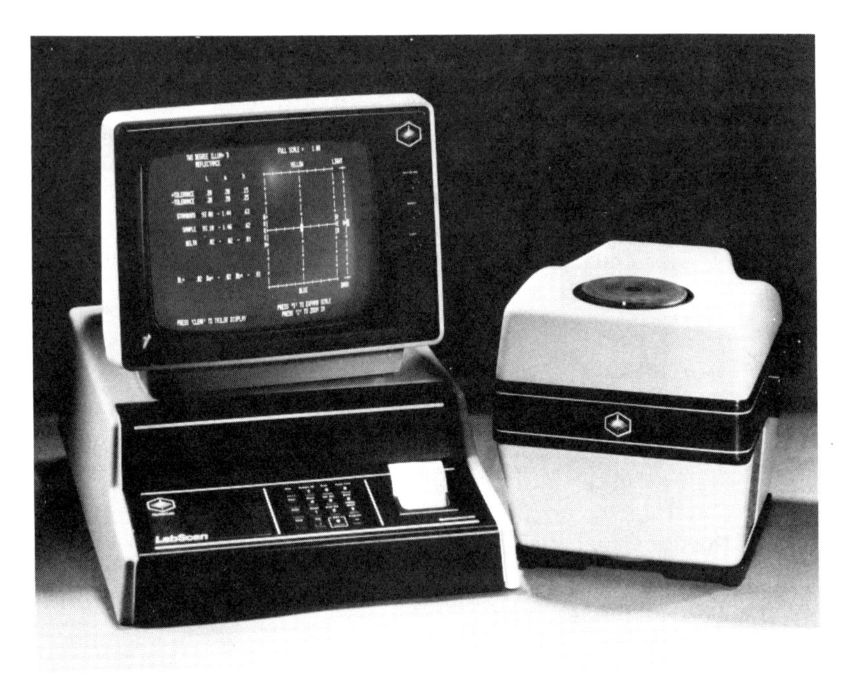

FIG. 11.5. HunterLab's Labscan spectrocolorimeter. (Courtesy HunterLab.)

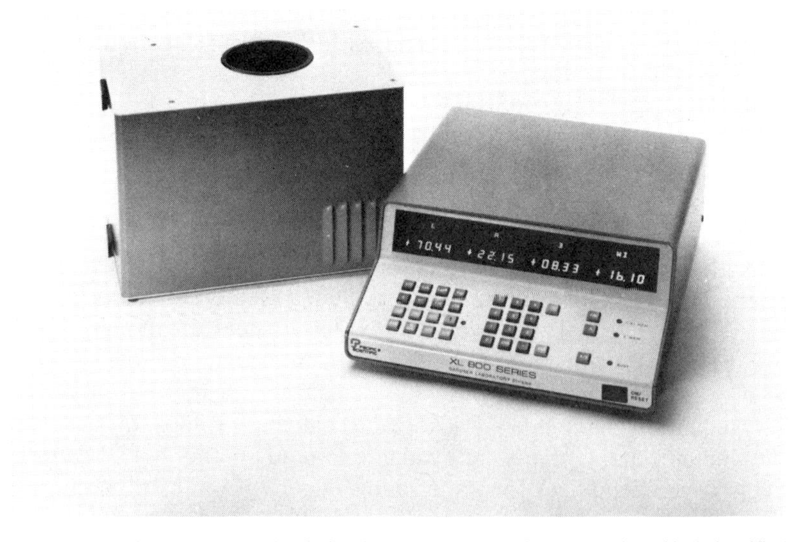

FIG. 11.6. Gardner XL-845 Tristimulus colorimeter. (Courtesy of Pacifc Scientific.)

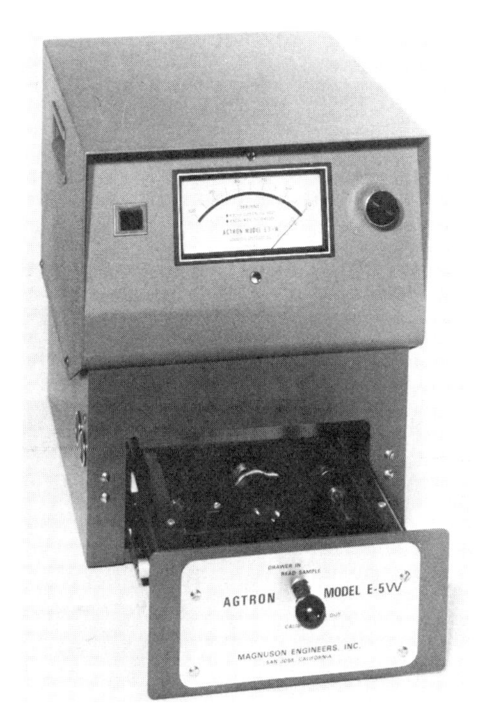

FIG. 11.7. Model E-5W Agtron colorimeter. (Courtesy Filper/Magnuson Corporation.)

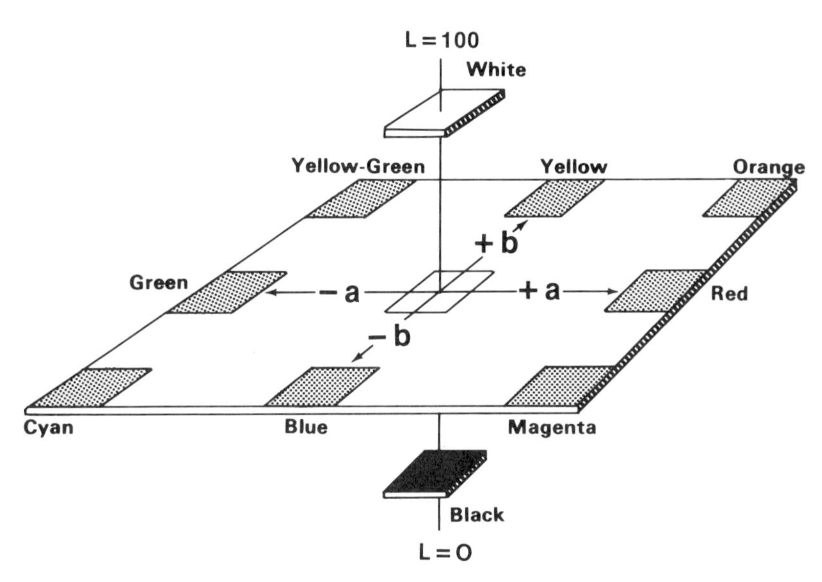

FIG. 11.8. *L, a, b* color solid.

tions. The selection of the appropriate color instrument for a fruit processing operation depends on many factors; (1) the range of products and materials to be measured, (2) the importance of color in the overall quality of the product, (3) the appropriateness of color values generated to product quality, (4) the sampling schedule and speed of the analysis, and (5) the skills of the person who will be operating the instrument.

The consumer is the ultimate judge of product quality. The significance of any color measurement is based on its ability to predict consumer response. Visual evaluation of a product is thus another method of measuring color. Unfortunately, visual evaluation is not consistent. Color perception of a product can vary depending on the light source and evaluator as well as the frame-of-mind and fatigue of the evaluator. Plastic standards for certain products, such as canned peach halves, help fix the frame of reference.

Instruments have been designed to sort apples (Reid 1976), oranges (Chuma *et al.* 1981), and other fruits (Krivioshiev and Chalukova 1981). Future instruments may be able to determine internal attributes of fruits such as soluble solids content and acidity (Dull *et al.* 1980).

The remainder of this section is devoted to color aspects of specific fruits and their products. The pigments color changes during handling and processing, and color measurement for each type of fruit is discussed.

Citrus Fruits

The external color of citrus fruits is primarily a function of the relative concentrations of chlorophylls and carotenoids present in the flavedo. As the fruit ripens, chlorophyll disappears. Carotenoids increase upon ripening in oranges but tend to decrease upon ripening in other citrus fruits. It is not the color of the peel, however, but the color of the pulp that is of interest to processors. The pigments are concentrated in cellular organelles called plastids or chromatophores, the shape of which vary with species. In addition, pigments occur in the cell sac of the peel of mature limes and in the cell wall of oranges and tangerines.

Factors in citrus color quality are color consistency and nonenzymic browning. Color consistency is difficult to maintain during a season because the carotenoid concentrations vary from batch to batch. This problem is easily remedied by blending juices from different batches to achieve the desired color. Nonenzymic browning occurs in canned and dried products. Browning is influenced by moisture content, acidity, temperature, and oxygen concentration. Darkening is observed during refrigeration of pasteurized citrus juices and is apparently related to the disappearance of ascorbic acid. In addition to Maillard browning resulting from the interaction of reducing sugars and amino acids, some darkening may be due to oxidation of ascorbic acid to dehydroascorbic acid, which can then react with α-amino acids to form objectionable brown pigments. Removel of oxygen lowers the susceptibility of fruit to browning due to ascorbic acid oxidation. Even in the absence of oxygen, ascorbic acid is converted to furfural, which can also lead to darkening but at a slower rate. Citrus concentrates should be kept frozen to prevent browning reactions.

A special colorimeter for measuring the color of citrus products was developed by Richard Hunter. A special color scale has been developed for use with this instrument. Although it has been applied to most citrus products, the prime use of the colorimeter has been with orange juice (Francis and Clydesdale 1975).

Oranges. The major pigments in ripe oranges are three xanthophylls—cryptoxantin (red), antheraxanthin (yellow), and violaxanthin (yellow) (Stewart 1977; Nordby and Nagy 1980). Carotenes (orange) are present in lower concentrations, but make an important contribution to the coloration of the fruit (Gross 1977). Although over 60% of the carotenoids are present in the peel (Curl and Bailey 1956), the relative concentrations of cryptoxanthin and carotenes are higher in the pulp than in the peel (Gross 1977). The "blood" orange, an exception to the normal pigmentation found in oranges, contains anthocyanins, primarily cyanidin-3-glucoside, which give the fruit a red coloration (Chandler 1958).

Total carotenoid concentration increases with maturity. It is also influenced by cultivar, cultural practices, location, rootstock, and weather conditions. (Nordby and Nagy 1980). Such apparently trivial factors as the side of the tree can influence coloration. Stewart (1977) noted that juice produced from oranges grown on the north side of trees (in the northern hemisphere) produced a deeper colored juice. Color development in orange peel results from chlorophyll degradation and carotenoid biosynthesis. Very little chlorophyll is found in unripe pulp, and thus pulp color development is dependent on carotenoid biosynthesis (Gross 1977). Although carotenoid biosynthesis patterns are similar in both peel and pulp, changes in peel chlorophyll will markedly change peel color with little or no effect on pulp color.

Growth regulators affect the coloration of oranges. Some orange cultivars will "regreen" upon reaching maturity, a process characterized by reappearance of chlorophyll in the peel but not in the pulp. Gibberellic acid promotes chlorophyll accumulation, while ethylene treatment results in the inhibition of chlorophyll synthesis (El-Zeftawi and Garrett 1978). Ethylene is used in the degreening of fruit that is not fully mature to improve peel color with little change in pulp color (McCornack and

Wardowski 1977). Treatment with CPTA (2-14-chlorophenylthiol-triethylamine hydrochloride) results in an accumulation in both pulp and peel of the red pigment lycopene, which is not normally found in oranges.

Carotenoids appear to be relatively stable during the processing and storage of juice and other orange products, although they are susceptible to oxidation in dried products (Chichester and McFeeters 1971). Nonenzymic browning is the major color quality problem in orange products. Browning increases with an increase in °Brix, temperature (Kanner *et al.* 1982), and pH (Bruemmer and Bowers 1977). Browning has been noted in canned juice, freeze-dried crystals, and aseptically packaged juice. Browning occurs more readily in glass than plastic containers for aseptically packaged juice (Johnson and Toledo 1975). Browning observed in the products of the blood orange is due in part to polymerization of the anthocyanins present.

The USDA has developed a series of color standards for orange juice. The sample is compared visually with a series of color standards to determine a color grade. A special citrus colorimeter used extensively in the orange juice industry has a color scale giving values of citrus red and citrus yellow instead of the more familiar *a* and *b* terms (Francis and Clydesdale 1975). Although the cirtus colorimeter is the most common in the industry, other instruments are as effective in measuring orange juice color (Eagerman 1978). Color degradation due to browning has been followed using values of the CIE color system (Robertson and Reeves 1981). These values relate to visual browning but not USDA grading standards.

Orange products also can be evaluated for color on the basis of pigment content. Instrumental grading of fresh fruit based on the absence of chlorophyll (measured by reflectance spectrophotometry at 680 nm) provided correlation coefficients greater than 0.90 with visual grading (Chuma *et al.* 1981). Most measurements of carotenoids are based on the absorbance of alcohol extracts at 450 nm (measuring carotenes) or a ratio of values at 450 nm and 420 nm (yellow xanthophylls). A method based on the sum of absorbances at 465, 443, and 425 nm correlates strongly with the citrus colorimeter.

Grapefruit. Carotenoids are the major pigments of grapefruit pulp. The colorless compounds phytoene and phytofluene constitute approximately 75% of the carotenoids of the pulp of ripe white grapefruit cultivars. Lycopene is the major carotenoid in pink grapefruit flesh, with β-carotene present in lower concentrations (Cruse *et al.* 1979). Carotenoids develop in the peel before chylorophyll begins to disappear. The carotenoid concentrations then decline as chlorophyll decreases (Gross 1977). During ripening the lycopene concentration in the pulp decreases and β-carotene increases. The best coloration of pink grapefruit is observed about midseason. Lycopene concentrations can be enhanced by preharvest treatment with CPTA (Maier and Yokoyama 1977).

Browning reactions are the major color quality problem in grapefruit products. Maillard browning of sections and canned juice (Nagy and Shaw 1980), in conjunction with darkening due to ascorbic acid reactants, leads to an unacceptable product. Degradation of lycopene in pink grapefruit leads to a "muddy brown" color not associated with classical Maillard browning. Good correlations with visual color evaluation have been obtained for grapefruit juice samples using the citrus color scale; *L a b* scale (Huggart *et al.* 1977), and a spectrophotometric method using absorbance at 504, 470, and 445 nm.

Lemons and Limes. The two major carotenoids in lemon pulp are β-carotene and cryptoxanthin (Gross 1977). Nonenzymic browning of dried lemon crystals can be controlled at low temperatures. Lowering the water activity reduces browning, but nitrogen packaging does not affect the rate of browning (Kopelman *et al.* 1977A). Chlorophyll is the major pigment in limes.

Pome Fruits

Browning reactions are responsible for most color changes in pome fruits during processing. Although the peel is incorporated into certain products, peel color usually has little impact on final product color. Peel color is important, however, in the grading of fresh apples. The flesh of pears and apples is white, thus browning is readily noticeable. These fruits possess active enzymic browning systems, but nonenzymic browning is a problem in certain products.

Apples. The peel color of ripe apples varies with cultivar. Red coloration is due to anthocyanins, green to chlorophylls, and yellow to carotenoids and flavonoids. Although many of the carotenoid pigments are present in the peel while it is still green, yellowing of apples is generally thought to be the result of loss of chlorophyll (Gorski and Creasy 1977). In 'Golden Delicious' apples, an apocarotenol is found in the peel of the ripe fruit (Gross and Eckhard 1978).

Fertilizer treatment affects both skin and flesh color of apples. An increase in nitrogen fertilizer decreases red color development in the peel (Williams and Billingsley 1974). In certain cultivars (e.g. 'Starkrimson', which sometimes produces unacceptably dark red apples), $CaNO_3$ is beneficial. Ethephon and etherel promote red color development. Best red color development results from cold nights and warm days which favors enhanced anthocyanin synthesis. Increases in nitrogen fertilization, tree vigor, and crop load all result in greener flesh (Olsen and Ketchie 1978).

Development of browning of apple flesh during storage appears to be more a factor of the physiological condition of the fruit upon harvest than the storage temperature.

Apple tissue possesses very active enzyme systems that catalyze browning reactions. Chlorogenic acid appears to be the main substrate. Enzymic browning can be controlled by heat inactivation of the enzymes, treatment with sulfur dioxide or ascorbic acid, storage at low temperatures, and deaeration. Addition of chloride as either the sodium or calcium salt inhibits enzymic browning of crushed apples. Acidification with either malic or citric acid also decreases enzyme activity. Some enzymic browning, although considered objectionable in products such as applesauce and apple slices, is responsible for the desirable amber-to-brown coloration of clarified apple juice, apple butter, and vinegar.

Deaeration is perhaps the best method of preventing enzymic oxidation. Apple slices may be protected by immersing them in water or brine for up to 30 min at 40°C. Internal oxygen will then be used up by respiration. Care must be exercised not to continue this process too long or fermentation will commence. Drawing a vacuum on water-immersed slices is an alternative procedure. Deaeration is difficult to achieve in processes involving crushing because the reaction happens so quickly. Conducting crushing operations under a nitrogen blanket or applying ascorbic acid as a spray reduce browning problems.

Nonenzymic browning is a greater problem in dehydrated products. At low moisture concentrations, the apparent activation energy of nonenzymic browning increases as the moisture level decreases. Heat damage predisposes dried products to nonenzymic browning (Resnik and Chirife 1979). This means that the lower the moisture concentration obtained wtih minimal heat damage, the longer the anticipated shelf life is for a dried apple product.

Pears. Coloration of pears is similar to that of yellow and green apples. Chlorophyll accounts for the green color, and carotenoids become apparent as the chlorophyll disappears. If pears are too immature when harvested, a yellow coloration is observed in the canned product.

Browning is the major problem associated with processing. Application of $CaCl_2$ either preharvest or postharvest leads to decreased internal browning of fresh pears (Zerbini and Sozzi 1980). Browning in pear juice concentrate is the result of both nonenzymic and enzymic browning. Nonenzymic browning appears to predominate, however, as removal of amino acids results in less deterioration upon storage than removal of polyphenols (Cornwell and Wrolstad 1981).

Pink discoloration of canned pears is the result of a conversion of leucoanthocyanins (colorless) to anthocyanins upon heating. The leucoanthocyanin concentration depends primarily on cultural conditions. Pears with anthocyanins in the skin are more susceptible to discoloration than those without, and increased nitrogen fertilization increases the possibility of discoloration (Czerkaskyj 1970). It has been suggested that the pink pigment is a tin-cyanidin complex because the intensity of the color is related to the insoluble tin content in the pears. The reaction apparently occurs during the heating process, and addition of stannous ions to purée before heating reduces discoloration. Sulfur dioxide inhibits the formation of the pink color. Cultivar selection appears to be the best method of preventing this effect.

Stone Fruits

Unlike pome fruits, stone fruits are brightly colored with similar pigmentation in the skin and flesh. Although browning is still a problem in some products of stone fruits, the visual impact of the carotenoids in peaches and apricots or the anthocyanins in plums and cherries is the primary basis for color grading.

Peaches. The yellow-orange coloration of peaches is due to carotenoids, primarily xanthophylls. A reaction pathway in the ripening fruit proceeds from β-carotene to β-cryptoxanthin to further oxygenated xanthophylls (Katayama *et al.* 1971). The reddish blush on the skin of some cultivars is due to anthocyanins, primarily cyanidin-3-glycoside.

Ethephon has been used as a spray to advance the date of harvest maturity in peaches. Improvement in color of the raw product has been noticed in ethephon-treated fruit (Sharma and Dhuria 1978), but the final product appears to be just as susceptible to enzymic browning upon freezing and thawing. The major factor in enzymic browning in fresh peaches is the chlorogenic acid concentration. Polyphenol oxidase activity in fresh peaches is dependent on maturity and pH (Fig. 11.9). Jen and Kahler (1974) found that even at 3°C the enzymic browning systems are very active. This would explain why rapid browning occurs in unpasteurized peach juice at refrigerator temperatures. Polyphenol oxidase catalyzes the converstion of catechol to quinones in tissue that has been bruised or cut. In the intact fruit the enzyme and its substrate are protected from each other by subcellular organization. Once cellular membranes are disrupted, however, the polyphenol oxidase and catechol are united with oxygen under favorable conditions. This reaction leads to other reactions that ultimately lead to the brown discoloration.

Careful selection of cultivars that are low in the enzyme, substrate, or both is important, but minimizing bruising during handling is the best protection against browning. Another step in processing that is extremely

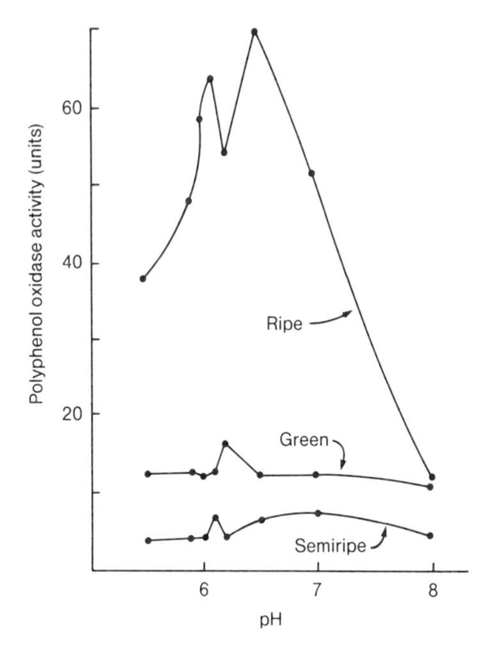

FIG. 11.9. Effect of maturity of 'Redhaven' peaches and pH on polyphenol oxidase activity. (From Jen and Kahler 1974.)

critical is lye peeling. Severe browning can occur within 15 min after peeling. Prompt removal of lye from the fruit by washing and acidification to pH 4.0 or below with 1% citric or malic acid may be advisable, but rapid handling of the fruit in the processing line is the most effective way to prevent browning.

Development of a deep red to purple coloration in canned peaches is undesirable, although a certain amount of anthocyanin pigment is considered desirable in the pit cavity of canned freestone peaches. Excess red coloration is the result of conversion of leucoanthocyanins to anthocyanins upon heating. Certain cultivars of peaches contain more leucoanthocyanins than others. The discoloration problem is more common when the weather is warmer than usual. A white deposit in the pit cavity is a warning of potential problems (Fig. 11.10). Upon heating, the deposit will change to a deep red or purple. Careful selection of cultivar, removal of the inner pit cavity, immersion peeling, thorough exhausting of the cans before sealing, and lowering the processing time reduce the probability of discoloration. Addition of SO_2 and use of C-enamel cans is effective but also leads to formation of the objectionable black compound iron sulfide. Treatments with EDTA, stannous chloride, or dehydroacetic acid were not effective in preventing discoloration (Van Blaricom and Hair

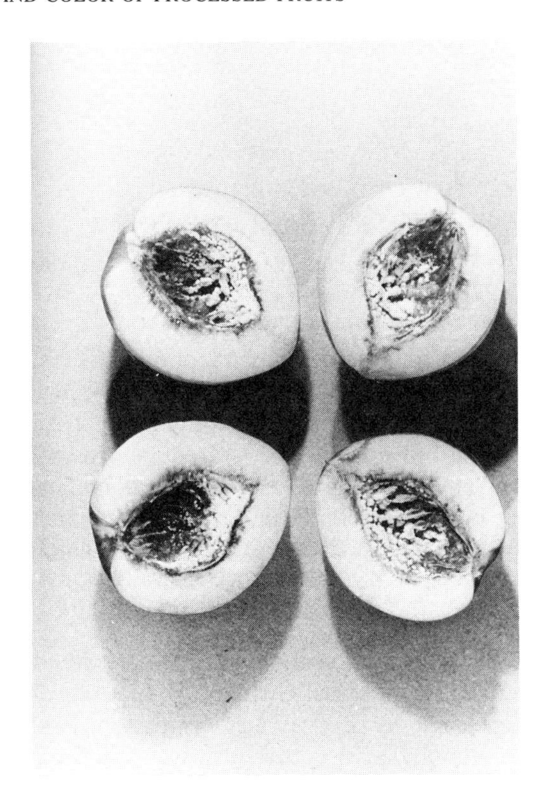

FIG. 11.10. White deposit in a fresh peach will turn a deep red to purple upon heating.

1972). Total carotenoids decrease during canning by as much as 42% (Kader *et al.* 1982) and during subsequent storage of peaches. Disappearance of β-carotene occurs more rapidly than disappearance of other pigments (Aczel 1977). Quality control measurements of the color of canned peaches employ either colorimeters or reflectance spectrophotometers.

Browning reactions are a major quality concern in dried and frozen peaches. Addition of SO_2 to packages of dried peaces effectively retards the browning process. Packages that do not permit escape of SO_2 and replacement of oxygen with nitrogen greatly improve product shelf life (Bolin *et al.* 1976). Freezing peaches in heavy syrup reduces the susceptibility of the product to rapid enzymic browning upon thawing. Ascorbic acid applied to the surface of the fruit is also helpful. Peaches frozen in 60% sugar syrup containing 0.2% ascorbic acid are shelf-stable up to a year at −18°C (0°F). A partial blanch prior to freezing is effective against browning but at a sacrifice to the characteristic flavor. Addition of SO_2 to frozen or dried product can also produce off-flavors.

The reason enzymic browning is such a difficult problem in frozen fruit tissue, particularly peaches, is membrane disruption. During the freezing process, as in bruising, the cellular membranes are disrupted by the formation of ice crystals. The enzyme reactions proceed at frozen temperatures but very slowly. As the temperature is raised by thawing, more disruption occurs and the reaction rates increase.

Apricots. The major pigment in ripe apricots is β-carotene although some xanthophylls are also present (Katayama *et al.* 1971). Red flecks in the apricot skin are due to cyanidin-3-glucoside. Enzymic browning in unprocessed fruit was found to be dependent on the substrate concentration of *o*-dioxy phenol primarily. Nonenzymic browning reactions occur in dried fruit. A sharp increase in the accumulation of Amadori products was noted when the moisture level was reduced from 15 to 12% (Fig. 11.11). Lee *et al.* (1979) observed a marked accumulation of carbonyls before noticeable brown discoloration. Carbonyl accumulation was inversely proportional to the disappearance of glucose and amino acids. Of the amino acids studied, alanine was most susceptible to reaction and glutamic acid was least susceptible.

Plums. Anthocyanins, primarily glycosides of cyanidin, are the predominant pigments of plum skins. Cyanidin-3-glucoside is the sole anthocyanin in 'Lamarck' plums (Bobbio and Scamparini 1982). Carotenoids, primarily the xanthophylls, are responsible for flesh pigmentation of plums and are rapidly degraded upon heat treatment in juice production.

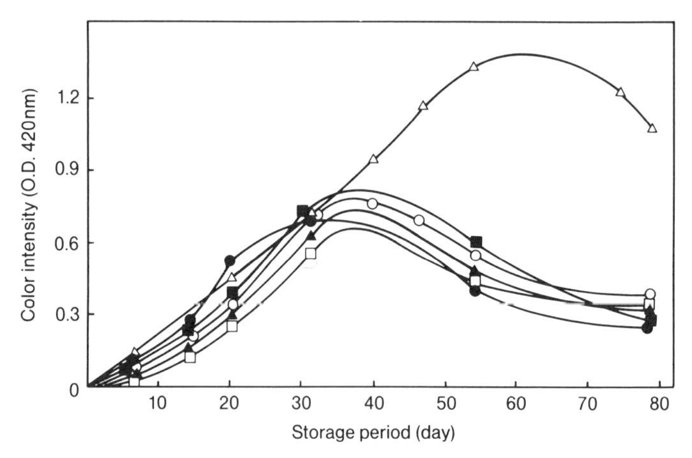

FIG. 11.11. Effect of moisture concentration on formation of Maillard browning reaction products during storage of apricots (△—12% moisture; ○—15%; ●18%; ■—20%; ▲—23%; □—26%). (From Lee et al. 1979.)

Carotenoid destruction is also observed upon browning of dried plum pulp (Moutounet 1976).

Cherries. Sweet cherries contain monoglucosides of cyanidin and peonidin, while sour cherries contain more complex combinations. Although most cherry anthocyanins are nonacylated, two acylated pigments have been isolated from black cherries.

Discoloration of cherries is observed shortly after bruising. Immersion of the fruit under water delays discoloration and cracking. The addition of ascorbic acid or calcium chloride is not effective in retarding browning. Removal of oxygen by vacuum is not effective either. Enzymatic degradation of anthocyanins in cherry juice can be controlled by a 1-min blanch of the cherries at 85°C prior to processing (Tressler *et al.* 1980). Pigment stability is greater in juice concentrate than in single-strength juice. Removal of oxygen from the product will reduce discoloration problems. Frozen cherry color is relatively stable during storage at temperatures of −18°C (0°F) or below (Polsello and Bonzini 1977).

Berry Fruits

Berry fruits are distinguished by their rich appealing coloration—reds, blue, and purples. The pigments primarily responsible for berry color are anthocyanins. These are relatively stable during frozen storage but are degraded upon heating. In addition to the classical enzymic and nonenzymic browning products, the degradation compounds of anthocyanins are polymers, which are also brown in color.

Blueberries. The anthocyanin composition of the various species of blueberries has been reviewed by Timberlake and Bridle (1982). As many as 15 different anthocyanins have been identified in ripe berries, all monoglucosides of five different aglycones. The blue coloration is not a property of the anthocyanin itself but of a metal complex with aluminum (Nakayama and Powers 1972). Extraction of the pigment from the berries results in a bright red coloration (Shewfelt and Ahmed 1978).

Color sorting of blueberries for maturity may be feasible. Good success has been observed in relating anthocyanin content to maturity of 'Wolcott' berries, but different cultivars require different settings. Solar drying of blueberries results in polymerization of the pigments, which is not as serious in a dark product like blueberries as it might be in other products. Explosion-puff drying results in much less pigment degradation (Sullivan *et al.* 1982). Explosion-puffed blueberries are rehydrated and used in pies and other baked goods and thus must closely approximate the color of fresh berries. Refrigeration and low pH (2.0–2.5) are recommended to preserve blueberry color in fresh juice. Storage in clear bottles in the light also accelerates pigment degradation.

Strawberries. Color is the single most important quality attribute in the grading of frozen strawberries. Little pigment degradation occurs during frozen storage, thus judicious sorting of the raw product is essential for high quality. Browning is the biggest problem in strawberry preserves. The key to control of browning in strawberry preserves is storage at low temperatures. Sistrunk and Morris (1978) reported that up to 50% green fruit could be used in preserves from highly colored cultivars of strawberries without adversely affecting the color or flavor of the product.

Cranberries. The pigments in cranberries can be divided into the anthocyanins and other flavonoids (Puski and Francis 1967). The anthocyanins present are the galactosides, arabanosides, and glucosides of cyanidin and peonidin. The total anthocyanin technique, first developed for cranberry products (Fuleki and Francis 1968), has been modified to a more rapid quality control technique (Deubert 1978). Each fruit product has a unique spectral profile based on its pigment composition. An example of a spectral profile of cranberry juice cocktail is shown in Fig. 11.12. The peak in the range of 520 nm decreases with storage at 38°C (100°F) for 1 year and is absent in juice from white berries. Transmission at 520 nm corresponds to red coloration, while transmission at 415 nm corresponds to browning pigments (Francis and Clydesdale 1975).

Cranberry anthocyanins have been extracted for use as colorants in beverage mixes (Chiriboga and Francis 1973). Increased nitrogen fertilization results in decreased pigment accumulation. Anthocyanin concentration in extracted juice is increased by freezing and thawing (Sapers *et al.* 1983). Browning accompanied by an increase in hydroxy-methylfurfural concentration is a major color quality problem in cranberry products.

Grapes. Anthocyanins have been characterized for vinifera grapes (Wulf and Nagel 1978) and muscadines (Ballinger *et al.* 1973). Malvidin-3-glucoside is the major pigment of vinifera grapes, and 3,5-diglucosides are found in muscadines. Anthocyanin production in grapes is increased by girdling and ethephon application, but it is decreased by shading the vines or the application of too much nitrogen fertilizer (Kliewer 1977). Polyphenol oxidase has been implicated as the most important factor in the postharvest color degradation of grapes to be used in juice processing and catechol is its major substrate (Cash *et al.* 1976). A rapid quality control method based on spectral evaluation of anthocyanin concentration has been developed for grape products by Watada and Abbott (1975).

Clarification of grape juice is very important due to the formation of muddy brown polymers. 'Concord' juice possesses much better pigmen-

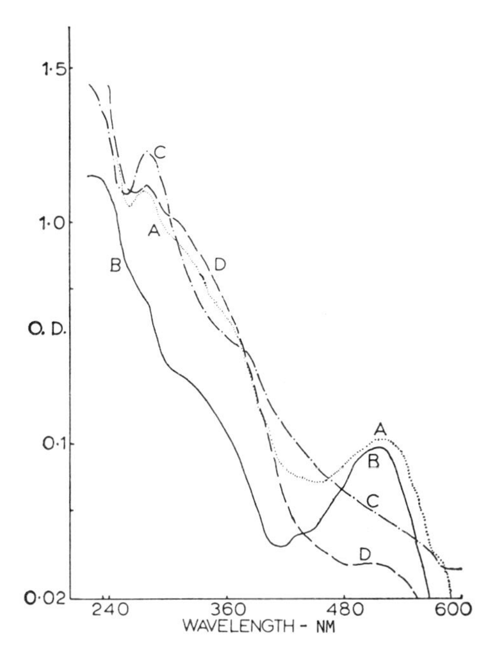

FIG. 11.12. Transmission spectra of cranberry juice cocktail. A—prepared from red berries and stored 1 year at 3°C; B—freshly prepared from red berries; C—prepared from red berries and stored 1 year at 38°C; and D—prepared from white berries and stored 1 year at 3°C. (From Francis and Clydesdale 1975.)

tation and color stability than other cultivars. Addition of $CaSO_4$ and $SnCl_2$ during processing results in greater color stability of 'Concord' juice (Sistrunk and Gascoigne 1983). During extraction, temperature is critical in obtaining the proper coloration. The solubility of both the anthocyanins and tannins increase as the temperature increases. A temperature of 60°–63°C (140°–145°F) optimizes anthocyanin extraction with a minimum of tannins. Time and temperature of extraction may be varied during the season to obtain similar color in juice from grapes of differing maturity.

Muscadine products are particularly susceptible to pigment degradation due to the greater presence of 3,5-diglucosides, which are more unstable than corresponding monoglucosides (Flora and Nakayama 1981). Juices from muscadine cultivars containing higher total anthocyanins and higher percentages of malvidin glycosides are more stable to heat preservation (Flora 1978). Low-temperature storage and SO_2 treatment provide the best protection for muscadine products (Flora and Nakayama 1981).

Color is a very important attribute in the quality of wines. Good wine color depends on full color development in grapes and optimal extraction prior to initiating fermentation. Color development in wines during aging is a complex process that involves three major types of reactions. Degradation of anthocyanins results in a color change from red to colorless to yellow. Condensation of anthocyanins with tannins stabilize the wine color. The orange character of wines results from tannin polymerization. The addition of SO_2 to prevent acetic acid formation can bleach out the anthocyanin color. For more detail on wine color development see Ribereau-Gayon (1982).

Cane Fruits. Cyanidin-3-glucoside is the major pigment of red raspberries (Torre and Barritt 1977). Cyanidin-3-rutinoside and cyanidin-3-xylosylrutinoside are important components of black raspberries. Blackberries, when frozen, change color from black to red. Jennings and Carmichael (1979) noted that this problem decreased with increasing maturity. They suggested that a decrease in pH is responsible for a blue-to-red color change in the product. An alternate explanation is a breakdown of metal–anthocyanin complexes.

Subtropical and Tropical Fruits

Few generalizations can be made on the color of tropical and subtropical fruits. All fruits in this group are susceptible to chilling injury and resulting color defects.

Bananas. Most of the pigmentation of bananas is in the peel. The predominant pigment of the plantains is chlorophyll, where xanthophylls characterize the color of ripe banana peels (Gross *et al.* 1976). Lutein, α-carotene, and β-carotene are the major pigments in the pulp, although all are present in very low concentrations. Both enzymic and nonenzymic browning reactions lead to color defects in banana products. Blanching of banana halves for 5 min and treatment with SO_2 (200 ppm) results in high-quality intermediate-moisture products.

Mangoes. Ripening mangoes change from green to yellow to red, losing chlorophyll while producing carotene and other carotenoids. An anthocyanin is present but is apparently not a major contributor to mango color. The change in color closely parallels changes in sugar concentrations, suggesting the possibility of instrumental color grading.

Failure to remove the peel before preparation of frozen mango purée results in a darker, but acceptable product. As in other fruit products, browning during storage, particularly of dried mango products, is the major color problem.

Pineapples. A major color problem in the postharvest handling of pineapples prior to processing is internal browning or endogenous brown spot. This condition is a physiological disorder resulting from chilling injury (Rohrbach and Paull 1982). The disorder is characterized by water spots, which enlarge and begin to turn brown. In extreme stages the whole internal tissue becomes brown. Holding of the fruit at 37°C for 24 hr prior to refrigeration minimizes this problem.

REFERENCES

ACZEL, A. 1977. Change of the carotenoid contents of yellow peach during the canning process and storage. Lebensm. Wiss. Technol. *10,* 120.

AHMED, A. A., WATROUS, G. H., HARGROVE, G. L. and DIMICK, P. S. 1976. Effect of fluorescent light on flavor and ascorbic acid content in refrigerated orange juice and drinks. J. Milk Food Technol. *39,* 332–336.

AHMED, E. M., DENNISON, R. A., DOUGHERTY, R. H. and SHAW, P. E. 1978A. Flavor and odor thresholds in water of selected orange juice components. J. Agric. Food Chem. *26,* 187–191.

AHMED, E. M., DENNISON, R. A., DOUGHERTY, R. H. and SHAW, P. E. 1978B. Effect of nonvolatile orange juice components, acid, sugar, and pectin on the flavor threshold of *d*-limonene in water. J. Agric. Food Chem. *26,* 192–194.

BALLINGER, W. E., MANESS, E. P., NESBITT, W. B. and CARROLL, D. E. 1973. Anthocyanins of black grapes of 10 clones of *Vitis rotundifolia,* MICHX. J. Food Sci. *38,* 909–910.

BERRY, R. E. and VELDHUIS, M. K. 1977. Processing of oranges, grapefruit, and tangerines. *In* Citrus Science and Technology, Vol. 2. S. Nagy, E. Shaw, and M. K. Veldhuis (Editors). AVI Publishing Co., Westport, CT.

BLANPIED, G. D. and BLAK, V. A. 1976. Relationship between ethylene level and varietal flavor in 'Delicious' apples. HortScience *11,* 596–597.

BOBBIO, F. O. and SCAMPARINI, A. R. P. 1982. Carbohydrates, organic acids and anthocyanin of *Eugenia jambolana* Lamarck. Ind. Aliment. *21,* 296–298.

BOGGESS, T. S. and HEATON, E. K. 1973. Canning southeastern freestone peaches. Univ. Ga. Coll. Agric. Exp. Stn. Res. Rep. *159.*

BOLIN, H. R. and SALUNKHE, D. K. 1971. Physiochemical and volatile flavor changes occurring in fruit juices during concentration and foam-mat drying. J. Food Sci. *36,* 665–668.

BOLIN, H. R., STAFFORD, A. E. and YANASE, K. 1976. Colour stability of dried peaches. Food Prod. Devel. *10,* 74–76.

BRECHT, J. K., KADER, A. A., HEINTZ, C. M. and NORONA, R. C. 1982. Controlled atmosphere and ethylene effects on quality of California canning apricots and clingstone peaches. J. Food Sci. *47,* 432–436.

BRUEMMER, J. H. and BOWERS, A. P. 1977. Storage stability of orange syrups. Proc. Fl. State Hort. Soc. *90,* 183–185.

BUTTERY, R. G. 1981. Vegetable and fruit flavors. *In* Flavor Research: Recent Advances. R. Teranishi, R. A. Flath, and H. Sugisawa (Editors). Marcel Dekker, New York.

CASH, J. N., SISTRUNK, W. A. and STUTTE, C. A. 1976. Characteristics of Concord grape polyphenoloxidase involved in juice color loss. J. Food Sci. *41,* 1398–1402.

CHAKRAWAR, V. R. 1978. Effect of some growth regulators on the ascorbic acid content and bitterness of the orange juice. Indian Food Packer *32,* 70–72.

CHANDLER, B. V. 1958. Anthocyanins of blood oranges. Nature *182,* 933.

CHANDLER, B. V. 1977. One of the "101 most interesting problems in food science"—bitterness in orange juice, a case history. Food Technol. (Australia) *29,* 303–311.

CHANDLER, B. V. and JOHNSON, R. L. 1977. Cellulose acetate as a selective sorbent for limonin in orange juice. J. Sci. Food Agric. *28,* 875–884.

CHICHESTER, C. O. and McFEETERS, R. 1971. Pigment degeneration during processing and storage. *In* The Biochemistry of Fruits and Their Products, Vol. 2. A. C. Hulme (Editor). Academic Press, New York.

CHIRIBOGA, C. D. and FRANCIS, F. J. 1973. Ion exchange purified anthocyanin pigments as a colorant for cranberry juice cocktail. J. Food Sci. *38,* 464–467.

CHUMA, Y., MORITA, K., and McCLURE, W. F. 1981. Application of light reflectance properties of Satsuma oranges to automatic grading in the packinghouse line. J. Fac. Agric. Kyushu Univ. *26,* 45–55.

CORNWELL, C. J. and WROLSTAD, R. E. 1981. Causes of browning in pear juice concentrate during storage. J. Food Sci. *46,* 515–518.

CRUSE, R. R., LIME, B. J. and HENSZ, R. A. 1979. Pigmentation and color comparison of Ruby Red and Star Ruby grapefruit juice. J. Agric. Food Chem. *27,* 641–642.

CURL, A. L. and BAILEY, G. F. 1956. Orange carotenoids. J. Agric. Food Chem. *4,* 156–162.

CZERKASKYJ, A. 1970. Pink discoloration in canned Williams' Bon Chretien pears. J. Food Sci. *35,* 608–611.

DEMOLE, E., ENGGIST, P., and OHLOFF, G. 1982. 1-*p*-Methene-8-thiol: A powerful flavor impact constituent of grapefruit juice (*Citrus paradisi* MacFayden). Helv. Chim. Acta *65,* 1785–1794.

DEUBERT, K. H. 1978. A rapid method for the extraction and quantitation of total anthocyanin of cranberry fruit. J. Agric. Food Chem. *26,* 1452–1453.

DIMICK, P. S. and HOSKIN, J. 1983. Review of apple flavor—state of the art. Crit. Rev. Food Sci. Nutr. *18,* 387–409.

DOUGHERTY, M. H., TING, S. V., ATTAWAY, J. A. and MOORE, E. L. 1977. Grapefruit juice quality improvement studies. Proc. Fl. State Hort. Soc. *90,* 165–172.

DULL, G. G. 1971. The pineapple: General. *In* The Biochemistry of Fruits and Their Products, Vol. 2. A. C. Hulme (Editor). Academic Press, New York.

DULL, G. G., BIRTH, G. S. and MAGEE, J. B. 1980. Nondestructive evaluation of internal quality. HortScience *15,* 60–63.

EAGERMAN, B. A. 1978. Orange juice color measurement using general purpose tristimulus colorimeters. J. Food Sci. *43,* 428–430.

EL-ZEFTAWI, B. M. and GARRETT, R. G. 1978. Effects of ethephon, GA and light exclusion on rind pigments, plastid ultrastructure and juice quality of Valencia oranges. J. Hort. Sci. *53,* 215–223.

ENNIS, D. M., KEEPING, L., CHIN-TING, J. and ROSS, N. 1979. Consumer evaluation of the inter-relationships between the sensory components of commercial orange juices and drinks. J. Food Sci. *44,* 1011–1012, 1016.

FELLERS, P. J. and BUSLIG, B. S. 1975. Relation of processing, variety and maturity to flavor quality and particle size distribution in Florida orange juices. I. Flavor considerations. Proc. Fl. State Hort. Soc. *88,* 350–353.

FLATH, R. A., BLACK, D. R., GUADAGNI, D. G., McFADDEN, W. H. and SCHULTZ, T. H. 1967. Identification and organoleptic evaluation of compounds in Delicious apple essence. J. Agr. Food Chem. *15,* 29–35.

FLATH, R. A., FORREY, R. R. and TERANISHI, R. 1969. High resolution vapor analysis for fruit variety and fruit product comparisons. J. Food Sci. *34,* 382–386.

FLORA, L. F. 1978. Influence of heat, cultivar and maturity on the anthocyanidin-3,5-diglucosides of muscadine grapes. J. Food Sci. *43,* 1819–1821.

FLORA, L. F. and NAKAYAMA, T. O. M. 1981. Quality characteristics of muscadine grape products. *In* Quality of Selected Fruits and Vegetables of North America. R. Teranishi and H. Barrera-Benitez (Editors). Amer. Chemical Society, Washington, DC.
FRANCIS, F. J. and CLYDESDALE, F. M. 1975. Food Colorimetry: Theory and Applications. AVI Publishing Co., Westport, CT.
FULEKI, T. and FRANCIS, F. J. 1968. Quantitative methods for anthocyanins. 2. Determination of total anthocyanin and degradation index for cranberry juice. J. Food Sci. *33,* 78–83.
GAWLER, J. H. 1962. Constituents of canned Malayan pineapple juice. I. Amino acids, non-volatile acids, sugars, volatile carbonyl compounds and volatile acids. J. Sci. Food Agric. *13,* 57–61.
GOODWIN, T. W. and GOAD, L. J. 1970. Carotenoids and triterpenoids. *In* The Biochemistry of Fruits and Their Products, Vol. 1. A. C. Hulme (Editor). Academic Press, New York.
GORDON, H. T. and BAUERNFEIND, J. C. 1982. Carotenoids as food colorants. Crit. Rev. Food Sci. Nutr. *18,* 59–97.
GORSKI, P. M. and CREASY, L. L. 1977. Color development in 'Golden Delicious' apples. J. Amer. Soc. Hort. Sci. *102,* 73–75.
GROSS, J. 1977. Carotenoid pigments in citrus. *In* Citrus Science and Technology, Vol. 1. S. Nagy, P. E. Shaw, and M. K. Veldhuis (Editors). AVI Publishing Co., Westport, CT.
GROSS, J. and ECKHARDT, G. 1978. A natural apocarotenol from the peel of the ripe Golden Delicious apple. Phytochemistry *17,* 1803–1804.
GROSS, J., CARMON, M., LIFSHITZ, A, and COSTES, C. 1976. Carotenoids of banana pulp, peel and leaves. Lebensm. Wiss. Technol. *9,* 211–214.
GUADAGNI, D. G., BOMBEN, J. L. and HARRIS, J. G. 1971A. Recovery and evaluation of aroma development in apple peels. J. Sci. Food Agric. *22,* 115–119.
GUADAGNI, D. G., BOMBEN, J. L. and HUDSON, J. S. 1971B. Factors influencing the development of aroma in apple peels. J. Sci. Food Agric. *22,* 110–115.
GUADAGNI, D. G., MAIER, V. P. and TURNBAUGH, J. G. 1973. Effect of some citrus juice constituents on taste thresholds for limonin and naringen bitterness. J. Sci. Food Agric. *24,* 1277–1288.
HEATON, E. K., BOGGESS, T. S., SHEWFELT, A. L., LI, K. C. and WOODROOF, J. G. 1973. The production and utilization of peach pulp, juice drink and concentrates. Univ. Ga. Coll. Agric. Exp. Stan. Res. Bull. *136.*
HEIKAL, H. A., EL-DASHLOUTY, M. S., EL-SIDAWI, M. H. and EL-WAKEIL, F. A. 1977. Results on studies on the preservation of lemon and orange juice concentrates. Confructa *22,* 15–21.
HODGE, J. E. and OSMAN, E. M. 1976. Carbohydrates. *In* Food Chemistry. O. R. Fennema (Editor). Marcel Dekker, New York.
HOROWITZ, R. M. and GENTILI, B. 1977. Flavonoid constituents of *Citrus. In* Citrus Science and Technology, Vol. 1. S. Nagy, P. E. Shaw, and M. K. Veldhuis (Editors). AVI Publishing Co., Westport, CT.
HOWARD, G. E. and HOFFMAN, A. 1967. A study of the volatile flavouring constituents of canned Malayan pineapple. J. Sci. Food Agric. *18,* 106–110.
HUGGART, R. L., PETRUS, D. R. and BUSLIG, B. S. 1977. Color aspects of Florida commercial grapefruit juices. Proc. Fl. State Hort. Soc. *90,* 173–175.
HULME, A. C. 1971. The mango. *In* The Biochemistry of Fruits and Their Products, Vol. 2. A. C. Hulme (Editor). Academic PRESS, New York.
HUNTER, R. S. 1975. The Measurement of Appearance. John Wiley & Sons, New York.
HUNTER, G. L. K., BUCEK, W. A. and RADFORD, T. 1974. Volatile components of canned Alphonso mango. J. Food Sci. *39,* 900–903.

ISMAIL, H. H., TUCKNOTT, O. G. and WILLIAMS, A. A. 1980. The collection and concentration of aroma components of soft fruit using Porapak Q. J. Sci. Food Agric. *31,* 262–266.

JEN, J. J. and KAHLER, K. R. 1974. Characterization of polyphenol oxidase in peaches grown in the Southeast. HortScience *9,* 590–591.

JENNINGS, D. L. and CARMICHAEL, E. 1979. Colour changes in frozen blackberries. Hort. Res. *19,* 15–24.

JENNINGS, W. G. and SEVENANTS, M. R. 1964. Volatile esters of Bartlett pear III. J. Food Sci. *29,* 158–163.

JEZEK, E. and SMYRL, T. G. 1980. Volatile changes accompanying dehydration of apples by the osmovac process. Can. Inst. Food Sci. Technol. J. *13,* 43–44.

JOHNSON, J. and CLYDESDALE, F. M. 1982. Perceived sweetness and redness in colored sucrose solutions. J. Food Sci. *47,* 747–752.

JOHNSON, R. L. and CHANDLER, B. V. 1981. A pilot-plant cellulose acetate gel bead column for the removal of limonin from citrus juices. J. Sci. Food Agric. *32,* 1183–1190.

JOHNSON, R. L. and TOLEDO, R. T. 1975. Storage stability of 55° Brix orange juice concentrate aseptically packaged in plastic and glass containers. J. Food Sci. *40,* 433–434.

KADER, A. A., HEINTZ, C. M. and CHORDAS, A. 1982. Postharvest quality of fresh and canned clingstone peaches as influenced by genotypes and maturity of harvest. J. Amer. Soc. Hort. Sci. *107,* 947–951.

KALLIO, H. 1976. Development of volatile aroma compounds in artic bramble, *Rubus articus L.* J. Food Sci. *41,* 563–566.

KANNER, J., FISHBEIN, J., SHALOM, P., HAREL, S. and BEN-GERA, I. 1982. Storage stability of orange juice concentrate packaged aseptically. J. Food Sci. *47,* 429–431, 436.

KATAYAMA, T., NAKAYAMA, T. O. M., LEE, T. H. and CHICHESTER, C. O. 1971. Carotenoid transformations in ripening apricots and peaches. J. Food Sci. *36,* 804–806.

KEALY, K. S. and KINSELLA, J. E. 1979. Orange juice quality with an emphasis on flavor components. Crit. Rev. Food Sci. Nutr. *11,* 1–40.

KEFFORD, J. F. and CHANDLER, B. V. 1970. The Chemical Constituents of Citrus Fruits. Academic Press, New York.

KLIEWER, W. M. 1977. Influence of temperature, solar radiation and nitrogen on coloration and composition of Emperor grapes. Amer. J. Enol. Vitic. *28,* 96–103.

KOPELMAN, I. J., MEYDAV, S. and WEINBERG, S. 1977A. Storage studies of freeze-dried lemon crystals. J. Food Technol. *12,* 403–410.

KOPELMAN, I. J., MEYDAV, S. and WILMERSDORF, P. 1977B. Freeze-drying encapsulation of water-soluble citrus aroma. J. Food Technol. *12,* 65–72.

KRIVOSHIEV, G. and CHALUKOVA, R. 1981. Spectral transmittance characteristics of peaches, apricots, pears, plums, strawberries, tomatoes and peppers. Acta Aliment. *10,* 45–60.

LaBELLE, R. L. 1981. Apple quality characteristics as related to various processed products. *In* Quality of Selected Fruits and Vegetables of North America. R. Teranishi and H. Barrera-Benitez (Editors). Amer. Chemical Society, Washington, DC.

LEBERMANN, K. W., NELSON, A. I. and STEINBERG, M. P. 1968. Post-harvest changes of broccoli stored in modified atmospheres. Food Technol. *22,* 487–518.

LEE, C. M., LEE, T.-C. and CHICHESTER, C. O. 1979. Kinetics of the production of biologically active Maillard browned products in apricot and glucose-L-tryptophan. J. Agric. Food Chem. *27,* 478–482.

LEONARD, S., PANGBORN, R. M. and TOMBROPOULOS, D. 1976. Influence of processing variables on flavor stability of pear purée. J. Food Sci. *41,* 840–844.

LI, K. C., BOGGESS, T. S. and HEATON, E. K. 1972. Relationship of sensory rating with tannin components of canned peaches. J. Food Sci. *37,* 177–178.

LIDSTER, P. D., LIGHTFOOT, H. T. and McRAE, K. B. 1983. Production and regeneration of principal volatiles in apples stored in modified atmospheres and air. J. Food Sci. *48,* 400–402, 410.

LUH, B. S. 1980. Nectars, pulpy juices and fruit juice blends. *In* Fruit and Vegetable Juice Processing Technology, 3rd ed. P. E. Nelson and D. K. Tressler (Editors). AVI Publishing Co., Westport, CT.

LUND, E. D. and BRYAN, W. L. 1976. Commercial orange essence: Comparision of composition and methods of analysis. J. Food Sci. *42,* 385–388.

MacLEOD, W. D. and BUIGUES, N. M. 1964. Sesquiterpenes. I. Nootkatone, a new grapefruit flavor constitutent. J. Food Sci. *29,* 565–568.

MacLEOD, A. J. and DE TROCONIS, N. G. 1982. Volatile flavour components of mango fruit. Phytochemistry *21,* 2523–2526.

MAIER, V. P., BENNETT, R. D. and HASEGAWA, S. 1977. Limonin and other limonoids. *In* Citrus Science and Technology, Vol. 2. S. Nagy, P. E. Shaw, and S. Hasegawa (Editors). AVI Publishing Co., Westport, CT.

MAIER, V. P. and YOKOYAMA, H. 1977. The concept of bioregulation of plant composition and its application to quality improvement of citrus fruits. *In* Citrus Science and Technology, Vol. 1. S. Nagy, P. E. Shaw, and M. K. Veldhuis (Editors). AVI Publishing Co., Westport, CT.

MARKAKIS, P. 1982. Anthocyanins as food additives. *In* Anthocyanins as Food Colors. P. Markakis (Editor). Academic Press, New York.

MARRIOTT, J. 1980. Bananas—physiology and biochemistry of storage and ripening for optimum quality. Crit. Rev. Food Sci. Nutr. *13,* 41–88.

MATSUURA, T., BAXTER, A. G. and SOURIRAJAN, S. 1975. Reverse osmosis recovery of flavor components from apple juice waters. J. Food Sci. *40,* 1039–1046.

McCORNACK, A. A. and WARDOWSKI, W. F. 1977. Degreening Florida citrus fruit: Procedures and physiology. Proc. Intern. Soc. Citriculture *1,* 211–215.

McINTOSH, C. A., MANSELL, R. L. and ROUSEFF, R. L. 1982. Distribution of limonin in the fruit tissues of nine grapefruit cultivars. J. Agric. Food Chem. *39,* 689–692.

MORRIS, J. R., KATTAN, A. A., BUESCHER, R. W., NELSON, G. S. and BAUGHMAN, A. J. 1976. Post-harvest quality of machine-harvested strawberries. Arkansas Farm Res. *25* (1) 13.

MORRIS, J. R., SPAYD, S. F., CAWTHON, D. L., KATTAN, A. A. and NELSON, G. S. 1980. Influence of hand picking prior to optimum and late timing of machine harvesting on yield and quality of A-5344 strawberries. J. Amer. Soc. Hort. Sci. *105,* 72–74.

MOSHONAS, M. G. and SHAW, P. E. 1978. Compounds new to essential orange oil from fruit treated with abscission chemicals. J. Agric. Food Chem. *26,* 1288–1290.

MOSHONAS, M. G., SHAW, P. E. and SIMS, D. A. 1976. Abscission agent effects on orange juice flavor. J. Food Sci. *41,* 809–811.

MOUTOUNET, M. 1976. Carotenoids in d'Ente variety plums and in prunes made therefrom. Ann. Technol. Agric. *25,* 73–84.

MOYER, J. C. and AITKEN, H. C. 1980. Apple juice. *In* Fruit and Vegetable Juice Processing Technology, 3rd ed. P. E. Nelson and D. K. Tressler (Editors). AVI Publishing Co., Westport, CT.

NAGY, S. and SHAW, P. E. 1980. Processing of grapefruit. *In* Fruit and Vegetable Juice Processing Technology, 3rd ed. P. E. Nelson and D. K. Tressler (Editors). AVI Publishing Co., Westport, CT.

NAKAYAMA, T. O. M. and POWERS, J. J. 1972. Absorption spectra of anthocyanin *in vivo*. *In* The Chemistry of Plant Pigments. C. O. Chichester (Editor). Academic Press, New York.

NELSON, G. S., MORRIS, J. R., KATTAN, A. A. and SHELBY, K. R. 1978. Mechanical system for harvesting and handling strawberries for processing. Trans. ASAE *21*, 442–450.

NORDBY, H. E. and NAGY, S. 1980. Processing of oranges and tangerines. *In* Fruit and Vegetable Juice Processing Technology 3rd ed. P. E. Nelson and D. K. Tressler (Editors). AVI Publishing Co., Westport, CT.

NURSTEN, H. E. 1970. Volatile compounds: The aroma of fruits. *In* The Biochemistry of Fruits and Their Products, Vol. 1. A. C. Hulme (Editor). Academic Press, New York.

O'BRIEN, M., FRIDLEY, R. B. and CLAYPOOL, L. L. 1978. Food losses in harvest and handling systems for fruits and vegetables. Trans. ASAE *21*, 386–390.

OLSEN, K. L. and KETCHIE, D. O. 1978. Coloring and dessert quality of Red Delicious strains. Proc. Washington State Hort. Assn. *74*, 80–84.

PANASIUK, O., TALLEY, F. B. and SAPERS, G. M. 1980. Correlation between aroma and volatile composition of McIntosh apples. J. Food Sci. *45*, 989–991.

PAPANICOLAOU, D., RIGAUD, J., SAAVAGEOT, F., DUBOIS, P. and SIMATOS, D. 1978. Behaviour of some volatile compounds during storage of orange juice powder with low and intermediate moisture contents. J. Food Technol. *13*, 511–519.

PARLIMENT, T. H. and SCARPELLINO, R. 1977. Organoleptic techniques in chromatographic food flavor analysis. J. Agric. Food Chem. *25*, 97–99.

PEDERSON, C. S. 1980. Grapejuice. *In* Fruit and Vegetable Juice Processing Technology, 3rd ed. P. E. Nelson and D. K. Tressler (Editors). AVI Publishing Co., Westport, CT.

POLESELLO, A. and BONZINI, C. 1977. Observations on pigments of sweet cherries and on pigment stability during frozen storage. 1. Anthocyanin composition. Confructa *22*, 170–175.

PUSKI, G. and FRANCIS, F. J. 1967. Flavonol glycosides in cranberries. J. Food Sci. *32*, 527–530.

PYYSALO, T., HONKANEN, E. and HIRVI, T. 1979. Volatiles of wild strawberries, *Fragaria vesca* L., compared to those of cultivated berries, *Fragaria* × *ananassa* cv Senga Sengana. J. Agric. Food Chem. *27*, 19–22.

PYYSALO, T., SUIHKO, M., and HONKANEN, E. 1977. Odour thresholds of the major volatiles identified in cloudberry (*Rubus chamaemorus* L.) and Artic bramble (*Rubus arcticus*, L.). Lebensm. Wiss. Technol. *10*, 36–39.

RAMANA, K. V. R., GOVINDARAJAN, V. S. and RANGANNA, S. 1981. Citrus fruits—varieties, chemistry, technology, and quality evaluation. Part 1. Varieties, production, handling, and storage. Crit. Rev. Food Sci. Nutr. *15*, 353–431.

RAMIREZ, J. M. and KREZDORN, A. H. 1975. Effect of date of harvest and spot picking on yield and quality of grapefruit. Proc. Fl. State Hort. Soc. *88*, 40–44.

RAMIREZ-MARTINEZ, J. R., LEVI, A., PADUA, H. and BAKAL, A. 1977. Astringency in an intermediate moisture banana product. J. Food Sci. *42*, 1201–1203, 1217.

REID, W. S. 1976. Optical detection of apple skin, bruise, flesh, stem and calyx. J. Agric. Eng. Res. *21*, 291–295.

RESNIK, S. and CHIRIFE, J. 1979. Effect of moisture content and temperature on some aspects of nonenzymatic browning in dehydrated apple. J. Food Sci. *44*, 601–605.

RIBÉREAU-GAYON, P. 1982. The anthocyanins of grapes and wines. *In* Anthocyanins as Food Colors. P. E. Markakis (Editor). Academic Press, New York.

RICHARDSON, T. 1976. Enzymes. *In* Food Chemistry. O. R. Fennema (Editor). Marcel Dekker, New York.

ROBERTSON, G. L. and REERES, M. J. 1981. Relationship between colour and brown pigment concentration in orange juices subjected to storage temperature abuse. J. Food Technol. *16,* 535–541.

ROE, B. and BRUEMMER, J. H. 1977. Treatment requirements for debittering and fortifying grapefruit and stable storage of the product. Proc. Fl. State Hort. Soc. *90,* 180–182.

ROHRBACH, K. G. and PAULL, R. E. 1982. Incidence and severity of chilling induced internal browning of waxed 'Smooth Cayenne' pineapple. J. Amer. Soc. Hort. Sci. *107,* 453–457. .

ROMANI, R. J. and JENNINGS, W. G. 1971. Stone fruits. *In* The Biochemistry of Fruits and Their Products, Vol. 2. A. C. Hulme (Editor). Academic Press, New York.

SÁNCHEZ NIEVA, F. and HERNANDEZ, I. 1977. Studies on the freezing of Red Spanish and Smooth Cayenne pineapples. J. Agric. Univ. Puerto Rico *61,* 354–360.

SÁNCHEZ NIEVA, F., MERCADO, M., and BUESO, C. 1980. Effect of the stage of development at harvest on the texture, flavor, quality and yields of frozen green bananas. J. Agric. Univ. Puerto Rico *64,* 275–282.

SAPERS, G. M., JONES, S. B., and MAHER, G.T. 1983. Factors affecting the recovery of juice and anthocyanin from cranberries. J. Amer. Soc. Hort. Sci. *108,* 246–249.

SARAVACOS, G. D., MOYER, J. C. , and WOOSTER, G. D. 1969. Stripping of high-boiling aroma compounds from aqueous solutions. N.Y. State Agric. Exp. Stn. Res. Circ. *21.*

SCHAMP, N. and DIRINCK, P. 1982. The use of headspace concentration on Tenax for objective flavor quality evaluation of fresh fruits (strawberry and apple). *In* Chemistry of Foods and Beverages: Recent Developments. G. Charalambous and G. Inglett (Editors). Academic Press, New York.

SCHREIER, P. 1980. Quantitative composition of volatile constituents in cultivated strawberries, *Fragaria ananassa* cvs. Senga Sengana, Senga Litessa and Senga Gourmella. J. Sci. Food Agric. *31,* 487–494.

SCHREIER, P., DRAWERT, F. and JUNKER, A. 1976. Identification of volatile constituents from grapes. J. Agric. Food Chem. *24,* 331–336.

SCHREIER, P., DRAWERT, F. and MICK W. 1978. The quantitative composition of natural and technologically changed aromas of plants. V. The influence of HIST-heating on the constituents of apple juice. Lebensm. Wiss. Technol. *11,* 116–121.

SHARMA, M. R. and DHURIA, H.S. 1978. Ethephon for colouring peach fruits. Indian Hort. *23* (1) 10, 18.

SHAW, P. E. 1977A. Essential oils. *In* Citrus Science and Technology, Vol. 1. S. Nagy, P. E. Shaw, and M. K. Veldhuis (Editors). AVI Publishing Co., Westport, CT.

SHAW, P. E. 1977B. Aqueous essences. *In* Citrus Science and Technology, Vol. 1. S. Nagy, P. E. Shaw, and M. K. Veldhuis (Editors). AVI Publishing Co., Westport, CT.

SHEWFELT, R. L. and AHMED, E. M. 1978. Enhancement of powdered soft drink mixes with anthocyanin extracts. J. Food Sci. *43,* 435–438.

SHRIKHANDE, A. J. 1976. Anthocyanins in foods. Crit. Rev. Food Sci. Nutr. *7,* 193–217.

SILVERSTEIN, R. M. 1971. The pineapple: Flavor. *In* The Biochemistry of Fruits and Their Products, Vol. 23. A. C. Hulme (Editor). Academic Press, New York.

SISTRUNK, W. A. 1976. Effect of extraction temperature on quality attributes of Concord grape juice. Arkansas Farm Res. *25* (1) 12.

SISTRUNK, W. A. and GASCOIGNE, H. L. 1983. Stability of color in 'Concord' grape juice and expression of color. J. Food Sci. *48,* 430–433, 440.

SISTRUNK, W. A. and MORRIS, J. R. 1978. Storage stability of strawberry products manufactured from mechanically harvested strawberries. J. Amer. Soc. Hort. Sci. *103,* 616–620.

SISTRUNK, W. A. and ROM, R. C. 1976. Quality attributes of peaches for processing. Arkansas Farm Res. *25* (3) 11.

SMOOT, J. J. and MELVIN. C. F. 1975. Market quality for citrus fruits dropped to the ground for harvesting. Proc. Fl. State Hort. Soc. *88,* 276–280.

SOMERS, T. C. 1971. The polymeric nature of wine pigments. Phytochemistry *10,* 2175–2186.

SOUTY, M. and REICH, M. 1978. Effects of processing on aroma compounds of peaches. Ann. Technol. Agric. *27,* 837–848.

STEWART, I. 1977. Provitamin A and carotenoid content of citrus juices. J. Agric. Food Chem. *25,* 1132–1137.

SULLIVAN, J. F., CRAIG, J. C., DEKAZOS, E. D., LEIBY, S. M. and KONSTANCE, R. P. 1982. Dehydrated blueberries by the continuous explosion-puffing process. J. Food Sci. *47,* 445–448.

TATUM, J. H., NAGY, S. and BERRY, R. E. 1975. Degradation products formed in canned single-strength orange juice during storage. J. Food Sci. *40,* 707–709.

TIMBERLAKE, C. F. and BRIDLE, P. 1982. Distribution of anthocyanins in food plants. *In* Anthocyanins as food colors. P. Markakis (Editor). Academic Press, New York.

TING, S. V. and ATTAWAY, J. A. 1971. Citrus fruits. *In* The Biochemistry of Fruits and Their Products, Vol. 2. A. C. Hulme (Editor). Academic Press, New York.

TING, S. V. and McALLISTER, J. W. 1977. Quality of Florida canned grapefruit juice in supermarket stores of the United States. Proc. Fl. State Hort. Soc. *90,* 170–172.

TORRE, L. C. and BARRITT, B. H. 1977. Quantitative evaluation of *Rubus* fruit anthocyanin pigments. J. Food Sci. *42,* 488–490.

TRESSLER, D. K., CHARLEY, V. L. S. and LUH, B. S. 1980. Cherry, berry and other miscellaneous fruit juices. *In* Fruit and Vegetable Juice Processing Technology, 3rd ed. P. E. Nelson and D. K. Tressler (Editors). AVI Publishing Co., Westport, CT.

VAN BLARICOM, L. O. and HAIR, B. L. 1972. Some experiments in controlling the discoloration of canned peaches. South Carolina Agric. Exp. Stn. Tech. Bull. *1043.*

VANDERCOOK, C. E. 1977. Organic acids. *In* Citrus Science and Technology, Vol. 1. S. Nagy, R. E. Shaw, and M. K. Veldhuis (Editors). AVI Publ. Co., Westport, CT.

VERSTEEG, C., MARTENS, L. J. H., ROMBOUTS, F. M., VORAGEN, A. G. J. and PILNIK. W. 1977. Enzymatic hydrolysis of naringen in grapefruit juice. Lebensm. Wiss. Technol. *10,* 268–272.

WATADA, A. E. and ABBOTT, J. A. 1975. Objective method of estimating anthocyanin content for determining color grade of grapes. J. Food Sci. *40,* 1278–1279.

WELCH, R. C., JOHNSTON, J. C. and HUNTER, G. L. K. 1982. Volatile constituents of the muscadine grape (*Vitis rotundifolia*). J. Agric. Food Chem. *30,* 681–684.

WILLIAMS, A. A., TUCKNOTT, O. G. and LEWIS, M. J. 1977. 4-methoxyallyl-benzene: An important aroma component of apples. J. Sci. Food Agric. *28,* 185.

WILLIAMS, M. W. and BILLINGSLEY, H. D. 1974. Effect of nitrogen fertilizer on yield, size, and color of 'Golden Delicious' apple. J. Amer. Soc. Hort. Sci. *99,* 144–145.

WILLS, R. B. H. and SCOTT, K. J. 1976. Influence of nitrogen and phosphorous fertilizers on the relation between apple volatiles and storage breakdown. J. Hort. Sci. *51,* 177–179.

WILSON, C. W. and SHAW, P. E. 1978. Quantitative composition of cold-pressed grapefruit oil. J. Agric. Food Chem. *26,* 1432–1434.

WOODROOF, J. G., HEATON, E. K., BOGGESS, T. S. and LI, K. C. 1965. Development of peach products. Univ. Ga. Dept. Food Sci., Res. Rep. *12.*

WOODROOF, J. G. 1980. Production history of fruit and vegetable beverages. *In* Beverages: Carbonated and Noncarbonated. J. G. Woodroof and G. F. Phillips (Editors). AVI Publishing Co., Westport, CT.

WOODRUFF, R. E., DEWEY, D. H. and SELL, H. M. 1960. Chemical changes of Jersey and Rubel blueberry fruit associated with ripening and deterioration. Proc. Amer. Soc. Hort. Sci. *75,* 387–401.

WULF, L. W. and NAGEL, C. W. 1978. High-pressure liquid chromatographic separation of anthocyanins of *Vitis vinifera*. Amer. J. Enol. Vitic. *29*, 42–49.

YAMASHITA, I., NEMOTO, Y., and YOSHIKAWA, S. 1975. Studies on flavor development in strawberries. 1. Formation of volatile esters in strawberries. Agri. Biol. Chem. *39*, 2303–2307.

ZERBINI, P. E. and SOZZI, A. 1980. The influence of postharvest calcium treatments on the internal browning of Passe Crassane pears. Acta Hort. *92*, 332–333.

12

Composition and Nutritive Value of Raw and Processed Fruits

C. T. Young and J. S. L. How

More than any other food group, fruits introduce a variety of color, taste, and texture to meals and snacks. Fruits and fruit products add bouquet to meals. Some contain considerable amounts of several vitamins, and others are major contributors of essential minerals. They aid digestion and are important in body development and tone. In addition to their nutritional qualities, fruits have unique and desirable functions in the diet that warrant more consideration by those who do the family food planning.

Fruits usually are low in calories and, with the exception of avocados and olives, contain practically no fat. Fresh fruits have a natural laxative effect because of the cellulose and organic acids present. Furthermore, upon digestion, they yield an alkaline residue or ash that neutralizes the acid residue formed by meat, eggs, and other protein-rich foods. Certainly, fruits rich in vitamins A and C should be included in every diet. Canned and frozen fruits are important sources of nutrients.

Citrus fruits are especially valuable for their ascorbic acid (vitamin C) content. Berries and melons are also fairly good sources, and many other fresh fruits contribute significant amounts of this highly unstable vitamin.

The vitamin precursor carotene, which is converted in the body to vitamin A, is common in fruits. Apricots, peaches, cantaloupes, and ba-

Commercial Fruit Processing, 2nd Edition

ISBN 0-87055-502-2

nanas supply appreciable amounts. Although tree fruits in general contain only small quantities of the B-complex vitamins, several dried fruits and citrus products contribute some thiamin (B_1) to the diet.

Certain vitamin-like factors in the form of bioflavonoids are suggested to have actions similar to those of vitamin C, such as the strengthening of capillary walls against breakage or leakage of fluids. Besides citrus fruits, apricots, cantaloupes, cherries, grapes, and papayas are rich sources of bioflavonoids.

Potassium is generally plentiful in tree fruits and is usually combined with organic acids. Large amounts of calcium are present in dried fruits, and moderate quantities are present in oranges, raspberries, and strawberries. Iron is found in significant amounts in dates, figs, and bananas and is plentiful in peaches, prunes, raisins, and apricots.

Dietary fiber in the form of poorly digested carbohydrate compounds has been implicated to have a positive effect on certain physiological disorders. Certain fruits such as figs and lemons, which are high in fiber, can be used to supplement a low-fiber diet.

Since fruits have a characteristically high water content and are low in fat and protein, they represent good sources of iron and other minerals without increasing the total caloric consumption.

The quality of fruits is determined basically by their chemical ingredients. Specific compounds and elements are needed to nourish the myriad of individual cells that make up the human body. At least 45 nutrient materials are needed by human cells. Each of these essential materials must be present in the diet and readily converted to bodily use. Absence of any of them leads to illness and eventually to death. No single fruit provides all 45 nutrients.

Until about 1800, the balance of nutrients in fruits was given little scientific thought. For centuries humans knew only of their need for enough food. To satisfy this quantitative need, they collected wild fruits and vegetables, caught fish, and killed animals. Later, they learned to process food in ways that slowed the rate at which it spoiled so that food products could be stored for subsequent use. Advances in analytical chemistry gave birth to scientific nutrition, which is concerned with the chemical constituents of food and the complex processes by which food is utilized.

Carbohydrates and fats furnish most of the calories the body needs for heat and energy. The carbohydrates, sugars, and starches in fruits are easily broken down during metabolism in reactions that eventually give off carbon dioxide and water and release energy for use by the body. Even more energy is released by the metabolism of fats, and some also is provided by proteins.

The essential nutrients include oxygen and water, which are not usually considered foods. Oxygen is essential because it is used in the combustion of food in the body. Water is equally essential because it carries nutrients through the body.

The other essential nutrients are still conveniently classified into five main groups: carbohydrates, fats, proteins, minerals, and vitamins. The carbohydrates include sugars and starches from which the important nutrient glucose is derived, either directly or through chemical conversion inside the body. Fats provide calorie-yielding triglycerides and linoleic acid, which was identified as an essential nutrient in 1932.

From proteins, the body derives nitrogen and the amino acids it needs for building and repairing tissues. Eight amino acids must be provided directly in the diet (children need a ninth, histidine). These are considered essential nutrients because they cannot be manufactured by the body; other amino acids are also necessary for good health, but they can be manufactured within the body from other food ingredients.

Of the 45 essential nutrients, 17 are minerals. Calcium, chlorine, iron, magnesium, phosphorus, potassium, sodium, and sulfur are required in amounts that range from 1 μg to 1 g/day. Trace elements include chromium, cobalt, copper, fluorine, iodine, manganese, molybdenum, selenium, and zinc.

Even the minute amounts of trace elements required seem large compared with the amounts of some vitamins believed necessary. Thirteen vitamins—A, C, D, E, K, and the eight members of the vitamin B complex—are believed essential to human health; all play a vital part in body chemistry.

Table 12.1 contains a summary of many of the nutrients, their function, recommended daily allowance (RDA), and several fruit sources that are highest in each nutrient. Table 12.2 lists the caloric and nutritional value of the leading fruits produced in the United States. Additional data on the composition of fruits appear in Tables 6.3–6.7, 6.10, 7.1, 7.6, 10.1, and 10.2.

As consumers become more aware of what they eat, they demand more knowledge of the nutritional value of fruits in all forms, and the FDA is requiring that such information be put on the label of processed fruit products. Commercial fruit processors should be aware of the ways that harvesting, handling, storing, and processing by canning, dehydration, freezing, brining, preserving, and other methods affect the composition and nutrition of the fruit. Losses in moisture, vitamins, sugars, and starches are not uncommon during the harvesting, handling, and storing of fruits; these losses may adversely affect the quality and nutritive value of fruit products. Commercial processing of fruits by various methods

Table 12.1. Recommended Daily Allowance of Nutrients and Their Fruit Sources

Nutrient	Function	RDA[a]	Examples of fruit sources
Protein (includes enzymes and amino acids)	1. Required for growth, maintenance and repair of body tissues 2. Forms an important part of enzymes, hormones, and body fluids 3. Supplies energy	65 g	Low; dried apricots and dried figs are highest
Carbohydrates (starches, sugars and celluloses)	1. Starches and sugars are major sources of energy for internal and external work and to maintain body temperature 2. Celluloses furnish bulk in diet 3. Help use body fat efficiently	None established	Most all fruits, especially bananas, currants, dates, and raisins
Fats (oils)	1. Supply food energy in compact form (weight-for-weight supplies twice as much energy as carbohydrates) 2. Some supply essential fatty acids 3. Help body use certain other nutrients	None established	Very low, except in avocados and olives
Water	1. Important part of all cells and fluids in body 2. Carrier of nutrients to and waste from cells in the body 3. Aids in digestion and absorption of food 4. Helps to regulate body temperature		Fresh fruits are usually high; dried fruits are low
Vitamin A (carotene)	1. Important for normal growth in children 2. Necessary for good vision 3. Helps keep skin smooth	5000 IU	Dried apricots, cantaloups, peaches, orange juice, and watermelons

(continued)

Table 12.1. (*Continued*)

Nutrient	Function	RDA[a]	Examples of fruit sources
	4. Helps keep lining of mouth, nose, throat, and digestive tract healthy and resistant to infection		
Vitamin C (ascorbic acid)	1. Essential for healthy teeth, gums, and bones 2. Builds strong body cells and blood vessels 3. Aids in healing wounds	60 mg	Citrus juices, cantaloupes, and strawberries
Thiamin (vitamin B_1)	1. Necessary for proper function of heart and nervous system 2. Helps body cells obtain energy from food 3. Promotes good appetite and digestion	1.5 mg	Avocados and orange juice
Riboflavin (vitamin B_2)	1. Necessary for healthy skin 2. Helps prevent sensitivity of the eyes to light 3. Essential for building and maintaining body tissues	1.7 mg	Avocados and dried peaches
Niacin (niacinamide)	1. Necessary for converting food to energy 2. Helps to maintain health of skin, tongue, digestive tract, and nervous system	20 mg	Dried apricots and dried peaches
Calcium	1. Helps build bone and teeth 2. Aids in clotting of blood 3. Helps nerves, muscles, and heart to function properly	1000 mg	Blackberries, currants, dried figs, and rhubarb

(*continued*)

Table 12.1. *(Continued)*

Nutrient	Function	RDA[a]	Examples of fruit sources
Iron	1. Combines with protein to make hemoglobin, the red substance of blood which carries oxygen from the lungs to muscles, brain, and other parts of the body 2. Helps cells use oxygen	18 mg	Dried apricots, dried figs, dried peaches, and prunes
Vitamin B_6	1. Important for healthy teeth and gums, the health of blood vessels, red blood cells, and the nervous system	2 mg	Bananas and raisins
Vitamin B_{12}	1. Helps prevent certain forms of anemia 2. Contributes to health of nervous system, and proper growth in children	6 μg	Low in fruits
Folacin (folic acid)	1. Helps prevent certain forms of anemia and is important in maintaining functions of the intestinal tract	0.4 mg	Dried dates, blackberries, and avocados
Vitamin D (includes D_2 = calciferol and D_3 = irradiated dehydrocholesterol)	1. Necessary for strong teeth and bones 2. Helps the body utilize calcium and phosphorus	400 IU	Not present in fruits
Vitamin E (tocopherols)	1. Essential for the functioning of red blood cells 2. Protects essential fatty acids	30 IU	Olives
Pantothenic acid	1. Necessary for the body's use of carbohydrates, fats, and protein	10 mg	Oranges

(continued)

Table 12.1. (*Continued*)

Nutrient	Function	RDA[a]	Examples of fruit sources
Biotin	1. Essential for the functioning of many body systems and use of food for energy	0.3 mg	Bananas and strawberries
Iodine	1. Maintains proper function of the thyroid gland	150 μg	Variable, depending upon iodine content of soil
Phosphorus	1. Builds bones and teeth (with other minerals) 2. Important in a number of body systems involving fats, carbohydrates, salts, and enzymes	1000 mg	Dried peaches, dried raisins, and dried figs

[a]Established as the standard Recommended Daily Allowance (RDA) for purposes of nutrition labeling by the FDA (Federal Register, Mar. 30, 1972).

changes the composition of the product, which is reflected in the nutritional value. Some of the losses in sugars, acids, vitamin C, and minerals are compensated for by the addition of these nutrients during processing.

When fruits are peeled, sliced, diced, or pressed, oxidation is stimulated; when they are blanched, pasteurized, or sterilized by heat, heat-sensitive constituents are affected; and when they are extracted by vacuum, pressing, centrifuging, or solvent (or a combination of these) their composition changes. Exposure to heat, air, light, irradiation, free water, or traces of certain metals may alter the nutritional qualities of fruits. High temperatures, fluctuating temperatures, and changes in moisture content (as experienced in regular canning, dehydration, freeze drying, spray drying, foam-mat drying, forced drying, or vacuum drying) alter the chemical composition of products and change their nutritional qualities to some extent.

Changes in the composition and nutritional qualities of cherries, as discussed in Chapter 2, Chapter 9, and Chapter 10, illustrate the differences between a fresh and processed fruit product. Cherries are bleached (and leached) in solutions of sulfur dioxide, then given a secondary bleaching in hypochlorite solution, and sometimes in a solution of hydrogen peroxide, to eliminate iron salts that could affect the final color.

Table 12.2. Caloric and Nutritive Value of 10 Fruits

Fruit[a]	Caloric Value[b] (kcal/100 g)	Nutritional value[b] per 100 g
1. Oranges	49	A good source of potassium (200 mg); an excellent source of vitamin C (50 mg) and bioflavonoids
2. Grapes	69	A very rich source of chromium; good source of potassium (158 mg); dried grapes (raisins) contain about 4 times the solids and calories of fresh fruit
3. Apples	56	Fair source of potassium and vitamin C; dried apples have almost 5 times the solids and caloric value of fresh apples
4. Grapefruit	41	A good source of vitamin C (38 mg) and bioflavonoids; a fair source of potassium; pink and red grapefruit contain moderate amounts of vitamin A (ca. 440 IU)
5. Peaches	38	A good source of potassium (202 mg); an excellent source of vitamin A (1330 IU); dried peaches contain about 7 times the solids and calories of the fresh fruit
6. Pears	61	A fair source of potassium and vitamin C; dried pears contain about 4.5 times the solids and calories of fresh fruit
7. Lemons	25	A good source of vitamin C (46 mg) and bioflavonoids; a fair source of potassium
8. Prunes	66	A good source of potassium, but contains only negligible amounts of vitamin C; dried prunes contain about 4 times the solids and calories of fresh fruit
9. Strawberries	37	A fair source of iron (1.0 mg) and potassium (164 mg); an excellent source of vitamin C (59 mg)
10. Cherries	58	An excellent source of vitamin A (1000 IU), potassium (291 mg), and bioflavonoids; a fair source of vitamin C

Source: Adapted from Anon (1980, 1983).
[a]Fruits listed in order of their U.S. production tonnage in 1980.
[b]Caloric and nutritive values based on fresh raw fruit.

After the bleaching agents (together with soluble sugars, water-soluble vitamins, minerals, flavors, and colors) are leached out, other colors, flavors, spices, and sugars are added. By this time, maraschino cherries, as they appear in confections, fruit cakes, or ice cream, are far different from the fresh fruit. In similar fashion, sliced apples are fortified, glazed, and made into butter; berries, peaches, pears, and other fruits are made into preserves and jams with loss of heat-sensitive volatiles and nutrients.

ANALYSIS OF RAW AND PROCESSED FRUITS

The analysis of fresh fruit for water, protein, carbohydrates, minerals, and fiber is not often required, nor does the determination of such constituents present any special difficulties. Only certain special determinations are, therefore, described in this section.

From the legal point of view, fresh fruits may need to be analyzed for preservatives such as diphenyl, orthophenylphenol, boric acid, thiourea, and mineral oil. It should be noted that traces of boric acid are present naturally in fruits. Dried fruits should be examined for sulfur dioxide, mineral oil, and trace metals.

Fruits for canning contain much less soluble solids than the syrup. During processing, therefore, the juice of the fruit dilutes the syrup. The density of the final syrup (referred to as the "cutout" syrup) is in consequence considerably lower than that of the added syrup; for example, an added syrup of 40° Brix will usually cut out at 22°–25°. The cutout syrup strength depends on several factors—the strength of the added syrup, the type of fruit, its ripeness, and the ratio of fruit to syrup. Similarly, the drained weight varies considerably depending on these same factors and the processing conditions. Syrup strength can be readily checked by means of a refractometer or a Brix hydrometer.

Canned fruits are normally examined for vacuum, pH, added color, and metallic contamination. Complaints occasionally arise alleging that canned grape products may contain glass. These fragments invariably consist of crystals of argol (cream of tartar derived from the grapes). Crystals of calcium DL-tartrate in cans of French cherries have been reported. The amount of calcium necessary to cause the formation of these crystals suggests that the cherries had been in contact with concrete tanks or other sources of lime at some stage.

Canned fruits are good sources of vitamins; canning results in little or only moderate loss of vitamins. Allowing for the preparation steps involved, there is little difference between raw foods and canned foods with respect to caloric, protein, carbohydrate, and fat content. Research has been done on the breeding, harvesting, and handling of fruit cultivars especially suitable for high-quality canned products (Anon. 1971B).

The following sections describe the various nutritional components of fruits and relevant analytical methods. Most of the analytical quantitation of these nutritional components was done before 1960; research since 1960 has been primarily directed toward the physiology of the fruits. Compositional data often are difficult to compare because researchers use a variety of sampling techniques. The Association of Official Analytical Chemists (AOAC 1970) has excellent well-documented procedures for sampling of fruits and fruit products.

COMPOSITION OF FRUITS

The edible portion of most types of fresh fruits contains 75–95% water. Fruits are low in protein but, in general, contain substantial carbohydrates. The latter may include varying proportions of dextrose, fructose, sucrose, and starch according to the type of fruit and its maturity. The principal acids in fruits are citric, tartaric, and malic. The total acidity often decreases during ripening and storage. The pH of fruits is usually from 2.5 to 4.5. Other constituents of fruits include cellulose and woody fibers, mineral salts, pectin, gums, tannins, pigments, and volatiles. From the nutritional point of view, certain fruits are valuable sources of vitamins A and C.

Since dried fruits contain only about 13–22% water, the other main constituents are present in proportionately greater amounts than in the corresponding fresh fruit. Some dried fruit products contain over 60% total sugars. Dried fruits are almost devoid of vitamin C, but prunes and dried apricots represent useful sources of vitamin A.

The composition and nutritive value on many fruits are given in Table 12.3. Details of the more important chemical constituents are discussed in the following sections.

PROTEINS

The protein constituents of fruits, although occurring in low concentrations, are of primary importance as components of nuclear and cytoplasmic structures involved in determining and maintaining cellular organization, and also as enzymes involved in metabolism during growth, development, maturation, and the postharvest life of the fruit.

Normally, protein content is determined by digestion of the fruit or fruit product with concentrated sulfuric acid and measurement of the ammonia liberated; multiplying the percentage of nitrogen by 6.25 gives the percentage of protein (AOAC method 22.052).

Studies on proteins in fruits have been impeded to some extent by their inherently low concentration and interference from other constituents, principally phenolic compounds and organic acids. These problems have been largely overcome by the development of new reagents and analytical techniques, so that the separation and isolation of proteins, specific enzymes, and organelles from fruit tissues are comparable in quality to those possible with other types of tissues.

Disc electrophoresis (Ornstein and Davis 1962), which employs small cylindrical columns of polyacrylamide gel as a medium, has limited capabilities as a quantitative method but has several advantages that make it an excellent tool for application to problems such as those encountered in fruit biochemistry. The method is sensitive, requiring less than 100 μg protein, and the technique is capable of separating up to thirty components.

Enzymes

Thatcher (1915) first reported the presence of enzymes in apples and their relation to the ripening process. He observed esterase, protease, and polyphenol oxidase activity in preparations from apple cortical tissue. In addition, he obtained evidence of diastase, emulsin, lipase, and protease activity in appleseed. He concluded from his studies that "the changes taking place (during ripening) in the apple were not simple respiratory changes, but probably in large part were internal enzymatic activities."

Besides their role in the ripening process, certain enzymes may be antinutritional factors. Ascorbic acid oxidase, which is present in pumpkin and peaches, causes the destruction of vitamin C. Choline esterase inhibitor, present in oranges and apples, is thought to interfere with transmission of nerve impulse.

Amino Acids

Little research has been done on the amino acids of fruit, probably because fruits are low in nitrogen and of little nutritional significance as a protein source. However, the amino acid content influences the processing of certain fruit, for example, the fermentation of grapes to form wine and the browning reactions of citrus products and dehydrated fruits.

Study of enzyme systems in fruits is difficult because of the low pH and the presence of polyphenols. The free amino acids are regarded as in metabolic equilibrium with the processes of protein synthesis and degradation, and the mechanisms involved are similar to those in other plant tissues. Tercelj (1965) has shown the presence of purines, nucleosides, and nucleotides in wine, and Burroughs (1957) found such compounds in apple juice. Hexosamines have been reported in grape juice and wines, and Ribereau-Gayon and Peynaud (1959) reported appreciable amounts of peptides in grapes.

CARBOHYDRATES

Sugars, either in the free state or as derivatives, play an important role in flavor. Flavor is influenced by the balance between sugar and acid. Some flavor constituents are glycosides. The attractive red pigments or many fruits are sugar derivatives of anthocyanins. Texture is influenced

Table 12.3.

Water (g)	Food energy (kcal)	Protein 6.25 N (g)	Total lipid (g)	Total carbohydrate (g)	Fiber (g)	Ash (g)	Calcium (mg)	Iron (mg)	Magnesium (mg)	Phosphorus (mg)	Potassium (mg)
Acerola or barbados cherry (*Malphighia glabra*), raw											
91.4	32	0.40	0.30	7.7	0.40	0.20	12	0.20	18	11	146
Apples (*Malus domestica* Borkhi), raw											
83.9	59	0.19	0.36	15.3	0.77	0.26	7	0.18	5	7	115
raw, without skin											
84.5	57	0.15	0.31	14.8	0.54	0.24	4	0.07	3	7	113
without skin, cooked, boiled											
85.5	53	0.26	0.36	13.6	0.54	0.28	5	0.19	3	8	88
without skin, cooked, microwave											
84.6	56	0.28	0.42	14.4	0.54	0.26	5	0.17	3	8	93
canned, sweetened, sliced, heated											
82.3	67	0.18	0.43	16.8	0.54	0.27	4	0.24	3	6	70
dehydrated (low moisture), sulfured, cooked											
79.4	74	0.28	0.12	19.9	0.87	0.33	4	0.43	5	12	136
dried, sulfured, uncooked											
31.8	243	0.93	0.32	65.9	2.87	1.10	14	1.40	16	38	450
dried, sulfured, cooked, without added sugar											
84.1	57	0.22	0.07	15.3	0.67	0.26	3	0.33	4	9	105
frozen, unsweetened, unheated											
86.9	48	0.28	0.32	12.3	0.54	0.24	4	0.18	3	8	77
frozen, unsweetened, heated											
87.2	47	0.29	0.33	12.0	0.54	0.23	5	0.19	3	8	76
juice, canned or bottled											
87.9	47	0.06	0.11	11.7	0.21	0.22	7	0.37	3	7	119
juice, frozen concentrate, undiluted											
57.0	166	0.51	0.37	41.0	—	1.12	20	0.91	17	25	448
Applesauce, canned, unsweetened											
88.3	43	0.17	0.05	11.3	0.53	0.15	3	0.12	3	7	75
canned, sweetened											
79.6	76	0.18	0.18	19.9	0.46	0.14	4	0.35	3	7	61
Apricots (*Prunus armeniaca*), raw											
86.4	48	1.40	0.39	11.1	0.60	0.75	14	0.54	8	19	296
canned, water pack, with skin											
92.4	27	0.71	0.16	6.4	0.42	0.38	8	0.32	7	13	192
canned, water pack, without skin											
93.4	22	0.69	0.03	5.5	0.37	0.38	8	0.54	9	16	154
canned, heavy-syrup pack, with skin											
77.6	83	0.53	0.08	21.5	0.40	0.37	9	0.30	7	12	140
frozen, sweetened											
73.3	98	0.70	0.10	25.1	0.60	0.80	10	0.90	9	19	229
nectar, canned											
84.9	56	0.37	0.09	14.4	0.19	0.29	7	0.38	5	9	114

Composition and Nutritive Value of Fruits

Sodium (mg)	Zinc (mg)	Copper (mg)	Manganese (mg)	Ascorbic acid (mg)	Thiamin (mg)	Riboflavin (mg)	Niacin (mg)	Pantothenic acid (mg)	Vitamin B_6 (mg)	Folacin (μg)	Vitamin A (IU)
7	—	—	—	1687	0.02	0.06	0.40	0.31	0.01	—	767
0	0.04	0.04	0.05	5.7	0.02	0.01	0.08	0.06	0.05	2.8	53
0	0.04	0.03	0.02	4.0	0.02	0.01	0.09	0.06	0.05	0.4	44
1	0.04	0.04	0.12	0.2	0.02	0.01	0.09	0.05	0.04	0.6	44
1	0.04	0.05	0.14	0.3	0.02	0.01	0.06	0.05	0.05	0.6	40
3	0.05	0.05	0.16	0.2	0.01	0.01	0.08	0.03	0.04	0.1	56
26	0.06	0.06	0.03	0.6	0.01	0.03	0.14	0.08	0.05	0.1	19
87	0.20	0.19	0.09	3.9	0.00	0.16	0.93	—	0.12	—	0
20	0.05	0.04	0.02	1.0	0.01	0.02	0.13	0.06	0.05	0	17
3	0.05	0.06	0.17	0.1	0.01	0.01	0.04	0.05	0.03	0.7	34
3	0.05	0.06	0.15	0.4	0.01	0.01	0.04	0.06	0.03	0.6	20
3	0.03	0.02	0.11	0.9	0.02	0.02	0.10	—	0.03	0.1	1
25	0.13	0.05	0.22	2.1	0.01	0.05	0.14	0.22	0.12	1.0	—
2	0.03	0.03	0.08	1.2	0.01	0.02	0.19	0.10	0.03	0.6	29
3	0.04	0.04	0.08	1.7	0.01	0.03	0.19	0.05	0.03	0.6	11
1	0.26	0.09	0.08	10	0.03	0.04	0.60	0.24	0.05	8.6	2612
3	0.11	0.08	0.05	3.4	0.02	0.02	0.40	0.09	0.05	1.7	1293
11	0.11	0.07	0.05	1.8	0.02	0.02	0.44	0.09	0.05	1.7	1810
4	0.11	0.08	0.05	3.1	0.02	0.02	0.38	0.09	0.05	1.7	1230
4	0.10	0.06	0.05	9.0	0.02	0.04	0.80	0.20	0.06	—	1680
3	0.09	0.07	—	0.6	0.01	0.01	0.26	—	—	1.3	1316

(*continued*)

Table 12.

Water (g)	Food energy (kcal)	Protein 6.25 N (g)	Total lipid (g)	Total carbohydrate (g)	Fiber (g)	Ash (g)	Calcium (mg)	Iron (mg)	Magnesium (mg)	Phosphorus (mg)	Po- si- (m
Avocados (*Persea americana*), raw, all commercial varieties											
74.3	161	1.98	15.3	7.4	2.11	1.04	11	1.02	39	41	5
Bananas (*Musa paradisiaca*), raw											
74.3	92	1.03	0.48	23.4	0.50	0.80	6	0.31	29	20	3
dehydrated, or banana powder											
3.0	346	3.89	1.81	88.3	1.88	3.02	22	1.15	108	74	14
Blackberries (*Rubus fructicosus*), raw											
85.6	52	0.72	0.39	12.8	4.10	0.48	32	0.57	20	21	
canned, heavy-syrup pack											
75.1	92	1.31	0.14	23.1	2.60	0.39	21	0.65	17	14	
frozen, unsweetened											
82.2	64	1.18	0.43	15.7	2.70	0.51	29	0.80	22	30	
Blueberries (*Vaccinium australe*), raw											
84.6	56	0.67	0.38	14.1	1.30	0.21	6	0.17	5	10	
canned, heavy-syrup pack											
76.8	88	0.65	0.33	22.1	0.90	0.17	5	0.33	4	10	
frozen, unsweetened											
86.6	51	0.42	0.64	12.2	1.50	0.18	8	0.18	5	11	
Boysenberries (*Rubus loganobaccus*), frozen, unsweetened											
85.9	50	1.10	0.26	12.2	2.70	0.54	27	0.85	16	27	
Breadfruit (*Artocarpus altilis*), raw											
70.7	103	1.07	0.23	27.1	1.48	0.93	17	0.54	25	30	
Carambola or jalea (*Averrhoa carambola*), raw											
90.9	33	0.54	0.35	7.8	0.92	0.37	4	0.26	9	16	
Carissa or natal plum (*Carissa grandiflora*), raw											
84.2	62	0.50	1.30	13.6	0.90	0.40	11	1.31	16	7	
Cherimoya or chirimoya (*Annona cherimola*), raw											
73.5	94	1.30	0.40	24.0	2.20	0.80	23	0.50	—	40	
Cherries, sour, red (*Prunus cerasus*), raw											
86.1	50	1.00	0.30	12.2	0.20	0.40	16	0.32	9	15	
sour, red, canned, light-syrup pack											
79.6	75	0.74	0.10	19.3	0.10	0.25	10	1.32	6	10	
sour, red, frozen, unsweetened											
87.2	46	0.92	0.44	11.0	0.30	0.42	13	0.53	9	16	
sweet (*Prunus avium*), raw											
80.8	72	1.20	0.96	16.6	0.40	0.53	15	0.39	11	19	
sweet, canned, juice pack											
85.0	54	0.91	0.02	13.8	0.22	0.31	14	0.58	12	22	
sweet, canned, heavy-syrup pack											
77.6	83	0.60	0.15	21.3	0.31	0.38	9	0.35	9	18	
sweet, frozen, sweetened											
75.5	89	1.15	0.13	22.4	0.40	0.83	12	0.35	10	16	

ntinued)

Zinc (mg)	Copper (mg)	Manganese (mg)	Ascorbic acid (mg)	Thiamin (mg)	Riboflavin (mg)	Niacin (mg)	Pantothenic acid (mg)	Vitamin B_6 (mg)	Folacin (μg)	Vitamin A (IU)
0.42	0.26	0.23	7.9	0.11	0.12	1.92	0.97	0.28	61.9	612
0.16	0.10	0.15	9.1	0.04	0.10	0.54	0.26	0.58	19.1	81
0.61	0.39	0.57	7.0	0.18	0.24	2.80	—	—	—	305
0.27	0.14	1.29	21	0.03	0.04	0.40	0.24	0.06	—	165
0.18	0.13	0.70	2.8	0.03	0.04	0.29	0.15	0.04	26.5	219
0.25	0.12	1.22	3.1	0.03	0.05	1.21	0.15	0.06	34	114
0.11	0.06	0.28	13	0.05	0.05	0.36	0.09	0.04	6.4	100
0.07	0.05	0.20	1.1	0.03	0.05	0.11	0.09	0.04	1.6	64
0.07	0.03	0.15	2.5	0.03	0.04	0.52	0.12	0.06	6.7	81
0.22	0.08	0.55	3.1	0.05	0.04	0.77	0.25	0.06	63.3	67
0.12	0.08	0.06	29.0	0.11	0.03	0.90	0.46	—	—	40
0.11	0.12	0.08	21.2	0.03	0.03	0.41	—	—	—	493
—	0.21	—	38.0	0.04	0.06	0.20	—	—	—	40
—	—	—	9.0	0.10	0.11	1.30	—	—	—	10
0.10	0.10	0.11	10	0.03	0.04	0.40	0.14	0.04	7.5	1283
0.07	0.07	0.07	2	0.02	0.04	0.17	0.10	0.04	7.7	726
0.10	0.09	0.06	1.7	0.04	0.03	0.14	0.18	0.07	4.5	870
0.06	0.10	0.09	7.0	0.05	0.06	0.40	0.13	0.04	4.2	214
0.10	0.07	0.06	2.5	0.02	0.02	0.41	—	—	—	125
0.10	0.14	0.06	3.6	0.02	0.04	0.40	—	0.03	—	154
0.04	0.02	0.11	1.0	0.03	0.05	0.18	—	—	—	189

(*continued*)

Table 12

	Water (g)	Food energy (kcal)	Protein 6.25 N (g)	Total lipid (g)	Total carbohydrate (g)	Fiber (g)	Ash (g)	Calcium (mg)	Iron (mg)	Magnesium (mg)	Phosphorus (mg)	Pc si (
Crabapples (*Malus pumila* × *Malus baccata*), raw	78.9	76	0.40	0.30	20.0	0.60	0.42	18	0.36	7	15	
Cranberries (*Vaccinium macrocarpon*), raw	86.5	49	0.39	0.20	12.7	1.20	0.19	7	0.20	5	9	
Cranberry juice cocktail, bottled	85.0	58	0.03	0.05	14.9	—	0.04	3	0.16	3	1	
Cranberry sauce, canned, sweetened	60.6	151	0.20	0.15	38.9	0.30	0.10	4	0.22	3	6	
Currants, European black (*Ribes nigrum*), raw	82.0	63	1.40	0.41	15.4	2.40	0.86	55	1.54	24	59	
red and white (*Ribes rubrum*), raw	84.0	56	1.40	0.20	13.8	3.40	0.66	33	1.00	13	44	
Zante dried	19.2	283	4.08	0.27	74.1	1.57	2.36	86	3.26	41	125	
Custard-apple or bullock's heart (*Anona reticulata*), raw	71.5	101	1.70	0.60	25.2	3.40	1.00	30	0.71	18	21	
Damsons or damson plum (*Prunus institia*), raw	69.8	34	0.5	Tr	12.3	—	—	21	0.37	10	15	
Dates, domestic (*Phoenix dactylifera*), natural and dry	22.5	275	1.97	0.45	73.5	2.20	1.58	32	1.15	35	40	
Elderberries (*Sambucus canadensis* and others), raw	79.8	73	0.66	0.50	18.4	7.00	0.64	38	1.60	—	39	
Figs (*Ficus carica*), raw	79.1	74	0.75	0.30	19.2	1.20	0.66	35	0.37	17	14	
canned, heavy-syrup pack	76.3	88	0.38	0.10	22.9	0.55	0.29	27	0.28	10	10	
dried, uncooked	28.4	255	3.05	1.17	65.4	4.80	2.01	144	2.23	59	68	
dried, cooked	69.8	108	1.29	0.49	27.6	2.02	0.85	61	0.94	25	29	
Fruit cocktail, canned, light-syrup pack	84.4	57	0.40	0.07	14.9	0.45	0.23	6	0.29	5	11	
Fruit salad, canned, light-syrup pack	84.2	58	0.34	0.07	15.1	0.61	0.21	7	0.29	5	9	
Gooseberries (*Ribes grossularia*), raw	87.9	44	0.88	0.58	10.2	1.90	0.49	25	0.31	10	27	
Grapefruit, pink and red, all areas (*Citrus paradisi*), raw	91.4	30	0.55	0.10	7.7	0.20	0.29	11	0.12	8	9	
raw, white, all areas	90.5	33	0.69	0.10	8.4	0.20	0.33	12	0.06	9	8	
juice, raw	90.0	39	0.50	0.10	9.2	—	0.20	9	0.20	12	15	

ontinued)

n)	Zinc (mg)	Copper (mg)	Manganese (mg)	Ascorbic acid (mg)	Thiamin (mg)	Riboflavin (mg)	Niacin (mg)	Pantothenic acid (mg)	Vitamin B_6 (mg)	Folacin (μg)	Vitamin A (IU)
1	—	0.07	0.12	8.0	0.03	0.02	0.10	—	—	—	40
1	0.13	0.06	0.16	13.5	0.03	0.02	0.10	0.22	0.07	1.7	46
4	0.02	0.01	0.16	42.6	0.01	0.02	0.05	0.07	—	0.2	—
9	0.05	0.02	0.06	2.0	0.02	0.02	0.10	—	0.01	—	20
2	0.27	0.09	0.26	181	0.05	0.05	0.30	0.40	0.07	—	230
	0.23	0.11	0.19	41	0.04	0.05	0.10	0.06	0.07	—	120
	0.66	0.47	0.47	4.7	0.16	0.14	1.62	0.04	0.30	10.2	73
4	—	—	—	19.2	0.08	0.10	0.50	0.14	0.22	—	—
2	—	0.07	—	—	—	—	—	—	—	—	—
3	0.29	0.29	0.30	0	0.09	0.10	2.20	0.78	0.19	12.6	50
	—	—	—	36.0	0.07	0.06	0.50	0.14	0.23	—	600
	0.15	0.07	0.13	2.0	0.06	0.05	0.40	0.30	0.11	—	142
	0.11	0.11	0.08	1.0	0.02	0.04	0.43	0.07	—	—	37
	0.51	0.31	0.39	0.8	0.07	0.09	0.69	0.44	0.22	7.5	133
	0.21	0.13	0.16	4.4	0.01	0.11	0.64	0.13	0.13	1.0	159
	0.09	0.07	—	1.9	0.02	0.02	0.38	0.06	0.05	—	208
	0.07	0.07	—	2.5	0.01	0.02	0.36	—	0.03	—	429
	0.12	0.07	0.14	27.7	0.04	0.03	0.30	0.29	0.08	—	290
	0.07	0.04	0.01	38.1	0.03	0.02	0.19	0.28	0.04	12.2	259
	0.07	0.05	0.01	33.3	0.04	0.02	0.27	0.28	0.04	10.0	10
	0.05	0.03	0.02	38.0	0.04	0.02	0.20	—	—	—	—

(continued)

Table 1

Water (g)	Food energy (kcal)	Protein 6.25 N (g)	Total lipid (g)	Total carbohydrate (g)	Fiber (g)	Ash (g)	Calcium (mg)	Iron (mg)	Magnesium (mg)	Phosphorus (mg)
juice, frozen concentrate, undiluted										
62.0	146	1.97	0.48	34.6	—	0.99	27	0.49	38	49
Grapes, American type, slip skin (*Vitus vinifera*), raw										
81.3	63	0.63	0.35	17.2	0.76	0.57	14	0.29	5	10
European type, adherent skin, raw										
80.6	71	0.66	0.58	17.8	0.45	0.44	11	0.26	6	13
juice, canned or bottled										
84.1	61	0.56	0.08	15.0	—	0.29	9	0.24	10	11
juice, frozen concentrate, sweetened, undiluted										
54.4	179	0.65	0.31	44.4	—	0.27	13	0.36	15	15
Greengages or greengage plum (*Prunus domestica*), raw										
78.2	48	0.08	Tr	14.4	—	—	17	0.37	8	23
Groundcherries or cape gooseberries (*Physalis pubescens*), raw										
85.4	53	1.90	0.70	11.2	2.80	0.80	9	1.00	—	40
Guavas, common (*Psidium guajava*), raw										
86.1	51	0.82	0.60	11.9	5.60	0.60	20	0.31	10	25
Jackfruit (*Artocarpus heterophyllus*), raw										
73.2	94	1.47	0.30	24.0	1.00	1.00	34	0.60	37	36
Java plum or jambolana plum (*Syzygium cumuni*), raw										
83.1	60	0.72	0.23	15.6	0.27	0.36	19	0.19	15	17
Jujube or chinese date (*Ziziphus jujuba*), raw										
77.9	79	1.20	0.20	20.2	1.40	0.51	21	0.48	10	23
Kiwifruit or chinese gooseberry (*Actinida sinensis*), raw										
83.1	61	0.99	0.44	14.9	1.10	0.64	26	0.41	30	40
Kumquats (*Citrus japonica*), raw										
81.7	63	0.90	0.10	16.4	3.70	0.87	44	0.39	13	19
Lemons (*Citrus medica* cv. limon), raw, without peel										
89.0	29	1.10	0.30	9.3	0.40	0.30	26	0.60	—	16
raw, with peel										
87.4	20	1.20	0.30	10.7	—	0.40	61	0.70	12	15
juice, raw										
90.7	25	0.38	0	8.6	—	0.26	7	0.03	6	6
juice, canned or bottled										
92.5	21	0.40	0.29	6.48	—	0.36	11	0.13	8	9
juice, frozen, single-strength										
92.4	22	0.46	0.32	6.50	—	0.33	8	0.12	8	8
Limes (*Citrus aurantifolia*), raw										
88.3	30	0.70	0.20	10.5	0.50	0.30	33	0.60	—	18
juice, raw										
90.2	27	0.44	0.10	9.0	—	0.24	9	0.03	6	7
Loganberries (*Rubus loganobaccus*), frozen										
84.6	55	1.52	0.31	13.0	—	0.54	26	0.64	21	26

ontinued)

Zinc (mg)	Copper (mg)	Manganese (mg)	Ascorbic acid (mg)	Thiamin (mg)	Riboflavin (mg)	Niacin (mg)	Pantothenic acid (mg)	Vitamin B_6 (mg)	Folacin (μg)	Vitamin A (IU)
0.18	0.12	0.07	120	0.14	0.08	0.77	0.67	0.15	12.8	31
0.04	0.04	0.72	4.0	0.09	0.06	0.30	0.02	0.11	3.9	100
0.05	0.09	0.06	10.8	0.09	0.06	0.30	0.02	0.11	3.9	73
0.05	0.03	0.36	0.1	0.03	0.04	0.26	0.04	0.06	2.6	8
0.13	0.05	0.62	83.1	0.05	0.09	0.43	0.08	0.15	4.4	27
—	0.08	—	—	—	—	—	—	—	—	—
—	—	—	11.0	0.11	0.04	2.80	—	—	—	720
0.23	0.10	0.14	183	0.05	0.05	1.20	0.15	0.14	—	792
0.42	0.19	0.20	6.7	0.03	—	0.40	—	0.11	—	297
—	—	—	14.3	0.01	0.01	0.26	—	0.04	—	3
0.05	0.07	0.08	69.0	0.02	0.04	0.90	—	0.08	—	40
—	—	—	98.0	0.02	0.05	0.50	—	—	—	175
0.08	0.11	0.09	37.4	0.08	0.10	—	—	—	—	302
0.06	0.04	—	53.0	0.04	0.02	0.10	0.19	0.08	10.6	29
0.10	0.26	—	77.0	0.05	0.04	0.20	0.23	0.11	—	30
0.05	0.03	0.01	46.0	0.03	0.01	0.10	0.10	0.05	12.9	20
0.06	0.04	0.02	24.8	0.04	0.01	0.20	0.09	0.04	10.1	15
0.05	0.03	0.03	31.5	0.06	0.01	0.14	0.12	0.06	9.5	13
0.11	0.06	—	29.1	0.03	0.02	0.20	0.22	—	8.2	10
0.06	0.03	0.01	29.3	0.02	0.01	0.10	0.14	0.04	—	10
0.34	0.12	1.25	15.3	0.05	0.03	0.84	0.24	0.06	25.7	35

(continued)

Table 12

Water (g)	Food energy (kcal)	Pro- tein 6.25 N (g)	Total lipid (g)	Total carbo- hy- drate (g)	Fiber (g)	Ash (g)	Cal- cium (mg)	Iron (mg)	Mag- ne- sium (mg)	Phos- pho- rus (mg)	Po- si- (m
Longans or lungan (*Euphoria longan*), raw											
82.8	60	1.31	0.10	15.1	0.40	0.70	1	0.13	10	21	
Loquats (*Eritobotrya japonica*), raw											
86.7	47	0.43	0.20	12.1	0.50	0.50	16	0.28	13	27	
Mangos (*Mangifera indica*), raw											
81.7	65	0.51	0.27	17.0	0.84	0.50	10	0.13	9	11	
Melons, cantaloupe (*Curcumis melo*), raw											
89.8	35	0.88	0.28	8.4	0.36	0.71	11	0.21	11	17	
honeydew, raw											
89.7	35	0.46	0.10	9.2	0.60	0.60	6	0.07	7	10	
Mulberries (*Morus nigra*), raw											
87.7	43	1.44	0.39	9.8	0.96	0.69	39	1.85	18	38	
Nectarines (*Prunus persica* cv. nectarina), raw											
86.3	49	0.94	0.46	11.8	0.40	0.54	5	0.15	8	16	
Olives (*Olea europaea*), in brine											
76.5	106	0.9	11.00	4.4	—	—	61	1.03	22	17	
in brine, with stones											
61.1	85	0.7	8.80	3.5	—	—	49	0.83	17	13	
Oranges, all commercial varieties (*Citrus aurantium* var dulcis), raw											
86.8	47	0.94	0.12	11.8	0.43	0.44	40	0.10	10	14	
juice, raw											
88.3	45	0.70	0.20	10.4	0.10	0.40	11	0.20	11	17	
juice, canned											
89.0	42	0.59	0.14	9.8	0.10	0.41	8	0.44	1	14	
juice, frozen concentrate, undiluted											
57.8	159	2.39	0.21	38.2	0.18	1.38	32	0.35	34	57	
peel, raw											
72.5	—	1.50	0.20	25.0	—	0.80	161	0.80	22	21	
Mandarin- or mandarins (*Citrus reticulata*), canned											
81.0	64	0.5	Tr	18.4	—	—	8	0.19	7	7	
Orange–grapefruit juice, canned											
88.6	43	0.60	0.10	10.3	—	0.39	8	0.46	10	14	
Papayas (*Carica papaya*), raw											
88.8	39	0.61	0.14	9.8	0.77	0.61	24	0.10	10	5	
Passion fruit, purple or grenadilla (*Passiflora edulis*), raw											
72.9	97	2.20	0.70	23.4	11.0	0.80	12	1.60	29	68	
juice, purple, raw											
85.6	51	0.39	0.05	13.6	0.04	0.34	4	0.24	—	13	
Peaches (*Prunus persica vulgaris*), raw											
87.7	43	0.70	0.09	11.1	0.64	0.46	5	0.11	7	12	
canned, juice pack											
87.5	44	0.63	0.03	11.6	0.25	0.27	6	0.27	7	17	

tinued)

Zinc (mg)	Copper (mg)	Manganese (mg)	Ascorbic acid (mg)	Thiamin (mg)	Riboflavin (mg)	Niacin (mg)	Pantothenic acid (mg)	Vitamin B_6 (mg)	Folacin (μg)	Vitamin A (IU)
0.05	0.17	0.05	84.0	0.03	0.14	0.30	—	—	—	—
0.05	0.04	0.15	1.0	0.02	0.02	0.18	—	—	—	1528
0.04	0.11	0.03	27.7	0.06	0.06	0.58	0.16	0.13	—	3894
0.16	0.04	0.05	42.2	0.04	0.02	0.57	0.13	0.12	17.0	3224
—	0.04	0.02	24.8	0.08	0.02	0.60	0.21	0.06	—	40
—	—	—	36.4	0.03	0.10	0.62	—	—	—	25
0.09	0.07	0.04	5.4	0.02	0.04	0.99	0.16	0.02	3.7	736
—	0.23	—	—	—	—	—	—	—	—	300
—	0.18	—	—	—	—	—	—	—	—	300
0.07	0.04	0.02	53.2	0.09	0.04	0.28	0.25	0.06	30.3	205
0.05	0.04	0.01	50.0	0.09	0.03	0.40	0.19	0.04	—	200
0.07	0.06	0.01	34.4	0.06	0.03	0.31	0.15	0.09	—	175
0.18	0.16	0.05	138	0.28	0.06	0.72	0.56	0.16	155.3	276
0.07	0.08	0.02	29.1	0.06	0.03	0.34	0.14	0.02	—	119
—	—	—	136	0.12	0.09	0.90	0.49	0.18	—	420
—	0.03	—	—	—	—	—	—	—	—	—
0.07	0.02	0.01	61.8	0.03	0.03	0.34	0.22	0.02	—	2014
—	—	—	30.0	—	0.13	1.50	—	—	—	700
—	—	—	29.8	—	0.13	1.46	—	—	—	717
0.14	0.07	0.05	6.6	0.02	0.04	0.99	0.17	0.02	3.4	535
0.11	0.05	—	3.6	0.01	0.02	0.58	—	—	—	381

(*continued*)

Table 1

	Water (g)	Food energy (kcal)	Protein 6.25 N (g)	Total lipid (g)	Total carbohydrate (g)	Fiber (g)	Ash (g)	Calcium (mg)	Iron (mg)	Magnesium (mg)	Phosphorus (mg)
canned, light-syrup pack	84.7	54	0.45	0.03	14.6	0.30	0.25	3	0.36	5	11
dehydrated (low-moisture), sulfured, uncooked	7.5	325	4.89	1.03	83.2	3.97	3.39	38	5.51	57	162
dried, sulfured, uncooked	31.8	239	3.61	0.76	61.3	2.93	2.50	28	4.06	42	119
frozen, sliced, sweetened	74.7	94	0.63	0.13	24.0	0.40	0.53	3	0.37	5	11
nectar, canned	85.6	54	0.27	0.02	13.9	0.14	0.15	5	0.19	4	6
Pears (*Pyrus communis*), raw	83.8	59	0.39	0.40	15.1	1.40	0.28	11	0.25	6	11
canned, juice pack	86.5	50	0.34	0.07	12.9	0.49	0.19	9	0.29	7	12
canned, heavy-syrup pack	80.4	74	0.20	0.13	19.2	0.59	0.15	5	0.22	4	7
dried, sulfured, uncooked	26.7	262	1.87	0.63	69.7	5.69	1.11	34	2.10	33	59
nectar, canned	84.0	60	0.11	0.01	15.8	0.31	0.10	5	0.26	3	3
Persimmons, Japanese or kaki (*Diospyros kaki*), raw	80.3	70	0.58	0.19	18.6	1.48	0.33	8	0.15	9	17
Japanese, dried	23.0	274	1.38	0.59	73.4	3.62	1.59	25	0.74	31	81
Pineapple (*Ananas comosus*), raw	86.5	49	0.39	0.43	12.4	0.54	0.29	7	0.37	14	7
canned, juice pack	83.5	60	0.42	0.08	15.7	0.35	0.30	14	0.28	14	6
juice, canned	85.5	56	0.32	0.08	13.8	0.10	0.30	17	0.26	13	8
juice, frozen concentrate, undiluted	53.1	179	1.30	0.10	44.3	0.30	1.20	39	0.90	35	28
Plums (*Prunus domestica*), raw	85.2	55	0.79	0.62	13.0	0.60	0.39	4	0.10	7	10
canned, purple, juice pack	84.0	58	0.51	0.02	15.2	0.26	0.30	10	0.34	8	15
Prunes, dehydrated (low-moisture) or dried prune-plums, uncooked	4.0	339	3.70	0.73	89.1	2.90	2.50	72	3.52	64	112
dried, uncooked	32.4	239	2.61	0.52	62.7	2.04	1.76	51	2.48	45	79
juice, canned	81.2	71	0.61	0.03	17.4	0.01	0.68	12	1.18	14	25

ntinued)

Zinc (mg)	Copper (mg)	Manganese (mg)	Ascorbic acid (mg)	Thiamin (mg)	Riboflavin (mg)	Niacin (mg)	Pantothenic acid (mg)	Vitamin B_6 (mg)	Folacin (μg)	Vitamin A (IU)
0.09	0.05	0.05	2.4	0.01	0.02	0.59	0.05	0.02	3.3	354
0.78	0.49	0.41	10.6	0.04	0.11	4.82	0.52	0.16	6.6	1417
0.57	0.36	0.30	4.8	0.01	0.21	4.38	—	0.07	—	2163
0.05	0.02	0.03	94.2	0.01	0.04	0.65	0.13	0.02	—	284
0.08	0.07	0.02	5.3	0.01	0.01	0.29	—	—	—	258
0.12	0.11	0.08	4.0	0.02	0.04	0.10	0.07	0.02	7.3	20
0.09	0.05	—	1.6	0.01	0.01	0.20	—	—	—	6
0.08	0.05	0.03	1.1	0.01	0.02	0.24	0.02	0.01	1.2	0
0.39	0.37	0.33	7.0	0.01	0.14	1.37	—	—	—	3
0.07	0.07	0.03	1.1	0.01	0.01	0.13	—	—	—	1
0.11	0.11	0.36	7.5	0.03	0.02	0.10	—	—	7.5	2167
0.42	0.44	1.39	0	—	0.03	0.18	—	—	—	558
0.08	0.11	1.65	15.4	0.09	0.04	0.42	0.16	0.09	10.6	23
0.10	0.09	—	9.5	0.09	0.02	0.28	—	—	—	38
0.11	0.09	0.99	10.7	0.06	0.02	0.26	0.10	0.10	23.1	5
0.40	0.31	3.44	42.0	0.23	0.06	0.90	0.43	0.26	—	50
0.10	0.04	0.05	9.5	0.04	0.10	0.50	0.18	0.08	2.2	323
0.11	0.05	—	2.8	0.02	0.06	0.47	—	—	—	1009
0.75	0.61	0.31	0	0.12	0.16	2.99	0.42	0.74	1.9	1762
0.53	0.43	0.22	3.3	0.08	0.16	1.96	0.46	0.26	3.7	1987
0.21	0.07	0.15	4.1	0.02	0.07	0.78	—	—	0.4	3

(continued)

Table 12

Water (g)	Food energy (kcal)	Protein 6.25 *N* (g)	Total lipid (g)	Total carbohydrate (g)	Fiber (g)	Ash (g)	Calcium (mg)	Iron (mg)	Magnesium (mg)	Phosphorus (mg)	Po si (r
Quinces (*Cydonia oblonga*), raw											
83.8	57	0.40	0.10	15.3	1.70	0.40	11	0.70	8	17	
Raisins, golden seedless (dried *Vinus vinifera*)											
15.0	302	3.39	0.46	79.5	1.43	1.66	53	1.79	35	115	
seeded											
16.6	296	2.52	0.54	78.5	0.67	1.89	28	2.59	30	75	
Raspberries (*Rubus idaeus*), raw											
86.6	49	0.91	0.55	11.6	3.00	0.40	22	0.57	18	12	
frozen, red, sweetened											
72.8	103	0.70	0.16	26.2	2.21	0.24	15	0.65	13	17	
Strawberries (*Fragaria vesca*), raw											
91.6	30	0.61	0.37	7.0	0.53	0.43	14	0.38	10	19	
frozen, unsweetened											
90.0	35	0.43	0.11	9.1	0.79	0.37	16	0.75	11	13	
frozen, sweetened, sliced											
73.2	96	0.53	0.13	25.9	0.62	0.24	11	0.59	7	13	
Sultanas, (*Vinus vinifera*), raw, dried											
18.3	249	1.7	Tr	71.7	—	—	53	1.82	35	9	
Tangerines, (*Citrus reticulata* var nobilis), raw											
87.6	44	0.63	0.19	11.2	0.33	0.39	14	0.10	12	10	
Watermelon (*Citrullus lanatus*), raw											
91.5	32	0.62	0.43	7.2	0.30	0.26	8	0.17	11	9	

Source: Adapted from Gebhardt *et al.* (1982) and McCance and Widdowson (19

by structural polysaccharides. Finally, ascorbic acid (vitamin C) is commonly considered a sugar derivative and occurs widely and sometimes abundantly in fruits.

The polysaccharides in plant cell walls are highly complex. During extraction by any of the usual methods, degradation occurs and subunits are produced. The degradation is caused either by enzymatic or chemical attack. Polysaccharides of the cell wall can form compounds with noncarbohydrate materials. Lignin is present in the cellulose fraction from cranberry pulp, and complexes of polysaccharides with lignin may well be present. Similarly, the association of polysaccharides with amino acids, particularly hydroxyproline, has been reported in corn pericarp (Boundy *et al.* 1967) and in fruit pulp of *Parkia biglobosa* (Lanza *et al.* 1962).

Dietary Fiber

Dietary fiber is the indigestible carbohydrate components present commonly in fruit and vegetable products. The various components of dietary *ntinued)*

Zinc (mg)	Copper (mg)	Manganese (mg)	Ascorbic acid (mg)	Thiamin (mg)	Riboflavin (mg)	Niacin (mg)	Pantothenic acid (mg)	Vitamin B_6 (mg)	Folacin (μg)	Vitamin A (IU)
—	0.13	—	15.0	0.02	0.03	0.20	0.08	0.04	—	40
0.32	0.36	0.31	3.2	0.01	0.19	1.14	0.14	0.32	3.3	44
0.18	0.30	0.27	5.4	0.11	0.18	1.11	—	0.19	3.3	0
0.46	0.07	1.01	25.0	0.03	0.09	0.90	0.24	0.06	—	130
0.18	0.10	0.65	16.5	0.02	0.04	0.23	0.15	0.03	26.0	60
0.13	0.05	0.29	56.7	0.02	0.07	0.23	0.34	0.06	17.7	27
0.13	0.05	0.29	41.2	0.02	0.04	0.46	0.11	0.03	16.8	45
0.06	0.02	0.25	41.4	0.02	0.05	0.40	0.11	0.03	14.9	24
—	0.35	—	—	—	—	—	—	—	—	—
—	0.03	0.03	30.8	0.11	0.02	0.16	0.20	0.07	20.4	920
0.07	0.03	0.04	9.6	0.08	0.02	0.20	0.21	0.14	2.2	366

fiber consist of cellulose, hemicellulose, lignin, and pectic substances, which are derived mainly from the cell wall and the exocarp (skin) of fruits.

The role of dietary fiber in human nutrition and the relationship between dietary fiber and certain diseases have attracted considerable attention. Epidemiological studies have implicated links between reduced dietary fiber consumption and increased incidences of certain disorders such as cardiovascular diseases, diverticulosis, and cancer of the colon. A more obvious effect of dietary fiber is the relieving of constipation by increasing the water-holding capacity of the stools. Consumption of fruits that are high in dietary fiber enhance the physiological well-being of humans.

Vidal-ValVerde *et al.* (1982) determined the dietary fiber and its various components in 21 Spanish fruits. The contents of the various constituents were as follows for lemon, apple, peach, pear, apricot, and cherimoya: 0.2–2.75% for dietary fiber, 0.06–1.8% for cellulose, 0.00–0.86% for hemicellulose, 0.06–0.5% for lignin, and 0.12–1.28% for pectic substances.

Pectic Substances

Pectic substances are complex, colloidal carbohydrate derivatives that occur in, or are prepared from, plants and contain a large proportion of anhydrogalacturonic acid units, which exist in a chainlike combination. The carboxyl groups of polygalacturonic acids may be partly esterified by one or more bases. Protopectin is the water-insoluble parent pectic substance occuring in plants; upon restricted hydrolysis, it yields pectinic acids.

Pectinic acids are colloidal polygalacturonic acids containing more than a negligible proportion of methyl ester groups. Pectinic acids, under suitable conditions, are capable of forming gels (jellies) with sugar and acid, or, if suitably low in methoxyl content, with certain metallic ions. The salts of pectinic acids are either normal or acid pectinates.

Pectin designates are water-soluble pectinic acids of varying methyl ester content and degree of neutralization. They are capable of forming gels with sugar and acid under suitable conditions.

Pectic acid is the term applied to pectic substances composed mostly of colloidal polygalacturonic acids and essentially free from methyl ester groups. The salts of pectic acid are either normal or acid pectates.

The pectin content of fruits may be readily measured (AOAC 1970). Pentosans may also be present in fruits and yield five-carbon sugars on hydrolysis. These are formed in increasing quantities as the liquefaction of plant tissue proceeds. There is a close association between cellulose and pentosans in fruit tissue. The following percentages for pentosan content have been reported: juniper berries, 6.0; raspberries, 2.7; elderberries, 1.2; grapes, 1.6; blackberries, 1.2; strawberries, 0.9; cranberries, 0.75; bilberries, 0.8; gooseberries, 0.5; and currants, 0.4.

LIPIDS

The fruits of some plant species (olive and avocados) accumulate large lipid reserves during their development. Consequently, these fruits have constituted an important source of dietary fats. The physiology of several nonoleaginous fruits (apples, pears, bananas) has been studied in great detail, and interest has developed in the lipid metabolism of these organs. Some of these lipids comprise the essential oils for flavoring of drinks and other foods.

VITAMINS

The main nutritional contribution of fruits and their processed products is their supply of the antiscorbutic vitamin (L-ascorbic acid or vitamin C). While different fruits and vegetables vary substantially in their ascorbic acid content, as a group they are the major source of this vitamin for all primates.

Several fruits—including apricots, peaches, melons, and cherries—are also good sources of β-carotene (provitamin A). Many fruits (apricots, gooseberries, black currants, figs, and citrus fruits) contain moderate amounts of pantothenic acid and biotin. Nicotinic acid, folic acid, thiamin, and riboflavin also occur in small amounts in fruits, but their concentrations are low compared with those in other animal or vegetable foods. Fruits have little or no vitamin D, tocopherols, or vitamin B.

The natural vitamin C content of apples depends on cultivar, stage of ripeness, growing environment, season, fruit acidity, and storage conditions. Levels in different cultivars vary from a trace to 34.0 mg vitamin C/100 g fresh fruit. The peel contains three to five times as much ascorbic acid as the pulp. Highly colored fruit and fruit exposed to greater light intensity have relatively higher levels of vitamin C than less colored fruit or fruit grown in the shade. The application of ascorbic acid to apple skin has shown some indication of increasing fruit color.

Fruits can be divided into two main groups: those that show discoloration on cutting or injury, such as the apple and banana, and those that do not, such as the lemon and the orange. Tissues that discolor may be low in ascorbic acid and/or have highly active phenolases. The unrestrained action of phenolase on an orthophenolic or flavonoid substrate in the presence of oxygen causes the formation of orthoquinone compounds. Although the phenolic components of individual fruits vary, the dominant forms include chlorogenic acid, leucoanthocyanins, catechin, and epicatechin. The role of the various polyphenolic compounds in the oxidation by phenolase also varies. Some of the polyphenols are substrates, some are inhibitors, and some are neither. The phenolase systems of the apple, mushroom, and potato exist in multiple forms, and have individual characteristics.

As long as adequate ascorbic acid is present, the orthoquinone compounds are reduced back to the orthophenolic forms and browning does not occur. When the ascorbic acid content of a fruit is exhausted, the orthoquinones no longer serve as an intermediate in this reaction. Polymerization occurs at this stage and browning becomes irreversible. The browning phenomenon in fruit and the role of ascorbic acid is reviewed by Bauernfeind and Pinkert (1970). Most cultivars of peaches, apricots, pears, plums, nectarines, bananas, and apples are low in vitamin C content, discolor readily, and develop a concomitant off-flavor when the tissue is exposed or when the frozen product is thawed. This enzymatic oxidation can be prevented by adding ascorbic acid. Certain cultivars of strawberries, cherries, and pineapple with intermediate natural levels of ascorbic acid can also benefit from the addition of ascorbic acid. In the case of porous fruit or fruit with oxygen contained within the tissue (e.g., apples and pears), the ascorbic acid must penetrate the tissue. This can be accomplished either by using thin slices or a vacuum treatment.

As an antioxidant, ascorbic acid has the following advantages: (1) it is a natural constituent of many fruits, (2) it is not detectable by odor or taste, (3) it is easily detected by recognized chemical and biological tests, (4) it is economical to use, and (5) it enhances the nutritional value of the product to which it is added. Added ascorbic acid (100–300 mg/lb) can protect the color of cherries stored frozen for 6 months, during thawing, and up to 6 hours thereafter. Solutions containing 0.5% ascorbic acid impart good browning control to heated apple tissue, without adverse effects. As a prefreezing treatment, fruit slices may be immersed in 0.5–7% ascorbic acid solutions.

Ascorbic acid-treated peaches and apricots, which were frozen in syrup, stored for 5 months, and then thawed, were found to possess 100% of the initial vitamin C activity; hence, the treatment improved the nutritive value of frozen fruit. Ascorbic acid treatment will not, however, upgrade original fruit quality or cover up poor processing techniques. Partial thawing and refreezing of frozen fruit products destroys some of the ascorbic acid so that less is available for protection of flavor and color on final defrosting.

In apple juice production, enzymatic browning can be delayed by addition of ascorbic acid sprayed on the apple fruit pulp during grinding. This treatment produces natural or opalescent juices that have substantially the flavor and color characteristics of fresh fruit. The addition of ascorbic acid to apple, pear, and grape juice and to tomato pulp improves both their color and flavor. Apple cultivars show considerable differences in the rate of loss of added ascorbic acid.

In the production of conventional cider, or conventional clarified apple juice, ascorbic acid is usually added in the processing just prior to sealing the cans. At this stage, most of the fruit browning process is already at the irreversible stage and only a partial reversal of color is obtained. The added ascorbic acid, however, is available to react with headspace oxygen during storage and provides a nutritional advantage to the product.

In summary, there are several advantages to adding ascorbic acid to fruit juices during processing. Ascorbic acid prevents or retards undesirable enzymatic browning oxidation during processing and storage and therefore improves organoleptic and color qualities. If a sufficiently high and uniform residual level of vitamin C results, it is possible to make a vitamin C claim for the juice. If vitamin C is used in beverages in combination with SO_2, the amount of SO_2 necessary may be substantially reduced (or eliminated) if adequate heat processing is used to control bacterial growth.

VOLATILE FLAVOR COMPOUNDS

Aroma and taste are the major components of flavor, but appearance, behavior on manipulation, mouthfeel, and even the sounds emitted on chewing, all play a part. Aroma is the subjective sensation of the olfactory epithelium, situated high up in the nasal cavity and reached normally only by gases or vapors. Only volatile substances provide an olfactory response but, because of the extraordinary sensitivity of the nose to some compounds, a very low degree of volatility may suffice.

For understanding of the aroma of a fruit, it is necessary to know (1) the nature of the constituents present; (2) the quality of the aroma of each, if any; (3) the quantity of each present; and (4) the intensity of the aroma of each. Significant changes in volatile consti tuents occur during the development of the fruit and during its storage and processing because of natural metabolic reactions (see Chapter 11).

ORGANIC ACIDS

The acidic properties of organic acids are due to the presence in their molecular structure of the carboxylic group in the free state. Other fruit constituents may have acidic properties without possessing this group in the free state; this is especially the case for phenols and ascorbic acid in which acidity is due to two phenol groups.

Many fruits are particularly rich in organic acids, which are usually dissolved in the water of the cell, either free or combined as salts, esters, glycosides, etc. Sometimes, the concentration is sufficiently high to cause crystallization (e.g., calcium oxalate in many young fruits; potassium bitartrate in grapes). Organic acids are an important source of respiratory energy in the plant cell.

Among the methods available for measuring the acidity of fruit solutions are titration (AOAC 22.058) and use of an electrode (AOAC 22.059).

PHENOLIC COMPOUNDS

Phenolic compounds are distributed widely in plants and are particularly prominent in fruits where they are important in determining color and flavor, for example in wines. Elaborate biosynthetic pathways exist for the production of phenolic compounds in plants. They are secondary products and apparently play no vital role in metabolism.

MOLD SUPPRESSANTS

Diphenyl is an effective antimold agent, which may be present in citrus fruits up to a maximum of 100 ppm. Diphenyl is incorporated into the paper wrappers or the lining papers of fruit packing boxes so that the vapor inhibits mold growth on the surface of the fruit. Diphenyl can be determined spectrophotometrically using cyclohexane as the solvent. It can be removed from wrapping materials by soaking the paper in solvent (Rajzman 1960) and from fruits by steam distillation, but the presence of citrus oils tends to cause interference.

Orthophenylphenol is effective both as a bactericide and as a fungicide and may be present in some fruits up to certain prescribed maximum limits. Mold growth on citrus fruits can be suppressed by dipping in a solution of the sodium salt. This treatment appears to give a residue of about 60 ppm on the whole fruit. During storage the sodium salt is converted into the free phenol. In view of its lower volatility, orthophenylphenol is probably less effective than diphenyl when incorporated into wrapping papers. Noncitrus fruits are dipped in or sprayed with an alkaline solution of the phenol and are then sometimes washed with water. Most of the residue remains on the outside of the fruit and may not be ingested if the fruit is peeled before consumption. Orthophenylphenol can be determined spectrophotometrically using cyclohexane as the solvent (Harvey and Penketh 1957).

If both diphenyl and orthophenylphenol are present, the fruit should first be distilled from alkali so that the former comes over but the latter remains behind. After acidifying and continuing the steam distillation, the phenol comes over.

Thiourea has been used as a rot and mold suppressant in oranges (see Chapter 5).

ANALYTICAL SOURCES

Food firms continually look to independent laboratories for support of new product research and development activities. Greater emphasis on laboratory capabilities has resulted from the FDA's new nutrient-labeling guidelines.

A directory of 100 independent laboratories and their analytical capabilities was published in *Food Product Development* (Anon. 1979). An abbreviated listing of these laboratories is shown in Table 12.4.

Table 12.4. Laboratories Offering Analytical Services on Fruits

U.S. region	Laboratory name	City, state
Central	Accra Laboratories, Inc.	Cleveland, OH
	Analytical Biochemistry Labs	Columbia, MO
	Analytical & Biological Labs	Garden City, MI
	Beckart Labs, Inc.	Morton Grove, IL
	Bio-Technical Resources	Manitowoc, WI
	Commerical Testing Laboratory	Colfax, WI
	Contech Laboratories, Pet, Inc.	Greenville, IL
	Crobaugh Div. Herron Labs	Cleveland, OH
	Daty Laboratories	North Kansas City, MO
	Harris Laboratories, Inc.	Lincoln, NE
	Hill Top Research, Inc.	Miamiville, OH
	Ingram Laboratories, Inc.	Minneapolis, MN
	Ingredient Control Laboratorics	St. Louis, MO
	Index Testing Laboratories	Park Forest South, IL
	Langston Laboratories	Leawood, KS
	Medallion Laboratories	Minneapolis, MN
	Milwaukee Food Laboratories	Germantown, WI
	Minnesota Valley Testing Labs	New Ulm, MN
	Nain Laboratories	Columbus, OH
	Northview Laboratories, Inc.	Northbrook, IL
	Nebraska Testing Laboratories	Omaha, NE
	Raltech Scientific Service	Madison, WI
	Rosner/Rumyon Laboratories	Chicago, IL
	Scientific Associates, Inc.	St. Louis, MO
	Seaway Industrial Labs, Inc.	Hammond, IN
	Shuman Chemical Laboratory	Battle Ground, IN
	Sommer-Frey Laboratories	Milwaukee, WI
	Springborn Inst. for Bioresearch	Spenceville, OH
	Suburban Laboratories	Hillside, IL
	Swift R&D Center	Oakbrook, IL
	Younger Laboratories	St. Louis, MO
Northeast	American Standard Testing Bur.	New York, NY
	Bacto Free	Corona, NY
	Bolaffi International, Ltd.	Newton, MA
	Buffalo Testing Laboratories	Buffalo, NY
	Certified Laboratories, Inc.	Corona, NY
	Corenco Quality Assurance Lab	Tewksbury, MA
	Eastern Laboratory Service	York, PA
	Erie Testing Laboratories	Erie, PA
	Fehmerling Associates	Bridgetown, NJ
	Fitelson Laboratories, Inc.	New York, NY
	Food Research Laboratories	Boston, MA
	Food Tek	Parsippany, NJ
	Industrial Testing Labs	New York, NY
	Lancaster Laboratories, Inc.	Lancaster, PA
	LuPuck Laboratories, Inc.	Watertown, MA

(*continued*)

Table 12.4 (*Continued*)

U.S. region	Laboratory name	City, state
	Leberco Laboratories	Roselle Park, NJ
	Microbac Labs, Inc.	Pittsburgh, PA
	E. Everett Meschter Associates	Ambler, PA
South	ABC Research Corp.	Gainesville, FL
	Bioresearch Labs, Inc.	San Antonio, TX
	Biosystems, Inc.	Atlanta, GA
	Central Analytical Labs, Inc.	Metairie, LA
	Graham Associates	Washington, DC
	Hazelton Laboratories	Vienna, VA
	Industrial Laboratories	Fort Worth, TX
	Institute for Research	Houston, TX
	O. D. Kurtz Assoc., Inc.	Baltimore, MD
	MacMillan Research	Marietta, GA
	Product Protection Agency	Arlington, TX
	Stewart Laboratories	Knoxville, TN
	Strasburger & Siegel, Inc.	Baltimore, MD
	Testron Labs, Inc.	Atlanta, GA
	Woodson-Tenent Labs	Memphis, TN
	W-W Laboratories, Inc.	Baltimore, MD
	Webb Food Lab, Inc.	Raleigh, NC
	Applied Rsch. Labs. of Fla.	Hileah, FL
West	A&L Agricultural Laboratories	Modesto, CA
	A.N.C. Research Lab.	Oakland, CA
	Anresco, Inc.	San Francisco, CA
	Bio Technics	Los Angeles, CA
	California Analytical Labs	Sacramento, CA
	Columbia Laboratories, Inc.	Corbett, OR
	Curtis & Tompkins, Ltd.	San Francisco, CA
	Daylin Laboratories	Los Angeles, CA
	Food Quality Analysts	Portland, OR
	Foremost Research Center	Dublin, CA
	Geo. W. Gooch Lab., Ltd.	Los Angeles, CA
	Hibbs Labs	Boise, ID
	Industrial Labs, Co.	Denver, CO
	LFE Environmental	Richmond, CA
	Mei-Charlton, Inc.	Portland, OR
	Michelson Laboratories, Inc.	Los Angeles, CA
	Morse Laboratories	Sacramento, CA
	Shankman Laboratories	Los Angeles, CA
	Stoner-Mcintosh Labs, Inc.	Santa Clara, CA
	Tech S. Corp.	Berkeley, CA
	Truesdail Laboratories, Inc.	Los Angeles, CA
	Unilab Research	Berkeley, CA
Canada	Bio Research Laboratories	Semmerville, Que.
	Can Test Ltd.	Vancouver, B.C.

(*continued*)

Table 12.4 (*Continued*)

U.S. region	Laboratory name	City, state
	Cambrian Processes Ltd.	Mississauga, Ont.
	Cor-El Food Corp.	Langley, B.C.
	Diversified Rsch. Labs, Ltd.	Toronto, Ont.
	Industrial Laboratories of Canada	Tilsonburg, Ont.
	Paul Melnychyn Consultants	Hudson, Que.
	Research Foods Ltd.	Downsview, Ont.
	Standard Biological Labs	Mississauga, Ont.
	Toronto Research Laboratories	Toronto, Ont.
	Wood Laboratory Ltd.	Vancouver, B.C.

Source: Anon. (1979).

REFERENCES

ANON. 1971A. Nutritive value of foods. U.S. Dept. Agric. Home Gard. Bull. *72*.

ANON. 1971B. Forums in focus. Research on canned foods. Food Technol. *25*(6), 79–80.

ANON. 1979. Directory of independent laboratory capabilities. Food Prod. Devel. *13*, 24–29.

ANON. 1980. U.S. Production of the leading fruit crops. U.S. Dept. Agric., Agric. Stat. 1980, p. 202.

ANON. 1983. Fruits. *In* Food and Nutrition Encyclopedia, 1st Edition, pp. 1012–1045, Ensminger, A. H., Ensminger, M. E., Kolando, J. E., and Robson, J. R. K. (editors). Pegasus, Press, Clovis, CA.

AOAC. 1970. Official Methods of Analysis of the AOAC. 11th ed. Assn. of Official Analytical Chemists, Washington, DC.

BAUERNFEIND, J. C. and PINKERT, D. M. 1970. Food processing with added ascorbic acid. Adv. Food Res. *18*, 219–315.

BREWSTER, L. and JACOBSON, M. F. 1978. The Changing American Diet. Center for Science and Public Interest, Washington, DC.

BUNDY, J. H. WALL, J. S., TURNER, J. E., WOYCHIK, J. H., and DIMLER, R. J. 1967. A mucopolysaccharide containing hydroxyproline from corn pericarp. J. Biol. Chem. *242*, 2410–2415.

BURROUGHS, L. F. 1957. Amino acids of apple juice and ciders. J. Sci. Food Agric. *8*, 122–131.

GEBHARDT, S. E., CUTRUFELLI, R. and MATTHEWS, R. H. 1982. Composition of foods. Fruits and fruit juices. U.S. Dept. Agric., Agric. 8–9.

HARVEY, D. and PENKETH, G. E. 1957. Determination of small amounts of o-phenylphenol. Analyst *82*, 498–503.

HULME, A. C. 1970. The Biochemistry of Fruits and Their Products. Vol. 1. Academic Press, New York.

HULME, A. C. 1971. The Biochemistry of Fruits and Their Products. Vol. 2. Academic Press, New York.

LANZA, M. and REGLI, P. 1962. Chemical study of the grain of *Digitaria iburua*. Med. Trop. *22*, 471–476.

LANZA, M., REGLI, P. and BUSSON, F. 1962. Chemical study of the fruit pulp of *Parkia biglobosa*. Med. Trop. *22,* 377–384.
LUND, E. D. and SMOOT, J. 1982. Dietary fiber content of some tropical fruits and vegetables. J. Agric. Chem. *30,* 1123–1127.
McCANCE, R. A., and WIDDOWSON, E. M. 1960. The Composition of Foods. **HMSD,** London.
NAS–NRC. 1980. Recommended Dietary Allowances. 9th ed., rev. National Academy of Sciences, National Research Council, Washington, D.C.
ORNSTEIN, L. and DAVIS, B. J. 1962. Disc Electrophoresis. Distillation Products Industries, Rochester, NY.
POMERANZ, Y. and MELOAN, C. E. 1971. Food Analysis: Theory and Practice. AVI Publishing Co., Westport, CT.
RAJZMAN, A. 1960. A method for the microdetermination of biphenyl in paper wrappers. Analyst *85,* 116–121.
RIBEREAU-GAYON, J. and PEYNAUD, E. 1959. Amino acids of grape musts and higher alcohols of wines. C. R. Congr. Soc. *248,* 247–251.
TERCELJ, D. 1965. Nitrogenous compounds in Slovenian wine. Ann. Technol. Agr. *14,* 307–319.
THATCHER, R. W. 1915. Enzymes of apples and their relation to ripening process. J. Agric. Res. *5,* 103–116.
VIDAL-VALVERDE, C., HERRANZ, J., BLANCO, I. and ROJAS-HIDALGO, E. 1982. Dietary fiber in Spanish fruits. J. Food Sci. *47,* 1840–1845.
ZYREN, J., ELKINS, E. R., DUDEK, J. A. and HAGEN, R. E. 1983. Fiber content of some selected fruits and processed vegetables, fruits and fruit juices as served. J. Food Sci. *48,* 600–603.

13

Grades and Standards for Raw and Processed Fruits

J. G. Woodroof

The federal government, each state, and many cities and counties have grades and standards for operating fruit processing plants, and grades for processed fruits. These may be obtained by writing to the Department of Agriculture of the respective states, cities or counties, or by writing to the appropriate federal agency in Washington, D.C.

The Grades and Standards discussed below are those promulgated by the federal agencies and pertain to processing fruits for interstate commerce. The regulations are so technically and legally involved that only an outline is given in this chapter.

FDA REGULATIONS

The U.S. Food and Drug Administration promulgates Standards of Identity, Quality and Fill for many fruit products, as summarized in Table 13.1. These Standards are constantly revised and processors should be sure to procure the latest information on the following: good manufacturing practice regulations; color additive amendments; frozen foods code; label statements for dietetic foods; nutritive and non-nutritive sweeteners; new quality statement; proposed recommended consumer size packages; sealing and loading recommendations for corrugated and solid fiber boxes; and general regulations covered by Standards of Identity.

Commercial Fruit Processing, 2nd Edition

ISBN 0-87055-502-2

Table 13.1. Fruit Products for Which FDA Standards of Identity, Quality, and Fill Have Been Issued

CANNED FRUITS
- Applesauce
- Apricots, artifically sweetened
- Apricots, canned with rum
- Berries (blackberries, blueberries, boysenberries, dewberries, gooseberries, huckleberries, loganberries, strawberries, youngberries)
- Cherries, artificially sweetened
- Figs, artificially sweetened
- Figs, preserved
- Fruit cocktail, artifically sweetened
- Grapes, seedless
- Grapefruit
- Peaches, artificially sweetened
- Peaches, canned with rum
- Pears, artificially sweetened
- Pears, canned with rum
- Pineapple, artificially sweetened
- Plums
- Prunes

FRUIT JUICES
- Canned lemon juice
- Frozen concentrate for lemonade
- Canned grapefruit juice
- Orange juice and orange juice products:
  - Canned orange juice
  - Frozen orange juice
  - Pasteurized orange juice
  - Canned orange juice from concentrate
  - Frozen concentrated orange juice
  - Frozen reduced-acid orange juice
  - Canned concentrated orange juice
- Concentrated orange juice for manufacturing
- Concentrated orange juice with preservatives
- Canned pineapple juice
- Canned prune juice

MISCELLANEOUS PRODUCTS
- Frozen strawberries (pending)
- Fruit butters
- Fruit jelly (apple, apricot, blackberry, black raspberry, boysenberry, cherry, crabapple, cranberry, damson plum, dewberry, fig, gooseberry, grape, grapefruit, greengage plum, guava, loganberry, orange, peach, pineapple, plum, pomegranate, prickly pear, quince, raspberry, red currant, strawberry, youngberry)
- Artificially sweetened fruit jelly
- Fruit preserves and jam
- Artificially sweetened preserves and jams
- Frozen cherry pie

STANDARDS PENDING[a]
- Artificially sweetened cranberry juice cocktail
- Canned fruit nectars
- Lemonade
- Colored lemonade
- Cranberry juice cocktail

STANDARDS TERMINATED[b]
- Canned mandarin orange
- Canned pineapple–grapefruit drink
- Canned tropical fruit salad
- Citrus marmalade
- Grape juice concentrate
- Limeade
- Table olives
- Quick frozen blueberries

[a]Fruit products for which standards were pending as of May 15, 1982.
[b]Fruit products for which earlier standards have been terminated.

Information on these and other processing regulations may be obtained by writing to one of the FDA regional offices:

Region I. 585 Commercial St., Boston, MA 02109
Region II. 850 Third Ave., Brooklyn, NY 11232
Region III. 1204 U.S. Customhouse, Philadelphia, PA 19106
Region IV. 60 8th St., N.E., Atlanta, GA 30309

Region V. 1222 Post Office Bldg., Chicago, IL 60607
Region VI. 3032 Bryan St., Dallas, TX 75204
Region VII. 1009 Cherry St., Kansas City, MO 64106
Region VIII. 513 U.S. Customhouse, Denver, CO 80202
Region IX. Room 518, Federal Office Bldg., San Francisco, CA
Region X. Room 5003, Federal Office Bldg., Seattle, WA 98104

Gertrude C. Gabuten, Office of Consumer Affairs, Food and Drug Administration, describes the purpose and scope of FDA regulations as follows:

All foods, including fruits and fruit products, shipped in interstate commerce are subject to the requirements of the Food, Drug, and Cosmetic Act. The Act requires that foods be prepared from clean, wholesome raw materials, processed, packed and held under sanitary conditions, and properly labeled.

Standards of identity, quality, and fill of container have been promulgated for a number of canned fruits and fruit juices. The specific standards should be consulted by distributors of canned fruits that are shipped in interstate commerce or anyone intending to ship these products into the United States. The Standards of Identity for canned fruits and fruit juices can be found in Title 21, Parts 100 to 169 of the *Code of Federal Regulations*.

Fill-of-container standards have been promulgated for a number of canned fruits and fruit juices, but in packing any other canned fruit the container must be well filled with the fruit and with only enough packing medium added to fill the interstices; otherwise the container may be deceptive and prohibited by the Act.

The standards of fill for canned peaches, pears, apricots, and cherries require that the fill of the solid food component be the maximum practicable quantity that can be sealed in the container and processed by heat without crushing or breaking such component.

Standards for canned fruit cocktail, grapefruit, and plums specify minimum drained weights for the solid food component, expressed as a percentage of the water capacity of the container. These drained weight requirements are as follows: fruit cocktail—50%; grapefruit—50%; whole plums—50%; and plum halves—55%. An additional 90% fill requirement based on the total capacity of the container for the solid food and the liquid packing medium has been established for plums. This 90% fill requirement also applies to applesauce in metal containers (85% fill for applesauce in glass), crushed pinapple, and pineapple juice.

Minimum drained weight requirements for canned pineapple provide that "crushed pineapple" may be labeled as "heavy pack" or "solid pack." The standard of quality requires a minimum drained weight for crushed pineapple of 63% of the net weight.

All fruit used for canning or juice should be mature and sound, that is, free from insect infestation, moldiness, or other forms of decomposition.

Some canned fruits fall into the category of low-acid foods (with a pH value greater than 4.6). Registration is required for both U.S. establishments and those in other countries that export such foods to the United States. Besides registering each plant, the processors are required to file processing information (cooking times and temperature, etc.) for each low-acid canned food product. Full text of the low-acid canned food regulations is in the *Code of Federal Regulations,* Title 21, Parts 108 and 113.

U.S. grade standards for most fresh fruits are under the jurisdiction of the U.S. Department of Agriculture (USDA), Washington, DC 20250. The local county Extension Service, which is a part of USDA, may be able to provide one with additional information.

Grades of some fruits may be either federal, state, or private. For example, in Florida, there is a citrus code of legislative acts administered and enforced by the Florida Citrus Commission. The Florida Department of Citrus establishes and amends regulations pursuant to the citrus code. The regulations cover such subjects as maturity, grades, containers, packing, inspection, use of labels, sampling and testing, and records of citrus fruit dealers. Florida citrus is further regulated by a federal marketing order administered by the U.S. Secretary of Agriculture. Grades and sizes for interstate shipment are recommended from time to time by the Growers Administrative Committee and Shippers Advisory Committee appointed by the Secretary.

California's agricultural code also governs the marketing of grapefruit, prescribing, for example, the minimum standards of maturity and other qualities of fruit. California and Arizona also have a federal marketing order that applies to all grapefruit grown in Arizona and in Imperial County and part of Riverside County in California. The order is administered by a committee that makes recommendations to the U.S. Secretary of Agriculture who is empowered to issue regulations under its provisions. The grade, size, and quality of grapefruit to be shipped are controlled.

A list of FDA consumer affairs offices is given in Table 13.2.

Imports

While the legal requirements that must be met are the same for imported and domestic products, the enforcement procedures are necessarily different.

Imported products regulated by the FDA are subject to inspection at the time of entry through U.S. Customs. Shipments found not to comply with the laws and regulations are subject to detention. They must be brought into compliance, destroyed, or re-exported.

At the discretion of the FDA, an importer may be permitted to try to bring an illegal importation into compliance with the law before the final decision is made as to whether it may be admitted. Any sorting, reprocessing, or relabeling must be supervised by an FDA investigator at the expense of the importer.

Both foreign shippers and importers in the United States should realize that conditional release of an illegal importation to bring it into compliance is not a right but a privilege. Abuse of the privilege, such as repeated shipments of the same illegal article, may result in denial of the privilege in the case of subsequent importations.

Exports

Many U.S. producers are interested in exporting some of their products. If the item is intended for export only, meets the specifications of the foreign purchaser, is not in conflict with the laws of the country to which it is to be shipped, and is properly labeled it is exempt from the adulteration and misbranding provisions of the Act [Sec. 801(d)]. This

Table 13.2. FDA Consumer Affairs Offices

California
1521 W. Pico Blvd.
Los Angeles, CA 90015
(213) 688-4395

50 United Nations Plaza
Room 524
San Francisco, CA 94102
(415) 556-2682

Colorado
500 U.S. Customhouse
19th and California Sts.
Denver, CO 80202
(303) 837-4915

Florida
6501 N.W. 36th St.,
Room 200
Miami, FL 33166
(305) 526-2920

P.O. Box 118
Orlando FL 32802
(305) 855-0900

Georgia
1182 W. Peachtree
St., N.W.
Atlanta, GA 30309
(404) 881-7355

Illinois
433 West Van Buren St.
1222 Main Post Office
Bldg.
Chicago, IL 60607
(312) 353-7126

Indiana
975 North Pennsylvania
Room 693
Indianapolis, IN 46204
(317) 269-6500

Louisiana
4298 Elysian Fields Ave.
New Orleans, LA 70122
(504) 589-2420

Maryland
900 Madison Avenue
Baltimore, MD 21201
(301) 962-3731

Massachusetts
585 Commercial St.
Boston, MA 01209
(617) 223-5857

Michigan
1560 East Jefferson Ave.
Detroit, MI 48207
(313) 226-6260

Minnesota
240 Hennepin Avenue
Minneapolis, MN 55401
(612) 725-2121

Missouri
1009 Cherry Street
Kansas City, MO 64106
(816) 374-3817

808 North Collins St.
Lacledes Landing
St. Louis, MO 63102
(314) 425-5021

Nebraska
1619 Howard St.
Omaha, NE 68102
(402) 221-4675

New Jersey
20 Evergreen Place
East Orange, NJ 07018
(201) 645-6365, 645-3265

New York
850 Third Avenue
Brooklyn, NY 11232
(718) 965-5754, 965-5043

599 Delaware Avenue
Buffalo, NY 14202
(716) 846-4483, 846-5452

Ohio
1141 Central Parkway
Cincinnati, OH 45202
(513) 684-3501

601 Rockwell Ave.,
Rm 463
Cleveland, OH 44114
(216) 522-4844

Pennsylvania
Room 900, U.S.
Customhouse
2nd and Chestnut Streets
Philadelphia, PA 19106
(215) 597-0837

7 Parkway Center
Suite 645
Pittsburgh, PA 15220
(412) 644-2858

Puerto Rico
P.O. Box S4427
Old San Juan Station
San Juan, PR 00905
(809) 753-4264

Tennessee
297 Plus Park Blvd.
Nashville, TN 37217
(615) 251-7127

Texas
1200 Main Tower
Building
Room 15-45
Dallas, TX 75202
(214) 767-5433

1440 N. Loop
Suite 250
Houston, TX 77009
(713) 229-3533

727 E. Durango,
Rm B-406
San Antonio, TX 78206
(512) 229-6737

Virginia
701 W. Broad St.
Room 309
Falls Church, VA 22046
(703) 285-2578

Washington
909 First Avenue
Federal Office Bldg.,
Rm 5009
Seattle, WA 98174
(206) 442-5258

Table 13.3. Fruit Products for which USDA Grade Standards were in Effect, December 1980

FRESH FRUITS	PROCESSING GRADES
Anise, sweet	Apples
Apples	Berries
Apricots	Blueberries
Avocados	Cherries, red sour for manufacture
Blueberries	Cherries, sweet for canning or freezing
Cherries	Cherries, sweet for export sulfur brining
Cranberries	Cranberries
Dewberries and blackberries	Grapes, American (Eastern type) bunch
Grapes, American (Eastern type) bunch	Grapes, for processing and freezing
Grapes, juice	Peaches, fresh
Grapes, table	Pears
Grapefruit (California and Arizona)	Raspberries
Grapefruit (Florida)	Strawberries, growers stock
Grapefruit (Texas and states other than Florida, California, and Arizona)	Strawberries, washed and sorted for freezing
Lemons	PROCESSED FRUITS
Limes	Apples, canned
Nectarines	Apples, dried
Oranges (California and Arizona)	Apples, dehydrated (low moisture)
Oranges (Texas and states other than Florida, California, and Arizona)	Apples, frozen
Oranges and tangelos (Florida)	Apple butter
Peaches	Apple juice, canned
Pears, summer and fall	Apple juice, concentrated frozen
Pears, winter	Applesauce, canned
Pineapples	Apricots, canned
Plums and prunes, fresh	Apricots, dehydrated
Raspberries	Apricots, dried
Strawberries	Apricots, frozen
Tangerines	Apricots, solid-pack, canned
Tangerines (Florida)	Berries, frozen

(continued)

exemption does not apply to "new drugs" that have not been approved as safe and effective, and certain devices.

USDA STANDARDS FOR GRADES

The U.S. Department of Agriculture promulgates quality grade standards for fresh and processed fruits. USDA grade standards are used extensively in marketing farm products. They are the basis of the federal and federal–state grading services, and manufacturers and packers often use them in quality control work. Use of the standards is voluntary in

Table 13.3. (*Continued*)

Blackberries	Limeade, concentrate, frozen
Blueberries, canned	Orange juice, canned
Blueberries, frozen	Orange juice, dehydrated
Cherries, red tart pitted, canned	Orange juice, concentrated, canned
Cherries, red tart pitted, frozen	Orange juice, concentrated, frozen
Cherries, sweet canned	Orange juice, concentrated for manufacturing
Cherries, sweet frozen	Orange juice, pasteurized
Cranberry sauce, canned	Orange juice, from concentrate
Cranberries, frozen	Orange marmalade
Dates	Peaches, dehydrated (low moisture)
Figs, Kadota, canned	Peaches, dried
Figs, dried	Peaches, frozen
Fruit cocktail, canned	Peaches, clingstone, canned
Fruit jelly	Peaches, freestone, canned
Fruit preserves (or jams)	Pears, canned
Fruits for salad, canned	Pears, dried
Grape juice, canned	Pineapple, canned
Grape juice, concentrate, sweetened, frozen	Pineapple, frozen
Grapefruit, canned	Pineapple juice, canned
Grapefruit, frozen	Plums, canned
Grapefruit and orange for salad, canned	Plums, frozen
Grapefruit juice	Prunes, dehydrated (low moisture)
Grapefruit juice, concentrated, frozen	Prunes, dried, canned
Grapefruit juice, dehydrated	Prunes, dried
Grapefruit juice and orange juice, canned	Raisins, processed
Grapefruit juice and orange juice	Raspberries, canned
Grapefruit juice for manufacturing	Raspberries, frozen
Grapes, canned	Strawberries, frozen
Lemon juice, canned	Tangerine juice, canned
Lemon juice, concentrated for manufacturing	Tangerine juice, concentrated for manufacturing, canned
Lemonade, concentrate, frozen	Cherries, sulfured

Source: Agricultural Handbook *533*. Revised April 1981.

most cases, but some states and communities require the use of certain standards in their areas.

All fresh and processed fruits and fruit products for which USDA grade standards were in effect as of December 1980 are listed in Table 13.3. The list is subject to revision. Copies of individual grade standards may be obtained from Fruit and Vegetable Division, A.M.S., U.S. Department of Agriculture, Washington, DC 20250.

Standards for grades are designed to aid processors in packing better and more uniform products; to enable processors, distributors, and retailers to market their products in a more orderly and efficient manner; to aid loan agencies in arriving at equitable loan values; to assist in adjusting damage claims; and to permit inspection and certification as to quality, condition, and grade by federal, federal–state, or state inspection services.

Compliance with the provisions of these grade standards shall not excuse failure to comply with the provisions of the U.S. Food, Drug and Cosmetic Act or with applicable state laws and regulations.

General requirements are adherence to standard fill of containers, maximum headspace, fill weights for canned fruits, and method of drained weights. Use of USDA Standards (for Grades, on a continuous basis, is voluntary on the part of the processor; in contrast, compliance with the FDA) Standards of Identity is legally required.

Standards for Grades vary with each fruit and product, but all include a definition of the product; style of pack (whole, halves, quarters, slices, dices, pieces, mashed); and grade (U.S. Grade A or U.S. Fancy, U.S. Grade B or U.S. Choice, U.S. Grade C or U.S. Standard, U.S. Grade D, and Substandard).

A scoring chart is provided for each product assigning points for color, size and symmetry, defects, character, and other qualities for determining the grade of the finished product.

PROPOSED ASSURANCE FOR PESTICIDE USE

A written statement that fruit processors should obtain from growers and other raw product suppliers—to ensure that only registered pesticide chemicals are used, and that no illegal or unsafe residues remain on the crop—may be adopted. The National Food Processors Association recommends that a statement be drafted in accordance with the following language:

> The grower agrees to use in the production, transportation, and storage of the crop only those pesticide chemicals that are registered under the Federal Insecticide, Fungicide, and Rodenticide Act, in strict accordance with the procedures recommended by the United States Department of Agriculture. The grower warrants that the crop, when delivered to the buyer, will contain no residue of any pesticide chemical which is unsafe within the meaning of the law.
>
> It should be understood, however, that no form of guarantee from the grower can operate to relieve the processor from his liability under the U.S. Food, Drug, and Cosmetic Act. If the processor sells processed fruits that are adulterated or misbranded—including foods that contain unsafe pesticides, or other food additives—he remains subject to criminal prosecution, injunction, and seizure of the product whether or not he has a guarantee from his grower. The written statement recommended above serves primarily as a means of informing the grower of his responsibility, and of assuring the processor that his raw product contains no hazardous residues of pesticide chemicals. In addition, the statement may be regarded by the FDA as evidence of the processor's good faith efforts to comply with the law (Anon. 1973A).

FDA OUTLOOK

The Food and Drug Administration emphasizes a cooperative approach between the agency and industry in controlling the quality, including the safety, of the food supply.

Probably no one questions the basic principle that the quality of the food is the responsibility of the purveyor thereof. He may share this responsibility with his supplier, but he cannot duck it, nor will a government agency assume such responsibility. In practice, with respect to processed foods, the responsibility tends to concentrate primarily on the processor.

The practical result of this intermediate position results in several courses of action:

1. An attempt will be made to communicate clearly what is expected of the industry. This has already been done in part through food standards, additive regulations, and the general regulation on current Good Manufacturing Practice (GMP). The regulatory status of ingredients will be more specifically communicated. Labeling policies will be updated and clarified.
2. Detection of deliberate violations will be assisted by a much augmented program of sampling and analysis of products at the retail level.
3. Emphasis of inspection will shift from operations to the control of operations. Inspectors will be showing a heightened interest in how the processor guards against inadvertent violations.
4. More emphasis can be expected on inadvertent and even unknown contaminants.

There is in process a series of regulations setting forth nutritional guidelines. The program set forth will be voluntary, even though at least in the case of the guidelines there is probably legal authority for a mandatory regulation.

An activity in the economic area is the revision of Standards of Identity to bring them into accord with proposals of the Codex Alimentarius where this is considered desirable. For better or for worse, economic considerations must always take second place to safety considerations in the work of FDA. There is a large program involved in re-examining the safety of items on the GRAS list. There has been a request for industry participation in furnishing information on uses of these substances, along with levels of use and total amounts consumed.

Each processor undoubtedly is asking what he should be doing to improve his relationship with FDA. Here are some suggestions:

1. Be sure of raw materials, ingredients, and supplies. When the processor takes title to purchased materials, he also takes responsibility for

them if they are chemically or microbiologically contaminated. Assurance that his purchased materials are free of such contamination constitutes an important first step.

2. Identify the processing steps that make or break the quality of the product, including its safety. Be sure that an effective feedback control system is operating at each such point.

3. Know the nutritive value of the product, particularly formulated products. A combination of FDA guidance and competitive pressures may soon generate strong motivation for nutritional labeling. If the product is not reasonably in accord with label claims, the FDA will take action against the misbranding. For many processors, this examination of nutritive values may lead to process changes.

4. Find a way to declare all ingredients, even on standardized foods. Although FDA is not authorized to require this, it strongly urges it, and there is a strong consumer demand for it. Legislation making it mandatory may come.

5. Check for inadvertent additives. Materials can get into foods from a variety of sources such as surface coatings, lubricants, detergents, and others. If you are using materials that can contribute small residues, make sure that they are cleared for safety.

6. Think about your environment. Increasing attention will be paid by all government agencies to the effect of any proposed actions on the environment. Present activities should also be reviewed because sooner or later some government agency will be looking at them.

7. Run a clean plant. Infestation with insects and rodents is still one of the leading causes of seizure. The increased emphasis on food-borne disease and environmental contaminants does not imply abandonment of traditional activities (Wodicka 1972).

Government control over the food industry is almost certain to set new records. Washington officials will have greater responsibility for the basic safety and cleanliness of manufactured foods.

Whether or not this trend is a healthy one has yet to be proven. The immediate point is that the acquisition of more authority by government not only is growing, but much of it is taking place at the instigation—even the urging—of the food industry itself. Today's prerogatives of management are fast becoming tomorrow's government demands.

What we are going to see in the years ahead is a continuing proliferation of government agencies (in Washington, in state capitals, in county courthouses, and in city halls) that claim, in one way or another, to "protect" consumers from the wickedness of merchants and manufacturers (Baker 1972).

FDA's new food-labeling regulations constitute one of the most far-reaching sets of regulations issued to an already much-regulated industry.

The industry is told that when *any* mention of a product's nutritional value is made on the package label, *all* new regulations must be followed. This involves the problem of the type-size requirement on the package information label. Many existing packages will have to be redesigned to accommodate specified nutritional information.

The FDA is developing microbiological standards for most food products. Such items as frozen cream pies are under immediate consideration; others are in the offing.

Numerous States are in varying degrees of development of their particular versions of regulations pertaining to open dating, unit pricing, and nutritional labeling.

FOOD ADDITIVES

Food additives are any ingredients or substances added to food, intentionally or otherwise, which are not "generally recognized as safe" (GRAS) and which have not had prior approval. Many substances are of concern, such as detergents for product washing or cleaning, germicides, antibiotics, any substances transmitted from a container to its contents, flavorings, colorings, oils, and processing ingredients for products which receive preprocessing treatment with these chemicals.

A material used in the production of containers and packages is subject to the definition if it may be reasonably expected to become a component, or to affect the characteristics, directly or indirectly, of foods packed in the container. If there is no migration of a packaging component from the package to the food, it does not become a component of the food and thus is not a food additive. A substance that does not become a component of the food, but that is used, for example, in preparation of an ingredient of the food to give a different flavor, texture, or other characteristic in the food, may be a food additive.

Fruit processors must themselves be sure about additives because the responsibility cannot be shifted to ingredient suppliers or others by use of guarantees, though such guarantees serve a useful purpose in emphasizing the duties of all concerned.

Processors should keep records on the product, the ingredients, any factors which may affect the product, and should relate these to package codes. The latter should be changed at frequent intervals so that, if necessary, any lot can be segregated as to source of raw materials, ingredients used, or other operation variables.

Processors should use only ingredients of food grade and purity, obtained from reliable suppliers, because substances, in addition to the desired ingredient, could be added at this source (Anon. 1973A).

"Safety" in a fruit product means that there is a reasonable certainty in the minds of competent scientists that the substance is not harmful under the intended conditions of use. It is impossible in the present state of scientific knowledge to establish with complete certainty the absolute harmlessness of the use of any substance.

This applies to processed fruits and fruit juices, including all commercially processed fruits, citrus, berries, and mixtures; salads, juices and juice punches, concentrates, dilutions, "ades," and drink substitutes made therefrom:

> Factors considered in determining the safety of an additive to processed fruits are (1) the probable consumption of the substance and of any substance formed in or on the food because of its use; (2) the cumulative effect of the substance in the diet; (3) safety factors, in the opinion of experts qualified to evaluate the safety of the ingredient.
>
> Fruit products to which additives are usually combined include beverages, carbonated and noncarbonated, alcoholic and nonalcoholic; condiments and relishes including plain seasoning, sauces and spreads, olives, pickles, and relishes; confections and frostings, including flavored sugars; fresh fruits and fruit juices, including melons and berries; fruit and water ices, including frozen fruit; jams and jellies, preserves, butters, and sweet spreads; processed fruits and fruit juices, including commercially processed fruits, citrus, berries and mixes, salads, juices and juice punches, concentrates, dilutions, "ades" and drink substitutes made therefrom; sugar substitutes, including granulated, liquid, and tablet; sweet sauces, toppings, including syrups of berries, maple, and corn; curing and pickling agents to improve color, texture and flavor-firming agents, to improve texture; flavoring agents and adjuvants to improve taste and aroma; nonnutritive sweeteners having less than 2% of the caloric value of sucrose; nutritive sweeteners having more than 2% of the caloric value of sucrose; texturizers, which affect the feel or appearance of the food [U.S. Code of Federal Regulations. Revised April 1, 1984. Food and Drugs *21*, Chapter 1, subchapter 8, pp. 5–8].

SANITATION REQUIREMENTS FOR USDA PURCHASES

The following sanitation requirements are prescribed for plants operating under the purchase contracts awarded by the Fruit and Vegetable Marketing Service, U.S. Department of Agriculture for canned, dried, dehydrated, and frozen fruits and their products.

Premises. Premises shall be clean, well drained, and free from any material or condition that creates rodent and insect harborage. All roadways and parking areas adjacent to the plant buildings creating a dust or dirt problem shall be hard-surfaced or treated in a manner to prevent a possible source of product contamination.

Buildings. Floors, walls, and ceilings shall be constructed of such materials that they can be kept clean and maintained in a sanitary manner. All windows, doors, and similar openings shall be protected by appropriate screening or other devices where flies, insects, or birds are a problem.

Syrup and Brine Rooms. All syrup tanks shall be adequately covered or syrup rooms shall be completely screened. Syrup, brine, sugar, salt, or other ingredients used in food products shall be kept clean and sanitary.

Ventilation. Ventilation shall be adequate for the operation that is to be performed and sufficient to control excessive condensation, mold growth, and objectionable odors.

Lighting. Lighting in all buildings and specific areas shall be sufficient and adequate for the operation that is to be performed. Light bulbs and fixtures suspended over food tanks or inspection belts shall be of the safety type or otherwise constructed to prevent product contamination.

Water Supply. A sufficient quantity of potable water shall be readily accessible for processing and cleanup. The water supply shall be approved by the state or local health authority.

Waste Disposal. Liquid waste from washing, processing, cooking, cooling, and cleaning or from toilet or hand-washing facilities shall be conveyed to a public sewer or shall be disposed of by methods or systems that will not create unsanitary conditions. Solid waste, including garbage, trash, rubbish, paper, cartons, etc., shall be removed from the premises at regular intervals and at such frequencies as to preclude the development of insect-attracting or breeding conditions, odors, or other nuisances. Receptables for wet waste collection within the plant shall be kept covered at all times and removed and cleaned at regular intervals.

Equipment, Utensils, and Food Containers. Equipment shall be maintained in a sanitary condition regardless of type, construction, or design, and shall be kept in a proper state of repair. Containers, utensils, pans, and buckets used for the storage or transportation of partially processed food ingredients shall not be nested unless rewashed before each use; and containers that are used for holding partially processed food ingredients shall not be stacked in such manner as to permit contamination of the partially processed food ingredients. Containers in which the food is packed shall be clean.

General Housekeeping. Thorough cleaning of all equipment and facilities shall be accomplished at such frequencies as to keep the plant in a sanitary condition. Frequent cleaning of floors and rest rooms shall be conducted during plant operations and adequate waste receptacles shall be provided at all times.

Toilet Facilities. Toilets, washrooms, and rest rooms shall be totally enclosed, well lighted, and ventilated to the outside. They shall be adequately screened and equipped with self-closing doors. Toiler facilities shall be adequate, operational, conveniently located, and in compliance with city and state codes. Handwashing facilities with soap, running water, and drying facilities shall be provided in convenient locations.

Storage Facilities. Storage facilities for raw materials, packing and packaging materials, and finished products shall be maintained in an orderly, clean, and sanitary manner.

Control of Insects, Birds, and Animals. Every practical precaution shall be taken to exclude dogs, cats, birds, and vermin (including rodents and insects) from the plant. Insecticides and rodenticides, if used, shall be only those suitable for the purpose, shall be used in accordance with approved methods, and shall be handled and stored in a safe manner.

Plant Operation. All operations after the final washing operation shall be constructed so as to prevent contamination from the elements and other hazards. Methods used in the processing, manufacture, handling, and storage of foods shall be such as to prevent contamination of the product.

Plant Personnel. No person affected with any communicable disease in a transmissible stage or infected lesions of the hands, arms, or face shall be permitted to have direct contact with the food or food products being processed. Employees having direct contact with food products or food components shall comply with the following:

- Wear clean suitable outer garments.
- Have adequate hair nets or head covering.
- Refrain from spitting, gum chewing, and all uses of tobacco except in designated areas.
- Take all necessary precautions to prevent the contamination of processed products and ingredients thereof with any foreign substance (including but not limited to perspiration, hair, cosmetics, and medications).
- Wash hands immediately prior to starting work and each resumption of work after each absence (Anon. 1971A).

GLOVES FOR HAND AND PRODUCT SAFETY

Use of gloves for product protection and employee hand safety is well established in the canning industry.

The canning industry uses gloves for a variety of operations. Over 80% of the plants use them for hand sorting of products; about 50% for cutting, peeling, and packaging; 32% for product washing; and 28% for freezing. Gloves are also worn during maintenance in 56% of the canneries, material handling in 52%, shipping in 36%, and research and development in 6%.

Employees' hands are exposed to acids and caustics in about 50% of the plants, solvents in 30%, fats in 24%, and chemicals in 22%. Thus, gloves are necessary to protect workers.

Among liquid-proof gloves, the unlined rubber style is the most popular, being used in 64% of the plants. Neoprene-coated fabric gloves are worn in 36%, unlined neoprene in 28%, flock-lined rubber in 22%, vinyl-coated fabric in 16%, flock-lined neoprene in 14%, and unlined vinyl in 12%.

Heat-sealed vinyl gloves lead the disposables. They are used in 18% of the canneries, compared with 6% for the molded vinyl type.

NUTRITIONAL LABELING

The most significant change in food-labeling practices since food labeling began is that of nutritional labeling. Processed fruits are among the foods coming under the nutritional labeling regulation, which became effective in 1973.

The basic regulation requires nutrient labeling for all food to which nutrients, other than those naturally found in the food, have been added and for all foods for which claims of nutritional value or information about nutritional value are included on product labels or in advertising. Complete nutrient labeling also is required when any reference in labeling or advertising is made to protein, fat, carbohydrate, or caloric content of a food product. Specific labeling requirements exist for special dietary foods and infant foods.

A point that must be emphasized is that food labeling is a voluntary action on the part of the food manufacturer.

In working out the format nutritional labeling must follow, a real effort was made to keep it simplified. The items must appear in the following order: serving size, servings per container, caloric content, protein content, carbohydrate content, fat content, percentage of U.S. Recommended Daily Allowances of protein, vitamins and minerals. Whenever nutrition labeling is used it must follow this format. It is hoped that consumers will use this information to compare the nutritional value of products, for instance, the vitamin C content of two cans of tomato juice.

By keeping the format simplified and by making the placement and wording of the new labels as uniform as possible, consumers should be able to find the information they desire with little difficulty.

Along with the introduction of nutrition labeling came the introduction of the U.S. Recommended Daily Allowances (RDA). These replaced the old Minimum Daily Requirements (MDR) that were based on the amount of a nutrient required to prevent a deficiency. The system had been outdated for some time because several generally well-known vitamins were not even included. The U.S. Recommended Daily Allowances for the normal adults are listed in Table 12.1.

In addition to establishing the RDA as the official measurements, the regulation also sets forth these five prohibitions:

1. With certain exceptions, it prohibits any claims or promotional suggestion that products intended to supplement diets are sufficient in themselves to prevent, treat, or cure disease.
2. It prohibits any implication that a diet of ordinary foods cannot supply adequate nutrients.
3. It prohibits all claims that inadequate or insufficient diet is due to the soil in which a food is grown.
4. It prohibits all claims that transportation, storage, or cooking of foods may result in inadequate or deficient diet.
5. It prohibits nutritional claims for non-nutritive ingredients such as rutin, other bioflavonoids, para-aminobenzoic acid, inositol, and similar ingredients and prohibits their combination with essential nutrients.

These regulations do not prohibit a manufacturer from making a claim that his product has a greater nutritive value than another, if he has adequate scientific data to support such claim. It also does not prohibit a claim that a particular food has a higher nutrient content because of the soil in which it is grown if that claim is backed up by scientific data. A manufacturer can also identify vitamins present naturally and those that are added.

Vitamins and Minerals. The regulation draws a clear distinction between ordinary foods, special dietary foods intended for diet supplementation, and drugs intended for the treatment of diseases.

In general, if a product contains less than 50% of the RDA, it is not a dietary supplement and only nutritional labeling is required.

If it contains from 50 to 150% of the RDA, it is a dietary supplement and must meet the standard.

If it exceeds 150% of the RDA, then it cannot be sold as a food or a dietary supplement, but must be labeled and marketed as a drug.

Flavorings. Before the nutritional labeling regulation was adopted, foods had to state whether flavorings were used, but the guidelines were not clear. The new regulation seeks to clarify the former condition. An example best illustrates this. If vanilla pudding contains only natural flavor, it would be called simply "vanilla pudding." If it contains a natural flavor that predominates with an added artificial flavor, it would then be called "vanilla flavored pudding." If both natural and artificial flavorings are used, and the artificial flavoring predominates, the name would be "artificially flavored vanilla pudding." If only artificial flavor is used, it would also be "artificially flavored vanilla pudding."

Imitation Foods. The term "imitation" is required only when a food is nutritionally inferior to a food product for which it is a substitute. The regulation also sets up a mechanism so that a new product similar to an established food product and nutritionally equivalent could be marketed without the use of the word "imitation." A new name must be established that is fully descriptive and informative to the consumer.

Nutritional Quality Guidelines. The regulation seeks to establish principles to be followed in describing desirable nutritional quality characteristics appropriate for given classes of food. A product that complies with the guideline may include on the label the statement that it "provides nutrients in amounts appropriate for this class of food as determined by the U.S. Government" (Vollertsen 1973).

Advertising Guidelines. Several guidelines for the treatment of nutritional information in advertising are included in the regulation.

All specific facts presented with regard to nutrition or the nutritional performance of a product must be completely accurate and documentable.

Advertising for products that have no specific nutritional value should not imply nutritional benefits, even in the most general way (such as by saying, "It's good for you"). Only fun, snack, taste, etc., values should be claimed.

Advertising for products that supply only a small part of daily nutritional requirements should be advertised primarily on the basis of nonnutritional benefits. General statements on nutrition education, however, can be used with such products as long as they do not imply that the products have high nutritive value.

Advertising for products that have a strong nutritional reason for being consumed should only make claims pertaining to the nutrients contained in relatively high amounts (e.g., ⅓ of the MDR).

Implications of physical benefits should be restricted to those that would actually result from normal usage of the product. Implication of superhuman or extreme recuperative powers should be avoided, as should the use of superpersonalities in the fields of physical prowess resulting from the use of a single product. Claims such as superpowered and supercharged should also be avoided.

Comparisons that emphasize nutritional value should be made only against products that are recognized sources of particular nutrients.

Efforts should be made to use easily understood, meaningful terminology, such as RDA. Advertising should also give the consumer a basis on which to compare products nutritionally.

Whenever feasible, advertising should attempt to educate consumers. Particular attention should be paid to building educational value into nutritional claims for children's products (Nesheim 1971).

NEW REGULATIONS FOR U.S. WINE LABELS

Regulations enacted in 1982 by the federal government require more precise information on the labels of U.S.-produced wines (Anon. 1983).

The new regulations raise the minimum percentage of a single grape variety required for a wine to carry a varietal designation. Multivarietal labeling requires that both the name and percentage of each grape variety in the blend be listed.

Every bottle of varietal wine is required to list an appelation of origin on the label—a state, county, or viticultural area. The viticultural area designates a significant grape-growing region whose geography and boundaries are recognized and strictly defined by the federal government. For "estate bottled" to be used on the label, both the winery and the vineyards that are the source of the grapes must be located in the same viticultural area.

REFERENCES

ANON. 1971A. 56th annual compilation of basic references for the canning, freezing, preserving and allied industries. *In* The Almanac. Edward E. Judge & Sons, Westminster, MD.

ANON. 1971B. Glove usage for product and hand safety. Canner/Packer *140*(10) 11.

ANON. 1973A. 58th annual compilation of the basic references for the canning, freezing, preserving and allied industries. *In* The Almanac. Edward E. Judge & Sons, Westminster, MD.

ANON. 1973B. Running the gauntlet of regulation. Quick Frozen Food *35*,(8) 18, 23.

ANON. 1983. Regulations changes for U.S. wine labels. Amer. Fruit Grower *103*,5, 31.

BAKER, G. 1972. More controls ahead. Food Eng. *44*(1) 62–63.

NESHEIM, R. O. 1971. Industry response to the nutrition challenge. Food Technol.*25*, 605, 608.

VOLLERTSEN, N. 1973. Nutritional labeling. Atlanta J. *91*(172) 1F, 20F.

WODICKA, V. O. 1972. Message from F.D.A. Food Eng. *44*(1) 58–61.

14

Storage Life of Canned, Frozen, Dehydrated, and Preserved Fruits

J. G. Woodroof

Modern methods of preparation, packaging, distribution, and sale of fruits and fruit products in recent years have vastly increased the variety, availability, and consumer acceptance of these items to the point that they are commonplace and are a part of the American diet. Processed fruits retain their healthfulness and aesthetic qualities over periods of months and years, but all deteriorate with age. It is of interest to the manufacturer, and all who handle or store and use these items, to know how to handle and preserve them in the best manner. An initial complication is the fact that the manufacturing dating practices and procedures are as diverse and numerous as the items themselves.

Since health hazards caused by food-borne diseases from processed fruits are rare, major attention can be directed to retaining their aesthetic and nutritional qualities in storage. Manufacturers preserve fruits in an endeavor to reduce the rate at which such deterioration and alteration occur. Even fully processed fruits deteriorate in a readily detectable manner, and aesthetic qualities—flavor, color, texture, appearance, and nutritional qualities—can be measured.

In addition to the normal and gradual deterioration that occurs in processed fruits, mishandling by the fruit processor, warehouseman, fruit

Commercial Fruit Processing, 2nd Edition

ISBN 0-87055-502-2

service establishment, and enduser may damage fruits and reduce their quality.

Manufacturers generally employ diverse systems of coding and self-monitoring for quality maintenance; however, individual consumers would not benefit from knowing the actual age of food items. But, to prevent quality losses in processed fruits due to aging, systems regulating their orderly rotational flow, on a first-in, first-out basis, should be used. These systems should also include temperature and humidity standards. Postmanufacturing handlers and consumers lack both awareness and technical sophistication regarding the viability of processed fruits. Traditional existing federal and state laws regulate only food manufacturers. To assure wholesome food quality, these laws should be extended to include warehousemen, food retailers, transporters, and distributors (Hoofnagle 1971A).

The most important causes of quality deterioration of canned fruits are very slow chemical changes that take place during storage. They are responsible for changes in nutritive value, flavor, color, and texture of the product; however, none of the chemical products formed are harmful to health. Air trapped within the container can cause surface darkening, development of off-flavors, and loss of vitamin C. Contact of raw materials with utensils or equipment and dissolution of minute quantities of iron and copper in the food may result in a grayish or black discoloration.

Processed fruits are rarely nutritionally equal to raw fruits. Nutrients are destroyed during processing because of their sensitivity to pH, oxygen, light, heat, or a combination of these. Trace elements and enzymes catalyze certain catabolic reactions. Canned fruits and fruit juices are often fortified with ascorbic acid to increase the vitamin C content and to offset that which is lost during storage. Vitamin C in prune juice is lost due to the antagonistic effect of anthocyanins.

Lowering of sensory qualities and nutrient levels of processed fruits occurs during all stages of processing, in channels of trade, in storage, and in the home. Since about 1965, there has been an increased interest in the quality of processed fruit on the part of consumers. Consumer interest has resulted in stricter labeling and demand for control of nutrient levels of foods, culminating in the establishment of nutrition-labeling guidelines by the federal government.

The sensory quality of citrus fruits, particularly orange juice, deteriorates more rapidly than does the vitamin C content. Orange juice concentrate will retain sensory quality more than 2 years at 0°F, and will lose less than 10% even at temperatures as high as 40°F.

At storage temperatures of 40°F or lower fruit juices suffer very small losses of vitamin C in 2 years. When held at 80°F canned fruit juices lose about 25% of vitamin C in 12 months, and 50% in 23 months; under tropical conditions (98°F) loss is 50% in 4 months.

Other nutrients in processed fruits, such as carotene, are more stable than vitamin C, and no special storage conditions are needed for this vitamin. Low-temperature storage of canned, frozen, dried, and preserved fruits is used mainly for retention of aesthetic qualities. Most dried fruits, concentrates, and powders are quite acid and are packed under vacuum in opaque containers, which offer triple protection. Such products have a shelf life of 1 year at cool (48°F) storage conditions (Kramer 1974).

SHELF LIFE OF FOODS

The period between manufacture and retail purchase of a food product during which the product is of satisfactory quality is commonly referred to as its "shelf life." All food products have a finite and variable shelf life. Food companies generally know the shelf life of their products under different environmental conditions. Storage studies are part of every product development program, whether it includes a new product, a product improvement, or simply a change in type or specification of an ingredient.

Attempts are made to maximize shelf life consistent with costs and patterns of handling and use of distributors, retailers, and consumers. Manufacturers avoid sources of bacterial contamination and follow prescribed procedures for pasteurization of perishable foods. Special ingredients are used as preservatives, antioxidants, emulsifiers, stabilizers, and chelating agents to retard chemical and physical changes in foods during storage. Any deficiency in shelf life will result in consumer complaints, which eventually will seriously affect acceptance and sales of brandname products.

Foods can be divided into two main categories—perishable (including semiperishable) and shelf stable.

Perishable foods are those that can be held for only a short time unless stored under refrigerated or freezing conditions to inhibit growth of spoilage microorganisms and action of enzymes. Semiperishable foods are those that are more stable because of natural inhibitors, or those that have received some type of mild preservation treatment for greater tolerance to environmental conditions and abuse during distribution and handling. Obviously, perishable foods are a special and sensitive group when one considers storage and shelf life.

Nonperishable or shelf-stable foods are not affected by microorganisms, because such foods are either preserved by heat sterilization, are formulated as dry mixes, or processed to give low water activity. Many unprocessed foods such as nuts, honey, and maple syrup are shelf stable because of low water activity.

Bacteria, molds, yeast, and enzymatic breakdown are the main causes of deterioration of perishable foods (see Chapter 5). Usually, bacterial deterioration occurs well before chemical or physical changes. Bacteria grow under a wide range of environmental conditions and are responsible for a variety of changes in food products. Some bacteria are associated with production of acid; others cause off-flavors or phase separation , which adversely affect the acceptability of food. Molds and yeast cause visible signs of deterioration. Enzymatic activity can be derived from microbial growth or arise from natural systems such as bruising or darkening in apples.

In the food distribution and supply system, the maintenance of chilling or freezing temperatures is an absolute requirement for perishable foods such as fresh juices. For semiperishable foods, the requirements for low temperature and short shelf life are less critical. Semiperishable foods have a shelf life of 30–90 days under ideal conditions, depending on the food, whereas some perishable foods can be held only 5–7 days even under ideal conditions.

Changes that occur in foods during storage include the following: oxidation, hydrolysis, and reversion of fats; breakdown of protein structure, which can cause textural changes as well as loss in water-binding and functional properties, such as whipping, gelation, and thickening; browning reactions, which involve sugars and amino groups found in proteins; oxidation of pigments; and chemical deterioration of certain micronutrients.

Under extreme conditions of prolonged storage or excessively high temperatures, interactions can take place even between the food and the containers, for example can corrosion. These reactions and breakdowns are manifested by visible changes in the food—fading of color, caking, browning, staling, softening, flocculation, cloud formation, or gelation. There would also be signs of staleness, and most likely off-flavors and -odors as well as nontypical texture and appearance.

As a result of improved methods for heat sterilization, vitamins in raw foods are largely retained during processing. Losses during prolonged storage are limited to the more vulnerable vitamins. In general, a 15–30% decrease in ascorbic acid may be expected if canned foods are stored for 1 year at 80°F, whereas only 5–15% is lost at 65°F, which is more representative of average storage conditions in food warehouses in temperate climates. Depending on the pH of the food, thiamin losses are of similar order of magnitude or somewhat larger. Carotene, riboflavin, and niacin levels decrease slightly on storage.

Various chemical and biological reactions take place in foods, depending upon the temperature encountered during storage. Protection against oxygen can be provided by a package system, by avoidance of pro-oxidants, or by use of antioxidants. Protection against moisture gain or loss can be accomplished by use of packaging materials with low moisture permeability.

In shelf-stable foods, the rate of simple chemical reactions increases as the temperature is raised; this relationship can be quantified in the van't Hoff and Arrhenius equations. As a rough guide, the rate of reaction is doubled for every 10°C (18°F) rise in temperature. The effect of temperature change on a reaction in a food system is often expressed as the "temperature quotient," defined as the ratio of the rate of change at one temperature to the rate at a specified lower or higher temperature. In biochemical systems, it has been common to give the temperature quotient for a 10°C temperature difference, which is designated at Q_{10}.

Temperature and level of bacterial contamination are the most important factors that determine shelf life for perishable foods. Most bacteria multiply very rapidly in the temperature range of 70°–100°F. At temperatures outside this range, the rate of growth is reduced. Under ideal growth conditions, bacteria can have a generation time as rapid as 20 min. At this rate, a single bacterium could increase in number to 2,097,152 within 7 hr.

Most perishable foods clearly demonstrate highly disagreeable signs of spoilage before they become a danger to health. Nevertheless, there is always the potential danger that overaged or abused perishable foods can cause food poisoning if the spoilage is not detected and the food is consumed without heating (Anon. 1974).

Determining Shelf Life

It is highly desirable to know the shelf life of canned fruits and fruit products at different storage time/temperature conditions, without actually storing samples under all environments. It is possible to get a fairly accurate indication by a single test at fairly high temperature. An accelerated test may be run at 100°F utilizing the principle that reactions are reduced one-half by each drop of 10°C (18°F) in temperature. After determining the number of days for deterioration in flavor, color, texture, consistency, and nutritive qualities to take place, this number may be doubled for storage at 82°F, doubled again for storage at 64°F, doubled the third time at 46°F, and doubled the fourth time for storage at 28°F. This rule does not follow in frozen products.

A major canner supplied the following information for a survey by Hoofnagle (1971A):

> When a product is developed, one part of the research is to store samples under various conditions of time, temperature, and humidity in several containers. At specified time intervals the product is examined for quality factors. The product is analyzed for metal pickup and the container itself is tested for changes. All current canned products packed in conventional cans have a corrosive service life of at least 36 months.
>
> Factory packs are made on each new product and each major formula change. A 150–200 can sample is divided: 100 cans for storage at 70°F for 36 months, 50 cans for storage at 100°F for 18 months, and 12–24 cans for storage at 120°F for 12 weeks.
>
> The 12-week storage at 120°F provides an acceleration of 10:1 over 70°F storage. We consider 12 weeks at 120°F to approximate 120 weeks' storage at 70°F. This rate is estimated from the tin and iron pickup in the product, which varies depending on the activity of the product. The 120°F packs provide an early indication of corrosion and product reaction on the container.
>
> The 18-month storage at 100°F provides an acceleration of 2 or 3 to 1 over the 70°F packs. Any product successfully passing the 18 months' storage at 100°F will usually be the equivalent of more than 3 years storage at 70°F.
>
> The 100°F storage at 18 months allows us to evaluate the effects of abusive storage in the trade and provides an accelerated means of predicting the corrosion effects of the product on the container and the stability of product quality. The 70°F storage at 36 months allows us to determine corrosion service life, product quality, and color under close to normal conditions. The 70°F packs are the ones used to follow flavor changes.

STABILITY OF CANNED FRUITS AND JUICES

Canned fruits are processed with sufficient heat to make them commercially sterile so that they can be safely shipped and stored without spoilage.

Canning of fruits involves a series of steps after harvesting and delivery to the cannery: washing, sorting, peeling, coring, blanching or preheating, filling containers, sealing, cooking, cooling, labeling, and storing. Most of these steps extend stability by cleaning and otherwise destroying microorganisms; other steps tend to shorten stability by subdividing and exposing the product. The order of some of the steps may be altered for a variety of reasons.

Juice for canning requires some form of extraction from the fruit followed by straining before filling into containers and processing by heating. These operations tend to shorten the life of the product because of the air entrapped into the product. Some acid juices are merely hot-filled and sealed.

The stability of canned fruits and juices may be extended by lowering the pH by the addition of acids or blending with more acid fruits or juices; or by raising the degree Brix by the addition of dry or liquid sugar. Other ways of extending the shelf life of canned fruits and juices are by packing under vacuum in nitrogen gas or by adding antioxidants, such as ascorbic acid.

The cooking time and temperature necessary to produce commercially sterile canned fruits or juices depend both on the product and size of the containers. To destroy pathogenic bacteria, processing conditions have been established in specific, precise detail by the National Food Processors Association in cooperation with can manufacturers and canners themselves. Most fruits are acid-type foods; these have a pH below 4.5 and usually require lower temperature, often about 212°F, than low-acid (above pH 4.5) foods such as most vegetables.

The changes that canned fruits undergo in sensory quality and nutritive value may result from reactions within the food, from interactions between fruit and the container, or, with fruit in glass, from surface reactions stimulated by light passing through the container wall. Each canned fruit has a definite storage potential. This is usually adequate to ensure acceptable quality of fruits on the retail market for at least 1 year. Common storage temperatures for canned fruits seldom average above 70°F.

Retention of physical quality and palatability are the most important criteria for determining the storage life of canned fruits. Gradual softening occurs with prolonged storage at temperatures above 80°F, but freezing causes even greater changes. For this reason canned fruits should not be allowed to freeze (Cecil and Woodroof 1963).

No special low-temperature conditions are necessary for storage and transportation of canned fruits and juices, but 80°F or lower is common in commercial channels. While such high-temperature storage and transportation of canned fruits and juices do not induce any health hazards, aesthetic qualities do change. These are time and temperature dependent and are usually accompanied by nutritional losses. Many canned fruits and juices can withstand storage for 1 year before expert taste panels can detect changes. High-temperature storage is considered abnormal in the industry, and most companies are turning to storage temperatures of 50°F or lower.

The order of breakdown in quality of canned fruits and juices is (1) flavor, (2) color, (3) texture, and (4) finally nutritive losses. The critical storage temperature for most fruits and juices is 80°–84°F, and temperatures higher than this should be limited to a few weeks.

Fruit juice manufacturers and dealers are increasingly aware of the effects of storage upon quality. Several types of changes occur at above freezing temperatures: destruction of anthocyanin pigment; formation of brown color compounds by reaction between sugars and amino acids; possible intermolecular oxidation; destruction of ascorbic acid, and others, all of which may affect the flavor of the product.

Lee *et al.* (1950) studied the stability of raw and pasteurized juices and concentrates prepared from apples; strawberries; red, purple and black raspberries; cherries; rhubarb; and plums. They investigated the influence of storage temperatures (0°, 18°, 34°, and 70°F) and the effect of added ascorbic acid upon changes in flavor, color, and ascorbic acid content of raw and pasteurized juices and concentrates.

The flavor of all the pasteurized juices and concentrates was changed during storage at 70°F; there was evidence of change in strawberry juice at 34°F. The addition of ascorbic acid to fortify the products did not noticeably prevent or definitely retard deterioration of flavor in any of the pasteurized juices. The flavor scores of pasteurized unfortified juices made from cherry, prune, purple raspberry, Latham red raspberry, and apple were somewhat higher than those of the corresponding raw juices. Raw apple juice highly fortified with ascorbic acid retained the characteristic fresh fruit flavor. Unfortified raw apple juice stored at 0° and 18°F developed varying degrees of musty flavor.

Definite loss of flavor in fruit juice containing anthocyanin is accompanied by destruction of this pigment. Change in color is observed by a decrease in optical density between 500 and 530 nm or by an increase in optical density, caused by browning, between 400 and 430 nm.

Ascorbic acid added to apple juice retarded development of brown color. It also retarded the development of brown color in strawberry juice but hastened destruction of red color. Ascorbic acid was retained well by the juices even after storage for 11 months at 70°F, and analyses for ascorbic acid gave no indication of the marked deterioration in color and flavor that occurred at 70°F.

STABILITY OF FROZEN FRUITS AND FRUIT PRODUCTS

If frozen fruits, fruit juices, purees, or concentrates are prepared and maintained in good sanitary conditions, and kept at 0°F, they are stable for 1–5 years. However, allowed to partially or entirely thaw, they deteriorate as perishable items within a few days or even hours.

The major quality features of frozen fruits, from promotional and consumer viewpoints, reside in their aesthetic appeal. These are the first apparent quality loss areas and serve as natural warning factors. Flavor and color are most sensitive to gross temperature fluctuations; nutrient losses occur, if at all, later on in the chain of changes. Storage of "frozen" fruits (0°F) differs from "chilled" (50°F), "cold" (40°F), and "refrigerated" (32°F) fruits only in the temperature used and resulting longer storage life. Frozen fruits have practically all of the water bound as ice and should remain in this state until ready to use.

The preservative benefits of freezing are generally greater than those of refrigeration owing to the lower temperatures employed, which slow down chemical changes, arrest enzyme activity, arrest growth of microorganisms, control rodents and insects, and preserve nutritional values (Hoofnagle 1971B).

While 0°F is the acceptable temperature for storage of frozen fruits for 1 year or longer, −10° or −20°F is preferred by many packers to protect an esthetic quality for a longer time, and to assure that a maximum temperature of 0°F is maintained at all times, especially during transportation and handling.

Since fluctuating temperatures are in reality a continuous thawing and refreezing process, it results in a "pumping action" in the product and package due to changes in the vapor pressure. This action results in migration of moisture, resulting in "cavity ice" within the package and deterioration of the texture and appearance of the product.

Most frozen fruits prepared in the recommended manner have a "built-in" storage life of 1 year or more at 0°F; if the temperature is raised to 10°F, the time is reduced; if the temperature is lower, the time is extended. The quality of various frozen fruit products stored at 0°F for 1 year is summarized in Table 14.1.

The superior quality of frozen fruits is not ensured automatically by the mere act of freezing the product and storing it at low temperature. There have been, and still are, flagrant temperature abuses in the distribution system from freezer to consumer in which the product is exposed to ambient temperatures for a time, resulting in rapid, partial thawing. Though these incidents represent the exception rather than the rule, they produce unfortunate results. It is neither temperature nor time alone that is harmful, but the integrated effect of both.

Regardless of the initial quality of a product when it enters the distribution system quality attributes can be measured from time to time and compared with control lots from the same sample held constantly at −10°F or lower, at which changes are negligible. Data in Table 14.2 show the relationship between temperature and approximate average time required to cause a definite organoleptically measurable change in a typical frozen fruit. Although the actual storage time associated with a given loss in quality varies among different fruits and products, the general exponential relation between temperature and time to produce a given degree of quality change holds. According to Table 14.2, for every 5°F that the temperature is increased, the rate or speed of quality loss increases 2 to 2½ times. The quality change brought about by these time-temperature combinations is essentially the same. From these data it is clear that a week at 20°F is no worse than 6 months at 0°F. In fact, neither of these time-temperature combinations alone would cause noticeable quality

Table 14.1. Desirability of Some Frozen Fruit Products after Storage at 0° for 1 Year

Fruit	Treatment	Form Packed[1]	Desirability
Apple	Cooked	Butter, solid pack	Excellent
	Cooked	Sauce, solid pack	Excellent
	Steamed	Stew, semisolid pack	Good
	Steamed	Slices, loose pack	Fair
	Boiling syrup	Slices, solid pack	Excellent
	None	Slices	Poor, discolored
Apricot	Steamed	Halves, semisolid pack	Good
Blackberry	Steamed	Whole, semisolid pack	Good, color loss
	Boiling syrup	Whole, solid pack	Excellent
	None	Whole	Fair, bitter
Blueberry	Steamed	Whole, semisolid pack	Good
	Boiling syrup	Whole, solid pack	Excellent
	None	Whole, loose pack	Poor, flavorless
Cherry	Steamed	Seeded, loose pack	Fair, color loss
	Boiling syrup	Seeded, semisolid pack	Excellent
Peach			
firm flesh	Steamed	Slices, solid pack	Good
	Boiling syrup	Slices, solid pack	Very good
soft flesh	Steamed	Slices, solid pack	Poor, flavorless
	Boiling syrup	Slices, solid pack	Fair, flavorless
Plum	Steamed	Seeded, semisolid pack	Fair, flavorless
	Boiling syrup	Seeded, semisolid pack	Good
	Steamed	Seeded, solid pack	Excellent
Raspberry	Steamed	Whole, semisolid pack	Fair, color loss
	Boiling syrup	Whole, solid pack	Excellent
	None	Whole, loose pack	Fair, bitter

Source: Woodroof and Atkinson (1945).
[1]The amount of sugar added to the various fruits varied, but all lots of the same fruit received the same amount.

Table 14.2. Time–Temperature Relationship That Causes Equivalent Quality Changes in Some Frozen Fruits

Temp. (°F)	Time
0	1 year
5	5 months
10	2 months
15	1 month
20	2 weeks
25	1 week
30	3 days

Source: Guadagni (1961).

change, but the two experiences combined would have the same effect as 1 year at 0°F or 2 weeks at 20°F.

Table 14.2 presents a simple straightforward relation between steady temperature-time and a definite quality change. In handling frozen foods, however, there is rarely, if ever, a steady temperature but rather a series of variable temperature excursions. By this procedure it is possible to evaluate the relationship between products stored at one place for a period and transferred to another. It can also be used to determine the degree of damage in products that are unintentionally partially thawed in transit or while waiting on the docks. Surveys have shown, however, that a major portion of the degradation of frozen foods occurs while they are in the hands of consumers (Guadagni 1961).

The ultimate quality of a frozen fruit product is determined by the care and control that is exercised in all phases of its handling, processing, freezing, and distribution. Exceptional care in one link of the frozen fruit chain will not make up for a weakness in another.

Should fruits that are unavoidably thawed be refrozen? The answer is "yes" if the temperature does not rise above 32°F, but all the aesthetic qualities—texture, color, flavor, aroma—will greatly deteriorate, depending upon the rate at which thawing and refreezing occur. Vitamins will be least affected. Since thawing and freezing of unopened packages will not generate a health hazard, it is recommended that such products be used immediately in bakery goods, ice cream, wine, or alcoholic products.

Failure to maintain a constant 0°F temperature, from the manufacturer and through handling to the home or end-user, results in cumulative effects. Qualities lost gradually, or by a single abuse, cannot be regained, and no subsequent treatment can restore the initial qualities.

Adequate packaging, with at least one liquid-tight, oxygen-moisture-vapor barrier is necessary to prevent loss of moisture (weight); volatile flavors; absorption of flavors from other products; entrance of oxygen-causing browning (in peaches, apricots, cherries) and staling (of citrus, apples, peaches); microbial contamination; and prevention of spillage when the fruit or juice is thawed. Since most fruits are "runny" before and after freezing, packages should be rigid, and may be made of fiberboard, aluminum cans, tin-coated cans, plastic, or laminated material. Vacuum packaging greatly extends the shelf life of frozen fruits and juices.

STABILITY OF DRIED FRUIT AND FRUIT POWDERS

The basis of preservation of fruits by drying is to reduce the moisture content to levels at which microorganisms cannot act to spoil the product.

When the moisture is not sufficiently low to do this, the fruits are partially dried, pasteurized, and packaged without the entrance of organisms. Some fruits, such as dates, are packaged then pasteurized.

A dehydrated product remains stable only as long as it is protected from water, air, sunlight, and contaminants. Packaging, therefore, is very important. The two major considerations in packaging are the exclusion of moisture and oxygen. Metal cans, plastic bags, and laminated bags and boxes effectively limit the passage of these.

Air may be eliminated by packing under vacuum, replacing the air with an inert gas like nitrogen or carbon dioxide, or treating the product with an antioxidant. The antioxidant is eventually used up in its fight with oxygen, so there is a practical limit to how long this protection can be provided.

In practically all fruit-drying processes, removal of the last portions of water becomes increasingly costly and time-consuming. It is advisable to stop the process when cost and quality reach an optimum. The point at which each type of product becomes stable involves a complex relationship between water content and the nature of the fruit being dried.

One way to reach the desired low-moisture content in the product is to package it with an in-package dessicant (a chemical having an extremely high affinity for water). The desiccant or drying agent is placed in a porous bag to keep it apart from the food, and then placed inside the container, which is sealed. In this way, orange juice powder containing 3% moisture when packaged can be brought down to less than 1% after a few weeks in storage.

With dried apricots, increasing the storage temperature accelerates oxygen consumption, carbon dioxide production, disappearance of sulfur dioxide, and darkening.

Molds and yeasts grow slowly and cause deterioration of dried fruits at 40°F unless relative humidity (RH) is very low (below approximately 50%). At 32°F, RH is less important than at higher temperatures.

STABILITY OF AVOCADO AND GUAVA PUREES

Avocado Puree

Latin Americans traditionally prepare avocados for domestic use in the form of *guacamole,* a mixture of mashed avocado, lime juice, salt, and other ingredients. In recent years the product has been prepared for commercial uses; when frozen, avocado puree retains flavor and texture for 1 year.

Freeze-drying has been found to be a satisfactory means of preserving the product, when the moisture content is about 5.8% and an equilibrium humidity of 32% at room temperature is maintained. The dried product developed an off-flavor when exposed to the atmosphere. This was partially corrected by the addition of 0.075% of butylated hydroxyanisole (BHA), by packaging under vacuum, and storage under refrigeration (Gomez and Bates 1970).

Guava Puree

Guava concentrates (2.5- to 3-fold) that retain natural flavor, aroma, and color can be made from puree by low-temperature evaporation at low pressure. However, this method affords little heat destruction of contaminating microorganisms. Many of the flora, particularly molds and yeasts, can grow at the relatively low soluble-solids content of such concentrates, causing spoilage even at refrigerated temperatures (45°F). Additional treatment is required to prolong the storage life of guava puree concentrate. Although freezing would provide the best means of retaining quality, the cost of handling and transportation from the tropics would be prohibitive. Heat processing would not be desirable because the prolonged time needed to heat a viscous concentrate would permit destruction of the natural flavor and color. A practical alternative is the use of chemical preservatives to prevent spoilage under refrigeration.

Brekke *et al.* (1970) found that guave puree concentrate, containing 1000 ppm potassium sorbate, showed no gross signs of spoilage during storage for 5 months at 45°F. During this period the initial number of yeasts decrease steadily to an insignificant level. Flavor and aroma were good and did not deteriorate appreciably until the fourth month in storage. Browning of concentrate was apparent after 2 months, and became more apparent as storage continued. The carotenoid level remained constant during the 5-month storage period.

STORAGE OF DRIED FRUIT

Refrigeration is beneficial in augmenting drying as a means of preserving fruits. The optimum conditions for holding most dried fruits are about 55% RH and just above their freezing temperature. Since the sugar content of these products is rather high, the freezing point varies from about 22°–26°F. Refrigerated storage helps to retain natural flavor, ascorbic acid, carotene, and sulfur dioxide, and to control browning, insects, rancidity, and molding. Except for insect control, low humidity is more important than low temperatures for storage of dried fruit.

While most dried fruits are adversely affected by softening and injured by freezing, dates are held best for freezing. Before storage, dried fruits should be brought to the desired moisture content and packed in moisture-proof containers. The latter not only ensures a constant moisture content while in storage, but prevents injury from moisture condensation upon removal from storage.

Sun drying has been progressively replaced by machine dehydrators for prunes, apples, figs, and bleached raisins, but regular raisins, apricots, peaches, and pears are still sun dried, and this influences the time and method of storage.

Kiln driers, cabinet driers, and tunnels with truck and trays are conventional dehydrators for fruits. Continuous-belt conveyor driers are used for apple slices and shredded coconut. Drum driers are used for drying bananas and fruit puree; and freeze driers have been used for drying strawberries, bananas, and blueberries. All of these have been treated to inactivate enzymes and/or microorganisms and some are aseptic at the time of packaging, which greatly extends their shelf life.

Fumigation with methyl bromide is often used to prevent insect infestation during storage and before packaging. In addition to fumigation, antimycotic agents (fungistats) are used to stabilize moist prunes and figs against mold growth at 30–35% moisture. Sorbic acid and sorbate salts are used as dips or sprays to prevent molding; sulfur dioxide or sulfite salts are used to preserve fruits during drying from color changes and browning, and to ward off insects. Storage of products at semitropical or summer temperatures requires residual sulfite to prevent darkening and flavor bittering, and to make the dried fruit a less favorable medium for growth of microorganisms. Ionizing radiation also prevents growth of all kinds of organisms.

Darkening and loss of flavor are the major types of deterioration of dried fruits in storage. Typically, their storage life at 70°F is about four times that at 90°F (Hoofnagle 1971B).

Raisins

At 32°–40°F and 50–60% RH, sugaring in raisins is prevented for 1 year, provided the moisture content of the dried fruit is not unusually high. Raisins contain 15–20% moisture, and for extremely long storage the lowest possible moisture content should be maintained.

Insects in raisins are now controlled by fumigation with methylbromide or phosphine. Storage in a low-oxygen atmosphere has been found to destroy insects in grain and naval oranges has without the use of toxic chemicals. Use of low-oxygen atmosphere is a less hazardous control method and does not increase pollutants in the atmosphere. Raisins stored in low-oxygen atmosphere for 1 year at 18.50 or 27°C were found equal or superior in flavor quality and acceptability to similar products stored in normal atmosphere for the same length of time. It was concluded that an insect-controlling, low-oxygen atmosphere could be used instead of fumigation, without changing the natural flavor characteristics of the product, except those normally occurring during storage at the same temperature (Guadagni *et al.* 1981).

Figs

Figs may be held for a year at 32°–40°F at 50–60% RH. A temperature of 55°F or lower prevents darkening for more than 5 months; low humidity controls sugaring.

Prunes

Prunes may be held for 1 year at 32°–40°F at 50–60% RH. For storage of 4–5 months, a relative humidity of 75–80% is not detrimental.

Apples

At 32°–40°F and a relative humidity of 55–65% dried apples retain excellent color and texture for more than 1 year. A relative humidity of 70–80% is not objectionable at 32°F, but at 40°F enough moisture may be gained to cause the fruit to mold within 8 months. Browning develops gradually at temperatures of 40°F and above.

Pears

Storage conditions for dried pears are the same as those for dried apples.

Peaches

Sun-dried freestone peaches are harder to store than most dried fruits. Therefore, the temperature should be held close to 32°F at 55–65% RH. At 40°F and a moderate humidity the moisture pickup is sifficient to cause molding and browning to develop rapidly.

Clingstone peaches (dehydrated after steam scalding) should be stored at 32°–40°F and at 55–75% RH. Sun-dried peaches will tolerate a slightly higher humidity than dehydrated peaches.

Apricots

Dried apricots kept in refrigerated storage at 32°–4040°F with a 55–65% RH remain in excellent condition for more than 1 year. At 40°F with moderate humidity there is enough gain in moisture content to cause molding.

Dates (Sucrose or Hard Type)

For storage of 6 months or less dates may be held at 32°F with 70–75% RH. But for longer storage they should be stored at 24°–26°F. Usually, it is more convenient to store them at 0°F; at this temperature they can be stored for over 1 year.

Soft or invert-sugar type dates may be held for 6 months at 28°–32°F, but if stored for 9–12 months they should be stored at 0°–10°F. Uncured dates should be stored at 0°–10°F.

Fruit Powders

Fruit powders are prepared by vacuum, spray drying, puff drying, or foam-mat drying, contain as low as 1.25% moisture, and are usually packed in tin or glass containers, with an antioxidant, under vacuum or with nitrogen. The stability of fruit powders varies widely but most remain stable more than 1 year without refrigeration. The most important requirement is that the relative humidity be maintained at 60% or lower to protect the cans (ASHRAE 1971).

Dehydrated Applesauce

Dehydrated applesauce is not a new product, but the explosion-puffing process used in its preparation is a recent innovation. Puffed-dried apple pieces have a partially porous structure (as do freezedried) and a partially compact structure (as do air-dried). Comparative porosity (surface connected pores as a percentage of total piece volume) is 88.2% for freeze-dried pieces, 61.5% for puffed, and 10.7% for air-dried.

The various processing techniques used in preparing dehydrated apple products have a pronounced effect on their composition, structure, physical state, and storage stability. Storage tests of amorphous, glassy-structured apple juice products showed no significant difference in storage stability among products packed in air, N_2, or under vacuum. However, apple nuggets, when reconstituted to applesauce, showed markedly improved keeping qualities when packed in inert gas compared with air-packed products.

Eisenhardt *et al.* (1969) stored dried applesauce powder made from explosion-puffed apple pieces by comminuting the dried pieces and blending them with sucrose and malic acid. Initial moisture content of the powder was 1.4% and SO_2 content was 175 ppm. Results showed that N_2 packing prolonged shelf life and allowed room-temperature (73°F) storage for at least 9 months. Air-packed samples stored at 38° and 73°F developed off-flavors in 5 months. Severe caking occurred in all samples stored at 100°F but not at 73°F and below. Color changes took place in all samples, ranging from darkening at 100°F to lightening at 73°F and below.

Color of the 73°F N_2-packed samples was essentially the same as that of samples at 0°F after 9 months.

STORAGE OF FRUIT PRESERVES

Preserves, jams, and jellies are about 65% sugar and about 30% water, are usually packed in glass jars, and may be vacuum-sealed. They contain very few minerals, vitamins, proteins, and fats, but are high in natural flavors and colors. It is the latter, along with texture, that it is hoped to retain during storage for 1–2 years.

Cecil and Woodroof (1963) included four kinds of preserves—apricot, grape, peace and plum—in a 7-year study of storage of foods at 100°, 70°, 32°, 0°, and −20°F. The foods were stored constantly at these temperatures for 1 and 2 years and some of them were alternated to other temperatures for storage. The results are summarized below.

Appearance

Major changes affecting appearance included darkening and masking of natural colors; syneresis; and growth of tartrate crystals in grape jam or increase of a light, relatively soft and smooth-textured precipitate (sugar separation) in plum jam. Changes in 7 years of storage at 32°F or below were limited mostly to crystal formation. Tartrate crystals appeared in grape jam within the first year, and continued to grow slowly in size throughout the storage period. The soft-crystal phenomenon was observed in plum jam during the fourth year.

Syneresis (separation of syrup) occurred in each kind of jam, increasing with storage beyond the fourth year at 32°F, the third year at 47°F, or the first 18 months at 70°F. Syneresis was moderate after 6 months at 100°F, with a pronounced separation of syrup occurring in these samples in 1 year.

Color

Only slight changes occured in the color of jams stored 7 years at 0° or −20°F, although grape jam faded and "blued" to a moderate extent. Changes at 32°F consisted of slight browning of peace and apricot jams after the fourth year, with considerable fading and moderate darkening of plum and grape jams during the sixth year. Storage for 3 years at 47°F was roughly equivalent to 5 years at 32°F for each of the jams.

Each of the jams darkened slightly within 6 months at 100°F or 1 year at 70°F. Grape and plum jams lost natural colors by darkening; the resulting blue-brown in grape and dull brown in plum jam were unacceptable in 6 months at 100°F, or 2 years at 70°F.

Aroma

Fruit aromas tended to disappear, leaving the jams bland and "lacking" after the third year at 47°F or the fifth year at 32°F, but there was very little change in jams at 32°F and below. While retention of fresh aroma was generally better in apricot and peach jams than in grape and plum jams, each lost practically all of the fresh fruit aroma, and each smelled "burned" or caramelized, after 6 months at 100°F or 18 months at 70°F.

Texture

Texture changes were minor in apricot and peach jams, although firmness increased with increased syneresis in all jams at higher temperatures. The fiber and skins of apricot and plum jams also became tougher with storage at 70° or 100°F. The most serious changes were graining of grape jam after the first year, and of most of the plum jam at 47°F and below during the fourth year. Crystals continued to increase in size and amount in both of these jams after the periods noted.

Flavor

Natural fruit flavors were lost slowly, becoming noticeably weaker after the third year at 32°F or the fifth year at 0°F and lower. Flavor became slightly strong after 3 years at 47°F and slightly burned after 2 years at 70°F or 6 months at 100°F. Strong flavors were less after 2 years at 70°F or 6 months at 100°F. Strong flavors were less objectionable in peach or apricot jams than in grape and plum products, because the natural flavors of peach and apricot were more pronounced and lost less readily.

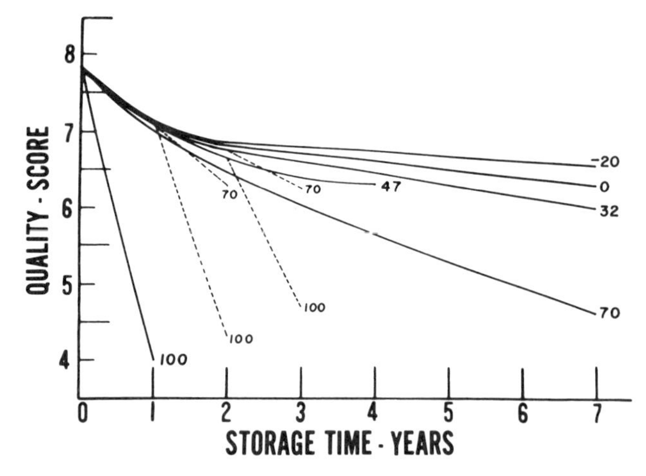

FIG. 14.1. Evaluation of storage quality of peach jam.

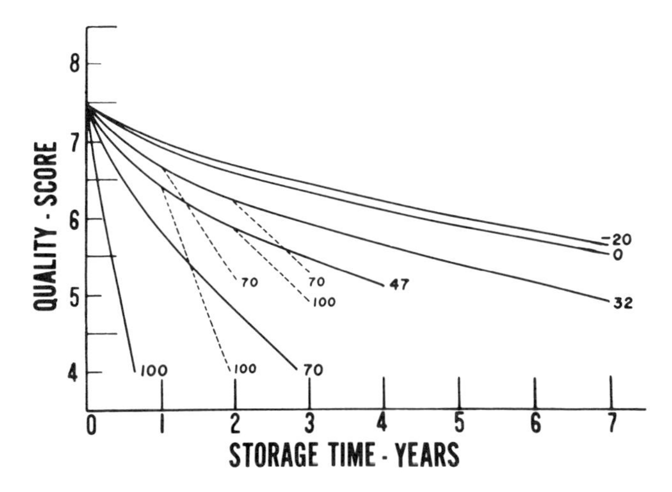

FIG. 14.2 Evaluation of storage quality of grape jam.

Summary

Based on tendencies for crystallization, storage of plum jam should be limited to 4 years, and of grape jam to 2 years, at any temperature. In order to prevent discoloration, jams should not be stored at 100°F for more than 6 months. However, at 32°F or lower, jams may be safely stored for 7 years.

Data in Figs. 14.1, 14.2, and 14.3 show the stability of peach jam, grape jam, and plum jam during long-time storage. Part of the samples that were stored at −40°, 0°, 32°, and 47°F were moved to 70° and 100°F after 1 and 2

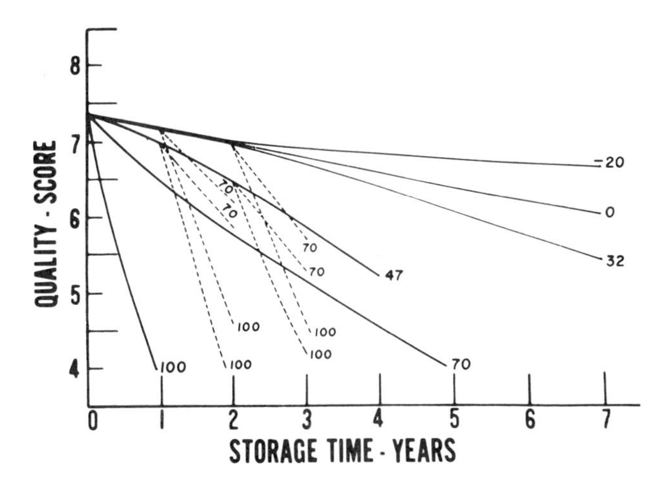

FIG. 14.3. Evaluation of storage quality of plum jam.

years storage. It is noteworthy that the rate of deterioration in quality was about the same at 100° and 70°F irrespective of whether the jams previously had been stored at −20°, 0°, 32°, or 47°F.

CONTAINERS

The durability of cans, and metal lids of glass bottles and jars, is a primary factor in determining the storage life of canned fruits and juices.

Cecil (1971A) tested cans for leaks by noting the emission of bubbles from cans immersed for 2 min in water at 110°F. Estimates of questionable leakers were based on examination of the interior of can surfaces and oxygen contents; cans having red rust on inside seam surfaces or a higher oxygen content than companion nonleaking cans were considered questionable leakers.

Cans were also tested for leaks by analysis of the headspace. Upon removal from storage rooms, cans were left 5 days at 73°F to equalize the temperature of the contents, then examined for evidence of imperfect seams or gross physical damage. Each was then punctured with a gas-sampling device to remove headspace gas samples for analysis of oxygen, nitrogen, carbon dioxide, and carbon monoxide contents by a gas chromatographic procedure. Headspace oxygen decreased from an initial 19.7% to 19.0% in 12 months at 68°F and to 16.5% in 12 months at 104°F. Corresponding values for carbon dioxide were 0.35 and 1.76%; for carbon monoxide 1.13 and 1.86%.

Maximum storage may depend as much on the container as the product. Cans containing acid foods, including most fruits and juices, may show can failure. Acid may eventually perforate the can and cause leaking, or hydrogen gas may form and swell the can ends. This is rather unusual. Such cans, while not a health hazard, should be discarded by the consumer or returned to retail stores.

Can swells may also appear during storage owing to overfilling or filling at lower than usual temperature. This type of swell is not directly related to time of storage. Some foods packed in glass containers may suffer color quality losses due to exposure to light.

Care should be used in selecting the proper can lacquer for the respective fruits. Can corrosion may produce bleaching of the fruit color. This effect is sometimes considered advantageous, as in the case of applesauce. Here, the slight amounts of tin that dissolve keep the product light in color; the same product packed in glass containers tends to darken in color, which is usually considered a loss in quality. In other instances, bleaching of the product is decidedly objectionable, such as in the case of canned cherries.

While manufacturers produce cans designed to be suitable for specific fruits or juices, individual canners or processors decide on the particular type of can or bottle, with metal lid, for their use, sometimes allowing their choices to be dictated by desire for economy. Some cans, therefore, are more likely to result in product degradation sooner than others, even though their contents remain edible and do not constitute a health hazard.

Sterilized canned fruits and juices have a long shelf life, but it is not economical for the goods to occupy warehouse and store shelves for long periods of time. The dates of manufacture, of storage, of shelf display, and consumer rotation are all important. Canners currently code date the ends of each can produced, and thus have adequate control over the history of their products up to a point. Storage of these products should be below 80°F at all times and preferably below 70°F. There is a growing demand for storage of all canned fruits in dry caves or in a warehouse cooled to 65°F or lower for long-time storage.

There are no uniform state or federal laws requiring open dating of canned fruits; even where canners follow the practice this in itself cannot regulate or assure proper handling with respect to quality maintenance. Code dating is effective when used as a tool by a concerned and profit-motivated industry to prevent old stocks from reaching the consumer.

The shelf-life of canned fruits varies widely from company to company. The overall quality is time- and temperature-dependent, and there are no distinct storage requirements for all processors to follow.

Manufacturers of tin cans, aluminum cans, glasses, and flexible containers are constantly testing and improving the type, shape, and kind of containers for specific fruits and juices; and fruit processors should keep abreast of these development.

Some manufacturers of glass and flexible containers make slightly different styles of containers each year for jellies, preserves, and other products. "New products" require distinctive kinds and styles of containers. Closures for fruit and juice containers also are changed constantly.

Information on containers for fruits and fruit juices may be obtained from one or more of the following:

- Alcoa Closure Systems International, Division of Aluminum Company of America, Richmond, IN 47374
- American Frozen Food Institute, Burlingame, CA 94010
- Brockway Containers, Inc., Brockway, PA 15824
- Can Manufacturers Institute, 1625 Massachusetts Ave., Washington, DC 20036
- Custom Metalcraft Processing Equipment, Springfield, MO 16580
- Florida Citrus Processors Institute, Winter Haven, FL 33880
- Food Machinery Corporation, Foods Products Division, 2000 Market St., Philadelphia PA 19103
- Food Processors Institute, 1133 20th St., Washington, DC 20036
- Glass Packers Institute, 2000 L. St., Washington, DC 20006

- Menasha Corporation, P.O. Box 508, Watertown, WI 53094
- Michigan Frozen Food Packers, South Haven, MI 49090
- Michigan Food Processors Association, Shelby, MI 49555
- National Cherry Growers, Inc., Corvallis, OR 97330
- National Food Processors Assoc., 1133 20th St., Washington, DC 20036
- New Jersey Food Processors, Egg Harbor, NJ 08215
- Northwest Food Processors, Portland, OR 97201
- Ohio Food Processors Assoc., Worthington, OH 43085
- Packaging Machinery Manufacturers Institute, Washington, DC 20006
- Penn. Food Processors Assoc., Harrisburg, PA 17101
- Pineapple Growers Assoc., 1902 Financial Plaza, Honolulu, HI 98613
- Plastics, Inc., St. Paul, MN 55164
- Processed Apples Institute, Hardwood Station, Littleton, MA 01460
- Reynolds Aluminum Company, Can Division, Richmond, VA 25237
- Texas Food Processors Assoc., Box 341, College Station, TX 77841
- U.S. Caps and Closure, Inc., 7107 Higging Road, Chicago, IL 60656
- Wisconsin Canners and Freezers, 110 E. Main St., Madison, WI 53703

Caramel-Caused Corrosion

Corrosion problems in canned beverages are of concern to the canner-packer, container manufacturer, and steel maker. The role of sugar breakdown products in detinning reactions on unenameled tinplate containers has been observed for many years. Caramel, which is a complex, dark-colored material resulting from heating or "burning" sugar under a variety of conditions, has been studied in connection with corrosion.

Three basic methods may result in production of caramel: (1) dry heating; (2) melting and refluxing; and (3) heating sugar solutions in the presence of acidic and basic catalysts. Disaccharides or monosaccharides are dehydrated and fragmented by heat into simpler but reactive compounds. Rearrangements, condensations, and polymerizations occur, resulting in high molecular weight dark substances and small quantities of such compounds as triose sugars, acetaldehyde, pyruvic aldehyde, acetal, crotonaldehyde, and many others. It is clear from the literature that caramels differ considerably in chemical characteristics that are not well understood.

Commercial caramels vary markedly on their attack rates on pure tin. All caramels in dilute acids attack tin more rapidly than do the acids alone adjusted to the same pH level. Small quantities of pyruvic aldehyde in dilute acid systems have been found to be extremely corrosive to tin. Variation among caramels in the amount of this compound or its labile precursors, glyceryl aldehyde or dihydroxyacetone, probable accounts for the differences in corrosivity encountered. Pyruvic aldehyde, which is so corrosive to tin, has little effect on steel. On the other hand, acetal and crotonaldehyde added to malic acid buffer solutions effectively inhibit steel corrosion but have no effect on tin. These compounds or their condensation products may well be responsible for the inhibition of steel corrosion in caramel systems.

Willey and Reese (1969) showed that some caramels allow rapid attack on poor steels yet inhibit corrosion on average and superior steels. The difference probably lies in variation in the inhibiting components of the caramels, some of which are selective with respect to surface adsorption on the steel. The findings suggest that the corrosion characteristics of caramel may be an important factor determining corrosivity of colas.

LIGHT DETERIORATION OF FOOD

Light accelerates the development of rancidity of fats and oils and lowers the vitamin C content of foods. Practically all foods darken in the presence of light.

Light is essentially an electromagnetic oscillation of wavelengths between 4000 and 7000 Å. It is radiation energy that is transferred to the radiated object and may cause photochemical reactions. Infrared radiation occurs above 7000 Å; wavelengths shorter than 4000 Å are called ultraviolet. Neither of these portions of the spectrum is perceived as light by the human eye. Ultraviolet radiation is responsible for many cases of discoloration and product spoilage.

The intensity of light is a significant factor in causing discoloration. Artificial sources of light have varied intensity. A stronger flow of light and a stronger bleaching effect is produced by fluorescent lights as contrasted to ordinary incandescent light bulbs. If the two sources of light are of the same strength, light radiation will be equalized. An additional factor of importance is that incandescent bulbs radiate heat and may cause discoloration due to a temperature increase. There is also a considerable difference between fluorescent lights and incandescent bulbs in their ability to reproduce the natural colors of the product. This is due to differences in spectral characteristics.

The length of time an object is illuminated is a significant factor in discoloration. In all cases, the period of illumination should be as brief as possible. Many supermarkets turn off all showcase lights at night to reduce excess product fading.

Distance between products and the light source is also an important factor in discoloration. If the distance is reduced by one-half, the effect of the lighting becomes four times stronger. The effect is reciprocally proportional to the square of the distance.

Other factors of importance include humidity, oxygen content, and specific wavelength of the light. Discoloration may even occur in complete darkness, but will not be as rapid and intense as when a product is exposed to light.

There are three basic classes of packaging materials relative to light transmission—transparent, opaque, and translucent. Glass and clear plastic films are transparent; aluminum foil and tin are opaque; and paper is translucent.

Plastic films vary considerably in their capacity to transmit ultraviolet and infrared radiation. The transmission rate of ultraviolet light (2537 Å) is highest through polyethylene and lowest through polyester film. Cellophane transmits almost as much ultraviolet as does polyethylene. If necessary, a film that has a high transmission may be modified by tinting with various dyes or adding certain chemicals.

The infrared transmission of various plastics may also be modified by suitable tinting. Dark colors absorb heat more readily than light ones. Aluminum foil is an excellent heat reflector and tends to keep products cooler than ordinary materials. Also, it protects products from ultraviolet radiation.

Many food substances are adversely affected by light and oxygen. In order to prevent light deterioration, opaque or amber-colored containers are generally used. The degree of protection depends upon the extent to which glass excludes light of the wavelengths to which the contents are sensitive. Amber glass used commercially for blown bottles (2 mm thick) excludes practically all light of wave lengths lower than 4500 Å. The corresponding figure for green glass is about 3500 Å. As the thickness and glass distribution vary, light filtration is affected.

The question as to which specific wavelengths of light cause the most damage to food products has never been fully resolved. Some researchers state that the entire light spectrum is damaging. Others feel that only the ultraviolet portion is responsible for food spoilage. Flavor and odor changes are usually brought about by ultraviolet radiation between 3250 and 4600 Å.

STACK BURN

The following information on stack burn is taken from Shallenberger and Pederson (1980):

> Stack burn is a form of deterioration occurring occasionally in all types of processed foods. When stack burn occurs, foods must often be sold at reduced prices, or be discarded.
>
> The most familiar form of stack burn occurs in the warehouse where cases of foods are stacked close together after processing, leaving little room for ventilation and cooling. The cases in the center of the stack may then hold their heat for several days;

the product discolors, and develops a disagreeable flavor. The can itself may be attacked. It is now recognized that stack burn has many ramifications beyond discoloration and off-flavor development due to improper cooling and warehousing. Prolonged storage periods at relatively low temperatures may produce similar effects.

Adequate control of stack burn requires identification of the various avenues stack burning may take. Classification of some deteriorative chemical reactions which may take place in canned foods resulting in stack burn is as follows:

- I. Browning reactions
  - 1. Sugar, amino acid reactions
  - 2. Caramelization reactions
  - 3. Oxidative reactions
    - a. Enzyme catalyzed
    - b. Nonenzymatic
- II. Natural pigment reactions
  - 1. Reactions involving chlorophyll
  - 2. Reactions involving the anthocyanins

These reactions may occur singly, or stack burn may involve two or more of these reactions occurring simultaneously. To illustrate these reactions in canned foods, it is more convenient to treat first the color aspects of these reactions. Later, their flavor implications will be discussed.

The browning reactions are characterized by the development of pink, yellow, brown, and even black pigments, usually undesirable in canned foods, but not always. Sugar—amino acid reactions have received the most attention. The "pinking" of applesauce in the stacks and the browning of sauerkraut are probably due to sugar-amino acid reactions. The color of maple syrup is due to sugar-amino acid reactions, and in very dark and bitter maple syrup, the reaction has gone too far. While not pertinent to this discussion, the colors of bread crusts, of dried cereals, coffee, and toast are also attributed to these reactions and serve as illustrations.

Carmelization reactions, while included in the classification, seldom, if ever, are involved in stack burn. Carmelization is the heat degradation and polymerization of pure chemical substances resulting in the formation of "caramels." However, stack burnt products are often described as being "carmelized."

Oxidative browning reactions may or may not be enzyme catalyzed. These reactions are usually associated with the action of oxygen on the tannins in the food. The pinking of apple slices exposed to the air and prior to blanching, and the red color which develops in peeled potatoes are examples of enzyme catalyzed oxidative browning. The enzyme is not entirely necessary, however. The reaction will proceed in blanched material in the presence of oxygen, although at a much slower rate. Sterilization, with inadequate deaeration, resulted in the problem of "black neck" in bottle catsup. This form of stack burn is largely unknown today. When identified as nonenzymatic oxidative browning, adequate deaeration procedures eliminated the problem.

The natural plant pigments may be subject to chemical change indicative of stack burn. Upon heating or prolonged storage chlorophyll readily converts to pheophytin, which is brown. The anthocyanins, which are red, blue, or purple, decompose with heat to yield anthocyanidins, which are also brown. Anthocyanin degradation may be observed in strawberry preserves and jams, or other fruit products.

Flavor reactions are concomitant with developing color in the browning reactions. Many of the browning reaction intermediates possess flavor. During sugar—amino reactions, flavorful compounds are generated by amino acid decomposition and sugar fragmentation. Diacetyl, a major component of butter flavor, is generated in sugar—amino acid reactions also. While the brown pigment itself is invariably bitter, a great variety of flavors may be produced. The flavors forming depend on the nature of the amino acids reacting with the sugar. This knowledge led to the finding that the formation of pyrrolidone-carboxylic acid (PCA) from the tasteless amino acid, glutamine, may be an index of stack burn prior to the onset of color formation.

To determine the course of stack burn in a given processed food quantitative measures are required. This may not be a simple matter, for the course which the defect has taken is not always easily determined. Overall measures may involve the estimation of the brown color intensity. Absorbance measures of extracts at wavelengths just over 400 mμ have been used. But these brown pigments do not form true solutions and do not follow established spectrophotometric laws. Three-dimensional reflectance measures have also been used, but occasionally too much color develops to be measured even in this way. Furfural, a sugar dehydration product, forms as an intermediate in color formation and has been used as an index of stack burn.

Previous knowledge of the chemical make-up of many foods is required to suggest the course which stack burn has taken. If processed foods are known to contain, in the raw state, considerable amounts of sugars and amino acids, these may be suspected. Demonstration requires measures of changes in the amino acids and sugars and correlation of these changes with the pigment developed.

Tracing oxidative browning may require measures of oxygen up-take. This may be difficult since the amount of oxygen absorbed during pigment formation may be less than that required for the most sensitive oxygen measures. Measures of changes in easily oxidizable substances, particularly of the vitamin ascorbic acid or the amino acid tyrosine, may be an index of oxidative browning. Discoloration due to natural pigment changes may be followed by measuring changes in the chlorophylls and anthocyanins. Absorbance measurements at 540 mμ may indicate anthocyanin destruction.

In addition to inferior colors and flavors resulting from stack burn, the nutritional value of a food may be impaired. It has been demonstrated that the ten amino acids essential to the diet participate in amino acid—sugar browning reactions resulting in loss of biological activity. This is especially true of lysine, which may be destroyed even while still combined in protein form. Thiamin, or vitamin B_1, may react with sugars since it has an active amine group, and the sugar-amino acid browning reaction partly explains the heat sensitivity of this nutrient.

When oxidative browning occurs, one may be assured that the easily oxidizable vitamins A and C have been partly destroyed. In some foods, oxidative stack burn does not become apparent until nearly all of the vitamin C has been oxidized.

Stack burn may be controlled. Beginning with the raw product, the cultural and storage practices which lead to the excess accumulation of known precursors to stack burn should be avoided. Careful preprocessing procedures, such as adequate blanching, exclusion of air, and deaeration procedures minimize both enzyme catalyzed and nonenzymatic oxidative reactions. Additives such as sulfur dioxide retard sugar-amino acid reactions, while ascorbic acid retards oxidative reactions. Unduly high processing temperatures and prolonged heating will increase the possibility of stack burn. Proper water cooling after sterilization followed by efficient air cooling will minimize storage changes. Since prolonged storage at normal temperature also results in stack burn, low temperature storage is desirable.

PRODUCT CODING

"Open dating" legislation requires marking "pull dates," "inspection dates," and "packing dates" on a wide variety of food products. Current consumer attitudes also require such age indications on products that change in acceptability with time.

A plant registration number is now required by a few states and for plants under USDA inspection. It may be required universally in the future for all food plants to aid regulatory and legal proceedings.

Product recalls are totally dependent on detailed production information. Product recalls are not the only reason for production codes. Investigators can trace complaints and eliminate their causes when the plant and time a product was made are known. The circumstances can be studied with the aid of production records. Distribution and inventory patterns are also aided by legible codes. And proper product rotation (first in, first out) is facilitated.

Coding each individual package involves costs for material, equipment, and maintenance, as well as numerous new technical problems. Problems include transfer or application of symbols on hot, soft, frozen, wet, brittle, and moving surfaces. As a last step in the development of codes, systems, and equipment, fail-safe devices will be needed to assure positive or foolproof coding.

Proper coding allows a manufacturer to study and control shelflife of a product during marketing. Occasional lots, which appear acdeptable during initial quality control tests, may show product defects during marketing. Proper coding allows their segregation and recall, if necessary.

Codes provide a means of tracing consumer complaints to establish their validity. They also offer legal protection from false claims and let a manufacturer discover faults due to improper production, distribution, or retailing. Codes can be used to provide data regarding the market turnover, sales volume, or movement of new products. Some food products must be coded to comply with federal, state, or municipal regulations (Anon. 1972).

FDA AND CONSUMER PROTECTION

The U.S. Food and Drug Administration has the mission of protecting consumers in certain defined areas—adequacy of controls exercised by the processor during processing and safety of the finished product on the market. Since responsibility for quality, including safety, is with the processor, his interest and that of FDA lie in the same direction. It is possible to inspect processing practices without inspecting the quality of products, but hardly possible to inspect for quality without knowing the processing procedures. A revised FDA field program consists of a greatly expanded testing program on finished products picked up at retail.

Food additives are not permitted in the food supply until after their safety has been demonstrated to the FDA. A major thrust is to document and, if necessary, further test the safety of all "generally recognized as safe" (GRAS) products. Testing is being done on both old and new additives.

FDA activity is being stepped up on (1) screening foods for harmful additives that might cause cancer such as cyclamates; (2) examining foods for water pollutants such as mercury, DDT, and other residues; and (3) tightening the bacterial limits in foods to include all pathogenic organisms. There is a growing movement in this direction on an international level and tentative count limits have already been arrived at for some of the more vulnerable products (Wodicka 1972).

REFERENCES

ANON. 1972. Tips on product coding. Food Eng. *44*(12) 63–64.

ANON. 1974. Shelf life of foods: A scientific status summary—IFT Expert Panel on Food Safety and Nutrition. J. Food Sci.*39,* 861.

ASHRAE. 1971. Guide and Data Book, Applications. pp. 399, 425. American Society of Heating, Refrigeration, and Air-Conditioning Engineers.

BREKKE, J. E. TONAKI, K. I., CAVELETTE, C. G. and FRANK, H. A. 1970. Stability of guava puree concentrate during refrigerated storage. J. Food Sci. *35,* 469–471.

CECIL, S. R. 1971A. Storage stability of civil defense shelter rations. Univ. Ga. Tech. Rep. *153-IX-FS*.

CECIL, S. R. 1971B. Accelerated storage tests of cereal ration. Univ. Ga. Tech. Rep. *153-X-FS*.

CECIL, S. R. and WOODROOF, J. G. 1962. Long-term storage of military rations. Ga. Exp. Stn. Tech. Bull. *N.S. 25*.

CECIL, S. R. and WOODROOF, J. G. 1963. The stability of canned foods in long term storage. Food Technol. *17*(5) 131–138.

EISENHARDT, N. H., CALHOU, M. J., TALLY, F. B. and DELLAMONICA, E. S. 1969. Storage properties of dehydrated applesauce made from explosive puffed pieces. Food Technol. *23,* 159–162.

GOMEZ, R. F. and BATES, R. P. 1970. Storage stability of freeze-dried avocado puree and guacamole. J. Food Sci. *35,* 472–475.

GUADAGNI, D. G. 1961. Time-temperature experience of food quality. ASHRAE, *3*(4) 66–69, 83.

GUADAGNI, D. G., STOREY, C. L. and SODERSTROM, E. L. 1981. Effect of controlled atmosphere on flavor stability of raisins. J. Food Sci. *43,* 1726–1728.

HOOFNAGLE, W. S. 1971A. Food Stability Survey, Vol. 1. Rutgers Univ., New Brunswick, NJ.

HOOFNAGLE, W. S. 1971B. Food Stability Survey, Vol. 2. Rutgers Univ., New Brunswick, NJ.

KRAMER, A. 1974. Storage retention of nutrients. Food Technol. *28,* 50–58.

LEE, F. A., ROBINSON, W. B., HENING, J. C. and PEDERSON, C. S. 1950. Low temperature preservation of fruit juices and fruit juice concentrates. N.Y. State Agric. Exp. Stn. Bull. *743*.

SACHAROW, S. 1969. Light as a factor in food deterioration. Food Prod. Devel. *3*(5) 67–68, 72.

SHALLENBERGER, R. S. and PEDERSON, C. S. 1960. Stack burn. N.Y. State Exp. Stn. Farm. Res. *26*(3) 4–5.

WILLEY, A. R. and REESE, R. E. 1969. Effect of beverage caramel on steel dissolution rates in dilute phosphoric acid. Proc. 6th Annu. Meet. Soc. Soft Drink Technol., Washington, DC.

WOODICKA, V. O. 1972. Revised FDA field inspection program. Frozen Food Age *20*(9) 23–24.

WOODROOF, J. G. and ATKERSON, I. S. 1945. Freezing cooked foods. Food Ind. *17,* 1041–1042, 1136–1138, 1179–1180, 1264–1265.

15

Plant Sanitation and Waste Disposal

N. J. Moon and J. G. Woodroof

PLANT SANITATION

There is evidence on all sides that plant owners are taking a closer look at their sanitation facilities in an effort to improve and update them. This is especially true in older plants where sanitation planning had only casual attention from engineers.

The influence of plant design sanitation seems clearly indicated by the fact that the cleanest plants are usually those of the largest packers, which are almost invariably the best designed. Smaller companies, on the other hand, seem to have a haphazard approach to sanitation simply because it was not considered in planning the total operation. Often buildings have been converted to canneries from some other use; in such cases, it is difficult to achieve an efficient sanitation setup because this usually requires changes in basic structure, in equipment layout, in production flow, and other expensive alterations.

Steam and high humidity in a conventional wood or metal frame building create an enormous problem of cleaning beams, ledges, piping, and all overhead exposed structures where dirt, insects, bats, and even birds may lurk.

Given enough space, sanitation becomes relatively simple even if it has not been considered in the design. One plant in the Central Valley of

Commercial Fruit Processing, 2nd Edition

ISBN 0-87055-502-2

California had been a massive warehouse before it was converted to a cannery. Although cleaning equipment was not particularly well placed, the overall impression was one of a sanitary plant simply because of space—wide aisles and high ceilings made it easier to avoid the clutter and mess of a small building.

Because wooden floors, which harbor bacteria, do not support forklift trucks, wood flooring started to disappear about 1945 in processing plants. The new concrete floors were loudly hailed as the ultimate material until it was noted that some fruit acids and other solutions dissolved the concrete. For several decades, various materials, ranging from asphalt to plastic coatings, have been used to protect and sanitize the concrete. One of the most durable floors is made of brick tile set over concrete with a corrosion-resistant resin.

Packers have become more conscientious about cleaning up the external plant environment. This is true even in rural areas because civic and environmental organizations are beginning to demand that processors maintain such areas in reasonable condition. Outside space also is being utilized more and more to eliminate extraneous vegetable material that was once brought into the plant. For example, western green pea processors a few years ago set up viner stations with cleaning and washing equipment. Much of the raw product is now already cleaned, washed, and cooled and is generally in fine condition before it reaches the plant. Often precleaning and washing equipment for handling raw products has been installed on slabs outside the plant where pods, sticks, vines, pebbles, snakes, insects, and other objectionable materials are sorted out. This makes the jobs of plant sanitarian and quality control much easier. Raw product awaiting processing may be treated to control ripening and reduce bacterial growth and stored in temperature- and humidity-controlled rooms.

Many canning and packing plants appear to have the most difficulty in maintaining proper sanitation in the preparation department. Spillage, drippage, overflow, and discarded portions of the food under preparation call for more efficient means of control and disposal. Between cleanup efforts, such spillage attracts flies and other germ-carrying insects. This department is often especially vulnerable to such invasion because of its proximity to the outside receiving area.

In recent years, packers have been trying to overcome the mistakes of floor design by making tremendous efforts to facilitate cleanup, especially in processing areas. A few years ago it was considered quite modern to build a big square or rectangular plant with parallel rows of processing lines stretching nearly from one end to another. But it was soon found that one type of building did not suit all products nor did it accommodate newer types of equipment—like special cookers, sorters, and graders—that take up tremendous room and must be built in place.

New types of sanitary pumps and conveyors make it possible and sometimes desirable to lift the product off the floor in a multilevel operation. Often, this has resulted in much of the equipment being placed above the main floor level, so that the floor surface is accessible for easy clean-up. Washers, cleaners, blanchers, graders, and inspection belts are often mounted on steel scaffold structures high above the floor so that product droppage and food residue can be easily cleaned up under the stilt legs.

Plants are now being designed around a particular production line, sometimes with three different floor elevations, so the floor level is dropped down with the flow of product. More plants are beginning to use all stainless steel for equipment that touches the product.

A feeling of cleanliness should surround workers in a processing plant. Rest room facilities and cafeteria have improved noticeably. Many managers make a fetish of keeping things looking freshly painted, believing that if the plant looks clean, the employees will keep it clean. The most important aspect of sanitation is the specific delegation of individual responsibility for every phase of sanitation. This is even more important than modern equipment, regular cleaning schedules, or inspection. When a worker or workers in a given area know that the responsibility for sanitation in their area rests squarely on their shoulders, they are sanitation-conscious all day long—not just at closing time or just prior to inspection (Anon. 1967).

LEGAL RESPONSIBILITY

Champion (1967) has written the following:

> Everyday compliance with FDA and other regulatory agencies is a universal problem of critical importance which requires a well-organized program of total sanitation.
>
> First, we must understand the principal aspect of the Food, Drug, and Cosmetic Act . . . what they are and how they work.
>
> Second, we must know and recognize the potential contamination and infestation problems inherent in our industry and in our own plants.
>
> Third, we must know how to cope with both actual and potential contamination problems, and develop and maintain a total program of effective, preventive sanitation pest control.
>
> Finally, we must accept the cost obligations by allocating enough money and sufficient manpower to establish and maintain a continuing, permanent program of preventive sanitation, maintenance, and pest control.

The principal enforcement tools of the FDA, in relation to the everyday operation of food processing plants, are citations, prosecutions, injunctions, and seizures under the Federal Food, Drug and Cosmetic Act, based primarily on those sections of the Act that concern product contamination, unsanitary conditions, and poisonous or deleterious substances.

Contamination of raw materials or finished products by insects, rodents, birds, mold, bacteria, microorganisms or extraneous matter is considered adulteration and is subject to seizure and other FDA action. Further, it is not necessary for the FDA to find or prove that a product is actually contaminated. The agency can act on the premise that contamination could or may occur under unsanitary operating conditions or with unsanitary environment.

Processors are legally obligated to control pests and vermin, maintain pest-free sanitary premises and equipment, and prevent contamination of products and raw materials. It is also the processor's legal responsibility to prevent contamination by the very same insecticides, rodenticides, fumigants, and other pesticides used for control. Interpretation of the laws and regulations governing the use of pesticides in food plants is made largely by inspectors on the scene. If, in their judgment, there is a reasonable possibility of contamination, they will collect samples of the pesticides and suspected products and, in general, collect facts and data to establish contamination, or the likelihood of contamination.

The maximum penalties for convictions under the Federal Food, Drug, and Cosmetic Act can be severe. Federal courts are empowered to levy fines from $1000 up to $10,000/count. Furthermore, the responsible official of the company may also be fined as an individual and is also subject to imprisonment for 1 to 3 years, or both.

Although the maximum legal penalties are rarely imposed, the "hidden" penalties can be disastrous. Adverse publicity almost always goes hand-in-hand with seizures, prosecutions, convictions, or injunctions. Loss of customers, loss of prestige, loss of good will, economic losses through default, decrees of condemnation and destruction, expenses of conversion, decharacterizing, and reconditioning—these are the very real and costly penalties incurred because of poor plant sanitation.

Important Components

Pest Control. A very high percentage of FDA seizures, citations, and criminal prosecutions are the result of filth contamination by rodents and insects. Therefore, preventive pest control is a highly important component of sanitation in fruit processing plants.

Chlorination. Most canneries have been using chlorination in process water and cooling water since about 1940. The use and storage of hypochlorites presents many problems, including inaccuracies of dosage due to deterioration of hypochlorite strength in storage, high cost of available chlorine, and employee safety in handling.

Many advantages can be realized through the use of chlorine gas. Chlorination with gas enables the system to be placed under vacuum, involves no gas pressure lines, and permits immediate, automatic chlorine shutoff in the event of mishap, loss of water supply, or lack of chlorine. A remote-mounted chlorine ejector is placed at the point of chlorine application, eliminating hazardous pressurized chlorine solution lines. Several remote metering tubes, ranging in maximum capacity from 10 to 25 lb/day (PPD), can be connected with pressure cookers, cooling canal, or atmospheric cooker-cooler. Through the use of air-operated thermostat solenoid valves, set to open at specific temperatures, the entire chlorination process is automatic. When the valve opens, water passes through an ejector, creating a vacuum that in turn, operates the chlorinator (Anon. 1971E).

Eliminating Smoke and Offensive Odors. A catalytic oxidation system has been developed that eliminates smoke and offensive odors by converting exhaust contaminants into harmless carbon dioxide and water vapor by chemical reaction at low temperatures (Celentane Bros. 1971). The system consists of a preheat burner, prepiped valve train, combustion air blower, inlet transition duct, and automatic controls.

Organic odors and fumes from food processing or manufacturing can be controlled by raising the process exhaust to 1400°F for 0.5 sec. Three types of equipment are available for installing in the exhaust:

1. *Afterburner*—a straightforward device consisting of a reflector-lined steel shield. Normally, it is mounted vertically with process gases entering at the bottom where the burner (and combustion blower, if used) is located. The process gases combine with the combustion products, raising the mixture to the required temperature. The hot mixture then passes through the afterburner and is exhausted to the atmosphere. Length of the afterburner determines the retention time at the purification temperature.
2. *Afterburner with Energy Recovery*—an afterburner with the addition of a heat exchanger. The hot gases leave the afterburner and pass through a heat exchanger unit having thin-wall metal tubes. The incoming process gases, to the afterburner, pass on the other side of the metal tubes. The efficiency of the afterburner and heat recovery is normally about 40%, provided the unit does not corrode and there is no buildup of oils or other materials.
3. *Thermal Regenerative System*—the same as No. 2 plus a heat-retaining, inert ceramic. This system slowly passes hot gases through the heat-retaining material for a few minutes, then gas flow reverses to use

the stored energy to preheat the incoming gases. The thermal recovery efficiency is related primarily to the surface area of the heat transfer media, although velocity of the gases and the makeup of the process gas stream influence the ultimate efficiency. The preheated gases, as they pass through the equipment, are raised to the purification temperature. The time that the gases are held at the purification temperature depends directly on the volume of the process gases and equipment dimensions.

Each of these three systems performs essentially the same fume and odor control function; and each claims particular advantages over the others. Fuel availability and usage is a key decision factor in selection of equipment. Timing and consideration of the effect pollution control regulations will have on overall plant profitability also must be considered (Mueller 1971).

WASTE DISPOSAL

Our society has become increasingly concerned with the environmental and social costs of the wastes generated during food production. Public concern, both in the United States and other countries, about the destruction of environmental quality has led to legislation that severely curtails past practices. New attitudes must be developed by every sector of agriculture to learn to live with these new constraints (Soderquist 1976).

In response to these concerns, new, more efficient processing techniques, which result in maximum benefit to all with minimum trade-offs, have been developed. Professional organizations such as the Institute of Food Technologists have made waste reduction and efficiency primary goals for new research (Table 15.1). New technologies emphasize reducing the waste load produced, as well as the alternatives and various disposal strategies available (Moon 1980).

The most common treatment of solid fruit wastes has involved disintegration by comminuting and removal of solid matter by screening, flocculation, settling, or centrifuging. Screening may be continuous or in stages, and there are many types of screens (stationary, vibrating, rotary). In most cases, such equipment is custom-built for each plant and modified for specific products, such as peaches, pears, apples, or berries. If the screening is effective and if the residue is not contaminated with spray materials, the residue may be further dewatered, dried, and used for livestock feed or fertilizer; the liquid portion has been recirculated, treated, and recycled and used for irrigation or treated to reduce the organic content so it can be disposed of in a flowing stream.

The second most common method of treating fruit waste is biological disintegration, a natural process in which microorganisms break down

Table 15.1. 1982 IFT Research Recommendations[a]

1.	Ensure food safety
* 2.	Reduce energy use and increase conservation
3.	Maintain and enhance nutritional quality
* 4.	Reduce food losses
5.	Physical chemical properties of foods
6.	Innovation and technological development
* 7.	Underutilized resources—wastes
* 8.	New sources of ingredients—wastes
9.	Microbes for nutrient production
*10.	Biomass conversion to energy

Source: Marion (1982).
[a]Asterisks indicate research needed in academia.

organic material. The resulting sludge is separated from the liquid portion by settling or centrifuge.

Biological disintegration of vegetable material takes place naturally when the temperature, acidity, oxygen supply, and proper organisms are present for sufficient time. The process can be facilitated by addition of inorganic nutrients to plant effluents. Previously, many fruit processors were allowed to dispose of plant waste into flowing streams, relying on the natural stream biologial process to degrade the wastes. However, many processing plants produced a waste load in great excess of stream capacity to oxidize it, with subsequent harmful effects on normal aquatic life. This practice has now ceased due to stricter state and federal laws in the United States and other countries, although standards can vary widely from country to country.

New Federal Regulatory System

The Water Pollution Control Act, as amended by the Clean Water Act of 1977 (Public Law 92-500), is a radically different approach to water quality control in the United States. The law prohibits the discharge of any pollutant by any person from a point source, except in compliance with requirements respecting effluent limitations and standards, standards of performance, and discharge permits.

The law sets forth as a national goal, though not as a requirement, the elimination by 1985 of the discharge of pollutants into the nation's navigable waters, and as an interim goal, the achievement, wherever possible by July 1, 1983, of water quality adequate for the protection and propagation of fish, shellfish, and wildlife, and for water-based recreation.

The amended law replaces the former federal water quality enforcement mechanisms with a new regulatory scheme. Its keystone is a national permit system—the National Pollutant Discharge Elimination System (NPDES).

Effluent Limitations. Initially, the Act required industry to meet effluent limitations attainable by application of the "best practicable control technology" currently available. Industries discharging to municipal systems must comply with pretreatment requirements and with any applicable effluent standards or prohibitions governing toxic pollutants.

The Act also directed the EPA to issue regulations setting effluent limitations based on the "best available technology economically achievable" for different categories and classes of point sources of pollution. If the EPA Administrator finds that the elimination of discharges of all pollutants is technologically and economically achievable for a category or class, the effluent limitations may so require. Industries discharging to municipal systems must comply with pretreatment requirements and with applicable toxic pollutant effluent limitations or prohibitions.

If the Administrator finds that discharges from one or more point sources, with the application of effluent limitations, would not be adequate to protect water quality for public water supply, agricultural, and industrial uses, and the protection and propagation of a balanced population of shellfish, fish, and wildlife, and to allow water-based recreation, he is directed to establish, after a public hearing, effluent limitations for those point sources designed to attain or to maintain adequate water quality. A person affected by a water quality-related effluent limitation may at a public hearing seek to demonstrate that there is no reasonable relationship between the economic and social costs and the benefits to be obtained; if he does so, the effluent limitation will be adjusted as it applies to that person.

Discharges into the territorial sea, the contiguous zone, and ocean water beyond the contiguous zone are subject to special guidelines for determining the degradation of those waters. Interim regulations for this purpose, as well as for the evaluation of ocean-dumping permit applications under the Marine Protection Research and Sanctuaries Act (Public Law 92–532), were published in the *Federal Register* on January 11, 1977.

New sources of pollution in 27 industrial categories must meet federal standards of performance that reflect the greatest degree of effluent reduction achievable through application of the best available demonstrated control technology processes, operating methods, or other alternatives. Among the 27 industry groups to which the new source standards of performance will be applied are these processing groups: meet products and rendering; dairy products; canned and preserved fruits and vegetables; canned and preserved seafood; and sugar.

Plants that discharge their wastes into municipal treatment systems are not required to secure permits under the NPDES. However, pretreatment standards are applied to pollutants determined not to be susceptible to treatment by the municipal facility or those that would interfere with its operation.

Economic Considerations. In some segments of the food processing industry, the measures necessary to achieve compliance with water quality requirements introduce added production costs at a time when many Americans are deeply concerned about the cost of food. The EPA is directed to include economic considerations in decisions on effluent requirements. The 1977 deadline for the achievement of the "best practicable control technology" currently available and the regulations issued for the achievement of the "best available technology economically achievable" are related to costs.

The age of equipment and facilities, the process employed, engineering aspects of the application of various control techniques, process changes, and non-water quality environmental impact, including energy requirements, are factors to be taken into account in the assessment of the "best practicable" and "best available" technology. The EPA administrators may consider other factors, as they deem appropriate.

The law now also includes a provision for waste classification into three control groups: toxic compounds, conventional pollutants, and nonconventional pollutants. Greater emphasis is now being placed on discharge of toxic substances (heavy metals, etc.) and thus is of less concern to food processors. Most processors fit into the second category by producing conventional pollutants. For these pollutants, biochemical oxygen demand (BOD_5), suspended solids (TSS), pH, and fecal coliforms are regulated.

Permit Procedures. The responsibility for issuance of permits to discharge waste into streams under the EPA standards resides with the states. The permits specify what substances can be discharged into streams and in what amounts. If some of the discharge requirements are not being met at the time the permit is issued, the permit must also contain a cleanup schedule. The final deadline for industrial water cleanup using "the best available technology" was July 1, 1977.

After a permit is granted, any new or increased discharges must be reported. Polluters must also monitor the amount of flow and pollutant content of all discharges averaging more than 50,000 gal./day, keeping detailed records for at least 3 years and reporting at least once a year to the state. In addition, EPA guidelines provide for state inspection of pollution sources. Permits to municipalities must require secondary treatment of sewage.

The guidelines also make broad provisions for public participation in setting limitations on discharges. When a state receives an application for a discharge, it must first tentatively decide if the permit will be issued and propose effluent limitations and a compliance schedule. The state must then publish this information in the vicinity of the proposed discharge and receive comments from interested parties for at least 30 days following the notice. Information should be available to anyone who requests it.

Once issued, any permit may be revoked, suspended, or modified if there is a change in conditions that requires a reduction or elimination of the discharge. The state is empowered to set civil penalties for violation of permit requirements and criminal fines for "willful or negligent violations" (Van Luven *et al.* 1972).

Penalties. The law provides strong penalties for violations. The violation of the provisions named above, or any condition or limitation in an EPA-issued or state-issued permit implementing any of those provisions, is subject to a civil penalty not to exceed $10,000/day; a willful or negligent violation is punishable by a fine of not less than $2500 nor more than $25,000/day, by imprisonment for not more than 1 year, or by both. A fine of $50,000/day or 2 years imprisonment, or both, is the maximum penalty for a second conviction.

Waste Reduction through Increased Efficiency

The most useful approach in waste reduction and increased efficiency is to change the production or processing of food. Postharvest technology, the first step in maintenance of food value, has increased the quality of foods reaching the consumer. These changes can include mechanical harvesting, cleaning, and controlled-atmosphere storage of products before they reach the processing plant.

In every sector of the food processing industry changes in processing schemes have resulted in more product reaching the consumer with less waste. Each particular food industry generates specific types and amounts of wastes. For example, the fruit and vegetable industry generates much more solid waste than does the dairy industry (Table 15.2). Any changes in processing must address the problem of each industry.

Waste Characterization. The first stage in the optimization of waste reduction schemes is to identify and characterize the wastes (solid and liquid) produced. This has been the subject of numerous reports in the scientific literature (Bough 1973; Moon 1980; Flora and Lane 1978). Data in Table 15.2 show the relatively higher amount of solid waste that vegetable processing produces compared with dairy or meat processing. Data in Table 15.3 characterize the waste load of various processed fruits and indicate the problem of suspended organic matter in the waste water.

Table 15.2. Solid and Liquid Wastes[a] Generated during Processing of Food

Processed food	Total solid waste (g/kg)	Liquid volume (m^3/kg)	BOD_5 (mg/kg)[b]
Vegetable			
Kale	16	0.004	11,000
Spinach	20		11,000
Mustard greens	16		10,000
Turnip greens	15		9,000
Collards	13		8,000
Potatoes	66	0.012	44,000
Peppers (lye peel)	65	0.020	33,000
Tomatoes (lye peel)	14	0.010	
Dairy			
Cheese whey		9.000	270,000
Skim milk		0.070	1,500
Ice cream		0.080	3,000
Meat			
Red	0.440	25.000	14,000
Poultry	0.270	50.000	13,000
Eggs	0.111		

Source: Moon (1980).
[a]Waste loads calculated per unit weight of product.
[b]BOD: biological oxidation demands.

From such information on individual processing plant trials, it is possible to identify the waste load produced in processing various vegetables. The magnitude of the problem is only apparent when the volume of wastes produced is considered.

The weight of tree fruits peeled by chemical methods is estimated at 1 million tons/year. Since each ton of apricots, peaches, or pears processed produce about 12 lb of BOD and 9 lb of suspended solids in rinse water, the total liquid waste is high. Solid waste from peels and cores can also be high. For example, over 1800 million lb of apples are canned, dried, or frozen in the United States annually. The amount of peels, cores, or trimmings and culls at processing plants varies widely with the maturity, size, and grade of apples, as well as the operating conditions. It has been estimated that 500 million lb of apple peels and cores accumulate as waste.

Processing Changes. The most feasible way to reduce wastes associated with food processing is adoption of practices and technologies that result in significant recovery of a salable product. Again it is useful to compare technologies used in different industries (Table 15.4). The most significant change in fruit and vegetable processing has been lye peeling

Table 15.3. Wastes from Processing Fruits

		Waste water		BOD		Suspended solids		Solid residuals	
	Raw tons (1000 tons)	(1000 gal per ton)	(Million gal)	(lb per ton)	(Million lb)	(lb per ton)	(Million lb)	(lb per ton)	(1000 tons)
Apple	1,000	5.0	5,000	40	40	5	5	600	320
Apricot	120	5.0	600	60	7	12	1	360	21
Cherry	190	2.0	400	20	4	6	1	300	27
Citrus	7,800	3.0	23,000	4	31	7	55	880	3390
Peach	1,100	4.0	4,400	60	66	10	11	500	270
Pear	400	4.0	1,600	70	28	20	8	660	120
Pineapple	1,000	0.5	500	20	20	8	8	900	450
Other fruit	400	8.0	3,200	20	8	10	4	—	80
Total	12,250		38,700		200		90		4680

Table 15.4. Technologies That Result in Increased Yield and Decreased Waste Production

Industry	Subclass	Process	Increased yield
Meat	Beef	Protein-enzyme tenderization	39%[a]
Dairy	Cheese	Ultrafiltration	16–20%
Fruit and vegetables	Peaches	Lye peeling	Marginal
	Pears	Lye peeling	Marginal
	Potatoes	Lye peeling	12–40%
	Pimiento peppers	Lye peeling	30%
	Apples	Lye peeling	6%

Source: Moon (1980).

[a]The increase in yield quoted here is an increase in meat acceptable for dry heat cooking from 29% in a conventional carcass to 68% in a Pro Ten carcass.

or dry-caustic peeling. In this method the surface of the product is conditioned with infrared energy at 1650°F, sprayed with 20% lye at 170°F for 50–100 sec, and subjected to infrared heat for 2–5 min, followed by a holding period of 5 min. The peel is removed by rotating rubber rolls and can be used as an ingredient of cattle feed. The surface of the product is rinsed with a minimum amount of water and weak acid to neutralize the lye (Graham *et al.* 1969).

The advantage of this process from a waste point of view is that it produces a thick sludge, which can be disposed of as a solid waste. It also results in a significant increase in product yield. The process is disadvantageous, however, because it results in a liquid rinse of pH > 10, which must be disposed of through neutralization or aeration.

Waste Accommodation

While efforts to increase production and processing yields have been made, most attention has been paid to developing new methods to accommodate wastes. These have been in two major areas: (1) application of product recovery technologies to food waste streams and (2) anaerobic and aerobic digestion methods.

Recovery of By-Products from Liquid Wastes. The recovery of useful by-products from food processing waste streams has been limited by the variability of wastes from season to season and from product to product. For by-product recovery to be successful, the waste water must not vary in composition and must contain a by-product that has some value or special use. Often a nondestructive recovery technique must be employed for maximum by-product value. Technologies developed for pure chemical separations have been applied to treatment of liquid food processing wastes for by-product recovery. The techniques found useful have included ultrafiltration, reverse osmosis, electrodialysis, and ion exchange.

For example ion exchange procedures have been tested for recovery of amino acids, potassium, organic acids, phosphates, and proteins from potato wastewaters. In this process (Heisler *et al.* 1972), the dilute wastes were first concentrated to 2.5% solids by inplant water use modifications, then the concentrate was subjected to ion exchange (Schwartz *et al.* 1972), which recovered inorganic cations and precipitated proteins. A second ion exchange recovered proteins and a third recovered organic acids. The protein recovered could be used as an animal feed, and the cations as a fertilizer. A 1971 economic analysis of these processes indicated that this ccheme was not economically practical at that time. The best alternative was effluent drying of potato process wastes and utilization as animal feed.

Pressure filtration of waste solutions to remove water only (reverse osmosis) or to harvest large-molecular-weight components such as proteins (ultrafiltration) has also been used to recover by-products from wastes. Whey proteins have been successfully harvested by ultrafiltration. This protein concentrate could be used as it is, or dried for food emlulations. Reverse osmosis is usually more difficult to employ because of membrane fouling, but has been developed for concentrating citrus centrifuge effluents (Berry and Casals 1971). Electrodialysis has been suggested as a useful method for waste treatment. In this process, ionic species can be removed from solution with anion- and cation-selective membranes and an imposed voltage. The process has been described for reducing the ionic content of cheese whey proteins, but has found little use so far in the fruit industry.

While these examples of nondestructive recovery techniques yield products that have undergone limited destruction, they can be expensive to operate and maintain. Therefore, other less costly recovery techniques have been often employed. For example, flocculating or precipitating agents can be used to recover specific large-molecular-weight products such as proteins from meat-packing effluent. They could have potential use in the fruit processing industry. However, precipitation depends on pH and the ionic species present, and these factors must be considered in any flocculation or chemical interaction method. The use of the by-product is dependent on the inherent value of the substrate.

In some instances the by-product recovered can be used directly in a human food, or be further processed to yield important components such as flavors. However, in many instances the harvested by-product has little value in human foods but can be used for animal feeds. This is an excellent use for some human food wastes because food is recycled and maintained in the food system and may replace some human foodstuffs normally fed to animals. Feeding waste to animals is also common (Moon 1980).

Recovering Solids by Spray Drying. Modern spray-drying techniques, now being used to convert many liquid products into marketable powders, are a valuable ally of industry in the battle against pollution. They are also an important method of reclamating useful materials.

Spray drying is a means of converting a solution or suspension into a dry product. The fluid is reduced to a fine spray and mixed with a stream of heated air to produce a dried powder. As the liquid portion of the pollutant is evaporated in the spray drier, a residual powder is collected and packaged. This may prove to be a commercially valuable product instead of a waste material.

Atomization is the most important single step in spray drying. Three common methods are used to atomize the fluid product into the drying air:

1. Two-fluid nozzle system uses air and the liquid for atomization. It involves large volumes of relatively high-pressure air.
2. Spinning disk or centrifugal atomization wheels are used in many spray driers. The spinning disks may revolve from 8000 to 18,000 rpm. The degree of atomization is varied by increasing or decreasing the speed of the wheel and by increasing and decreasing the rate of product feed into the wheel and by increasing and decreasing the rate of product feed into the wheel. The spray angle of a centrifugal atomizing wheel is always 180° as the centrifugal force of the wheel throws the product out in a straight line.
3. High-pressure nozzles, operating at up to 8000 psi, usually consist of two basic pieces—the orifice through which the product is swirled, and the spinner whirl chamber which gives the product a rotational motion. These are matched sets, and at a given pressure thousands of combinations can be made to meter different flow rates and give spray angles from 25° to 120°.

The viscosity of the fluid to be sprayed is of great importance. As viscosity increases, atomization becomes more difficult. Maximum upper limits for the high-pressure atomizing nozzle are 10,000–12,000 cp. Maximum upper limits for atomizing on the centrifugal atomizer are 4000–5000 cp.

Inlet air temperature can approach 1000°F with normal exit temperature between 200° and 300°F (Haak 1971).

Reduction of Salt Content by Ion Exchange. Brine fermentation is unique among food processing operations using substantial quantities of salt because most of the salt is separated from the final product and discarded to sanitary sewers when the incremental sodium chloride load which it contributes does not substantially increase the total dissolved solids level in the effluent from the treatment plant. In other areas, the total dissolved solids in a treatment plant effluent become excessive due to lack of dilution of the processing brines with sanitary sewage or industrial wastes low in sodium chloride. One such area where potential saline pollution of receiving waters is of concern is in the Central Valley of California. Here are located a number of olive processing plants, usually in small towns, that generate substantial volumes of liquid waste containing sodium chloride and sodium hydroxide. The bulk of the salt used in olive and cucumber processing is contained in the storage brines for freshly harvested olives. These relatively low-volume, high-salt content brines can be managed by ponding or by reconditioning and reuse. It is the large-volume, low-salt content processing and rinsing brines that present a more challenging waste management problem (Ralls *et al.* 1971).

New processing technology such as in-the-jar fermentation of olives and salt-free storage of olives may provide solutions to part of the potential saline pollution from olive production. However, it is still often necessary to use sodium hydroxide to hydrolyze bitter olive constituents, so the problem of handling large-volume, low-salt content processing waste still must be solved.

Ion exchange is the most promising method currently available to treat saline wastes that contain dissolved organic compounds as well as inorganic salts. Of five methods, the Aqua-Ion process has the lowest cost, and is a small-scale operation. The treatment consists of passage of waste effluent over a mixed bed of cation and anion exchange resins. The cation exchanger is in the calcium form and is a sulfonated polystyrene resin (Duolite C-20). The anion exchanger is the hydroxyl form (Duolite A-102-DD). The polar constituents of the waste react with the exchangers as follows:

$$\text{(cation) } R_2Ca + 2\ NaCl \rightarrow 2\ RNa + CaCl2$$
$$\text{(anion) } 2\ ROH + CaCl_2 \rightarrow 2\ RCl + Ca\ (OH)_2$$

Depending on the solution concentration, the calcium hydroxide formed during the removal of sodium chloride will stay in solution or will precipitate. The precipitated calcium hydroxide can be removed by inexpensive filters. The slurry enters the filter distributor at the top and flows down through the individual hoses. The clear liquid passes through cloth filters and runs down the outside of the hose to the collecting point of the Uni-Flow filter. The sludge moves along inside the hose and is discharged periodically at the bottom.

The product of the ion exchange operation is a solution of calcium hydroxide and organic material. Part of the organic material originally present in the waste is converted to insoluble organocalcium salts, which are removed by filtration. The calcium hydroxide can be removed from the ion exchange effluent by carbonation and filtration of the resulting calcium carbonate or by ion exchange of the calcium for magnesium. Formation of insoluble magnesium hydroxide to remove calcium hydroxide is feasible in locations where either the waste water or the water supply contains high levels of magnesium. In regions that have high bicarbonate hardness, the effluent from the ion exchange unit can be blended with hard water to produce cold lime softening.

The resin must be regenerated to convert it into usable form for further sodium hydroxide removal. This is accomplished with a solution or suspension of calcium hydroxide in the saline waste water. The regenerant effluent is saturated with calcium hydroxide and contains the salts and part of the organic compounds originally present in the saline waste waters. The regenerant is recycled many times in order to increase the sodium chloride concentration to a level that makes salt recovery or reuse attractive economically. The regenerated resin is rinsed with tap water to remove residual calcium hydroxide and is then ready for treatment of saline waste water. Tests have shown that sodium chloride content of processing water can be reduced from about 600 to 145 ppm. When brine of approximately 6000-ppm sodium chloride was passed through the ion exchange unit, the desalted product had an average salinity of 790 ppm as sodium chloride.

Through the ion exchange process the sodium chloride present in the original olive processing water can be concentrated into one-tenth the original volume. At the same time a volume of desalinated water equal to the volume of the treated olive processing brine can be produced for possible reuse (Ralls *et al.* 1971).

Solid Waste Recovery. Waste seeds of peaches, nectarines, plums, apricots, olives, and cherries, and waste shells of almonds, walnuts, and other nuts, estimated at 150,000 tons in California alone, can be made into commercial charcoal. In the past, dumping at sea was virtually the only alternative to cut-and-cover burial for such waste material; sometimes, it was used for direct firing of furnaces.

A $2 million plant to convert waste seeds and shells into briquets is located in Elk Grove, California. A key element in the process is the pyrolysis furnace, which is widely used for clear disposal/reclamation in sewage treatment plants. It has four separate sections operating at different controlled temperatures. An internal mechanism, operating continuously at temperatures up to 1500°F, transports material between sections of the furnace. The capacity of the plant is 35,000 tons of high-quality briquets per year.

Typical infeed of the raw material to the charring furnace averages

25,000–30,000 lb/hr of a blend comprising 60% hard and soft almond shells, 30% walnut shells, and 10% peach pits. "Hard" shells include peach pits, apricot shells, black and English walnut shells, one variety of almonds, and olive pits. These average 70–75% of the overall infeed to the char process. Average moisture content of the raw materials is from 10 to 15%, while peach pits average 15%.

The function of the first unit in the conversion process is to drive off volatiles from the infeed of shells and pits in a controlled charring process that takes place within an oxygen-poor atmosphere. The prevents ignition of the product while effectively reducing its volume by a ratio of 5 to 1.

The products proceed through three more hearths. Temperature on the fourth hearth is raised to 1500°F, and the product is raked outward to a single discharge outlet. At this point the product is water-quenched by a series of sprays that reduce the temperature to 200°F, in a horizontal distance of 20 ft. The water evaporates as it contacts the hot material.

The quenched product, "char," passes to a vibrating screen separator where the fine material falls through a screen into a storage hopper, while the larger material is carried by a 10-in. horizontal screw conveyor to a hammer mill. The ground material is fed into a bucket elevator that carries it to the storage bin.

On entering the bricket plant, char feeds into a double-shafted paddle mixer, or "pug mill," where a starch binder is introduced in slurry form. Mixed with steam, the starch cooks to a gelatinous state in which it is blended with the char in a proportion of about 95% char and 5% starch. The paddle mixer moves the blended char–starch mixture about 20 ft where it is discharged into the briquet press.

Wet briquets (containing about 30% moisture) come off the forming press at a rate of 7 tons/hr. They feed into a continuous chain briquet drier, which is heated by seven natural gas burners at 1.3 million Btu's each. The briquet storage hopper supplies an automated packaging line, which fills 5-, 10-, or 20-lb bags at an average of 25–30/min.

Product quality is directly related to the degree of control maintained in the charring furnace. A deviation of more than 20°F from specified temperatures can materially affect the hearth dwell time, and reduce the ultimate yield from a given input.

This process eliminates smoke from burning shell waste, and provides a year-round supply of fuel for barbecue- and hibachi-style cooking (Anon. 1971B).

Many other processes for solid waste utilization have been recommended and investigated. One popular alternative is the use of wastes in animal feed (Moon 1980). This certainly is feasible if water can be removed to give about 30% dry matter.

Conjector Centrifuges. Liquid and solid wastes must often be separated before further by-product recovery can occur. A vertical, constant-speed, cone-shaped, continuous centrifuge (Fig. 15.1) provides a means of separating a solid phase from a liquid phase waste at a rate of over 70 tons/hr/unit (Anon. 1972A). A wide choice of nonclogging screens permits controlled classification of fine from coarse, and valuable from less valuable solids.

Three dynamic forces are continuously applied to the material being processed: (1) the separating action of the high centrifugal force (1800 × *g*) as the conical bowl assembly rotates at high speed; (2) a filtering action of the replaceable perforated plate on which the solids are deposited and through which the mother liquor passes; (3) a metering action of the differential volute, which controls the flow of solids downward at a constant rate.

Because the rotating assembly is conical, the centrifugal forcc becomes greater as the diameter increases. As a result, the solids are subjected to maximum centrifugal force just before they are discharged from the widest section of the bowl at the bottom, assuring maximum dryness. At the same time, the major portion of the mother liquor is separated at the smaller diameter near the top, with resulting minimum power consump-

FIG. 15.1. Exterior view of conjecttor continuous screening centrifuge. (Courtesy Sharples Corp.)

tion. This efficient application of the centrifugal force is a major factor in the overall processing effectiveness of the unit.

As the slurry is fed into the top of the rotor and brought up to speed, the solids are uniformly deposited on the "filter" surface. Because of the specially selected angle of the rotor, the solids are subjected to minimum mechanical action as they pass over the gradually increasing screen area. In effect, they float on the filter surface, and the volute acts to meter the flow of solids so that each relative turn permits a regulated quantity of solids to flow through. This action, combined with uniform distribution over a filter area, provides maximum performance with minimum degradation of the shape or size of solids.

Continuous screening centrifuges come in four sizes, and numerous modifications are possible to dewater slurry during the processing of fruits, vegetables, and other foods.

PRACTICAL WASTE MANAGEMENT

Because the quality and characteristics of fruit processing wastes are diverse, the most effective control systems needs to be "custom built" to the specific situation. Fruit processing wastes vary from liquid wastes to liquid slurries, from continuous to variable flows and concentrations, from nutrient deficient to nutrient rich, and from seasonal to year-round loads. Geographical location, availability of land, and proximity to ground and surface waters are factors to be considered in designing waste management systems.

The first step in any waste management program is to inventory the amount of waste produced in each stage of processing. The amount of waste entering the waste water should be minimized to reduce the organic load on the treatment facilities. Because of the narrow limits of BOD load legally discharable into streams, the discharge of both solid and liquid waste must be reduced drastically.

Rose *et al.* (1971) have outlined several steps in evaluating the problems and procedures of utilizing fruit wastes:

1. Determine the volume and composition of the wastes and measure their pollution strengths and magnitudes. This, in turn, permits an evaluation of (a) potentials of possible treatments for the waste; (b) potentials for use of the waste in light of current market practices, whether as feed or industrial derivatives; (c) potentials for isolation, reduction, or retention of the waste at the point of harvest; (d) evaluation of possible methods for disposal of waste to the soil; and (e) determination of methods for reduction of discrete wastes within the process lines and reduction of leachates in all transport phenomena.

2. Investigate the economic feasibility of handling wastes consolidated from several plants where potentials for agricultural or industrial derivatives exist.

3. Investigate possible markets for new industrial products in light of changing needs: fertilizers, mulches and soil conditioners, chemical and nutritional derivatives, feedstuffs and feed derivatives.

4. Consider establishing centers for research in solid waste utilization where industry may collectively support investigations and potentials in waste utilization.

The wastes generated by processing of fruits are primarily solid (i.e., peels, stems, pits, and culls) or organic matter in suspension or solution in the processing waste water. Each individual plant should make an inventory of the in/out balances of its waste. That is, where does the waste come from during processing and where does it go.

Some information is available on the wastes load generated and the methods used to dispose of these wastes.

Data in Tables 15.5 and 15.6 show the distribution of liquid waste disposal methods used in fruit processing. Many plants use more than one disposal method. Generally the removal of pollution is less in a holding than in an aerated treatment pond because of aerobic oxidation being inherently more efficient and faster than anaerobic. Some processors use pond treatment in addition to land application.

Citrus processing is concentrated largely in Florida, California, Texas, and Arizona. More than 1300 sq miles in Florida are planted to citrus, with a yield of greater than 10 million tons; 52 processing plants convert about 82% of the citrus crop into single-strength and concentrated juice. The industry has made substantial progress in solving its waste problem through recovery of such valuable by-products as peel, pulp, citrus molasses, citrus flavors, citric acid, pectin, and essential oils. However, the combined operations of the processing plants discharge about 130 million

Table 15.5. Distribution of Liquid Waste by Fruit

	Percentage of plants using			
Product	City systems	Holding ponds	Treatment ponds	Irrigation
Citrus	12	—	10	24
Peach	83	3	11	11
Apple	30	10	40	30
Pear	92	4	8	8
Other fruit	67	10	16	20

Table 15.6. Distribution of Liquid Waste by Location

Discharge to	Percentage of plants		
	Urban	Nonurban	Total
City systems	100	0	47
Ponds	9	28	16
Spray irrigation	0	55	31
Surface water	0	9	5

gal/day of waste water, with a 5-day BOD loading equivalent to a population of 2 million people.

Unfortunately in many instances the economics of by-product recovery and utilization is not profitable and plants are forced to use traditional methods of disposal including land fill for solid wastes and anaerobic/aerobic digestion similar to sewage treatment for liquid wastes.

Liquid Waste Treatment

Initially, suspended solids are usually removed from waste waters by screening or sedimentation. The pressure-air floatation system is effective in removing suspended solids from waste water. The principle behind this system is simple: when waste water under high pressure is suddenly released into an open tank at atmospheric pressure, the extra volume of air is released in the form of bubbles, which adhere to the particles of suspended matter in the liquid and carry the particles to the surface, forming a layer of debris that can be removed by skimming mechanism.

Another effective way of removing suspended solids from liquid waste streams is through the use of vibrating screens. Screens of various mesh size have been tested, but the basic idea remains the same—to trap the solids as water flows through the screens. A two-deck screen system is considerably superior to a single-deck screen, and circular screens with 48-mesh or finer remove more solids than rectangular screens in the 20-mesh range.

After solids are removed at this stage, further dissolved wastes can be removed by trickling filters. In many ways, a plastic-filled trickling filter system appears to be the best way of reducing the pollution load of waste water. Addition of nitrogen to the waste water has improved the performance of these filters and eliminated odors that were present when filters were operated without nitrogen. In one study, a trickling filter without nitrogen removed 190 lb of biochemical oxygen demand (BOD)/1000 ft^3/day when operated at the standard flow rate; with the addition of nitrogen, the removal rate was 450 lb of BOD/1000 ft^3/day.

After filtration, the waste water is sent to a pond or a lagoon to further reduce the organic load. The design and construction of waste treatment systems are complex and individualized for each processing plant. It is beyond the scope of this chapter to itemize such design or indicate appropriate solutions to individual processing problems.

There is considerable engineering information on the disposal of effluent wastes upon soil. Soil type, pollution load, application rate, soil cover, and other conditions vary widely. The results have depended upon the experience of the participants in experimental projects. There is a dearth of information on the potential effects of such applications on the character of the soil. It is evident, however, that in many operations effluent loads are being applied at maximum levels, resulting in some degree of runoff and development of undesirable odors. A compendium of engineering information for land disposal systems is needed.

Waste Evaluation. The concentration and characterization of pollution loads in most fruit and process wastes often are measured by tests of chemical oxygen demand (COD) or biochemical oxygen demand (BOD). To obtain reliable values in these tests, trained personnel must supervise their application.

The COD test can be performed in several hours, whereas the BOD test requires 5 days, each after procedures have been standardized. These tests are used to determine the acceptability of effluents disposed into natural waterways and, in part, to determine sewer charges.

The BOD test is designed to determine the amount of organic matter in wastes by measuring the ability of bacteria present or added to the sample to degrade it. During their metabolism, they utilize much of the oxygen available and the analyst essentially determines how much is left after a 5-day incubation (hence the term BOD_5).

The COD test attempts to measure the same process chemically. An excess of a strong oxidizing agent is added to the wastes, and then the amount of residual agent is determined after a short digestion period. This process is faster and for most food processing wastes provides a sufficiently accurate measure of waste load.

Other measures of waste load commonly used are the amount of solids that readily settle out and those that can be removed by some means of filtration.

Information Needed by Management. Data on the costs of liquid waste treatment and disposal are not available in a form that can guide individuals in management who must make final investment decisions. A board of directors must have information on the available alternatives for meeting requirements for adequate treatment and disposal of wastes. They need to know capital requirements, reliability of results from the capital to be invested, the cost effects on the product, and the competitive advantage of alternative possible treatment systems.

To reach decisions about waste treatment choices, plant managements need an organized compendium of relevant information. Such a compendium should include the following:

1. A listing of the laws and ordinances affecting most processors, with a summary covering major restrictions.
2. A procedure for determining the cost of water supply and water handling and distribution, including its handling as waste and its recovery.
3. Procedures for evaluating the effectiveness of various methods of screening of wastes to remove collodial, suspended, and smaller particulate matter; an assessment of the relative merits of these procedures for specific applications and their capital and operational costs.
4. Procedures for determining costs of municipal sewage service, including specific situation information with factor data for application to operations where the service is available.
5. Procedures for evaluating land disposal systems, including engineering factors, limitations, operational capabilities, and capital and operational costs.
6. Procedures for determining costs of solid waste removal, utilization, and disposal.
7. Recommendations for monitoring systems or procedures for control of pollution loads in wastes and effluents.

The compendium should include, first, a concise summary of the problems facing a processing plant, relating these to economic and environmental problems and current external controls on how waste may be treated and discharged from the property. It should include evaluation of trends in enforcement by such external controls. A second section should include summary graphs of current waste disposal possibilities for application by management to mixes of suspended and soluble materials in the wastes. The graphs could relate the following:

1. Variables on the required work for different mixes of dissolved and suspended solids plus nutrients.
2. Relation between any given variable and investment and operation costs.
3. Relations between variables for various mixes and for different product wastes.
4. Relations between investment and operating costs against variable

production levels for available methods of treatment, including municipal, land, and biological degradation systems.

The relationship of the composition of effluent to its biodegradability is not clearly established; for example, the rates of degradation of effluents bearing different types of organic solids, such as different types of vegetable starches, or different types and concentrations of fruit acids, cannot be accurately predicted. However, as much data as possible should be provided regarding measures to attain complete and rapid biodegradation of the wastes from a particular plant to the point of enabling economic treatment and recycling/reuse of the water. New engineering concepts may be feasible in the use of waste energy from stack gases or thermal processes applied to specific wastes or to biodegrading systems. Some information of this type has been developed. The efforts at treatment of wastes with unicellular microorganisms indicate the need for better understanding of the requirements to attain greater rates and more complete degradation (Rose *et al.* 1971).

Innovative Waste Treatment—An Example

A New York winery has developed a waste treatment system that is effective, economical, and adapted to other types of wastes. Constant aeration purifies the effluent, reducing solids in the waste water to 20 ppm and dissolving enough oxygen into the water to reduce BOD by 97%. The flow of waste begins at the press with most solids removed by screens. The remaining waste water is measured, analyzed, and doctored to improve the treatment in the aeration ponds (Anon. 1972B).

During pressing season, waste water flows from the press plant into the screenhouse, where a vibrating screen removes the solids that are not removed by pressing. The solids are transported by conveyor into a storage hopper and later used as fertilizer. Waste water and solids too small to be retained on the screen flow into an entrance structure. Waste water from other parts of the processing plant join a 1000-ft main drainage line and also flow into the entrance structure, which houses equipment to (1) measure flow rate in gallons per day (up to 120,000 gal./day in peak periods, (2) measure water pH, and (3) add nutrients. For bacterial growth to occur, the water must have a near-neutral pH. If the wastes become too acid, as they periodically tend to, automatic equipment is activated to pump caustic soda into the water. Because winery wastes are virtually devoid of phosphates and nitrogen, which are essential for bacterial growth, these nutrients are pumped into the waste water in small amounts as needed.

From the entrance structure, waste water flows into four 10-ft-deep aeration ponds that encourage bacterial growth. Bacteria feed on the dissolved organic chemicals that come from the processing plant. The two primary ponds are 80 ft in diameter; the two secondary ponds are 88 ft in diameters. From the waste material and natural decomposition, activated sludge is formed.

For proper bacterial growth and to keep the solids in suspension, the ponds are constantly aerated with four 10-hp aerators. The units, with 5-ft-diameter impellers, produce a high flow for complete solids suspension. The amount of daily flow determines how many ponds, and in what sequence, are needed for the operation.

Overflow from the ponds travels to two 12-ft-diameter final clarifiers. As water flows into the clarifier, solids (activated sludge) settle to the bottom and the clear water at the top flows over small surface troughs into the filter house. As the solids settle in the clarifier, they are removed. On the bottom of the clarifier is a large rotating arm with suction cups on the underside that slowly sweep the bottom and gently pick up the activated sludge that has settled. The sludge is then returned to the head of the plant for seeding of new bacterial growth.

Recycling of the activated sludge is continuous until such time, determined by testing, that the solids accumulation is too great. A concentration of 2500 ppm is maintained for maximum efficiency. When this level is reached, part of the returned solids are transferred to an aerobic digester where they are aerated by a 10-hp surface aerator. When solids level in the digester must be reduced, digested solids are pumped from the pond to a tank truck and distributed over the soil as mulch.

Meanwhile, overflow from the clarifier passes into the filter house and flows through a bed of sand for final filtering. From the filter house, the water flows into the receiving waters. The final effluent entering the receiving stream has an average of 20 ppm solids and a pH of 7.2–8.

DISPOSAL OF CONTAINERS

The problems involved in disposal of containers, and their impact on the environment, depend in large part on the materials from which they are made. Three basic types of steel, as well as aluminum, are used in can manufacture today.

Blackplate is steel without any coating applied by the steel mills and is used primarily in the production of steel pails.

Tinplate, the major metal used by can manufacturers today, consists of steel with a very thin coating of tin applied to both sides of the sheet or coil. This material is used in all food cans and to a great extent in beer and beverage cans.

TFS (tin-free steel) is steel with a microscopic thickness of chromium. It was first used in beer cans and later in soft drink containers. Some TFS is used in beverage ends today, but since the material cannot be readily soldered, a new method of can making had to be developed. The use of TFS for can bodies was made possible by the advent of the bonded side seam and welded side seam.

Aluminum found its place in the can-making business first in easy-open beer and beverage ends and later in the full-panel removal end. Aluminum is also used in making two-piece beer and soft drink cans.

The rusting characteristics of these metals vary depending upon their chemical properties and the environment in which they are placed. A standard tin can, for example, will decompose in 6 months to 1 year if incinerated when initially disposed. The burning takes off the organic coatings and most of the protective layer of tin. When not burned, estimates show that a tin can may last from 6 to 10 years before being completely oxidized.

An aluminum can takes about three times as long as a tin can to degrade.

The disposability of metal cans could be substantially enhanced by their destruction, recyling, and/or use by new applications. Of these, recycling seems to present the most beneficial opportunity. Such an approach is economically sound, conserves resources, and gives the highest utility in terms of total consumption.

A recycling program would have as its objective the total collection of all used steel container waste (household, commercial, or litter) and recycling of that waste to one or more steel companies. A device might be devised for the home that would transform cans into little particles. In this form, the can metal would be easier to transport for recycling and might even be applicable to lawns, gardens, and croplands as an iron supplement.

In the area of new applications for existing metal cans, crushed cans could be used as aggregate for poured concrete and shredded cans used for reinforcing concrete. Like the steel rods used in building highways, the all-steel can would not rust or react with the concrete. The beverage can may become a building block, a toy, or part of a partition in the family room. Perhaps someone will come up with an idea that will transform the used can into a valuable commodity, thus eliminating altogether the disposability dilemma.

The destruction or reduction of cans to a concentrated form would enhance disposability since cans would be disintegrated, shredded, or compacted either in the home or at a minicipal dump. Package planning design should lend itself to one or a combination of these three major approaches to can disposal.

The simplest product to control, whether it be aimed at disposal, litter, or pollution, is the one in its most pure form; that is, a can made of (1) all steel with no tin coating; (2) all steel with tin coating; or (3) all aluminum. Ideally, a solution will be reached to provide the ultimate can design for disposability as well as control of litter and pollution.

Some progress has been made on the development of new coatings that will dissolve and aid in the deterioration of the container once the product has been consumed. Also, much work is being done on catalytic agents as additives to inks, coatings, or basic steel that would hasten oxidation.

The possibilities for accelerating the degradation process are more numerous on the exterior of the can than the interior. Interior coatings are selected with great care to ensure that no reactions can occur to affect flavor, shelf life, premature perforation, etc., even with highly corrosive liquids. Exterior coatings are selected to protect label inks against scuffing and the metal against premature corrosion, and to provide a slick surface for rapid transfer of empty cans in the filling operation. Therefore, exterior coatings could be more easily adapted for accelerated degradability than interior coatings.

Although the environment of city dumps and litter areas may be quite diverse, they usually have two conditions in common—rain and sunlight. Exterior coating systems are under development that provide a degree of water sensitivity. They must be sufficiently impervious to moisture to allow normal handling (washing and drying, wet filling, normal refrigerator storage, etc.) but must react to prolonged moisture attack or repeated wet/dry cycling. The coating should crack, peel, dissolve, or in some way expose the protected metal surface to atmospheric oxidation after the containers are discarded. Other coating systems being investigated are subject to degradation from ultraviolet radiation from sunlight (Anon. 1970).

Steel Can Recovery

Los Angeles By-Products Co., Los Angeles, California, has been a pioneer in salvaging steel cans from mixed refuse in sanitary landfills. This company designed and manufactureed a portable sorter that consists of conveyors and magnetic separating equipment. The unit is portable, and power is provided by a diesel generator contained in the unit. Refuse is fed onto the conveyor system by a Traxcavator equipped with a large bucket. The mixed refuse is passed through a magnetic separator, which extracts the ferrous metal—mostly tin cans. Ferrous metal is delivered to a secondary magnet for final separation, and then loaded directly into trucks or trailers for transportation to the processing plant. The remaining solid waste is deposited on the ground near the "working face" of the landfill. At this point, the landfill operator spreads the refuse, which is then covered in a normal manner. The collection system handles all ferrous metal that can be put in a 40-gal. container. One sorter operates efficiently when it receives from 350 to 400 tons of refuse/8-hr day. A thorough study of the economics of such a salvaging system showed that the capital investment required for a sorter, Traxcavator, and trucking equipment for delivery to the processing plant was approximately $150,000 (Anon. 1971B) in the early 1970s.

Concerned citizens in many cities throughout the United States are separating cans from household garbage and carrying them to collecting centers established by can manufacturers and the steel and aluminum industries. They recovered an estimated 800 million cans in 1971, and more in 1972.

However, magnetic extraction of steel cans as a component of municipal and regional trash-collecting systems is a better way to salvage and recycle can materials. Numerous cities have established such systems (Table 15.7).

Several different systems are employed to produce reusable ferrous materials. In Oakland, California cans are extracted from household refuse at the landfill site. In St. Louis and Los Gatos, California, the systems are capable of removing cans before the remainder of the garbage is incinerated. In Amarillo, Chicago, Atlanta, and Stickney, Illinois, cans are retrieved after incineration. In Franklin, Ohio, cans are removed from a slurry that is formed by pulverizing the garbage and mixing it with water (Anon. 1973).

A problem in recovery of iron, steel, or tins from cans is that several metals are involved, and the proportions vary with different products. Melted-down cans produce a mixture of the main metal with molten metals from other sources. This results in a product of unknown composition and properties—one batch will vary from another in strength, hardness, resistance to rust, and suitability for cans. Thus, recycled steel from cans may be lower in quality than that used originally.

There are industry-wide trends to replace steel–tin cans for fruits and fruit products with aluminum and plastic containers.

Degradable Plastics

What were claimed to be the first truly degradable plastics appeared in late 1972, under the trade name Ecolyte. Degradable plastics are similar to other polystyrenes, polyethylenes, ethylenes, vinyls, and polypropylenes as far as in-use properties are concerned. They can thus be used for the same packaging applications and will provide the same product protection. The difference is in the after-use property: they are completely degradable when discarded outdoors.

Table 15.7. Cities Operating Steel Can Recovery Systems

Location	Separation system	Amounts collected[a] Garbage (tons/day)	Amounts collected[a] Cans (tons/day)	Annual can recovery[a] (million)	Markets
Amarillo, TX	After incineration	200	12	50	Copper mines
Atlanta, GA	After incineration	700	16	100	Ferroalloys
Chicago, IL	After incineration	4000	100	730	Copper mines
Franklin, OH	Slurry system	60	5	30	Steel making
Houston, TX	Dry separation at a transfer station	450	20–25	104–130	Copper mines
Los Gatos, CA	After shredding, before incineration	300	20	120	Copper mines
Madison, WI	After shredding	250	7–8	38–41	Steel making Copper mines
Martinez (Contra Costa County), CA	Portable separator at landfill	500	20	80	Copper mines
Melrose Park, IL	After incineration	400	16	83	Copper mines
New Castle County, DE	After shredding	1200	60–96	312–500	Detinners Steel making
Oakland, CA	Portable separator at landfill	600	40	182	Copper mines
Pompano Beach, FL	After shredding	200	10	60	To be established
Sacramento, CA	Portable separator at landfill	250	12	74	Copper mines
St. Louis, MO	After shredding, before incineration	1000	50	260	Pilot program
St. Petersburg, FL	Cans segregated by householders before magnetic separation	na	na	3	Detinners
Stickney, IL	After incineration	250	10	84	Steel making Copper mines
Tampa, FL	After incineration	750	20	104	Steel making Copper mines

Source: Anon. (1973).
[a]Estimated values.

These plastics have the property of degradability built into them during their initial polymeric synthesis. They are programmed "at birth" to self-destruct when exposed to ultraviolet radiation from the sun. As long as they are kept away from ultraviolet light, they remain stable. They may be displayed in store windows, since the window glass removes enough of the sun's UV rays to prevent the degradation process from occurring. They contain no additives that can be extracted to contaminate the contents of packages.

The very small particles that remain after these plastics have degraded become part of the natural soil, susceptible to biological degradation. The time required for a degradable plastic package to be reduced to particulate matter varies from a few days to 6 months depending on the physical properties of the specific plastic, the wall thickness of the finished package, and the amount of sunlight to which it is exposed (Anon. 1972C).

Recycling Centers

The first permanent, full-time glass and tin recycling effort, known as the Metropolitan Bottle and Can Recycling Center, was located in Midway Center, St. Paul, Minnesota. All types of clean glass containers and all types of metal containers were acceptable for industrial recycling. Through a revenue-sharing plan it was hoped that the operation would be self-supporting; in case it was not, Coca Cola Bottling Midwest agreed to underwrite the operation (Anon. 1971C).

Brooks Products, Holland, Michigan, pioneered the first use of a recycled glass and asphalt formulation as a road-paving material. The "glassphalt" formula, worked out in 1968 at the University of Missouri, contained ground glass, asphalt, crushed limestone, and rock aggregates, and was used to resurface pavement 1½ in. thick. Glassphalt paving offers one way to recycle glass in a direct, meaningful way (Anon. 1971D).

Canada Dry announced in November 1971 the opening of four recycling and reclamation centers in New York. All types of bottles and cans were accepted, on a part-time basis, and the profits derived from the sale were donated to Boy Scouts and other youth organizations. Similar centers for reclamation have been set up in other sections of the country.

One of the first aluminum can reclamation projects was begun in San Diego, in 1970 by Aluminum Company of America. The company paid $150,000 for nearly 30 million cans for recycling, the first year. The "Yes we can" program was set up in the Dallas/Fort Worth area and other cities in 1971.

Recovery and Recycling of Aluminum

The amount of aluminum recycled is impressive (Testin *et al.* 1972) and is rapidly increasing each year. Of the 5,383,000 tons of domestic aluminum supply in 1969, slightly less than 20% was provided from domestic secondary recovery and imported scrap. During that year new scrap, old scrap, and imports included 2,215,000 tons, of which about 48% (1,056,000 tons) was recovered. Of the old scrap (1,334,000 tons), only about 13% was recycled. Most old scrap came from packaging materials. Consumer durables accounted for 197,000 tons of old scrap, of which 13% was recycled. Practically all new scrap was recycled.

Clearly, if recycling of aluminum is to be increased in the future, the recycling of consumer-oriented scrap must play a major role. There are several long-term advantages to recycling aluminum. The metal can be recycled almost indefinitely (excluding melt losses) at energy expenditures of less than 5% of that required to make aluminum from virgin sources. Recycling aluminum requires far lower capital investment than building new reduction facilities. Old scrap represents a "domestic ore" of ever-increasing quantity, reducing the reliance on virgin ore. Recycling reduces the drain on natural resources and can make aluminum competitive in otherwise economically marginal situations. Finally, recycled metal re-enters the process stream near the end of the production process, thus eliminating the expensive environmental control operations involved in primary production.

Since 87% of old scrap is lost and the largest single factor in this loss is packaging scrap, this is a place to begin if one wishes to make inroads into that portion of scrap that has not, historically, been recycled. More specifically, cans make up about half of the industry shipments in the packaging area, and comprise the most rapidly growing segment of the expanding aluminum packaging field. Thus, Reynolds aluminum-can reclamation program, which had its beginning in 1967, sought to open up a new form of scrap for recycling—scrap of a character that had not heretofore been recycled. Two basic problems were encountered: (1) how to set up a reverse distribution system for a large quantity of scrap that existed in numerous low-volume end points around the nation, and (2) how to prepare, ship, and utilize can scrap most efficiently.

In the recycling of aluminum cans, the ultimate source of scrap is the individual consumer. Therefore, it is necessary to find a way to motivate consumers' to bring scrap cans to some sort of centralized collection point. This motivation is best provided by a cash payment for cans and other clean aluminum household scrap collected.

Testin *et al.* (1972) have described the early Reynolds reclamation efforts:

> Once we established that we could motivate the public to bring substantial quantities of material to us for cash payment, we began developing other points in our reverse distribution network and, simultaneously, began developing the technology for preparation and use of the scrap. Reynolds now has a network of over 800 collection points

in more than 30 states cooperating with our program. At all of these, the public can redeem all their aluminum beverage cans for $0.10/lb. These collection points are primarily operated by beer and soft drink distributors. The cans collected at these points are either hauled to central locations through normal back-hauling procedures available in the beverage industry or are picked up. Most of this material ultimately ends up at one of 14 can reclamation centers or 5 mobile units (can reclamation centers mounted in trucks) that we now have in operation servicing areas where cans are being redeemed. At these reclamation centers the cans are magnetically separated, shredded, and loaded onto rail cars for shipment to smelting plants in Alabama or Virginia. At the smelting plants, the can scrap is remelted in reclamation type furnaces. Most of this scrap is now being put back into can stock. We do maintain flexibility so that this material can be used whenever this scrap product fits the company needs.

The aluminum can reclamation program is now an economically viable long-term operation for Reynolds Metals Company and we anticipate its continued expansion in the future. Other aluminum companies now have similar programs.

The aluminum industry redeemed over 35 million lb in 1971. This industry figure is 8½ times more than the aluminum packaging materials redeemed in 1969. It is estimated that the aluminum industry will reclaim 70 million lb in 1972 through these programs and by 1976 will be reclaiming 200 million lb annually. In 1971, the aluminum industry redeemed approximately 12% of all aluminum beverage cans manufactured in the United States. In 1972, this redemption percentage jumped to 18%, despite the rapid rise in production of aluminum cans.

We realize that regardless of the success of this program, a large part of the old scrap will remain in the junk heap or garbage dump unless additional steps are taken. We are stepping up our efforts to concentrate and segregate aluminum consumer scrap at the source. Collection programs at schools, ball parks, military bases, and bars where aluminum cans, rigid foil containers, and other consume scrap is expected to concentrate are being worked into our overall aluminum recycling network.

REFERENCES

ANON. Undated. The Metal Can, Its Past, Present and Future. Can Manufacturers Institute, Washington, DC.

ANON. 1967. Sanitation milestones—Past and present. Canner/Packer *136*(10) 23–24.

ANON. 1970. Can manufacturers shift into high gear in anti-litter drive. Soft Drink Ind.*50*(4) 19–20.

ANON. 1971A. Treating liquid waste in canning. Canner/Packer *140*(6) 17–18.

ANON. 1971B. Turning waste materials into profits. Canner/Packer *140*(7) 16–17.

ANON. 1971C. Guidelines for glass recycling. Form *0798*. Owens-Illinois Co., Toledo, OH.

ANON. 1971D. Recycling glass. Glass Container Manufacturers Institute, New York.

ANON. 1971E. Sanitation through chlorination. Canner/Packer *140*(2) 15–17.

ANON. 1972A. Sharples conjector continuous screening centrifuges. Sharples Div. of Pennsalt, Philadelphia. Bull. *1288*.

ANON. 1972B. Winery innovates waste treatment. Food Eng. *44*(6) 73–75.

ANON. 1972C. Degradable plastic. Soft Drinks Trade J. *26*(10) 510.

ANON. 1973. Magnetic Separation of Steel Cans—A Key to Solid Waste Management. American Iron Steel Institute, New York.

BERRY, R. R. and CASALS, P. 1971. Use of reverse osmosis process for concentrating citrus centrifuge effluent solids: Food processing waste. *ARS 72*. U.S. Agric. Res. Serv.

BOUGH, W. A. 1973. Composition and waste load of unit effluents from a commercial leafy greens canning operation. J. Milk Food Technol. *36,* 547–553.

CELENTANE BROS. 1971. Eliminates smoke and offensive odors. Food Eng. *43*(5) 119.

CHAMPION, E. 1967. Total sanitation. Canner/Packer *136*(10) 25–26.

EIDSNESS, F. A., GOODSON, J. B. and SMITH, J. J., JR. 1971. Biological treatment of citrus processing wastewaters. 2nd Natl. Symp. Food Process. Natl. Canners Assoc., Washington, DC.

FLORA, L. F. and LANE, R. P. 1978. Processing trails with muscadine grapes. Ga. Agric. Exp. Stn. Res. Rep. 289.

GOLUEKE, C. G. 1977. Biological Reclamation of Solid Wastes. Rodale Press, Emmaus, PA.

GRAHAM, R. P., HUXELL, C. C., HART, M. R., WEAVER, M. L., and MORGAN, A. I., Jr. 1969. Dry caustic peeling of potatoes. Food Technol. *23*(2) 195.

HAAK, M. P. 1971. Recover solids by spray drying. Pollut. Eng. *3*(4) 33–35.

HEISLER, E. G., SICILIANS, J. and KRULICK, S. 1972. Potato starch factory waste effluents. Development of a process for recovery of amino acids, proteins and potassium. J. Sci. Food Agric.*23,* 745–762.

MARION, W. W. 1982. Needs and capabilities in food research in academia. Food Technol. *36,* 89–91.

MOON, N. J. 1980. Maximizing efficiencies in the food system: A review of alternatives for waste abatement. J. Food Protect. *42,* 231–238.

MUELLER, J. H. 1971. What it costs to control odors. Food Eng. *42*(4) 62–65.

NICKEL, W. J., LEIN, C. L. and DOMINICK, D. D. 1969. A primer on water treatment. *CWA 12*. U.S. Dep. Inter., Fed. Water Control Admin.

RALLS, J. W., MERCER, W. A., GRAHAM, R. P., HART, M. R., and MAAGDENBERG, H. J. 1971. Dry caustic peeling of tree fruits to reduce liquid waste volume and strength. 2nd Natl. Symp. on Food Processing Wastes, Natl. Canners Assoc., Washington, DC.

ROSE, W. W., MERCER, W. A., KATSUYAMA, A., STERNBERG, R. W., BRAUNER, G. V., OLSON, N. A. and WECKEL, K. G. 1971. Production and disposal practices for liquid waste from cannery and freezing plants. 2nd Natl. Symp. Food Process. Natl. Canners Assoc., Washington, DC.

SCHWARTZ, J. H., KRULICK, S. and PORTER, W. L. 1972. Potato starch factor waste effluents. III. Recovery of organic acids and phosphate. J. Sci. Food Agric. 23, 977–985.

SODERQUIST, M. R. 1976. Waste Management in the food processing industry. J. Environ. Qual. *1,* 81–85.

STEIN, M. 1973. The new regulatory system in federal water quality law. Food Technol. *27,* 66–72.

TESTIN, R. F., BOLLING, R. N. and ABBATE, S. T. 1972. Recovery and recycling of aluminum. Materials Eng. Congr., Cleveland.

VAN LUVEN, A. L., VAN SOEST, R. and MCCORMICK, B. J. 1972. High rate complete (water) treatment. Ind. Waste Conf., Toronto. (Unpublished)

16

Fruit Consumption Trends and Prospects

J. E. Epperson

In recent years significant changes in the consumption of fruit in the United States have resulted from shifts in the prices of fresh and processed fruits, consumer income, population, and tastes and preferences. Tastes and preferences have been influenced significantly since 1970 by major changes in our society.

The number of women entering the U.S. labor force has increased dramatically; by 1982, over 43% of the work force was composed of women. This phenomenon, in turn, has precipitated a notable change in the eating habits of American consumers. They seem to prepare fewer meals at home, while eating away from home continues to become more popular. In addition, greater public awareness of health and nutrition has caused a change in the diet of the average consumer. Thus, the per capita consumption of fruits and vegetables experienced a surge in the mid-1970s and continues upward.

In an advanced society, two factors limit the amount of food that the average consumer will eat: income and the biological capacity to ingest food. The consumer has the task of allocating disposable income among various commodities and services that bring the greatest amount of satisfaction or utility. Implicitly, then, disposable income must be allocated between food and nonfood items. Because only so much food is needed or desired, producers and processors of various food commodities compete

Commercial Fruit Processing, 2nd Edition

ISBN 0-87055-502-2

for parts of an overall food market. Fruit processors, then, compete for a *market share*. Those who are more adept at anticipating or influencing consumer demand and are able to produce accordingly will be the successful processors of tomorrow.

The purpose of this chapter is to explore trends in consumer demand for processed fruits since the early 1960s. This discussion includes an examination of the factors that have influenced consumption patterns and projections of the likely directions of consumer demand for processed fruits over the next decade. The evaluation of trends in consumer demand are based largely on per capita consumption estimates over time. These estimates reveal the direction of consumer demand through comparison of consumption data over time. Year-to-year fluctuations in per capita consumption estimates often are not indicative of changes in consumer demand but rather reflect variations in supply that are due for the most part to favorable or unfavorable weather.

Table 16.1 depicts the allocation of per capita personal consumption expenditures in 1981 for various food categories. Total domestic food expenditures of $1249/person represented about 14% of total personal consumption expenditures. Per capita purchases of fresh and processed fruits of $80.45, not including melons or wine, amounted to 6.4% of all farm food expenditures and 0.9% of total personal consumption expenditures. The fruit share of the consumer food dollar in Table 16.1 is that portion for which fruit growers and processors compete. In addition, fruit processors compete with producers and processors in other segments of the food industry for a share of the consumer food dollar. However, the

Table 16.1 Per Capita Expenditures for Domestic Farm Food Products, 1981

Commodity group	($)	(%)
Meat products	371.81	29.8
Dairy products	177.23	14.2
Poultry and eggs	86.03	6.9
Fruits[a]	80.45	6.4
Melons	5.72	0.4
Vegetables	178.71	14.3
Grain mill products	40.83	3.3
Bakery products	123.31	9.9
Miscellaneous	185.25	14.8
All farm food expenditures	1249.34	100.0
Total personal consumption expenditures	8911.50	

Source: USDA (1982).
[a]Does not include wine.

scope of this chapter pertains primarily to the relative changes and the reasons for such changes in per capita consumption of fresh versus processed fruit with emphasis on relative changes among processed fruits.

CONSUMPTION TRENDS

Total fruit consumption per person in the United States since 1960 has trended downward and upward periodically. Table 16.2 illustrates the decline in per capita fruit consumption in the early 1960s. From 1963 through 1970 per capita fruit consumption trended upward primarily due to gains by processed fruit. From 1970 through 1974, per capita fruit consumption again fell substantially. However, after 1974, fruit consumption per person again increased. All of these most recent gains, however, have been due to significant increases in the per capita consumption of fresh fruit. When wine consumption is included, though, overall fruit consumption per person improves substantially. In fact, with wine included, gains in the per capita consumption of processed fruit from 1964 through 1975 were much more impressive than without wine. However, even with wine included in the processed fruit category, the growth in per capita consumption of fresh fruit compares favorably with that of processed fruit beginning in 1979.

Table 16.2. Per Capita U.S. Fruit Consumption (lb)[a], 1960–1981

Year	Processed[b]	Fresh	Melons[c]	Total[d]
1960	50.3	89.1	28.2	167.6
1962	49.8	80.4	24.9	155.1
1964	46.6	76.7	24.8	148.1
1966	49.4	79.2	24.0	152.6
1968	51.2	76.3	24.9	152.4
1970	55.6	78.9	25.1	159.6
1972	55.0	74.9	23.8	153.7
1974	52.1	76.4	21.0	149.5
1976	55.6	82.8	22.3	160.7
1978	55.2	79.0	23.8	158.0
1980	55.4	85.7	20.9	162.0
1981	55.0	87.3	23.0	165.3

Source: USDA (1982).
[a]Retail weight equivalent.
[b]Does not include wine.
[c]Includes home garden consumption.
[d]Includes frozen citrus juices in product weight instead of single strength equivalent.

Fruit Product Categories

Table 16.3 shows per capita consumption of fruit by category and the average annual rate of change from 1960 through 1981. The total quantities in Table 16.3 differ from those in Table 16.2 for two reasons: wine is included in Table 16.3 and is not included in Table 16.2; and the quantity of juice concentrate is given in single-strength equivalent in Table 16.3 and is included as product weight in Table 16.2. The greatest increase in per capita consumption occurred in the category with an average annual rate of change of 4.1%. This upward trend showed no signs of leveling off as of 1981. Chilled juice had the next highest average annual rate of change (3.1%) from 1960 through 1981. Per capita consumption of chilled juice, however, appears to have peaked in recent years. The average annual rate of change of juice concentrate was 2.9% over the total period; again, however, per capita consumption for this category seems to have peaked in recent years. An average annual rate of change of 1.8% for canned juice is also notable. In addition, the change in per capita consumption for canned juice from 16.67 pounds in 1980 to 19.13 pounds in 1981, an increase of nearly 15%, is quite substantial. Per capita consumption of canned fruit continues to decline, whereas that of frozen and dried fruits has been relatively stable in recent years.

Particular Fruit Products

In this section per capita consumption estimates are presented for individual processed fruit items in each product category. Estimates are presented for 1960 through 1982 in most cases.

Table 16.4 depicts per capita consumption of each canned fruit commodity; Table 16.5 of citrus fruit juices; Table 16.6 of noncitrus fruit juices; Table 16.7 of frozen fruit items; and Table 16.8 of the various dried fruits.

Figure 16.1 displays in graph form the annual per capita consumption of the leading item in each processed fruit category. The per capita consumption of canned peaches was fairly steady through the 1960s but trended down through the 1970s. No rebound is indicated for this commodity. Consumption of frozen orange juice, the leading processed fruit commodity, has leveled in recent years after a dramatic rise between 1966 and 1976. The per capita consumption of apple juice has trended upward throughout the period from 1960 to 1982. Gains in the per capita consumption of apple juice have been substantial with an average annual rate of increase of nearly 10%. This upward trend shows no sign of weakening in the near term.

Per capita consumption of frozen strawberries has been rather stable, ranging from a low of 1.13 lb/person in 1974 to a high of 1.56 lb/person in

Table 16.3. U.S. Per Capita Consumption (lb)[a] of Fruit Products, 1960–1981

Year	Fresh fruit	Melons	Canned fruit[b]	Canned juice	Frozen fruit	Juice concentrate	Chilled juice	Dried fruit	Wine	Total fruit
1960	89.1	23.2	23.0	12.86	3.50	18.00	2.12	3.07	7.57	182.42
1962	80.4	20.9	23.2	11.56	3.76	19.93	2.27	3.00	7.53	172.60
1964	76.7	20.8	23.4	11.01	3.59	13.73	1.36	2.89	8.08	161.61
1966	79.2	20.0	23.4	11.61	3.55	15.20	3.18	3.01	8.13	167.28
1968	76.3	20.9	22.3	11.66	3.84	21.30	4.20	2.82	8.92	172.24
1970	78.9	21.1	23.7	14.51	3.33	22.20	4.69	2.68	10.93	182.04
1972	74.9	19.8	21.6	13.86	3.65	29.67	5.18	2.01	13.48	184.15
1974	76.4	17.0	19.6	13.00	2.77	31.39	5.16	2.40	13.77	181.49
1976	82.8	18.3	18.9	14.47	3.09	35.52	6.09	2.60	14.61	196.38
1978	79.0	19.8	18.1	16.46	3.29	30.63	6.05	2.08	16.33	191.79
1980	85.7	16.9	17.6	16.67	3.07	34.13	5.85	2.37	17.67	200.01
1981	87.3	19.0	16.6	19.13	2.95	33.43	4.15	2.40	18.33	203.29
Average annual rate of change, 1960–1981 (%)	−0.1	−0.9	−1.5	1.8	−0.8	2.9	3.1	−1.1	4.1	0.5

Source: USDA (1982) and Wine Institute.
[a]Retail weight equivalent.
[b]Includes chilled citrus sections.

Table 16.4. U.S. Per Capita Consumption (lb)[a] of Canned Fruits, 1960–1982

Year	Apples & apple-sauce	Apricots	Berries	Cherries	Cran-berries	Salad & cocktail	Peaches	Pears	Pine-apples	Plums & prunes	Olives	Citrus sections[b]	Total[c]
1960	3.4	1.1	0.2	1.1	0.6	2.7	6.1	2.0	3.2	0.3	0.8	1.4	23.0
1962	3.4	0.9	0.2	1.2	0.8	2.8	6.4	2.1	2.8	0.4	0.8	1.3	23.2
1964	3.7	1.0	0.1	1.3	0.7	2.6	6.6	1.6	3.2	0.3	1.0	1.2	23.4
1966	3.3	1.1	0.2	1.0	0.8	3.0	6.2	1.9	3.1	0.4	0.8	1.5	23.4
1968	3.5	0.9	0.1	0.7	0.9	2.8	5.7	1.4	3.7	0.3	0.7	1.5	22.3
1970	3.7	1.0	0.1	0.9	0.9	3.2	5.9	2.0	3.3	0.3	1.1	1.3	23.7
1972	3.5	0.7	0.1	0.7	0.8	2.6	5.7	2.0	3.4	0.2	0.7	1.1	21.6
1974	3.1	0.6	0.1	0.7	0.9	2.7	4.9	1.7	2.6	0.2	0.9	1.1	19.6
1976	2.3	0.6	0.1	0.7	0.7	2.7	4.9	2.1	2.7	0.2	1.0	0.9	18.9
1978	2.6	0.4	0.1	0.7	0.8	2.8	4.2	1.8	3.0	0.2	0.6	0.9	18.1
1980	2.4	0.5	0.1	0.9	0.8	2.5	3.9	1.8	3.0	0.1	0.8	0.8	17.6
1981	2.1	0.4	0.1	0.8	0.7	2.3	3.6	1.6	3.0	0.2	1.0	0.9	16.6
1982	2.0	0.4	0.1	0.5	0.7	2.2	3.7	1.8	—	0.2	0.8	0.6	—

Source: USDA (1982, 1983A).
[a]Retail weight equivalent.
[b]Includes chilled citrus sections.
[c]Includes figs.

Table 16.5. U.S. Per Capita Consumption (lb)[a] of Citrus Fruit Juices, 1960–1982

Year	Canned orange	Canned grapefruit	Canned orange–grapefruit	Canned lemon & lime	Canned citrus concentrate	Total canned[b]	Chilled orange	Chilled grapefruit	Total chilled	Frozen orange concentrate[c]	Frozen grapefruit & other concentrate[c]	Total frozen concentrate[c]
1960	2.12	1.51	0.51	0.13	1.35	5.69	2.10	0.02	2.12	16.07	1.93	18.00
1962	1.92	1.48	0.47	0.13	0.95	5.01	2.19	0.08	2.27	18.29	1.69	19.98
1964	1.17	1.09	0.30	0.11	1.16	3.87	1.29	0.07	1.36	12.20	1.58	13.78
1966	1.53	1.73	0.34	0.10	0.94	4.66	3.04	0.14	3.18	13.92	1.28	15.20
1968	1.19	2.22	0.32	0.10	0.81	4.65	3.96	0.24	4.20	20.13	1.17	21.30
1970	1.74	2.98	0.33	0.10	1.32	6.48	4.35	0.34	4.69	20.83	1.37	22.20
1972	1.46	3.27	0.25	0.10	1.61	6.70	4.56	0.62	5.18	27.85	1.82	29.67
1974	1.46	3.51	0.21	0.10	1.01	6.29	4.64	0.52	5.16	29.64	1.75	31.39
1976	1.36	3.35	0.31	0.08	1.84	6.94	5.36	0.73	6.09	34.54	0.98	35.52
1978	1.72	3.51	0.16	0.07	1.49	6.95	5.30	0.75	6.05	27.64	3.04	30.68
1980	1.99	2.94	0.09	0.05	1.97	7.04	5.20	0.65	5.85	31.90	2.28	34.18
1982	1.76	2.08	0.04	0.03	0.82	4.73	3.19	0.31	3.50	33.49	3.68	37.17

Source: USDA (1982, 1983A).
[a]Retail weight equivalent.
[b]Includes tangerine juice.
[c]Single strength equivalent.

Table 16.6 U.S. Per Capita Consumption (lb)[a] of Noncitrus Fruit Juices, 1960–1982

Year	Apple	Fruit nectars	Grape	Pineapple[b]	Prune	Total
1960	0.89	1.06	0.76	3.40	1.06	7.17
1962	1.05	0.52	0.65	3.27	1.06	6.55
1964	1.49	0.28	0.65	3.61	1.11	7.14
1966	1.17	0.40	0.63	3.65	1.10	6.95
1968	1.69	0.37	0.55	3.65	0.75	7.01
1970	2.67	0.70	0.58	2.97	1.11	8.03
1972	2.62	0.56	0.54	2.77	0.67	7.16
1974	2.54	0.53	0.67	2.25	0.72	6.71
1976	3.32	0.77	0.56	1.88	1.00	7.53
1978	4.26	0.76	0.92	2.63	0.94	9.51
1980	4.77	0.68	0.65	2.67	0.86	9.63
1982	7.16	0.52	0.64	—	0.79	—

Source: USDA (1982, 1983A).
[a]Retail weight equivalent.
[b]Single-strength juice and concentrate.

1963. The trend in per capita consumption of raisins and currants has also been rather flat since 1960. However, in 1982, the latest year reported, per capita consumption of raisins and currants was at the highest level reported, 1.70 pounds, for the 23-year period. Wine, which is a category by itself, is also represented in Fig. 16.1 for purposes of comparison. The phenomenal growth in the per capita consumption of wine has already been discussed.

Examination of Tables 16.4–16.8 shows that per capita consumption of frozen citrus concentrate and canned apple juice have had the largest and most consistent increases. Per capita consumption of most other products has remained fairly constant or trended downward, with the exception of canned grapefruit juice, chilled orange juice, and raisins and currants.

FACTORS INFLUENCING CONSUMPTION

Consumption of processed fruit commodities is influenced by variables that determine the quantities supplied and the quantities demanded from year to year. Given the advanced state of the economy, the abundance of natural resources, and the advanced agricultural technology in the United States, it can be assumed that ample fruit supplies at reasonable costs are available except for fluctuations due to weather variations. Therefore, this section is devoted to a discussion of the factors that affect the demand

Table 16.7. U.S. Per Capita Consumption (lb)[a] of Frozen Fruits, 1960–1982

Year	Blackberries	Raspberries	Strawberries	Blueberries	Apples	Apricots	Cherries	Peaches	Miscellaneous[b]	Total[c]
1960	0.14	0.21	1.28	0.10	0.40	0.07	0.71	0.24	0.23	3.50
1962	0.14	0.17	1.42	0.19	0.32	0.06	0.74	0.30	0.31	3.76
1964	0.12	0.17	1.31	0.18	0.44	0.06	0.62	0.24	0.38	3.59
1966	0.07	0.15	1.40	0.15	0.39	0.10	0.74	0.30	0.22	3.55
1968	0.17	0.18	1.42	0.25	0.49	0.08	0.53	0.29	0.31	3.84
1970	0.11	0.16	1.18	0.21	0.48	0.06	0.61	0.26	0.20	3.33
1972	0.11	0.12	1.35	0.18	0.67	0.04	0.64	0.31	0.17	3.65
1974	0.06	0.09	1.13	0.14	0.33	0.06	0.50	0.28	0.14	2.77
1976	0.12	0.13	1.29	0.13	0.39	0.06	0.68	0.13	0.11	3.09
1978	0.10	0.10	1.38	0.11	0.39	0.07	0.64	0.27	0.18	3.29
1980	0.02	0.08	1.39	0.19	0.35	0.06	0.49	0.27	0.19	3.07
1982	0.09	0.07	1.15	0.12	0.43	0.06	0.62	0.24	0.16	2.97

Source: USDA (1982, 1983A).
[a]Retail weight equivalent.
[b]Includes grapes and pulp.
[c]Includes berries not listed separately.

Table 16.8. U.S. Per Capita Consumption (lb)[a] of Dried Fruits, 1960–1982

Year	Apples	Apricots	Dates	Figs	Prunes	Raisins and currants	Total[b]
1960	1.10	0.07	0.45	0.34	0.62	1.42	3.07
1962	0.12	0.05	0.36	0.26	0.68	1.47	3.00
1964	0.09	0.06	0.31	0.27	0.67	1.45	2.89
1966	0.15	0.06	0.31	0.27	0.54	1.64	3.01
1968	0.11	0.06	0.27	0.25	0.66	1.44	2.82
1970	0.11	0.05	0.27	0.22	0.68	1.34	2.68
1972	0.08	0.05	0.28	0.12	0.49	0.96	2.01
1974	0.11	0.03	0.25	0.16	0.51	1.33	2.40
1976	0.14	0.06	0.42	0.18	0.52	1.25	2.60
1978	0.13	0.04	0.32	0.17	0.42	0.98	2.08
1980	0.10	0.03	0.14	0.14	0.44	1.50	2.37
1982	0.17	0.06	0.22	0.13	0.45	1.70	2.76

Source: USDA (1982, 1983A).
[a]Retail weight equivalent.
[b]Includes peaches and pears.

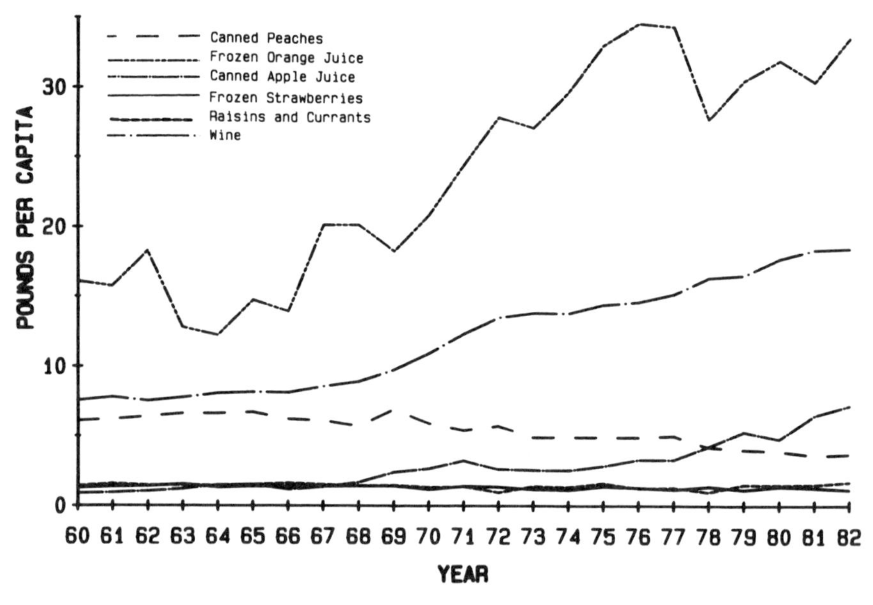

FIG. 16.1. U.S. per capita consumption of leading processed fruit commodities, retail weight equivalent, 1960–1982. (Orange juice, single-strength equivalent.)

for processed fruit commodities over time. Such factors include population, price, income, technology, and tastes and preferences.

Population

In a developed economy like that of the United States, the total quantity of food consumed is largely a function of population because the per capita consumption of food in general is virtually constant over time. For example, the annual per capita consumption of all food from 1960 through 1981 varied by only 3.4%, from a low of 1367.8 lb in 1974 to a high of 1415.1 lb in 1979 (Table 16.9). The components of the quantity of food consumed, however, are more variable. For instance, the per capita consumption of fruit consumed during the same years varied by 18.4%, from a low of 120.2 lb in 1963 to a high of 142.3 lb in 1981. Fruit averaged 9.5% of total food consumed from 1960 through 1981.

Since population is such an important determinant of fruit consumption, population projections can be useful for planning purposes. However, because of several behavioral assumptions that must be made in order to make population projections, the Bureau of the Census makes three projections, each based on different sets of assumptions. The assumptions pertain to the birth rate, life expectancy, and net immigration. The

Table 16.9. Per Capita Consumption[a] of All Food and Fruit, 1960–81

Year	All foods (lb)	Fruit[b] (lb)	Percentage of fruit for all food
1960	1,400.1	139.4	10.0
1962	1,374.6	130.2	9.5
1964	1,375.2	123.3	9.0
1966	1,377.2	128.6	9.3
1968	1,394.4	127.5	9.1
1970	1,397.0	134.5	9.6
1972	1,398.8	129.9	9.3
1974	1,367.8	128.4	9.4
1976	1,403.7	138.5	9.9
1978	1,395.0	134.2	9.6
1980	1,406.9	141.2	10.0
1981	1,399.8	142.3	10.2

Source: USDA (1982).
[a]Retail weight equivalent.
[b]Does not include melons; includes frozen citrus juices in product weight instead of single-strength equivalent.

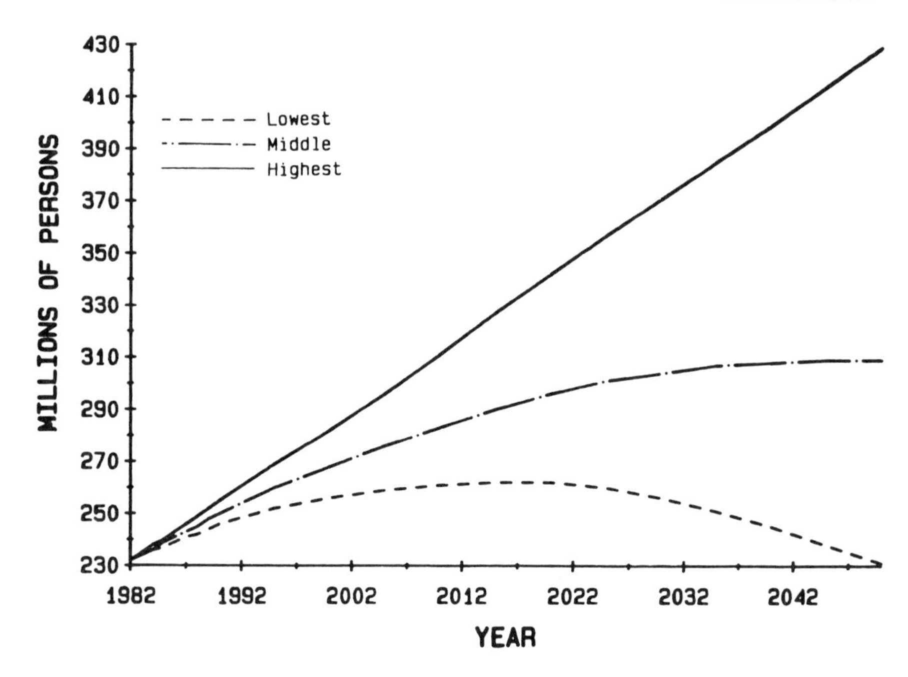

FIG. 16.2. U.S. population projections, 1982–2050. See text for assumptions underlying each projection. (From USDC 1982B.)

three population projections through the year 2050 for the United States are shown in Fig. 16.2.

The lowest population projection is based on the assumptions of 1.6 lifetime births per woman, life expectancy of 76.7 years in 2050, and yearly net immigration of 250,000 people. The middle projection is based on 1.9 lifetime births per woman, life expectancy of 79.6 years in 2050, and yearly net immigration of 450,000 people. The highest projection was made given 2.3 lifetime births per woman, life expectancy of 83.3 years in 2050, and yearly net immigration of 750,000 people.

According to these projections, population by the year 2050 could be anywhere from 231 million, a loss of a million people from 1982, to 429 million. This represents quite a range of possibilities. Even so, the highest population projection is conservative relative to past projections.

Until about 1970, agriculture outlook work was based on a population growth rate of approximately 1.2% a year. The highest estimate in Fig. 16.2 represents an average annual growth rate of nearly 0.91% from 1982 through 2050, considerably less than the previous rate used for long-range planning. The middle projection represents an average annual growth rate of just over 0.42%. Thus, significant changes have occurred in the assumptions behind population projections over the past 15 years. For example, lifetime births per woman have fallen dramatically. This phenomenon is reflected in a steadily declining average annual rate of population growth. From 1960 through 1982 the rate was just over 1.14%; from 1970 through 1982 the rate was about 1.03%; from 1975 through 1982 the rate had fallen to near 1.02%; and from 1980 through 1982 the rate was down to just under 0.94%.

A slowdown in population growth would have some impact on per capita consumption by changing the age distribution of the population. However, any possible increase in per capita consumption would not be sufficient to offset the lower population growth, exerting downward pressure on the total demand for fruit.

Table 16.10 shows population projections from the middle series by age group for selected years from 1982 through 2050. Significant changes occur for all age groups. The proportion of people in the under 14 and 15–29 age groups falls sharply from 1982 through 2050. The proportion of the population in the 30–44 age category increases sharply through 1990 and declines in like manner through 2050. The 45–64 age group declines slightly in proportion to total population through 1990 and then rebounds markedly through 2025. The most dramatic change of all is in the 65 and

Table 16.10. Population Projections[a] of the U.S. by Age Group (Middle Series)

Year	Age groups					Total
	Under 14	15–29	30–44	45–64	65 and Over	
1982						
Millions	52	62	47	44	27	232
(%)	22.4	26.7	20.3	19.0	11.6	100.0
1990						
Millions	55	57	60	46	32	250
(%)	22.0	22.8	24.0	18.4	12.8	100.0
2000						
Millions	56	53	63	61	35	268
(%)	20.9	19.8	23.5	22.8	13.0	100.0
2025						
Millions	55	55	60	72	59	301
(%)	18.3	18.3	19.9	23.9	19.6	100.0
2050						
Millions	54	56	58	74	67	309
(%)	17.5	18.1	18.8	23.9	21.7	100.0

Source: USDC (1982B).
[a]Includes armed forces overseas.

over category. The proportion of population in this category rises continuously from 11.6% in 1982 to 21.7% in 2050, an 87% increase in 68 years. This kind of change in the makeup of the population without doubt will precipitate a mammoth evolution in the food industry. The fruit processing sector can expect changes in all facets of the industry—packaging, processing methods, product form, and product type.

Processed Fruit Prices

The prices of processed fruits and vegetables increased significantly from 1960 through 1981. However, the most profound price increases occurred from the early 1970s through 1981. This was a period of high inflation in the United States. From 1967 through 1981 the general price level in the United States rose 189.1%. During the same period the price of fresh fruit rose 178.1%, while the price of processed fruit rose 158.5%, noticeably less than the general rate of inflation.

Figures 16.3 and 16.4 show the indices for consumption and price of, respectively, fresh and processed fruit from 1960 through 1981. In each figure, the index for the base year (1967) is 100. Figure 16.3 shows an inverse relationship between deflated or real price and per capita consumption of fresh fruit. This relationship is based largely on year-to-year

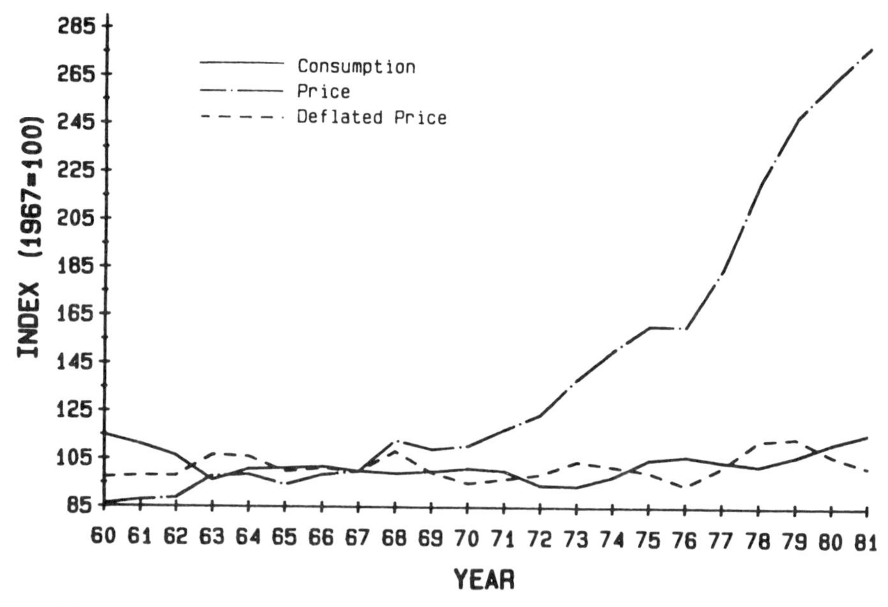

FIG. 16.3. Indices of U.S. per capita consumption and retail prices of fresh fruit, 1960–1981. (From USDA 1982.)

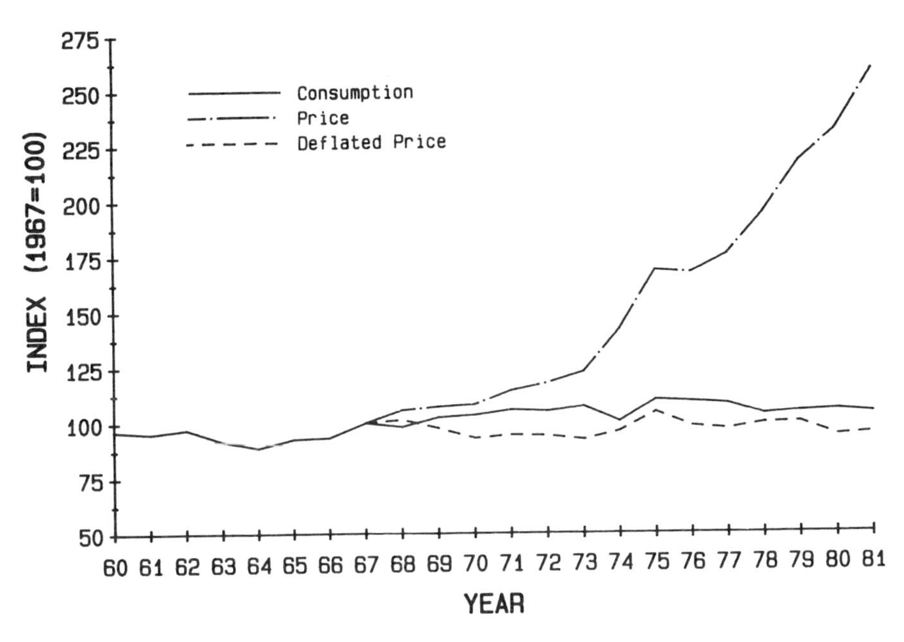

FIG. 16.4. Indices of U.S. per capita consumption and retail prices of processed fruit, 1960–1981. (From USDA 1982.)

fluctuations and, thus in part, represents price response to availability of supplies. Figure 16.4 does not reveal a consistent relationship between per capita consumption and the real price of processed fruit. Moreover, the real price of processed fruit does not seem to have an influence on the consumption of fresh fruit; nor does the real price of fresh fruit appear to have an impact on the consumption of processed fruit.

The recent marked increase in per capita consumption of fresh fruit does not seem to have adversely affected the per capita consumption of processed fruit, which is holding fairly steady (Table 16.2). This, however, does not preclude substantial substitution among individual fresh and processed fruit items.

Overall, the data reveal a supplemental relationship between fresh and processed fruit. This supplemental relationship seems enhanced by the apparent efficiency of the fruit processing industry as reflected in the lower real prices of processed fruit since 1967. Table 16.11 lists prices in January for selected processed fruit items for 1960 through 1983. (Prices for wine are average annual prices). Table 16.12 is the same as Table 16.11 except that all prices in Table 16.12 are deflated by the general consumer price index. Deflating prices in this manner allows one to more easily see relative changes in prices of individual processed fruit commodities.

Table 16.11. Retail Prices in Dollars for Selected Processed Fruit Items (Baltimore, Maryland), January, 1960–1983

Year	Applesauce, canned (16 oz)	Apricots, canned (30 oz)	Fruit cocktail, canned (17 oz)	Peaches, canned (16 oz)	Pears, canned (29 oz)	Pineapple, sliced, canned (20 oz)	Pineapple juice, canned (46 oz)	Grapefruit juice, canned (46 oz)	Orange juice, canned (46 oz)	Orange juice concentrate, frozen (6 oz)	Straw-berries, frozen (10 oz)	Wine[a] (gal)
1960	0.16	0.46	0.26	0.24	0.40	0.40	0.31	0.31	0.42	0.20	0.24	4.75
1962	0.15	0.39	0.26	0.23	0.42	0.43	0.30	0.33	0.44	0.22	0.26	4.92
1964	0.17	0.36	0.25	0.22	0.44	0.42	0.33	0.38	0.54	0.29	0.24	4.87
1966	0.16	0.40	0.28	0.23	0.48	0.43	0.32	0.39	0.42	0.18	0.30	4.90
1968	0.18	0.49	0.25	0.24	0.56	0.44	0.31	0.37	0.36	0.19	0.31	5.86
1970	0.18	0.47	0.27	0.25	0.46	0.48	0.33	0.41	0.46	0.23	0.31	6.83
1972	0.18	0.42	0.30	0.27	0.53	0.44	0.36	0.54	0.49	0.25	0.32	7.67
1974	0.32	0.60	0.36	0.35	0.57	0.54	0.40	0.53	0.53	0.26	0.41	8.82
1976	0.30	0.83	0.45	0.42	0.70	0.57	0.68	0.57	0.63	0.32	0.53	9.66
1978	0.40	0.82	0.49	0.52	0.79	0.69	0.84	0.71	0.80	0.49	0.63	11.01
1980	0.35	0.97	0.53	0.53	0.96	0.70	1.01	0.91	1.10	0.56	0.76	12.91
1982	0.42	1.27	0.64	0.64	1.04	0.81	1.13	1.07	1.19	0.59	0.94	14.23
1983	0.51	1.44	0.70	0.72	1.09	0.86	1.19	1.14	1.37	0.61	0.98	—

Source: Maryland Department of Agriculture, Marketing Services Section, and the Wine Institute.
[a]Average value as computed from aggregate consumer expenditures and gallons consumed in the United States.

Table 16.12. Deflated[a] Retail Prices in Dollars for Selected Processed Fruit Items (Baltimore, Maryland), January, 1960–1983

Year	Applesauce, canned (16 oz)	Apricots, canned (30 oz)	Fruit cocktail, canned (17 oz)	Peaches, canned (16 oz)	Pears, canned (29 oz)	Pineapple, sliced, canned (20 oz)	Pineapple juice, canned (46 oz)	Grapefruit juice, canned (46 oz)	Orange juice, canned (46 oz)	Orange juice concentrate, frozen (6 oz)	Straw-berries, frozen (10 oz)	Wine (gal)
1960	0.18	0.52	0.30	0.27	0.46	0.46	0.35	0.35	0.48	0.23	0.27	5.36
1962	0.17	0.43	0.29	0.26	0.47	0.48	0.33	0.37	0.49	0.24	0.29	5.43
1964	0.18	0.39	0.27	0.24	0.48	0.45	0.36	0.41	0.58	0.31	0.26	5.24
1966	0.17	0.42	0.29	0.24	0.50	0.45	0.34	0.41	0.44	0.19	0.31	5.04
1968	0.18	0.48	0.25	0.24	0.55	0.43	0.30	0.36	0.35	0.19	0.30	5.62
1970	0.16	0.41	0.24	0.22	0.41	0.42	0.29	0.36	0.41	0.20	0.27	5.87
1972	0.15	0.34	0.24	0.22	0.43	0.36	0.29	0.44	0.40	0.20	0.26	6.12
1974	0.23	0.43	0.26	0.25	0.41	0.39	0.29	0.38	0.38	0.19	0.29	5.97
1976	0.18	0.50	0.27	0.25	0.42	0.34	0.41	0.34	0.38	0.19	0.32	5.67
1978	0.21	0.44	0.26	0.28	0.42	0.37	0.45	0.38	0.43	0.26	0.34	5.63
1980	0.15	0.42	0.23	0.23	0.41	0.30	0.43	0.39	0.47	0.24	0.33	5.23
1982	0.15	0.45	0.23	0.23	0.37	0.29	0.40	0.38	0.42	0.21	0.33	4.92
1983	0.17	0.49	0.24	0.25	0.37	0.29	0.41	0.39	0.47	0.21	0.33	—

Source: Table 16.11 and USDC (1960–1983).

[a]Deflated by the January Consumer Price Index, 1967 = 100, except wine; wine was deflated by the Annual Consumer Price Index, 1967 = 100.

Table 16.12 seems to indicate that wine, canned fruit cocktail, pears, and sliced pineapple became less expensive from 1960 through 1982, whereas canned pineapple juice and frozen strawberries appear to have become more expensive. Although the remaining commodities in Table 16.12 show real price variability, no noticeable trends, up or down, are evident.

For illustration purposes, the data on per capita consumption and prices for canned orange juice and wine are graphed in Figs. 16.5 and 16.6. In periods of high inflation (1970–1981) the real or deflated price is more relevant to observe than the current price (in this case, inflated price).

The price-quantity relationship for canned orange juice is difficult to isolate since many factors are involved in the determination of quantity consumed. The inverse relationship between price and quantity is clearly seen in only four years from 1960 through 1981 (Fig. 16.5). These years, 1964–1965 and 1968–1969, are revealing because there were severe shocks largely due to adverse weather. A cursory view of the data indicates that in general a relative change in the price of canned orange juice is accompanied by a larger relative change in consumption. This is to be expected for a commodity that has many substitutes, indicating a relatively high price elasticity of demand for canned orange juice.

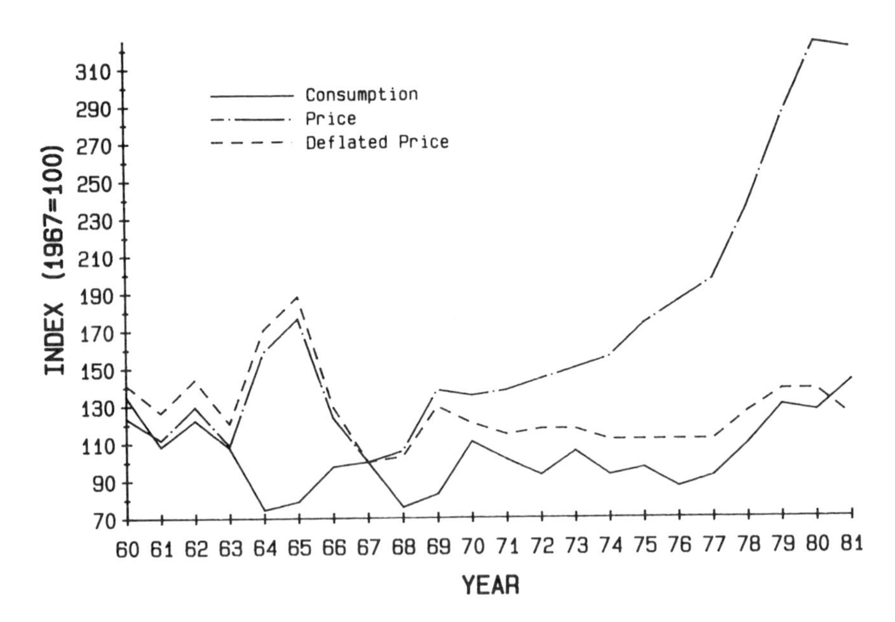

FIG. 16.5. Indices of U.S. per capita consumption and retail prices for canned orange juice, 1960–1981. (From Tables 16.5, 16.11, and 16.12.)

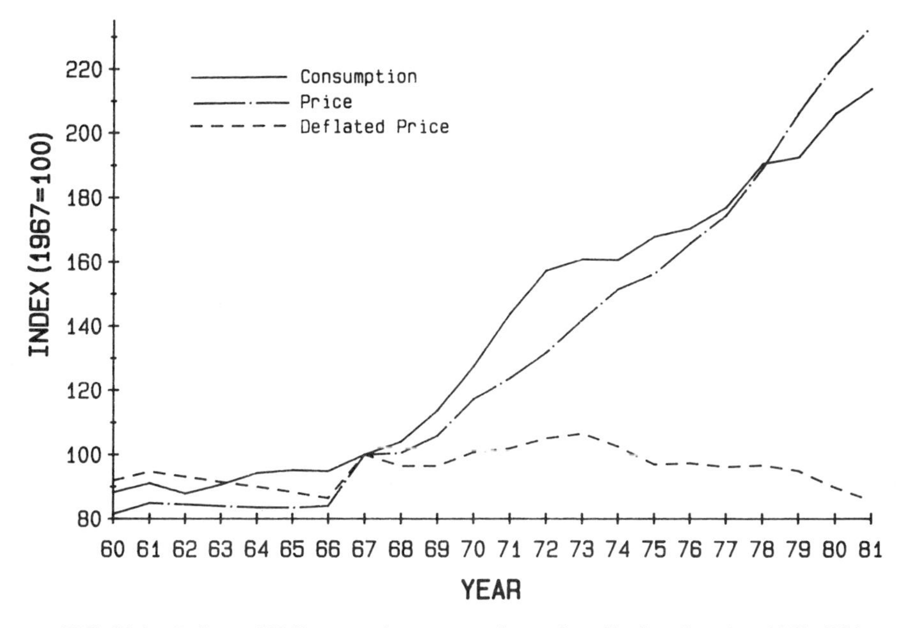

FIG. 16.6. Indices of U.S. per capita consumption and retail prices for wine, 1960–1981. (From Tables 16.3, 16.11, and 16.12.)

The inverse price–quantity relationship for wine is even more obscure (Fig. 16.6). The per capita consumption of wine has trended upward throughout the period from 1960 through 1981; and the real (deflated) price of wine peaked in 1973. Examination of the data reveals that the magnitude of relative changes in real price are accompanied by relative changes in consumption of a similar magnitude, all else equal. Thus, substitutes for wine are apparent, but not nearly to the extent of that for canned orange juice. In the long run the falling real price of wine since 1973 appears to be a major factor in the marked upward trend in the per capita consumption of wine.

Income Influences

As consumer income expands, family food budgets normally increase but at a lower rate than income. The increase in food spending may go for increased quantity, better quality, or different forms. The 1977–1978 Household Food Consumption Survey revealed that the per capita value of food purchased increased uniformly as household income rose. At a household income of under \$3000, per capita food expenditures were about 23% below the national average of \$1142. For a household income of \$30,000 or more, per capita food expenditures were about 37% above the U.S. average. In the $3000–$4000 income category, food expenditures accounted for 49% of the household budget. In the $25,000–$30,000 income category, food purchases encompassed just under 16% of the household budget.

As incomes rise, families are better able to indulge their individual tastes and preferences and consume a wider variety of foods. Rising incomes of U.S. consumers have contributed to increased demand for convenience foods and rapid expansion of the foodservice industry.

From 1960 through 1981 per capita disposable personal income in current dollars rose 353% (Table 16.13). This represents an average annual rate of increase of 7.46%. However, real or deflated per capita disposable personal income rose by only about 48% during the same period, or 1.87% per annum. Real income represents true purchasing power of the consumer.

Although inflation was a major problem in the United States from 1970 through 1981 and real purchasing power declined in 1974 through 1975 and 1979 through 1981, the trend in per capita consumption of processed fruit remained flat through these troubled years. Per capita consumption of processed fruit, then, does not appear to be responsive to changes in real per capita income over time, reflecting a zero income elasticity of demand as measured through time-series data.

Cross-section data (Table 16.14) reveal that per capita consumption of processed fruit in general is relatively low at extremely low levels of household income (under $3000/year) and relatively high at extremely

Table 16.13. U.S. Per Capita Disposable Personal Income in Current and Deflated Dollars, 1960–1981

Year	Current disposable personal income	Deflated disposable personal income[a]
1960	1947.94	2196.10
1962	2073.95	2289.13
1964	2295.92	2471.39
1966	2612.87	2688.14
1968	2956.74	2837.56
1970	3390.00	2914.88
1972	3860.51	3081.01
1974	4667.34	3160.01
1976	5478.71	3213.32
1978	6621.70	3388.79
1980	8010.87	3245.90
1981	8830.06	3241.58

Source: USDA (1982).
[a]Deflated by the Consumer Price Index, 1967 = 100.

Table 16.14. Weekly Quantity of Fruit Products Consumed per Household by U.S. Household Income, 1977–1978

Annual household income ($)	Average household size (no.)	Fresh fruit (lb)	Jelly and jam (lb)	Canned fruit (lb)	Frozen fruit (lb)	Canned juice (lb)	Frozen juice (lb)	Fresh juice (lb)	Dried fruit (lb)	Wine (lb)	Total processed (lb)	Fresh and processed (lb)
All households	2.76	7.87	0.29	0.73	0.02	1.32	0.65	0.79	0.07	0.28	4.15	12.02
Under 3000	1.77	3.95	0.16	0.37	[a]	1.04	0.25	0.51	0.03	0.10	2.46	6.41
3000–3999	1.84	4.68	0.22	0.43	0.02	1.02	0.31	0.88	0.05	0.09	3.02	7.70
4000–4999	2.20	5.22	0.25	0.47	0.02	1.20	0.33	0.87	0.05	0.08	3.27	8.49
5000–5999	2.31	6.87	0.26	0.60	0.02	1.33	0.45	0.84	0.05	0.09	3.64	10.51
6000–6999	2.42	6.74	0.30	0.57	[a]	1.19	0.43	0.89	0.05	0.08	3.51	10.25
7000–7999	2.60	7.41	0.24	0.70	0.02	1.04	0.50	0.73	0.06	0.10	3.39	10.80
8000–9999	2.64	7.36	0.26	0.74	0.03	1.28	0.58	0.87	0.06	0.14	3.96	11.32
10,000–11,999	2.83	8.04	0.29	0.74	0.03	1.32	0.56	0.81	0.07	0.17	3.99	12.03
12,000–14,999	2.99	8.57	0.33	0.81	0.02	1.22	0.68	0.76	0.08	0.19	4.09	12.66
15,000–19,999	3.19	8.53	0.29	0.86	0.02	1.40	0.76	0.84	0.08	0.33	4.58	13.11
20,000–24,999	3.12	8.96	0.32	0.76	0.02	1.41	0.86	0.75	0.09	0.40	4.61	13.57
25,000–29,999	3.15	9.83	0.30	0.85	0.02	1.48	0.90	0.79	0.08	0.66	5.08	14.91
30,000 and over	3.19	11.35	0.35	0.87	0.05	1.53	1.19	1.07	0.09	0.77	5.92	17.27

Source: USDA (1983B).
[a]Less than 0.01.

high levels of household income ($30,000 and over/year). There does not appear to be, however, any relationship between per capita consumption of processed fruit in general and household income between the two extreme categories of household income. This finding tends to support the time-series examination that revealed a zero income elasticity of demand for processed fruit. The findings for fresh fruit are similar.

The lack of responsiveness of per capita consumption of processed fruit to changes in income in the United States in recent years seems to indicate that income is sufficiently high to allow virtually all desired purchases of processed fruit by U.S. households.

Changing Technology

Thus far we have seen that population is the major factor influencing food consumption, including processed fruit. We have seen that price and income no longer dictate marked changes in per capita consumption of processed fruit because the prices of these products are low enough and U.S. household income has risen enough that income in general is no longer a constraining influence on processed fruit consumption.

Aside from the necessity of sophisticated advertising, new technology is essential for significant improvement in per capita consumption of processed fruit. New technology may result in new tastes, improved flavor, greater convenience, or lower cost and subsequent adjustments in the pattern of consumption reflecting consumer reaction to these changes.

Per capita consumption of several processed fruit items has increased noticeably; but overall per capita consumption of processed fruit has been flat in recent years. The processed fruit commodities recently gaining in importance—various citrus juices, apple juice, raisins, and wine—are similar in that they are near in form to fresh fruit except wine. At the same time, per capita consumption of fresh fruit has improved markedly in recent years. Due to sophisticated advertising and promotion and declining real prices, wine has become fashionable on middle-class American menus at home and away from home. In addition, Americans have become health conscious, becoming more aware of the factors that enhance or lessen good health. The nutritional importance of increased consumption of fresh fruits and vegetables has received massive publicity in recent years.

The implication for the processed fruit industry is clear. Aside from promoting the nutritive benefits of processed fruits, research and development should be directed toward changing the form and packaging of processed fruits to more nearly resemble fresh fruit products. Product development and complimentary promotion efforts should seek to deter the distinction between fresh and processed fruits.

Other Influences

There are a number of other changing factors of the U.S. economy that are now and could continue influencing processed fruit consumption patterns. A major influence is the increasing number of women entering the labor force. This trend is recognized as an important influence on consumption habits and is having a dual impact: more convenience foods are desired at home and more meals are eaten away from home.

The number of working women has increased markedly over the years (Table 16.15). In 1982, 43.3% of the total labor force were women compared with 32.3% in 1960. The percentage of all women of working age who were in the labor force increased from 37.8% in 1960 to 52.6% in 1982. This trend is likely to continue but at a declining rate. In fact, the average annual rate of increase in the percentage of total women in the labor force has been declining since the late 1960s.

The increasing number of single adult households and households with

Table 16.15. Women in the U.S. Labor Force, Selected Years, 1900–1982

Year	Women in labor force (1000)	Total labor force (%)	All women of working age (%)
1900	5,114	18.1	20.4
1910	7,889	20.9	25.2
1920	8,430	20.4	23.3
1930	10,679	22.0	24.3
1940	12,845	24.3	25.4
1945	19,270	29.6	35.7
1950	18,412	28.8	33.9
1955	20,584	30.2	35.7
1960	23,272	32.3	37.8
1965	26,232	34.0	39.3
1970	31,560	36.7	43.4
1975	36,998	39.9	46.3
1980	45,487	42.5	51.3
1982	47,755	43.3	52.6

Source: Economic Report of the President, 1973; USDA (1981, 1982A); U.S. Dept. of Labor (1983).

Note: Data from 1900 to 1940 are from decennial censuses and refer to a single date; beginning 1945, data are annual averages. For 1900 to 1945, data include women 14 years of age and over; beginning 1950, data include women 16 years of age and over. Labor force data for 1900 to 1930 refer to gainfully employed workers.

both husband and wife working has spurred rapid growth of the foodservice industry. Real expenditures in 1967 dollars for food away from home rose from $16.0 billion in 1960 to $31.6 billion in 1981, an increase of 97%. Real purchases in 1967 dollars of food used at home rose almost 41% during this period, from $63.4 billion in 1960 to $89.2 billion in 1981. The share of total food expenditures spent for away-from-home food purchases increased from 20% in 1960 to 26% in 1981. However, the average annual rate of increase has been declining since 1973.

The tremendous growth of the foodservice industry in conjunction with a greater awareness of nutrition spawned the now-popular salad and fruit bars, which first appeared in restaurants. Salad and fruit bars are now spreading throughout the foodservice industry—school, government, prison, hospital, and corporate eating facilities. Originally, salad and fruit bars were stocked mostly with fresh fruits and vegetables; but recently, a greater assortment of processed fruit items has been appearing on fruit and salad bars.

The impact of synthetics and substitutes on fruit product markets, especially fruit juices, will likely continue in the years ahead. Fruit products have long been subjected to significant competition from synthetics and substitutes. Further penetration of this market by substitutes is a continuing threat. There is also commercial activity to develop fruit analogs having characteristic fruit structure, color, and flavor. The dangers posed by synthetics and substitutes can be minimized through continued efficient fruit processing and sophisticated advertising and promotion that project the desirability of natural goodness and nutrition.

For many years the U.S. population became increasingly urbanized. However, the 1980 Census revealed a reversal of this trend. The urbanization trend could be counted on to enhance the consumption of processed fruit since city inhabitants consume more processed fruit than other Americans (Table 16.16). It appears that the changing mix of population segments in the United States is no longer a positive stimulus to consumption of processed fruit.

Distinct regional differences also exist in fruit consumption patterns (Table 16.17). These regional differences have not changed at least in the last 15 years. The implication for the processed fruit industry is to target the north-central and southern regions for effective promotion and advertising to increase consumption.

CONSUMPTION PROSPECTS

The total demand for fruit is expected to increase in the years ahead primarily because of population growth, increasing awareness of nutrition as it relates positively to health, and the exploding popularity of salad and

Table 16.16. Weekly Quantity of Fruit Products Consumed per Household by Population Segment in the United States, 1977–1978

	Population segment		
	Central city	Suburban	Nonmetropolitan
Average household size (No.)	2.59	2.86	2.82
Fresh fruit (lb)	7.85	8.45	7.25
Jelly and jam (lb)	0.22	0.30	0.32
Canned fruit (lb)	0.61	0.79	0.77
Frozen fruit (lb)	0.02	0.02	0.03
Canned juice (lb)	1.40	1.23	1.34
Frozen juice (lb)	0.59	0.74	0.62
Fresh juice (lb)	1.13	0.89	0.39
Dried fruit (lb)	0.07	0.08	0.07
Wine (lb)	0.32	0.36	0.14
Total processed (lb)	4.36	4.41	3.68
Fresh and processed (lb)	12.21	12.86	10.93

Source: USDA (1983B).

fruit bars in the foodservice industry. From 1980 to 1995, processed fruit consumption should increase almost 14.3% assuming that population grows according to the middle-series projection and assuming per capita consumption of processed fruit does not change.

Based on recent patterns of per capita consumption, the consumption of frozen fruit juice concentrate, chilled juice, and dried fruit will increase in line with the increase in population, about 14.3% from 1980 to 1995. During the same period the consumption of canned fruit could fall nearly 21% due to a reduction in per capita consumption to about 12 lb in 1995. The consumption of canned juice and wine could climb significantly from 1980 to 1995. Based on recent increases in per capita consumption and

Table 16.17. Weekly Quantity of Fruit Products Consumed per Household by U.S. Region, 1977–1978

Region	Average household size (No.)	Fresh (lb)	Processed (lb)	Total (lb)
Northeast	2.77	7.98	5.09	13.07
North Central	2.87	7.87	3.76	11.63
South	2.71	6.90	3.57	10.47
West	2.69	9.37	4.43	13.80

Source: USDA (1983C–F).

expected population growth, consumption of canned juice could increase almost 58% and wine, nearly 57%. Per capita consumption of canned juice and wine in 1995 could be just over 24 lb each.

The projections that show increases in consumption are considered conservative. The conservative estimate of the overall increase in per capita consumption of processed fruit from 1980 to 1995 is about 8.9%. Increases beyond this estimate will depend on how well the food processing industry capitalizes on American consumers' increasing awareness of the relationship between nutrition and health and on the tremendous growth of the foodservice industry in conjunction with the amazing popularity of salad and fruit bars.

For optimum results in obtaining a greater share of the American food dollar, the food processing industry should develop new and convenient packaging and new forms of processed fruits that will appear more like fresh fruits. Such developments should be augmented with sophisticated advertising and promotion, thus obscuring the apparent difference between fresh and processed fruits. In this way competition in terms of price, quality, convenience, appearance, and form will be among alternative forms of fruit without the added difficulties posed by irrelevant differences between fresh and processed fruits. It is imperative that the differences in consumer perceptions of fresh and processed fruits be eliminated to the greatest extent possible.

REFERENCES

USDA. 1982. Food consumption, prices, and expenditures 1960–1981. U.S. Dept. Agric. Econ. Res. Serv., Stat. Bull. *694*.

USDA. 1983A. Fruit outlook and situation. U.S. Dept. Agric. Econ. Res. Serv. *TFS-227*.

USDA. 1983B. Food consumption: Households in the United States, seasons and year 1977–1978. U.S. Dept. Agric. Human Nutr. Inf. Serv., NFCS Rep. *H-6*.

USDA. 1983C. Food consumption: Households in the Northeast, seasons and year 1977–1978. U.S. Dept. Agric. Human Nutr. Inf. Serv., NFCS Rep. *H-7*.

USDA. 1983D. Food consumption: Households in the North Central, seasons and year 1977–1978. U.S. Dept. Agric. Human Nutr. Inf. Serv., NFCS Rep. *H-8*.

USDA. 1983E. Food consumption: Households in the South, seasons and year 1977–1978. U.S. Dep. Agric., Human Nutr. Inf. Serv., NFCS Rep. *H-9*.

USDA. 1983F. Food Consumption: Households in the West, seasons and year 1977–1978. U.S. Dept. Agric. Human Nutr. Inf. Serv. NFCS Rep. *H-10*.

USDC. 1960–1983. Survey of Current Business. U.S. Dept. Commerce, Bur. Econ. Anal.

USDC. 1981. Population characteristics. U.S. Dept. Commerce, Bur. Census, Ser. *P-20* (363) June.

USDC. 1982A. Population characteristics. U.S. Dept. Commerce, Bur. Census, Ser. *P-20* (374) Sept.

USDC. 1982B. Population estimates and projections. U.S. Dept. Commerce, Bur. Census, Ser. *P-25* (922) Oct.

U.S. Dept. of Labor. 1983. Employment and earnings. U.S. Dept. Labor, Bur. Labor Stat. *30* (1) Jan.

Index

F

G

H

I

J

L

M

N

P

R

S

T

U

V

W